# Handbook of Bacterial Adhesion

# Handbook of Bacterial Adhesion

*Principles, Methods, and Applications*

Edited by

## Yuehuei H. An, MD

*Department of Orthopaedic Surgery,*
*Medical University of South Carolina, Charleston, SC*

and

## Richard J. Friedman, MD

*Department of Orthopaedic Surgery,*
*Medical University of South Carolina, Charleston, SC*

Humana Press  Totowa, NJ

QR
96
.8
.H36
2000

Library of Congress Cataloging in Publication Data

Handbook of bacterial adhesion : principles, methods, and applications / edited by
Yuehuei H. An, Richard J. Friedman.
      p. ; cm.
   Includes bibliographical references and index.
   ISBN 0-89603-794-0 (alk. paper)
   1. Bacteria--Adhesion--Handbooks, manuals, etc. 2. Biomedical
materials--Microbiology--Handbooks, manuals, etc. 3. Biofilms--Handbooks, manuals,
etc. I. An, Yuehuei H. II. Friedman, Richard J.
   [DNLM: 1. Bacterial Adhesion--physiology. 2. Biocompatible Materials. 3.
Biofilms--growth & development. QW 52 H2354 2000]
   QR96.8 .H36 2000
   571.2'93--dc21

                                                                99-057867

*To Kay Q. Kang, MD*

*Without her love, inspiration, and support,*
*this book would not have been possible*

**Yuehuei H. An**, MD


*To my wife Vivian, and my daughters Arielle and Leah,*
*for their patience, understanding, love, and support*

**Richard J. Friedman**, MD

# PREFACE

Research on bacterial adhesion and its significance is a major field involving many different aspects of nature and human life, such as marine science, soil and plant ecology, the food industry, and most importantly, the biomedical field. The adhesion of bacteria to human tissue surfaces and implanted biomaterial surfaces is an important step in the pathogenesis of infection.

*Handbook of Bacterial Adhesion: Principles, Methods, and Applications* is an outgrowth of the editors' own quest for information on laboratory techniques for studying bacterial adhesion to biomaterials, bone, and other tissues and, more importantly, a response to significant needs in the research community.

This book is designed to be an experimental guide for biomedical scientists, biomaterials scientists, students, laboratory technicians, or anyone who plans to conduct bacterial adhesion studies. More specifically, it is intended for all those researchers facing the challenge of implant infections in such devices as orthopedic prostheses, cardiovascular devices or catheters, cerebrospinal fluid shunts or extradural catheters, thoracic or abdominal catheters, portosystemic shunts or bile stents, urological catheters or stents, plastic surgical implants, oral or maxillofacial implants, contraceptive implants, or even contact lenses. It also covers research methods for the study of bacterial adhesion to tissues such as teeth, respiratory mucosa, intestinal mucosa, and the urinary tract. In short, it constitutes a handbook for biomechanical and bioengineering researchers and students at all levels.

*Handbook of Bacterial Adhesion: Principles, Methods, and Applications* is the first inclusive and organized reference book on how to conduct studies on bacterial adhesion to biomaterials and tissues, a topic that has not been covered adequately by existing works. The book also complements other reference titles on bacterial adhesion. The book has six parts: Part I (6 chapters), Mechanisms of Microbial Adhesion and Biofilm Formation; Part II (6 chapters), General Considerations for Studying Microbial Adhesion and Biofilm; Part III (7 chapters), Techniques for Studying Microbial Adhesion and Biofilm; Part IV (7 chapters), Studying Microbial Adhesion to Biomaterials; Part V (8 chapters), Studying Microbial Adhesion to Host Tissue; and Part VI (5 chapters), Strategies for Prevention of Microbial Adhesion. Since yeasts are also a major factor in implant and/or tissue infections, the book includes a chapter covering Candida adhesion and related infections (Chapter 33).

*Handbook of Bacterial Adhesion: Principles, Methods, and Applications* is designed to be concise as well as inclusive, and more practical than theoretical. The text is simple and straightforward. A large number of diagrams, tables, line drawings, and photographs is used to help readers better understand the content. Full bibliographies at the end of each chapter guide readers to more detailed information. Although a work of this length cannot discuss every aspect of bacterial adhesion that has been studied over the years, it is hoped that all the major principles, methods, and applications have been included.

*Yuehuei H. An, MD*

# TABLE OF CONTENTS

# THE EDITORS

**Yuehuei H. (Huey) An,** MD, graduated from the Harbin Medical University, Harbin, Northeast China in 1983 and was trained in orthopaedic surgery at the Beijing Ji Shui Tan Hospital (Residency), and in hand surgery at Sydney Hospital (Clinical Fellow), Sydney, Australia. In 1991, Dr. An joined with Dr. Richard J. Friedman in the Department of Orthopaedic Surgery at the Medical University of South Carolina to establish the MUSC Orthopaedic Research Laboratory, which is now a multifunctional orthopaedic research center. Dr. An has published more than 70 scientific papers and book chapters and more than 60 research abstracts and edited 3 reference books. His first book *Animal Models in Orthopaedic Research*, a major contribution to orthopaedic research, was published by CRC Press in 1998. His second book, *Mechanical Testing of Bone and the Bone-Implant Interface,* a 700-page handbook, was also published by CRC Press in late 1999. He enjoys art and created most of the line drawings used in his books and papers. He is an active member of eight academic societies, including the *Orthopaedic Research Society, Society for Biomaterials, American Society of Biomechanics,* and *Tissue Engineering Society.* Dr. An's current research interests include bone and cartilage repair using tissue engineering techniques, bone or soft tissue ingrowth to implant surfaces, bone biomechanics, bacterial adhesion and prosthetic infection, and animal models for orthopaedic applications.

**Richard J. Friedman,** MD, FRCSC, graduated from the University of Toronto School of Medicine in 1980 and was trained in orthopaedic surgery at Johns Hopkins University and Harvard University. Dr. Friedman is the founder of the Orthopaedic Research Laboratory at the Medical University of South Carolina. He has published more than 100 scientific papers and book chapters and more than 200 research abstracts. In 1994, he edited *Arthroplasty of the Shoulder*, published by Thieme Medical Publishers, New York. He is the guest editor of the 29th volume of Orthopaedic Clinics of North America in July 1998 on the subject of shoulder arthroplasty. He has been invited as a guest speaker for many national and international academic meetings. He has served as a chairman or member of more than thirty academic or professional committees or meetings and is a member of more than twenty societies or associations. Dr. Friedman's current research interests include bacterial adhesion and prosthetic infection, orthopaedic biomaterials, bone and cartilage repair using tissue engineering techniques, bone ingrowth, shoulder and elbow surgery, and total joint replacement.

# CONTRIBUTORS

**Colin G. Adair,** PhD, *Pharmaceutical Devices Group, School of Pharmacy, The Queen's University of Belfast, United Kingdom*

**Donald G. Ahearn,** PhD, *Professor, Biology Department, Georgia State University, Atlanta, GA, USA*

**M. John Albert,** PhD, MRCPath, *Senior Research Microbiologist, Laboratory Sciences Division, International Centre for Diarrhoeal Disease Research, Bangladesh (ICDDRB), Dhaka, Bangladesh*

**Yuehuei H. An,** MD, *Associate Professor, Orthopaedic Research Laboratory, Department of Orthopaedic Surgery, Medical University of South Carolina, Charleston, SC, USA*

**Roland Andersson,** MD, PhD, *Professor of Surgery, Department of Surgery, Lund University Hospital, Lund, Sweden*

**Jeffrey O. Anglen,** MD, *Associate Professor, Department of Orthopaedic Surgery, University of Missouri-Columbia, Columbia, Missouri, USA*

**Carla R. Arciola,** MD, PhD, *Professor, Laboratory for Biocompatibility Research on Implant Materials, Istituti Ortopedici Rizzoli, Bologna, Italy*

**Reza Ardehali,** BSc, *Artificial Heart Research Laboratory, University of Utah, Salt Lake City, UT*

**Masahiro Asaka,** MD, PhD, *Professor, Department of Internal Medicine, Hokkaido University School of Medicine, Sapporo, Japan*

**Alan J. Barton,** MD, MSc, *Resident, Department of Pediatrics and Communicable Diseases, University of Michigan Hospitals, Ann Arbor, Michigan, USA*

**Katrin Bartscht,** MSc, *Institut für Medizinische Mikrobiologie und Immunologie, Universitäts-Krankenhaus Eppendorf, University of Hamburg, Germany*

**Brian K. Blair,** BSc, *Medical Student, Medical University of South Carolina, Charleston, SC, USA*

**Thomas Boland,** PhD, *Assistant Professor, Department of Bioengineering, Clemson University, Clemson, SC, USA*

**Robert A. Burne,** PhD, *Department of Microbiology and Immunology, and Center for Oral Biology, University of Rochester, School of Medicine and Dentistry, Rochester, NY, USA*

**Lisa B. Byers,** BSc, *Pharmaceutical Devices Group, School of Pharmacy, The Queen's University of Belfast, United Kingdom*

**Douglas E. Caldwell,** PhD, *Department of Applied Microbiology and Food Science, University of Saskatchewan, Saskatoon, Saskatchewan, Canada*

**Yi-Ywan M. Chen,** PhD, *Assistant Professor, Department of Microbiology and Immunology, and Center for Oral Biology, University of Rochester, School of Medicine and Dentistry, Rochester, NY, USA*

**Gordon D. Christensen,** MD, *Associate Chief of Staff for Research and Development, Harry S. Truman Memorial Veterans' Hospital, Columbia, Missouri; Professor, Departments of Internal Medicine and Molecular Microbiology and Immunology, University of Missouri-Columbia, Columbia, Missouri, USA*

**Aaron R. Clapp,** PhD, *Department of Chemical Engineering & NSF Engineering Research Center for Particle Science and Technology, University of Florida, Gainesville, FL, USA*

**Harry S. Courtney,** PhD, *Research Microbiologist, Veterans Affairs Medical Center, Assistant Professor of Medicine, Department of Medicine, University of Tennessee, Memphis, TN, USA*

**James B. Dale,** MD, *ACOS for Education, Veterans Affairs Medical Center, Professor of Medicine and Chief of Infectious Diseases, Department of Medicine, University of Tennessee, Memphis, TN, USA*

**Helen M. Dalton,** MSc, *Senior Research Fellow, School of Microbiology and Immunology, The University of New South Wales, Sydney, Australia*

**Rabih O. Darouiche,** MD, *Associate Professor, Department of Medicine, Baylor College of Medicine and VAMC Infectious Disease Section, Houston, TX, USA*

**Sophie de Bentzmann,** PhD, *INSERM U 514, Reims University, Reims, France*

**Richard B. Dickinson,** PhD, *Assistant Professor, Department of Chemical Engineering, University of Florida, Gainesville, FL, USA*

**Kurt Dierickx,** *Catholic University of Leuven, Research Group for Microbial Adhesion, Department of Periodontology, School of Dentistry, Oral Pathology and Maxillofacial Surgery, Leuven, Belgium*

**Sabine Dobinsky,** MD, *Institut für Medizinische Mikrobiologie und Immunologie, Universitäts-Krankenhaus Eppendorf, University of Hamburg, Germany*

**Ronald J. Doyle,** PhD, *Professor, Department of Microbiology, University of Louisville Health Sciences Center, Louisville, KY, USA*

**W. Michael Dunne, Jr.,** PhD, *Division Head of Microbiology, Department of Pathology, Henry Ford Hospital, Detroit, MI, USA*

**Charles E. Edmiston, Jr.,** PhD, *Associate Professor, Department of Surgery, Medical College of Wisconsin, Milwaukee, USA*

**Arjuna N. B. Ellepola,** BDS, *Lecturer, Department of Oral Medicine and Periodontology, University of Peradeniya, Peradeniya, Sri Lanka and Oral Bio-Sciences, The Prince Philip Dental Hospital, University of Hong Kong, Hong Kong*

**David A. Elliott,** PhD, *Graduate Student, Department of Pathology, Johns Hopkins School of Medicine, Baltimore, MD, USA*

**Theresa A. Fassel,** PhD, *Technical Research Associate, Core Electron Microscope Unit, The Scripps Research Institute, LaJolla, CA, USA*

**Richard J. Friedman,** MD, FRCS(C), *Professor, Department of Orthopaedic Surgery, Medical University of South Carolina, Charleston, SC, USA*

**Taku Fujiwara,** DDS, PhD, *Assistant Professor, Department of Oral Microbiology, Osaka University, Faculty of Dentistry, Osaka, Japan*

**Manal M. Gabriel,** DDS, PhD, *Research Associate, Biology Department, Georgia State University, Atlanta, GA, USA*

**Barry J. Gainor,** MD, *Professor, Department of Orthopaedic Surgery, University of Missouri-Columbia, Columbia, Missouri, USA*

**Thomas A. Gardiner,** PhD, *Department of Ophthamology, The Queen's University of Belfast, United Kingdom*

**George Georgiou,** PhD, *Professor, Department of Chemical Engineering, University of Texas at Austin, Austin, TX, USA*

**Michael P. Goheen,** MSc, *Department of Pathology, Indiana University School of Medicine, Indianapolis, IN, USA*

**Sean P. Gorman,** PhD, *Professor, Pharmaceutical Devices Group, School of Pharmacy, The Queen's University of Belfast, United Kingdom*

**Travis Grant,** BSc(Hons), PhD, *Research Officer, Department of Microbiology and Immunology, University of Melbourne and the Department of Microbiology, Royal Children's Hospital, Melbourne, Victoria, Australia*

**Shigeyuki Hamada,** DDS, PhD, *Professor, Department of Oral Microbiology, Osaka University, Faculty of Dentistry, Osaka, Japan*

**David L. Hasty,** PhD, *Career Scientist, Veterans Affairs Medical Center, Professor of Anatomy and Neurobiology, Department of Anatomy and Neurobiology, University of Tennessee, Memphis, TN, USA*

**Shunji Hayashi,** MD, PhD, *Instructor, Department of Microbiology, Jichi Medical School, Tochigi-ken, Japan*

**Yoshikazu Hirai,** MD, PhD, *Professor, Department of Microbiology, Jichi Medical School, Tochigi-ken, Japan*

**Matthias A. Horstkotte,** MD, *Institut für Medizinische Mikrobiologie und Immunologie, Universitäts-Krankenhaus Eppendorf, University of Hamburg, Germany*

**Bernd Jansen,** MD, PhD, *Professor, Department of Hygiene and Environmental Medicine, University of Mainz, Mainz, Germany*

**David S. Jones,** PhD, *Pharmaceutical Devices Group, School of Pharmacy, The Queen's University of Belfast, United Kingdom*

**M. Bernice Kaack,** PhD, *Research Scientist, Tulane Regional Primate Research Center, Covington, LA, USA*

**Subramanian Karthikeyan,** *Department of Applied Microbiology and Food Science, University of Saskatchewan, Saskatoon, Saskatchewan, Canada*

**Shigetada Kawabata,** DDS, PhD, *Associate Professor, Department of Oral Microbiology, Osaka University, Faculty of Dentistry, Osaka, Japan*

**Kathrin Kiel,** MSc, *Institut für Medizinische Mikrobiologie und Immunologie, Universitäts-Krankenhaus Eppendorf, University of Hamburg, Germany*

**Johannes K.-M. Knobloch,** MD, *Institut für Medizinische Mikrobiologie und Immunologie, Universitäts-Krankenhaus Eppendorf, University of Hamburg, Germany*

**Wolfgang Kohnen,** PhD, *Senior Researcher, Department of Hygiene and Environmental Medicine, University of Mainz, Mainz, Germany*

**Darren R. Korber,** PhD, *Department of Applied Microbiology and Food Science, University of Saskatchewan, Saskatoon, Saskatchewan, Canada*

**Robert A. Latour,** PhD, *Associate Professor, Department of Bioengineering, Clemson University, Clemson, SC, USA*

**Franklin D. Lowy,** PhD, *Professor of Medicine, Montefiore Medical Center, Albert Einstein College of Medicine, Bronx, NY, USA*

**Dietrich Mack,** MD, *Associate Professor, Institut für Medizinische Mikrobiologie und Immunologie, Universitäts-Krankenhaus Eppendorf, University of Hamburg, Germany*

**Paul E. March,** PhD, *Senior Lecturer, School of Microbiology and Immunology, The University of New South Wales, Sydney, Australia*

**Roger E. Marchant,** PhD, *Professor, Department of Biomedical Engineering, Case Western Reserve University, Cleveland, OH, USA*

**Kylie L. Martin,** BSc, *Research Specialist, Orthopaedic Research Laboratory, Department of Orthopaedic Surgery, Medical University of South Carolina, Charleston, SC, USA*

**Katharine Merritt,** PhD, *Senior Scientist, Division of Life Sciences, Office of Science and Technology, Food and Drug Administration, Gaithersburg, MD, USA*

**Syed F. Mohammad,** PhD, *Professor, Department of Pathology, University of Utah, Salt Lake City, UT, USA*

**Lucio Montanaro,** MD, PhD, *Professor, Laboratory for Biocompatibility Research on Implant Materials, Istituti Ortopedici Rizzoli, Bologna, Italy*

**Keiji Oguma,** MD, PhD, *Professor, Department of Bacteriology, Okayama University Medical School, Okayama, Japan*

**William G. Pitt,** PhD, *Professor, Chemical Engineering Department, Brigham Young University, Provo, UT, USA*

**Maria Cristina Plotkowski,** PhD, *Associate Professor, Department of Microbiology and Immunology, State University of Rio de Janeiro, Brazil*

**Edith Puchelle,** PhD, *INSERM U 514, Reims University, Reims, France*

**Marc Quirynen,** *Professor, Catholic University of Leuven, Research Group for Microbial Adhesion, Department of Periodontology, School of Dentistry, Oral Pathology and Maxillofacial Surgery, Leuven, Belgium*

**Issam I. Raad,** MD, *Associate Professor, Infection Control, Department of Medical Specialties, University of Texas M.D. Anderson Cancer Center, Houston, USA*

**Anneta P. Razatos,** PhD, *Assistant Proffessor, Department of Chemical Engineering, Arizona State University, Tempe, AZ, USA*

**Gregor Reid,** PhD, MBA, *Associate Professor, Department of Microbiology and Immunology, University of Western Ontario Health Sciences Center, London, Canada*

**James A. Roberts,** MD, *Professor of Urology, Tulane University School of Medicine, New Orleans, LA, and Senior Research Scientist, Tulane Regional Primate Research Center, Covington, LA, USA*

**Roy Robins-Browne,** MB, BCh, PhD, *Professor of Microbiology and Immunology, University of Melbourne; Director of Microbiological Research, Royal Children's Hospital, Melbourne, Victoria, Australia*

**Lakshman P. Samaranayake,** BDS, DDS, FRCPath, *Chair Professor of Oral Microbiology, The Prince Philip Dental Hospital, University of Hong Kong, Hong Kong*

**Peter Schäfer,** MD, *Institut für Medizinische Mikrobiologie und Immunologie, Universitäts-Krankenhaus Eppendorf, University of Hamburg, Germany*

**David Silverhus,** MD, *Surgical Microbiology Research Laboratory, Department of Surgery, Medical College of Wisconsin, Milwaukee, WI, USA*

**W. Andrew Simpson,** PhD, *Director, Orthopaedic and Infectious Diseases Research Laboratory, Harry S. Truman Memorial Veterans' Hospital, Columbia, Missouri; Associate Professor, Departments of Internal Medicine and Molecular Microbiology and Immunology, University of Missouri-Columbia, Columbia, MO, USA*

**Chris H. Sissons,** PhD, *Senior Research Fellow, Dental Research Group, Department of Pathology, Wellington School of Medicine, University of Otago, Wellington South, New Zealand*

**Piotr Skowronski,** *Undergraduate Student, College of Charleston, Charleston, SC, USA*

**Mark S. Smeltzer,** PhD, *Associate Professor, Department of Microbiology and Immunology, University of Arkansas for Medical Sciences, Little Rock, Arkansas, AR, USA*

**Doron Steinberg,** PhD, *Senior Lecturer, Department of Oral Biology, Faculty of Dentistry, Hebrew University-Hadassah, Jerusalem, Israel*

**Fred J. Stutzenberger,** PhD, *Professor, Department of Microbiology and Molecular Biology, Clemson University, Clemson, SC, USA*

**Toshiro Sugiyama,** MD, PhD, *Assistant Professor, Department of Internal Medicine, Hokkaido University School of Medicine, Sapporo, Japan*

**Stephen E. Truesdail,** PhD, *Department of Chemical Engineering & NSF Engineering Research Center for Particle Science and Technology, University of Florida, Gainesville, FL, USA*

**David Uyeno,** MD, *Surgical Microbiology Research Laboratory, Department of Surgery, Medical College of Wisconsin, Milwaukee, WI, USA*

**Katanchalee Vacheethasanee,** PhD, *PhD Student, Department of Biomedical Engineering, Case Western Reserve University, Cleveland, OH, USA*

**Daniel van Steenberghe,** *Catholic University of Leuven, Research Group for Microbial Adhesion, Department of Periodontology, School of Dentistry, Oral Pathology and Maxillofacial Surgery, Leuven, Belgium*

**Pierre E. Vaudaux,** PhD, *Professor and Chief, Antibiotics and Microbial Research Laboratory, Department of Medicine, Division of Infectious Diseases, University Hospital, Geneva, Switzerland*

**Gideon M. Wolfaardt,** PhD, *Department of Microbiology, University of Stellenbosch, Stellenbosch, South Africa*

**Lisa Wong,** MSc, *Research Fellow, Dental Research Group, Department of Pathology, Wellington School of Medicine, University of Otago, Wellington South, New Zealand*

**Kenji Yokota,** PhD, *Instructor, Department of Bacteriology, Okayama University Medical School, Okayama, Japan*

**Jian-Lin Yu,** MD, PhD, *Infectious Disease Division, Massachusetts General Hospital, Harvard Medical School, Boston, MA, USA*

# PART I

## MECHANISMS OF MICROBIAL ADHESION AND BIOFILM FORMATION

# Mechanisms of Bacterial Adhesion and Pathogenesis of Implant and Tissue Infections

**Yuehuei H. An,[1] Richard B. Dickinson,[2] and Ronald J. Doyle[3]**

[1]*Department of Orthopaedic Surgery, Medical University of SC, Charleston, SC, USA*
[2]*Department of Chemical Engineering, University of Florida, Gainesville, FL, USA*
[3]*Department of Microbiology, University of Louisville, Louisville, KY, USA*

## I. INTRODUCTION

The research of bacterial adhesion and its significance is a large field covering different aspects of nature and human life, such as marine science, soil and plant ecology, food industry, and most importantly, the biomedical field.[3,23,101] Adhesion of bacteria to human tissue surfaces and implanted biomaterial surfaces is an important step in the pathogenesis of infection.[3,50]

Like tissue cells growing in in vitro culture, bacteria prefer to grow on available surfaces rather than in the surrounding aqueous phase, as first noted by Zobell in 1943.[173] Stable adhesion of a bacterium on a surface requires the following events: transport to the vicinity of the substrate (tissue cells or biomaterial surfaces), attachment to the substrate, then by molecular interactions the resistance of detachment in the presence of any dislodging force. Adhesion of bacteria to solid surfaces has been described as a two-phase process including an initial, instantaneous, and reversible physical phase (phase 1) and a time-dependent and irreversible molecular and cellular phase (phase 2).[111,112]

The exact mechanisms by which implant or foreign body infections occur still remain unclear. The mechanism of bacterial adhesion is a very complicated topic. The excellent reviews of the mechanisms of bacterial adhesion by Dankert et al.,[25] and the mechanisms of prosthetic infection by Dougherty,[39,40] and Gristina et al.,[60,65] will be very helpful to the reader. It is known that certain strains of bacteria, particularly *Staphylococcus epidermidis*, secrete a layer of slime after adhering to the implant surface, making themselves less accessible to the host defense system[3,58] and significantly decreasing antibiotic susceptibility.[62,63,119,124,144] These bacteria can remain quietly on the material's surface for a long period of time until the environment allows them to overgrow, such as with decreased host immune function or poor tissue ingrowth around the prosthesis, and a clinical infection then occurs.

Microbial adherence to tissue cells or mucosal surfaces is the initial step in the development of most infectious conditions.[22] Specific interactions between microbial

*Handbook of Bacterial Adhesion: Principles, Methods, and Applications*
Edited by: Y. H. An and R. J. Friedman © Humana Press Inc., Totowa, NJ

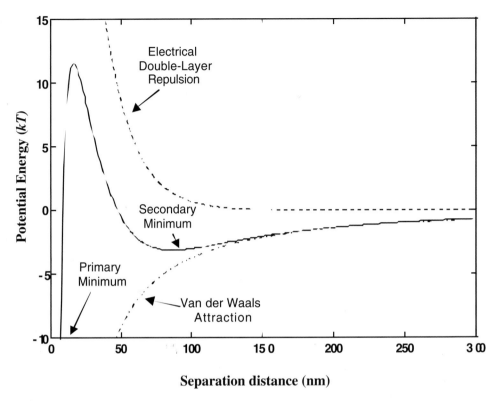

**Figure 1.** A hypothetical plot of the free energy of interaction, *f(z)*, between a colloidal particle and a surface vs separation distance based on DLVO theory for a particle and substratum with like charge. The dashed line is the contribution from electrostatic double layer interaction, and the dash-dot line is the contribution from van der Waals attraction. The primary minimum (not shown) reflects a balance between short range repulsion forces and short range van der Waals attractive forces at molecular contact, and the secondary minimum reflects a balance between longer range electrostatic repulsion and van der Waals attraction at a point away from contact. The characteristic time for attachment scales exponentially with the height of the interceding energy barrier.

surface ligands or adhesins and host receptors influence the distribution of microbes in the sites of infection. By adherence and adhesion, microbes gain access to host tissues, upset the integrity of the cells of the organ surface and cause tissue infection. The induction of mucosal epithelial inflammation is one aspect of this process. Bacterial attachment to mucosal surfaces activates the production and release of pro-inflammatory cytokines which can cause both local and systemic inflammation.[69,147,152]

Over the years, significant work has been done to investigate the process of bacterial adhesion to tissue and biomaterial surfaces, but still many questions remain unanswered.[25,111]

## II. MECHANISMS OF BACTERIAL ADHESION

### A. Physicochemical Interactions (Phase 1)

A common approach to predicting or interpreting bacterial attachment data has been to apply concepts developed in the colloid literature.[14,98,142,160] The long-range forces that are typically considered to govern the rate of deposition of a charged colloidal particle to a surface include long-range van der Waals forces and electrical double layer forces, which collectively form the basis for the well-known DLVO theory (after Derjaguin, Landau, Verway, and Overbeek[31,163]). van der Waals forces are generally attractive and result from induced dipole interactions between molecules in the colloidal particle and molecules in the substrate. Electrical double layer forces result from the overlap of counter-ion clouds near charged surfaces and the change in free energy as the surfaces are moved closer or farther apart. The result is an attractive force for like-charged surfaces and a repulsive force for oppositely charged surfaces.

Theoretical expressions based on DLVO theory for quantitative prediction of force and potential energy as a function of separation distance have been developed and validated for ideal systems where the assumptions of homogeneity, uniformly-charged, molecularly-flat surfaces are good approximations. However, these assumptions are likely to be strongly violated for real bacteria, as discussed below; therefore the actual nature of the long-range forces may be substantially different. Nonetheless, we discuss DLVO theory here only to provide a useful conceptual framework for understanding the relationship between long-range interaction forces and attachment of a colloidal particle to a surface.

Figure 1 is a hypothetical plot of the interaction energy, $f(z)$, as a function of separation distance, $z$, for a charged colloidal particle in a weak electrolyte near a flat plate accounting for contributions from the electrical double-layer repulsion and van der Waals attraction. The force on the particle at any separation distance, $F(z)$, is proportional to the gradient of the potential energy, i.e., $F(z) = -df(z)/dz$. For like-charged surfaces, characteristic features of this plot are two minima separated by a maximum. The two minima correspond to positions of mechanical equilibrium: the secondary minimum results from a balance between long-range attractive and repulsive forces away from contact, and the primary minimum results from a balance between short range attraction and repulsive forces usually considered to be at the point of contact. In this view, the energy of primary minimum is the change in free energy upon the formation of a new interface between the particle and surface. Therefore, a thermodynamically stable primary minimum exists only if the free energy of the new interface is less than that of the two individual surfaces in contact with the medium.

The rate at which particles will reach the primary minimum is determined by how quickly Brownian thermal fluctuations will provide the energy necessary to cross the energy barrier. It can be shown theoretically[131] that the rate at which particles spontaneously cross the energy barrier and attach decreases exponentially with the height of the energy barrier, analogous to an Arrhenius reaction rate expression. Another viewpoint is that initial deposition corresponds to the cell reaching the secondary minimum to become weakly associated with the surface before the time-dependent, long-range molecular binding begin to create irreversible interactions (phase 2),[112] which is in essence a deepening of the secondary minimum over time.

Because of the heterogeneous macromolecular structure of the cell surface, there are a number of serious limitations in applying theoretical expressions based on DLVO theory to make quantitative predictions of bacterial attachment. Several examples are listed below.

1. The concept of "separation distance" requires a clear delineation between "surface" and "medium"; however, the reference point to identify the separation distance is not obvious. For example, the charges involved in any electrostatic interations may be diffusely dispersed in the cell glycocalyx instead of in one flat layer.
2. The characteristic length scale of the forces (<1 nm for electrical double layer interactions in physiological electrolyte solution) is often smaller than the characteristic size of the macromolecules in the interface between the bacterium and the surface.
3. The steric or bridging interactions of macromolecules in the interface (e.g., the glycocalyx) may dominate any DLVO forces.[28]
4. The parameters required such as zeta potential and Hamaker constant can be difficult to estimate and reflect "average" properties of the cell surface, whereas macromolecules involved in the initial contact may have substantially different properties or interaction forces.[14] This is especially true for adhesin-mediated bacterial attachment, where the initial attachment may be due to the formation of a few specific higher-energy macro-molecular bonds.[34] Similarly, non-specific hydrogen bonding or hydrophobic interactions between cell surface appendages and sites on the substrate may also result in a long-range bridging interaction, which explains the common observation that short-range hydro-phobic forces can be involved in determining the rate of bacterial deposition to surfaces.[169]
5. DLVO expressions that predict force versus separation distance assume that cell-substrate interaction is at thermodynamic equilibrium at any separation distance. However, as a cell approaches a surface, a significant amount of time may be required for macromolecules in the interface to bind to the substrate or to rearrange to lower free–energy configurations.[92,93]

For these reasons, it is not surprising that application of DLVO theory to bacterial attachment studies has yielded mixed results, even in predicting of qualitative trends. More detailed and realistic theories that account for the macromolecular structure of the cell surface and the discrete binding interactions or steric repulsion due the macromolecules in the interface are needed. (For an example of such an approach refer to the work by Dickinson[33] for a model that accounts for both macromolecular bridging and DVLO forces in particle deposition.) However, advancement of such theories will require more sophisticated force-measurement techniques, such as those described in Chapters 18 and 19.

It was reported that atomic force microscopy (AFM) can be used to analyze the initial events in bacterial adhesion with unprecedented resolution.[133] Interactions between the cantilever tip and confluent monolayers of isogenic strains of *Escherichia coli* mutants exhibiting subtle differences in cell surface composition were measured. It was shown that the adhesion force is affected by the length of core lipopolysaccharide molecules on the *E. coli* cell surface and by the production of the capsular polysaccharide, colanic acid. Furthermore, by modifying the atomic-force microscope tip a method was developed for determining whether bacteria are attracted or repelled by virtually any biomaterial of interest. This information will be critical for the design of materials that are resistant to bacterial adhesion.[133]

## B. Molecular and Cellular Interactions (Phase 2)

### 1. Adhesion to Solid Surface

In the second phase of adhesion, molecular reactions between bacterial surface structures and substratum surfaces become predominant. It implies a firmer adhesion of bacteria to a surface by the bridging function of bacterial surface polymeric structures, which include capsules, fimbriae or pili, and slime. In fact, the functional part of these structures should be adhesins which are part of these structures, especially when the substrata are host tissues. To understand the structure, composition, and functions of these surface elements, the cell wall description of Gram-negative bacteria by Wicken will be useful.[168] The bacterial cell wall has three regions, namely an outer membrane, a peptidodoglycan monolayer, and the inner plasma membrane. The two membranes may show continuity at various attachment points and may also be bridged by the basal bodies of flagellae.

Bacterial capsules, firmly adherent as a discrete covering layer with a distinct margin on the bacterial cells (at the outside of the cell wall), are separated from extracellular slime. They can be seen in stained or unstained preparations as a clear zone surrounding the bacterial cell. They occur in both Gram-negative and Gram-positive bacteria. Most capsules are composed of polysaccharides and proteins.[100,151,158] In a number of studies it has been suggested that cell surface polysaccharides and proteins act as bacterial adhesins.[53,72,100] From a surface hydrophobicity point of view, different encapsulated coagulase-negative staphylococci (CNS) have different surface hydrophobicity and their hydrophobicity and adhesion appears to be reduced by proteolytic enzyme treatments, indicating the involvement of proteins. The presence of hydrophilic capsules reduced adhesion.[70-72] It is suggested that the capsules of both *Staphylococcus aureus* and *S. epidermidis* are important factors in the pathogenesis of these bacteria. [81]

Fibrillae are the more amorphous surface adhesive structures which appear to lack the regular filamentous forms of fimbriae. Fibrillar structures has been observed at the surface of various streptococci and they are thought to contribute to the better adhesion onto hydroxyapatite substrata.[129] Sex pili are proteinaceous appendages which occur on donor strains of bacteria and whose formation is determined by the presence of conjugative plasmids.[149,168] Many different types have been described on the basis of their plasmid determinants, diameter (8.5–9.5 nm), length (2–20 nm), and attachment of specific RNA and DNA phages. Flagella are composed of many polypeptides forming a filamentous hook (diameter: 20 nm) and a complex basal body which interacts with both outer and inner membranes.[9,149]

Fimbriae (or pili) are a group of rigid, straight, filamentous appendages on a bacterial surface and are often no more than 4 to 7 nm in diameter and from 0.2 up to 20 nm in length. The filamentous nature of fimbriae is often demonstrated by electron microscopy. They arise from proteins in the outer surface of the outer membrane and may extend for considerable distances beyond the outer membrane. On the bacterial surface usually several hundred to one thousand fimbriae are present.[13,47] Bacterial fimbriae are polymers, composed primarily of identical protein subunits called pilin.[86] They are most prominent in gram-negative bacteria and are believed to be a major adhesive structure on bacterial surface. For detailed description of different types of fimbriae, one may consult the reviews by Ørskov and Krogfelt.[99,123]

Fimbriae may mediate bacterial adhesion by adhesins associated with fimbriae, such as F71, F8, and F13 adhesins,[73,117] or by adhesive subunits (adhesins), such as the K88 fimbriae[29] and SS142 fimbriae.[115] The nature of adhesins at large appeared more complex when bacteria were found to be adhesive but did not express fimbriae. It was possible to isolate adhesins (nonfimbrial adhesins) from these bacteria which did not form electron microscopically recognizable structures. Fimbriae mediate adhesion also by fimbriae-dependent surface hydrophobicity, since fimbriated bacteria are often more hydrophobic than nonfimbriated ones and hydrophobicity is dependent on the types of fimbriae present.[106] Bacterial fimbriae are believed to be related to virulence. Gram-negative bacteria possessing fimbriae are more infectious than their nonfimbriated variants.[66,110]

Slime is defined as an extracellular substance produced by the bacteria which may be partially free from the bacteria after dispersion in a liquid medium.[46] Capsules and slime are both subclasses of extracellular polymeric substances and are usually polysaccharides. An accumulated biomass of bacteria and their extracellular materials (slime) on a solid surface is called a biofilm. The slime produced by CNS is a loose hydrogel of polysaccharides associated through ionic interactions.[61]

Bacterial strains not producing slime are less adherent and less pathogenic.[18,19,26,89] Slime is thought to be important for intercell connection during surface colonization.[76,77] Bacteria washed to remove slime showed á similar adhesion onto fluorinated poly(ethylenepropylene) (FEP) comparing to nonwashed bacteria.[72] The current concept is that the production of slime will be especially important for events after the initial phase of adhesion, including colonization of various surfaces, protection against phagocytosis, interference with cellular immune response, and reduction of antibiotic effects. CNS which do not quickly adhere to the surfaces are rapidly killed by the immune system. Slime-forming CNS are less susceptible to vancomycin and other antibiotics following adhesion to biomaterials than bacteria grown in culture.[119] Such antibiotic resistance may be partly due to the slow growth rate of CNS in biofilm or to the limited transport of nutrients, metabolites and oxygen to and from the biofilm surface. Biofilm on biomaterial surfaces also protects bacteria by sequestering them from defense systems, including inhibition of PMN chemotaxis, phagocytosis and oxidative metabolism; depression of immunoglobulin synthesis; mononuclear lympho-proliferative response; killer cell cytotoxicity; and antibiotic action. Chronic infection occurs when a bacterial inoculum reaches critical size and overcomes the local host defenses.

Specific adhesion can be defined as the selective binding between bacterial adhesin (a specific molecular component on bacterial surface) and substratum receptor (a specific component on material surface or tissue surface), and is less affected by many common environmental factors such as electrolytes, pH, or temperature. Specific adhesion of bacteria is believed to be important in the pathogenesis of prosthetic infections.[155,156] Morris and McBride found that saliva-coated HA has two specific receptors for *S. sanguis*[118] and this bacterium has distinct sialic acid lectins (adhesins) which are responsible for the adhesion onto saliva-coated HA.[104,118] Specific adhesion was also proposed by Bayer et al. in 1983 by studying the process of *Clostridium thermocellum* adhesion to cellulose.[8] The adhesion was determined to be selective for cellulose as the observed adhesion was not significantly affected by various parameters including salts, pH, temperature, detergent, or soluble sugar, and seemed to be due to the specific interaction between the cellulose-binding factor (CBF) of the bacteria and surface of the cellulose substrata.[8] A capsular polysaccharide adhesin from *S. epidermidis* strain RP62A was isolated by Tojo et al. in 1988.[156] The adhesin was composed of a complex mix of monosaccharides, bound well to silastic catheter tubing, inhibited adhesion of strain RP62A to the catheter, and elicited antibodies that both blocked adherence and stabilized an extracellular structure that appeared to be a capsule. Timmerman et al.[155] also found a proteinous adhesin of a strain of *S. epidermidis* (strain 354) which mediated the adhesion of this bacterium to polystyrene.

## 2. Adhesins and Adhesion to Host Tissue

The past few decades has resulted in a reasonable understanding of adhesion of some bacterial pathogens. Early studies were phenomenological, in that it was useful to determine if a particular bacterium could hemagglutinate, or bind to coated particles. Later studies revealed that adhesins were located on defined structures, such as fimbriae or outer membranes. Subsequently, some adhesins were isolated and characterized. Adhesion studies now are at the molecular level, with attempts to clone, sequence and study regulation of adhesin expression. Some efforts are being made to develop adhesin-based vaccines and to employ adhesin inhibitors in therapy.

There are several generalizations that can be made about bacterial adhesion and adhesins:

1. All bacteria seem to possess the ability to make multiple adhesins. The need for multiple adhesins is understood when considering the tendency bacteria have to adhere. Phase variations in bacteria enable a bacterium to escape from one environment and become adherent in another. Some bacteria produce multiple adhesins at the same time, raising the probability of successful colonization of a surface.

2. Bacterial adhesins penetrate into the medium environment from a scaffold. Bacterial fimbriae are scaffolded onto the outer membrane, with adhesins often found on the very tips of the fimbriae. The fimbriae are particularly well-suited to interact with various soluble or insoluble substrata. The extension of fimbriae away from the cell surface helps the bacterium overcome repulsive forces. Sometimes adhesins are located on fibrillae, short, stubby appendages emanating from the cell surface. Fibrillae are anchored in either the cell wall or in the cytoplasmic membrane only to penetrate the wall. Fibrillae may carry lectins or they may recognize protein sequences. Fibrillae are more common in Gram-positive than in Gram-negative bacteria. "Bald" bacteria, devoid of surface appendages, such as fimbriae or fibrillae, have a low tendency to adhere to various surfaces. *Bacillus subtilis* is an example if a bacterium with little tendency to bind to animal tissues. Its spore however, binds to various surfaces because the spores are hydrophobic.[44]

3. Adhesion does not dictate tropism. *Escherichia coli* is no better able to adhere to intestinal cells than it is to buccal cells, yet its natural environment is the intestine. Most *E. coli* in fact, can adhere to any mannosylated surface. Tropism is dictated by nutritional requirements, secretions, nearest neighbors, and other factors as well as by adhesion.

4. Adhesion is the preferred way of bacterial life. Biofilms are unable to develop in the absence of adhesion. Adhesion seems to commonly (but not always) endow a bacterium with greater resistance to enzymes (lysozyme, etc), antibodies, antibiotics, and disinfectants. In some cases, the adherent bacterium is the superior competitor for nutrients.[170]

Table 1 outlines some of the best-studied mechanisms of bacterial adhesion. All of them may be involved in biofilm development. Interactions between a bacterium and a substratum may involve the hydrophobic effect[43] between hydrophobin adhesins and receptors. The hydrophobic effect is said to be nonspecific because any hydrophobin adhesin will interact with any hydrophobic receptor. It must be kept in mind, however, that the specificity determinants of the hydrophobic effect are poorly understood. Future research may reveal that complementary adhesin–receptor complexes may have subtle stereo-chemical restraints, conferring specificity. The hydrophobic effect reduces the repulsive forces between interacting species, and probably contributes to most bacterium–substratum

**Table 1. Some Proposed Mechanisms of Microbial Adhesion***

| Adhesin | Receptor | Characteristics |
|---------|----------|-----------------|
| Hydrophobin | Hydrophobin | Nonspecific, temperature-dependent, high entropy change |
| Lectin | Glycoconjugate | Specific for certain carbohydrates, reversible |
| MSCRAMM | Proteins (ECMs) | Membrane or wall-bound proteins, protein or glyco-conjugate receptors |
| Protein | RGD sequence | RGD-containing peptides inhibit |
| Polyelectrolyte | Polyelectrolyte | Salts tend to inhibit |

*Derived from Dawson and Ellen 1990;[27] Doyle and Rosenberg 1990;[43] Hasty et al 1992;[68] Ofek & Doyle 1994,[122] and Patti et al 1994.[125]

complexes. The hydrophobic effect is much more pronounced at 37°C than at 0–4°C, making it possible to assess the extent of the effect in an adhesive event. Interactions between a bacterium and a surface depending solely on the hydrophobic effect are not known. Even the binding of bacteria to oil droplets involves multiple interacting components, witnessed by the tendency of bacteria to adhere next to each other on oil droplets.[43] Missing components in research on hydrophobins are studies on the molecular bases of adhesin expression and presentation. In protein hydrophobins there may be clefts, akin to clefts for lectin saccharide binding sites of lectins. In many lectin binding sites, hydrophobic amino acids surround the carbohydrate-specific site rendering that site stable.

As regards lectins in adhesion, there are numerous specificities known (Table 2). The most common appears to be mannose-specific. Many Enterobacteriaceae possess mannose lectins in their fimbriae.[122] In *E. coli*, the lectin prefers hydrophobic mannosides, whereas in salmonellae, there appears to be no tendency to bind hydrophobic mannosides. Recent work suggests that a truncated Fim H (the lectin of type 1 fimbriae of *E. coli*) is a good candidate for a vaccine. Thus far, vaccines based on lectin adhesins are not promising. Many of the lectins are not immunogenic and strain variability is common. The role of lectin adhesins in biofilms is unknown. Some biofilm bacteria may express lectins, although specific saccharides do not often disrupt the biofilm structures. It is likely the initial adhesive events, leading to biofilm formation, can be inhibited by saccharides but once the biofilm is formed, the saccharides are unable to dissociate the bacteria from their environment. Detachment is known to be more difficult, compared to adhesion, even when *E. coli* binds to tissue culture cells.[122]

Lectin adhesins have not only been described in enteric bacteria, but also in members of the genera *Pseudomonas, Vibrio, Fusobacterium, Helicobacter, Haemophilus, Staphylococcus, Streptococcus, Actinomyces* and numerous others. Specificities are known for Gal, Gal-1,4Gal, GalNAc, GlcNAc, Fuc, sialic acids and α-1,6 glucan. *Streptococcus sobrinus* has an interesting lectin in that the lectin binds to 7 to 10 internally-linked α-1,6 glucose residues.[45] The α-1,6 glucans are derived from extracellular streptococcal glucosyltransferases acting on sucrose. *S. sobrinus* is one of the few bacteria that can make its own adhesin (glucan-binding lectin or GBL) and substratum (α-1,6 glucan). This combination may play a prominent role in the development of dental plaques. Some oral bacteria use lectins to coaggregate, forming large masses of cells in subgingival spaces. It is not clear that these masses constitute a biofilm, even though

## Table 2. Some Common Adhesins of Bacterial Pathogens

| Bacterium | Adhesin | Cell location | Specificity |
|---|---|---|---|
| *Aeromonas* sp. | Protein | Surface array | Hydrophobin |
| *E. coli* | Type 1 (lectin) | Fimbrium | Mannose |
| | Type P (lectin) | Fimbrium | Gal-1,4 Gal |
| | Type G (lectin) | Fimbrium | GlcNAc |
| | Type S (lectin) | Fimbrium | Sialic acid |
| | LPS | Outer membrane | Unknown |
| *Mycoplasma pneumoniae* | PI protein | Cytoplasmic membrane | Sialic acids, sulfated glycolipids |
| *Neisseria gonorrhoeae* | PII protein | Outer membrane | Glycolipids |
| *Pseudomonas aeruginosa* | Lectins | Cytoplasmic origin | Fuc,Gal, hydrophobic saccharides |
| *Staph. aureus* | MSCRAMM | Cell wall | Collagen |
| | MSCRAMM | Cell wall | Laminin |
| | MSCRAMM | Cell wall | Vitronectin |
| *Staph. saprophyticus* | Lectin | Cell wall | GalGlcNAc |
| *Strept. sanguis* | Fim (lectin) | Fibrillin | Sialic acid |
| | Hydrophobin | Cell wall | Hydrophobin |
| *Strept. sobrinus* | Lectin | Cell wall | $(Glc-1,6)_{6-10}$ |
| *Strept. pyogenes* | M-protein | Cytoplasmic membrane, extending into cell periphery | Fibronectin |
| | LTA | | Fibronectin |

Results derived from Altwegg and Geiss, 1989;[1] Drake et al., 1988;[45] Hasty et al., 1992;[68] Jacques 1996;[78] Kahane et al., 1985;[87] Krogfelt, 1991;[99] Ma et al., 1995;[108] Nesbitt et al., 1982;[120] Ofek and Doyle, 1994;[122] Patti et al., 1994[128]

Abbreviations:
   Fuc, fucose
   Gal, galactose
   Glc, glucose
   MSCRAMM, microbial surface component recognizing adhesive matrix molecules;
   NAc, N-acetylglucosamine
   LPS, lipopolysaccharide
   LTA, lipoteichoic acid

natural biofilms usually consist of mixed cellular communities. It is more likely the coaggregating bacteria are able to resist phagocytosis.[42]

Adhesins must be surface-exposed in order to serve an adhesive function (Table 2). One bacterium, *P. aeruginosa*, synthesizes two internal lectins called PA-I and PA-II. These lectins are synthesized only when carbon in the medium becomes limiting or when the cell density becomes high, suggesting a possible quorum-sensing type of regulation.

When the cells are no longer able to maintain a protonmotive force (PMF), a portion of the population lyses, releasing the internal lectins. The released lectins can weakly bind to nonlysed bacteria, enabling the bacteria to adhere to glycoconjugate-containing substrata.[166] This type of adhesion requires the lysis of some bacteria in order to assist the adhesion of survivors. Whether such a mechanism of adhesion is involved in the adhesion of *P. aeruginosa* to lung tissues of cystic fibrosis patients is unknown, although *P. aeruginosa* biofilms are common in such patients. If the lectins do indeed contribute to biofilm development in cystic fibrosis (or other situations involving *P. aeruginosa*), it would be interesting to determine if lectin inhibitors could modify the course of the infection/biofilm.

Some staphylococci are able to adhere to various ECMs (Table 2). The staphylococcal adhesins (MSCRAMMs) or microbial surface components recognizing adhesive matrix molecules.[127] These adhesins are able to bind fibronectin, collagen, laminin, vitronectin, heparin, etc. One clone may express one or more adhesins. The adhesins are not lectins in that neutral sugars or saccharides do not inhibit. Although MSCRAMMs have been studied the most in staphylococci, other bacteria are known to possess similar adhesive proteins. The MSCRAMMs are anchored in the cytoplasmic membrane or tightly bound to the cell wall. If wall bound, the adhesins may be released by bacteriolytic enzymes, such as lysostaphin, lysozyme or mutanolysin. Sometimes a single MSCRAMM may complex with several ECMs. The general redundancy of MSCRAMMs, especially in staphylococci, may ensure successful adhesion of the bacterium. It is likely MSCRAMMs can contribute to biofilm formation, as ECMs are readily abundant in tissues where biofilms are normally found. There are several cloned and sequenced MSCRAMMs, and regions defining adhesin activity, wall binding motifs or membrane spanning sequences, identified. A recent review on the molecular biology of MSCRAMMs is available.[54]

Some bacteria produce cell surface proteins capable of recognizing RGD (arginine-glycine-aspartic acid) sequences. The RGD-binding proteins serve as adhesins, and RGD-containing peptides are generally good inhibitors of adhesion. Similarly, some bacteria express RGD-containing proteins, while some animal cell proteins are complementary to the sequence.[127] Fibronectin is known to possess an RGD sequence recognizable by group A streptococci, *Treponema denticola* and a few other pathogens. RGD-containing peptides are not known to modify biofilm masses. although it is likely an RGD peptide could have prevented adhesion leading to biofilm development involving some bacteria.

Finally, some bacteria are not able to express known lectins, MSCRAMMs or RGD-binding proteins. These bacteria are found more commonly in the nonanimal cell environments and their adhesive capacity may depend on the composite effects of charge interactions. Of particular importance is the adhesion of bacteria to ion-exchange resins, glass beads and ceramic supports.[52] The density of adherent bacteria bound to solid supports is much greater than the numbers of bacteria achievable in suspension. The high density, combined with metabolism enhanced by adhesion, has resulted in increased production of commercial products. Bacterial adhesion to solid supports may lead to biofilm, providing the adherent bacteria secrete polymeric matrix materials. A somewhat neglected area of study is biofilm physiology. How do bacteria in biofilms respond to feast and famine, ions, pH fluctuations, nearest neighbors, cold or heat shocks, osmotic changes, etc. More is known about the physiology of adherent bacterial masses from an industrial viewpoint than from a medical or environmental viewpoint.

As an example of a neglected area in biofilm research, it is known that some bacteria must obligatorily shed (turnover) their cell wall during growth.[41] The extent of this turnover can be as high as 50% of the wall material per generation. Turned-over wall components in biofilms include peptidoglycan fragments and teichoic acids. These components could supply nutrients for unrelated bacteria which are members of the biofilm community or the turnover components could bind to cells capable of competing with such glyconjugates. In Gram-positive bacteria, the turnover products cannot be reutilized and are simply shed into the surrounding medium. In Gram-negative bacteria, the turnover products may cause turgor on the outer membrane, resulting in the formation of blebs or membrane vesicles.[171] Membrane vesicles therefore contain muramyl peptides and lipopolysaccharide which can act synergistically to cause inflammation. Beveridge et al.[10] have suggested that membrane vesicles in biofilms may be predatory, as the vesicles also contain autolysins and periplasmic hydrolases. It is predicted that vesicle research in biofilm will soon see an increased activity.

## III. BACTERIAL ADHESION AND SUBSEQUENT IMPLANT OR TISSUE INFECTIONS

### A. Bacterial Adhesion and Subsequent Implant Infections

Based on several reports, staphylococci are the most important pathogens of implant infection. *S. epidermidis*, which was once recognized as a nonpathogen, is the cause of a large percentage of late or chronic infections, whereas *S. aureus* remains a common pathogen for prosthetic infections, especially those occuring relatively early on.[4,17,88] Our previous review shows a roughly equal incidence of *S. epidermidis* and *S. aureus* causing prosthetic hip joint infections, accounting for 50 to 60% of all infections since 1980.[2] This estimate agrees with the calculation by Sanderson.[139] A trend of increasing CNS infections is noted, rising from 13% in the 1970's to 25% in 1980's and 33% in 1990's.

Staphylococci are members of the *Micrococcaceae* family, characterized as Gram-positive, nonmotile, catalase-positive, coagulase-negative, aerobic or faculatively anaerobic cocci. Strains are distinguished by coagulase and mannitol fermentation test. *S. aureus* is coagulase-positive and *S. epidermidis* is negative. Coagulase-negative staphylococci (CNS) are a normal component of the skin flora, and *S. epidermidis* is the most common species and most predominant.[107] CNS are widely recognized as significant pathogens in patients with infections associated with orthopedic prosthesis or implants, prosthetic heart valves, vascular prostheses, cerebrospinal fluid shunts, urinary tract catheters, peritoneal dialysis catheters, and others. *S. aureus* causes more severe and more rapid infection than *S. epidermidis* and its effects may therefore be more clinically obvious at an earlier stage after surgery. *S. epidermidis* is less virulent and the clinical features less severe than that with *S. aureus*.

### 1. Bacterial Colonization of Biomaterial Surfaces

Bacterial adhesion to biomaterial surfaces is the initial step in the pathogenesis of prosthetic infections.[59] Generally, bacteria come from two sources. The most obvious one is direct contamination of the wound and prosthesis surfaces during surgery, when bacteria have a chance to reach these surfaces from the patient's skin and air.

The second type of contamination is hematogenous or lymphatic seeding from infections elsewhere in the body.[51,57,109] Theoretically, bacteria can reach the prosthesis as early as the time of operation or as late as few years after when an infection, such as periodontal abscess or urinary tract infection, occurs elsewhere in the body. Bartzokas et al.[7] reported four cases of prosthetic infection caused by strains of *S. sanguis*. For each patient the strain of the *S. sanguis* isolated from the mouth (severe periodontal disease and caries) was indistinguishable from that isolated from the infected prosthesis. In a recent report, two cases of hip arthroplasties were infected by *Mycobacterium tuberculosis* 18 months and 14 years after the surgery.[159] The primary source was pulmonary tuberculosis. It is very clear from these two reports that the infection was caused by hematogenous bacterial seeding.

Bacteria attach to material surfaces by the actions of physical forces (phase 1). If the local environment is suitable to bacteria, such as abnormal tissue integration and a weak host defense, bacteria will remain viable on the material surface and complete the second phase (phase 2) of adhesion by secreting exopolysaccharides.

### 2. Tissue Integration and Bacterial Invasion

Tissue biofilms are glycoproteinaceous-conditioning films consisting of albumin, collagen, fibronectin, vitronectin and other proteins.[145] Plasma proteins rapidly coat any biomaterial introduced into the body and modify the extent of bacterial adhesion. The expression "race for the surface" refers to a contest between tissue cell integration of, and bacterial adhesion to, an available implant surface.[64,65] Once a biofilm is established on the surface, it is not easily traumatized or altered. If tissue integration occurs first at the surface, the biomaterial becomes relatively resistant to subsequent bacterial colonization. If bacterial adhesion occurs first, then host cells can seldom displace the primary colonizers on that portion of the surface, establishing conditions for eventual infection.[24]

Thus, the susceptibility of biomaterials to infection is a function not only of the number and type of bacteria but also the time needed for tissue integration on the implanted surface versus the time needed for adhesion of bacteria to the same surface. When prosthetic loosening occurs due of technical or mechanical reasons, the local tissue integration is damaged by the process of loosening, and the local environment may become susceptible to bacterial duplication or hematogenous bacterial seeding and subsequent infection.

Many strains of CNS elaborate copious amounts of slime after adhesion to a surface. This slime/bacteria complex probably enhances the subsequent growth of bacteria on the surface of biomaterials and inhibits recovery of the bacteria during routine diagnostic procedures. The biofilm provides a favorable surface colonization by either the bacteria or eukaryotic cells, whichever makes contact first.[11,157] On biomaterials such as metallic alloys, binding sites are further modified by ionic and glycoproteinaceous constituents from the host environment. In addition to these specific interactions, exposed surfaces may also act as catalytic surfaces for close-range molecular and cellular activities. These interactions can be dramatically changed by variations in local pH, inflammation and tissue damage caused by such factors as surgery, trauma and infection.

Merritt et al.[113] reported that the presence of bacterial growth can have a significant effect on the corrosion of stainless steel, and raised the question that we should rethink the role of infection in complications following the use of metallic implants. Significant destruction of a hydroxyapatite (HA) coating after exposure to bacteria has also been reported, and this may lead to a better understanding of prosthetic loosening.[5,90] These findings raised questions and concerns about the damage to implant surfaces caused by bacteria and the effects to the host tissue by the subsequent harmful products produced from the interaction between bacteria and material surface. These effects can change the local physiological environment, such as pH value and chemical compositions of tissue fluid, and consequently stimulate bacterial duplication or aggravate existing infections. Due to their unique characteristics, the adherent bacteria can stay on the material surface for a long period of time because of their behavioral changes resulting in resistance to antibiotics[5,62,63,124,162] and the human immune system.[36,150,161]

### 3. Effects of Implants

The effects of an implant on the incidence of bacterial colonization and subsequent infection have been summarized by Dougherty[39,40] and include a foreign body reactivity leading to local tissue damage and inflammation, harmful effects of the implant to local host defenses and the effects of trapping and sequestering of bacteria. Foreign body reaction has long been recognized as a very important infection-promoting factor[80] and this concept has been verified both experimentally and clinically. The determinants of implant reactivity include chemical composition of the implant material,[20] surface characteristics such as surface configuration or particle size,[94] implantation site, and mechanical interactions with host tissues.[114]

Particulate wear debris is another important factor which can stimulate inflammation and facilitate prosthetic loosening. If sufficient stimulus exists, inflammation will occur.[32] The products of inflammation will trigger the local defense system to release tissue toxic enzymes and oxygen-free radicals. The later will further damage the local tissues. If precolonized or hematogenous seeded bacteria are present, infection may occur.

Contact between neutrophils and the implant surface can impair leukocyte bactericidal capacity.[114,172] Using a tissue cage model, Zimmerli et al.[172] demonstrated that, when compared to neutrophils from peripheral blood, polymorphonuclear neutrophils from sterile tissue cages showed decreased phagocytic and microbicidal activity. The mechanism of this effect from the implant is not clear, but some evidence exists suggesting that the local release of lysosomal enzymes and oxygen free radicals from leukocytes triggered by contact with the implant surface may damage leukocytes themselves,[6] or that certain metal ions like nickel or cobalt can interfere with bacterial phagocytosis by neutrophils.[132]

The variety and complexity of implant surface configurations provide bacteria with harbors which protect them from the impact of host defense systems and antibiotics. Porous and multifilament surfaces are examples of this effect, and these surfaces have a much higher implant-site infection rate.[84,103]

### 4. Effects of Host Endogenous Factors

The susceptibility of the human body to infection is high after implantation surgery since both the insertion of a foreign body (implant) and the tissue inflammation caused by the surgical procedure provide a favorable local environment for bacterial colonization. If the immune system functions normally and the local tissues are in a healthy state that allows for normal healing, there will be no bacterial colonization, no bacterial aggregation and no infection. Certain individuals are more predisposed to prosthetic infection, such as those with rheumatoid arthritis,[17,51,103,109] previous joint surgery,[15] previous joint sepsis,[84,91] remote infection at the time of surgery,[15] diabetes mellitus,[15] or those with an immune deficiency. These patients are especially susceptible to hematogenous infections.[57]

## B. Bacterial Adhesion and Subsequent Tissue Infections

The bacterial requirements for pathogenicity, as summarized by Smith,[147] include infection of the mucous surfaces (adhesion), entering the host through these surfaces, multiplication in the environment of the host, interference with host defense system, and damaging the host tissue.

Most bacterial infections are initiated by the adherence of microorganisms to host tissues. Bacteria usually attack the susceptible animal or human at mucosal surfaces of the oral cavity, respiratory, gastrointestinal, or genitourinary tract. To colonize these surfaces they must penetrate a number of nonspecific defense barriers including cleansing mechanisms such as sneezing, coughing, peristalsis and fluid flow. Some micro-organisms escape recognition by soluble immune or nonimmune molecules, and bind to the mucosal surfaces via specialized molecules exposed on their surface (adhesins) which recognize and interact with complementary molecules (receptors) on the surface of specific host cells. This is the key step in the pathogenesis of infectious diseases which is currently the subject of intensive investigation. Adherence, however, is also a virulence factor through which microbes gain access to host tissues, upset the integrity of the mucosal barrier, and cause disease. The induction of mucosal inflammation is one aspect of this process. Bacterial attachment to mucosal surfaces activates the production of pro-inflammatory cytokines that cause both local and systemic inflammation. Here the mechanism and the role of adherence in different bacterial infections are briefly discussed.

### 1. Oral Infections[67,83,97,102,167]

Adherence to a surface is a key element for colonization of the human oral cavity by the more than 500 bacteria recorded from oral samples.[167] In the oral cavity, there are three surfaces available: teeth, epithelial mucosa, and the nascent surface created as each new bacterial cell binds to existing dental plaque. Oral bacteria exhibit specificity for their respective colonization sites. Such specificity is directed by adhesin–receptor pairs on genetically distinct cells. Colonization is successful when adherent cells grow and metabolically participate in the oral bacterial community.

Streptococci express arrays of adhesins on cell surfaces that facilitate adherence to oral substrates. A consequence of this binding ability is that streptococci adhere simultaneously to a spectrum of substrates, including salivary glycoproteins, extracellular matrix and serum components, host cells, and other microbial cells. Adhesion facilitates colonization and may be a precursor to tissue invasion and immune modulation. Many of the streptococcal adhesins and virulence-related factors are cell-wall-associated proteins.[82,83]

Bacterial fimbriae have been shown to play an important role in the interaction between bacteria and host cells or among bacterial cells.[67] The fimbrial structure of *Porphyromonas gingivalis* is composed of 41-kDa fimbrillin proteins which exhibit a wide variety of biological activities including immunogenicity, binding to various host proteins, stimulation of cytokine production, and promotion of bone resorption. *Actinobacillus actinomycetemcomitans* also possesses fimbriae; however, little is known about their biological properties. Fimbriae of *Prevotella intermedia* are shown to induce hemagglutination, while those of *P. loescheii* are found to cause coaggregation with other bacteria. Fimbriae from Gram-positive oral bacteria such as oral *Actinomyces* sp. and *S. sanguis* may participate in coaggregation, binding to saliva-coated hydroxyapatite or surface glycoprotein of epithelial cells. Taken together, fimbriae are key components in cell-to-surface and cell-to-cell adherence of oral bacteria and pathogenesis of some oral and systemic diseases.[67]

*P. gingivalis* is a major etiological agent in the initiation and progression of severe forms of periodontal disease.[102] Colonization of the subgingival region is facilitated by the ability to adhere to available substrates such as adsorbed salivary molecules, matrix proteins, epithelial cells, and bacteria within an existing biofilm on tooth or epithelial surfaces. The binding may be mediated by various regions of *P. gingivalis* fimbrillin, the structural subunit of the major fimbriae. *P. gingivalis* is an asaccharolytic organism, with a requirement for hemin and peptides for growth. At least three hemagglutinins and five proteinases are produced to satisfy these requirements. Many of the virulence properties of *P. gingivalis* appear to be consequent to its adaptations to obtain hemin and peptides. Thus, hemagglutinins participate in adherence interactions with host cells, while proteinases contribute to inactivation of the effector molecules of the immune response and to tissue destruction. In addition to direct assault on tissue, *P. gingivalis* can also modulate eucaryotic cell signal transduction pathways, directing its uptake by gingival epithelial cells. Within this privileged location, *P. gingivalis* can replicate and impinge upon components of the innate host defense. A variety of surface molecules stimulate production of cytokines and other participants in the immune response. *P. gingivalis* may also undertake a stealth role whereby pivotal immune mediators are selectively inactivated.[102]

Most oral bacteria exhibit coaggregation, cell-to-cell recognition of genetically distinct cell types.[96,97] Many interactions appear to be mediated by a lectin on one cell type that interacts with a complementary carbohydrate receptor on the other cell type. A lactose-sensitive adhesin has been isolated from *Prevotella loescheii* PK1295. Other adhesins have been identified. One *S. sanguis* adhesin is a lipoprotein that appears to have a dual function of recognizing both a bacterial carbohydrate receptor and a receptor in human saliva. Carbohydrate receptors for some adhesins have been purified from several streptococci, and they specifically block the coaggregations with the streptococcal partners that express the complementary adhesins. Coaggregation offers an explanation for the temporally related accretion of dental plaque and bacterial recognition of mucosal surfaces.

Bacterial adhesion to oral surfaces have been reviewed extensively.[48,79,105,116,146] Also *see* Chapters 6, 23, and 33 in this volume.

## 2. Respiratory Infections[130,136]

Normally, the mucosa of the nasooropharynx, trachea, and major bronchi is colonized with aerobic and anaerobic microbes.[136] Thus epithelial cells coexist with the microbial flora and are not overgrown with it. Also, the physiologic functions of the mucosa (protective barrier, mucociliary clearance, and air humidification and warming) are not impeded. A balance is maintained during health in which epithelial cell integrity — a function of proper nutrition, available secretory immunoglobulins and glycoproteins, and ciliary motion — resists the microbe's attempt to attach via specialized receptors (pili) or by proteolytic destruction of local proteins. When colonization is excessive and aspiration of more microbes into the lower airway occurs, infection is likely. Certain bacteria such as *Streptococcus pneumoniae* and *Hemophilus influenzae*, which are associated with chronic bronchitis, illustrate a mechanism in which the host–microbial balance may be upset by selective impairment of a host protein, secretory IgA1. Alternatively, cilotoxic microbes (mycoplasma) can favor colonization of bacteria when mucosal clearance mechanisms are impaired. Last, mucosal integrity can be breached by noxious gases or inflammation that may allow bacteria entry into the submucosal that provides a nidus for infection.[136]

In pathological conditions, the mucociliary clearance may be severely reduced, and mucus-associated bacteria may multiply and infect the underlying epithelium. Only a few bacteria have been shown to adhere to ciliary membranes of functionally active ciliated cells. Therefore, the first way in which most of the respiratory pathogens associate with the airway epithelium is likely to be by their adhesion to mucus. Some bacteria also secrete products that may affect ciliary function, cause cell death or epithelial disruption. Respiratory pathogens that do not bind to normal ciliated cells may readily adhere to injured cells or exposed extracellular matrix. Also following injury, epithelial respiratory cells in the process of migration, in order to repair the wounds, may present receptors to which bacteria adhere. The adhesion to all of these epithelial receptors may contribute to the chronicity of many bacterial respiratory infections.[130]

Methods of studying bacterial adhesion to respiratory mucosa have been summarized by Plotkowski et al. in this volume (*see* Chapter 34).

### 3. Gastric Infections[12,35,148]

Attachment of the bacterium to polarized gastric epithelial cells causes damage to microvilli and stimulates actin polymerization, which is associated with adherence pedestal formation.[21,148] *Helicobacter pylori* can directly contribute to the injury of gastric epithelial cells by the elaboration of cytotoxic factors. The first toxin identified from *H. pylori*, known as vacuolating cytotoxin, induces vacuole formation in eukaryotic cells. Elaborated enzymes by *H. pylori* may also contribute directly to epithelial cell injury. Ammonia produced through urease activity may be toxic to gastric epithelial cells. *H. pylori* protease and lipase degrade gastric mucus and disrupt the phospholipid-rich layer at the apical epithelial cell surface, allowing for cell injury from back diffusion of gastric acid. This cell injury may lead to cell death, resulting from induction of apoptosis. *H. pylori*, through direct pathogenic mechanisms, contributes significantly to the gastric mucosal injury associated with this infection, and may enhance the susceptibility of gastric epithelial cells to carcinogenic conversion.[148]

*H. pylori* adherence, the production of a vacuolating cytotoxin and bacterial enzymes all contribute to epithelial damage. Recruitment and activation of immune cells in the underlying mucosa involves *H. pylori* chemotaxins, epithelial-derived chemotactic peptides (chemokines) such as IL-8 and GRO-$\alpha$, and pro-inflammatory cytokines liberated by mononuclear phagocytes (TNF-$\alpha$, IL-1 and IL-6) as part of nonspecific immunity. Antigen-specific cellular immunity results in a predominant Th1 lymphocyte response with an increase in IFN-gamma secreting T-helper cells, whilst humoral responses lead to the production of anti-*H. pylori* antibodies and complement activation. Molecular mimicry of host structures by *H. pylori*, with the generation of specific immunity directed against self-antigens may also contribute to host injury. Progress in molecular biology has revealed considerable genomic diversity amongst *H. pylori* strains, with cag+ bacteria being associated with increased chemokine and cytokine responses and more severe degrees of gastric inflammation. Strain hetereogeneity may contribute towards the wide spectrum of disease manifestations encountered in clinical practice.[12]

*See* Chapter 36 for methods on studying bacterial adhesion to gastric mucosa. The readers are also referred to the review by Evans and Evans.[49]

### 4. Intestinal Infections[37,56,95,140,141,153]

In general, bacteria causing gastrointestinal infection need to penetrate the mucous layer before attaching themselves to epithelial and other absorptive cells in the intestine. This attachment is usually mediated by fimbriae or pilus structures although other cell surface components of bacteria may also take part in the process. Adherent bacteria colonize intestinal epithelium by multiplication and initiation of a series of biochemical reactions inside the target cell through signal transduction mechanisms (with or without the help of toxins). Alternatively, adherent bacteria induce extensive rearrangement of the cytoskeletal structure of the epithelial cell thereby making more intimate contact with the cell or even forcing their entry into it. This is followed by bacterial multiplication and intercellular spread leading to eventual death of the target cell.[56]

Attachment is not only a mechanism of tissue targeting but also a first step in the pathogenesis of many infections. The attaching bacteria engage in a "crosstalk" with the host cells through the mutual exchange of signals and responses. Enteropathogenic *E. coli* (EPEC) induce cytoskeletal rearrangements in epithelial cells and cause attaching/effacing lesions. *Shigella* sp. and *Listeria* sp. invade the cells and cause actin polymerization. Bacteria have the ability to trigger mucosal inflammation through activation of cells in the mucosal lining. Receptors for bacterial adhesins bind their ligands with a high degree of specificity and that ligand-receptor interactions trigger transmembrane signaling events that cause cell activation. Receptors for microbial ligands thus appear to also fulfill the same criteria as those used to define receptors for other classes of ligands such as hormones, growth factors, and cytokines.[153]

EPEC, first described in the 1940's and 1950's, remain an important cause of severe infantile diarrhea and hemorrhagic colitis. EPEC do not produce enterotoxins and are not invasive; instead their virulence depends upon exploitation of host cell signaling pathways and the host cell cytoskeleton both as a means of colonizing mucosal surfaces of the small intestine and causing diarrhea. Bacteria-induced signal transduction activates the receptor that allows tenacious adherence of the bacteria to the host cell surface. Both type IV fimbriae and a type III secretion apparatus play principal roles in interactions between the bacteria and host cells. Following initial mucosal attachment, EPEC secrete "signaling" proteins and express a surface adhesin, intimin, to produce attaching/effacing lesions in the enterocyte brush border membrane characterized by localized destruction of brush border microvilli, intimate bacterial adhesion and cytoskeletal reorganization and accretion beneath attached bacteria. The pathophysiology of EPEC diarrhea is also complex and probably results from a combination of epithelial cell responses including both electrolyte secretion and structural damage.[37,95]

The pathogenesis of shigellosis and bacillary dysentery is characterized by the capacity of the causative microorganism, *Shigella*, to invade the epithelial cells that compose the mucosal surface of the colon.[140,141] The invasive process encompasses the entering of bacteria into epithelial cells which involves activation of signaling pathways that elicit a macropinocitic event. Upon contact with the cell surface, *S. flexneri* activates a Mxi/Spa secretory apparatus, through which Ipa invasins are secreted. Two of the invasins, IpaB

(62 kDa) and IpaC (42 kDa), form a complex which is itself able to activate entry via its interaction with the host cell membrane. This interaction elicits major rearrangements of the host cell cytoskeleton, essentially the polymerization of actin filaments that form bundles supporting the membrane projections which achieve bacterial entry. Active recruitment of the proto-oncogene pp 60c-src has been demonstrated at the entry site with consequent phosphorylation of cortactin. Also, the small GTPase Rho is controlling the cascade of signals that allows elongation of actin filaments from initial nucleation foci underneath the cell membrane. The regulatory signals involved as well as the proteins recruited indicate that *Shigella* induces the formation of an adherence plaque at the cell surface in order to achieve entry. Once intracellular, the bacterium lyses its phagocytic vacuole, escapes into the cytoplasm and starts moving the inducing polar, directed polymerization of actin on its surface. In the context of polarized epithelial cells, bacteria then reach the intermediate junction and engage their components, particularly the cadherins, to form a protrusion which is actively internalized by the adjacent cell. Bacteria then lyse the two membranes, reach the cytoplasmic compartment again, and resume actin-driven movement.

Suggested further readings on bacterial adhesion to intestinal mucosa are the articles by Dean-Nystrom[30] and Donnenberg and Nataro.[38]

### 5. Urinary Tract Infections[22,135,137,138]

Urinary tract infections are caused by a variety of Gram-negative bacteria that ascend into the urinary tract and establish a population of ≥105 bacteria/mL of urine.[22] Bacterial adherence to the uroepithelium is recognized as an important mechanism in the initiation and pathogenesis of urinary tract infections. The uropathogens originate predominantly in the intestinal tract and initially colonize the periurethral region and ascend into the bladder, resulting in symptomatic or asymptomatic bacteriuria. Thereafter, depending on host factors and bacterial virulence factors, the organisms may further ascend and give rise to pyelonephritis. Considerable progress has been made in identifying bacterial adhesins and in demonstrating bacterial receptor sites on uroepithelial surfaces. Several studies have identified natural antiadherence mechanisms in humans as well as possible increased susceptibility to urinary tract infections when these mechanisms are defective and when receptor density on uroepithelial cells is altered.[135]

Initially, bacterial adhesion occurs via bacterial fimbriae in the case of Gram-negative bacteria, while Gram-positive bacteria adhere more frequently via extracellular polysaccharides. Urinary tract infections occur most frequently because of adherence via P fimbriae of uropathogenic *E. coli*. P fimbriae are important in cystitis as well, while type 1 fimbriae (mannose-sensitive fimbriae) and nonfimbrial adhesins may also be responsible for its initiation. The usual initiating mechanism in bacterial infections involves bacterial adhesion to specific molecules on cell surfaces or secreted mucosal components, followed by invasive disease. The tip proteins of P fimbriae of *E. coli* lead to the initiation of urinary tract infection.[138] Adhering bacteria and adhesin-positive P fimbriae (binding to glycolipid receptors) stimulated cells to secret significantly more IL-6 than non-adhering bacteria or adhesin-negative P fimbriae. This is a receptor-mediated transmembrane signaling, which results in the induction of cytokines and subsequent tissue inflammation.[22]

Chapter 22 and 37 describe methods for studying bacterial adhesion to urinary tract and catheters.

## 6. Septic Arthritis and Osteomyelitis[75,126,154,165]

*S. aureus* strains isolated from patients with septic arthritis or osteomyelitis possess a collagen receptor present in two forms, which contains either two or three copies of a 187-amino-acid repeat motif. Collagen receptor-positive strains adhered to both collagen substrata and cartilage in a time-dependent process. Collagen receptor-specific antibodies blocked bacterial adherence, as did pre-incubation of the substrate with a recombinant form of the receptor protein. Furthermore, polystyrene beads coated with the collagen receptor bound collagen and attached to cartilage. Taken together, these results suggest that the collagen receptor is both necessary and sufficient to mediate bacterial adherence to cartilage in a process that constitutes an important part of the pathogenic mechanism in septic arthritis.[154]

Hudson et al.[75] investigated the ability of *S. aureus* to associate with chick osteoblasts in culture and demonstrated internalization of bacteria by the osteoblasts. Two strains of *S. aureus* were examined that were ingested by osteoblasts to different extents, suggesting strain differences in uptake. Initial association of *S. aureus* strains with osteoblasts was independent of the presence of matrix collagen produced by the osteoblasts. Internalization of bacteria required live osteoblasts, but not live *S. aureus*, indicating osteoblasts are active in ingesting the organisms. The bacteria were not killed by the osteoblasts, since viable bacteria were cultured several hours after ingestion. Bacterial internalization may be an important step in the pathogenesis of osteomyelitis.

The surface-adherent mode of bacterial growth has been shown to play a pivotal role in the persistent nature of infections involving retained foreign bodies, biomaterials, or dead bone (e.g., osteomyelitis). Bone and implant materials provide a surface environment that promotes a type of bacterial growth characterized by an enhanced antibiotic resistance. Antibiotic resistance was found to vary with mode of bacterial growth. For the staphylococcal subtypes, adherent growth on bone was associated with the most antibiotic resistance.[165]

## 7. Pathogenesis of Candida Infections[16,74,121]

*Candida* infections of the skin and superficial mucosal sites are the result of an interplay between fungal virulence and host defenses. Epidermal proliferation and T lymphocyte immune responses are expressed by the host to combat fungal invasion, but inflammatory responses and nonspecific inhibitors also probably play a role. *Candida albicans* can express at least three types of surface adhesion molecules to colonize epithelial surfaces (protein–protein interactions, lectin-like interactions, and incompletely defined interactions in which the adhesive ligand is as yet unidentified), plus an aspartyl proteinase enzyme able to facilitate initial penetration of keratinized cells. Deeper penetration of keratinized epithelia is assisted by hypha formation, and *C. albicans* hyphae may use contact sensing (thigmotropism) as a guiding mechanism. Pathogenesis requires differential expression of virulence factors at each new stage of the process: a propensity for rapid alteration of the expressed phenotype in *C. albicans* may therefore be a significant factor in establishing the comparatively high pathogenic potential of this species.[74,121]

*Candida albicans* is frequently isolated from the mouth. Oral candidiasis presents clinically in many forms, reflecting the ability of the yeast to colonize different oral surfaces and the variety of factors which predispose the host to *Candida* colonization and subsequent infection. Colonization of the oral cavity appears to be facilitated by several specific adherence interactions between *C. albicans* and oral surfaces, which enable the yeast to resist host clearance mechanisms. *Candida* has been shown to adhere to complement receptors, various extracellular matrix proteins, and specific sugar residues displayed on host or bacterial surfaces in the oral cavity. Oral candidiasis results from yeast overgrowth and penetration of the oral tissues when the host physical and immunological defenses have been compromised. Tissue invasion may be assisted by secreted hydrolytic enzymes, hyphal formation, and contact sensing.[16]

*C. albicans* adhesion to tissues has been reviewed by Segal and Sandovsky-Losica[143] and by Samaranayake and Ellepola in this book (*see* Chapter 38).

*8. Other Infections*

Group A streptococci are nonmotile and have no structures that would enable them to penetrate submucosally into the pharynx. Reed et al.[134] have found that these bacteria adhere to host pharyngeal mucosal cells called Langerhans cells that are motile and could transport them into deeper tissues, causing pharyngitis.

Infective endocarditis begins with adherence of microorganisms to cardiac tissues.[85] These tissues have often been previously damaged, creating a thrombotic lesion consisting of platelets and fibrin. Circulating microorganisms localize to this lesion. The tissue specificity of endocarditis likely results from interactions between cell-surface determinants on the endocardium, platelet, and microorganism. Interference with these binding events may offer a means of modifying the course of the infection. *See* Chapter 35 for methods for studying bacterial adhesion to endothelial cells.

In eye infections, gonococci possess long range adhesins in the form of pili permitting initial contact with conjunctival cells. Subsequently, sticky surface protein (Protein II) bonds the *Gonococcus* close to the host cell surface. Damage is mediated both by the intracellular uptake and the introduction of pores in the cell membrane. *Pseudomonas* can only attach to damaged cells but once the eye is invaded a wide range of enzymes and toxins leads to rapid tissue destruction. The mechanisms by which *Chlamydia trachomatis* induce trachoma are ill-understood but involve intense antigenic stimulation.[164]

Lyme disease affects several major organ systems and leads to chronic illness. The pathogenic organism, *Borrelia burgdorferi*, shows preference for cell surfaces and tissues which may explain the paucity of isolations but also displays characteristic nonspecificity in its adherence to eukaryotic cells. This lack of specificity may explain its capacity to reside and injure vastly different tissues.[55]

## REFERENCES

1. Altwegg M, Geiss HK: Aeromonas as a human pathogen. *Crit Rev Microbiol* 16:253–86, 1989
2. An YH, Friedman RJ: Prevention of sepsis in total joint arthroplasty. *J Hosp Infect* 33:93–108, 1996
3. An YH, Friedman RJ, Draughn RA, et al: Bacterial adhesion to biomaterial surfaces. In: Wise DE, et al, ed. *Human Biomaterials Applications.* Humana Press, Totowa, NJ, 1996:19–57
4. Andrews HJ, Arden GP, Hart GM, et al: Deep infection after total hip replacement. *J Bone Joint Surg [Br]* 63-B:53–7, 1981

5. Arizono T, Oga M, Sugioka Y: Increased resistance of bacteria after adherence to polym-ethyl methacrylate. An *in vitro* study. *Acta Orthop Scand* 63:661–4, 1992

6. Barnett GR, Tannock GA, Paul JA, et al: An improved membrane-filtration enzyme immunoassay for the rapid serological diagnosis of viral infections. *J Virol Methods* 20:323–32, 1988

7. Bartzokas CA, Johnson R, Jane M, et al: Relation between mouth and haematogenous infection in total joint replacements. *Br Med J* 309:506–8, 1994

8. Bayer EA, Kenig R, Lamed R: Adherence of *Clostridium thermocellum* to cellulose. *J Bacteriol* 156:818–27, 1983

9. Beveridge TJ: Ultrastructure, chemistry, and function of the bacterial wall. *Int Rev Cytol* 72:229–317, 1981

10. Beveridge TJ, Makin SA, Kadurugamuwa JL, et al: Interactions between biofilms and the environment. *FEMS Microbiol Rev* 20:291–303, 1997

11. Blenkinsopp SA, Costerton W: Understanding bacterial biofilm. *Tebtech* 9:138–42, 1991

12. Bodger K, Crabtree JE: *Helicobacter pylori* and gastric inflammation. *Br Med Bull* 54:139–50, 1998

13. Brinton CC, Jr.: The structure, function, synthesis and genetic control of bacterial pili and a molecular model for DNA and RNA transport in gram negative bacteria. *Trans N Y Acad Sci* 27:1003–54, 1965

14. Busscher HJ, Cowan MM, van der Mei HC: On the relative importance of specific and non-specific approaches to oral microbial adhesion. *FEMS Microbiol Rev* 8:199–209, 1992

15. Canner GC, Steinberg ME, Heppenstall RB, et al: The infected hip after total hip arthro-plasty. *J Bone Joint Surg [Am]* 66:1393–9, 1984

16. Cannon RD, Holmes AR, Mason AB, et al: Oral Candida: clearance, colonization, or candidiasis? *J Dent Res* 74:1152–61, 1995

17. Charnley J: Postoperative infection after total hip replacement with special reference to air contamination in the operating room. *Clin Orthop* 87:167–87, 1972

18. Christensen GD, Simpson WA, Bisno AL, et al: Adherence of slime-producing strains of *Staphylococcus epidermidis* to smooth surfaces. *Infect Immun* 37:318–26, 1982

19. Christensen GD, Simpson WA, Bisno AL, et al: Experimental foreign body infections in mice challenged with slime-producing *Staphylococcus epidermidis*. *Infect Immun* 40:407–10, 1983

20. Chu CC, Williams DF: Effects of physical configuration and chemical structure of suture materials on bacterial adhesion. A possible link to wound infection. *Am J Surg* 147:197–204, 1984

21. Clyne M, Drumm B: Adherence of *Helicobacter pylori* to the gastric mucosa. *Can J Gastroenterol* 11:243–8, 1997

22. Connell H, Hedlund M, Agace W, et al: Bacterial attachment to uro-epithelial cells: mechanisms and consequences. *Adv Dent Res* 11:50–8, 1997

23. Costerton JW, Lewandowski Z, Caldwell DE, et al: Microbial biofilms. *Annu Rev Microbiol* 49:711–45, 1995

24. Costerton JW, Marrie TJ, K.J. C: Phenomena of bacterial adhesion. In: Savage DC, Fletcher M, eds. *Bacterial Adhesion. Mechanisms and Physiological Significance*. Plenum, New York, 1985:1–43

25. Dankert J, Hogt AH, Feijen J: Biomedical polymers: Bacterial adhesion, colonization, and infection. *CRC Crit Rev Biocompat* 2:219–301, 1986

26. Davenport DS, Massanari RM, Pfaller MA, et al: Usefulness of a test for slime production as a marker for clinically significant infections with coagulase-negative staphylococci. *J Infect Dis* 153:332–9, 1986

27. Dawson JR, Ellen RP: Tip-oriented adherence of *Treponema denticola* to fibronectin. *Infect Immun* 58:3924–8, 1990

28. de Gennes PG: Model polymers at interfaces. In: Bongrad P, ed. *Physical Basis of Cell-Cell Adhesion*. CRC Press, Boca Raton, FL, 1988:

29. De Graaf FK: Fimbrial structures of enterotoxigenic *E. coli. Antonie Van Leeuwenhoek* 54:395–404, 1988

30. Dean-Nystrom EA: Identification of intestinal receptors for enterotoxigenic *Escherichia coli. Methods Enzymol* 253:315–24, 1995

31. Derjaguin BV, Landau LD: Theory of the stability of strongly charged lyophobic sols and of the adhesion of strongly charged particles in solutions of electolytes. *Acta Physica Chimica, USSR* 14:633, 1941

32. DiCarlo EF, Bullough PG: The biologic responses to orthopaedic implants and their wear debris. *Clin Mater* 9:235–60, 1992

33. Dickinson RB: A dynamic model for attachment of a Brownian particle mediated by discrete macromolecular bonds. *J Colloid Interface Sci* 190:142–51, 1997

34. Dickinson RB, Nagel JA, McDevitt D, et al: Quantitative comparison of clumping factor- and coagulase-mediated *Staphylococcus aureus* adhesion to surface-bound fibrinogen under flow. *Infect Immun* 63:3143–50, 1995

35. Dixon MF: Pathophysiology of *Helicobacter pylori* infection. *Scand J Gastroenterol Suppl* 201:7–10, 1994

36. Dobbins JJ, Seligson D, Raff MJ: Bacterial colonization of orthopedic fixation devices in the absence of clinical infection. *J Infect Dis* 158:203–5, 1988

37. Donnenberg MS, Kaper JB, Finlay BB: Interactions between enteropathogenic *Escherichia coli* and host epithelial cells. *Trends Microbiol* 5:109–14, 1997

38. Donnenberg MS, Nataro JP: Methods for studying adhesion of diarrheagenic *Escherichia coli. Methods Enzymol* 253:324–36, 1995

39. Dougherty SH: Pathobiology of infection in prosthetic devices [see comments]. *Rev Infect Dis* 10:1102–17, 1988

40. Dougherty SH, Simmons RL: Endogenous factors contributing to prosthetic device infections. *Infect Dis Clin North Am* 3:199–209, 1989

41. Doyle RJ, Chaloupka J, Vinter V: Turnover of cell walls in microorganisms. *Microbiol Rev* 52:554–67, 1988

42. Doyle RJ, de La Barca AM, Buck G: Lectin interactions in oral microbial adhesion. In: Busscher HJ, Evans LV, eds. *Oral Biofilms and Plaque Control*. Harwood Publishers, Amsterdam, 1998:111–24.

43. Doyle RJ, Rosenberg M: *Microbial Cell Surface Hydrophobicity*. American Society for Microbiology, Washington, DC, 1990

44. Doyle RJ, Singh S, Nedjat-Haiem F: Hydrophobic characteristics of *Bacillus* spores. *Curr Microbiol* 10:329–92, 1984

45. Drake D, Taylor KG, Bleiweis AS, et al: Specificity of the glucan-binding lectin of *Streptococcus cricetus. Infect Immun* 56:1864–72, 1988

46. Duguid JO: The demonstration of bacterial capsules and slime. *J Pathol Bacteriol* 63:673–85, 1951

47. Duguid JO, Smith IW, Dempster G, et al: Nonflagellar filamentous appendages ("fimbriae") and haemagglutinating activity in *Bacterium coli. J Pathol Bacteriol* 70:335–48, 1955

48. Ellen RP, Lepine G, Nghiem PM: *In vitro* models that support adhesion specificity in biofilms of oral bacteria. *Adv Dent Res* 11:33–42, 1997

49. Evans DG, Evans DJ, Jr.: Adhesion properties of *Helicobacter pylori. Methods Enzymol* 253:336–60, 1995

50. Fitzgerald RH, Jr.: Infections of hip prostheses and artificial joints. *Infect Dis Clin North Am* 3:329–38, 1989

51. Fitzgerald RH, Jr., Nolan DR, Ilstrup DM, et al: Deep wound sepsis following total hip arthroplasty. *J Bone Joint Surg [Am]* 59:847–55, 1977

52. Fletcher M: Effect of solid surfaces on the activity of attached bacteria. In: Savage DC, Fletcher M, eds. *Bacterial Adhesion: Mechanisms and Physiological Significance.* Plenum Press, New York, 1985:339–62

53. Fletcher M, Floodgate GD: An electron-microscopic demonstration of as acidic polysaccharide involved in the adhesion of a marine bacterium to solid surfaces. *J Gen Microbiol* 74:325–34, 1973

54. Foster TJ, Hook M: Surface protein adhesins of *Staphylococcus aureus. Trends Microbiol* 6:484–8, 1998

55. Garcia-Monco JC, Benach JL: The pathogenesis of Lyme disease. *Rheum Dis Clin North Am* 15:711–26, 1989

56. Ghose AC: Adherence and colonisation properties of *Vibrio cholerae* and diarrhoeagenic *Escherichia coli. Indian J Med Res* 104:38–51, 1996

57. Gillespie WJ: Infection in total joint replacement. *Infect Dis Clin North Am* 4:465–84, 1990

58. Gray ED, Peters G, Verstegen M, et al: Effect of extracellular slime substance from *Staphylococcus epidermidis* on the human cellular immune response. *Lancet* 1:365–7, 1984

59. Gristina AG, Costerton JW: Bacterial adherence to biomaterials and tissue. The significance of its role in clinical sepsis. *J Bone Joint Surg [Am]* 67:264–73, 1985

60. Gristina AG, Giridhar G, Gabriel BL, et al: Cell biology and molecular mechanisms in artificial device infections. *Int J Artif Organs* 16:755–63, 1993

61. Gristina AG, Hobgood CD, Barth E: Biomaterial specificity, molecular mechanisms, and clinical relevance of *S. epidermidis* and *S. aureus* infections in surgery. In: Pulverer G, Quie PG, Peters G, eds. *Pathogenesis and Clinical Significance of Coagulase-negative Staphylococci.* Fisher Verlag, Stuttgart, 1987:143–57

62. Gristina AG, Hobgood CD, Webb LX, et al: Adhesive colonization of biomaterials and antibiotic resistance. *Biomaterials* 8:423–6, 1987

63. Gristina AG, Jennings RA, Naylor PT, et al: Comparative in vitro antibiotic resistance of surface-colonizing coagulase-negative staphylococci. *Antimicrob Agents Chemother* 33:813–6, 1989

64. Gristina AG, Naylor PT, Myrvik QN: Musculoskeletal infection, microbial adhesion, and antibiotic resistance. *Infect Dis Clin North Am* 4:391–408, 1990

65. Gristina AG, Naylor PT, Myrvik QN: Mechanisms of musculoskeletal sepsis. *Orthop Clin North Am* 22:363–71, 1991

66. Hacker J: Role of fimbrial adhesins in the pathogenesis of *Escherichia coli* infections. *Can J Microbiol* 38:720–7, 1992

67. Hamada S, Amano A, Kimura S, et al: The importance of fimbriae in the virulence and ecology of some oral bacteria. *Oral Microbiol Immunol* 13:129–38, 1998

68. Hasty DL, Ofek I, Courtney HS, et al: Multiple adhesins of streptococci. *Infect Immun* 60:2147–52, 1992

69. Hecht G, Savkovic SD: Review article: Effector role of epithelia in inflammation — interaction with bacteria. *Aliment Pharmacol Ther* 11 Suppl 3:64-8; discussion 8–9, 1997

70. Hogt AH, Dankert J, de Vries JA, et al: Adhesion of coagulase-negative staphylococci to biomaterials. *J Gen Microbiol* 129:2959–68, 1983

71. Hogt AH, Dankert J, Feijen J: Adhesion of *Staphylococcus epidermidis* and *Staphylococcus saprophyticus* to a hydrophobic biomaterial. *J Gen Microbiol* 131:2485–91, 1985

72. Hogt AH, Dankert J, Hulstaert CE, et al: Cell surface characteristics of coagulase-negative staphylococci and their adherence to fluorinated poly(ethylenepropylene). *Infect Immun* 51:294–301, 1986

73. Hoschutzky H, Lottspeich F, Jann K: Isolation and characterization of the alpha-galactosyl-1,4-beta-galactosyl-specific adhesin (P adhesin) from fimbriated *Escherichia coli. Infect Immun* 57:76–81, 1989

74. Hostetter MK: Adhesins and ligands involved in the interaction of *Candida* sp. with epithelial and endothelial surfaces. *Clin Microbiol Rev* 7:29–42, 1994

75. Hudson MC, Ramp WK, Nicholson NC, et al: Internalization of *Staphylococcus aureus* by cultured osteoblasts. *Microb Pathog* 19:409–19, 1995

76. Hussain M, Collins C, Hastings JG, et al: Radiochemical assay to measure the biofilm produced by coagulase-negative staphylococci on solid surfaces and its use to quantitate the effects of various antibacterial compounds on the formation of the biofilm. *J Med Microbiol* 37:62–9, 1992

77. Hussain M, Hastings JG, White PJ: Isolation and composition of the extracellular slime made by coagulase-negative staphylococci in a chemically defined medium. *J Infect Dis* 163:534–41, 1991

78. Jacques M: Role of lipo-oligosaccharides and lipopolysaccharides in bacterial adherence [see comments]. *Trends Microbiol* 4:408–9, 1996

79. Jacques N: Molecular biological techniques and their use to study streptococci in dental caries. *Aust Dent J* 43:87–98, 1998

80. James RC, MacLeod CJ: Induction of staphylococcal infections in mice with small inocula introduced on sutures. *Br J Exp Pathol* 42:266–77, 1961

81. Jann K, Jann B: Capsules of *Escherichia coli*, expression and biological significance. *Can J Microbiol* 38:705–10, 1992

82. Jenkinson HF: Genetic analysis of adherence by oral streptococci. *J Ind Microbiol* 15:186–92, 1995

83. Jenkinson HF, Lamont RJ: Streptococcal adhesion and colonization. *Crit Rev Oral Biol Med* 8:175–200, 1997

84. Jerry GJ Jr, Rand JA, Ilstrup D: Old sepsis prior to total knee arthroplasty. *Clin Orthop*:135–40, 1988

85. Johnson CM: Adherence events in the pathogenesis of infective endocarditis. *Infect Dis Clin North Am* 7:21–36, 1993

86. Jones GW, Isaacson RE: Proteinaceous bacterial adhesins and their receptors. *Crit Rev Microbiol* 10:229–60, 1983

87. Kahane I, Tucker S, Baseman JB: Detection of *Mycoplasma pneumoniae* adhesin (P1) in the nonhemadsorbing population of virulent *Mycoplasma pneumoniae*. *Infect Immun* 49:457–8, 1985

88. Kamme C, Lindberg L: Aerobic and anaerobic bacteria in deep infections after total hip arthroplasty: differential diagnosis between infectious and non-infectious loosening. *Clin Orthop* 154:201–7, 1981

89. Khardori N, Rosenbaum B, Bodey GP: Evaluation of *in vitro* markers for clinically significant infections with coagulase-negative staphylococci. *Clin Res* 35:20A, 1987

90. Kieswetter K, Merritt K, Myers R: Effects of infection on hydroxyapatite coating. *Trans Soc Biomater* 16:220, 1993

91. Kim YY, Ko CU, Ahn JY, et al: Charnley low friction arthroplasty in tuberculosis of the hip. An eight to 13-year follow-up. *J Bone Joint Surg [Br]* 70:756–60, 1988

92. Klein J: Forces between mica surfaces bearing layers of adsorbed polystyrene in cyclohexane. *Nature* 288:248–9, 1980

93. Klein J, Luckham P: Forces between two adsorbed polyethylene oxide layers immersed in a good aqueous solvent. *Nature* 300:429–31, 1982

94. Klock JC, Bainton DF: Degranulation and abnormal bactericidal function of granulocytes procured by reversible adhesion to nylon wool. *Blood* 48:149–61, 1976

95. Knutton S: Cellular responses to enteropathogenic *Escherichia coli* infection. *Biosci Rep* 15:469–79, 1995

96. Kolenbrander PE: Coaggregations among oral bacteria. *Methods Enzymol* 253:385–97, 1995

97. Kolenbrander PE, Ganeshkumar N, Cassels FJ, et al: Coaggregation: specific adherence among human oral plaque bacteria. *Faseb J* 7:406–13, 1993

98. Krekeler C, Ziehr H, Klein J: Physical methods for characterization of microbial surfaces. *Experientia* 45:1047–55, 1989

99. Krogfelt KA: Bacterial adhesion: genetics, biogenesis, and role in pathogenesis of fimbrial adhesins of *Escherichia coli*. *Rev Infect Dis* 13:721–35, 1991

100. Kroncke KD, Orskov I, Orskov F, et al: Electron microscopic study of coexpression of adhesive protein capsules and polysaccharide capsules in *Escherichia coli*. *Infect Immun* 58:2710–4, 1990

101. Kumar CG, Anand SK: Significance of microbial biofilms in food industry: a review. *Int J Food Microbiol* 42:9–27, 1998

102. Lamont RJ, Jenkinson HF: Life below the gum line: pathogenic mechanisms of porphyromonas gingivalis. *Microbiol Mol Biol Rev* 62:1244–63, 1998

103. Lidwell OM: Air, antibiotics and sepsis in replacement joints. *J Hosp Infect* 11(Suppl C):18–40, 1988

104. Liljemark WF, Bloomquist CG: Isolation of a protein-containing cell surface component from *Streptococcus sanguis* which affects its adherence to saliva-coated hydroxyapatite. *Infect Immun* 34:428–34, 1981

105. Liljemark WF, Bloomquist CG, Reilly BE, et al: Growth dynamics in a natural biofilm and its impact on oral disease management. *Adv Dent Res* 11:14–23, 1997

106. Lindahl M, Faris A, Wadstrom T, et al: A new test based on "salting out" to measure relative surface hydrophobicity of bacterial cells. *Biochim Biophys Acta* 677:471–6, 1981

107. Lowy FD, Hammer SM: *Staphylococcus epidermidis* infections. *Ann Intern Med* 99:834–9, 1983

108. Ma Y, Lassiter MO, Banas JA, et al: Multiple glucan-binding proteins of *Streptococcus sobrinus*. *J Bacteriol* 178:1572–7, 1996

109. Maderazo EG, Judson S, Pasternak H: Late infections of total joint prostheses. A review and recommendations for prevention. *Clin Orthop* 229:131–42, 1988

110. Marrie TJ, Costerton JW: Scanning electron microscopic study of uropathogen adherence to a plastic surface. *Appl Environ Microbiol* 45:1018–24, 1983

111. Marshall KC: Mechanisms of bacterial adhesion at solid-water interfaces. In: Savage DC, Fletcher M, eds. *Bacterial Adhesion. Mechanisms and Physiological Significance*. Plenum Press, New York, 1985:133–61

112. Marshall KC, Stout R, Mitchell R: Mechanism of initial events in the sorption of marine bacteria to surfaces. *J Gen Microbiol* 68:337–48, 1971

113. Merritt K, Brown SA, Payer JH, et al: Influence of bacteria on corrosion of metallic biomaterials. *Trans Soc Biomater* 14:106, 1991

114. Merritt K, Shafer JW, Brown SA: Implant site infection rates with porous and dense materials. *J Biomed Mater Res* 13:101–8, 1979

115. Mett H, Kloetzlen L, Vosbeck K: Properties of pili from *Escherichia coli* SS142 that mediate mannose-resistant adhesion to mammalian cells. *J Bacteriol* 153:1038–44, 1983

116. Meyer DH, Fives-Taylor PM: Adhesion of oral bacteria to soft tissue. *Methods Enzymol* 253:373–85, 1995

117. Moch T, Hoschutzky H, Hacker J, et al: Isolation and characterization of the α-sialyl-β-2,3-galactosyl-specific adhesin from fimbriated *Escherichia coli*. *Proc Natl Acad Sci U S A* 84:3462–6, 1987

118. Morris EJ, McBride BC: Adherence of *Streptococcus sanguis* to saliva-coated hydroxy-apatite: evidence for two binding sites. *Infect Immun* 43:656–63, 1984

119. Naylor PT, Myrvik QN, Gristina A: Antibiotic resistance of biomaterial-adherent coagulase-negative and coagulase-positive staphylococci. *Clin Orthop* 261:126–33, 1990

120. Nesbitt WE, Doyle RJ, Taylor KG: Hydrophobic interactions and the adherence of *Streptococcus sanguis* to hydroxylapatite. *Infect Immun* 38:637–44, 1982

121. Odds FC: Pathogenesis of *Candida* infections. *J Am Acad Dermatol* 31:S2–5, 1994
122. Ofek I, Doyle RJ: *Bacterial Adhesion to Cells and Tissues.* Chapman & Hall, New York, 1994:578
123. Ørskov I, Ørskov F: Serologic classification of fimbriae. *Curr Top Microbiol Immun* 151:71–90, 1990
124. Pascual A, Ramirez de Arellano E, Martinez Martinez L, et al: Effect of polyurethane catheters and bacterial biofilms on the *in vitro* activity of antimicrobials against *Staphylococcus epidermidis. J Hosp Infect* 24:211–8, 1993
125. Patti JM, Allen BL, McGavin MJ, et al: MSCRAMM-mediated adherence of micro-organisms to host tissues. *Annu Rev Microbiol* 48:585–617, 1994
126. Patti JM, Bremell T, Krajewska-Pietrasik D, et al: The *Staphylococcus aureus* collagen adhesin is a virulence determinant in experimental septic arthritis. *Infect Immun* 62:152–61, 1994
127. Patti JM, Hook M: Microbial adhesins recognizing extracellular matrix macromolecules. *Curr Opin Cell Biol* 6:752–8, 1994
128. Pegden RS, Larson MA, Grant RJ, et al: Adherence of the Gram-positive bacterium *Ruminococcus albus* to cellulose and identification of a novel form of cellulose-binding protein which belongs to the Pil family of proteins. *J Bacteriol* 180:5921–7, 1998
129. Phillips GN, Jr., Flicker PF, Cohen C, et al: Streptococcal M protein: alpha-helical coiled-coil structure and arrangement on the cell surface. *Proc Natl Acad Sci U S A* 78:4689–93, 1981
130. Plotkowski MC, Bajolet-Laudinat O, Puchelle E: Cellular and molecular mechanisms of bacterial adhesion to respiratory mucosa. *Eur Respir J* 6:903–16, 1993
131. Prieve DC, Ruckenstein E: Effect of London forces upon the rate of deposition of Brownian particles. *AIChE J* 20:1178–87, 1974
132. Rae T: The action of cobalt, nickel and chromium on phagocytosis and bacterial killing by human polymorphonuclear leucocytes; its relevance to infection after total joint arthroplasty. *Biomaterials* 4:175–80, 1983
133. Razatos A, Ong YL, Sharma MM, et al: Molecular determinants of bacterial adhesion monitored by atomic force microscopy. *Proc Natl Acad Sci U S A* 95:11059–64, 1998
134. Reed WP, Metzler C, Albright E: Streptococcal adherence to Langerhans cells: a possible step in the pathogenesis of streptococcal pharyngitis. *Clin Immunol Immunopathol* 70:28–31, 1994
135. Reid G, Sobel JD: Bacterial adherence in the pathogenesis of urinary tract infection: a review. *Rev Infect Dis* 9:470–87, 1987
136. Reynolds HY: Bacterial adherence to respiratory tract mucosa—a dynamic interaction leading to colonization. *Semin Respir Infect* 2:8–19, 1987
137. Roberts JA: Bacterial adherence and urinary tract infection. *South Med J* 80:347–51, 1987
138. Roberts JA: Tropism in bacterial infections: urinary tract infections. *J Urol* 156:1552–9, 1996
139. Sanderson PJ: The choice between prophylactic agents for orthopaedic surgery. *J Hosp Infect* 11(Suppl C):57–67, 1988
140. Sansonetti PJ: Molecular and cellular mechanisms of invasion of the intestinal barrier by enteric pathogens. The paradigm of *Shigella. Folia Microbiol* 43:239–46, 1998
141. Sansonetti PJ, Egile C: Molecular bases of epithelial cell invasion by *Shigella flexneri. Antonie Van Leeuwenhoek* 74:191–7, 1998
142. Schamhart DH, de Boer EC, Kurth KH: Interaction between bacteria and the lumenal bladder surface: modulation by pentosan polysulfate, an experimental and theoretical approach with clinical implication. *World J Urol* 12:27–37, 1994
143. Segal E, Sandovsky-Losica H: Adhesion and interaction of *Candida albicans* with mammalian tissues *in vitro* and *in vivo. Methods Enzymol* 253:439–52, 1995

151. Sutherland IW: Microbial exopolysaccharides — their role in microbial adhesion in aqueous systems. *Crit Rev Microbiol* 10:173–201, 1983

152. Svanborg C, Agace W, Hedges S, et al: Bacterial adherence and epithelial cell cytokine production. *Zentralbl Bakteriol* 278:359–64, 1993

153. Svanborg C, Hedlund M, Connell H, et al: Bacterial adherence and mucosal cytokine responses. Receptors and transmembrane signaling. *Ann N Y Acad Sci* 797:177–90, 1996

154. Switalski LM, Patti JM, Butcher W, et al: A collagen receptor on *Staphylococcus aureus* strains isolated from patients with septic arthritis mediates adhesion to cartilage. *Mol Microbiol* 7:99–107, 1993

155. Timmerman CP, Fleer A, Besnier JM, et al: Characterization of a proteinaceous adhesin of *Staphylococcus epidermidis* which mediates attachment to polystyrene. *Infect Immun* 59:4187–92, 1991

156. Tojo M, Yamashita N, Goldmann DA, et al: Isolation and characterization of a capsular polysaccharide adhesin from *Staphylococcus epidermidis*. *J Infect Dis* 157:713–22, 1988

157. Tollefson DF, Bandyk DF, Kaebnick HW, et al: Surface biofilm disruption. Enhanced recovery of microorganisms from vascular prostheses. *Arch Surg* 122:38–43, 1987

158. Troy FAd: The chemistry and biosynthesis of selected bacterial capsular polymers. *Annu Rev Microbiol* 33:519–60, 1979

159. Ueng WN, Shih CH, Hseuh S: Pulmonary tuberculosis as a source of infection after total hip arthroplasty. A report of two cases. *Int Orthop* 19:55–9, 1995

160. van Loosdrecht MC, Norde W, Zehnder AJ: Physical chemical description of bacterial adhesion. *J Biomater Appl* 5:91–106, 1990

161. Vaudaux PE, Zulian G, Huggler E, et al: Attachment of *Staphylococcus aureus* to polymethylmethacrylate increases its resistance to phagocytosis in foreign body infection. *Infect Immun* 50:472–7, 1985

162. Vergères P, Blaser J: Amikacin, ceftazidime, and flucloxacillin against suspended and adherent *Pseudomonas aeruginosa* and *Staphylococcus epidermidis* in an *in vitro* model of infection. *J Infect Dis* 165:281–9, 1992

163. Verway EJW, Overbeek GTG: *Theory of Stability of Lyophobic Colloids*. Elsevier, Amsterdam, 1948

164. Watt PJ: Pathogenic mechanisms of organisms virulent to the eye. *Trans Ophthalmol Soc U K* 105:26–31, 1986

165. Webb LX, Holman J, de Araujo B, et al: Antibiotic resistance in staphylococci adherent to cortical bone. *J Orthop Trauma* 8:28–33, 1994

166. Wentworth JS, Austin FE, Garber N, et al: Cytoplasmic lectins contribute to the adhesion of *Psuedomonas aeruginosa*. *Biofouling* 4:99–104, 1991

167. Whittaker CJ, Klier CM, Kolenbrander PE: Mechanisms of adhesion by oral bacteria. *Ann Rev Microbiol* 50:513–52, 1996

168. Wicken AJ: Bacterial cell walls and surfaces. In: Savage DC, Fletcher M, eds. *Mechanisms of Bacterial Adhesion at Solid-Water Interfaces*. Plenum Press, New York, 1985:45–70

169. Wiencek KM, Fletcher M: Bacterial adhesion to hydroxyl- and methyl-terminated alkanethiol self-assembled monolayers. *J Bacteriol* 177:1959–66, 1995

170. Zafriri D, Oron Y, Eisenstein BI, et al: Growth advantage and enhanced toxicity of *Escherichia coli* adherent to tissue culture cells due to restricted diffusion of products secreted by the cells. *J Clin Invest* 79:1210–6, 1987

171. Zhou L, Srisatjaluk R, Justus DE, et al: On the origin of membrane vesicles in Gram-negative bacteria. *FEMS Microbiol Lett* 163:223–8, 1998

172. Zimmerli W, Waldvogel FA, Vaudaux P, et al: Pathogenesis of foreign body infection: description and characteristics of an animal model. *J Infect Dis* 146:487–97, 1982

173. Zobell CE: The effect of solid surfaces upon bacterial activity. *J Bacteriol* 46:39–56, 1943

# 2

# Molecular Basis of Bacterial Adhesion

**Thomas Boland,[1] Robert A. Latour,[1] and Fred J. Stutzenberger[2]**

*[1]Department of Bioengineering, and [2]Department of Microbiology and Molecular Biology, Clemson University, Clemson, SC, USA*

## I. INTRODUCTION

Bacterial infections are responsible for a broad spectrum of human illnesses and medical device complications. For example, urinary tract infections caused by *Escherichia coli* affect over 7 million people annually and are among the most common infectious diseases acquired by humans.[39] Enteropathogenic *E. coli* (EPEC) and shiga toxin-producing *E. coli* (STIC) are diarrhoegenic pathogens causing serious health problems in both industrialized and developing countries.[26,15] *Helicobacter pylori* have been found to be a main factor in the development of gastric and duodenal ulcers and are believed to be a causitive factor of gastic cancer.[34] *Staphylococcus aureus* and *Staphylococcus epidermidis* are major causes of infections associated with wounds, indwelling catheters, and cardio-vascular and orthopedic implant devices.[1,19,24,25,35,49,56,59]

Bacteria have a strong tendency to attach to surfaces. Attached cells will form a colony (biofilm) consisting of prokaryotic cells, surrounded by a matrix of biomolecules secreted by the cells. Although the structure and functions of biofilms are as varying as the type of bacteria, the same four step process is always followed in the creation of the biofilm.[21,57]

During the first step, a series of small molecules (initially water and salt ions) will adsorb to the surface. Subsequently, the substrate surface will be covered with a single layer of small organic molecules or proteins that are present in the medium. The mixture of water, ions and proteins is called conditioning film and is always present before the first microorganisms arrive at the surface.[17,50]

The second step is characterized by the initially reversible adsorption of micro-organisms to the conditioning film. The microbes arrive by Brownian motion, gravitation, diffusion, or intrinsic motility. They may also adhere to each other forming microbial aggregates before adsorbing to the conditioning film. Since the microorganisms adhere to the conditioning film and not the surface itself, the strength of the initial biofilm depends on the structure of the conditioning film.[8,29]

The initially reversible adsorption becomes irreversible, mainly through the secretion of exopolymeric substances by the adsorbed microorganisms in step three.[18,40] These substances will incorporate in the conditioning film and strengthen its cohesiveness. In a

*Handbook of Bacterial Adhesion: Principles, Methods, and Applications*
Edited by: Y. H. An and R. J. Friedman © Humana Press Inc., Totowa, NJ

few cases, an entirely microbially derived conditioning film has been observed.[7] Once bound to the surface, many bacterial strains, including *S. aureus* and *S. epidermidis*, additionally form a polysaccharide based biofilm surrounding the bacterial colony. Biofilm effectively inhibits phagocytosis and makes the contained bacteria impervious to antibiotics, thus making implant removal an essential part of treatment once an infection is established.[12,24,25,35,56]

Finally, the number of microorganisms in the biofilm accumulates mainly through *in situ* cell growth. The final structure and composition of the biofilm is determined by these initial events. Other aspects such as the influence of surface active compounds secreted by the microorganisms,[41] the hydrodynamic environment,[48,56] the surface roughness,[5,46] the available nutrients,[6,22] and the attraction and adhesion to other microorganisms from the surrounding medium[4,20,32,33] are thought to be of secondary importance regarding the final outcome of the biofilm.[9]

Bacterial adhesion leading to infection can be divided into three distinct categories: specific adhesion to host cell surface molecules, specific adhesion to extracellular matrix and blood plasma derived molecules, and adhesion to biomaterial surfaces of medical devices. In this chapter, an overview of the current understanding of the molecular basis of bacterial adhesion as it pertains to each of the three categories of bacterial adhesion is presented, followed by modeling of bacterial adhesion based upon the general principles governing molecular adhesion. Particular emphasis is given to interactions between the initially arriving microorganisms and the conditioning film at the molecular level.

## II. MICROBIAL ADHESION
## TO EXTRACELLULAR MATRIX MOLECULES

The interactions of arriving microorganisms with a conditioning film on a surface are usually mediated by specific binding events between adhesins on the microbe surface and receptors of the extracellular matrix (ECM). Receptor binding may subsequently activate a series of complex signal transduction cascades in the host cell, which may be either inhibitive or beneficial to bacterial invasion. In several bacterial species, including *E. coli, Pseudomonas aeruginosa, Vibrio cholerae*, and *Salmonella enteritidis*, adhesins are presented at the tips of complex cell-surface structures which extend from the outer cell membrane called pili or fimbriae.[52] Pili are classified as P, type 1, type IV, and curli, each with distinct structural organization and assembly mechanisms. Alternatively, nonpilus adhesins may be directly presented from the bacterial cell surface as well.[52]

The molecular basis for bacterial adhesion to ECM molecules has been widely studied, and found to occur through specific binding mechanisms involving both piliated and nonpilated bacterial adhesins. These processes involve integrins which are a family of heterodimeric ($\alpha\beta$) cell-surface receptors that recognize specific ECM submolecular structures.[43] While much remains to be learned of the specific molecular mechanisms involved, bacterial cells have been found to utilize many of these same cell wall receptors to specifically adhere to ECM molecules, including fibronectin, collagens, laminin, vitronectin, thrombospodin, elastin, bone sialoprotein and GAGs like heparin, heparan sulphate, and chondroitin sulphate.[34] Bacterial adhesins, which bind with ECM, are termed microbial surface components recognizing adhesive matrix molecules (MSCRAMM).[37,43,44]

One of the most widely studied systems of bacteria-ECM interaction is *S. aureus* binding to fibrinogen. Fibrinogen is specifically recognized by several host cell integrins, including the platelet integrin $\alpha_{IIb}\beta_3$.[43] Two of the most widely understood fibrinogen binding proteins expressed by *S. aureus* strains are called clumping factors A and B (ClfA and ClfB). Studies have indicated that ClfA binds exclusively with the $\gamma$-chain of fibrinogen, while ClfB binds to both $\alpha$- and $\beta$-chains.[19]

ClfA is a 933 residue protein which includes a 520 residue region that contains its fibrinogen binding domain. This binding domain is preceded by a very interesting molecular structure consisting of 154 repeats of a serine-aspartate dipeptide sequence.[19] At a physiologic pH of 7.4, the carboxylic acid side groups of the aspartic acid residues will be deprotonated (thus carrying a single negative charge on each residue) and the interdispersed hydroxyl side groups of the serine residues will be strongly hydrophilic. The repulsion of the sequential aspartic acid residues coupled with the hydrophilicity of the serine residues should thus provide a large electrostatic driving force to extend the adhesin outward from the bacterial cell surface, much like hairs standing up on one's head when electrostatically charged.

Studies suggest that ClfA binds to two distinct sites in the $\gamma$-chain of fibrinogen by molecular structures very similar to the fibrinogen-binding integrin, $\alpha_{IIb}\beta_3$. ClfA and the platelet receptor $\alpha_{IIb}\beta_3$ have been found to recognize the same 400-411 residue section at the extreme C-terminus of the fibrinogen $\gamma$-chain (residue sequence …HHLGGA-KQAGDV), and studies have shown that alteration of only the last four residues (… AGDV) is sufficient to inhibit ClfA binding.[37,19] The $\gamma$-chain binding site of ClfA has been mapped to a region of the $\alpha_{IIb}$ polypeptide of the $\alpha_{IIb}\beta_3$ integrin and both contain the $Ca^{2+}$ binding EF-hand motif found in many eukaryotic calcium ion binding proteins. The EF-hand motif consists of about 12 residues, with the proposed ClfA sequence being **DSDGN**VIYT**FTD**, which represents residue numbers 310-321 in the protein. The sequence letters indicated in bold are the residues specifically involved with cation binding. These residues form a coordination sphere for the divalent cation and are flanked by $\alpha$-helices which provide support structure for this motif to maintain proper functional conformation.

Studies have shown that high concentrations of calcium ions inhibits ClfA-fibrinogen binding in a manner similar to that observed for integrin-ligand binding.[19,43] Thus, calcium ion concentration serves as an important regulator of *S. aureus* binding to fibrinogen. The second fibrinogen binding site of ClfA, which is also the $\alpha_M\beta_2$ integrin binding site, involves fibrinogen $\gamma$-chain residues 190-202. Similar to $\alpha_M\beta_2$, this ClfA binding site also includes a cation-binding <u>m</u>etal <u>i</u>on-<u>d</u>ependent <u>a</u>dhesion <u>s</u>ite (MIDAS) motif.[19] The ClfB protein does not have an EF hand-like tertiary structure, however, it does possess a MIDAS-like motif, such that the binding of both ClfA and ClfB to their fibrinogen ligand sites are regulated by $Ca^{2+}$ concentration. The specific binding sequences in ClfB and their role in ligand binding have not yet been elucidated to the extent of ClfA.[19]

*S. aureus* also has been found to express MSCRAMM adhesins for both fibronectin (FnbpA and FnbpB) and collagen.[13,34] Studies have shown that there are at least two binding sites on fibronectin, one occurring in its N-terminus domain, and the other in its C-terminus domain.[34] Although the molecular details of these interactions have not been as widely reported in the literature as the fibinogen binding system, the actual structure of an *S. aureus* collagen-binding adhesin has been determined and is available in the Brookhaven Protein Data Base under protein code 1AMX.

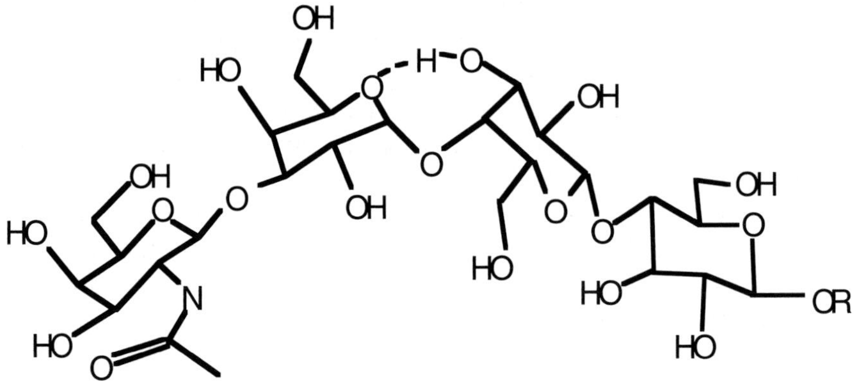

**Figure 1.** Molecular structure of the globotetraose-binding site of a class II PapG. Oxygen and hydroxyl groups indicated in bold are believed to form hydrogen bonds with the adhesin. Adaped from Striker et al.[53]

## III. HOST CELL ADHESION MECHANISMS

Bacterial adhesion to host cells of the urinary tract has been found to occur by specific molecular interactions between adhesins located on the distal tip of pili extending from the bacterial outer membrane and receptor molecular structures present on the host cell outer surface. Although the exact molecular structures involved for many of these interaction are yet to be determined, the specific binding mechanisms involved in a few systems have been relatively well characterized.

Uropathogenic *E. coli* has been shown to adhere to erythrocytes and uroepithelial cells of the kidney and urinary tract.[28,53] Studies have revealed that this specific binding event occurs between PapG adhesins located at the tip of P pili and Galα(1-4)Gal saccharide epitopes in the globo series of glycolipids. This saccharide structure has been determined to be linked by a β-glucose residue to a ceramide group anchoring the receptor in the host cell membrane. The receptor-binding domain of PapG has been determined to lie in the N-terminus of the protein. An example of the globotetraose-binding site of a class II PapG is presented in Figure 1. Studies involving sequential functional group replacement in Galα(1-4)Gal have revealed that PapG binds to this receptor, in part, by hydrogen binding with a series of five oxygen atoms located along the edge of the disaccharide surrounding a central hydrophobic core.[28]

Uropathogenic *E. coli* expressing type-1 pili tipped with the FimH adhesin (a 30-kDa protein[51]) have been shown to specifically bind to mannosylated integral membrane glycoproteins (uroplakins) presented from the luminal surface of bladder epithelial cells using a murine cystitis model.[39] This bacteria exhibits a very interesting but not yet fully understood mechanism to facilitate close bacteria–host cell interactions. Host cell contacting pili were found to be only 0.12 μm in length in contrast to the typical 1-2 μm length for non-contacting pili. This suggests a possible pili retraction mechanism to enhance tight bacterial–host cell apposition, with subsequent possible host-cell internalization of the *E. coli*[39] Sokurenko and coworkers[51] investigated the differences in the 300 residue sequence of FimH and their respective adhesive characteristics for fourteen *E. coli* strains. This study revealed that the FimH sequences where essentially homologous to one another except for 2 specific residue changes involving a swapping of arginine and serine residues at positions 70 and 78. The exchange of these two residues was found to result in distinct differences in bacterial adhesion. Although not well understood, this alteration is believed to influence the structure of the saccharide binding pocket in FimH for mannose binding.[51]

A different bacterial mechanism has been found to occur for the mucosal lining of the intestine leading to microvilli effacement and diarrhea. A four stage infection process has been suggested involving initial attachment of enteropathogenic *E. coli* bacteria (EPEC) to the microvilli enterocyte cell surface:

1. A nonpiliated adhesin mechanism;
2. Type III bacterial secretion of 80kDa proteins (*E. coli* secreted protein, EspE) which mediate cytoskeleton disruption and the formation of tyrosine-phosphorylated translocated intimin receptors (TIR) on the host-cell surface for intimin binding;
3. Intimin-binding mediated bacterial attachment to the intestinal mucosa; and
4. Bundle-forming pili (BFP) mediated bacterial colonization.[15,26]

Others have suggested that BFP serves as the adhesin controlling initial host cell contact as well as bacterial colonization.[2]

*S. aureus* can bind to endothelial cells through its fibrinogen binding clumping factors ClfA and ClfB. Adhesion studies found that the preferential attachment of *S. aureus* to umbilical vein endothelial cells is mediated by fibrinogen adsorbed from plasma. Antifibrinogen antibodies could block the binding, indicating the specificity. Cheung et al. found that fibrinogen acts as a bridging molecule, attaching to both endothelial and *S. aureus* cell-wall integrins with each of its two γ-chains.[11]

Finally, some bacteria use the integrin on endothelial cells to invade the host. Filamentous hemagglutinin (FHA), an adhesin formed as a 50 nm monomeric rigid rod of *Bordetella pertussis*, interacts with two classes of molecules on macrophages, galactose containing glycoconjugates and the $\alpha_M\beta_2$ integrin which binds to the Arg-Gly-Asp (RGD) sequence in FHA.[47] Intimin, the outer membrane protein of *E. coli* also binds specifically to $\alpha_M\beta_2$ integrins and is inhibited by RGD containing peptides.[23]

## IV. ADHESION TO BIOPOLYMERS AND BIOMATERIALS

### A. Natural Biopolymers

In natural heterogeneous microbial ecosystems (such as soil or an aquatic environment), adherence of a cell to a solid surface confers several competitive advantages. The ability to bind to a solid biopolymer (such as the cellulose fiber to which the filamentous bacterium, *Thermomonospora curvata*, has bound itself in Figure 2) provides the cells with a reliable constant carbon and energy source.[27] The adhesion not only brings its surface bound enzymes (cellulosomes[31]) into intimate contact with the substrate, but also affords it prime access to whatever soluble depolymerization products are released by their catalytic action. Cellulosomal organization and molecular structure of its complex components has been most extensively studied in the mesophilic anaerobe, *Clostridium cellulovorans*. Its cellulosome is composed of three major subunits: as caffolding protein, an endoglucanase, and an exoglucanase.[16] The binding of cellulolytic microbes such as *Clostridium cellulovorans* to cellulose-containing substrate surfaces is

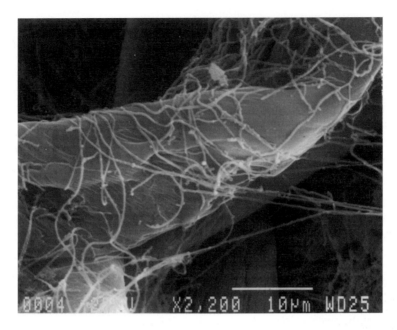

**Figure 2.** Filamentous bacterium *Thermonospora curvata* bound to cellulose fiber.

mediated by one or more of a heterologous group of cell surface-bound proteins containing cellulose binding domains (CBD). These CBDs have been classified into 10 families (I-X) on the basis of amino acid sequence homology.[55] The amino acid sequences of CBDs in *C. cellulovorans* and *C. josui* show high homology with those from other cellulolytic genera such as *Bacillus*. CBDs in this family contain several highly conserved amino acid sequences:[30]

1. Tryptophane-asparate-phenylalanine-asparagine-asparate-glycine-threonine
2. Isoleucine-alanine-alanine-isoleucine-proline-glutamine
3. Isoleucine-leucine-phenylalanine-valine-glycine

The cell surface-bound cellulosome in *Clostridium* species and in others is a complex of adherence and catalytic proteins. A major cellulosomal subunit (EngE) has been recently characterized.[54] EngE is anchored via a protein having triple-repeated surface layer homology (SLH) domains at the N-terminus; these domains appear to integrate into the lattice of the cell surface peptidoglycan-surface protein complex; they also bind with hydrophobic domains of the EngE. Therefore the cellulosic surface adhesion architecture in *C. cellulovorans* consists of catalytic units which have specific cellulose-binding domains, hydrophobic domains which act as linkers between the catalytic units and the SLH domains, and the SLH which integrates into the peptidoglycan-teichoic acid-lipoteichoic acid lattice which is the major structural component of the bacterial cell wall.[45] This attachment structure appears similar to that complexed with other surface-bound exoenzymes in related bacteria.[38]

### B. Synthetic Biomaterials

In vivo, coagulase-negative staphylococcal (CNS) (in particular *S. epidermidis*) and *S. aureus* are the leading causes of infection for body fluid-contacting medical devices, including intravascular catheters, cerebrospinal fluid shunts, prosthetic artificial heart valves, orthopedic devices, cardiac pacemakers, chronic peritoneal dialysis catheters, and vascular grafts.[19,25,35,49,58,59]

Serum proteins, such as albumin, fibronectin, fibrinogen and laminin, have been found to readily adsorb nonspecifically to biomaterial surfaces following body fluid exposure.[41] These adsorbed proteins form the conditioning film onto which specific bacterial adhesion takes place in mechanisms very similar to those that govern bacterial adhesion to extracellular matrix components.[1,34] An additional factor unique to biomaterials, however, is that the serum protein adsorption process may abnormally denaturate the proteins structure, thus potentially exposing binding sites not normally present in either the soluble or ECM form of the proteins. Fibrinogen is one of the major plasma proteins adsorbed onto implanted biomaterials and the adhesion of *S. aureus* to adsorbed fibrinogen and fibrin is an important initiator of infection[13,19,24] The specific adhesion mechanisms between bacteria and surface bound fibrinogen/fibrin are believed to be essentially the same as those which occur for bacteria adhesion to fibrinogen as an ECM component and have been addressed in the previous section.

Fibronectin is another predominant glycoprotein component of plasma which is incorporated into the fibrin matrix during blood clot formation and which also readily adsorbs to implant surfaces following blood contact.[59] This macromolecule is a dimeric glycoprotein composed of a series of domains comprising different combinations of type I, type II, and type III modules.[24] *S. aureus* has been shown to specifically bind to both soluble and adsorbed fibronectin, but has little affinity for fibrinogen incorporated into thrombi by itself without preexposure of the bacteria to soluble fibronectin.[59] Binding has been found to involve at least two *S. aureus* MSCRAMM fibronectin-binding proteins, known as FnbpA and FnbpB, and at least two binding sites on fibronectin involving both the N- and C-termini.[24,34] FnbpA and FnbpB are identical to one another apart from their N-terminal regions which is only 45% homologous.[13,24] In contrast to this, coagulase negative staphylococcus (CNS) does not bind soluble fibronectin; however, it exhibits significant affinity for the N-terminus domain of fibronectin incorporated into fibrin thrombi or immobilized on plastic surfaces.[34,59] These differences between soluble and bound fibronectin are believed to reflect conformational changes induced by adsorption; although soluble fibronectin has a globular structure, the two arms of the dimer are believed to unfold upon binding to expose previously hidden binding sites.[59]

Another pathway for bacterial infection involves the secretion of exopolymeric substances following initial bacterial adhesion to implant surfaces. *S. epidermidis*, for example, colonizes within a self-generated viscous biofilm largely composed of polysaccharides. This amorphous extracellular matrix substance enhances bacterial adhesion to biomaterials surfaces and provides the bacterial colony with protection from antibiotics and phagocytoses.[1,35,49,53] The process of colonization and biofilm production by *S. epidermidis* involves several key molecular interactions. *S. epidermidis* strains produce two similar insoluble polysaccharides, termed polysaccharide adhesins (PS/A) and polylsaccharide intercellular adhesins (PIA). PS/A production is generally correlated with bacterial adherence to biomaterial surfaces, apparently by nonspecific adsorption mechanisms, with subsequent biofilm formation and bacterial colonizaton mediated by PIA.[38] PS/A is a high-molecular-weight variably N-succinylated (65-100%) β-1,6-linked N

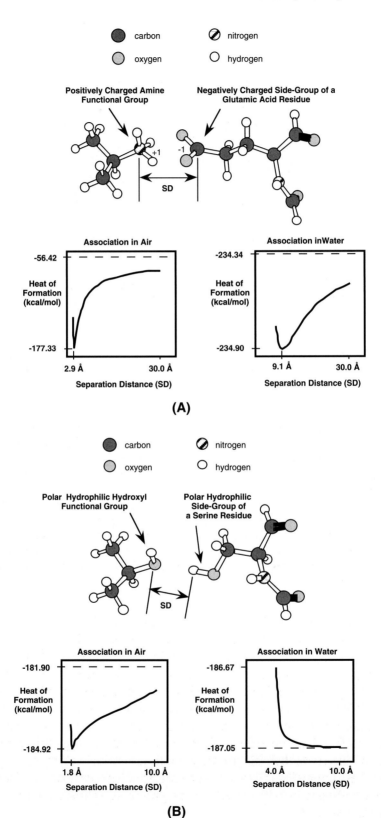

**(A)**

**(B)**

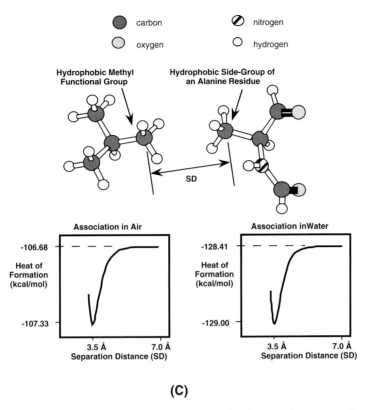

**(C)**

**Figure 3.** Molecular models illustrating intermolecular interactions as a function of chemical structure and environment. Plots present calculated heats of formation versus functional group separation distance (SD) in air and water. Dotted lines indicate energy at infinite separation. A) Ionic interaction between a positively charged amine functional group and a negatively charged carboxylic acid side-group of a glutamic acid residue. B) Polar interaction (hydrogen bonding) between a hydrophilic hydroxyl functional group and a hydrophilic hydroxyl side-group of a serine residue. C) Nonpolar interaction between a hydrophobic methyl functional group and a hydrophobic methyl side-group of an alanine residue. (Calculations conducted using semi-empirical quantum mechanical theory using MOPAC/PM3 with COSMO water simulation; CAChe Software, Oxford Molecular, Beaverton, OR.)

acetylated glucosamine.[38] NMR data suggests that PIA has an unbranched structure with β-1,6 interglycosidic linkages containing at least 130 residues. PIA is reportedly made up two primary types of polysaccharides, termed PS I (80%) and PS II (20%). PS I has 80 to 85% *N*-acetyl-D-glucosaminyl residues, with the remainder being non-*N*-acetylated positively charged D-glucosaminyl residues which are apparently randomly distributed along the chain. PS II is similar to PS I, however, it contains fewer non-*N*-acetylated side groups and a small number of phosphate and ester-linked succinyl residues, giving the overall chain a moderately anionic character.[25,35,49] The unbranched, long-chain structure of PS I and PS II, combined with their distributed anionic and cationic charges, is believed to be important for both intercellular adhesion and biofilm adherence, with van der Waals, electrostatic, and hydrogen bonding mechanisms providing molecular attraction between polysaccharide chains.[35]

Recent studies have found some evidence for cell-to-cell signaling in bacterial biofilms adsorbed to surfaces such as in the growth of *Pseudomonas aeruginosa* adsorbed onto glass.[14] Cell-to-cell signaling in bacterial biofilms may alter antibiotic resistance or adherence characteristics of other pathogens on medical implant surfaces.

## V. MODELING MICROBIAL ADHESION

Molecular adhesion involves noncovalent intermolecular interactions governed by either secondary bonding between induced dipolar and dipolar functional groups (also referred to as physical bonding or van der Waals bonding) or electrostatic interactions between charged functional groups.[10] Although individual secondary bonds and electrostatic interactions within an electrolyte solvent (i.e., physiologic saline) are relatively weak compared to primary chemical bonds, their combined effects can result in very strong binding between macromolecules. The reader is referred to Chapter 1 for a review of intermolecular interactions.

These three basic types of secondary bonding and ionic interactions are graphically presented in Figure 3, which presents molecular models of 3 sets of molecules illustrating hydrophobic, neutral hydrophilic and ionic interactions. Two energy *vs* distance plots are presented for each model; one representing the calculated system heat of formation versus molecular separation in a solvent free environment (i.e., air) and the other in an aqueous solution. The net adsorption enthalpy for each set of molecules represents the difference in the heats of formation between the molecules at infinite separation compared to the system heat of formation when they are closely associated. From these energy plots, the hydrophobic functional groups are shown to only weakly attract one another over a short range of only about 6 Å in both the air and aqueous environments. In an aqueous environment, however, additional effects due to the entropy associated with the orientation of water molecules surrounding the functional groups provide an even larger thermodynamic driving force to minimize the solvent accessible surface, thus tending to strongly bind hydrophobic groups together. In contrast to this, interaction between the polar hydrophilic hydroxyl groups is shown to provide a relatively strong short-range molecular attraction in air but an effective energy barrier to binding in an aqueous environment due to preferential adsorption of water. Finally, interactions between positive and negative charges in an ionic system show very long-range strong attraction in air with close ion group association in the bound state. However, in an aqueous environment the ionic functional groups show long-range but relatively weak attraction with the low energy binding state, predicting that the functional groups are separated by a distance approximately equal to two water molecules, thus representing a solvation layer about each charged group.

These results have recently been measured using atomic force microscopy (AFM). The interactions between uncharged glycolipids with varying head-groups and the organic surfaces of self-assembled monolayers (SAMs) were measured in deionized water. It was found that in the absence of ions, the water structure surrounding the organic layers influences the adhesion. The presence of $Ca^{2+}$ ions produced an enhanced attraction between sialic acid and the ganglioside GM1.[3]

## VI. CONCLUDING REMARKS

As discussed above, the critical first step in the development of bacterial infections of tissues and medical devices involves bacterial adhesion to host cells, ECM glycoproteins, or biomaterials surfaces. These binding mechanisms most often involve very specific submolecular interactions, many of which are still not well understood. Therapies which can prevent or disrupt these molecular mechanisms have great potential for development as alternatives to antibiotics for the prevention and treatment of infection, thus emphasizing the importance of research directed toward the elucidation of the molecular basis for bacterial adhesion.

# REFERENCES

1. An YH, Friedman RJ: Concise review of mechanisms of bacterial adhesion to biomaterials surfaces, *J Biomed Mater Res (Appl Biomater)* 43:338–48, 1998
2. Bieber D, Ramer SW, Wu C-Y, et al: Type IV pili, transient bacterial aggregates, and virulence of enteropathogenic *Escherichia coli. Science* 280:2114–8, 1998
3. Boland T, Dufrêne Y, Barger WR, et al: Characterization of hybrid lipid bilayers with atomic force microscopy. *Crit Rev Bioeng* In press, 1999
4. Bos R, van der Mei HC, Busscher HJ: A quantitative method to study co-adhesion of micro-organisms in a parallel plate flow chamber. II: Analysis of kinetics of co-adhesion. *J Microbiol Methods* 23:169–82, 1995
5. Boulangé-Petermann L, Rault J, Bellon-Fontaine MN: Adhesion of *Streptococcus thermophilus* to stainless steel with different surface topography and roughness. *Biofouling* 11:201–16, 1997
6. Bradshaw DJ, Homer KA, Marsh PD, et al: Metabolic cooperation in oral microbial communities during growth on mucin. *Microbiology* 140:3407–12, 1994
7. Bradshaw DJ, Marsh PD, Watson GK, et al: Effect of conditioning films on oral microbial biofilm development. *Biofouling* 11:217–26, 1997
8. Busscher HJ, van der Mei HC: Use of flow chamber devices and image analysis methods to study microbila adhesion. *Meth Enzymol* 253:455–77, 1995
9. Busscher HJ, van der Mei HC: Physico-chemical interactions in initial microbial adhesion and relevance for biofilm formation. *Adv Dent Res* 11:24–32, 1997
10. Callister WD Jr: Atomic structure and interatomic bonding. In: *Materials Science and Engineering. An Introduction.* 4th Ed, John Wiley & Sons, New York, 1997:19–26
11. Cheung AL, Krishnan M, Jaffe EA, et al: Fibrinogen acts as a bridging molecule in the adherence of *Staphylococcus aureus* to cultured human endothelial cells. *J Clin Invest* 87:2236–45, 1991
12. Costerton JW, Stewart PS, Greenberg EP: Bacterial biofilms: A common cause of persistent infections. *Science* 284:1318–22, 1999
13. Darouiche RO, Landon GC, Patti JM, et al: Role of *Staphylococcus aureus* surface adhesins in orthopaedic device infections: Are results model-dependent? *J Med Microbiol* 46:75–9, 1997
14. Davies DG, Parsek MR, Pearson JP, et al: The involvement of cell-to-cell signals in the development of a bacterial biofilm. *Science* 280:295–8, 1998
15. Deibel C, Krämer S, Chakraborty T, et al: EspE, a novel secreted protein of attaching and effecing bacteria, is directly translocated into infected host cells, where it appears as a tyrosine-phosporylated 90 kDa protein. *Mol Microbiol* 28:463–74, 1998
16. Doi RH, Goldstein MA, Park JS, et al: Structure and function of the subunits of the Clostridium cellulovorans cellulosome, In: Shimada K, Hoshino S, Ohmiya K, et al., eds. *Genetics, Biochemistry and Ecology of Lignocellulose Degradation.* Uni Publishers, Tokyo, 1994:43–52
17. Dristina AG: Biomaterial-centered infection, microbial vs tissue integration. *Science* 237:1588–97, 1987
18. Dufrêne YF, Boonaert CPJ, Rouxhet PG: Adhesion of *Azospirilium brasilense*, role of proteins at the cell-support interface. *Colloids Surfaces B Biointerfaces* 7:113–28, 1996
19. Eidhin DN, Perkins S, Francois P, et al: Clumping factor B (ClfB), a new surface-located fibrinogen-binding adhesin of *Staphylococcus aureus. Mol Microbiol* 30:245–57, 1998

20. Ellen RP, Veisman H, Buivids IA, et al: Kinetics of lactose-reversible co-adhesion of *Actinomyces naeslundii* WVU 398A and *Streptococcus oralis* 34 on the surface of hexadecane droplets. *Oral Microbiol Immunol* 9:364–71, 1994

21. Escher A, Characklis WG: Modeling the initial events in biofilm accumulation. In: Characklis WG, Marshall KC, eds. *Biofilms*. John Wiley and Sons, New York, 1990:445–86

22. Fletcher M, Lessmann JM, Loeb GI: Bacterial surface adhesives and biofilm matrix polymers of marine and freshwater bacteria. *Biofouling* 4:129–40, 1991

23. Frankel G, Lider O, Herhkoviz R, et al: The cell-binding domain of intimin from entero-pathogenic *Escherichia coli* binds to $\beta_1$ integrins. *J Biol Chem* 271:20359–64, 1996

24. Greene C, McDevitt D, Francois P, et al: Adhesion properties of mutants of *Staphylococcus aureus* defective in fibronectin-binding proteins and studies on the expression of fnb genes. *Mol Microbiol* 17:1143–52, 1995

25. Heilmann C, Schweitzer O, Gerke C, et al: Molecular basis of intercellular adhesion in the biofilm-forming *Staphylococcus epidermidis*. *Mol Microbiol* 20:1083–91, 1996

26. Hicks S, Frankel G, Kaper JB, et al: Role of intimin and bundle-forming pili in enteropatho-genic *Escherichia coli* adhesion to pediatric intestinal tissue *in vitro*. *Infect Immun* 66:1570–8, 1998

27. Hostalka F, Moultrie A, Stutzenberger F: Influence of carbon source on cell surface topology of *Thermomonospora curvata*. *J Bacteriol* 174:7048–52, 1992

28. Hultgren SJ, Jones CH, Normark S: Bacterial adhesins and their assembly. In: Neidhardt FC, ed. *Escherichia Coli and Salmonella. Cellular and Molecular Biology*. 2nd ed, Vol 2, ASM Press, Washington, DC, 1996:2750–56

29. Husscher HJ, Doornbusch GI, van der Mei HC: Adhesion of mutans streptococci to glass with and without salivary coating as studied in a parallel plate flow chamber. *J Dent Res* 71:491–500, 1992

30. Ichi-ishi S, Sheweita S, Doi RH: Characterization of EngF from Clostridium cellulovorans and identification of a novel cellulose binding domain. *Appl Environ Microbiol* 64:1086–90, 1998

31. Lamed R, Bayer EA: The cellulosome concept: exocellular and extracellular enzyme factor centers for efficient binding and cellulolysis. In: Aubert J-P, Beguin P, Millet J, eds. *Biochemistry and Genetics of Cellulose Degradation*. Academic Press, San Diego, 1988:101–16

32. Lamont RJ, Rosan B: Adherence of mutant streptococci to other oral bacteria. *Infect Immun* 58:1738–43, 1990

33. Liljemark WF, Bloomquist CG, Coulter MC, et al: Utilization of a continuos streptococcal surface to measure interbacterial adherence *in vitro* and *in vivo*. *J Dent Res* 67:1455–60, 1988

34. Ljungh Å, Moran AP, Wadström T: Interactions of bacterial adhesins with extracellular matrix and plasma proteins: pathogenic implications and therapeutic possibilities. *FEMS Immunol Med Microbiol* 16:117–26, 1996

35. Mack D, Fischer W, Krokotsch A, et al: The intercellular adhesin involved in biofilm accumulation of *Staphylococcus epidermidis* is a linear $\beta$-1,6-linked glucosaminoglycan: Purificaton and structural analysis. *J Bacteriol* 178:175–83, 1996

36. Matuschek M, Burchhart G, Sahm K, et al: Pullalanase of *Thermoanaerobacterium thermosturigenes* EMI (*Clostridium thermosturigenes*): molecular analysis of the gene, composite structure of the enzyme, and a common model for its attachment to the cell surface. *J Bacteriol* 176:3295–302, 1998

37. McDevitt D, Nanavaty T, House-Pompeo K, et al: Characterization of the interaction between the *Staphylococcus aureus* clumping factor (ClfA) and fibrinogen. *Eur J Biochem* 247:416–24, 1997

38. McKenney D, Hübner J, Miller E, et al: The ica locus of *Staphlococcus epidermidis* encodes production of the capsular polysaccharide/adhesin. *Infect Immun* 66:4711–20, 1998

39. Mulvey MA, Lopez-Boado YS, Wilson CL, et al: Induction and evasion of host defenses by type 1-piliated uropathogenic *Escherichia coli*. *Science* 282:1494–7, 1998

40. Neu TR, Marshall KC: Bacterial polymers: Physico-chemical aspects of their interaction at interfaces. *J Biomater Appl* 5:107–33, 1990

41. Neu TR: Significance of bacterial surface-active compounds in interaction of bacteria with interfaces. *Microbiol Rev* 60:151–65, 1996

42. Norde W: Driving forces for protein adsorption at solid surfaces. In: Maalmsten M, ed. *Biopolymers at Interfaces.* Marcel Dekker, NY, 1998:27–54

43. O'Connell DP, Nanavaty T, McDevitt D, et al: The fibrinogen-binding MSCRAMM (clumping factor) of *Staphylococcus aureus* has a $Ca^{2+}$-dependent inhibitory site. *J Biol Chem* 273:6821–29, 1998

44. Patti JM, Allen BL, McGavin MJ, et al: MSCRAMM-mediated adherence of micro-organisms to host tissues. *Ann Rev Microbiol* 48:585–617, 1994

45. Prescott LM, Harley JP, Klein DA: Procaryotic cell structure and function. In: Prescott LM, Harley JP, Klein DA, eds. *Microbiology.* 4th edition, McGraw-Hill, Boston, 1999:52

46. Quirynen M, Bollen CM: The influence of surface roughness and surface-free energy on supra- and subgingival plaque formation in man. A review of the literature. *J Clin Periodontol* 22:1–14, 1995

47. Relman D, Tuomanen E, Falkow S, et al: Recognition of a bacterial adhesion by an integrin: macrophage CR3 ($\alpha_M\beta_2$, $CD_{11}/CD_{18}$) binds filamentous hemagglutinin of *Bordella pertussis. Cell* 61, 1375–82, 1990

48. Rittman BE: Detachment from biofilms. In: Characklis WG, Wilderer PA, eds. *Structure and Function of Biofilms.* John Wiley and Sons, Chichester, 1989:49–58

49. Rupp ME, Ulphani JS, Fey PD, et al: Characterization of the importance of polysaccharide intercellular adhesin/hemagglutinin of *Staphylococcus epidermidis* in the pathogensis of biomaterial-based infection in a mouse foreign body infection model. *Infect Immun* 67:2627–32, 1999

50. Schneider RP, Marshall KC: Retention of the Gram-negative marine bacterium SW8 on surfaces - effects of microbial physiology, substratum nature and conditioning films. *Colloids Surfaces B Biointerfaces* 2:387–96, 1994

51. Sokurenko EV, Courtney HS, Maslow J, et al: Quantitative differences in adhesiveness of type 1 fimbriated *Escherichia coli* due to structural differences in fimH genes. *J Bacteriol* 177:3680–6, 1995

52. Soto GE and Hultgren SJ: Minireview. Bacterial adhesins: Common themes and variations in architecture and assembly. *J Bacteriol* 181:1059–71, 1999

53. Striker R, Nilsson U, Stonecipher A, et al: Structural requirements for the glycolipid receptor of human uropathogenic *Escherichia coli. Mol Microbiol* 16:1021–9,1995

54. Tamaru Y, Doi RH: Three surface layer homology domains at the N terminus of the *Clostridium cellulovorans* major cellulosomal subunit EngE. *J Bacteriol* 181:3270–6, 1999

55. Tomme P, Warren RA, Miller RC, et al: Cellulose binding domains: classification and properties. In: Saddler JN, Penner MH, eds. *Enzymatic Degradation of Insoluble Carbohydrates.* American Chemical Society, Washington, DC, 1995:42–161

56. Vacheethasanee K, Temenoff JS, Higashi JM, et al: Bacterial surface properties of clinically isolated *Staphylococcus epidermidis* strains determine adhesion on polyethylene. *J Biomed Mater Res* 42:425–32, 1998

57. Van Loosdrecht MC, Lyklema J, Norde W, et al: Influence of interfaces on microbial activity. *Microbiol Rev* 54:75–87, 1990

58. Vaudaux PE, Franzçois P, Proctor RA, et al: Use of adhesion-defective mutants of *Staphlococcus aureus* to define the role of specific plasma proteins in promoting bacterial adhesion to canine arteriovenous shunts. *Infect Immun* 63:585–90, 1995

59. Weigand PV, Timmis KN, Chhatwal GS: Role of fibronectin in staphylococcal colonization of fibrin thrombi and plastic surfaces. *J Med Microbiol* 38:90–5, 199

# 3

# Molecular Genetics
# of Bacterial Adhesion and Biofouling

## Helen M. Dalton and Paul E. March

*School of Microbiology and Immunology, University of New South Wales,*
*Sydney, Australia*

## I. INTRODUCTION

It has been estimated that in natural environments ninety nine percent of all bacteria exist in biofilms, or at least reside at surfaces.[6] To date, bacterial attachment or adhesion to both natural and artificial surfaces has been examined in many different contexts. These include attachment to inert (solid and agar) and animate (eukaryotic cells) surfaces. In the vast majority of cases little is known about how bacteria attach to these surfaces. Recent advances, particularly in the area of bacteria adhering to eukaryotic cells, have pointed to the importance of surface organelles and receptors in the attachment process. Following attachment, growth on solid surfaces results in the formation of a biofilm, and in most cases this contributes to biofouling. Advances in fluorescent microscopy and reporter gene-fusion technology have provided a means of detecting altered gene expression as it occurs during microbial growth at surfaces and within biofilms (*see* Chapter 12).

Contemporary research has focused on elucidation of the process of adhesion of bacteria to human, animal and plant cells. These studies also provide a conceptual framework for understanding how bacteria may attach to surfaces, leading to paradigms that explain initial bacterial–surface interactions. Adhesion to animal cells (human and otherwise) proceeds via a mechanism that requires receptor–ligand interactions and altered expression of target cell genes.[11] Factors on the bacterial cell surface for attachment include fimbrial and afimbrial adhesins. This chapter will highlight recent advances made towards revealing the genes and molecules necessary for bacterial adhesion to surfaces and cells. The oral cavity offers an accessible model system for studying global concepts of bacterial adhesion and it is in this environment that many principles of biofilm formation on implanted inert devices have evolved. Whittaker et al.,[29] and Jenkinson and Lamont[17] have provided comprehensive contemporary reviews on this topic.

## II. ATTACHMENT

### A. Flagella and Fimbriae as Antennae

Flagella have been extensively studied as the locomotory apparatus of bacterial cells. Like fimbriae, flagella (Fig. 1a) are large complex protein assemblages spanning the

*Handbook of Bacterial Adhesion: Principles, Methods, and Applications*
Edited by: Y. H. An and R. J. Friedman © Humana Press Inc., Totowa, NJ

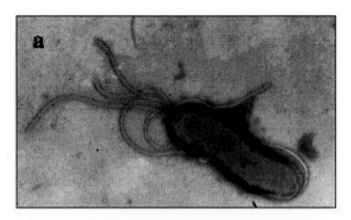

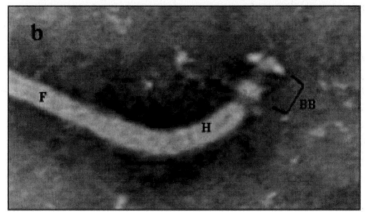

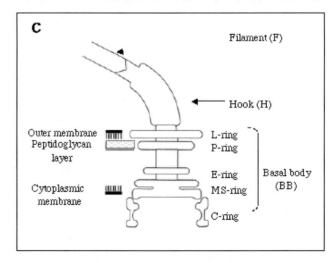

**Figure 1.** (a) Transmission electron micrograph of a negatively stained bacterium with sheathed bipolar flagella (*Helicobacter* sp). Magnification ×12000. (b) Transmission electron micrograph of sheathed filament (F), hook (H) and basal body (BB) of these flagella revealed after Teepol treatment. The rings of the basal body correspond to the MS, P and L rings (S.J. Danon, personal communication). Magnification ×150000. (c) Diagram of some key structures of a flagellum. Based on the *Caulobacter crescentus* flagellum.[23] Micrographs courtesy of S.J. Danon and J. O'Rourke.

bacterial cell wall. Protein subunits exposed in the hook and filament of flagella (Fig. 1b,c) may be ideally positioned to mediate adhesion to either animate or inert surfaces. Interestingly, research into the induction of lateral flagella of marine *Vibrio* sp. suggests that during surface colonization flagella may function as mechanosensors instead of adhesins.[18] These particular bacteria exist in the marine environment as planktonic rod-shaped cells, 2 μm in length, and containing a single polar flagellum (swimmer cells). They also colonize both animate and inert surfaces in the marine environment. In laboratory contrived systems surface attachment leads to conversion of the swimmer cell into a swarmer cell, more than 30 μm in length and possessing many lateral flagella. This alteration of the bacterial cell morphology allows the swarmer cell to efficiently colonize the surface with which it has interacted. The polar flagellum of the swimmer cell derives energy for rotation from sodium ion transport, whereas the lateral flagella utilize proton transport.[3] Specific inhibition of rotation of the polar flagellum by agents that block sodium ion channels results in production of lateral flagella.[18] Together these observations suggest that when a swimmer cell approaches a surface, the rotation of the polar flagellum will be negatively affected. The decrease in rotation (or sodium ion flux) provides a signal to up-regulate surface-specific genes (in this case lateral flagella). In this way the polar flagellum acts as a mechanosensor sensitive to the proximity of a surface. In aquatic and marine environments there are many swimming bacteria that also have the ability to colonize surfaces. It is anticipated that future investigations will reveal other bacteria that commonly employ flagella as mechanosensors during surface colonization in natural ecosystems.

In *Pseudomonas aeruginosa* flagellar function is directly coupled to adhesion to mucin. Mutations in the *fliF* gene lack the MS ring (the flagellum's base), are nonmotile and nonadhesive.[1] In addition, the expression of both adhesin and flagella are dependent on transcriptional activators (FleQ and FleR) that function in concert with the alternative sigma factor RpoN. The reason for this co-regulation has not been established and it is possible that flagellar protein components are the actual adhesins in this system.[2] The role of flagellar genes in early stages of attachment to inert surfaces has been investigated by transposon mutagenesis coupled with time-lapse video microscopy.[24] In that investigation it was suggested that *Pseudomonas aeruginosa flgK* mediates a role for flagella in initial cell-to-surface attachment. The same observation has been made for *Escherichia coli* where it was shown (using defined *fli*, *flh*, *mot* and *che* alleles) that motility, but not chemotaxis, was critical for biofilm development. Motility was specifically required for normal initial attachment and movement along a surface.[26]

Fimbriae (or pili) are filamentous structures composed of protein subunits found on a variety of bacterial cell surfaces (including *Escherichia coli*, *Haemophilus influenzae*, *Pseudomonas aeruginosa*, *Vibrio cholerae*, *Neisseria* sp., *Moraxella* sp.). The role of fimbriae in bacterial adhesion during pathogenesis has been the topic of a wide variety of investigations and has recently been reviewed by Finlay and Falkow.[11] Several important themes have emerged from these studies. First, specific fimbriae-dependent, bacterial–host cell interactions depend on protein resident either in the body of the filament, or at the tip. Second, fimbriae bind to specific receptors on the host and activate host cell genes and signal transduction pathways leading to enhanced adhesion or invasion. It is intriguing to speculate that fimbriae may have a broader role in attachment of bacteria to surfaces outside of the realm of bacterial pathogenesis. An investigation in *E. coli* has implicated a role for type I pili in initial attachment to inert surfaces (polyvinylchloride plastic).[26] This is in contrast to a second study, which indicated that fimbriae are not involved in attachment of *E. coli* to urinary catheters.[28] A role for fimbriae in attachment of the marine bacterium *Hyphomonas* has been investigated. Using transmission electron microscopy of synchronized populations, it has been shown that the timing of fimbriae production correlates with the adhesion stage of the developmental cycle of *Hyphomonas.*[27]

### B. The Role of Specific Bacterial Cell-Wall Protein Receptors

It is certainly possible that receptor–ligand interactions could be a very important aspect of bacterial attachment to both inert as well as living surfaces. Such a process could provide specificity determinants that allow bacteria to selectively colonize a particular environment or surface. Many contemporary studies in this area have focused on bacterial colonization of animate surfaces. Studies most relevant to colonization of inert surfaces are discussed here.

Streptococci express an array of cell surface components that are important for adhesion to host cells. It has been proposed that the type of adhesin used will depend upon the strain, environmental factors and the receptor expressed by the host cell.[13] Further, these authors suggest that, adhesion of streptococci to host cells proceeds through an initial reversible attachment followed by irreversible binding. They have subsequently provided evidence that suggests that the reversible step involves a hydrophobic interaction between the host cell and lipoteichoic acid of the bacterial cell wall. Additional results with the streptococcal M protein adhesin indicated that subsequent irreversible adhesion required exposed M protein.[7] The two-stage model for attachment (loose reversible binding followed by more secure irreversible attachment) is exactly analogous to what is likely to happen when bacteria colonize inert surfaces during biofilm formation.

An important class of adhesins includes factors that bind specifically to extracellular matrix (ECM) components. Fibronectin (Fn) is a major component of the ECM and a number of recent studies have focused on establishing the role of Fn-binding proteins in bacterial adhesion. Using a combination of biochemical and genetic approaches, Konkel et al.[19] have demonstrated that the gram-negative bacterium *Campylobacter jejuni* expresses a 37 kDa outer membrane protein. This protein binds to Fn. Interestingly, the amino acid sequence of the protein (CadF) is 52% similar to the root adhesin protein from *Pseudomonas fluorescens*. This finding suggests that similar outer membrane proteins act as adhesins in diverse environments. Further experimentation is warranted to examine whether such adhesins are widespread in bacteria.

Greene et al.[12] employed a genetic approach to demonstrate that expression of both of the *Staphylococcus aureus* cell-wall associated Fn-binding proteins (FnBPA and FnBPB) were required for adhesion to Fn-coated coverslips. Mutated *S. aureus* harboring an insertion in either the *fnbA* or *fnbB* gene was not defective in the adhesion assay, but the *fnbA-fnbB* double mutant was completely defective in adhesion. If either wild type gene was supplied separately on a multicopy plasmid, Fn-mediated adhesion was restored. Furthermore, the double mutant was defective in adhesion to coverslips explanted after subcutaneous implantation of guinea pigs which indicates that the FnBPA-FnBPB/Fn-interaction is important in *S. aureus* attachment to biomaterials in vivo.[12] This demonstrates the potential of an adhesin to direct attachment to either a living surface or an inert surface so long as extracellular material is first deposited on the inert surface. It is likely that inert surfaces decorated with remnants of biological substances provide niches that are widespread in many diverse environments. This being the case, then it is also likely that cell-specific adhesins may be co-opted for the process of colonizing inert as well as living surfaces.

## III. GENES NECESSARY FOR BIOFILM FORMATION

What is the fate of the bacterial cell after attachment to a surface? The genes and molecules involved in this process have been reviewed recently with regard to bacterial–animal cell interaction[11] and bacterial–plant cell interaction.[5] In contrast, much less is known about genes specifically required for maintenance of bacterial growth on inert surfaces. Initially, a nude surface exposed to an environment is coated with adsorbed organic molecules followed by a bacterial biofilm. Subsequent biofouling occurs as a result of continual accumulation of biological material. The genetic control of biofilm formation and maintenance is a young science, and less is known about the complex interactions that occur during late stages of biofouling. One system that has recently provided important information in this regard is the formation of biofilms by *Staphylococcus epidermidis.*

*S. epidermidis* is a commensal of human skin; however, some strains can efficiently colonize solid surfaces and form biofilms. These biofilm-forming strains are a major cause of infections arising from medical implants in humans. Two distinct phases are necessary for biofilm formation: initial attachment and intercellular adhesion. Transposon mutagenesis was employed to obtain mutations blocked in each phase.[14] A mutation that blocked initial attachment was localized to a gene whose product is 61% identical to the amino acid sequence of the major autolysin of *S. aureus.*[16] It is intriguing to speculate that an autolysin may be important for attachment of a bacterium to a surface because autolysins are important for cell-wall turnover and cell division. Unfortunately the authors were unable to establish whether the autolysin had a specific versus an indirect role in the attachment process.

A transposon insertion that blocked the second stage of biofilm formation was localized to an operon involved in the synthesis of a polysaccharide known to be essential for biofilm formation.[15,20] The mutated strain was deficient in intercellular adhesion, the ability to form biofilms, and the synthesis of a cell surface polysaccharide. A clone of the unmutated operon restored all three phenotypes when transformed into the mutated strain. Most strikingly, following DNA transformation the same clone converted the non-biofilm forming species *Staphylococcus carnosus* to a biofilm-forming species. Importantly, this demonstrates the necessity of a specific molecule for the establishment of a biofilm.

Transposon mutagenesis has also been applied to genetic screens in *Pseudomonas aeruginosa* aimed at obtaining mutant strains that were defective in the initial stages of cell-to-surface attachment. A screen of 14,000 transposon mutants yielded 37 mutants that were unable to form a biofilm on inert surfaces, but which grew at wild-type rates in liquid medium. Initial DNA sequence analysis was performed on 24 of these isolates, with only three of these 24 mutants showing transposition within a known gene. Two of the three were involved in flagella synthesis and the third encoded the Clp protease.[24] Since 21 out of 24 loci had no known function, these results provide a strong indication that the biology of attachment to inert surfaces may involve factors that have yet to be characterized.

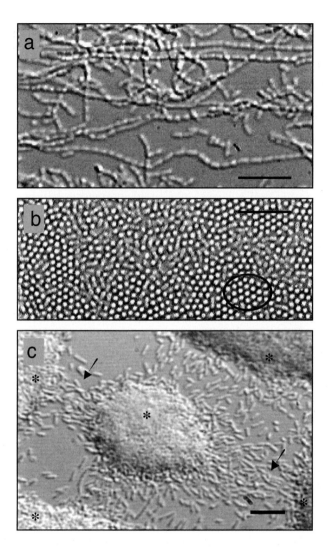

**Figure 2.** Differential interference contrast microscope images representing colonization features of biofilm development by the marine bacteria; (a) *Psychrobacter* sp. SW5 showing long chains of cells, (b) *Psychrobacter* sp. SW8 showing cells vertically packed into a honeycomb-like structure (oval) and (c) *Pseudoalteromonas* sp. S9 cells showing micro-colonies (*) and cells translocating between these colonies (arrows). Bar = 10 μm.

## IV. REGULATION OF GENE EXPRESSION IN BIOFILMS

As discussed in Chapter 12, methodological limitations have prohibited robust advances in understanding how genes are regulated when cells grow in biofilms. Recently however, there has been a considerable amount of interest in the role that extracellular signals may play in this process within biofilms. It is clear that intra- and interspecies cell–cell communication occurs in biofilms found in nature. The cell–cell interactions can be quite sophisticated as evidenced by microscopic examination of marine bacteria using laminar flow chambers and time-lapse video imaging. Such studies have shown that some bacteria form long chains of cells when attached to specific surfaces (Fig. 2a), whereas other strains pack into honeycomb-like structures (Fig. 2b) or aggregate into inter-connected microcolonies (Fig. 2c).[8,9] In the context of these observations it was anticipated that cell density dependent signaling pathways will regulate biofilm-specific gene expression. In fact, initial studies demonstrated that cell density regulated phenotypes were present in cells growing within biofilms[4] and that *N*-acyl homoserine lactone (AHL)-dependent regulation occurred in biofilms.[21] Davies et al.[10] provided the first definitive demonstration of the role of this type of signaling within biofilms when they showed that *Pseudomonas aeruginosa lasI* mutants (defective in AHL synthesis) were unable to form differentiated biofilms. However, exogenous addition of a synthetic signal molecule restored the wild-type biofilm development.[10]

## V. CONCLUSION

Growth and survival of microbial communities depends on adaptation to a series of changing environmental milieux. The ultimate goal of investigations into the molecular genetics of bacterial adhesion and growth at surfaces is to gain insight into the molecular mechanisms of biofilm formation and maintenance. In recent times profound progress in the elucidation of genes and molecules necessary for bacterial attachment to surfaces and subsequent biofilm formation has been made. It is anticipated that this information will lead to the development of strategies to prevent/minimize biofilm formation which result in, for example, reduced material corrosion and frictional drag, and prevention of tissue damage by pathogens. Future research will need to combine multidisciplinary approaches. An illustration of this is the recently developed *in situ* methods aimed at detecting alterations in bacterial gene expression at solid surfaces.[22,25] Such experimental analyses should mimic natural environmental conditions as much as possible. Furthermore, when drawing direct parallels between attachment of bacteria to living and non-living surfaces it is essential to establish relevance since it has been demonstrated that the host cell is not an inert surface, but plays an active role in the attachment process through the activation of specific genes[11].

**Acknowledgment:** The authors acknowledge support by a grant N00014-96-1-0668 from the US Office of Naval Research.

## REFERENCES

1. Arora SK, Ritchings BW, Almira EC, et al: Cloning and characterization of *Pseudomonas aeruginosa fliF* necessary for flagellar assembly and bacterial adherence to mucin. *Infect Immun* 64:2130–6, 1996
2. Arora SK, Ritchings BW, Almira EC, et al: A transcriptional activator, FleQ, regulates mucin adhesion and flagellar gene expression in *Pseudomonas aeruginosa* in a cascade manner. *J Bacteriol* 179:5574–8, 1997
3. Atsumi T, McCarter L, Imae Y: Polar and lateral flagellar motors of marine *Vibrio* are driven by different ion-motive forces. *Nature* 355:182–4, 1992
4. Baker B, Zambryski P, Staskawicz B, et al: Signaling in plant-microbe interactions. *Science* 276:726–33, 1997
5. Batchelor SE, Copper M, Chabra SR, et al: Cell density-regulated recovery of starved biofilm populations of ammonia-oxidizing bacteria. *Appl Environ Microbiol* 63:2281–6, 1997
6. Costerton JW, Cheng K-J, Geesey GG, et al: Bacterial biofilms in nature and disease. *Ann Rev Microbiol* 41:435–64, 1987

7. Courtney HS, Ofek I, Hasty DL: M protein mediated adhesion of M type 24 *Streptococcus pyogenes* stimulates release of interleukin-6 by Hep-2 tissue culture cells. *FEMS Microbiol Lett* 151:65–70, 1997

8. Dalton HM, Poulsen LK, Halasz P, et al: Substratum-induced morphological changes in a marine bacterium and their relevance to biofilm structure. *J Bacteriol* 176:6900–6, 1994

9. Dalton HM, Goodman AE, Marshall KC: Diversity in surface colonization behavior in marine bacteria. *J Indust Microbiol* 17:228–34, 1996

10. Davies DG, Parsek MR, Pearson JP, et al: The involvement of cell-to-cell signals in the development of a bacterial biofilm. *Science* 280:295–8, 1998

11. Finlay BB, Falkow S: Common themes in microbial pathogenicity revisited. *Microbiol Mol Biol Rev* 61:136–69, 1997

12. Greene C, Mcdevitt D, Francois P, et al: Adhesion properties of mutants of *Staphylococcus aureus* defective in fibronectin-binding proteins and studies on the expression of *fnb* genes. *Mol Microbiol* 17:1143–52, 1995

13. Hasty DL, Courtney HS: Group A streptococcal adhesion: all theories are correct. *Adv Exp Med Biol* 408:81–94, 1996

14. Heilmann C, Gerke C, Perdreau-Remington F, et al: Characterization of Tn917 insertion mutants of *Staphylococcus epidermidis* affected in biofilm formation. *Infect Immun* 64:277–82, 1996

15. Heilmann C, Schweitzer O, Gerke C, et al: Molecular basis of intercellular adhesion in the biofilm-forming *Staphylococcus epidermidis*. *Mol Microbiol* 20:1083–91, 1996

16. Heilmann C, Hussain M, Peters G, et al: Evidence for autolysin-mediated primary attachment of *Staphylococcus epidermidis* to a polystyrene surface. *Mol Microbiol* 24:1013–24, 1997

17. Jenkinson HF, Lamont RJ: Streptococcal adhesion and colonization. *Crit Rev Oral Biol* 8:175–200, 1997

18. Kawagishi I, Imagawa M, Imae Y, et al: The sodium-driven polar flagellar motor of marine *Vibrio* as the mechanosensor that regulates lateral flagellar expression. *Mol Microbiol* 20:693–9, 1996

19. Konkel ME, Garvis SG, Tipton SL, et al: Identification and molecular cloning of a gene encoding a fibronectin-binding protein (CadF) from *Campylobacter jejuni*. *Mol Microbiol* 24:953–63, 1997

20. Mack D, Fischer W, Krokotsch A, et al: The intercellular adhesin involved in biofilm accumulation of *Staphylococcus epidermidis* is a linear β-1,6-linked glucosaminoglycan: purification and structural analysis. *J Bacteriol* 178:175–83, 1996

21. McLean RJ, Whiteley M, Stickler DJ, et al: Evidence of autoinducer activity in naturally occurring biofilms. *FEMS Microbiol Lett* 154:259–63, 1997

22. Moller S, Sternberg C, Andersen JB, et al: *In situ* gene expression in mixed-culture biofilms: evidence of metabolic interactions between community members. *Appl Environ Microbiol* 64:721–32, 1998

23. Mohr CD, MacKichan JK, Shapiro L: A membrane-associated protein, FliX, is required for an early step in *Caulobacter* flagellar assembly. *J Bacteriol* 180:2175–85

24. O'Toole GA, Kolter R: Flagellar and twitching motility are necessary for *Pseudomonas aeruginosa* biofilm development. *Mol Microbiol* 30:295–304, 1998

25. Poulsen LK, Dalton HM, Angles ML, et al: Simultaneous determination of gene expression and bacterial identity in single cells in defined mixtures of pure cultures. *Appl Environ Microbiol* 63:3698–702, 1997

26. Pratt LA, Kolter R: Genetic analysis of *Escherichia coli* biofilm formation: roles of flagella, motility, chemotaxis and type I pili. *Mol Microbiol* 30:295–304, 1998

27. Quintero EJ, Busch K, Weiner RM: Spatial and temporal deposition of adhesive extracellular polysaccharide capsule and fimbriae by *Hyphomonas* strain MHS-3. *Appl Environ Microbiol* 64:1246–55, 1998

28. Reid G, van der Mei HC, Tieszer C, et al: Uropathogenic *Escherichia coli* adhere to urinary catheters without using fimbriae. *FEMS Immunol Med Microbiol* 16:159–62, 1996

29. Whittaker CJ, Klier CM, Kolenbrander PE: Mechanisms of adhesion by oral bacteria. *Ann Rev Microbiol* 50:513–52, 1996

# 4

# Factors Influencing Bacterial Adhesion

## Katharine Merritt[1] and Yuehuei H. An[2]

*[1]Division of Life Sciences, Food and Drug Administration, Gaithersburg, MD, USA*
*[2]Department of Orthopaedic Surgery, Medical University of South Carolina,*
*Charleston, SC, USA*

## I. INTRODUCTION

Bacterial adhesion is a very complicated process which is affected by many factors, including some characteristics of the bacteria, the chemical and physical nature of the target material surface, and factors in the bacterial suspension medium including the physical conditions of the medium and the presence of carbohydrates, proteins, serum proteins, or bactericidal substances. In addition, the methods used for material–bacteria interaction and for detection and/or quantitation of bacterial adherence will affect the results. A better understanding of the unique behavior of certain bacteria, the surface characteristics of the material, and the relevant environment would make it possible for one to control the adhesion process by changing these factors. This chapter will address the interaction of bacteria with material surfaces, especially the factors affecting bacterial adhesion to surfaces. Several review articles with similar topics have previously been published, including the ones authored by Neu and Marshall,[102] Merritt,[91] and An and Friedman.[6] This chapter will not address the wealth of information on bacterial coaggregation, adherence of one type of microorganism to another, or adherence of bacteria to mammalian cells or tissue in a fluid or in vivo environment. Rather, this chapter will attempt to address these interactions when they occur on a material surface, either in vivo or in vitro. Intact mammalian tissue will not be considered a material in this analysis. However, tissue that has been processed (engineered) as a material substrate will be considered a material.

## II. METHODS FOR PROVIDING MATERIAL–BACTERIA INTERACTION

For bacterial adhesion to occur, the bacteria must first interact with the surface of the material. The nature of the bacterial surface, the material surface, and the subsequent interactions will dictate whether or not bacterial adhesion and colonization occur. Thus it is critical to carefully describe the conditions under which the interaction took place, and to maintain the same conditions for initial interaction when comparisons of effects of material or of bacterial properties are to be made. Such detailed information is often not

*Handbook of Bacterial Adhesion: Principles, Methods, and Applications*
Edited by: Y. H. An and R. J. Friedman © Humana Press Inc., Totowa, NJ

readily available in published studies and inconsistencies in results may perhaps be attributable to minor differences in technique. The nature of the material surface will reflect the environment into which the material was placed. Insertion into water or saline will change the hydrophobicity/hydrophilicity characteristics of the material. Insertion into protein or carbohydrate-containing solutions will result in immediate coating of the material surface. Thus bacteria added to a solution already containing a material will see a different material surface than bacteria already in solution to which the material is then added. The initial interaction will be very different and the endpoint of adherence and proliferation may be altered. If it is not clear as to what was done, studies cannot be compared adequately. This does not imply that one technique is more biologically relevant, it is simply that they are different and need to be explicitly defined. If one is interested in the effect of organisms on an implant at the time of surgery, then the material needs to be added to the bacteria. If one is interested in the interaction of bacteria in the host on an implanted material, then the material should be preconditioned to mimic the appropriate biological coating before the bacteria are added.

Similarly, there are many methods for studying bacteria–material interaction. In some studies the bacteria and material are left together undisturbed for a designated time period (static test); in some cases there is intermittent mixing; in some cases there is constant agitation by rocking, spinning, or mixing with a magnetic stir bar; and sometimes elaborate devices which allow flow of the bacteria over the material are devised. (For example, a Robbins device or modified Robbins device is often used.) In some cases, a flow cell or loop to mimic blood flow is used. The issue of air–fluid interface is often raised. It is virtually impossible to conduct an in vitro experiment where the material and solution do not first encounter an air–material–fluid interface. With flow cells, it may be possible to avoid a bacteria–air–material interface by initiating solution flow over the material and then adding the bacteria to the solution reservoir. However, this results in a preconditioned material encountering the bacteria. There are no easy solutions to these problems of interaction and the most meaningful system that will address the particular question must be selected.

In addition, there is also the issue of the history of the contact of the bacteria with other surfaces. The bacteria must be grown in a vessel of some kind. The two most common materials for the vessels are borosilicate glass and polystyrene. Some individuals may use polycarbonate. However, there are subtle differences in the nature of these polymers, which may be designated tissue culture grade or bacteriological grade. The exact nature of the material used is often not indicated in published reports. A few studies have been done addressing this issue and there is probably no influence of the past history of bacteria–material contact on the adhesion, however it will forever remain a question.[91] It is advisable that bacteria used in an adhesion experiment be discarded and not used for other experiments. A known stock exposed to the identical environment should be used for subsequent experiments. However, it is also advisable that the stock solution be checked to be sure the adhesion characteristics have been maintained over time and a new stock prepared as indicated. For instance, *Staphylococcus epidermidis* that adheres to polystyrene may lose adherence characteristics over time and with repeated passaging. Thus the strain should be selected periodically from organisms that have adhered to the polystyrene tube or plate.

## III. METHODS OF DETECTION AND QUANTITATION

Subsequent chapters will deal at length with this issue. However, it is important to keep in mind that it is the initial contact of the microorganism with the surface of the material that will dictate whether this organism will stick and proliferate. The surface of the material will be altered by the initial adhesion of microorganisms or interaction with microorganisms and the nature of this surface will determine the fate of further organism-material interaction. Unfortunately, this initial interaction is rarely observed. It is the end result of sufficient organisms on the surface to be detectable by the selected method that is used to measure bacterial adhesion. This is really adhesion and proliferation.

## IV. ENVIRONMENT (THE MEDIUM)

Bacterial adhesion is determined by the properties of all three phases involved, i.e., the adhering bacteria, the substrate, and the suspending liquid medium.[3] Some factors in the general environment of the suspending liquid medium, such as the types of medium,[71] the shear stress of the flowing medium,[36,102] temperature,[75] time period of exposure, bacterial concentration, chemical treatment or the presence of antibiotics,[139] and surface tension of the medium,[3] will affect bacterial adhesion.

It was reported that marked differences in both the production of slime and adherence of *S. epidermidis* were observed when comparing four culture media.[71] Slime production was notably poor in used peritoneal dialysis fluid (PUD). Adherent growth was markedly increased in a chemically defined medium (HHW) and synthetic dialysis fluid (SDF) but was poor in tryptic soy broth (TSB) and PUD when air with 5% $CO_2$ was used. These findings emphasize the advantages in using chemically defined and biological fluids when studying slime production and adherence by *S. epidermidis*.

Under flowing conditions the flow pattern is an important factor in attachment of bacteria to a solid surface and under these conditions a shear stress is applied to bacterial cells attached to the surface. Adhesion is optimal under a shear stress of 6-8 $N/m^2$, but still occurs under shear forces up to 130 $N/m^2$.

It was found that adhesion of *Streptococcus faecium* to glass increased with increasing temperature (to 50°C), time period of exposure, and bacterial concentration. The equilibrium apparently was not reached even after incubation for 8 h or at a cell concentration of $3 \times 10^{10}$/mL.[110] This effect of temperature was also found by other researchers.[42] Various studies on the adhesion of *S. epidermidis* and *Staphylococcus aureus* have shown dependence on time and temperature. If the conditions are not correct for optimal growth, then proliferation may be considerably slower, but there does appear to be a saturation level that is reached between 18 h and 48 h in adequate growth conditions.

Limited studies with *S. epidermidis* indicated that the bacterial concentration plays a role in adhesion and biofilm formation. It is bacterial concentration (cfu/mL) that is important and not the available organisms, indicating a surface hit phenomenon. For example, a concentration of $10^6$ cfu in 1 mL would give greater adhesion than would $10^6$ cfu in 10 mL. The number of available organisms is the same, but the number at the surface in any given time is different.[91]

Fletcher and Marshall[44] and Satou et al.[123] found that the number of bacteria adhering to substrata surfaces increased with time until they reached a saturation level which was specific for each type of surface. The adhesion of *Staphylococcus pyogenes* to hexadecane was abolished by pretreating the organisms with trypsin and pepsin or HCl solutions.[27,104] Hogt et al. also found that treatment of *S. epidermidis* with pepsin or extraction with aqueous phenol resulted in a decreased adhesion to poly (tetrafluorethylene-co-hexa-fluoropropylene) (FEP).[69]

Concentrations of electrolytes (KCl or NaCl),[1,16,110] or $CO_2$,[33,71,139] and pH value[56,61,110] in the culture environment also influence bacterial adhesion and slime production. The presence of iron,[30,75] cadmium or zinc,[89] sugars,[31] or surfactants such as Tween 20 or 80[74] in the medium or bacterial suspension may also alter adhesion. The presence of antibiotics will also affect adhesion and proliferation. Cultures of *S. epidermidis* in subinhibitory concentrations (0.5 MIC) of cephalothin, clindamycin and vancomycin resulted in a 30–80% reduction in adhesion.[111] All of these factors may influence bacterial adhesion by either changing physical interactions in phase one of adhesion, or changing surface characteristics of bacteria or materials. This will be amplified in further sections.

Finally, Giridhar et al.[54] reported that the exposure of polymethylmethacrylate (PMMA) discs to extracellular slime extracted from strain RP12 greatly reduced adherence of strain RP12, SP2, SE-360, and *S. epidermidis* RP62A. The active component(s) was present in the >10 kD mol wt fraction obtained by Amicon YM10 ultrafiltration of crude slime; heat treatment of the fraction did not affect its inhibitory activity. When the bacteria and RP12 slime fractions were added simultaneously to the PMMA discs, the >10-kD mol wt fraction of slime competitively inhibited adherence of strain RP12 to PMMA discs.

## V. BACTERIAL CHARACTERISTICS

The nature of the bacterial species has a major impact on the adhesion to surfaces. Some bacteria (mostly the Gram-negative bacteria) adhere by virtue of protein-based structures such as pili, fimbriae, and flagellae. Most bacteria under discussion in this chapter are staphylococci and streptococci with adherence based on carbohydrate structures. However, the factors which affect adherence will be very different for adhesion by protein structures and by carbohydrate structures and may also be different within these mechanisms. It will be difficult to draw generalities and there needs to be attention to the specifics. For a given material surface, different bacterial species, and strains adhere differently, and this can be explained physicochemically since physicochemical characteristics of bacteria are different between species and strains.

### A. Bacterial Hydrophobicity

Surface hydrophobicity of bacteria is determined by cell surface components such as fimbriae,[51,65] polypeptides,[72] and 60,000- to 90,000-mol wt proteins of *Streptococcus sanguis*[51] prodigiosin and amphyipathic aminolipids of *Serratia marcescens*.[12,65] The hydrophobicity of bacteria varies according to bacterial species and is influenced by growth medium, bacterial age, and bacterial surface structure. The reviews by Krekeler et al.[76] and Dankert et al.[29] provide more details.

Hydrophobicity of bacteria can be examined by 1) contact angle measurements, such as the sessile drop method,[2,22] 2) evaluation of ability of bacteria to adhere to hexadecane, hydrocarbon or polystyrene,[35,111,119,120,136] 3) partitioning of bacteria in an aqueous two-phase system,[49,132,133] 4) the salt aggregation test,[35,81,85] 5) hydrophobic interaction chromatography.[35,85,98] 6) latex particle agglutination test,[35,79] or 7) direction of spreading.[122]

Surface hydrophobicity of bacteria is an important physical factor for adhesion, especially when the substrata surfaces are either hydrophilic or hydrophobic. Generally, bacteria with hydrophobic properties prefer materials with hydrophobic surfaces, those with hydrophilic characteristics prefer hydrophilic surfaces,[69,123,131] and hydrophobic bacteria adhere to a greater extent than hydrophilic bacteria.[133] Hogt et al.[69] found that one strain of *S. epidermidis* with hydrophobic characteristics showed a significantly higher adhesion to hydrophobic FEP than *S. saprophyticus*. The adhesion of *S. epidermidis* to the more hydrophilic cellulose acetate was always low. Treatment of *S. epidermidis* with pepsin or extraction with aqueous phenol yielded cells with a decreased hydrophobicity and resulted in decreased or abolished adhesion to FEP.[69,111] Similar changes of decreased hydrophobicity were produced by repeated subculture of oral streptococcus, which were accompanied by decreased adhesion to hydroxyapatite.[138] Satou et al.[123] also found *S. sanguis* strains with hydrophobic surfaces adhered more to hydrophobic glass slides than others with a less hydrophobic character. According to the recent study by Zita and Hermansson,[140] the cell surface hydrophobicity of single bacterium estimated by microsphere adhesion to cells also correlates well with adhesion of bacteria to hydrocarbons or hydrophobic interaction chromatography for a set of hydrophilic and hydrophobic bacteria.

However, the effect of hydrophobicity characteristics may be dependent on methods used to perform the study, such as the shear rate and the hydrophobicity or surface tension of the liquid phase. One study demonstrated that the effect of hydrophobicity was most important at low shear rate in protein-free solutions.[131] At higher shear rates or in other solutions, the effects of bacterial hydrophobicity were less important. In essence, adhesion is more extensive to hydrophilic substrata (i.e., substrata of relatively high surface tension) than to hydrophobic substrata, when the surface tension of the bacteria is larger than that of the suspending medium. When the surface tension of the suspending liquid is greater than that of the bacteria, the opposite pattern prevails.[3]

## B. Bacterial Surface Charge

The surface charge of bacteria may be another important physical factor for bacterial adhesion. It is involved in the initial step of bacterial colonization, also governed by long-range van der Waals forces.[29,70,73] Most particles acquire an electric charge in aqueous suspension due to the ionization of their surface groups. The surface charge attracts ions of opposite charge in the medium and results in the formation of an electric double layer. The surface charge is usually characterized by the isoelectric point[62] or the electrokinetic potential (or zeta potential) or electrophoretic mobility.[52,85,98,108,118,132] The surface charge of bacteria can also be characterized by colloid titrition[103] or electrostatic interaction chromatography.[113] Bacteria in aqueous suspension are always negatively charged.[70] A high surface charge is accompanied by a hydrophilic character of the bacteria, but a hydrophobic bacterium may still have a rather high surface charge.[70] The surface charge of bacteria varies according to bacterial species and is influenced by growth medium, bacterial age, and bacterial surface structure.[29]

Long-range electrostatic forces may influence the initial phase of bacterial adhesion onto solid surfaces. Several studies of the effect of bacterial surface charge on adhesion found that different surfaces were not significantly affected by the relative surface charge of bacteria.[1,63,70] However, other studies have called this into question.[52,85,132,133]

### C. Multiple Species or Strains

It is well known that bacteria interact in the environment and in the mammalian body. Each bacterium has its own niche in the ecosystem and mixed colonies and interactions are formed. When there is an imbalance in the controlling factors, then one bacterium may overgrow and crowd out the others. This observation has led to many studies on these mixed colonies. It was of interest to determine how the bacteria interacted on surfaces. There are many issues involved and only the influence on adherence will be discussed here since this will be a major topic in some of the other chapters in this book.

If adherence to materials were a major event in the pathogenesis of implant site infections in humans, then this might be modulated by pre-coating the material with less virulent organisms or portions of organisms which would use all the binding sites and make the material surface unacceptable to pathogenic organisms. However, studies with mixed organisms have demonstrated that the presence of bacteria on the surface of a material makes that surface even more attractive for other bacteria. When a *Pseudomonas* sp. biofilm was allowed to form on biomaterial surfaces and then the material was implanted into a host, the material became colonized with a mix of organisms, phagocytes, fibrin, and live and dead mammalian cells.[21] The presence of pre-adhered organisms greatly enhanced the biofilm mass formed in vivo. Similar findings were observed when *Pseudomonas*, *Proteus*, and *S. epidermidis* were used in in vitro and in vivo studies.[23,24] The presence of biofilm formed by one organism greatly enhanced the subsequent adherence of the same or another species. This was true whether the initial biofilm remained alive or was killed by autoclaving or by the presence of antibiotics. Thus the preconditioning with bacteria actually increased subsequent bacterial adherence and increased the infection rate. This has also been observed with *Streptococcus mutans* and *Candida albicans*.[18] The severity of infections is increased when there is a mix of orrganisms in the tissue,[10,90] especially when there is a mix of a Gram-positive and a Gram-negative species.[92]

### VI. MATERIAL SURFACES

The factors influencing bacterial adherence to biomaterial surfaces include chemical composition of the material,[13,106] surface charge,[70] hydrophobicity,[115,117] and surface roughness or physical configuration.[8,82] Also, the surface energy, empty binding sites, and hydrophobic/hydrophilic characteristics can be quickly altered by the adsorption or binding of serum proteins and formation of biofilms.[57,58] It is crucial in evaluating effects of modifications of material surfaces that the nature of the surface the bacteria encounter is known. As indicated before, prewetting the surface, especially with protein solutions, will decrease bacterial adherence. Other surface modifications which go unnoticed or unreported may also have a major effect.

## A. Surface Chemical Composition

Chu and Williams[26] examined the effects of physical configurations of suture materials on bacterial adhesion. In the group of absorbable sutures, the polydioxanone (PDS) sutures exhibited the smallest affinity toward the adherence of both *E. coli* and *S. aureus*, while Dexon sutures had the highest affinity toward these two bacteria. Studies by Sugarman and Musher[126] demonstrated that adherence of bacteria to gut was up to 100 times greater than to nylon and adherence to polyglycolic acid or silk was intermediate. According to Gristina et al.,[58] *S. epidermidis* preferentially adheres to polymers and *S. aureus* to metals. This result may explain why most *S. epidermidis* infections are associated with polymeric implants and *S. aureus* is usually the major pathogen implant site infections associated with metal implants.

If the surface chemicals are changed or modified, such as with a Pluronic surfactant coating,[19] poly(vinyl pyrrolidone) (PVP) coating,[46] antimicrobial peptide coating,[37] non-steroidal antiinflammatory drug coating,[40] cations absorbed solid surface,[55] or amine-containing organosilicon surfaces,[125] bacterial adhesion to these surfaces is discouraged. More recently, Oga et al.[105] found that the number of *S. epidermidis* adhered to sintered hydroxyapatite was higher than that to three other material surfaces. The effects of surface roughness were possibly involved in this study according to the manufacturing procedure for preparing the samples. A similar study by Prewett et al.[114] also indicates that the number of adherent bacteria which bind to a metal surface is dependent upon the strain of the microorganism and the type of metal. An effect of chemical composition on bacterial adhesion was also found in our recent study.[7]

## B. Surface Roughness

Surface roughness is a two dimensional parameter of a material surface measured by roughness measuring systems such as the Stylus system and commonly described as arithmetic average roughness (Ra). It is a distance measurement between the peaks and valleys on a material surface and does not represent the morphological configurations of the surface. Surface finish and smoothness are the easy-to-understand "alternative terms" of surface roughness.

Baker[11] found that roughening the surface of either glass or polystyrene with a grindstone greatly increased the rate of bacterial colonization in a river environment. Another study recommended searching for optimal surface smoothness for all intraoral and intrasulcular hard surfaces for reduction of bacteria colonization and plaque formation, since the results showed rough surfaces harbored 25 times more bacteria.[116] There are more reports on the effects of surface roughness on bacterial adhesion to central venous catheters,[130] enamel surfaces,[59] or teeth surfaces.[15] All of these studies indicate surface roughness influences bacteria adhesion. The causes for this phenomenon may include 1) a rough surface has a greater surface area, and 2) the depressions in the roughened surfaces provide more favorable sites for colonization.[11]

However, a recent study in our laboratory showed that cp-Ti (commercially pure titanium) surface roughnesses produced by 120 to 1200 grit sandpaper polishing had virtually no effect on the number of adhered *S. epidermidis* (VAS-11).[8] What caused this is a question for further studies such as analysis of surface area, or evaluation of surface configuration. A further point for consideration is whether the range of roughnesses studied was great enough.

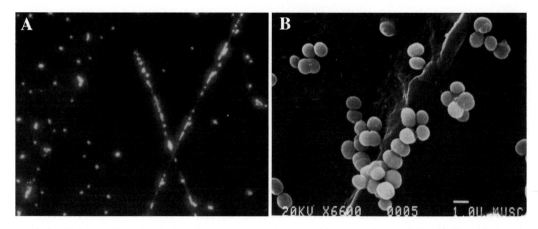

**Figure 1.** *S. aureus* adherence to material surfaces in PBS suspension at 37°C with agitation evaluated by epifluorescent microscope and SEM. Bacteria prefer to attach to grooves on cp-Ti surface (A) or ridges on UHMWPE) surface (B).

Since clinically different prostheses or implant devices have different surface roughnesses, which may play a role in bacterial adhesion and implant infection, more studies are needed to test the effects of a broader range of surface roughness.

### C. Surface Morphology or Configuration

Physical configuration of a material surface is different from surface roughness, and is rather complicated. It is a morphological description of the pattern of a material surface, such as a monofilament surface, a braided surface, a porous surface, or a grid-like surface, and it is a three dimensional parameter. Routinely, physical configurations are evaluated by scanning electron microscopy.

The irregularities of material surfaces promote bacterial adhesion, biofilm deposition, and accumulation of biliary sludge, while ultrasmooth surfaces do not allow bacterial adhesion and biofilm deposition.[87] Figure 1 shows that surface configurations of solid surfaces play an important role on the patterns of bacterial adhesion. Bacteria prefer to attach to grooves on cp-Ti surface or ridges on ultra high molecular weight polyethylene (UHMWPE) surfaces.

Merritt et al.[94] found that implant site infection rates are obviously different between porous and dense dental materials, with a much higher rate for porous material. This implies bacteria adhere and colonize the porous surface preferentially. This finding was confirmed in a recent study using <20 cfu of *S. aureus* on implanted segments of suture.[93] The bacterial adherence in one hour to these various suture materials was not markedly different since less than a fivefold difference in cfu in the solution of bacteria used for adherence resulted in the same number of organisms on the suture segment. However, the infection rate was significantly higher with the multifilament sutures. Thus, in addition to adherence mechanisms, there is a major difference in host response, which is yet to be elucidated.

Locci et al.[82] perfused intravenous catheters with coagulase-negative staphylococci (CNS) and found that the initial bacterial adhesion occurred at different irregularities of the inner surface of the catheters. With nonabsorbable sutures, the physical configuration of the suture contributed more to their ability to attract bacteria than the surface finish. Braided suture materials may have increased bacterial adherence compared to nonbraided ones, probably partially due to increased surface area.[126]

## D. Surface Hydrophobicity or Wettability

Metal surfaces have a high surface energy, are negatively charged, and hydrophilic as shown by water contact angles, while polymers such as UHMWPE or Teflon have low surface energy, are less electrostatically charged, and hydrophobic. The hydrophobicity of a material surface has been determined mainly by contact angle measurement.

Depending on the hydrophobicity of both bacteria and material surfaces, bacteria adhere differently to materials with different hydrophobicities.[43,69,123] Several groups reported that hydrophilic materials are more resistant to bacterial adhesion than hydrophobic materials.[69,84] Fletcher and Loeb[43] investigated the attachment of a marine *Pseudomonas* sp. to a variety of surfaces. Large numbers of bacteria attached to hydrophobic plastics with little or no surface charge (Teflon, polyethylene, polystyrene, and polyethylene terephthalate); moderate numbers attached to hydrophilic metals with a positive or neutral surface charge; and very few attached to hydrophilic, negatively charged substrata (glass, mica, oxidized plastics). Satou et al.[123] also studied the adhesion of two *S. sanguis* strains and two *S. mutans* strains to four surface-modified glass slides with different hydrophobicity. *S. sanguis* strains (with more hydrophobic surfaces) adhered more to hydrophobic glass slides than others. Coating substrata surfaces with proteins, such as bovine serum albumin (BSA), bovine glycoprotein, or fatty-acid free BSA, decreased surface hydrophobicity leading to an inhibited bacterial adhesion to the surfaces.[44] In a study of surface hydrophiliation and adherence of *S. mutans,* the presence of saliva made a difference and the effect of various other factors was obscured by this.

Bridgett et al.[19] studied the adherence of three clinical isolates of *S. epidermidis* to model polystyrene surfaces in vitro using epifluorescent image analysis. A series of 16 Pluronic surfactants (A-B-A block copolymers where A is poly(ethylene oxide) (PEO) and B is poly(propylene oxide) (PPO)) were used as hydrophilic surface modifiers for the model polystyrene surfaces. Substantial reductions (up to 97%) in bacterial adhesion levels were achieved with all copolymers tested, irrespective of the PPO or PEO block lengths. It appears likely that such treatments create a sterically stabilized surface with adsorbed PEO chains, conferring nonspecific anti-adhesive properties which can limit bacterial attachment. Similar effects of PEO were also found by other researchers.[60,109]

In a recent study in Dr. An's laboratory, PEO coated titanium surfaces inhibited the adherence of *S. epidermidis* (RP62A) by 95-98% (Table 1). Briefly two different PEOs, with molecular weight being 2 kDa and 12 kDa, were dissolved in dimethyl sulfoxide (5% PEO). The solution was used to coat 600 grit sandpaper polished cp-Ti discs. The discs were exposed to *S. epidermidis* (RP62A) PBS suspension for 1 h at 37°C with agitation. Then the discs were stained with propidium iodide and examined under a computerized epifluorescent microscope. This dramatic effect of PEO on staphylococcal adhesion is under further experimentation.

However, in an attempt to study the effect of hydrophobicity/hydrophilicity on bacterial adherence, titanium surfaces were treated with chemicals to form a "functionalized

**Table 1. Effect of PEO Coating on Adhesion of *S. epidermidis* to Titanium Surface**

|  | Bare cp-Ti | cp-Ti coated with 2 kDa PEO | cp-Ti coated with 12 kDa PEO |
|---|---|---|---|
| RP62A adherence | $339 \pm 31^a$ | $14.9 \pm 6.9$ | $5.6 \pm 9.9$ |
| Inhibition rate | — | 95.6% | 98.4% |

[a]The values are the average counts of 40 microscopic fields (×40 objective) taken from 4 identical sample discs. Each field is about 0.01627 mm$^2$.

**Table 2. Staphylococcal Adherence to Materials with Different Surface Hydrophobicity**

| Material | n | cp-Ti | Ti Alloy | CoCr | SS | PE |
|---|---|---|---|---|---|---|
| Surface roughness (μm) | 9 | $0.44 \pm 0.04$ | $0.51 \pm 0.05$ | $0.47 \pm 0.08$ | $0.51 \pm 0.04$ | $0.55 \pm 0.12$ |
| Water contact angle | 9 | $39 \pm 1$ | $34 \pm 2$ | $49 \pm 4$ | $50 \pm 8$ | $111 \pm 4$ |
| *S. aureus* adhered[a] | 40 | $137 \pm 12^a$ | $118 \pm 8$ | $116 \pm 16$ | $147 \pm 25$ | $38 \pm 6$ |
| *S. epidermidis* adhered | 40 | $194 \pm 12$ | $180 \pm 11$ | $178 \pm 11$ | $171 \pm 9$ | $47 \pm 13$ |

[a]Average number of bacterial counts per field under epifluorescence microscope (The values were calculated from 40 fields on 4 identical sample discs and each field is about 0.01627 mm$^2$)

surface" with ranges from highly hydrophobic to highly hydrophilic. These changes to the base material did not significantly alter the adherence of bacteria.[95] However, in the parallel study using cell culture,[127] there was a marked difference in mammalian cell adhesion and proliferation. Whether the surface hydrophobicity made no difference to the bacterial adherence or the bacteria penetrated the functionalized surface and saw the same base metal in all treatments remains unknown.

Another exception was found in An's laboratory on bacterial adhesion to hydrophobic and hydrophilic material surfaces (Table 2).[7] With nearly the same surface roughnesses, hydrophobic UHMWPE discs attracted much less bacteria compared to the hydrophilic metal surfaces.

## VII. SERUM OR TISSUE PROTEINS

Many proteins (serum or tissue proteins) have been studied for their effects on bacterial adhesion to material surfaces,[5,8,88,135] including albumin, fibronectin, fibrinogen, laminin, denatured collagen, and more. They promote or inhibit bacterial adhesion by altering binding to substrata surfaces, binding to bacterial surfaces, or by their presence in the liquid medium during the adhesion period. For the latter situation most of the proteins inhibited bacteria adhesion,[20,41] possibly affecting bacterial adhesion by their association with the bacterial cell surface, the material surface, or both. Most of the binding between bacteria and proteins is specific ligand–receptor-like interactions. Proteins may also change the adherent behavior of bacteria by changing bacterial surface physicochemical characteristics.[96,117,120] A comprehensive review on the interrelationships between protein surface adsorption and bacterial adhesion has been published recently by Daeschel and McGuire.[28]

## A. *Fibronectin*

Fibronectin (Fn), which is recognized for its ability to mediate surface adhesion of eukaryotic cells, has also been shown to bind to *S. aureus*.[77] Fn clearly promotes *S. aureus* adhesion to substratum surface.[78,86,134,135] Kuusela et al.[78] demonstrated a time-dependent and Fn concentration-dependent adhesion of *S. aureus* to Fn-coated coverslips. In vitro adhesion of *S. aureus* strain Wood 46 to explanted coverslips was inhibited in a dose-dependent manner by anti-Fn antibody.[134,135] The binding of $^{125}$I-Fn to a strain of *S. aureus* is specific, time-dependent, irreversible, and occurs with both live and heat-killed cells. Staphylococci may be saturated with Fn at a level which suggests the presence of specific receptors (staphylococcal Fn-binding molecules) on bacterial cells.[38,86,121] This Fn-binding molecule has been cloned in *E. coli* and purified.[45,48] The *S. aureus* binding domain of Fn was also found in the Fn molecule.[17,97]

Fibronectin plays an important role in foreign body infection.[124,134] In the presence of serum, the level of *S. aureus* adherence to explanted coverslips (from guinea pig, covered by Fn deposits) was 20 times higher than that of adherence to unimplanted coverslips.[135] Scheld et al.[124] also found that Fn exposed to the constituents of nonbacterial thrombotic endocarditis may mediate microbial adhesion of circulating organisms to initiate colonization during the early pathogenesis of infective endocarditis.[124] Although strain specific, adherence of clinical staphylococcal isolates to foreign surfaces was significantly increased by Fn, suggesting the possible contribution of these proteins to the pathogenesis of intravenous device infection.[68] Delmi et al.[32] found that incubation of either in vitro Fn-coated or explanted metallic coverslips with anti-Fn antibodies produced a significant decrease in staphylococcal adhesion. These results suggest that the presence of Fn on the surface of implanted metallic devices is an important determinant of colonization of orthopaedic biomaterials by staphylococci. Francois et al.[46] found that poly(vinyl pyrrolidone) (PVP)-coated Pellethane showed a strong reduction in either fibrinogen or fibronectin adsorption compared to all other PVP-free polyurethane central venous catheters (CVCs). This decreased protein adsorption led to a proportional reduction in protein-mediated adhesion of either *S. aureus* or *S. epidermidis* and in the binding of a monoclonal antibody directed against the cell-binding domain of fibronectin.

There are controversies regarding Fn effect on *S. epidermidis* adhesion to material surfaces. Herrmann et al.[68] found Fn markedly promoted adherence of all *S. aureus* strains but only four out of 19 strains of *S. epidermidis*. Naylor[101] found Fn can inhibit *S. epidermidis* adherence to cobalt–chrome alloy and poly(methyl methacrylate) surface by 90%. Pre-incubated with silicone catheters, fibrinogen and Fn inhibited the binding of *S. epidermidis*, while both of them enhanced the binding of *S. aureus*.[39] Our recent results showed that Fn, unlike albumin, has no effect on the adhesion of VAS-11 to cp-Ti surface.[8]

## B. *Collagen*

Collagen is an important factor in performance and colonization of vascular grafts and some other implanted devices, is an important matrix for colonization leading to infections of damaged host tissue, and is a common constituent of tissue engineered devices. In general the presence of collagen has effects similar to that of fibronectin and promotes the adherence of *S. aureus*.[135] It also promotes the binding of *S. mutans*.[128] The binding of *S. aureus* to collagen appears to be under bacterial genetic control.[53]

## C. Albumin

Albumin adsorbed on material surfaces has shown obvious inhibitory effects on bacterial adhesion to polymer,[70,111,112,115] ceramic,[50] and metal[5,8,9,88] surfaces. In a recent study, human serum albumin (HSA) inhibited *S. epidermidis* adhesion to cp-Ti surfaces by more than 95% after treatment of the cp-Ti sample with 200 mg/mL of human serum albumin at 37°C for 2 h. Most of the proteins reduced the adhesion through adsorption to substrata surface, while serum albumin also inhibited the adhesion by means of binding to the bacterial cells.[20] The mechanism of the inhibiting effect of albumin is not clear. Albumin may reduce bacterial adhesion by changing substratum surface hydrophobicity, because in the presence of dissolved and adsorbed bovine serum albumin (BSA) substrata surface became much less hydrophobic.[44,117]

As stated, HSA inhibited *S. epidermidis* adhesion to commercially pure titanium (cp-Ti) surfaces by more than 95% after adsorption of 200 mg/mL of human serum albumin at 37°C for 2 h.[8] Cp-Ti surfaces were then coated with BSA using a cross-linking agent, carbodiimide. Only 10% of the coated BSA decayed off the surface during the 20 d incubation period (at 37°C, in phosphate buffered saline, with intermittent agitation). The inhibition rate of the albumin coating on bacterial adherence remained high (more than 85 percent) throughout the length of the experiment (20 d).[39,88] In a recent study, a cross-linked albumin coating has been shown to reduce prosthetic infection rate in a rabbit model. Animals with albumin coated implants had a much lower infection rate (27%) than those with uncoated implants (62%). This finding may represent a new method for preventing prosthetic infection.[4]

## D. Fibrinogen

Fibrinogen is another important serum protein mediating bacterial adhesion to biomaterials and host tissues. Most studies showed adsorbed fibrinogen promotes adherence of bacteria, especially staphylococci, to biomaterials. Herrmann et al.[68] found that adsorbed fibrinogen had a promoting effect on *S. aureus* and CNS adhesion to polymethylmethacrylate (PMMA) coverslips. In the study by Hermann et al., fibrinogen markedly promoted adherence of all *S. aureus* strains but only a few coagulase-negative strains. The later finding was supported by the studies by Muller et al.[99] and by Paulsson et al.[112] Recently, Herrmann et al.[66] also studied the mediating role of fibrinogen/fibrin and platelet integrin on *S. aureus* adhesion to surface-bound platelets. Staphylococcal adherence to polymer catheters coated with fibrinogen was significantly increased compared with that to control catheters (pre-incubated in phosphate-buffered saline).[25] Fibrinogen bound to coverslips also increase streptococcal adhesion.[78] In another in vitro study, pretreatment of bacteria or both bacteria and polyethylene (PE) catheter surfaces with fibrinogen enhanced bacterial adherence suggesting the presence of ligands for fibrinogen on the staphylococcal cell surface.[20]

## E. Laminin

Laminin, a major component of basement membranes, was shown to bind some strains of *S. pyogenes*. Binding of [125]I-laminin to bacteria was time dependent and functionally irreversible. Laminin receptors (0–103 receptors/per cell) were isolated from these bacteria.[129] Laminin has a promoting effect on *S. aureus* and CNS adhesion to PMMA coverslips, but to a lesser extent compared to the effects of fibronectin and fibrinogen.[68]

The presence of laminin receptors in *S. aureus* (about 100 binding sites per cell) has been also reported. [83]

## F. Serum or Plasma

The adhesion of various CNS onto plasma-coated FEP was studied by Hogt et al.[70] The adhesion of all strains onto plasma FEP was much lower than onto the untreated control FEP surface. Pascual et al.[111] found that pre-incubation of Teflon catheters in human serum caused an 80 to 90% reduction of adhesion of *S. epidermidis*. Pre-incubation of *S. epidermidis* in serum similarly decreased adhesion. This effect of serum was mainly due to albumin, while IgG and fibronectin were less effective. Similar effects were also found when polymers were pre-incubated with plasma[34] or albumin.[112]

## G. Other Proteins or Factors

Fletcher found that adsorbed gelatin or pepsin impaired the attachment of a marine pseudomonad to polystyrene petri dishes.[41] The basic proteins such as histone and poly-L-lysine facilitated *S. mutans* adherence to hydroxyapatite discs and the acidic proteins such as phosvitin, β-lactoglobulin, or poly-L-glutamate inhibited adhesion.[117] *S. epidermidis* adhesion to Teflon catheters was significantly related to the degree of hydrophobicity of the strains. When hydrophobic groups were removed from the bacteria by pepsin treatment, adhesion was almost completely abolished.[111]

Saliva has a major effect on the binding of streptococci to materials. The study by Olsson et al. described in the section on surface hydrophilicity demonstrated promotion and inhibition of adherence of *S. mutans* by some saliva constituents.[107] Saliva-coated hydroxyapatite has been used as a model for adherence of organisms important in the use of dental materials and dental hygiene. Cell–cell interaction of *Propionibacterium* and *S. sanguis* has been demonstrated and may be an important factor in periodontal disease. *Bacteriodes* strains and *Actinomyces viscosus* have also been shown to adhere to these surfaces.[80] The presence of glycoproteins in the saliva was demonstrated to be of importance in the promotion of adherence of oral streptococci.[100] It is important when evaluating studies on adherence to materials to know what proteins or other factors are present and which are important in modeling the in vivo conditions encountered by the organisms studied. Once a material is in the biological system, its surface will be coated by biological material and it is no longer the surface that was studied so carefully in vitro.

In the in vivo condition, implants inserted not only encounter serum proteins but also different cells or tissue elements such as platelets. Wang et al. reported the mediating effect of platelets on *S. epidermidis* (RP62A) adhesion onto hydrophobic polyethylene surfaces.[137] Thrombospondin has a promoting effect on bacterial adhesion.[67] Franson et al. found that D-mannosamine inhibits in vitro coagulase-negative staphylococci adhesion to intravascular catheters.[47] Tamm Horsfall protein from pooled urine interferes with adhesion of *E. coli, P. aeruginosa,* and *S. epidermidis* to polymers by binding directly to these bacteria.[64] Baumgartner and Cooper demonstrated components of thrombus are important in mediating *S. aureus* adherence to polyurethane surfaces.[14]

## I. Tissue Proteins

Bacterial adhesion to material surfaces is a complex process. Anything that alters either the bacterial surface or the material surface will alter the interaction. Thus it is critical that all the variables in the assay system be identified. In this chapter we have attempted to identify the important variables. There are conflicting reports in the literature as to which are the important factors. Some proteins have a great effect on some bacteria in some studies but show little effect in others. Some physical characteristics such as flow and turbulence affect some interactions and not others. Unless all of the components of a bacteria–material interaction study are known, minor differences in the procedures may be unrecognized and yet cause major differences in results leading to confusion in interpreting the literature. The important variables need to be identified and controlled before a rational approach to controlling the interaction can be attempted.

## REFERENCES

1. Abbott A, Rutter PR, Berkeley RC: The influence of ionic strength, pH and a protein layer on the interaction between *Streptococcus mutans* and glass surfaces. *J Gen Microbiol* 129:439–45, 1983
2. Absolom DR: Measurement of surface properties of phagocytes, bacteria, and other particles. *Methods Enzymol* 132:16–95, 1986
3. Absolom DR, Lamberti FV, Policova Z, et al: Surface thermodynamics of bacterial adhesion. *Appl Environ Microbiol* 46:90–7, 1983
4. An YH, Bradley J, Powers DL, et al: The prevention of prosthetic infection using a cross-linked albumin coating in a rabbit model. *J Bone Joint Surg [Br]* 79:816–9, 1997
5. An YH, Friedman RJ: Prevention of sepsis in total joint arthroplasty. *J Hosp Infect* 33:93–108, 1996
6. An YH, Friedman RJ: Concise review of mechanisms of bacterial adhesion to biomaterial surfaces. *J Biomed Mater Res* 43:338–48, 1998
7. An YH, Friedman RJ, Draughn RA: Staphylococci adhesion to orthopaedic biomaterials. *Trans Soc Biomater* 16:148, 1993
8. An YH, Friedman RJ, Draughn RA: Rapid quantification of staphylococci adhered to titanium surfaces using image analyzed epifluorescence microscopy. *J Microbiol Meth* 24:29–40, 1995
9. An YH, Stuart GW, McDowell SJ, et al: Prevention of bacterial adherence to implant surfaces with a crosslinked albumin coating *in vitro*. *J Orthop Res* 14:846–9, 1996
10. Baddour LM, Meyer J, Henry B: Polymicrobial infective endocarditis in the 1980s. *Rev Infect Dis* 13:963–70, 1991
11. Baker AS, Greenham LW: Release of gentamicin from acrylic bone cement. Elution and diffusion studies. *J Bone Joint Surg [Am]* 70:1551–7, 1988
12. Bar-Ness R, Avrahamy N, Matsuyama T, et al: Increased cell surface hydrophobicity of a Serratia marcescens NS 38 mutant lacking wetting activity. *J Bacteriol* 170:4361–4, 1988
13. Barth E, Myrvik QM, Wagner W, et al: *In vitro* and *in vivo* comparative colonization of *Staphylococcus aureus* and *Staphylococcus epidermidis* on orthopaedic implant materials. *Biomaterials* 10:325–8, 1989
14. Baumgartner JN, Cooper SL: Influence of thrombus components in mediating *Staphylococcus aureus* adhesion to polyurethane surfaces. *J Biomed Mater Res* 40:660–70, 1998
15. Bollen CM, Papaioanno W, Van Eldere J, et al: The influence of abutment surface roughness on plaque accumulation and peri-implant mucositis. *Clin Oral Implants Res* 7:201–11, 1996
16. Bos R, van der Mei HC, Busscher HJ: Influence of ionic strength and substratum hydrophobicity on the co-adhesion of oral microbial pairs. *Microbiology* 142:2355–61, 1996
17. Bozzini S, Visai L, Pignatti P, et al: Multiple binding sites in fibronectin and the staphylococcal fibronectin receptor. *Eur J Biochem* 207:327–33, 1992
18. Branting C, Sund ML, Linder LE: The influence of *Streptococcus mutans* on adhesion of *Candida albicans* to acrylic surfaces *in vitro*. *Arch Oral Biol* 34:347–53, 1989

19. Bridgett MJ, Davies MC, Denyer SP: Control of staphylococcal adhesion to polystyrene surfaces by polymer surface modification with surfactants. *Biomaterials* 13:411–6, 1992

20. Brokke P, Dankert J, Carballo J, et al: Adherence of coagulase-negative staphylococci onto polyethylene catheters *in vitro* and *in vivo*: a study on the influence of various plasma proteins. *J Biomater Appl* 5:204–26, 1991

21. Buret A, Ward KH, Olson ME, et al: An *in vivo* model to study the pathobiology of infectious biofilms on biomaterial surfaces. *J Biomed Mater Res* 25:865–74, 1991

22. Busscher HJ, Weerkamp AH, van der Mei HC, et al: Measurement of the surface free energy of bacterial cell surfaces and its relevance for adhesion. *Appl Environ Microbiol* 48:980–3, 1984

23. Chang CC, Merritt K: Effect of *Staphylococcus epidermidis* on adherence of *Pseudomonas aeruginosa* and *Proteus mirabilis* to polymethyl methacrylate (PMMA) and gentamicin-containing PMMA. *J Orthop Res* 9:284–8, 1991

24. Chang CC, Merritt K: Infection at the site of implanted materials with and without preadhered bacteria. *J Orthop Res* 12:526–31, 1994

25. Cheung AL, Fischetti VA: The role of fibrinogen in staphylococcal adherence to catheters in vitro. *J Infect Dis* 161:1177–86, 1990

26. Chu CC, Williams DF: Effects of physical configuration and chemical structure of suture materials on bacterial adhesion. A possible link to wound infection. *Am J Surg* 147:197–204, 1984

27. Courtney HS, Ofek I, Simpson WA, et al: Human plasma fibronectin inhibits adherence of *Streptococcus pyogenes* to hexadecane. *Infect Immun* 47:341–3, 1985

28. Daeschel MA, McGuire J: Interrelationships between protein surface adsorption and bacterial adhesion. *Biotechnol Genet Eng Rev* 15:413–38, 1998

29. Dankert J, Hogt AH, Feijen J: Biomedical polymers: Bacterial adhesion, colonization, and infection. *CRC Crit Rev Biocompat* 2:219–301, 1986

30. Deighton M, Borland R: Regulation of slime production in *Staphylococcus epidermidis* by iron limitation. *Infect Immun* 61:4473–9, 1993

31. Deighton MA, Balkau B: Adherence measured by microtiter assay as a virulence marker for *Staphylococcus epidermidis* infections. *J Clin Microbiol* 28:2442–7, 1990

32. Delmi M, Vaudaux P, Lew DP, et al: Role of fibronectin in staphylococcal adhesion to metallic surfaces used as models of orthopaedic devices. *J Orthop Res* 12:432–8, 1994

33. Denyer SP, Davies MC, Evans JA, et al: Influence of carbon dioxide on the surface characteristics and adherence potential of coagulase-negative staphylococci. *J Clin Microbiol* 28:1813–7, 1990

34. Dickinson RB, Nagel JA, Proctor RA, et al: Quantitative comparison of shear-dependent *Staphylococcus aureus* adhesion to three polyurethane ionomer analogs with distinct surface properties. *J Biomed Mater Res* 36:152–62, 1997

35. Dillon JK, Fuerst JA, Hayward AC, et al: A comparison of five methods for assaying bacterial hydrohobicity. *J Microbiol Methods* 6:13–9, 1986

36. Duddriege JE, Cent CA, Laws JE: Effect of surface shear stress on the attachment of *Pseudomonas fluorescens* to stainless steel under defined flow conditions. *Biotech Bioeng* 24:153–64, 1982

37. Duran JA, Malvar A, Rodriguez-Ares MT, et al: Heparin inhibits *Pseudomonas* adherence to soft contact lenses. *Eye* 7:152–4, 1993

38. Espersen F, Clemmensen I: Isolation of a fibronectin-binding protein from *Staphylococcus aureus*. *Infect Immun* 37:526–31, 1982

39. Espersen F, Wilkinson BJ, Gahrn-Hansen B, et al: Attachment of staphylococci to silicone catheters *in vitro*. *Apmis* 98:471–8, 1990

40. Farber BF, Wolff AG: The use of nonsteroidal antiinflammatory drugs to prevent adherence of *Staphylococcus epidermidis* to medical polymers. *J Infect Dis* 166:861–5, 1992

41. Fletcher M: The effects of proteins on bacterial attachment to polystyrene. *J Gen Microbiol* 94:400–4, 1976

42. Fletcher M, Floodgate GD: An electron-microscopic demonstration of as acidic polysaccharide involved in the adhesion of a marine bacterium to solid surfaces. *J Gen Microbiol* 74:325–34, 1973

43. Fletcher M, Loeb GI: Influence of substratum characteristics on the attachment of a marine pseudomonad to solid surfaces. *Appl Environ Microbiol* 37:67–72, 1979

44. Fletcher M, Marshall KC: Bubble contact angle method for evaluating substratum interfacial characteristics and its relevance to bacterial attachment. *Appl Environ Microbiol* 44:184–92, 1982

45. Flock JI, Froman G, Jonsson K, et al: Cloning and expression of the gene for a fibronectin-binding protein from *Staphylococcus aureus*. *Embo J* 6:2351–7, 1987

46. Francois P, Vaudaux P, Nurdin N, et al: Physical and biological effects of a surface coating procedure on polyurethane catheters. *Biomaterials* 17:667–78, 1996

47. Franson TR, Sheth NK, Rose HD, et al: Quantitative adherence *in vitro* of coagulase-negative staphylococci to intravascular catheters: inhibition with D-mannosamine. *J Infect Dis* 149:116, 1984

48. Froman G, Switalski LM, Speziale P, et al: Isolation and characterization of a fibronectin receptor from *Staphylococcus aureus*. *J Biol Chem* 262:6564–71, 1987

49. Gerson DF, Scheer D: Cell surface energy, contact angles and phase partition. III. Adhesion of bacterial cells to hydrophobic surfaces. *Biochim Biophys Acta* 602:506–10, 1980

50. Gibbons RJ, Etherden I: Comparative hydrophobicities of oral bacteria and their adherence to salivary pellicles. *Infect Immun* 41:1190–6, 1983

51. Gibbons RJ, Etherden I, Skobe Z: Association of fimbriae with the hydrophobicity of *Streptococcus sanguis* FC-1 and adherence to salivary pellicles. *Infect Immun* 41:414–7, 1983

52. Gilbert P, Evans DJ, Evans E, et al: Surface characteristics and adhesion of *Escherichia coli* and *Staphylococcus epidermidis*. *J Appl Bacteriol* 71:72–7, 1991

53. Gillaspy AF, Lee CY, Sau S, et al: Factors affecting the collagen binding capacity of *Staphylococcus aureus*. *Infect Immun* 66:3170–8, 1998

54. Giridhar G, Kreger AS, Myrvik QN, et al: Inhibition of *Staphylococcus* adherence to biomaterials by extracellular slime of *S. epidermidis* RP12. *J Biomed Mater Res* 28:1289–94, 1994

55. Goldberg S, Doyle RJ, Rosenberg M: Mechanism of enhancement of microbial cell hydrophobicity by cationic polymers. *J Bacteriol* 172:5650–4, 1990

56. Gordon AS, Gerchakov SM, Udey LR: The effect of polarization on the attachment of marine bacteria to copper and platinum surfaces. *Can J Microbiol* 27:698–703, 1981

57. Gristina AG: Biomaterial-centered infection: microbial adhesion versus tissue integration. *Science* 237:1588–95, 1987

58. Gristina AG, Hobgood CD, Barth E: Biomaterial specificity, molecular mechanisms, and clinical relevance of *S. epidermidis* and *S. aureus* infections in surgery. In: Pulverer G, Quie PG, Peters G, eds. *Pathogenesis and Clinical Significance of Coagulase-negative Staphylococci*. Fisher Verlag, Starttgart, 1987:143–57

59. Gurgan S, Bolay S, Alacam R: In vitro adherence of bacteria to bleached or unbleached enamel surfaces. *J Oral Rehabil* 24:624–7, 1997

60. Han DK, Park KD, Kim YH: Sulfonated poly(ethylene oxide)-grafted polyurethane copolymer for biomedical applications. *J Biomater Sci Polym Ed* 9:163–74, 1998

61. Harber MJ, Mackenzie R, Asscher AW: A rapid bioluminescence method for quantifying bacterial adhesion to polystyrene. *J Gen Microbiol* 129:621–32, 1983

62. Harden VP, Harris JO: The isoelectric point of bacterial cells. *J Bacteriol* 65:269–71, 1953

63. Harkes G, Feijen J, Dankert J: Adhesion of *Escherichia coli* on to a series of poly(methacrylates) differing in charge and hydrophobicity. *Biomaterials* 12:853–60, 1991

64. Hawthorn L, Reid G: The effect of protein and urine on uropathogen adhesion to polymer substrata. *J Biomed Mater Res* 24:1325–32, 1990

65. Hejazi A, Falkiner FR: Serratia marcescens. *J Med Microbiol* 46:903–12, 1997

66. Herrmann M, Lai QJ, Albrecht RM, et al: Adhesion of *Staphylococcus aureus* to surface-bound platelets: role of fibrinogen/fibrin and platelet integrins. *J Infect Dis* 167:312–22, 1993

67. Herrmann M, Suchard SJ, Boxer LA, et al: Thrombospondin binds to *Staphylococcus aureus* and promotes staphylococcal adherence to surfaces. *Infect Immun* 59:279–88, 1991

68. Herrmann M, Vaudaux PE, Pittet D, et al: Fibronectin, fibrinogen, and laminin act as mediators of adherence of clinical staphylococcal isolates to foreign material. *J Infect Dis* 158: 693–701, 1988

69. Hogt AH, Dankert J, de Vries JA, et al: Adhesion of coagulase-negative staphylococci to biomaterials. *J Gen Microbiol* 129:2959–68, 1983

70. Hogt AH, Dankert J, Feijen J: Adhesion of *Staphylococcus epidermidis* and *Staphylococcus saprophyticus* to a hydrophobic biomaterial. *J Gen Microbiol* 131:2485–91, 1985

71. Hussain M, Wilcox MH, White PJ, et al: Importance of medium and atmosphere type to both slime production and adherence by coagulase-negative staphylococci. *J Hosp Infect* 20:173–84, 1992

72. Jenkinson HF: Cell-surface proteins of *Streptococcus sanguis* associated with cell hydrophobicity and coaggregation properties. *J Gen Microbiol* 132:1575–89, 1986

73. Jucker BA, Harms H, Zehnder AJ: Adhesion of the positively charged bacterium *Stenotrophomonas (Xanthomonas) maltophilia* 70401 to glass and Teflon. *J Bacteriol* 178: 5472–9, 1996

74. Klotz SA, Drutz DJ, Zajic JE: Factors governing adherence of *Candida* species to plastic surfaces. *Infect Immun* 50:97–101, 1985

75. Krajewska-Pietrasik D, Wykrota M, Sidorczyk Z: [The influence of temperature on culture of Staphylococcus aureus for cell adhesion to collagen]. *Med Dosw Mikrobiol* 49:123–30, 1997

76. Krekeler C, Ziehr H, Klein J: Physical methods for characterization of microbial surfaces. *Experientia* 45:1047–55, 1989

77. Kuusela P: Fibronectin binds to *Staphylococcus aureus*. *Nature* 276:718–20, 1978

78. Kuusela P, Vartio T, Vuento M, et al: Attachment of staphylococci and streptococci on fibronectin, fibronectin fragments, and fibrinogen bound to a solid phase. *Infect Immun* 50: 77–81, 1985

79. Lachica RV, Zink DL: Determination of plasmid-associated hydrophobicity of *Yersinia enterocolitica* by a latex particle agglutination test. *J Clin Microbiol* 19:660–3, 1984

80. Li J, Ellen RP: Relative adherence of *Bacteroides* species and strains to *Actinomyces viscosus* on saliva-coated hydroxyapatite. *J Dent Res* 68:1308–12, 1989

81. Lindahl M, Faris A, Wadstrom T, et al: A new test based on 'salting out' to measure relative surface hydrophobicity of bacterial cells. *Biochim Biophys Acta* 677:471–6, 1981

82. Locci R, Peters G, Pulverer G: Microbial colonization of prosthetic devices. I.Microtopographical characteristics of intravenous catheters as detected by scanning electron microscopy. *Zentralbl Bakteriol Mikrobiol Hyg [B]* 173:285–92, 1981

83. Lopes JD, dos Reis M, Brentani RR: Presence of laminin receptors in *Staphylococcus aureus*. *Science* 229:275–7, 1985

84. Ludwicka A, Jansen B, Wadstrom T, et al: Attachment of staphylococci to various synthetic polymers. *Zentralbl Bakteriol Mikrobiol Hyg [A]* 256:479–89, 1984

85. Mafu AA, Roy D, Goulet J, et al: Characterization of physicochemical forces involved in adhesion of *Listeria monocytogenes* to surfaces. *Appl Environ Microbiol* 57:1969–73, 1991

86. Maxe I, Ryden C, Wadstrom T, et al: Specific attachment of *Staphylococcus aureus* to immobilized fibronectin. *Infect Immun* 54:695–704, 1986

87. McAllister EW, Carey LC, Brady PG, et al: The role of polymeric surface smoothness of biliary stents in bacterial adherence, biofilm deposition, and stent occlusion. *Gastrointest Endosc* 39:422–5, 1993

88. McDowell SG, An YH, Draughn RA, et al: Application of a fluorescent redox dye for enumeration of metabolically active bacteria on albumin-coated titanium surfaces. *Lett Appl Microbiol* 21:1–4, 1995

89. McEldowney S: Effect of cadmium and zinc on attachment and detachment interactions of *Pseudomonas fluorescens* H2 with glass. *Appl Environ Microbiol* 60:2759–65, 1994

90. Merritt K: Factors increasing the risk of infection in patients with open fractures. *J Trauma* 28:823–7, 1988

91. Merritt K, Chang CC: Factors influencing bacterial adherence to biomaterials. *J Biomater Appl* 5:185–203, 1991

92. Merritt K, Dowd JD: Role of internal fixation in infection of open fractures: studies with *Staphylococcus aureus* and *Proteus mirabilis*. *J Orthop Res* 5:23–8, 1987

93. Merritt K, Hitchins VM, Neale AR: Tissue colonization from implantable biomaterials with low numbers of bacteria. *J Biomed Mater Res* 44:261–265, 1999

94. Merritt K, Shafer JW, Brown SA: Implant site infection rates with porous and dense materials. *J Biomed Mater Res* 13:101–8, 1979

95. Merritt K, Sukenik CN, Balachander N: Bacterial adherence to functionalized surfaces. *Adv Biomaterials* 9:321–6, 1990

96. Miorner H, Myhre E, Bjorck L, et al: Effect of specific binding of human albumin, fibrinogen, and immunoglobulin G on surface characteristics of bacterial strains as revealed by partition experiments in polymer phase systems. *Infect Immun* 29:879–85, 1980

97. Mosher DF, Proctor RA: Binding and factor XIIIa-mediated cross-linking of a 27-kilodalton fragment of fibronectin to *Staphylococcus aureus*. *Science* 209:927–9, 1980

98. Mozes N, Rouxhet PG: Methods for measuring hydrophobicity of microorganisms. *J Microbiol Meth* 6:99–112, 1987

99. Muller E, Takeda S, Goldmann DA, et al: Blood proteins do not promote adherence of coagulase-negative staphylococci to biomaterials. *Infect Immun* 59:3323–6, 1991

100. Murray PA, Prakobphol A, Lee T, et al: Adherence of oral streptococci to salivary glycoproteins. *Infect Immun* 60:31–8, 1992

101. Naylor PT, Ruch D, Brownlow C: Fibronectin binding to orthopedic biomaterials and its subsequent role in bacterial adherence. *Trans Orthop Res Soc* 14:561, 1989

102. Neu TR, Marshall KC: Bacterial polymers: physicochemical aspects of their interactions at interfaces. *J Biomater Appl* 5:107–33, 1990

103. Noda Y, Kanemasa Y: Determination of surface charge of some bacteria by colloid titration. *Physiol Chem Phys Med NMR* 16:263–74, 1984

104. Ofek I, Whitnack E, Beachey EH: Hydrophobic interactions of group A streptococci with hexadecane droplets. *J Bacteriol* 154:139–45, 1983

105. Oga M, Arizono T, Sugioka Y: Bacterial adherence to bioinert and bioactive materials studied *in vitro*. *Acta Orthop Scand* 64:273–6, 1993

106. Oga M, Sugioka Y, Hobgood CD, et al: Surgical biomaterials and differential colonization by *Staphylococcus epidermidis*. *Biomaterials* 9:285–9, 1988

107. Olsson J, Carlen A, Holmberg K: Inhibition of *Streptococcus mutans* adherence by means of surface hydrophilization. *J Dent Res* 69:1586–91, 1990

108. Olsson J, Krasse B: A method for studying adherence of oral streptococci to solid surfaces. *Scand J Dent Res* 84:20–8, 1976

109. Olsson J, van der Heijde Y, Holmberg K: Plaque formation *in vivo* and bacterial attachment *in vitro* on permanently hydrophobic and hydrophilic surfaces. *Caries Res* 26:428–33, 1992

110. Ørstavik D: Sorption of *Streptococcus faecium* to glass. *Acta Path Microbiol Scand* 85:38–46, 1977

111. Pascual A, Fleer A, Westerdaal NA, et al: Modulation of adherence of coagulase-negative staphylococci to Teflon catheters *in vitro*. *Eur J Clin Microbiol* 5:518–22, 1986

112. Paulsson M, Kober M, Freij-Larsson C, et al: Adhesion of staphylococci to chemically modified and native polymers, and the influence of preadsorbed fibronectin, vitronectin and fibrinogen. *Biomaterials* 14:845–53, 1993

113. Pedersen K: Electrostatic interaction chromatography, a method for assaying the relative surface charges of bacteria. *FEMS Microbiol Lett* 12:365–7, 1980

114. Prewett AB, Domenick JM, Tsang N, et al: Differential adherence of three clinical isolates of *Staphylococcus* to various metal surfaces. *Trans Soc Biomater* 14:105, 1991

115. Pringle JH, Fletcher M: Influence of substratum hydration and adsorbed macromolecules on bacterial attachment to surfaces. *Appl Environ Microbiol* 51:1321–5, 1986

116. Quirynen M, van der Mei HC, Bollen CM, et al: An *in vivo* study of the influence of the surface roughness of implants on the microbiology of supra- and subgingival plaque. *J Dent Res* 72:1304–9, 1993

117. Reynolds EC, Wong A: Effect of adsorbed protein on hydroxyapatite zeta potential and *Streptococcus mutans* adherence. *Infect Immun* 39:1285–90, 1983

118. Richmond DV, Fisher DJ: The electrophoretic mobility of microorganisms. *Adv Microb Physiol* 9:1–29, 1973

119. Rosenberg M: Bacterial adherence to polystyrene: a replica method of screening for bacterial hydrophobicity. *Appl Environ Microbiol* 42:375–7, 1981

120. Rosenberg M, Gutnick D, Rosenberg E: Adherence of bacteria to hydrocarbons: a simple method for measuring cell-surface hydrophobicity. *FEMS Microbiol Lett* 9:29–33, 1980

121. Ryden C, Rubin K, Speziale P, et al: Fibronectin receptors from *Staphylococcus aureus*. *J Biol Chem* 258:3396–401, 1983

122. Sar N: Direction of spreading (DOS): A simple method for measuring the hydrophobicity of bacterial lawns. *J Microbiol Methods* 6:211–9, 1987

123. Satou N, Satou J, Shintani H, et al: Adherence of streptococci to surface-modified glass. *J Gen Microbiol* 134:1299–305, 1988

124. Scheld WM, Strunk RW, Balian G, et al: Microbial adhesion to fibronectin *in vitro* correlates with production of endocarditis in rabbits. *Proc Soc Exp Biol Med* 180:474–82, 1985

125. Speier JL, Malek JR: Destruction of microorganisms by contact with solid surfaces. *J Colloid Interface Sci* 89:68–76, 19082

126. Sugarman B, Musher D: Adherence of bacteria to suture materials. *Proc Soc Exp Biol Med* 167:156–60, 1981

127. Sukenik CN, Balachander N, Culp LA, et al: Modulation of cell adhesion by modification of titanium surfaces with covalently attached self-assembled monolayers. *J Biomed Mater Res* 24:1307–23, 1990

128. Switalski LM, Butcher WG, Caufield PC, et al: Collagen mediates adhesion of *Streptococcus mutans* to human dentin. *Infect Immun* 61:4119–25, 1993

129. Switalski LM, Speziale P, Hook M, et al: Binding of *Streptococcus pyogenes* to laminin. *J Biol Chem* 259:3734–8, 1984

130. Tebbs SE, Sawyer A, Elliott TS: Influence of surface morphology on *in vitro* bacterial adherence to central venous catheters. *Br J Anaesth* 72:587–91, 1994

131. Vacheethasanee K, Temenoff JS, Higashi JM, et al: Bacterial surface properties of clinically isolated *Staphylococcus epidermidis* strains determine adhesion on polyethylene. *J Biomed Mater Res* 42:425–32, 1998

132. van Loosdrecht MC, Lyklema J, Norde W, et al: Electrophoretic mobility and hydrophobicity as a measured to predict the initial steps of bacterial adhesion. *Appl Environ Microbiol* 53:1898–901, 1987

133. van Loosdrecht MC, Lyklema J, Norde W, et al: The role of bacterial cell wall hydrophobicity in adhesion. *Appl Environ Microbiol* 53:1893–7, 1987

134. Vaudaux PE, Suzuki R, Waldvogel FA, et al: Foreign body infection: role of fibronectin as a ligand for the adherence of *Staphylococcus aureus. J Infect Dis* 150:546–53, 1984

135. Vaudaux PE, Waldvogel FA, Morgenthaler JJ, et al: Adsorption of fibronectin onto polymethylmethacrylate and promotion of *Staphylococcus aureus* adherence. *Infect Immun* 45:768–74, 1984

136. Verheyen CC, Dhert WJ, de Blieck-Hogervorst JM, et al: Adherence to a metal, polymer and composite by *Staphylococcus aureus* and *Staphylococcus epidermidis. Biomaterials* 14: 383–91, 1993

137. Wang IW, Anderson JM, Marchant RE: *Staphylococcus epidermidis* adhesion to hydro phobic biomedical polymer is mediated by platelets. *J Infect Dis* 167:329–36, 1993

138. Westergren G, Olsson J: Hydrophobicity and adherence of oral streptococci after repeated subculture in vitro. *Infect Immun* 40:432–5, 1983

139. Wilcox MH, Finch RG, Smith DG, et al: Effects of carbon dioxide and sub-lethal levels of antibiotics on adherence of coagulase-negative staphylococci to polystyrene and silicone rubber. *J Antimicrob Chemother* 27:577–87, 1991

140. Zita A, Hermansson M: Determination of bacterial cell surface hydrophobicity of single cells in cultures and in wastewater *in situ. FEMS Microbiol Lett* 152:299–306, 1997

# Nonspecific *Staphylococcus epidermidis* Adhesion

## Contributions of Biomaterial Hydrophobicity and Charge

**Katanchalee Vacheethasanee and Roger E. Marchant**

*Departments of Biomedical Engineering and Macromolecular Science,*
*Case Western Reserve University, Cleveland, OH, USA*

## I. INTRODUCTION

In this chapter, we examine the contribution of material physicochemical surface properties on nonspecific *Staphylococcus epidermidis* adhesion under dynamic flow conditions in phosphate buffered saline (PBS). These are conditions that are relevant to pre-implantation conditions in the absence of adsorbing biological components such as proteins. In addition, the applied shear stress provides a quantitative measure of the strength of the bacteria–material surface adhesion. Here, we describe the influence of material surface hydrophobicity on *S. epidermidis* RP62A adhesion in 132.7 mM PBS using four self assembled monolayer (SAM) surfaces bearing a net negative, positive, or neutral charge and five biomedical polymers. Materials with higher surface hydrophobicity show statistically higher bacterial adhesion at shear stresses of 0 to 32 dyne/cm$^2$, demonstrating the contribution of material surface hydrophobicity on nonspecific adhesion. However, this positive correlation is limited to materials with water contact angles of >70° at 0–8 dyne/cm$^2$, and >100° at higher shear stresses (16–32 dyne/cm$^2$). For materials with contact angles below these critical values ($\theta_{crit}$), *S. epidermidis* adhesion was found to be relatively independent of material surface hydrophobicity, indicating that a minimum number of bacteria can adhere to biomaterials independent of surface energy. This behavior is explained by a proposed model of nonspecific interactions based on DVLO forces, hydrodynamic effect and fimbriae formation. The contribution of material surface charge to adhesion was assessed by quantifying adhesion of *S. epidermidis* RP62A to charged surfaces in PBS of increasing ionic strength under shear stresses of 0–15 dyne/cm$^2$. Both repulsive and attractive interactions were found to be involved in *S. epidermidis* adhesion, but only the attractive interaction was found to significantly alter adhesion in buffered saline (132.7 mM ion concentration) at zero shear stress.

*Handbook of Bacterial Adhesion: Principles, Methods, and Applications*
Edited by: Y. H. An and R. J. Friedman © Humana Press Inc., Totowa, NJ

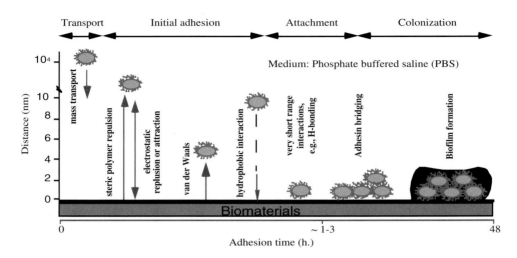

**Figure 1.** Physicochemical forces involved in nonspecific bacterial adhesion to biomaterials. The effective distances are based on using phosphate buffered saline as media.

## II. BACTERIAL ADHESION AND INFECTION

The extended use of cardiovascular and other implants is severely limited by device-centered infection, a serious complication which may lead to implant failure, prolonged hospitalization, possible amputation and even death.[15,30,34] In most cases, surgical removal and replacement of the implanted device is the only successful treatment. A major cause of device-centered infection is contamination by staphylococci, particularly strains of *S. epidermidis*, at or near the time of implantation via nonspecific adhesive interactions and subsequent colonization in vivo.[4,26] Nonspecific bacterial adhesion to a biomaterial has been proposed as a four-stage process: transport to the biomaterial surface, initial adhesion, attachment, and colonization (Fig. 1).[3,7] The first two steps (transport and initial adhesion) lead to physical contact between bacteria and biomaterial surfaces. The attachment stage is governed by both long- and short-range interactions. Long-range interactions position bacteria near the surface and include van der Waals forces, electrostatic forces and the hydrophobic effect.[14,45] As bacteria are transported sufficiently close to the surface (<2–3 nm), short-range chemical interactions, including hydrogen and covalent bonding between bacterial extracellular moieties and the device surface, predominate. Subsequent to, or concomitant with, attachment, specific and nonspecific interactions between bacterial surface structures and substrate surfaces induce an irreversibly firm adhesion in a time-dependent fashion (>1–3 h).[3,7,14] Capsules and fimbriae which act as bacterial adhesins bridge the bacteria to the material surface via nonspecific hydrophobic and electrostatic interactions and/or specific ligand–receptor interactions. Some strains of *S. epidermidis* also produce extracellular polysaccharides called slime, which may permanently bind bacteria to device surfaces.[7] In addition, slime may act as a protective biofilm, shielding bacteria from antibiotic therapy, physiologic shear, and possibly from host cell-mediated defenses.[5,8,13]

Hydrophobic and electrostatic interactions between the material surface and bacteria have been proposed to mediate both stages of nonspecific bacterial adhesion.[1] Thermodynamic analysis is commonly employed to predict the effect of hydrophobic interaction on bacterial adhesion to a solid surface in the absence of specific ligand–receptor binding.[17,18,42] The fundamental principle is that bacterial adhesion to a biomaterial surface becomes more thermodynamically favored as free energy of adhesion decreases. In the thermodynamic analysis, only physicochemical surface properties of the material, the bacteria, and bulk properties of the surrounding media are taken in account. Prevailing local hemodynamic conditions, which also play a role in bacterial adhesion, may limit the validity of these theory for predicting bacterial adhesion in vivo.[35]

Previously we have studied the effect of bacterial surface properties on *S. epidermidis* adhesion to polyethylene (PE) using a rotating disk system.[35] This system enables adhesion to be studied under a well-defined range of shear stress within the physiologic regime (0–32 dyne/cm$^2$). A positive correlation between bacterial surface hydrophobicity and adhesion was found under low shear stresses (0–8 dyne/cm$^2$). This correlation was eliminated at higher shear, suggesting that adhesion of *S. epidermidis* may be predicted by thermodynamic theory reasonably well only under conditions of low shear. *S. epidermidis* surface charge did not correlate with bacterial adhesion to PE. The experiments reported here were performed to evaluate the validity of thermodynamic analysis in predicting the contribution of material surface hydrophobicity to nonspecific *S. epidermidis* adhesion under dynamic flow. Additionally, the influence of biomaterial surface charge on *S. epidermidis* adhesion under shear stresses was investigated.

## III. MATERIALS AND METHODS

### A. Bacteria

The bacteria used in this study was the clinical isolate *S. epidermidis* RP62A (ATCC 35984) which is an encapsulated slime producer. Cultures of the RP62A strain were maintained on tryptic soy agar, a mixture of tryptic soy broth (TSB; Difco, Detroit, MI, USA) and Bacto agar (Difco) and stored at 4°C. Prior to each experiment, RP62A strains were inoculated in TSB and cultured at 37°C for 24 h in order to reach the stationary growth phase. The cultures were washed twice with phosphate buffered saline (PBS, pH 7.4; Sigma, St. Louis, MO, USA), suspended in PBS and maintained at ambient temperature. Experiments were conducted within 5 h of resuspension.

### B. Polymeric Biomaterials

Sterile sheets of NHLBI primary reference low density polyethylene film (PE; Abiomed, MA) and polydimethylsiloxane (PDMS; Thoratec Laboratories, Berkeley, CA, USA) were used as received. Extruded films of Pellethane 2363-55D, a poly(ether urethane), (PEU), supplied by Medtronic Inc. (Minneapolis, MN, USA) were polymerized from 4,4-methylene bisphenyldiisocyanate, poly(tetramethyleneglycol) (MW 1000) and 1,4-butanediol in a 4.5:1:3.5 weight ratio. The PEU contained butylated hydroxytoluene antioxidant and methylene bis-stearamide extrusion lubricant. Some of the PEU sheets were extracted in hot toluene for 48 h to remove extrusion wax.[43] Unmodified base poly(ether urethane urea) films (PEUU A') were provided by E. I. DuPont Nemours. PEUU A' was prepared from p-diphenyl-methanediisocyanate and poly(tetramethyleneoxide) (mol wt ~2000) in an approximately 1.6:1 capping ratio, and ethylenediamine was used as a chain extender. The PEUU A' was cast on Mylar sheets from a 20% solution in *N*, *N'*-dimethylacetamide. The air side surface was used for bacterial adhesion experiments as it provided a smoother surface than the Mylar side. The roughness of the Mylar side can lead to large surface aggregates of *S. epidermidis* RP62A.[42]

Biomaterial samples were prepared by cutting the polymer sheets into 17±0.05 mm diameter disks using a precision-machined and heat-treated stainless steel die and a manual hydraulic press. This procedure ensured that the sample disks were reproduced with high dimensional accuracy and smooth edges. PEU, extracted PEU and PEUU A' disks were sonicated for 20 min in MeOH, rinsed with deionized water (>18.2 MΩ, Millipore, Bedford, MA, USA) and air dried.

### C. Self-Assembled Monolayers

Four self assembled monolayers (SAM) composed of positively charged 3-aminopropyltriethoxysilane (APS), negatively charged carboxylic acid terminated SAM (CTS), hydrophobic octadecyltrichlorosilane (OTS) and trichlorovinylsilane (TVS) were prepared on 18 mm diameter glass coverslips and used as model surfaces. All glassware was cleaned by immersion in base solution overnight, followed by rinsing with deionized water and drying in an oven at 100°C overnight. The coverslips were sonicated in chloroform (HPLC grade, Fisher Scientific, Fair Lawn, NJ, USA), rinsed with a stream of chloroform, and then cleaned in an argon plasma system for 30 min.

OTS, APS, and TVS modified glass were prepared via silanation reactions. CTS was prepared by derivatization of the TVS surface. Surface modification using OTS (Aldrich Chemical, Milwaukee, WI, USA) was performed by immersing glow-discharged treated coverslips for 30 min in a 1% OTS solution in dicyclohexyl (Aldrich).[31] OTS and dicyclohexyl were both vacuum distilled before use. Silanation with APS (Aldrich) was performed by placing the coverslips for 2 min in a 0.1% APS solution in chloroform.[21] After silanation, OTS and APS modified coverslips were removed with Teflon tweezers and rinsed with a stream of chloroform. Silanation with TVS (Aldrich) was performed by placing the coverslips for 5 min in a 5% TVS solution in chloroform.[24] The silane solution was exchanged with chloroform before pulling the modified sample through the air–liquid interface to remove unreacted TVS, which can react with water in the air, forming a white film on the surface. After fluid exchange, the TVS modified glass was removed with Teflon tweezers and rinsed in chloroform. All SAMs were then sonicated in chloroform for 30–45 min, rinsed with chloroform, air dried, and stored in sealed fluoroware containers.

CTS surfaces were prepared via permanganate–periodate oxidation of the TVS modified surface using the procedure of Wasserman et al.[44] Briefly, TVS modified cover-slips were immersed in a 0.5 mM $KMnO_4$/9.5 mM $NaIO_4$/1.8 mM $K_2CO_3$ aqueous solution for 24 h. All chemicals used in the oxidation reaction were purchased from Aldrich Chemical. The samples were removed from the oxidant, sonicated in 3 M $NaHSO_3$ (Fisher Scientific) and 0.1N HCl (Fisher), and rinsed with water and ethanol. The samples were then dried in a 100°C oven and stored in sealed fluoroware containers. Bromination of vinyl groups in TVS was used to assess the extent of $KMnO_4$/$NaIO_4$ oxidation.[44] The TVS and CTS modofied coverslips were immersed for 2 h in a 2% (v/v) solution of bromine (Fisher Scientific) in chloroform. The materials were rinsed with chloroform and then sonicated in chloroform for 20 min. After sonication, samples were rinsed with ethanol and dried at 100°C.

## D. Surface Characterization

Advancing and receding water contact angles using the sessile drop method were measured in air with a contact angle goniometer (Model 100-00; Ramé-Hart, Mountain Lakes, NJ, USA) using deionized water. Measurements were conducted on at least three spots for each sample, and three samples of each substrate were used. Analysis of surface roughness (root mean square value) was performed on each material surface, using an atomic force microscope (Bioscope; Digital Instruments, Santa Barbara, CA).

X-ray photoelectron spectroscopy (XPS) was carried out using a Perkin Elmer PHI-5400 XPS system equipped with 400 W anode and 15 V X-ray source. The X-ray source was monochromatic Al $K_\alpha$ radiation (1487 eV) and the take-off angle was fixed at 45 degrees. A pass energy of 93.9 eV was employed to obtain survey scans from 0 to 1100 eV for the $C_{1s}$, $O_{1s}$, and $Br_{3d}$ regions. The spectra were referenced to the C1s peak in C-C binding energy of 285 eV. Atomic compositions were calculated with Phi analysis software (Physical Electronics, Inc.), which utilized peak areas and pre-programmed atomic sensitivity factors. The XPS atomic compositions were measured from at least 3 samples.

## E. Rotating Disk System

The rotating disk system (RDS) provides a well-defined and reproducible dynamic flow environment in which to study bacterial adhesion under steady state conditions, which has been described previously.[40-42] The shear stress ($\tau_{ss}$, dyne/cm$^2$) in a fluid laminar boundary layer at the surface of the disk varies linearly with the radial distance (r) from the center of the disk, such that

$$\tau_{SS} = 0.800 \eta r \left(\frac{\omega^3}{\nu}\right)^{1/2}, \tag{1}$$

where $\eta$ is the absolute viscosity of testing medium (poise), $\nu$ is the kinetic viscosity (stokes), and $\omega$ is the angular velocity (rad/s). In addition, the flux of bacteria (j, cfu bacteria/s·mm$^2$) to the disk is constant across the surface of the disk and is defined as:

$$j = 0.62 \, D^{2/3} \, \nu^{-1/6} \, \omega^{1/2} \, C_\infty, \tag{2}$$

where $D$ is the diffusivity of the bacteria through the medium and $C_\infty$ (cfu/mL) is the bulk concentration of bacteria in the test medium. $D$ is given by the relationship $D = KT/6\pi\eta b$, where $K$ is the Boltzmann constant, $T$ the absolute temperature (273°K), and $b$ the radius of the bacteria (0.5 μm). Bacterial adhesion was shown previously to be approximately shear stress independent at a shear stress greater than 15 dyne/cm$^2$.[16,35,40-42] Therefore, studies were conducted at rotation speed of 1100 rpm, providing a shear stress range of 0-65 dyne/cm$^2$. Bacterial adhesion was evaluated in a shear stress range of 0–33 dyne/cm$^2$. The effect of ionic strength in test media on bacterial adhesion was examined by rotating sample disks at 500 rpm, providing a shear stress range of 0–20 dyne/cm.$^2$ Bacterial adhesion was evaluated in a shear stress range of 0–15 dyne/cm$^2$.

## F. Bacterial Adhesion

Material samples were attached to 17 mm diameter stainless steel disks using cyanoacrylate adhesive and press fit into a polytetrafluorethylene (PTFE) holder. The disk assemblies were dried overnight, equilibrated in PBS for 1 h and then mounted on the rotating disk apparatus. The samples were positioned in a PTFE beaker containing the test media prior to disk rotation. All equipment used in the rotating disk experiments was steam-sterilized or presterilized and other precautions were taken in order to minimize adventitious bacterial contamination. To study the effect of changing ionic strength in the medium on bacterial adhesion, several PBS solutions (pH 7.4) were prepared at increasing total ion concentrations of 13.3, 132.7, 265.4, and 398.1 mM.

Each inoculum was injected into testing media to a final concentration of ~$10^8$ cfu/mL using a syringe with a sterile 30 gauge needle to minimize bacterial aggregation. These experimental parameters provided a bacterial flux of 34 and 54 cfu/mm$^2$·sec to the disk surface at the rotation speeds of 500 and 1100 rpm, respectively. Each rotation experiment was conducted for 1 h at 37°C, using a constant temperature water bath. Experiments were repeated no less than three times. After rotation, the test medium was immediately removed by fluid exchange with 150 mL of fresh PBS, followed by 100 mL of 1% paraformalde-hyde (PFA, Sigma). The bacteria adherent on sample disks were fixed by incubating in 1% PFA for 10 min, before bringing the disk surface through the air–water interface. Samples were then removed from the rotating disk assembly and stored in 1% PFA. Each sample was rinsed in deionized water and stained for 10 min with the fluorescent nucleic acid stain acridine orange (AO; Molecular Probes, Eugene, OR, USA) at a concentration of 50 µg/mL in sodium acetate buffer at pH 3.5 for 10 min.[23] Samples were rinsed with deionized water to remove excess stain and attached to clean glass microscope slides with a cyanoacrylate adhesive. After mounting with immersion oil, sample slides were stored in the dark at 20°C.

Bacteria adherent to the sample disks were imaged and enumerated using standard image-analyzed epifluorescent microscopy techniques.[10,32] Samples were imaged using an inverted optical microscope (Nikon Diaphot 200) with epifluorescent attachment and mercury arc lamp light source (Nikon) using the B-2A filter cube in the light path. Fluorescence images were captured using a software driven digital chilled CCD camera (Photometrics AT200, Tucson, AZ, USA) and then displayed and stored using the software package Metamorph (Universal Imaging, West Chester, PA, USA). Adhesion was quantified by direct enumeration of fluorescently stained bacteria on the disk surface at radial distances spaced 1 mm apart. For each of the radial distances evaluated, bacterial counts over ten 6250 µm$^2$ fields were summed and normalized to give the number of bacteria per square millimeter (N).

### G. Analysis of Bacterial Adhesion

Bacterial adhesion data were analyzed quantitatively by calculating an adhesive coefficient (AC) for each of the radial distances evaluated. The AC is defined as:

$$AC(\%) = [N/(j \cdot t)] \times 100, \tag{3}$$

where $N$ is the number of bacteria counted per square millimeter, $j$ is the bacterial flux to the rotating disk surface and $t$ is the duration of the rotation experiment in seconds. Note that $j \cdot t$ is the total number of bacteria transported to that surface area over the course of experiment. Therefore, the AC represents the percentage of bacteria that overcome the energy barrier opposing sufficiently close positioning to the surface to allow adhesion.

**Table 1. Atomic Composition of Test Materials Determined from XPS Analysis**

| Surface | Atomic composition[†] | | | |
| --- | --- | --- | --- | --- |
| | %C | %O | %Si | %other* |
| Glow discharge-treated glass | 6.2±0.6 | 68.4±0.1 | 25.4±0.6 | - |
| OTS | 53.0±0.3 | 31.5±0.7 | 15.6±0.5 | - |
| APS | 17.8±5.7 | 58.6±4.5 | 21.7±0.7 | 2.0±0.4 N |
| TVS | 16.5±1.6 | 58.6±1.7 | 24.9±0.3 | - |
| CTS | 18.6±2.7 | 58.9±2.7 | 22.2±1.1 | - |
| Brominated TVS | 30.6±2.5 | 43.9±1.8 | 20.1±0.6 | 5.5±0.2 Br |
| Brominated CTS | 25.7±3.2 | 52.4±2.1 | 21.6±1.1 | 0.26±0.0 Br |

\* Elements already existing on clean glass coverslips and less than 2% are not included.
[†] Mean ± standard error ($n = 3$).

The *AC* reflects bacterial "affinity" for the material surface under the specified experimental conditions. Given that the theoretical constraints on the system are well approximated, the *AC* should not exceed 100%, which would indicate that all bacteria transported to the surface attach to the surface.

Statistically significant differences ($p < 0.05$) between multiple means were determined by the method of analysis of variance (ANOVA). The statistical analysis was completed using Statistica 3.0b software (Statsoft Inc., Tulsa, OK, USA).

## IV. RESULTS

### A. *Surface Properties of Test Materials*

In this study, materials derived from self assembled monolayers (SAM) were used as model surfaces to investigate the influence of surface properties, including hydrophobicity and charge, on nonspecific *S. epidermidis* adhesion. Analysis of surface roughness by atomic force microscopy indicated root mean square roughness (RMS) values of 1.66 nm for OTS, 1.34 nm for APS, 1.88 nm for TVS, and 2.47 nm for CTS. Therefore, the SAM surfaces were relatively smooth compared with 1 μm diameter *S. epidermidis*. In addition, bacterial adhesion has been found to be independent of surface roughness up to 1.4 μm RMS.[20]

Atomic composition data of SAM surfaces, determined from XPS measurements, are summarized in Table 1. Self assembled monolayers of carbon-rich silane compounds were formed on a silicon-rich glass substrate. The monolayers are very thin (<1 nm); therefore, the XPS spectra obtained are dominated by the $Si_{2p}$ (103 eV) and $O_{1s}$ (532 eV) peaks attributable to the underlying glass. As seen in Table 1, approx 6% carbon (285 eV) was found on glow-discharge treated glass coverslips. OTS exhibited the highest carbon and the lowest oxygen and silicon contents among the surfaces, because it had the largest chain length and monolayer thickness. The other surfaces, APS, TVS and CTS, showed similar C, O and Si contents due to their similar thicknesses (1–3 carbon atom(s) per chain). The nitrogen ($N_{1s}$, 398 eV) peak was found in the APS spectra, but not in the others, and identifies the presence of the terminal amino groups in the APS. The percentages of C, O and Si were similar on the TVS and CTS surfaces; therefore, these surfaces could not be distinguished from the XPS survey scans.

**Table 2. Water Contact Angles on Test Materials**

| Surface | Contact angle* | |
| --- | --- | --- |
| | A[†] | R[‡] |
| OTS | 108.7±0.2 | 108.0±0.2 |
| PDMS | 106.7±0.5 | 80.4±1.5 |
| Pellethane | 97.9±0.6 | 96.0±1.0 |
| PE | 96.1±0.2 | 95.6±0.4 |
| TVS | 81.1±0.5 | 72.0±0.5 |
| PEUU A' | 69.6±0.6 | 58.2±0.6 |
| Extracted pellethane | 69.4±1.0 | 56.5±2.1 |
| APS | 40.0±0.3 | 38.3±0.1 |
| CTS | 17.5±0.2 | 15.9±0.5 |

\* Water-in-air sessile drop method; mean ± standard error ($n = 9$).
[†] Advancing contact angles.
[‡] Receding contact angles.

**Table 3. Correlation Coefficients and p-Values
for the Adhesive Coefficients of *S. epidermidis* to Materials
and Surface Hydrophobicity at Different Shear Stress**

| Shear stress (dyn/cm$^2$) | Correlation coefficient | p-Value |
| --- | --- | --- |
| 0 | 0.44 | 0.008 |
| 4 | 0.35 | 0.023 |
| 8 | 0.32 | 0.057 |
| 16 | 0.33 | 0.050 |

The extent of oxidation in CTS was assessed measuring bromination of remaining vinyl groups. TVS and CTS modified glass were placed in a 2% solution of bromine in chloroform for 2 h. Comparison of the amount of bromine ($Br_{3d}$, 70 eV) in TVS to CTS demonstrated that ~95% of vinyl groups had been oxidized. Wasserman et al. reported that, by using a similar oxidant solution, vinyl groups were converted only to carboxyl groups.[44] Therefore, the amount of carboxyl groups on the CTS surface was presumed to be close to 95%, even though some vinyl groups might have been converted to hydroxyl groups, another possible oxidization product, and not further oxidized to carboxylic groups.[25] With this surface density of carboxylic acid groups, the effect of negative charge on CTS surface on *S. epidermidis* adhesion should be detectable and comparable to that of the positively charged APS surface.

In addition to SAM surfaces, polymeric biomaterials also were included to determine the effect of hydrophobicity on *S. epidermidis* adhesion. Table 2 shows the results of the water contact angle measurements on both SAM and biomaterial surfaces. The advancing and receding contact angles are reported in order, from the highest to lowest surface hydrophobicity.

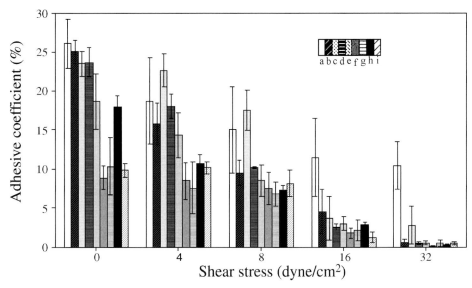

**Figure 2.** Adhesive coefficients of *S. epidermidis* RP62A as a function of shear stress to the following materials ordered from the highest to lowest surface hydrophobicity: OTS (a), PDMS (b), Pellethane (c), PE (d), TVS (e), PEUU (f), extracted Pellethane (g), APS (h), and CTS (i) in 137 mM PBS. Error bars reflect standard error, and are not visible if the SE is smaller than the symbol (*n* = 3).

## B. Effect of Hydrophobicity on S. epidermidis *Adhesion*

Figure 2 is a plot of adhesive coefficients (AC) of *S. epidermidis* RP62A to all material surfaces in order from the highest to lowest material hydrophobicity (left to right) as a function of shear stress in 132.7 mM PBS, approximately physiological electrolyte concentration.

Statistical analysis of bacterial adhesion data at zero shear stress is shown in Figure 3. To compare bacterial adhesion on the different surfaces, results are arranged from highest to lowest hydrophobicity based on contact angle measurement. ANOVA analysis was used to show whether bacterial adhesion was statistically different on two surfaces. One can ascertain whether differences in bacterial adhesion exist between the surfaces by looking at results in the vertical column on the left and intersecting with results for another surfaces from the right diagonal. At zero shear stress, all materials with advancing contact angle >80° show significantly greater *S. epidermidis* adhesion than those with advancing contact angle <70°, except for positively charged APS ($q_A$ = 40°). Nevertheless, OTS with the highest contact angle ($q_A$ = 108°) showed significantly higher bacterial adhesion than APS. Bacterial adhesion was significantly greater on the positively charged APS surface than the noncharged PEUU A' and extracted PEU surfaces which have much higher contact angles, demonstrating the ability of attractive electrostatic interactions to enhance adhesion of the net negatively charged *S. epidermidis* strain at zero shear stress

Table 3 shows correlation coefficients and p-values of correlation between the AC and material surface hydrophobicity. Positive correlations were found over the entire range of shear stresses studied (0–32 dyne/cm²), however, correlations only at 0 and 4 dyne/cm² were statistically significant. As seen by the decreasing correlation coefficients and

|          | OTS | PDMS | PEU | PE | TVS | PEUU A' | ex-PEU | APS | CTS |
|----------|-----|------|-----|----|-----|---------|--------|-----|-----|
|          | OTS |      |     |    |     |         |        |     |     |
| PDMS     |     | PDMS |     |    |     |         |        |     |     |
| PEU      |     |      | PEU |    |     |         |        |     |     |
| PE       |     |      |     | PE |     |         |        |     |     |
| TVS      | X   |      |     |    | TVS |         |        |     |     |
| PEUU A'  | X   | X    | X   | X  | X   | PEUU A' |        |     |     |
| ex-PEU†  | X   | X    | X   | X  | X   |         | ex-PEU |     |     |
| APS      | X   |      |     |    |     | X       | X      | APS |     |
| CTS      | X   | X    | X   | X  | X   |         | X      |     | CTS |

*Increasing hydrophobicity* (vertical axis) — *Increasing hydrophobicity* (diagonal axis)

† extracted pellethane

**Figure 3.** Statistical analysis of bacterial adhesion data at zero shear stress.

p-values with increasing shear stress, the sensitivity of bacterial adhesion to material hydrophobicity was limited by shear stress. At a shear stress of 4 dyne/cm$^2$, the ACs of bacteria to OTS and PEU were statistically greater than those to PEUU A' ($p < 0.03$), PEU ($p < 0.02$) and APS ($p < 0.04$). Even though no statistically significant positive correlations were observed at shear stresses >10 dyne/cm$^2$, the AC of bacteria to APS was statistically lower as compared with OTS ($p < 0.03$) and PEU ($p < 0.05$) at 8 dyne/cm$^2$. Adhesion to OTS was statistically greater than those to the other surfaces ($p < 0.05$) with the exception of PDMS at 16 dyne/cm$^2$, and OTS exhibited significantly greater bacterial adhesion than all other surfaces ($p < 0.001$) at 32 dyne/cm$^2$.

### C. Effect of Ionic Strength on S. epidermidis Adhesion

Figure 4 shows plots of AC of RP62A to (a) OTS, (b) TVS, (c) APS, and (d) CTS as a function of shear stress in PBS of varying ionic strength. As expected, no correlation between electrolyte concentration and adhesion to the nonionized OTS and TVS surfaces was seen over the entire shear range. At shear stresses of 0–32 dyne/cm$^2$, average bacterial adhesion to positively charged APS in 13.7 mM PBS was greater than at higher ionic strength, and the attractive electrostatic interaction dropped off quickly with increasing ion concentration. Due to counting variability, however, this trend was not statistically significant. For the negatively charged CTS surface, at zero shear stress *S. epidermidis* tended to show gradually greater adhesion as the ionic strength of the testing media was increased, and adhesion in 398.1 mM PBS is statistically greater than at 13.27 and 132.7 mM. The weak correlation with ionic strength was eliminated at higher shear stress.

## V. DISCUSSION

DLVO theory is a well-known construct which utilizes electrostatic and van der Waals forces to explain long range interactions between two macroscopic surfaces having the same sign surface charge (distance >150 nm).[19,29] The total interaction energy ($\Delta G_{total}$) represents the summation of the free energy of van der Waals attraction and electrostatic repulsion as a function of separation distance. Adhesion between two surfaces is more

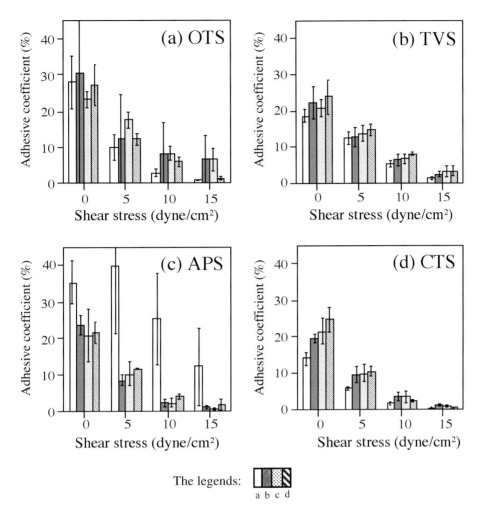

**Figure 4.** *S. epidermidis* RP62A adhesion to (A) OTS, (B) TVS, (C) APS, (D) CTS in 13.7 (a); 137 (b); 274 (c); and 411 (d), mM PBS. Error bars reflect standard error (SE), and are not visible if the SE is smaller than the symbol (*n* = 3).

favorable at primary (distance <1 nm) and secondary (distance ~5–20 nm) minima, with an unfavorable energy barrier between. Fitting this model to the current study, bacterial adhesion would be considered irreversible at the primary minimum and reversible at the secondary minimum.[12,29,36]

Typically, bacteria initially interact with a surface at the secondary minimum, where the bacteria are able to move laterally about the biomaterial surface.[39] The primary minimum, which leads to closer contact, is then reached by two pathways, either 1) by overcoming the energy barrier by attractive hydrophobic interaction; or 2) by bridging this separation distance by protruding their fibrillar adhesin structures.[29]

For both pathways (direct contact and bridging), the water layer(s) between bacteria and substrate must be removed, which can be predicted from thermodynamic analysis. The thermodynamically favored adherence of bacteria to surfaces is indicated by the negetive changes in Helmholtz free energy of adhesion ($\Delta G_{adh}$).[9,17,18] A thermodynamic model has been developed by Absolom et al. to predict the bacteria–substrate adherence considering only nonspecific interactions.[1,2] This model is based on calculating $\Delta G_{adh}$, as a function of biomaterial surface tension, in three distinct solutions. For the case in which bacterial surface tension is less than the media surface tension, then $\Delta G_{adh}$ increases with increasing material surface tension ($\gamma_{SV}$). On the other hand, if the bacterial surface tension is greater than the media surface tension, then $\Delta G_{adh}$ decreases with increasing $\gamma_{SV}$. In the last case, if bacterial surface tension is equal to the media surface tension, then $\Delta G_{adh}$ becomes zero and independent of $\gamma_{SV}$. In the system of aqueous medium, solid surface tension corresponds to surface hydrophibicity. For *S. epidermidis*, the surface tension (64.5–70.4 mN/m) is less than that of physiological saline buffer (73 mM/m).[2,11] Thus, using the thermodynamic model, increasing *S. epidermidis* adhesion is predicted with increasing material hydrophobicity. This prediction is in agreement with numerous studies in which the same strains of *S. epidermidis* were tested in static environments on different surfaces.[2,11]

Charged and noncharged SAMs covering a wide range of hydrophobicity (water contact angle = 18–109°) were selected as model surfaces to examine the validity of the thermo-dynamic analysis for predicting *S. epidermidis* RP62A adhesion. The SAM surfaces provided consistency in type, amount, and configuration of functional groups, with mini-mal changes in the intersample and intrasample surface roughness, as seen in the data obtained from XPS and AFM analysis. The high intersample and intrasample chemical and physical homogeneity of SAM surfaces minimized the contributions of other factors such as topography and deviations in surface wettability on adhesion. In addition, four polymer biomaterials were examined to test whether the effect of the hydrophobicity of SAMs on bacterial adhesion were quantitatively similar to that of the polymeric biomaterials.

A PBS solution of physiological ion concentration (132.7 mM) was used as the medium to determine the contribution of material surface hydrophobicity on nonspecific bacterial adhesion. In this study, the adherent *S. epidermidis* were fixed with 1% PFA solution before bringing through air–water interface, therefore, both reversible and irreversible adhesions were quantified. Over the entire range of shear studied (0–32 dyne/cm$^2$), surfaces with higher hydrophobicity tended to show greater *S. epidermidis* adhesion, and quantitative statistical differences in adhesion were found between materials having sufficiently different hydrophobicity, indicating a positive correlation between the surface hydrophobicity of materials and the adhesion of *S. epidermidis*. The highest correlation coefficients and p-values were found at zero shear and decreased as increasing shear stress was applied to the surfaces, demonstrating that the sensitivity of this positive correlation is suppressed by increasing the applied shear stress.

Figure 5 shows a plot of the average AC of *S. epidermidis* RP62A strains to all materials vs. advancing contact angles of the test materials in 132.7 mM PBS. Critical advancing contact angles ($\theta_{crit}$) were defined as the intersection of the two lines fitted by linear regression to adhesion curves, separating the shear range into two regions, which are relatively independent (region I) and dependent (region II) upon substrate hydrophobicity. At least three adhesion data sets represented each region. At zero shear stress, the AC of RP62A to APS was excluded from region I due to the observed enhancement of bacterial adhesion via attractive electrostatic interaction between positively charged APS and net negative bacteria surface.

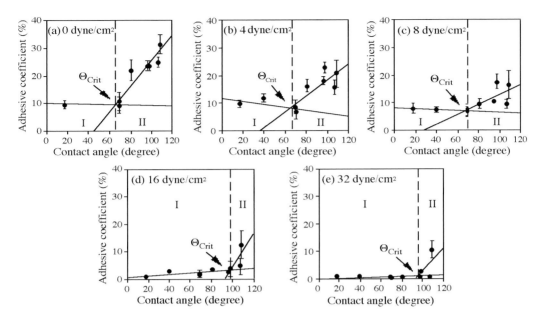

**Figure 5.** Average adhesive coefficient profile (*n* = 3 for each datum) of *S. epidermidis* RP62A, as a function of material contact angle data (*n* = 9), and shear stress. (a) 0, (b) 4, (c) 8, (d) 16 and (d) 32 dyne/cm$^2$. For region II in each plot, bacterial adhesion is strongly dependent on contact angle, but independent in region I. The transition between regions I and II, the critical contact angle ($\theta_{crit}$), is defined by the intersection of two lines fitted by linear regression to the adhesive coefficient profile. Error bars are standard errors (SE) and are not visible if the SE is smaller than the symbol (*n* = 3).

Bacterial adhesion is thermodynamically favored when the adhesion energy is negative and decreases with increasing surface energy. The data in region II, which are strongly dependent on material hydrophobicity, are in agreement with this theory. In contrast, the adhesion in region I appears to be relatively constant as material surface hydrophobicity increases. $\theta_{crit}$ defines a point above which the biomaterial surfaces have sufficiently low surface energy such that bacterial adhesion can be predicted using thermodynamic theories. At shear stresses of 0–8 dyne/cm$^2$ (Fig. 5a,b,c), $\theta_{crit}$ was observed at ~70°. At higher shear stresses (16–32 dyne/cm$^2$), $\theta_{crit}$ was found to shift to ~100°. Jansen et al.[20] also found that *S. epidermidis* adhesion to biomaterials increased with increasing material hydrophobicity, but remained constant when the surface hydrophobicity was sufficiently low to provide positive values of adhesion enthalpy. A similar bacterial adhesion profile, as a function of biomaterial surface tension, has been reported by Kiermitci-Gümüsderelioglu et al.[22] However, the adhesion of negatively charged *E. coli* was tested against only three biomaterials including polypropylene, Pellethane, and poly(hydroxyethylmethacrylate); therefore, no discussion of a $\theta_{crit}$ was developed. These studies suggest that, in vitro, a certain minimum number of bacteria will adhere even under apparently unfavorable free energy conditions.

Bacterial adhesion under apparent thermodynamically unfavorable conditions ($\Delta G_{adh} > 0$) has been reported previously[6,28,37] and found to be strain dependent.[27] Busscher et al. explained this observation as the result of protrusion of fibrillar surface structures or

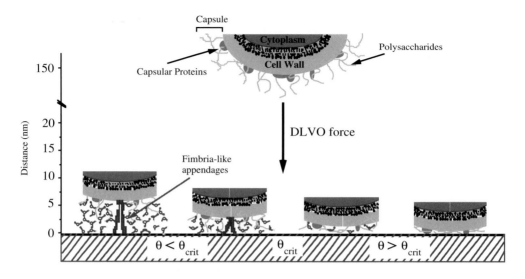

**Figure 6.** A proposed mechanism of nonspecific *S. epidermidis* adhesion to biomaterials of differing surface hydrophobicity in aqueous medium. The shaded regions represent biomaterial surfaces varying low to high water contact angles (left to right). *S. epidermidis* are positioned at the material surface within a secondary energy minimum as predicted by DLVO theory. Bacteria may then firmly adhere to the materials by either the hydrophobic effect, if $\theta < \theta_{crit}$, or by fimbria-like appendage formations, if $\theta > \theta_{crit}$.

by the extrusion of an intercellular glue, or both.[6] Therefore, this may indicate that the previous thermodynamic analysis is valid only for direct contact of bacteria with the surface. However, van Pelt et al. and Busscher et al. have observed that in vivo and in vitro bacterial adhesions were not qualitatively different, but reversible if $\Delta G_{adh} > 0$, indicating the mediation of material surface energy in binding strength of bacteria.[6,37] *S. epidermidis* surface proteins (SSP-1 and SSP-2) associated with fimbria-like polymers have been found that mediate bacteria adhesion to polystyrene.[38] The production of proteins and formation of adhesive patches have been observed after 1 h of adhesion, which is the testing duration of our studies.[9] Based on these findings, we propose a model, illustrated in Figure 6, to explain bacterial adhesion as a function of biomaterial surface tension. Water layers formed on the surfaces with contact angles $>\theta_{crit}$ can be removed, allowing attractive hydrophobic interactions to overcome repulsion and thus position bacteria at the primary minimum. This results in direct contact between bacteria and the material surface (<1 nm). The ease of water removal depends upon the hydrophobicity of both bacterial and material surfaces, resulting in more bacteria passing over the energy barrier and consequently direct attachment to the surface. When material hydrophilicity is sufficiently high to anchor the water layer(s), the hydrophobic effect is not strong enough to overcome the energy barrier and bacteria will remain at the separation distance of the secondary minimum. Bacteria may then conformationally change their fibrillar surface structures to reversibly adhere to the surface. The strength of fibrillar attachment may depend on surface energies of both bacteria and material, as reported by van Pelt et al.[6,37] However, under these conditions, hydrophobic interactions do not govern the position of bacteria at the secondary minimum, therefore, the number of adherent bacteria will be independent on material surface contact angle.

The shift of $\theta_{crit}$ from 70 to 100° at shear stress >16 dyne/cm$^2$ indicates that high shear stress can overcome the attractive hydrophobic interactions, resulting in the detachment of adherent bacteria. Materials with low surface energy can maintain high adhesion via stronger hydrophobic interaction. However, the AC at $\theta_{crit}$ drops dramatically for all materials, as shear stress increases to >16 dyne/cm$^2$.

Nonspecific electrostatic shielding in PBS solutions of increasing ionic strength was used to determine the effect of material surface charge on bacterial adhesion. Modulation of adhesion to both positively charged APS and negatively charged CTS surfaces due to electrostatic shielding demonstrates the influence of material surface charge on both repulsive and attractive electrostatic interactions. The results are in agreement with many previous studies conducted under static conditions.[20] Coverage of amino groups on the APS surface is expected to be ~100%, and coverage of carboxyl groups on the CTS surface was estimated to be ~95% by XPS analysis. Therefore, the adhesion data obtained from these two surfaces should be comparable. However, as seen in Figure 5, only the attractive interaction was found to significantly modulate *S. epidermidis* adhesion at physiological electrolyte concentration. For positively charged APS, the electrostatic interaction is attractive so that the energy barrier vanishes and *S. epidermidis* approach the primary minimum without difficulty,[29,33] resulting in greater adhesion than predicted based on material contact angle. For CTS surfaces, which carry a charge of the same sign as the RP62A strain, electrostatic interaction is repulsive and the resulting energy barrier may prevent the bacteria from positioning in the primary minimum. Electrostatic repulsion between CTS and bacteria was greater than those between uncharged surfaces and bacteria, resulting in the secondary minimum at further distance of separation. However, modulation of electrostatic repulsion as reflected in the bacterial adhesion was not found. This indicates that the separation distance of the secondary minimum of RP62A adhesion to CTS at physiological ionic strength is still in the range that reversible adhesion through fibrils may form.

## VI. SUMMARY

Material surface hydrophobicity plays an important role in nonspecific bacterial adhesion at shear stresses of 0–32 dyne/cm$^2$. However, dynamic flow conditions limit the effective strength of this correlation. Thermodynamic analysis was found to be valid only for materials with greater than 70° water contact angle. For materials with contact angles below 70°, *S. epidermidis* adhesion is relatively independent of material surface hydrophobicity, indicating that a minimum number of bacteria in vitro adhere to the biomaterial surfaces independent of surface energy. Both repulsive and attractive interactions were found to be involved in *S. epidermidis* adhesion, but only the attractive interaction was found to significantly alter adhesion in saline buffer at physiological ion concentrations (132.7 mM total ion concentration) at zero shear stress. In the previous studies, *S. epidermidis* strains with higher surface hydrophobicity showed significantly greater adhesion to PE in 137 mM PBS at low shear stresses (0–8 dyne/cm$^2$). At higher shear stress, this correlation between bacterial surface hydrophobicity and bacterial adhesion was eliminated. Also no correlation was found between *S. epidermidis* surface charge and bacterial adhesion to uncharged PE. Combining previous results with those of the current study, we suggest that modulation of material surface properties, including material hydrophobicity and charge, will be useful in controlling unwanted nonspecific bacterial adhesion. The range of relative hydrophobicity for the biomaterials tested is much wider than that of different bacteria. Nonionic surface modifications to biomaterials which enhance hydrophilicity should help suppress the contributions of both bacterial and substrate hydrophobicity to the nonspecific mechanism of adhesion, but may not prove sufficient to completely prevent device-centered infection. More sophisticated molecular designs of surface modifications, including highly hydrated long-chain molecules with a long range effective hydration distance and antibiotic modified surfaces, may be more effective approaches for developing infection-resistant biomaterials.

**Acknowledgments:** We thank Brian J. Lestini for editorial assistance and Madhu Raghavachari and Ping-fai Sit for performing AFM experiments. Assistance with bacterial adhesion experiments by Sona Sivikova is appreciated. This work is financially supported by NIH grant HL47300.

## REFERENCES

1. Absolom DR: The role of bacterial hydrophobicity in infection: Bacterial adhesion and phagocytic ingestion. *J Microbiol* 34:287–98, 1988
2. Absolom DR, Lamberti FF, Policova Z, et al: Surface thermodynamics of bacterial adhesion. *Appl Environ Microbiol* 46:90–7, 1998
3. An YH, Friedman RJ, Draughn RA, et al: Bacterial adhesion to biomaterial surfaces. In: Wise DL, Trantolo DJ, Altobelli DE, et al, eds. *Human Biomaterials Applications*. Humana Press, Totowa, NJ, 1996:19–57
4. Bandyk DF, Esses GE: Prosthetic graft infection. *Surg Clin North Am* 74:571–90, 1994
5. Bayston R, Penny SR: Excessive production of mucoid substance in staphylococcus SIIA: a possible factor in colonization of Holter shunts. *Dev Med Child Neurol Suppl* 27:25–8, 1972
6. Busscher HJ, Uyen MC, Van Pelt AWJ, et al: Kinetics of adhesion of the oral bacterium *Streptococcus sanguis* CH3 to polymers with different surface free energies. *Appl Environ Microbiol* 51:910–4, 1986
7. Christensen GD: The confusing and tenacious coagulase-negative staphylococci. *Adv Intern Med* 32: 177–92, 1987
8. Christensen GD: The 'sticky' problem of *Staphylococcus epidermidis* sepsis. *Hosp Pract* (Off Ed) 28:27–36, 1993
9. Esperson F, Wilkinson B, Gahrn-Hanson B, et al: Attachment of staphylococci to silicone catheters *in vitro*. *APMIS* 98:471–8, 1990
10. Evans-Hurrell JA, Adler J, Denyer S, et al: A method of the enumeration of bacterial adhesion to epithelial cells using image analysis. *FEMS Microbiol Lett* 107:77–82, 1993
11. Ferreiros CM, Carballo J, Craido MT, et al: Surface free energy and interaction of *Staphylococcus epidermidis* with biomaterials. *FEMS Microbiol Lett* 51:89–94, 1989
12. Fletcher M: Attachment of *Pseudomonas* fluorescence to glass and influence of electrolytes on bacterium-substratum separation distance. *J Bacteriol* 170:2027–30, 1988
13. Gray ED, Peters G, Verstegen M, et al: Effect of extracellular slime substance from *Staphylococcus epidermidis* on the human cellular immune response. *Lancet* 1:365–7, 1984
14. Gristina AG: Biomaterial-centered infection: Microbial adhesion versus tissue integration. *Science* 237:1588–95, 1987
15. Higashi JM, Marchant RE: Implant infections. In: Recum AFV, ed. *Handbook of Biomaterials Evaluation*. Taylor and Francis, Washington, DC, 1998:245–253
16. Higashi JM, Wang I-W, Shlaes DM, et al: Adhesion of *Staphylococcus epidermidis* and transposon mutant strains to hydrophobic polyethylene. *J Biomed Mater Res* 39:341–50, 1998

17. Hjerten S, Wadstrom T: What types of bonds are responsible for the adhesion of bacteria and viruses to native and artificial surface? In: Wadstrom T, Eliasson I, Holder IA, et al, eds. *Pathogenesis of Wound and Biomaterial-Associated Infections.* Springer-Verlag, London, 1990:225–232

18. Hogt AH, Dankert J, Feijen J: Adhesion of coagulase-negative staphylococci to biomaterials. *J Gen Microbiol* 129:2959–68, 1983.

19. Siedlecki CA, Eppell SJ, Marchant RE: Interactions of human von Willebrand factor with a hydrophobic self-assembled monolayer studied by atomic force microscopy. *J Biomed Mater Res* 28:971–80, 1994

20. Jansen B, Kohnen W: Prevention of biofilm formation by polymer modification. *J Ind Microbiol* 15:391–6, 1995

21. Jenney CR, Defife KM, Colton E, et al: Human monocyte/macrophage adhesion, macrophage motility, and IL-4-induced foreign body giant cell formation on silane-modified surfaces *in vitro. J Biomed Mater Res* 41:171–84, 1998

22. Kiermitci-Gumusderelioglu M, Pesmen A: Microbial adhesion to ionogenic PHEMA, PU and PP implants. *Biomaterials* 17:443–9, 1996

23. Kronvall G, Myhre E: Differential strain of bacteria in clinical specimens using acridine orange buffered at low pH. *Acta Path Microbiol Scand B* 85:249–54, 1997

24. McPherson TB, Shim HS, Park K: Grafting of PEO to glass, nitinol, and pyrolytic carbon surfaces by γ irradiation. *J Biomed Mater Res* 38:289–302, 1997

25. Higashi JM, Wang IW, Shlaes DM, et al: Adhesion of *Staphylococcus epidermidis* and transposon mutant strains to hydrophobic polyethylene. *J Biomed Mater Res* 39:341–50, 1998

26. O'Brien T, Collin J: Prosthetic vascular graft infection. *Br J Surg* 79:1262–7, 1992

27. Pratt-Terpstra IH, Weerpamp AH, Busscher HJ: On a relation between interfacial free energy-dependent and noninterfacial free energy-dependent adherence of oral streptococci to solid substrata. *Curr Microbiol* 16:311–3, 1988

28. Pratt-Terpstra IH, Weerkamp AH, Busscher HJ: Adhesion of oral streptococci from a flowing suspension to uncoated and albumin-coated surfaces. *J Gen Microbiol* 133:3199–206, 1985

29. Quirynen M, Marechal M, Busscher HJ, et al: The influence of surface free-energy on planimetric plaque growth in man. *J Dent Res* 68:796–9, 1989

30. Reilly LM, Altman H, Lusby RJ, et al: Late results following surgical management of vascular graft infection. *J Vasc Surg* 1:36–44, 1984

31. Siedlecki CA, Eppell SJ, Marchant RE: Interactions of human von Willebrand factor with a hydrophobic self-assembled monolayer studied by atomic force microscopy. *J Biomed Mater Res* 28:971–80, 1994

32. Sieracki ME, Johndon P, Sieburth JM: Detection, enumeration, and sizing of planktonic bacteria by image-analyzed epifluorescent microscopy. *Appl Environ Microbiol* 49:799–810, 1985

33. Skvarla J: A physio-chemical model of microbial adhesion. *J Chem Soc Faraday Trans* 89:2913–21, 1993

34. Sugarman B, Young EJ: Infections associated with prosthetic devices: magnitude of the problem. *Infect Dis Clin N Am* 3:187–98, 1989

35. Vacheethasanee K, Temenoff JS, Higashi JM, et al: Bacterial surface properties of clinically isolated *Staphylococcus epidermidis* strains determine adhesion on polyethylene. *J Biomed Mater Res* 42:425–32, 1998

36. Van Loosdrecht MCM, Zehnder AJB: Energetics of bacterial adhesion. *Experientia* 46:817–22, 1990

37. Van Pelt AWJ, Weerkamp AH, Uyen HC, et al: Adhesion of *Streptococcus sanguis* CH3 to polymers with different surface energies. *Appl Environ Microbiol* 49:1270–5, 1985

38. Veenstra GJ, Cremers FF, van Dijk H, et al: Ultrastructural organization and regulation of a biomaterial adhesin of *Staphylococcus epidermidis. J Bacteriol* 178:537–41, 1996

39. Vigeant MAS, Ford RM: Interactions between motile *Escherichia coli* and glass in media with various ionic strengths, as observed with a three-dimensional-tracking microscope. *Appl Environ Microbiol* 63:3437–79, 1997

40. Wang I-W, Anderson J, Marchant RE: *Staphylococcus epidermidis* adhesion to hydrophobic biomedical polymer is mediated by platelets. *J Infect Dis* 167:329–36, 1993

41. Wang I-W, Anderson J, Marchant RE: Platelet-mediated adhesion *of Staphylococcus epidermidis* to hydrophobic NHLBI reference polyethylene. *J Biomed Mater Res* 27:1119–28, 1993

42. Wang I-W, Danilich M, Anderson J, et al: Adhesion of *Staphylococcus epidermidis* to biomedical polymers: contributions of surface thermodynamics and hemodynamic shear conditions. *J Biomed Mater Res* 29:485–93, 1995

43. Waples LM, Olorundare OE, Goodman SL, et al: Platelet-polymer interactions: Morphologic and intracellular free calcium studies of individual human platelets. *J Biomed Mater Res* 32:311–3, 1996

44. Wasserman SR, Tao Y, Whitesides GM: Structure and reactivity of alkylsiloxane monolayers formed by reaction of alkyltrichlorosilanes on silicon substrates. *Langmuir* 5:1074–87, 1989

45. Wiencek KM, Fletcher M: Bacterial adhesion to hydroxyl- and methyl-terminated alkanethiol self-assembled monolayers. *J Bacteriol* 177:1959–66, 1995

# 6

# Effects of Surface Roughness and Free Energy on Oral Bacterial Adhesion

**Marc Quirynen, Kurt Dierickx, and Daniel van Steenberghe**

*Research Group for Microbial Adhesion, Department of Periodontology,*
*Oral Pathology and Maxillo-facial Surgery, School of Dentistry,*
*Catholic University of Leuven, Belgium*

## I. THE ORAL CAVITY, A UNIQUE ENVIRONMENT FOR STUDYING BIOFILM FORMATION

The oral cavity offers two interesting particularities which allow the evaluation of bacterial adhesion under a variety of conditions. From an ecological view point the oropharynx is considered an "open growth system" with an uninterrupted ingestion and removal of microorganisms and their nutrients. In order to colonize the oral cavity, bacteria must tackle the host defense mechanisms (including continuous shedding of mucosal membranes) and a variety of removal forces (e.g., friction by food intake, tongue and oral hygiene implements as well as the wash-out effect of the salivary and crevicular fluid outflow) (Fig. 1). Adherence to a surface is thus a key element for the colonization of the human oral cavity by the more than 500 bacterial taxa recorded from human oral samples which may cause inflammatory processes (e.g., gingivitis and periodontitis) (Fig. 1).

In the oral cavity bacteria can select from three distinct surfaces: the intra-oral hard surfaces (e.g., teeth, oral implants, dentures), the epithelial mucosae, and the nascent surface created when a new bacterium binds to an existing bacterial plaque (co-aggregation). Most research on intra-oral biofilm formation has been devoted to nonshedding intra-oral hard surfaces (teeth, oral implants, prostheses, etc.), because it is generally accepted that the constant renewal of the epithelial surfaces by shedding prevents the accumulation of large masses of microorganisms on the latter. However, several observations highlighted a significant intraspecies variation in the adhesion ability of pathogenic species to oral epithelia (dependent on the presence, composition and/or size of their capsule) (Fig. 2). Moreover, recent reports also seem to indicate an important intersubject variation in the rate of the initial bacterial adhesion to in vitro cultured epithelial cells.[32] These data are to be taken into consideration to understand the intersubject variation in susceptibility to different oral infections.

*Handbook of Bacterial Adhesion: Principles, Methods, and Applications*
Edited by: Y. H. An and R. J. Friedman © Humana Press Inc., Totowa, NJ

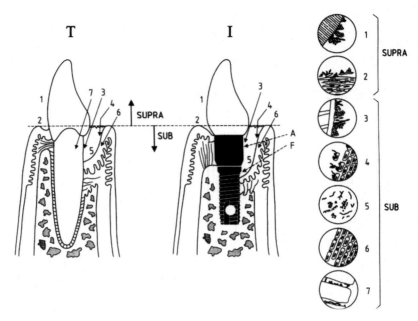

**Figure 1.** Schematic drawing representing a variety of intra-oral ecological sites to study biofilm formation with a distinction between the supra- and subgingival areas. A tooth (left) and a permucosal implant (right, consisting of an endosseous part, the fixture, F, and a transmucosal part, the abutment, A) form an unique ectodermal interruption which is reassured by a specific epithelial–connective tissue seal. In case of bacterial accumulation both gingivitis (the swelling of the gingiva) and periodontitis (the destruction of the periodontal tissues including the connective tissues, alveolar bone and periodontal ligament, the latter only present around teeth ) results in the formation of a periodontal pocket (right site for each abutment type) with a specific subgingival environment. Supragingivally bacteria can adhere to the hard surface (1) or, to a lower extent, to the desquamating oral epithelium (2). Subgingivally more niches become available for bacterial survival: (3) adhesion to the hard surface (root cementum or dentin in case of a tooth, variety of surfaces in case of an implant), (4) adhesion to the desquamating pocket epithelium, (5) swimming/floating in the crevicular fluid (the inflammatory exudate leaking through the epithelium into the pocket), (6) invasion into the epithelium with its large intercellular spaces which line the pocket, (7) invasion into the hard tissue (via the dentine tubules in case of a tooth or via penetration along the implant components in case of an implant). Adapted from ref.[30]

The biofilm on intra-oral hard surfaces is commonly called "plaque" because of its yellowish color, reminiscent of the mucosal plaques caused by syphilis. This accumulation of bacteria on teeth and/or permucosal implants forms a challenge to the periodontium because of the unique interruption of the ectodermal surfaces. Teeth and permucosal implants form a direct contact between the external environment and the bodily tissues through a seal of epithelial–connective tissue (Fig. 1). A bacterial accumulation on the supragingival area induces a swelling of the gingival margin due to an inflammatory host response reaction, which results in the creation of a small crevice between the gingiva and the tooth crown. Thereby an anaerobic environment (a subgingival area) is created which results in an overgrowth by anaerobic species that

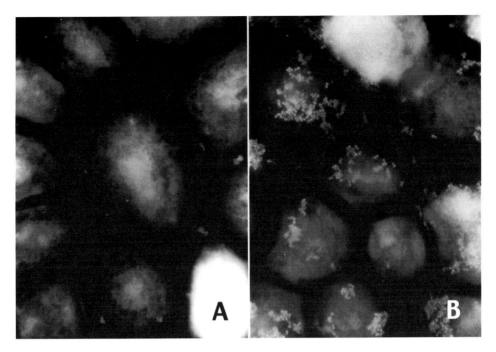

**Figure 2.** Fluorescent microscopic picture (enlargement ×1000) of two in vitro cultured monolayers form epithelial samples of one patient, after staining (LIVE/DEAD "Baclight" Bacterial Viability Kit; Molecular Probes, Eugene OR, USA) live cells and bacteria green and dead cells or bacteria red. Each monolayer had been seeded with a comparable number of different *Porphyromonas gingivalis* strains. The number of bacteria/cell is clearly lower in image A (capsule free strain) than in image B (capsule positive strain).

are considered even more pathogenic, resulting in the initiation of a vicious circle. With time, such an inflammatory process will lead to the creation of, at least in susceptible patients, a deep periodontal pocket resulting from direct (by bacteria) and indirect (due to the host response) tissue breakdown. These gingival pockets, with their subgingival area, offer a range of new possibilities for bacteria to survive in the oral environment (Fig. 1).

In the edentulous patient, dentures will offer other easily accessible surfaces to study biofilm formation. Whereas on the outer side of the denture removal forces retard a biofilm, bacteria are shielded against them at the mucosal surface of the denture.

This chapter will illustrate and discuss intra-oral observations on the impact of surface roughness and surface free energy on the biofilm formation. It will also link these data to reports from other areas. Modification of these variables may facilitate the prevention of oral diseases and be of clinical relevance.

## II. PHYSICOCHEMICAL INTERACTIONS BETWEEN BACTERIA AND INTRA-ORAL HARD SURFACES

Although, up to now, no completely satisfactory picture for bacterial adhesion to hard surfaces in an aquatic environment exists,[21] the following concept can help to understand most aspects of this adhesion process. Microbial adhesion in the oral cavity has been described (Fig. 3a) as a four-stage sequence.[6,7,36,41,42] This diagram of events clarifies the

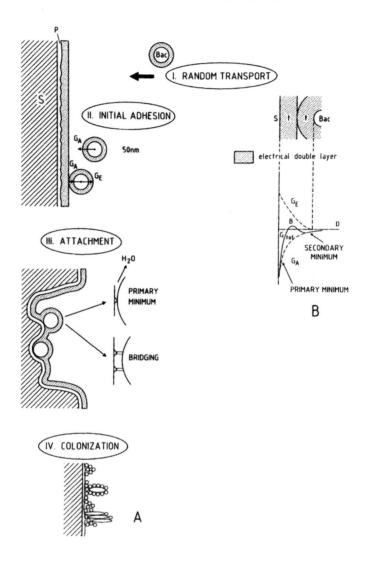

**Figure 3.** (a) Schematic representation of the dynamic plaque formation process as a four-stage sequence: I: random transport of bacterium (Bac) to the surface (S), the latter is covered with the aquired pellicle (P) and an electrical double layer (t), II: initial adhesion at secondary minimum (which often does not reach large negative values so that the adhesion is reversible), or directly at the primary minimum (with an irreversible binding) depending on the resultant of the van der Waals attractive force ($G_A$) and the electrostatic repulsive force ($G_E$), III: attachment of bacterium to the surface by specific interactions after bridging the separation gap or after passing the energy barrier, IV: colonization of the surface and biofilm formation (primarily by cell dividing and by bacterial intrageneric and/or inter-generic co-aggregation. (b) Long-range interaction between a negatively charged bacterium and a negatively charged surface according to the DLVO theory.[35] The Gibbs energy of interaction ($G_{tot}$) is calculated, in relation to the separation gap (D), as the summation of the van der Waals force ($G_A$) and the electrostatic interaction ($G_E$). Electrostatic interactions start when the electrical double layers overlap each other (t, thickness of the electrical double layer or Stern layer). Adapted from refs.[7,30,42] (Note that the size of the bacterium is too small in relation to the separation gap.)

hard-surface–bacteria interaction as well as the impact of surface characteristics on this interaction.

### Phase I: Transport to the Surface

The first stage involves the initial transport of a bacterium to the surface. Random contact may occur, for instance, through Brownian motion (average displacement of 40 µm/h), through sedimentation, through liquid flow (several orders of magnitude faster than diffusion), or through active bacterial movement (chemotactic activity).

### Phase II: Initial Adhesion

The second stage results in an initial reversible adhesion of the bacterium, initiated by the interaction between the bacterium and the surface from a certain distance (50 nm) via long and short-range forces (Fig. 3a).

*Long-range forces.* Bacteria may be considered as living colloidal particles, and as such they obey the laws of physical chemistry. If a colloidal particle approaches a surface, it interacts with that surface by means of 2 forces: the van der Waals forces (the first force to become active at distances even above 50 nm) and the electrostatic forces (at closer approach). Three types of van der Waals attractive forces ($G_A$) have been identified depending on the interacting molecules/atoms: the London dispersion forces, the Debye forces and the Kesson forces. Moreover, charged particles, in water, will be neutralized by a counter-charged layer that is diffusely distributed around the particle (the electrical double layer or Stern layer, Fig. 3b). When the double layer of a particle overlaps the double layer of the surface, an electrostatic interaction will take place. Since in the oral cavity most bacteria and the surfaces are negatively charged, the latter because of the spontaneous formation of the acquired pellicle (a glycoprotein rich layer that is formed immediately after introduction of a hard surface in the oral cavity), this electrostatic interaction is of a repulsive nature ($G_E$). The energy of this electrostatic interaction is determined by the zeta potential of the surface.[35] The distance at which this interaction appears is dependent on the thickness of the double layers, which themselves depend on the ionic charge of the surface and the ionic concentration of the suspension medium (for saliva the suspension ionic strength is considered to be medium). Derjaguin, Landau, Verwey, and Overbeek (DLVO) have postulated that, above a separation distance of 1 nm, the summation of the above mentioned 2 forces ($G_A$ and $G_E$) describes the total long-range interaction.[35] The total interaction energy (also called the total Gibbs energy, $G_{tot}$) for most bacteria in a suspension with a medium ionic strength (such as saliva) consists of a secondary minimum (where a reversible binding takes place: 5–20 nm from the surface), a positive maximum (an energy barrier to adhesion), and a steep primary minimum (located <2 nm away from the surface) where an irreversible adhesion is established (Fig. 3b). For bacteria in the mouth the secondary minimum does not frequently reach large negative values,[41] which results in a "weak" reversible adhesion (defined as a deposition to a surface in which the bacterium continues to exhibit Brownian motion and can readily be removed from the surface by mild shear or by the bacterium's own mobility). Surface irregularities, however, will protect bacteria against shear forces, so that an irreversible attachment can be established *(see below)*. On the contrary, on smooth surfaces, bacteria will easily detach. The positive maximum (B), that decreases with increasing ionic strength of the flow medium, is frequently low in the oral cavity (the smaller the particle the lower the height of the energy barrier B) so that a fraction of the bacteria may contain sufficient thermal energy to pass this barrier in order to reach the primary minimum (irreversible binding).

*Short-range interactions (Fig. 3a).* If a particle reaches the primary minimum (< 1nm from the surface), a group of short range forces (e.g., hydrogen bonding, ion pair formation, steric interaction, etc.) dominate the adhesive interaction and determine the strength of adhesion. Therefore, the DLVO theory is only able to predict whether primary minimum adhesion can occur; it cannot quantify the depth of this minimum. When a bacterium and a surface make direct contact, provided the water film present between the interacting surfaces can be removed, the interaction energy can be calculated from the assumption that the interfaces between bacterium–liquid (bl) and solid–liquid (sl) are replaced by a solid–bacterium (sb) interface. The change in the interfacial excess Gibbs energy upon adhesion is described by the formula:[1,3]

$$\Delta G_{adh} = \gamma_{sb} - \gamma_{sl} - \gamma_{bl}$$

in which the interfacial free energy of adhesion for bacteria ($\Delta G_{adh}$) is correlated with the solid–bacterium interfacial free energy ($\gamma_{sb}$), the solid-liquid interfacial free energy ($\gamma_{sl}$), and the bacterium–liquid interfacial free energy ($\gamma_{bl}$). This formula however, assumes that the effect of electric charges as well as specific biochemical interactions may be neglected. If $\Delta G_{adh}$ is negative (nature tends to minimize free energy), adhesion is thermodynamically favored and will proceed spontaneously. Bacteria initially adhering in the secondary minimum may also reach the surface by bridging this distance by protruding their fibrils, fimbriae, etc. (Fig. 3a). Because fimbriae have considerably smaller radii than the microbe itself, the electrostatic repulsion on these structures (which depends on their radius) will decrease, whereas the attractive van der Waals forces (which do not depend on the radius) remain constant so that for these structures the value of the energy barrier (B) decreases. For both situations (direct contact or bridging) the water film between the interacting surfaces has to be removed. This dehydrating capacity of bacteria occurs by hydrophobic groups associated with bacteria or their surface appendages. Hypothetically, the removal of interfacial water may be the main mechanism by which *"cell surface hydrophobicity"* and *"substratum surface hydrophobicity"* influence bacterial adhesion.[5,6,8] Sometimes, bacteria are forced to stay at a certain distance from the surface, not because of an energy barrier but because of steric hindrance between the surface coating polymers. Sometimes the electrostatic forces are so important that the thermodynamic concept becomes overruled.

### Phase III: Attachment

After initial adhesion a firm anchorage between bacteria and surface will be established by specific interactions (covalent, ionic, or hydrogen bonding) after direct contact with or bridging by true extracellular filamentous appendages (with a length of up to 10 nm). Such bonding is mediated by specific extracellular proteinaceous components of the organism (adhesins) and complementary receptors on the surface (e.g., pellicle) and is species-specific. Indeed, immediately following professional cleaning, a thin host-derived layer (the acquired pellicle) covers the tooth surface (as is the case for nearly every surface in the human being or in other environments). The pellicle in the oral cavity consists of mucins, glycoproteins, proline-rich proteins, histidine-rich proteins, enzymes like α-amylase, and other molecules. Some molecules from the pellicle (e.g., proline-rich proteins) evidently undergo a conformational change when they adsorb to the surface so that new receptors become available. *Actinomyces viscosus,* an early intra-oral colonizer for example, recognizes cryptic segments of the proline-rich proteins which are only available on adsorbed molecules.[12,13] This provides a microorganism with a mechanism for efficiently attaching to teeth and also offers a molecular explanation for their sharp tropisms for human teeth. It has been proven convenient to refer to such hidden receptors for bacterial adhesins as "cryptitopes" (cryptic = hidden, topos = place). On the other hand, the surface also has an impact on the developing pellicle.

### *Phase IV: Colonization*

When the firmly attached microorganisms start growing and newly formed cells remain attached, a biofilm may develop. From now on, new events are involved, because intrabacterial connections (co-aggregation) may occur. Each attached streptococcal and actinomyces strain binds specific salivary molecules. Thus from a common pool of salivary molecules, each strain of early colonizer may be coated with distinct molecules. Identical cells coated with a specific salivary molecule may also agglutinate, which would lead to a microconcentration and juxtapositioning of a particular strain. Alternatively, growth of a particular accreted strain would also lead to a microcolony coated with specific salivary molecules. Such events could dramatically alter the diversity of salivary molecules exposed to later colonizers.

In this concept of bacterial adhesion, both surface roughness and surface free energy of the solid substratum play an important role. On a rough surface bacteria are better protected against shear forces so that a change from reversible to irreversible bonding occurs more easily and probably more frequently. The substratum surface free energy becomes important when the water film between the interacting surfaces has to be removed before short range forces can be involved.

## III. SURFACE ROUGHNESS AND INTRA-ORAL BIOFILM FORMATION

Scanning electron microscopy clearly reveals that initial colonization of the intra-oral hard surfaces starts from surface irregularities such as cracks, grooves, perikymata, or abrasion defects and subsequently spreads out from these areas as a relatively even mono-layer of cells. With time, plaque areas develop at the irregularities which alternate with less extensively colonized surrounding areas.[18-20,24] Similar observations were recorded for the colonization of the mucosal surface of acrylic dentures.[22] Thus initial adhesion, especially supragingivally, preferentially starts at locations where bacteria are sheltered against shear forces, because time is given to the change from reversible to irreversible attachment (see adhesion model, Fig. 3a). Moreover, at surface irregularities and other sites of stagnation, bacteria, once attached, can survive longer because they are protected against removal forces[23] and even against oral hygiene measures.[25] Finally, one should keep in mind that a roughening of the surface also increases the area available for adhesion by a factor 2 to 3.

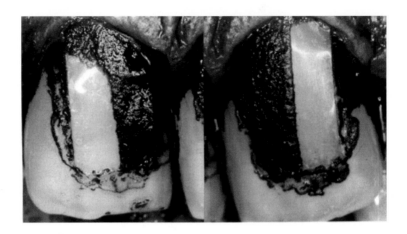

**Figure 4.** Photographs showing the clinical impact of surface roughness and surface free energy on *de novo* plaque formation. Two small strips were glued to the central upper incisors of a patient who refrained from oral hygiene for 3 d. Each strip was divided in two halves, a rough region ($R_a$ 2.0 µm) located mesially, and a smooth region ($R_a$ 0.1 µm) distally located. The left strip was cellulose acetate (medium sfe: 58 ergcm$^{-2}$) and the right strip Teflon (low sfe: 20 ergcm$^{-2}$). Plaque was disclosed with 0.5% neutral red solution. The smooth regions show the decrease in biofilm formation and the impact of the low surface free energy; the rough regions demonstrate the predominance of surface roughness, i.e., more plaque and no difference between the two materials. (Sample from the clinical study of Quirynen et al.[27]).

## A. Surface Roughness and Supragingival Area

Numerous in vivo studies have examined the effect of surface roughness on intra-oral biofilm formation. An overview of these studies[30] justifies the following general statements: 1) Rough surfaces accumulate and retain more bacteria (thickness, area, and colony forming units). 2) After several days of undisturbed biofilm formation, rough surfaces harbor a more mature plaque characterized by an increased proportion of motile organisms and spirochetes. 3) As a consequence, rough surfaces were more frequently surrounded by an inflamed gingiva. The significant contribution of surface roughness is illustrated in Figure 4.

## B. Surface Roughness and Percutaneous Surfaces

The impact of surface roughness on the subgingival biofilm formation could easily be examined using the abutments of 2 stage implants (Fig. 1). In an in vivo study plaque formation on standard ($R_a = 0.3$ µm) and roughened abutments ($R_a = 0.8$ µm) was evaluated after 3 mo of habitual oral hygiene.[28] Supragingivally, rough abutments harbored significantly more noncoccoid microorganisms (36 vs 19%), which is indicative of a more mature and pathogenic flora. Subgingivally, rough surfaces harbored 25-fold more bacteria, with a slightly lower density of coccoid organisms. Because bacteria have more possibilities of survival in the subgingival area as compared to the supragingival area (Fig. 1), it is clear that modifications in only one of these possibilities will have a less significant impact. Two more recent studies confirmed these observations and established a "threshold level" ($R_a = 0.2$ mm) for the surface roughness below which a further smoothening had no additional effect on the reduction in bacterial adhesion.[4,31] These observations were confirmed by Rimondini and co-workers[33] who examined initial biofilm formation (first 24 h) by scanning electron microscopy on intra-orally fixed titanium specimens with $R_a$ values ranging from 0.1 to 2.4 μm. Whereas smooth surfaces hosted comparable amounts of bacteria, the rough surfaces harbored significantly higher numbers.

## IV. SURFACE FREE ENERGY

Absolom calculated the change in $\Delta G_{adh}$ for the attachment of bacteria in suspension to substrata with different surface free energy (sfe) taking into account the sfe of the three interacting species, i.e., the sfe of the bacterium $\gamma_{bv}$, the sfe of the substratum $\gamma_{sv}$, and the surface tension of the suspending medium $\gamma_{lv}$.[1,2] When $\gamma_{lv}$ is greater than the surface free energy of the bacterium ($\gamma_{bv}$), then $\Delta G_{adh}$ becomes progressively less negative with increasing substratum surface free energy ($\gamma_{sv}$) predicting enhanced adhesion on the low energy (hydrophobic) substrata. On the other hand, when $\gamma_{lv} < \gamma_{bv}$ the opposite pattern of behavior is predicted, i.e., enhanced adhesion on the high energy (hydrophilic) substrata. For the rare cases in which $\gamma_{lv} = \gamma_{bv}$, $\Delta G_{adh}$ becomes equal to zero independently of the value of $\gamma_{sv}$. This model does not predict the number of bacteria that will adhere, it only predicts the relative extents (i.e., greater or lesser) of bacterial adhesion that are likely to be observed.[1,2] From this mathematic equation, two conclusions can be drawn: 1) Since most oral bacteria have a high $\gamma_{bv}$,[43] and because the saliva has a relative low $\gamma_{lv}$,[15] the situation $\gamma_{lv} < \gamma_{bv}$ will be frequently (for most bacteria) encountered so that one might conclude that the higher the substratum sfe the easier bacterial adhesion will occur (Fig. 3). 2) This formula suggests that bacteria with a low $\gamma_{bv}$ preferentially adhere to substrata with a low sfe, whereas bacteria with a high $\gamma_{bv}$ prefer high sfe substrata.

### A. Surface Free Energy and the Intra-Oral Biofilm Formation

Glantz was the first to recognize in vivo the correlation between substratum surface free energy and its retaining capacity for supragingival flora.[14] When he followed undisturbed supragingival biofilm formation on test surfaces of different free energies (mounted on a partial fixed bridge) he detected a "positive" correlation between substratum sfe and the weight of accumulated bacteria, measured at days 1, 3, and 7. In the dog, low sfe surfaces (like Teflon and Parafilm) were found to collect slightly less micro-organisms (within 2 h) than medium or high sfe surfaces such as dentine, enamel and glass.[40] Rölla and co-workers[34] demonstrated that the application of a silicone oil to teeth, which lowered their surface free energy, resulted in a significant reduction in biofilm formation.

Quirynen and co-workers[26,27] studied the influence of the substratum surface free energy on undisturbed biofilm formation in man over a 9-d period. Polymer films with different sfe were glued to tooth crowns (Fig. 4) in the immediate vicinity of the gingival margins to mimic natural plaque formation. Hydrophobic surfaces (e.g., Teflon) harbored 10 times less plaque than hydrophilic ones, with a retained adhesion strength. A microbiological examination of 3-d-old biofilms indicated that Teflon was preferentially colonized by bacteria with a low surface free energy, whereas the opposite was observed for surfaces with a higher sfe.[44] Moreover, strains of *Streptococcus sanguis* I isolated from Teflon were found to be significantly more hydrophobic than those isolated from higher energy surfaces. This suggests bacterial selection by, or adaptation to, the surfaces up to and even within the species level.[44]

The effect of substratum sfe on supra and subgingival biofilm maturation was investigated by comparing 3 mo-old plaque from implant abutments with either a high (titanium) or a low (Teflon) sfe. Low sfe substrata harbored a significantly less mature plaque supragingivally as well as subgingivally, characterized by a higher proportion of cocci and a lower proportion of motile organisms and spirochetes.[29]

This thermodynamic concept and thus of the influence of the substratum surface free energy for binding strength and facility of adhesion has a universal value such as for: 1) the adhesion of uropathogens to polymer materials;[17] 2) the colonization of vascular prostheses or prosthetic materials for abdominal wall reconstruction;[37] 3) the adhesion of catheter-associated bacteria;[16] 4) the attachment of freshwater bacteria to solid surfaces;[11] 5) the adhesion of mussels and barnacles to solid substrata;[9] 6) the binding strength of green alga to several surfaces;[10] 7) the attachment of insect residues to aircraft wings;[38] and 8) the adhesion of *Salmonella typhimurium* to soil particles.[39]

## V.  INTERACTION BETWEEN SURFACE ROUGHNESS AND SURFACE FREE ENERGY

The relative importance of both parameters (sfe and roughness) on supragingival biofilm formation has been examined in vivo by Quirynen and co-workers.[27] They followed undisturbed plaque formation on polymer strips with low and medium sfe on which one half was smooth (Ra = 0.1 μm) and the other half roughened (Ra > 2.0 μm). After 3 d of undisturbed plaque formation, significant intersubstrata differences were observed on the smooth regions, while the rough regions of the strips were nearly all completely covered with plaque (Fig. 4). Surface roughening resulted in a 4-fold increase in biofilm formation (extension as well as thickness) for both polymers. Surface roughness therefore seems an important factor in determining surface free energy for bacterial adhesion.

## VI. CONCLUSION

Surface characteristics (roughness and to a lesser degree surface free energy) are responsible for the majority of the intra-oral biofilm variability. These observations are applicable to nearly all other environments with nonshedding hard surfaces in a bioliquid (e.g., implanted membranes, larynx or cardiovascular prostheses, submarine surfaces, pipelines, etc.).

## REFERENCES

1. Absolom DR, Lamberti FV, Policova Z, et al: Surface thermodynamics of bacterial adhesion. *Appl Environ Microbiol* 46:90–7, 1983
2. Absolom DR: The role of bacterial hydrophobicity in infection: bacterial adhesion and phagocytic ingestion. *Can J Microbiol* 34:287–98, 1988
3. Bellon-Fontaine M-N, Mozes N, Van der Mei HC, et al: A comparison of thermodynamic approaches to predict the adhesion of dairy microorganisms to solid substrata. *Cell Biophys* 17:93–106, 1990
4. Bollen CML, Papaioannou W, Van Eldere J, et al: The influence of abutment surface roughness on plaque accumulation and peri-implant mucositis. *Clin Oral Implant Res* 7:201–11, 1996

5. Busscher HJ, Uyen, MH, Van Pelt, AWJ, et al: Kinetics of adhesion of the oral bacterium Streptococcus sanguis CH3 to polymers with different surface free energies. *Appl Environ Microbiol* 51:910–4, 1986

6. Busscher HJ, Weerkamp AH: Specific and non-specific interactions in bacterial adhesion to solid substrata. *FEMS Microbiol Rev* 46:165–73, 1987

7. Busscher HJ, Sjollema J, Van der Mei HC: Relative importance of surface free energy as a measiure of hydrophobicity in bacterial adhesion to solid surfaces. In: Doyle RJ, Rosenberg M, eds. *Microbial Cell Surface Hydrophobicity*. American Society for Microbiology, Washington, DC, 1990:335–59

8. Busscher HJ, Quirynen M, Van der Mei HC: Formation and prevention of dental plaque - a physico-chemical approach. In: Melo LF, ed. *Biofilms – Science and Technology*. Kluwer, Amsterdam, 1992:327–54

9. Crisp DJ, Walker G, Young GA, et al: Adhesion and substrate choice in mussels and barnacles. *J Colloid Interface Sci* 104:40–50, 1985

10. Fletcher RL, Baier RE: Influence of surface energy on the development of the green alga *Enteromorpha*. *Marine Biol Lett* 5:251–4, 1984

11. Fletcher M, Pringle JH: The effect of surface free energy and medium surface tension on bacterial attachment to solid surfaces. *J Colloid Interface Sci* 104:5–14, 1985

12. Gibbons RJ, Hay DI: Human salivary acidic proline-rich proteins and statherin promote the attachment of *Actinomyces viscosus* LY7 to apatitic surfaces. *Infect Immun* 56:439–45, 1988

13. Gibbons RJ, Hay DI: Adsorbed salivary proline-rich proteins as bacterial receptors on apatitic surfaces. In: Switalski LM, Hook M, Beachey E, eds. *Molecular Mechanisms of Microbial Adhesion*. Springer-Verlag, New York, 1988:143–69

14. Glantz P-O: On wettability and adhesivenesss. *Odontologisk Revy* 20 (suppl 17):1–132, 1969

15. Glantz P-O: The surface tension of saliva. *Odontologisk Revy* 21:119–27, 1970

16. Harkes G, Dankert J, Feijen J: Growth of uropathogenic *Escherichia coli* strains at solid surfaces. *J Biomater Sci, Polymer Ed* 3:403–18, 1992

17. Hawthorn L, Reid G: The effect of protein and urine on uropathogen adhesion to polymer substrata. *J Biomed Mater Res* 24:1325–32, 1990

18. Lie T: Ultrastructural study of early dental plaque formation. *J Periodont Res* 91–409, 1978

19. Lie T: Morphologic studies on dental plaque formation. *Acta Odontol Scand* 37:73–85, 1979

20. Lie T, Gusberti F: Replica study of plaque formation on human tooth surfaces. *Acta Odontol Scand* 37:65–72, 1979

21. Morra M, Cassinelli C: Bacterial adhesion to polymer surfaces: a critical review of surface thermodynamic approaches. *J Biomater Sci Polymer Ed* 9:55–74, 1997

22. Morris IJ, Wade WG, Aldred MJ, et al: The early bacterial colonization of acrylic palates in man. *J Oral Rehab* 14:13–21, 1987

23. Newman HN: Diet, attrition, plaque and dental disease. *Br Dent J* 136: 491–7, 1974

24. Nyvad B, Fejerskov O: Scanning electron microscopy of early microbial colonization of human enamel and root surfaces *in vivo*. *Scand J Dent Res* 95:287–96, 1987

25. Quirynen M: Anatomical and Inflammatory Factors Influence Bacterial Plaque Growth and Retention in Man. Thesis, Leuven, 1986

26. Quirynen M, Marechal M, Busscher HJ, et al: The influence of surface free energy on planimetric plaque growth in man. *J Dent Res* 68:796–9, 1989

27. Quirynen M, Marechal M, Busscher HJ, et al: The influence of surface free energy and surface roughness on early plaque formation. *J Clin Periodontol* 17:138–44, 1990

28. Quirynen M, van der Mei HC, Bollen CML, et al: An *in vivo* study of the influence of surface roughness of implants on the microbiology of supra- and subgingival plaque. *J Dent Res* 72:1304–9, 1993

29. Quirynen M, Van der Mei HC, Bollen CML, et al: The influence of surface free energy on supra- and subgingival plaque microbiology. An *in vivo* study on implants. *J Periodontol* 65:162–7, 1994

30. Quirynen M, Bollen CML: The influence of surface roughness and surface free energy on supra- and subgingival plaque formation in man. A review of the literature. *J Clin Periodontol* 22:1–14, 1995

31. Quirynen M, Bollen CML, Papaioannou W, et al: The influence of titanium abutments surface roughness on plaque accumulation and gingivitis. Short term observations. *Int J Oral Maxillofac Implants* 11:169–78, 1996

32. Quirynen M, Papaioannou W, van Steenbergen TMJ, et al: Adhesion of *Porphyromonas gingivalis* strains to cultured epithelial cells from patients with a history of chronic adult periodontitis and resistant patients. Submitted, 1999

33. Rimondini L, Farè S, Brambilla E, et al: The effect of surface roughness on early *in vivo* plaque colonization on titanium. *J Periodontol* 68:556–62, 1997

34. Rölla G, Ellingsen JE, Herlofson B: Enhancement and inhibition of dental plaque formation - some old and new concepts. *Biofouling* 3:175–81, 1991

35. Rutter PR, Vincent B: Physicochemical interactions of the substratum, microorganisms and the fluid phase. In: Marshall KC, ed. *Microbial Adhesion and Aggregation*. Springer Verlag, Berlin, 1984:21–38

36. Scheie AA: Mechanisms of dental plaque formation. *Adv Dent Res* 8:246–53, 1994

37. Schmitt DD, Bandyk DF, Pequet AJ, et al: Bacterial adherence to vascular prostheses. A determinant of graft infectivity. *J Vasc Surg* 3:732–40, 1986

38. Siochi EJ, Eiss NS, Gilliam DR, et al: A fundamental study of the sticking of insect residues to aircraft wings. *J Colloid Interface Sci* 115:346–56, 1987

39. Stenstrom TA: Bacterial hydrophobicity, an overall parameter for the measurement of adhesion potential to soil particles. *Appl Environ Microbiol* 55:142–7, 1989

40. Van Dijk J, Herkströter F, Busscher H, et al: Surface-free energy and bacterial adhesion, an *in vivo* study in beagle dogs. *J Clin Periodontol* 14:300–4, 1987

41. Van Loosdrecht MCM, Zehnder AJB: Energetics of bacterial adhesion. *Experientia* 46:817–22, 1990

42. Van Loosdrecht MCM, Lyklema J, Norde W, et al: Influence of interfaces on microbial activity. *Microbiol Rev* 54:75–87, 1990

43. Van Pelt AWJ, Van der Mei HC, Busscher HJ, et al: Surface free energies of oral streptococci. *FEMS Microbiol Lett* 25:279–82, 1984

44. Weerkamp AH, Quirynen M, Marechal M, et al: The role of surface free energy in the early *in vivo* formation of dental plaque on human enamel and polymeric substrata. *Microbial Ecol Health Dis* 2:11–18, 1989

# PART II

## GENERAL CONSIDERATIONS AND METHODS FOR STUDYING MICROBIAL ADHESION AND BIOFILM

# 7

# Basic Equipment and Microbiological Techniques for Studying Bacterial Adhesion

**Kylie L. Martin and Yuehuei H. An**

*Orthopaedic Research Laboratory, Medical University of South Carolina,*
*Charleston, SC, USA*

## I. INTRODUCTION

The knowledge of laboratory safety, major equipment, and basic microbiological techniques are essential for conducting studies on bacterial adhesion. This chapter is written for beginners in the field of microbiology and for nonmicrobiologists who want to set up a laboratory for conducting bacterial adhesion studies. For a beginner, it is a good idea to visit a nearby microbiology laboratory either before or after reading this chapter. The reader should also refer to the biological or microbiological safety manuals available at most research institutions and facilities.

## II. SAFETY

General laboratory safety principles should be applied when studying bacteria. One of the most important aspects is ensuring that all laboratory personnel have been adequately trained in the handling of potentially infectious materials as well as general laboratory safety procedures. The laboratory director is responsible for ensuring adequate training of personnel.

Everyone working in the laboratory should wear a lab coat and gloves, both of which should be removed before leaving the work area. These may be supplemented by such items as shoe covers, boots, respirators, face shields and safety glasses as required. For the purposes of studying bacterial adhesion, most laboratories would fall into Biohazard Safety Level 1 as defined by the United States Department of Health and Human Services.[31] Occupational Safety and Health (OSHA) regulations make it mandatory for an employer to provide a working environment free from recognized hazards that could cause injury, illness or death to an employee.[19]

All personnel should be aware that eating, drinking, smoking and applying makeup are hazardous in the laboratory setting. The laboratory should have a designated hand washing area and hands should be washed before leaving the work area, preferably with a germicidal soap. Mild germicidal soaps designed for frequent hand washing are available from laboratory suppliers. Frequent, thorough hand washing is one of the best ways of

*Handbook of Bacterial Adhesion: Principles, Methods, and Applications*
Edited by: Y. H. An and R. J. Friedman © Humana Press Inc., Totowa, NJ

preventing cross-infection in the laboratory. Additionally, all personnel should be aware that mouth pipetting is hazardous and they should not pipette any reagent or broth culture by mouth.

Other safety precautions to be followed include keeping work areas clean by spraying countertops with a 10% household bleach solution after working is completed. Inoculating loops are sterilized by flaming immediately after use, being careful not to cause spattering of the inoculant. The use of an electric incinerator, into which the loop is inserted to be heated and sterilized, will greatly reduce the incidence of sputtering.

An eyewash station and emergency "deluge" type shower should be installed in the laboratory to be used in case of chemical accidents. A fire extinguisher rated for use on both chemical and electrical fires should also be in close proximity to the laboratory, and all personnel made aware of its location and mode of operation. All safety equipment should be tested monthly to ensure its readiness for use if needed.

If radioisotopes are to be used, laboratory personnel using them must undergo training in safe use of the particular isotopes and be familiar with procedures to be followed in the event of an accident involving the radioactive materials. In addition, laboratory personnel must practice safe disposal methods for any radioactive waste generated and follow the established procedures for their particular institution.

## III. MAJOR EQUIPMENT

### A. Main Experimental Counter

#### 1. Centrifuge

A centrifuge capable of exerting a force of up to 3000 $g$ should be available. This force is necessary to deposit bacteria in a reasonable time. The centrifuge needs to be capable of handling multiple rotors, which may be needed for different types of tubes, plates, etc. For maximum microbiological safety, covered "safety cups" should be fitted to prevent spills. These safety cups may be as simple as using sealed centrifuge tubes or may consist of screw-capped buckets and sealed rotors.

#### 2. Stirrer/Hot Plate

A stirrer/hotplate is invaluable in the laboratory for preparation of solutions such as media, buffers and tryptic soy broth. The heating mode will be necessary for the preparation of agar solutions or broth in laboratories not using commercial agar plates or broth.

#### 3. Vacuum/Suction

Suction will be necessary to remove liquid media from plates or tubes. This can be provided by several methods including the use of a vacuum pump or the building's vacuum system. A trap will need to be installed in the vacuum line to collect the liquids being aspirated. A simple, effective trap consists of a sidearm Erlenmeyer flask fitted with a rubber stopper with two holes, one for inflow of the aspirated liquid and the other for the outflow of air. Household bleach should be added to the fluids in the trap at a concentration of 1 to 5% to sterilize the aspirated fluids, which can then be disposed of safely. Care should be taken not to overfill the reservoir as this will result in contamination of the suction system.

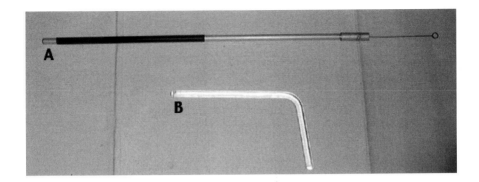

**Figure 1.** Images of an inoculating loop (A) and a "hockey stick" glass rod (B).

### 4. Sonicator

A sonicator or ultrasonic cleaner will be useful for critical cleaning of biomaterials. The item to be cleaned should be totally immersed in the cleaning solution and sonicated for 5 to 15 min. Care should be taken when cleaning contaminated items in an ultrasonic cleaner to ensure that aerosols are not created, thereby exposing laboratory personnel to potential pathogens. If there is any risk of this exposure, sonication should be performed in a biological safety cabinet.

Sonication can also be used as a means of removing attached cells from biomaterials for further analysis such as calcium levels, etc.

### 5. Inoculating Loops and Glass Spreaders

Inoculating loops have traditionally been made of inert metal wire, either platinum or a nickel–chromium alloy. Both types are available in several different gauges and can be chosen according to the individual user's preference. The wires can be cut to about 12 cm long, then twisted together forming a completely closed loop not more than 3 to 4 mm in diameter (Fig. 1A).[7] The twisted wires can then be inserted into a holder. Metal holders are preferable for both durability and ease of cleaning. Presterilized disposable plastic loops are readily available. These do not require flaming before use, which eliminates the hazards of open flames, sputtering and cross-contamination. They are available with different size loops for different applications.

Glass spreaders, or "hockey sticks", can be made using a 3 to 4 mm diameter borosilicate glass rod cut into 18 cm lengths and the ends smoothed by flaming. The rod can then be bent into an L shape by heating over a flame about 4 cm from one end (Fig. 1B).

### 6. Burners

In laboratories where wire loops are used a Bunsen burner may be used for flaming to sterilize immediately before use. The common type of ethanol glass burner is usually suitable. A burner may also be used to heat slides when performing acid-fast staining for the Gram stain.

### 7. Racks and Baskets

Racks and baskets will be needed for the transporting of items to be autoclaved and also for loading items into the autoclave. Metal racks may be used for most applications but polypropylene or polypropylene-coated racks are more suitable for test tubes as they

**Figure 2.** A rotator placed in a incubator for culturing bacteria in test tubes.

reduce the chances of breakage. Autoclavable plastic boxes specifically designed for test tubes are ideal when dealing with cultures. These can be used for supporting the tubes at the correct angle when pouring media into tubes.

## B. Incubator and Other Major Equipment

### 1. Multipurpose Incubator

Most microorganisms require a constant temperature (for many bacteria, 37°C) for optimal growth and this may be achieved by the use of an incubator. Incubators are able in various sizes and should be chosen with the requirements of the individual laboratory in mind. The bacteriological incubator is a dry air environment and care must be taken to ensure all experiments are terminated before they dry out. However, supplemental moisture can be supplied by placing a beaker of water in the incubator. This addition ensures a moist environment slowing dehydration as well as preventing random experimental results.

### 2. Rotator and Horizontal Shaker

A shaker or combination shaker/waterbath is useful for mixing and shaking cultures. This should be fitted with racks to hold tubes, bottles and flasks securely. The shaker should also have a lid to prevent the dispersal of aerosols. A rotator can be used to keep nonadhesion dependent bacteria in suspension during culture (Fig. 2). Frequently, these items will need to be placed in the incubator, so size will be a consideration when purchasing this piece of equipment. The advantage of using this device includes transfering heat in a quick and uniform fashion and providing aeration through agitation of the vessel, thus, culture growth is considerably accelerated.

### 3. Refrigerator and Freezer

Refrigerated storage is required for culture media, some specimens and other reagents such as thermolabile solutions, serums, and antibiotics. The refrigerator temperature should be set at 0 to 4°C and checked regularly as the quality of media and reagents may be adversely affected by storing at temperatures other than those specified. Some reagents will require storage below 0°C. For laboratory purposes, a self-defrosting freezer should

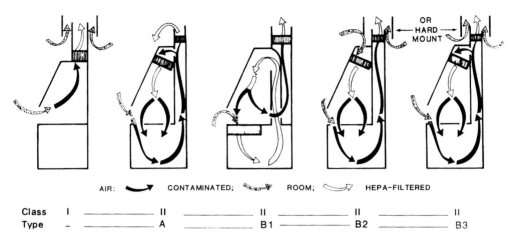

**Figure 3.** Different biological hoods for microbiological experiments (from ref.[8] with permission).

be avoided as the temperature cycling which occurs during the defrost cycle can have adverse effects on stored chemicals.

### 4. Biological Hood

A biological containment hood is necessary when studying pathogenic organisms. The hood should be of a type where no air escapes into the surrounding air. Such hoods are designated Class I, Class II, or Class III,[8] depending on the level of safety needed when working with particular types of pathogens. This is achieved by different designs: 1) For a Class I cabinet, the operator sits at the front and can see what he is doing through a glass screen. An exhaust fan located at the top of the cabinet provides a constant inflow of room air directly into the working area, preventing the escape of any aerosols released from cultures. The air passes through a HEPA filtering system and is then released into the room. These hoods need have the airflow checked regularly and the filters replaced when the airflow falls below the level stipulated by the manufacturer. A disadvantage of this type of cabinet is that items in the cabinet are subject to contamination by organisms in the room air (Fig. 3).[8] 2) In a Class II cabinet, the operator is seated at front and views the work area through a glass screen. Room air is drawn in at the front of the hood and down into a series of HEPA filters before flowing over the work area which is "bathed" in the filtered air. The continuous airflow is then drawn back over another series of filters before being exhausted. There are several types of Class II hoods (A, B, C, D) which all have different airflow patterns (Fig. 3). Class IIA is the most commonly used.[8] 3) The Class III cabinet is an airtight, enclosed system which is equipped with protective gloves. The workspace inside the cabinet is supplied with HEPA filtered air and the exhaust air is passed through two HEPA filters in series before being discharged outside the building.[8] Furthermore, these hoods also possess ultraviolet light capability permitting sterilization between usages. All biological safety cabinets should be recertified annually.

### 5. Peristaltic Pump

A peristaltic pump may be needed for the setup of flow systems for producing bacterial biofilm or for studying bacterial adhesion or biofilm formation under flow condition

**Figure 4.** An image of a peristaltic pump.

(Fig. 4). The pump may also be used when filter sterilizing large batches of liquid media and buffers.

### 6. Autoclave

An autoclave should be located in close proximity to the laboratory. The autoclave will be necessary for sterilization of glassware, pipettes, and other items before and after use to prevent cross-contamination of cultures. The autoclave also is a proven treatment method for rendering infectious material safe for disposal. Some buffers, such as phosphate buffered saline (PBS), may be sterilized by autoclaving.

## C. Radioactive Area

If no extra space is available, a part of the main experimental counter and a sink can be designated for the use of radioactive materials. The area may be clearly designated by the use of specially marked tape. For certain protocols, radiolabeling methods are very useful, including 1) prelabeling bacterial cells with radiotracers such as $^{14}$C-leucine,[28] $^{14}$C-glucose,[23] $^{3}$H-thymidine,[25] or $^{3}$H-palmitic acid,[13] 2) bacterial adhesion process, and 3) scintillation counting. It has been shown that the radiolabeling of bacteria is very useful for the study of bacterial adhesion to irregular material surfaces, such as surgical sutures, the inner-surface of catheters, or the surfaces of particles or spheres.[26,39] These methods also have the advantage of being very sensitive and very accurate and allowing rapid processing of a large number of samples. The disadvantage of radiolabeling is the need for specially designated laboratory space, specialized techniques for safe handling of radioactive materials and the potential risk to researchers.

## D. Disposable Materials

### 1. Glassware

For bacteriological work, rimless test tubes are available in a variety of sizes. These may be stoppered with a plug of cotton or with a specially designed stainless steel closure called the "Morton test tube closure." Other alternatives include rubber or plastic caps.These, however, are heat sensitive and subject to damage by overheating. Disposable plastic caps that can be autoclaved are also available.

## 2. Petri Dishes and Multi-Well Culture Plates

Petri dishes and multi-well culture plates may be used for static bacterial adhesion and biofilm formation experiments. Disposable plastic items are usually more convenient and often more economical than reusable glass because of the amount of time needed to adequately clean and sterilize the glassware. The plastic dishes are presterilized and can be stored easily for long periods. Both Petri dishes and multi-well culture plates are available in a range of sizes to suit the needs of the individual investigator. Petri dishes and multi-well culture plates are also useful for storage of biomaterial samples to be tested.

## 3. Culture Media

Bacteria will grow wherever conditions are suitable. The factors influencing their growth include availability of moisture, food and oxygen, as well as proper temperature and medium at correct (neutral) pH. The selection of an appropriate medium for the bacteria being studied will ensure that optimum growth is achieved. The most commonly used culture medium and agar are tryptic soy broth and trypticase soy agar, respectively. They are suitable for a wide range of bacteria. Agar is a convenient solidifying agent because of its properties: liquefying at 100°C and solidifying at 40°C. Most types of culture media are commercially available in powdered form or ready to use in tubes or plates of varying sizes.

## 4. Making Agar Plates and Tubes

Commercially prepared tubes and plates with several types of media are readily available from major suppliers (e.g., Fisher Scientific, Pittsburgh, PA). Other premixed media and supplements are available and can be used in the preparation of tubes and Petri dishes to be used for culturing bacteria. To prepare media, the powdered ingredients are mixed with water and heated until melted, taking care not to overheat as this may cause degradation of the media through destruction of growth factors and changes in pH. The water used should be distilled or deionized, as impurities in potable water may have an inhibitory effect on the growth of bacteria. Another consideration is the type of container used in preparation of the media. Glass is suitable for most applications. Metals other than stainless steel should be avoided for heating media because trace amounts will dissolve in the media and have a bactericidal effect. To pour plates, raise the lid slightly and pour in 15 mL of media. Dry the plates slightly open in an incubator, and store them in an inverted position in a refrigerator. To prepare tubes, add the required amount of media and prop the tube at an angle on the bench top while the media sets.

## 5. PBS and Liquid Media

PBS is widely used in many laboratories for rinsing, diluting and preparation of stains. It is also used in flow systems. PBS is easily prepared in large quantities and may be sterilized by autoclaving (for small batches) or, more commonly, by filtration methods. PBS may be mixed using readily available chemicals, purchased ready to use or in a concentrated form or in convenient tablet form (Sigma, St. Louis, MO). Liquid culture media may be purchased in ready to use or concentrated form. Additionally, these media are available in powdered form and may be mixed as needed, supplemented as required and filter sterilized or autoclaved as required.

### E. Waste Disposal

In cases where radiolabeling techniques have been used, special measures must be taken to dispose of infectious waste. Mixed waste containing conjugated tritium, technetium-99, carbon-14, and other radionuclides has been autoclaved safely at some institutions, but this method requires approval by the radiation safety officer of the institution.[8] Radioactive waste should be disposed of in accordance with the regulations of the institution in which the laboratory is located. Generally, solid and liquid wastes will need to be stored separately in approved containers and disposed according to regulations.

All bacterially contaminated material should be disposed of as biohazardous waste in plastic bags clearly labeled "Biohazard" and sealed securely. The bagged contaminated waste should be autoclaved for one hour at 120°C and may then be disposed of as general waste or in the manner required by the institution.

Toxic liquid waste should be collected in leakproof containers with secure closures and stored separately by type (solvents, acids, fixatives, etc.). The containers must be clearly labeled with the name of the chemical and its percentage as a part of the total waste volume. Large amounts should not be accumulated in the laboratory.

"Sharps" encompasses used needles, broken glass, surgical blades and pipettes, all of which should be deposited in a container which is clearly labeled, puncture-resistant and leak-proof. Pipettes and broken glassware may be disposed of in a cardboard box lined with a plastic bag to contain any spills. Used needles and surgical blades can be placed in disposable buckets made from high-strength, temperature-resistant plastic.

### F. Equipment for Materials Science Procedures

For experiments on bacterial adhesion to biomaterial surfaces, it is ideal to have the following items in the laboratory. A band saw may be used for cutting metal and plastic sheets to make samples. A diamond-wafering saw is desirable for cutting hard metals. A wheel polisher can be used to polish metal surfaces to obtain the desired finish. Nitrogen gas can be used for drying sample surfaces after washing with detergent and water. Stylus devices are used for measuring material surface roughness. Water contact angle devices are used for measuring material surface wettability.

## IV. EQUIPMENT FOR EVALUATING BACTERIAL ADHESION

### A. Light Microscopes

Light microscopy is a fundamental method for both bacterial enumeration and observation. A standard laboratory microscope with ×10, ×25, ×40, and ×100 objectives is ideal for a bacterial adhesion laboratory. Light microscopy is used to observe bacteria directly[40] or through a histologic section, as biofilm can be embedded in paraffin and cut into conventional histologic sections.[5] The substratum surfaces have to be translucent to use light microscopy. Normally bacteria are stained with dyes such as Gram stain, crystal violet, or carbol fuchsin. Some special staining methods allow the observation of rial surface structures such as capsules or appendages. Light microscopy has been combined with bacterial flow chamber or slide culture to observe living attached bacterial cells in real time.[11] Slide culture is composed of a microscope slide, a drop of bacterial culture medium containing the bacteria to be studied and a cover slip.[3] A continuous-flow slide culture is also available.[4] The advance of image analysis makes bacterial counting by light microscopy much faster and more efficient. This combination has been used to determine the numbers of cells attached, area coverage, and biovolume of attached cells, as well as the much more complicated evaluation of bacterial adhesion and growth in real time.[3]

## B. Special Equipment for Evaluating Bacterial Adhesion

### 1. Epifluorescent Microscope

Epifluorescent microscopes are rather expensive equipment available in most core facilities or image analysis laboratories. For smooth and opaque surfaces, like metal, plastic, or ceramic, epifluorescence microscopy based on fluorescence stained samples may offer the most reliability.[1,24] The main advantages of this technique include that: 1) it is very quick, reducing the time required to count a sample for visual counting by 85%,[35] making it especially suitable for large sets of samples; 2) it can reduce the possibility of operator bias; and 3) it makes direct observation and enumeration possible for attached bacteria on opaque surfaces. It should be realized that image-analyzed epifluorescence microscopy (IAEM) only counts the bright spots brought by stained bacteria which could be individual or a cluster of bacteria. In such cases, adjustments should be used for reliable data.[1]

### 2. Spectrophotometry

Spectrophotometry is based on the quantitative relation between optical density (commonly from crystal violet or Congo red staining) and colony counts being derived from standard curves prepared for each bacterial species. Theoretically there should be two basic measuring techniques; 1) bacteria attached to the substratum are examined directly in multi-well culture plates after staining using a spectrophotometer;[12,32,33] and 2) bacteria are washed off the substratum, stained in solution, and the solution examined using a spectrophotometer.

### 3. Coulter Counter

Kubitschek first used the Coulter counter in 1958 for the purpose of counting and measuring the size of bacteria.[18] This method measures the resistance of a conducting solution as a particle passes through an aperture and was originally used for counting blood cells. The Coulter counter method has several advantages over optical methods including; 1) the volume of each particle is measured accurately; 2) in the presence of a background of small particles, the signal-to-noise ratio is greater when volume is measured than when a cross-section is measured; 3) the signal-to-noise ratio is further improved by the action of the aperture as a filter to exclude large unwanted particles; and 4) the counter is simple to operate. Combined with the colony count method, particle counting is used to count the total number of bacteria while colony counting is used to determine the viable bacterial count.[5]

### 4. SEM and TEM

Scanning and transmission electron microscopy (SEM and TEM) have proved invaluable for examining the adhesion pattern of bacteria[22] and the structure of biofilm.[13,27]

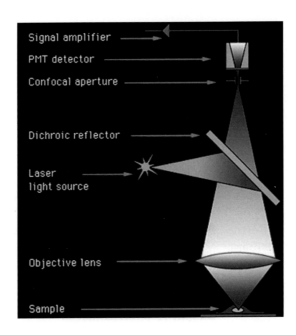

**Figure 5.** Basic principle of a CLSM system (Courtesy Bio-Rad, Hercules, CA).

Careful planning is needed before the preparation of samples for microscopic observation, since bacteria or bacterial biofilm are highly hydrated and thus readily deformed during preparation of specimens for electron microscopy or optical microscopes. For sample preparation for SEM or TEM, one can refer to the articles by Gristina and Costerton,[16] Ladd and Costerton,[20] and Diaz-Blanco et al.[9]

*5. Confocal Laser Scanning Microscopy*

Confocal laser scanning microscopy (CLSM) is a valuable newly developed method for morphological observation of adherent bacteria and biofilm.[14,15,21,29] As shown in Figure 5, a laser light source provides excitation illlumination which is reflected by a dichroic reflector into an objective lens and brought to a focus in the sample. This focused point is scanned in a horizontal (XY) plane. Emitted fluorescent light is collected by the same lens, passes back through the dichroic reflector and on to the confocal aperture. Only light from the plane of focus passes through the aperture into the detector. Thus, out of focus light is rejected and an optical section is generated.

It needs to be emphasized that CLSM has a bright future in the research of biofilm formation on surfaces. Because of its wide range of applications in biomedical research, it is becoming more and more available at many research facilities. Using CLSM, the structure of biofilm, such as the distribution of bacteria or the thickness of the biofilm, can be examined *in situ* and under the original hydrated conditions. Also as a feature of CLSM, samples can be sectioned optically to reveal the three-dimensional structure. Figure 6 shows a recent model of CLSM made by Bio-Rad (Bio-Rad Laboratories, Hercules, CA).

*6. Atomic Force Microscopy*

Atomic force microscopy (AFM) is emerging as an important alternative to electron microscopy as a technique for analyzing submicron details on biological surfaces.

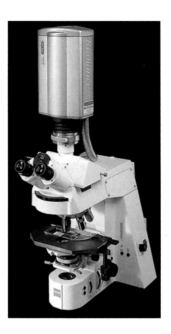

**Figure 6.** A recent model of CLSM system, Radiance Plus (Courtesy Bio-Rad, Hercules, CA).

Microbiological specimens such as viruses, bacteriophages, and ordered bacterial surface layers and membranes have played an important role in the development of atomic force microscopy in cellular and molecular biology.[10] With the development of atomic force microscopes, which allow imaging of similar native structures, AFM applications have widened to include straightforward surface structure analysis, such as polysaccharides,[17] analysis of surface elastic and inelastic properties, bonding force measurements between molecules, and micromanipulations of such individual molecules.[42]

Razatos et al.[30] showed that AFM could be used to analyze the initial events in bacterial adhesion with unprecedented resolution. Interactions between the cantilever tip and confluent monolayers of isogenic strains of *Escherichia coli* mutants exhibiting subtle differences in cell surface composition were measured. It was shown that the adhesion force is affected by the length of core lipopolysaccharide molecules on the *E. coli* cell surface and by the production of the capsular polysaccharide, colanic acid. Furthermore, by modifying the atomic force microscope tip they developed a method for determining whether bacteria are attracted or repelled by virtually any biomaterial of interest.

AFM has also been used for characterization of cell–biomaterial interaction at molecular level[34] and for evaluating surface characteristics of Gram-negative and Gram-positive bacteria.[42] For the latter application, bacterial images can be obtained rather easily with an atomic-force microscope (AFM) in the magnification range of 5,000 to 30,000 times without any pretreatment of the specimens for such observations as chemical fixation, dehydration or staining. The bacterial shapes or the presence of flagella can be clearly recognized in these magnification ranges.[42]

## 7. XPS, EDX, SIMS, ATR, and FT-IR

All of these methods were originally designed for characterizing elemental compositions of material surfaces. Tyler[41] reported three case studies which demonstrate that X-ray photoelectron spectroscopy (XPS) and secondary-ion mass spectroscopy (SIMS) are valuable techniques for studying all three surfaces involving in the process of cell or bacterial adhesion, the material surface, the biopolymer conditioning film, and the cell surface. XPS and SIMS are not only capable of providing an accurate analysis of the synthetic polymer surface but they are also sensitive to the composition and orientation of biomolecules. The potential for rapid characterization of cell surfaces with SIMS suggests that the application of these techniques may ultimately aid in answering the elusive question of how cells adhere to synthetic surfaces.

The collected information by XPS concerns only the outermost molecular layers, about 1 to 3 nm in thickness. Both working by irradiation of a sample surface with an X-ray beam which induces the ejection of electrons from the surface, XPS can provide an average elemental analysis of the whole surface, whereas EDX (energy-dispersive X-ray analysis) examines one particular point on the surface.[44] EDX and scanning electron microscopy are complementary and the instruments constitute one apparatus, so SEM helps to define a area of interest for elemental analysis by EDX. It is known by using XPS that the elements associated with biofilm encrustation to ureteral stents are calcium, magnesium, and phosphorus.[38]

Another noninvasive technique for collecting information on bacterial adhesion at molecular level is attenuated total reflectance (ATR) waveguides integrated within flow cells coupled with Fourier transform infrared spectroscopy (FT-IR).[2] IR waves can only penetrate less than 1 μm into bacterial biofilm. FT-IR has been used to characterize the biochemical contents of Lactobacillus biosurfactants, such as protein, polysaccharide, and phosphate.[43] Recently, Suci et al.[36] employed ATR/FT-IR to monitor bacterial colonization of a germanium substratum, transport of antibiotics to the biofilm-substratum interface, and interaction of biofilm components with the antibiotic in a flowing system. Using the same method, they also provided information on both transport of an antimicrobial agent to bacteria in the biofilm and interactions between antimicrobial agents and biofilm components.[37]

## V. BASIC MICROBIOLOGICAL TECHNIQUES

### A. Overnight Culture in Culture Tubes

Overnight bacteria culture means that the inoculation of bacteria is done in the afternoon and cultured until the next morning (about 18 h). Briefly, one colony of bacteria taken from an agar plate using an inoculating loop is added to 5 mL Tryptic Soy Broth (TSB) media in a 15- or 20-mL test tube. The tube is then placed on a rotator in a multi-purpose incubator and cultured for 18 h. For most bacteria, the culture will be confluent in 18 h.

### B. Plating Cultures for Isolated Colonies

For routine uses and also for stock cultures, making agar plates with isolated bacterial colonies is often necessary. The Petri dish lid should be held in a slanted position slightly above the plate to protect the agar surface. The sterilized inoculating loop is dipped into the bacterial solution and a pattern is traced on the surface of the nutrient agar. If the source of the bacteria is an existing agar plate, one bacterial colony should be taken using the loop for spreading (Fig. 7A). Care should be taken not to gouge the surface of the agar

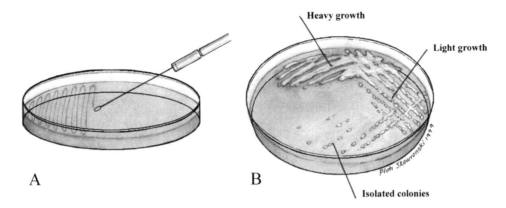

**Figure 7.** (A) shows plating in action using a inoculating loop; (B) shows the most common pattern of bacterial plating.

during streaking. There are several patterns that may be used in order to generate isolated colonies. The most commonly used pattern is shown in Figure 7B. The culture should be allowed to grow for 24–48 h. Petri dishes are usually inverted during incubation as this prevents the drying of the agar layer and also prevents condensation dripping on the agar surface.

If the source of bacteria is a suspension, a drop of the suspension (50–100 μL) can be pipetted onto the agar surface and then a "hockey stick" glass rod may be used for plating (Fig. 8A). A Petri dish turntable can be used for easier turning if the Petri dish (Fig. 8B). The usual method for sterilizing the glass rods is autoclaving. The number of glass rods needed for an experiment is based on the number of plates to be made. A quick way to sterilize a glass rod for immediate use is by placing the rod in 70% ethanol, flaming over a burner (or using an electric incinerator), and then placing it, flamed end up, in a clean container to cool down.

## C. Plating for Counting Bacterial Colony Forming Units

When plating for counting bacterial colony forming units, the microorganisms must first be introduced to the medium and spread as evenly as possible across the entire surface. This can be done using an inoculating loop or a "hockey stick" glass rod. The inoculating loop is sterilized by holding in the flame of a Bunsen burner or electric incinerator until it is red hot, and allowed to cool before use. If used while still hot it will cause sputtering of the bacterial sample, resulting in contamination of the user and surrounding areas. The loop may be cooled quickly by inserting into the agar at the periphery of the plate for 1 to 2 s before the spreading.

A colony forming unit (cfu) is a colony on a culture plate that is thought to have derived from a single bacterium (Fig. 9). For many bacteria, 24 to 48 h of incubation is enough to obtain countable colonies. For counting, plates are placed over an illuminated screen and examined with a magnifier. The plate cover is marked above each colony with a felt tip pen mounted on a hand-held digital counter. The colony count is obtained by dividing the average number of colonies per plate by the dilution factor used. This is the cfu/mL. Automatic colony counters are available for large work loads.

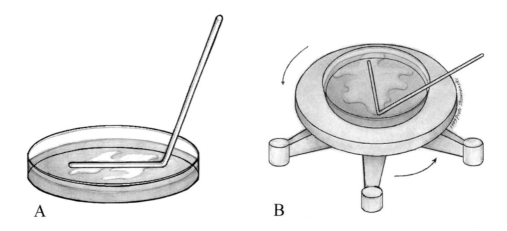

**Figure 8.** Plating with the "hockey stick" method.

To obtain bacterial colony forming units (50–300 colonies per plate), serial dilutions are needed. An appropriate diluent needs to be selected, depending on the bacteria under study. The most commonly used diluents, PBS and distilled (or deionized) water, are toxic in some cases, so 0.1% peptone water is the diluent of choice.[7] Start by preparing dilution blanks, pipetting 9 mL of diluent into sterile test tubes with screw caps. For bacteria in liquid suspensions, continue as follows. Mix the sample to ensure the bacteria are evenly distributed throughout, then remove a 1.0 mL sample and add to the first blank. Pipette up and down to mix, taking care not to create any bubbles. Remove 1 mL from the tube containing the diluted sample and add this to the next blank tube. Continue until the required number of dilutions is completed (Table 1).

### D. Membrane Filter Counts

This is another means of obtaining colony counts.[7] In this procedure, a known volume of the solution containing the bacteria is passed through a cellulose–ester filter that retains the organisms. The filter is then carefully removed from the filtering apparatus and placed on filter paper, which has been prewetted with the culture media. The filter is incubated in a petri dish or shallow box, sandwiched between layers of absorbent pads. The filters are printed with a grid pattern to aid counting, which is usually done using low power magnification. If the colonies are difficult to visualize, the filter may be stained with 0.01% methylene blue. The colonies will stain a darker blue than the filter. The colony count is expressed as membrane colony count per standard volume.

### E. Cultures for Stock Bacteria

The serial subculture method described by Collins and Lyne[7] is very useful for many laboratories doing bacterial adhesion studies. In this method, two tubes are inoculated and incubated until growth is established after which the cultures are stored refrigerated for 1 to 2 mo. One culture (A) is reserved as the stock and the other (B) is used for any laboratory purpose (Fig. 9). Subculturing should be done as seldom as possible, in order to prevent the appearance and possible subsequent selection of mutant strains.

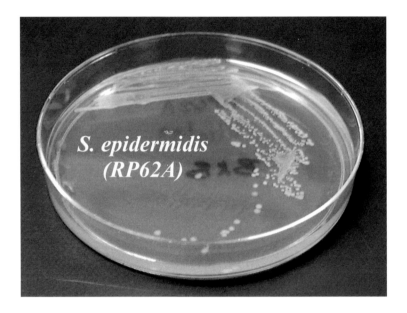

**Figure 9.** *S. epidermidis* colonies on an agar plate.

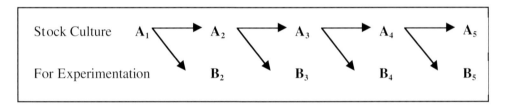

**Figure 10.** Illustration of subculture method.

**Table 1. Serial Dilutions of Bacterial Suspension to Obtain Bacterial Colony
Forming Units Which Are Countable (50–300 Colonies per Plate)**

| Tube no. | 1 | 2 | 3 | 4 | 5 | 6 | 7 | 8 |
|---|---|---|---|---|---|---|---|---|
| Dilution | 1:10 | 1:100 | 1:1000 | 1:10 000 | 1:100 000 | 1:1000 000 | 1:10 000 000 | 1:100 000 000 |
| Vol. of original | 0.1 | 0. 01 | 0.001 | 0.0001 | 0.00001 | 0.000001 | 0.0000001 | 0.00000001 |
| fluid/mL | $(10^{-1})$ | $(10^{-2})$ | $(10^{-3})$ | $(10^{-4})$ | $(10^{-5})$ | $(10^{-6})$ | $(10^{-7})$ | $(10^{-8})$ |

Freezing at −20°C has been used as a means of storing a stock culture for a short period.[7] Thick suspensions of bacteria in skim milk or water are frozen in 0.5 mL aliquots in screw-topped containers. Cryopreservation, storage at ultralow temperature in liquid nitrogen, is used for long term storage of bacteria. The liquid suspension of bacteria is pelleted by centrifugation and resuspended in growth medium containing 2 to 10% methanol or 5 to 15% dimethyl sulfoxide (DMSO). Glycerol may also be used as the cryoprotectant. Freezing is undertaken in a series of steps, first cooling to 4°C, then to –70°C for 2 h, then into liquid nitrogen.

Lyophilization (freeze-drying) is one of the most effective methods for long term preservation of bacterial cultures. It is particularly useful for long term maintenance of organisms with extremely precise growth requirements. A large number of bacterial species have been preserved with this technique and are still viable after several decades. Before freeze-drying, the bacteria are suspended in a mixture consisting of 1 part 30% glucose in nutrient broth and 3 parts horse serum. The glucose acts as a cryo-protectant and prevents damage to the bacteria during both the freezing process and the subsequent thawing. The glucose–serum solution should be autoclaved before use, and the cells should be frozen in small quantities of 3 to 5 mL.

## REFERENCES

1. An YH, Friedman RJ, Draughn RA: Rapid quantification of staphylococci adhered to titanium surfaces using image analyzed epifluorescence microscopy. *J Microbiol Meth* 24:29–40, 1995
2. Bryers JD, Hendricks S: Bacterial infection of biomaterials. Experimental protocol for *in vitro* adhesion studies. *Ann N Y Acad Sci* 831:127–37, 1997
3. Caldwell DE, Germida JJ: Evaluation of different imagery for visualizing and quantitating microbial growth. *Can J Microbiol* 31:35–44, 1985
4. Caldwell DE, Lawrence JR: Growth kinetics of *Pseudomonas fluorescens* microcolonies within the hydrodynamic boundary layers of surface microenvironments. *Microb Ecol* 12:299–312, 1986
5. Chang CC, Merritt K: Microbial adherence on poly(methyl methacrylate) (PMMA) surfaces. *J Biomed Mater Res* 26:197–207, 1992
6. Cappuccino JG, Sherman N: *Microbiology—A Laboratory Manual.* 5th Ed., Benjamin/Cummings, Menlo Park, CA, 1999
7. Collins CH, Lyne PM: *Microbiological Methods.* Butterworth, London, 1984
8. Committee on Hazardous Biological Substances in the Laboratory: *Biosafety in the Laboratory: Prudent Practices for the Handling and Disposal of Infectious Materials.* National Academy Press, Washington DC, 1989
9. Diaz-Blanco J, Clawson RC, Roberson SM, et al: Electron microscopic evaluation of bacterial adherence to polyvinyl chloride endotracheal tubes used in neonates. *Crit Care Med* 17:1335–40, 1989
10. Firtel M, Beveridge TJ: Scanning probe microscopy in microbiology. *Micron* 26:347–62, 1995
11. Fletcher M: Attachment of *Pseudomonas fluorescens* to glass and influence of electrolytes on bacterium-substratum separation distance. *J Bacteriol* 170:2027–30, 1988
12. Fletcher M: Methods for studying adhesion and attachment to surfaces. *Method Microbiol* 22:251–80, 1990
13. Ganderton L, Chawla J, Winters C, et al: Scanning electron microscopy of bacterial biofilms on indwelling bladder catheters. *Eur J Clin Microbiol Infect Dis* 11:789–96, 1992
14. Gorman SP, Adair CG, Mawhinney WM: Incidence and nature of peritoneal catheter biofilm determined by electron and confocal laser scanning microscopy. *Epidemiol Infect* 112:551–9, 1994
15. Gorman SP, Mawhinney WM, Adair CG, et al: Confocal laser scanning microscopy of peritoneal catheter surfaces. *J Med Microbiol* 38:411–7, 1993
16. Gristina AG, Costerton JW: Bacterial adherence and the glycocalyx and their role in musculoskeletal infection. *Orthop Clin North Am* 15:517–35, 1984

17. Kirby AR, Gunning AP, Morris VJ: Imaging polysaccharides by atomic force microscopy. *Biopolymers* 38:355–66, 1996
18. Kubitschek HE: Electronic counting and sizing of bacteria. *Nature* 182:234–5, 1958
19. Kuehne RW, Chatigny MA, Stainbrook BW, et al: Primary barriers and personal protective equipment in biomedical laboratories. In: Fleming DO, Richardson JH, Tulis JJ, et al., eds. *Laboratory Safety: Principles and Practices.* ASM Press, Washington DC, 1995
20. Ladd TI, Costerton JW: Methods for studying biofilm bacteria. *Meth Microbiol* 22:287–307, 1990
21. Lawrence JR, Korber DR, Hoyle BD, et al: Optical sectioning of microbial biofilms. *J Bacteriol* 173:6558–67, 1991
22. Locci R, Peters G, Pulverer G: Microbial colonization of prosthetic devices. I. Microtopographical characteristics of intravenous catheters as detected by scanning electron microscopy. *Zentralbl Bakteriol Mikrobiol Hyg [B]* 173:285–92, 1981
23. Mackowiak PA, Marling-Cason M: A comparative analysis of in vitro assays of bacterial adherence. *J Microbiol Meth* 2:147–58, 1984
24. McDowell SG, An YH, Draughn RA, et al: Application of a fluorescent redox dye for enumeration of metabolically active bacteria on albumin-coated titanium surfaces. *Lett Appl Microbiol* 21:1–4, 1995
25. Ørstavik D: Sorption of *Streptococcus faecium* to glass. *Acta Path Microbiol Scand* 85:38–46, 1977
26. Pascual A, Fleer A, Westerdaal NA, et al: Modulation of adherence of coagulase-negative staphylococci to Teflon catheters in vitro. *Eur J Clin Microbiol* 5:518–22, 1986
27. Peters G, Locci R, Pulverer G: Adherence and growth of coagulase-negative staphylococci on surfaces of intravenous catheters. *J Infect Dis* 146:479–82, 1982
28. Pringle JH, Fletcher M: Influence of substratum wettability on attachment of freshwater bacteria to solid surfaces. *Appl Environ Microbiol* 45:811–7, 1983
29. Qian Z, Stoodley P, Pitt WG: Effect of low-intensity ultrasound upon biofilm structure from confocal scanning laser microscopy observation. *Biomaterials* 17:1975–80, 1996
30. Razatos A, Ong YL, Sharma MM, et al: Molecular determinants of bacterial adhesion monitored by atomic force microscopy. *Proc Natl Acad Sci U S A* 95:11059–64, 1998
31. Richardson JH, Barkley WE: *Biosafety in Microbiological and Biomedical Laboratories.* U.S. Government Printing Office, Washington DC, 1988
32. Rutter PR, Abbott A: A study of the interaction between oral streptococci and hard surfaces. *J Gen Microbiol* 105:219–26, 1978
33. Shea C, Williamson JC: Rapid analysis of bacterial adhesion in a microplate assay. *Biotechniques* 8:610–1, 1990
34. Siedlecki CA, Marchant RE: Atomic force microscopy for characterization of the biomaterial interface. *Biomaterials* 19:441–54, 1998
35. Sieracki ME, Johnson PW, Sieburth JM: Detection, enumeration, and sizing of planktonic bacteria by image-analyzed epifluorescence microscopy. *Appl Environ Microbiol* 49:799–810, 1985
36. Suci PA, Mittelman MW, Yu FP, et al: Investigation of ciprofloxacin penetration into *Pseudomonas aeruginosa* biofilms. *Antimicrob Agents Chemother* 38:2125–33, 1994
37. Suci PA, Vrany JD, Mittelman MW: Investigation of interactions between antimicrobial agents and bacterial biofilms using attenuated total reflection Fourier transform infrared spectroscopy. *Biomaterials* 19:327–39, 1998
38. Tieszer C, Reid G, Denstedt J: XPS and SEM detection of surface changes on 64 ureteral stents after human usage. *J Biomed Mater Res* 43:321–30, 1998
39. Timmerman CP, Fleer A, Besnier JM, et al: Characterization of a proteinaceous adhesin of *Staphylococcus epidermidis* which mediates attachment to polystyrene. *Infect Immun* 59:4187–92, 1991

40. Trulear MG, Characklis WG: Dynamics of biofilm processes. *J WPCF* 54:1288–301, 1982
41. Tyler BJ: XPS and SIMS studies of surfaces important in biofilm formation. Three case studies. *Ann N Y Acad Sci* 831:114–26, 1997
42. Umeda A, Saito M, Amako K: Surface characteristics of Gram-negative and Gram-positive bacteria in an atomic force microscope image. *Microbiol Immunol* 42:159–64, 1998
43. Velraeds MM, van der Mei HC, Reid G, et al: Inhibition of initial adhesion of uropathogenic *Enterococcus faecalis* by biosurfactants from *Lactobacillus* isolates. *Appl Environ Microbiol* 62:1958–63, 1996
44. Wollin TA, Tieszer C, Riddell JV, et al: Bacterial biofilm formation, encrustation, and antibiotic adsorption to ureteral stents indwelling in humans. *J Endourol* 12:101–11, 1998

# 8

# General Considerations
# for Studying Bacterial Adhesion to Biomaterials

Yuehuei H. An and Piotr Skowronski

*Orthopaedic Research Laboratory, Medical University of South Carolina,
Charleston, SC, USA*

## I. INTRODUCTION

Bacterial adherence to biomaterial and tissue surfaces is an important step in the pathogenesis of prosthetic[37,94] and tissue[13] infection and the exact mechanism of prosthetic infection remains unclear. Before effective preventive or therapeutic methods can be achieved, the process, characteristics, or the mechanism of bacterial adhesion to biomaterials have to be studied. In this review, the experimental methods for bacterial adhesion to biomaterial surfaces are discussed.

Many in vitro models for bacterial adhesion to biomaterials have been developed.[26] They can be classified as the following: 1) Bacterial adherence onto commercially available flat surfaces such as plastic culture tubes,[57] petri dishes,[39,41] or tissue culture plates;[10,18,32] 2) Bacterial adherence to substrata placed into culture tubes,[73] petri dishes,[51] tissue culture plates,[4,45] or chemotaxis chamber;[61] 3) Bacterial adhesion tested under flow conditions, such as a flow cell perfusion model,[108] rotating disc apparatus,[1,85] parallel plate flow chamber,[84] radial flow chamber,[31,42,72] or Robbins device.[65,70] The sample surfaces to be tested are often built as a part of the circulating or perfusion system, such as the inner wall of tubings or a sample piece fixed inside the flow system.

When designing an in vitro model for bacterial adhesion study, the following factors should be taken into consideration, including 1) the selection of bacteria, 2) sample surface preparation, 3) bacterial adhesion experiments, and 4) the sample preparation for evaluation. The reviews by Fletcher,[40] Dankert et al.,[26] and Christensen et al.[22] are excellent references for additional information.

## II. SELECTION OF MICROORGANISMS

If the experimental question relates to bacterial adhesion to certain medical devices or implants, it is clearly important to select microorganisms representative of the clinical conditions. Common pathogens isolated from human prosthetic infections should be used for most of the infection models, such as *Staphylococcus aureus*, *Staphylococcus epidermidis*, *Escherichia coli*, *Proteus*, or some of the common anaerobics. They can be

*Handbook of Bacterial Adhesion: Principles, Methods, and Applications*
Edited by: Y. H. An and R. J. Friedman © Humana Press Inc., Totowa, NJ

**Table 1. Bacterial Selection for Adhesion Study
on Common Medical Devices**

| Devices | Species | Reference |
|---------|---------|-----------|
| Orthopedic implants | *S. epidermidis* | 2,4,7,8,20,27,58,67,102 |
| | *S. aureus* | 3,8,27,67,102 |
| Surgical implants, sutures | *S. epidermidis* | 28,71,103 |
| | *S. aureus* | 6,23,28 |
| | *P. aeruginosa* | 28,47,69 |
| | *E. coli* | 23,28 |
| Vascular prostheses | *S. epidermidis* | 15,48,64,88,90,109,110 |
| | *S. aureus* | 9,64,105,109,110 |
| | *E. coli* | 109,110 |
| Bile stents | *E. coli* | 54,95,100,108 |
| | Enterococci | 54,95,108 |
| | *Klebsiella* sp. | 54 |
| Urinary catheters, stents | *E. coli* | 43,44,75 |
| | *P. aeruginosa* | 36,43,44,75,80,81 |
| | Enterococci | 36,44 |
| | *Klebsiella* sp. | 36,44 |
| Oral biomaterials | Streptococci | 17,30,34,53,55,87 |
| | *P. gingivalis* | 34,46,55,91,106 |
| | Actinomyces | 34,46,91,106 |

obtained directly from clinical isolation, or ATCC (American Type Culture Collection, Rockville, MD) collections or other major collections or laboratories. In recent years, the commonly used bacteria for the studies of in vitro or in vivo prosthetic infection or related bacterial adhesion to biomaterial surfaces include rich slime producer *S. epidermidis* RP62A (ATCC-35984),[21,50,98] poor slime producer RP12 (ATCC 35983),[22] nonslime producer *S. epidermidis* RP62NA,[21,50] and fimbriae producers as *E. coli* (ATCC-25922),[19,29] *Pseudomonas aeruginosa*,[11,104] and *Candida parapsilosis*[14,63] which are also slime producers.

Based on the specific purpose of the project, bacteria need to be selected carefully in order to achieve optimum results (Table 1). Cultures of microorganisms may be obtained from the sources listed in Table 2. One should keep in mind that 1) significant surface change can occur with natural isolates upon culture in the laboratory; 2) the production of cell surface polymers can be influenced considerably by growth conditions; 3) centrifugation and washing may remove bacterial capsule or slime; and 4) adhesion of a given bacteria can be influenced by the presence of other species.[40]

### III. SAMPLE SURFACE PREPARATION

Generally, materials can be divided into two main classes.[40] They are 1) high-surface energy materials, which are also hydrophilic, frequently negatively charged, and usually

## Table 2. Common Sources of Microorganisms

| Source and address | Collections | Contact numbers |
| --- | --- | --- |
| ATCC (American Type Culture Collection), 10801 University Blvd., Manassas, VA 20110, USA | 60,000 μ-organisms & cells | www.atcc.org Tel: 01 703 365 2700 |
| Summit Pharmaceutical (an ATCC distributor) 2-9 Kanda Nishiki-cho, Tokyo, 101-0054 Japan | N/A | Tel: 81 (03) 3294 1619 Fax: 81 (03) 3294 1645 |
| BCCM™ (Belgian Coordinated Collections of Microorganisms), Laboratorium voor Microbiologie Universiteit Gent (RUG), K. L. Ledeganckstraat 35 B-9000 Gent, Belgium | 50,000 μ-organisms | www.belspo.be/bccm Tel: 32 9 264 5108 Fax: 32 9 264 5346 |
| Culture Collection, University of Göteborg Guldhedsgatan 10, S-413 46 Göteborg, Sweden | 40,000 μ-organisms | www.ccug.gu.se Tel: 46 31 342 4625 |
| VKM (All-Russian Collection of Microorganisms) 142292 Pushchino, Moscow Region, Russia | 20,000 μ-organisms | www.vkm.ru Tel: 7 (0)95 925 7448 |
| DSMZ (German Collection of Microorganisms & Cell Culture), Mascheroder Weg 1 b D - 38124 Braunschweig - Germany | 11,000 μ-organisms | www.dsmz.de Tel: 49 (0)531 26 16 319 Fax: 49 (0)531 26 16 444 |
| JCM (Japan Collection of Microorganisms) RIKEN (The Institute of Physical and Chemical Research), Wako, Saitama 351-0198, Japan | 6000 μ-organisms | www.jcm.riken.go.jp No phone orders Fax: 81 48 462 4617 |
| NCTC (National Collection of Type Cultures) Central Public Health Lab., 61 Colindale Ave., London NW9 5HT | 5000 μ-organisms Medical & veterinary | Tel: 44 0181 200 4400 Fax: 44 0181 200 7874 |
| NCIMB (National Collection of Industrial & Marine Bacteria), 23 St Machar Drive, Aberdeen, AB2 1RY, Scotland | 3800 Bacteria | Tel: 44 01224 273332 Fax: 44 01224 487658 |
| Czechoslovak Collection of Microorganisms Masaryk University Tvrdeho 14, 602 00  Brno, Czech Republic | 2000 μ-organisms | Tel: (425) 23407 Fax: (425) 755247 |
| MSDN (Microbial Strain Data Network) | N/A | panizzi.shef.ac.uk/msdn |
| MINC (Microbial Information Network of China) | N/A | m.ac.cn www.im.ac.cn |

inorganic materials such as metals, or minerals, and 2) low-energy surfaces, which are relatively hydrophobic, low in electrostatic charge, and are generally organic polymers such as plastics. Because of the complicated physicochemical characteristics of material surfaces and also those of bacteria, microorganisms will adhere to different material surfaces differentially. Selection of material surfaces should be dependent on 1) the purpose of the study, 2) the existing knowledge of certain material surfaces, and 3) the availability of required material surfaces.

Material surfaces can be simply commercially available flat surfaces such as the surface of a coverslip,[60] the inner wall of a plastic culture tube,[57] catheters,[107] stents,[108] the bottom of a petri dish,[41] or tissue culture plates.[18,32] The advantages of using these surfaces for studies of bacterial adhesion include 1) most of these surfaces are optically clear, smooth, and uniform; 2) attached bacterial layers are easily attained;[83,86] 3) standardized manufacturing practices and widespread availability allow for ready comparison of results from different laboratories; and 4) they are economical.

Sample surfaces can also be made according to the requirements of a specific study. The common method is cutting the stock material sheets or rods into pieces (square sheets or circular discs) of certain size. The cut sample surfaces can also be modified by using different surface treatments, such as sandpaper grinding, sandblasting, macromolecule or surfactant coating. Chemical treatments can make samples with different surface chemistry such as the four types of glass surfaces reported by Satou et al.[86] Another possible way to get a material surface for bacterial adhesion study is the available prostheses or implant devices. They can be used as manufactured or after certain modifications.

Any homemade sample surfaces or readily available surface which is thought to be dirty should be cleaned and/or chemically treated before use. Plastic or polymer samples are usually cleaned using detergent and distilled water to remove debris and grease. These materials have low energy surfaces and do not adsorb contaminants as readily as high energy surfaces. Metal samples are normally cleaned by detergent and water followed by specific types of chemical treatment, such as the passivation method.[5] Sometimes an appropriate sterilization, i.e., radio frequency glow discharge (RFGD),[56,96] is needed.

It has been noted that certain studies ignored the necessary surface characterization, which made comparison of results impossible. It is strongly recommended that basic surface characterization, such as surface roughness, physical configuration, hydrophobicity, or even chemical composition at certain levels, should be performed.

## VI. POSITIONING OF THE SURFACE

One of the major concerns of positioning a sample surface in bacteria suspension is whether or not the surfaces to be studied are in a consistent contact with the suspension. Flow of the suspension is very important, as it influences the flux of bacteria and physical forces brought to the material surfaces, and it should be also consistent to every surface of the sample. Whether the surface is vertical or horizontal is also important because sedimentation plays a part when a horizontal surface is used. To our knowledge, there are two ways of placing a sample surface in bacterial suspension:

1) Static or random-flow systems: flat sample discs or plates can be placed in a petri dishes,[51] or culture plate wells, horizontally or vertically.[4,45] Samples can be immersed in solution, suspended, hanging on stainless steel wires[73] or using an O-ring method.[79,101] Incubation of the samples within bacterial suspension can be static or agitated randomly.

2) Flow systems, such as the flow cell perfusion model,[108] rotating disc,[84,85] laminar flow system,[52,78] and radial flow chamber.[31,42] All of these systems provide an oriented fluid (containing bacteria) flow which brings bacteria to the material surfaces to be tested. The latter are often built as a part of the circulating or perfusion system, with the inner wall of tubing or a sample piece fixed inside the flow system. The major advantages of

flow systems are: (1) controlled shear and mass transport, (2) a high data density in time, and (3) no air–liquid interface passages over the adhering bacteria.[16] Real-time observation is possible for certain designs.[107] It is a true noninvasive method for observing biofilm formation. This technique requires a flow chamber dimension small enough to fit on a microscopic stage or a special stage to be made and transparent for focusing and viewing.

## V. RINSING AND DRYING PROCEDURE

Rinsing is a very important part of a bacterial adhesion study.[40,77] The purpose of rinsing is to remove the unattached and loosely attached bacteria from the substrata surface. Based on specific purpose and experimental design (such as design of adhesion devices and samples), rinsing and drying procedures can be very different. Therefore, standardization of a washing and drying protocol in a given laboratory is required.

Attention should be paid to the force, direction, and content of the rinsing fluid. The rinsing stream should be directed to the walls of a incubation bath or a container. A gentle rinsing can be achieved simply by changing fluid in the incubation bath, which has no strong jets of washing on the substrata surfaces.

Liquids commonly used for rinsing include sterile water, normal saline, and phosphate-buffered saline. One shortcoming of using these liquids is the effect of a moving air–liquid interface.[16,77] A moving air–liquid interface on a substrate surface has a strong shearing force which can detach adhered bacteria and move them elsewhere, leading to unevenly distributed bacteria on the surface. This is more obvious on hydrophobic surfaces. This effect changes the actual adhesion pattern and makes the surface uncountable. An effective method has been published recently to avoid this air–liquid interface effect by using ethanol as the final rinsing liquid since it evaporates much faster than water.[77] The substrate surface should remain immersed in liquid during the rinsing. If possible, the staining procedure can also be carried out before the sample is taken out of the liquid for drying. After rinsing, samples can be air dried or with the aid of a gentle flow of nitrogen air.

## VI. STAINING

To visualize adherent bacteria or biofilm, light, fluorescent, and confocal laser scanning microscopy (CLSM) are often utilized. To use these methods, the substrate must be flat or have flat portions on the surfaces. Scanning electron microscopy (SEM) has the advantage of being able to observe bacterial adhesion on rough or irregular surfaces.

### A. Stains for Light Microscopy

The Gram stain is possibly the most widely used stain in bacterial studies, and is used as a first step in identifying bacteria. The Gram stain distinguishes between Gram-positive and Gram-negative bacteria. Gram-positive bacteria retain the crystal (gentian) violet stain and after decolorizing in alcohol will appear blue or purple. Gram-negative bacteria appear red as the crystal violet is decolorized by alcohol, allowing counterstaining with red safranin. Commercially available Gram stain kits should contain everything needed to perform the stain. The Gram stain is performed on a smear which is heat-fixed to a slide. The staining protocol is as follows: the crystal violet stain is applied to the smear for 1 to

**Table 3. Common Bacterial Staining Protocols**

| Stains | Solution | Concentration | Bacteria fixation and staining procedure | Ref. no. |
|--------|----------|---------------|------------------------------------------|----------|
| AO | PBS, water | 0.005-0.5 w/v | 2–3 min | 40,41,74 |
| CTC | PBS, culture media | 5 mmol/L | No fixation, stain in incubation | 3,67 |
| DAPI | Water | 10 µM | 1.86% formaldehyde | 74 |
| Hoechst 33258 | PBS, water | 10 µg/mL | No fixation, stain in incubation | 67,74 |
| Hoechst 33258 | PBS, water | 10 µg/mL | 1.86% formaldehyde | 74 |
| PI | PBS, water | 0.05 mg/mL | 75% ethanol, stained after incubation | 3,67 |

2 min, after which time all cells are stained purple. Excess stain is rinsed off with water, then the slide is rinsed with iodine solution and incubated 1 to 2 min with fresh iodine solution. The iodine solution fixes the crystal violet stain by forming a dye–iodine complex and reducing the solubility of the stain. The slide is rinsed again with water and an acid–alcohol solution is applied to decolorize. Decolorization takes 5 to 10 s. At the end of this step, Gram-positive bacteria will be purple, while Gram-negative bacteria will be colorless. The slide is then counterstained with safranin O for one minute, which results in a pink or reddish stain in the Gram-negative bacteria.

Other popular staining methods for light microscopy include methylene blue stain and fuchsin stain.[24] For methylene blue staining, the sample or sample slide is stained for 1 min in a 0.5% aqueous solution of methylene blue. For fuchsin staining, the samples are stained 30 s with the following solution: Dissolve 1.0 g basic fuchsin in 100 mL 90% ethanol, stand for 24 h, filter, and add 900 mL distilled water.[24]

### B. Fluorescence Staining

Propidium iodide (PI) is the most popular fluorescence dye used for DNA staining and DNA content analysis with the aid of flow cytometry.[3,59] PI intercalates with DNA and RNA and this binding markedly enhances the fluorescence of the dye (excitation at 536 nm and emission at 617 nm).[62] Athough the most popular fluorescence dye for bacteria staining to be observed by fluorescence microscopy is acridine orange (AO),[40,41,76] in our laboratory the contrast of the acridine orange stained field was not as significant as that of PI. It is known that AO tends to stain the substrate surfaces, which can obscure the images of adherent bacteria. Also, acridine orange staining needs a rinsing procedure to remove the unbound part in order to reduce the brightness of the background. This may loosen or wash off more bacteria. By comparison, PI has the advantages of significant field contrast and no need for rinsing.

The fluorescent redox dye, 5-cyano-2,3-ditolyl tetrazolium chloride (CTC), has been used as an indicator of electron transport activity in Ehrlich ascites tumor cells[92] and for the direct visualization of respiring bacteria in environmental water samples.[82] When oxidized, CTC is colorless and does not fluoresce, but when reduced by electron transport activity the dye forms fluorescent CTC-formazan. CTC has been tested in the authors' laboratory for quantifying actively respiring bacteria, *S. aureus* and *S. epidermidis*, that adhere to comercially pure titanium surfaces coated with cross-linked albumin.[67] The adherent bacteria appeared bright red against a dark red background (excitation at 420 nm).[82]

Other popular DNA stains include Hoechst 33258, Hoechst 33342 (excitation at 345 nm and emission at 460 nm), and DAPI (4'6-diamidino-2-phenylindole) (excitation at 359 nm and emission at 461 nm).[40,49] In this laboratory, similar results have been obtained with these fluorescence stains. Based on the method of evaluation, many different staining protocols are available which are summarized in Table 3.

## C. Preparation for SEM

SEM is frequently used to observe the patterns of bacterial adhesion on material surfaces. One advantage of SEM is that it can be used to visualize rough or irregular surfaces. It can be used to count the number of attached bacteria[3] and also for evaluating the distribution of the bacteria. It can be applied to a wide range of material surfaces, such as metals, plastics, minerals, plants or tissues. Routinely, the samples are fixed in 2 to 3% buffered glutaraldhyde (cacodylate buffer, 0.067 M, pH 6.2), dehydrated in a graded series of acetone or ethanol, further treated in a critical point dryer and coated with gold in a sputter coater. If the substrate is a plastic or polymer, acetone should not be used for dehydration to prevent potential dissolution of the material in acetone solution.

## VII. REMOVING BACTERIA FROM SUBSTRATA SURFACES

Sometimes, the adherent bacteria need to be removed from the sample surfaces to facilitate enumeration, especially for rough or irregular surfaces. Methods for removing bacteria from substrate surfaces include homogenization, sonication, vortexing, and the use of surfactants (such as Tween 80, Triton X-100, or sodium periodate).[66] Most the studies of the effects of methodology on the enumeration of sedimentary bacteria agree that either sonication or homogenization gives maximum yield. Scraping can be used to remove large amounts of biofilm on biomaterial surfaces. Vortexing is especially useful for breaking down large pieces of biofilm. According to the comparative study by McDaniel and Capone,[66] sonication appears to be an efficient and safe way to remove bacteria from biomaterial surfaces.[12,99] It is less harmful to the organisms than chemical elution,[93] and has less potential for harm to the investigators than scraping the biomaterial. Sonication of explanted vascular prosthetic graft material can disrupt surface biofilms and increase the recovery of adherent microorganisms.[93,99]

After removal from sample surfaces, the number of bacteria can be counted using the standard plate count method. There is virtually always incomplete removal of the adherent bacteria and incomplete disruption of bacterial aggregates in the suspension, so numbers will be underestimated by this method.[40]

## REFERENCES

1. Abbott A, Rutter PR, Berkeley RC: The influence of ionic strength, pH and a protein layer on the interaction between *Streptococcus mutans* and glass surfaces. *J Gen Microbiol* 129:439–45, 1983
2. An YH, Bradley J, Powers DL, et al: The prevention of prosthetic infection using a cross-linked albumin coating in a rabbit model. *J Bone Joint Surg Br* 79:816–9, 1997
3. An YH, Friedman RJ, Draughn RA: Rapid quantification of staphylococci adhered to titanium surfaces using image analyzed epifluorescence microscopy. *J Microbiol Meth* 24:29–40, 1995
4. An YH, Stuart GW, McDowell SJ, et al: Prevention of bacterial adherence to implant surfaces with a crosslinked albumin coating *in vitro. J Orthop Res* 14:846-9, 1996

5. Anonymous: Designation: F 86-91. Surface preparation and marking of metallic surgical implants. *Annual Book of ASTM Standards.* Vol. 13.01. ASTM Publications, Philadelphia, 1997:6–8

6. Arciola CR, Cenni E, Caramazza R, et al: Seven surgical silicones retain *Staphylococcus aureus* differently *in vitro. Biomaterials* 16:681–4, 1995

7. Arciola CR, Montanaro L, Moroni A, et al: Hydroxyapatite-coated orthopaedic screws as infection resistant materials: *in vitro* study. *Biomaterials* 20:323–7, 1999

8. Barth E, Myrvik QM, Wagner W, et al: *In vitro* and *in vivo* comparative colonization of *Staphylococcus aureus* and *Staphylococcus epidermidis* on orthopaedic implant materials. *Biomaterials* 10:325–8, 1989

9. Baumgartner JN, Cooper SL: Bacterial adhesion on polyurethane surfaces conditioned with thrombus components. *ASAIO J* 42:M476–9, 1996

10. Bayston R, Rodgers J: Production of extra-cellular slime by *Staphylococcus epidermidis* during stationary phase of growth: its association with adherence to implantable devices. *J Clin Pathol* 43:866–70, 1990

11. Benson DE, Burns GL, Mohammad SF: Effects of plasma on adhesion of biofilm forming *Pseudomonas aeruginosa* and *Staphylococcus epidermidis* to fibrin substrate. *ASAIO J* 42:M655–60, 1996

12. Bergamini TM, Bandyk DF, Govostis D, et al: Identification of *Staphylococcus epidermidis* vascular graft infections: a comparison of culture techniques. *J Vasc Surg* 9:665–70, 1989

13. Bonten MJ, Weinstein RA: The role of colonization in the pathogenesis of nosocomial infections. *Infect Control Hosp Epidemiol* 17:193–200, 1996

14. Branchini ML, Pfaller MA, Rhine-Chalberg J, et al: Genotypic variation and slime production among blood and catheter isolates of *Candida parapsilosis. J Clin Microbiol* 32:452–6, 1994

15. Brunstedt MR, Sapatnekar S, Rubin KR, et al: Bacteria/blood/material interactions. I. Injected and preseeded slime-forming *Staphylococcus epidermidis* in flowing blood with biomaterials. *J Biomed Mater Res* 29:455–66, 1995

16. Busscher HJ, van der Mei HC: Use of flow chamber devices and image analysis methods to study microbial adhesion. *Methods Enzymol* 253:455–77, 1995

17. Cai S, Simionato MR, Mayer MP, et al: Effects of subinhibitory concentrations of chemical agents on hydrophobicity and *in vitro* adherence of *Streptococcus mutans* and *Streptococcus sanguis. Caries Res* 28:335–41, 1994

18. Carsenti-Etesse H, Entenza J, Durant J, et al: Efficacy of subinhibitory concentration of pefloxacin in preventing experimental *Staphylococcus aureus* foreign body infection in mice. *Drugs Exp Clin Res* 18:415–22, 1992

19. Ceri H, Olson ME, Stremick C, et al: The calgary biofilm device: new technology for rapid determination of antibiotic susceptibilities of bacterial biofilms. *J Clin Microbiol* 37:1771–6, 1999

20. Chang CC, Merritt K: Microbial adherence on poly(methyl methacrylate) (PMMA) surfaces. *J Biomed Mater Res* 26:197–207, 1992

21. Christensen GD, Baddour LM, Simpson WA: Phenotypic variation of *Staphylococcus epidermidis* slime production *in vitro* and *in vivo. Infect Immun* 55:2870-7, 1987

22. Christensen GD, Baldassarri L, Simpson WA: Methods for studying microbial colonization of plastics. *Methods Enzymol* 253:477–500, 1995

23. Chu CC, Williams DF: Effects of physical configuration and chemical structure of suture materials on bacterial adhesion. A possible link to wound infection. *Am J Surg* 147:197–204, 1984

24. Collins CH, Lyne PM: *Microbiological Methods.* Butterworth, London, 1984

25. Cowell BA, Willcox MD, Herbert B, et al: Effect of nutrient limitation on adhesion characteristics of *Pseudomonas aeruginosa. J Appl Microbiol* 86:944–54, 1999

26. Dankert J, Hogt AH, Feijen J: Biomedical polymers: Bacterial adhesion, colonization, and infection. *CRC Crit Rev Biocompat* 2:219–301, 1986

27. Delmi M, Vaudaux P, Lew DP, et al: Role of fibronectin in staphylococcal adhesion to metallic surfaces used as models of orthopaedic devices. *J Orthop Res* 12:432–8, 1994

28. Desai NP, Hossainy SF, Hubbell JA: Surface-immobilized polyethylene oxide for bacterial repellence. *Biomaterials* 13:417–20, 1992

29. Dix BA, Cohen PS, Laux DC, et al: Radiochemical method for evaluating the effect of antibiotics on *Escherichia coli* biofilms. *Antimicrob Agents Chemother* 32:770–2, 1988

30. Drake DR, Paul J, Keller JC: Primary bacterial colonization of implant surfaces. *Int J Oral Maxillofac Implants* 14:226–32, 1999

31. Duddriege JE, Cent CA, Laws JE: Effect of surface shear stress on the attachment of *Pseudomonas fluorescens* to stainless steel under defined flow conditions. *Biotech Bioeng* 24:153–64, 1982

32. Dunne WM, Jr., Burd EM: *In vitro* measurement of the adherence of *Staphylococcus epidermidis* to plastic by using cellular urease as a marker. *Appl Environ Microbiol* 57:863–6, 1991

33. Duran JA, Malvar A, Rodriguez-Ares MT, et al: Heparin inhibits *Pseudomonas* adherence to soft contact lenses. *Eye* 7:152–4, 1993

34. Ellen RP, Lepine G, Nghiem PM: *In vitro* models that support adhesion specificity in biofilms of oral bacteria. *Adv Dent Res* 11:33–42, 1997

35. Farber BF, Hsieh HC, Donnenfeld ED, et al: A novel antibiofilm technology for contact lens solutions. *Ophthalmology* 102:831–6, 1995

36. Farber BF, Wolff AG: Salicylic acid prevents the adherence of bacteria and yeast to silastic catheters. *J Biomed Mater Res* 27:599–602, 1993

37. Fitzgerald RH, Jr: Infections of hip prostheses and artificial joints. *Infect Dis Clin North Am* 3:329–38, 1989

38. Fleiszig SM, Evans DJ, Mowrey-McKee MF, et al: Factors affecting *Staphylococcus epidermidis* adhesion to contact lenses. *Optom Vis Sci* 73:590–4, 1996

39. Fletcher M: The effects of proteins on bacterial attachment to polystyrene. *J Gen Microbiol* 94:400–4, 1976

40. Fletcher M: Methods for studying adhesion and attachment to surfaces. *Method Microbiol* 22:251–80, 1990

41. Fletcher M, Loeb GI: Influence of substratum characteristics on the attachment of a marine pseudomonad to solid surfaces. *Appl Environ Microbiol* 37:67–72, 1979

42. Fowler HW, McKay AJ: The measurement of microbial adhesion. In: Berkeley RW, et al., ed. *Microbial Adhesion to Surfaces*. Ellis Horwood, Chichester, UK, 1980:143

43. Gabriel MM, Mayo MS, May LL, et al: *In vitro* evaluation of the efficacy of a silver-coated catheter. *Curr Microbiol* 33:1–5, 1996

44. Gabriel MM, Sawant AD, Simmons RB, et al: Effects of silver on adherence of bacteria to urinary catheters: *in vitro* studies. *Curr Microbiol* 30:17–22, 1995

45. Gristina AG, Jennings RA, Naylor PT, et al: Comparative *in vitro* antibiotic resistance of surface-colonizing coagulase-negative staphylococci. *Antimicrob Agents Chemother* 33:813–6, 1989

46. Haas R, Dortbudak O, Mensdorff-Pouilly N, et al: Elimination of bacteria on different implant surfaces through photosensitization and soft laser. An *in vitro* study. *Clin Oral Implants Res* 8:249–54, 1997

47. Habash MB, van der Mei HC, Reid G, et al: Adhesion of *Pseudomonas aeruginosa* to silicone rubber in a parallel plate flow chamber in the absence and presence of nutrient broth. *Microbiology* 143:2569–74, 1997

48. Harris JM, Martin LF: An *in vitro* study of the properties influencing *Staphylococcus epidermidis* adhesion to prosthetic vascular graft materials. *Ann Surg* 206:612–20, 1987

49. Haugland RP: *Handbook of Fluorescent Probes and Research Chemicals*. Molecular Probes, Inc., Eugene, OR, 1992

50. Henke PK, Bergamini TM, Garrison JR, et al: *Staphylococcus epidermidis* graft infection is associated with locally suppressed major histocompatibility complex class II and elevated MAC-1 expression. *Arch Surg* 132:894–902, 1997

51. Hogt AH, Dankert J, de Vries JA, et al: Adhesion of coagulase-negative staphylococci to biomaterials. *J Gen Microbiol* 129:2959–68, 1983

52. Hogt AH, Dankert J, Feijen J: Adhesion of coagulase-negative staphylococci to methacrylate polymers and copolymers. *J Biomed Mater Res* 20:533–45, 1986

53. Ichikawa T, Hirota K, Kanitani H, et al: *In vitro* adherence of *Streptococcus constellatus* to dense hydroxyapatite and titanium. *J Oral Rehabil* 25:125–7, 1998

54. Jansen B, Goodman LP, Ruiten D: Bacterial adherence to hydrophilic polymer-coated polyurethane stents. *Gastrointest Endosc* 39:670–3, 1993

55. Kato T, Kusakari H, Hoshino E: Bactericidal efficacy of carbon dioxide laser against bacteria-contaminated titanium implant and subsequent cellular adhesion to irradiated area. *Lasers Surg Med* 23:299–309, 1998

56. Kawahara D, Ong JL, Raikar GN, et al: Surface characterization of radio-frequency glow discharged and autoclaved titanium surfaces. *Int J Oral Maxillofac Implants* 11:435–42, 1996

57. Kefford B, Marshall KC: Adhesion of *Leptospira* at a solid-liquid interface: a model. *Arch Microbiol* 138:84–8, 1984

58. Konig DP, Perdreau-Remington F, Rutt J, et al: Slime production of *Staphylococcus epidermidis*: increased bacterial adherence and accumulation onto pure titanium. *Acta Orthop Scand* 69:523–6, 1998

59. Krishan A: Rapid flow cytofluorometric analysis of mammalian cell cycle by propidium iodide staining. *J Cell Biol* 66:188–93, 1975

60. Kuusela P, Vartio T, Vuento M, et al: Attachment of staphylococci and streptococci on fibronectin, fibronectin fragments, and fibrinogen bound to a solid phase. *Infect Immun* 50:77–81, 1985

61. Leake ES, Gristina AG, Wright MJ: Use of chemotaxis chambers for studying *in vitro* bacterial colonization of biomaterials. *J Clin Microbiol* 15:320–3, 1982

62. LePecq JB, Paoletti C: A fluorescent complex between ethidium bromide and nucleic acids. Physical-chemical characterization. *J Mol Biol* 27:87–106, 1967

63. Levin AS, Costa SF, Mussi NS, et al: *Candida parapsilosis* fungemia associated with implantable and semi-implantable central venous catheters and the hands of healthcare workers. *Diagn Microbiol Infect Dis* 30:243–9, 1998

64. Malangoni MA, Livingston DH, Peyton JC: The effect of protein binding on the adherence of staphylococci to prosthetic vascular grafts. *J Surg Res* 54:168–72, 1993

65. McCoy WF, Bryers JD, Robbins J, et al: Observations of fouling biofilm formation. *Can J Microbiol* 27:910–7, 1981

66. McDaniel JA, Capone DG: A comparison of procedures for the separation of aquatic bacteria from sediments for subsequent direct enumeration. *J Microbiol Methods* 3:291–302, 1985

67. McDowell SG, An YH, Draughn RA, et al: Application of a fluorescent redox dye for enumeration of metabolically active bacteria on albumin-coated titanium surfaces. *Lett Appl Microbiol* 21:1–4, 1995

68. Miller MJ, Ahearn DG: Adherence of Pseudomonas aeruginosa to hydrophilic contact lenses and other substrata. *J Clin Microbiol* 25:1392–7, 1987

69. Nickel JC, Heaton J, Morales A, et al: Bacterial biofilm in persistent penile prosthesis-associated infection. *J Urol* 135:586–8, 1986

70. Nickel JC, Ruseska I, Wright JB, et al: Tobramycin resistance of *Pseudomonas aeruginosa* cells growing as a biofilm on urinary catheter material. *Antimicrob Agents Chemother* 27:619–24, 1985

71. Nomura S, Lundberg F, Stollenwerk M, et al: Adhesion of staphylococci to polymers with and without immobilized heparin in cerebrospinal fluid. *J Biomed Mater Res* 38:35–42, 1997

72. Olson ME, Ruseska I, Costerton JW: Colonization of *N*-butyl-2-cyanoacrylate tissue adhesive by *Staphylococcus epidermidis*. *J Biomed Mater Res* 22:485–95, 1988

73. Olsson J, Krasse B: A method for studying adherence of oral streptococci to solid surfaces. *Scand J Dent Res* 84:20–8, 1976

74. Paul JH: Use of Hoechst dye 33258 and 3342 for enumeration of attached and planktonic bacteria. *Appl Environ Microbiol* 43:939–44, 1982

75. Petas A, Vuopio-Varkila J, Siitonen A, et al: Bacterial adherence to self-reinforced polyglycolic acid and self-reinforced polylactic acid 96 urological spiral stents *in vitro*. *Biomaterials* 19:677–81, 1998

76. Pettipher GL, Rodrigues UM: Semi-automated counting of bacteria and somatic cells in milk using epifluorescence microscopy and television image analysis. *J Appl Bacteriol* 53:323–9, 1982

77. Pitt WG, McBride MO, Barton AJ, et al: Air-water interface displaces adsorbed bacteria. *Biomaterials* 14:605–8, 1993

78. Pratt-Terpstra IH, Weerkamp AH, Busscher HJ: Adhesion of oral streptococci from a flowing suspension to uncoated and albumin-coated surfaces. *J Gen Microbiol* 133:3199–206, 1987

79. Pringle JH, Fletcher M: Influence of substratum hydration and adsorbed macromolecules on bacterial attachment to surfaces. *Appl Environ Microbiol* 51:1321–5, 1986

80. Reid G, Khoury AE, Preston CA, et al: Influence of dextrose dialysis solutions on adhesion of *Staphylococcus aureus* and *Pseudomonas aeruginosa* to three catheter surfaces. *Am J Nephrol* 14:37–40, 1994

81. Reid G, Sharma S, Advikolanu K, et al: Effects of ciprofloxacin, norfloxacin, and ofloxacin on in vitro adhesion and survival of *Pseudomonas aeruginosa* AK1 on urinary catheters. *Antimicrob Agents Chemother* 38:1490–5, 1994

82. Rodriguez GG, Phipps D, Ishiguro K, et al: Use of a fluorescent redox probe for direct visualization of actively respiring bacteria. *Appl Environ Microbiol* 58:1801–8, 1992

83. Rosenberg M: Bacterial adherence to polystyrene: a replica method of screening for bacterial hydrophobicity. *Appl Environ Microbiol* 42:375–7, 1981

84. Rutter P, Leech R: The deposition of *Streptococcus sanguis* NCTC 7868 from a flowing suspension. *J Gen Microbiol* 120:301–7, 1980

85. Rutter PR, Abbott A: A study of the interaction between oral streptococci and hard surfaces. *J Gen Microbiol* 105:219–26, 1978

86. Satou N, Satou J, Shintani H, et al: Adherence of streptococci to surface-modified glass. *J Gen Microbiol* 134:1299–305, 1988

87. Schilling KM, Doyle RJ: Bacterial adhesion to hydroxylapatite. *Methods Enzymol* 253:536–42, 1995

88. Schmitt DD, Edmiston CE, Krepel C, et al: Impact of postantibiotic effect on bacterial adherence to vascular prostheses. *J Surg Res* 48:373–8, 1990

89. Simmons PA, Tomlinson A, Seal DV: The role of *Pseudomonas aeruginosa* biofilm in the attachment of Acanthamoeba to four types of hydrogel contact lens materials. *Optom Vis Sci* 75:860–6, 1998

90. Siverhus DJ, Schmitt DD, Edmiston CE, et al: Adherence of mucin and non-mucin-producing staphylococci to preclotted and albumin-coated velour knitted vascular grafts. *Surgery* 107:613–9, 1990

91. Steinberg D, Sela MN, Klinger A, et al: Adhesion of periodontal bacteria to titanium, and titanium alloy powders. *Clin Oral Implants Res* 9:67–72, 1998

92. Stellmach J: Fluorescent redox dyes. 1. Production of fluorescent formazan by unstimulated and phorbol ester- or digitonin-stimulated Ehrlich ascites tumor cells. *Histochemistry* 80:137–43, 1984

93. Sugarman B: Adherence of bacteria to urinary catheters. *Urol Res* 10:37–40, 1982
94. Sugarman B, Young EJ: Infections associated with prosthetic devices: magnitude of the problem. *Infect Dis Clin North Am* 3:187–98, 1989
95. Sung JY, Shaffer EA, Lam K, et al: Hydrophobic bile salt inhibits bacterial adhesion on biliary stent material. *Dig Dis Sci* 39:999–1006, 1994
96. Swart KM, Keller JC, Wightman JP, et al: Short-term plasma-cleaning treatments enhance *in vitro* osteoblast attachment to titanium. *J Oral Implantol* 18:130–7, 1992
97. Taylor RL, Willcox MD, Williams TJ, et al: Modulation of bacterial adhesion to hydrogel contact lenses by albumin. *Optom Vis Sci* 75:23–9, 1998
98. Tojo M, Yamashita N, Goldmann DA, et al: Isolation and characterization of a capsular polysaccharide adhesin from *Staphylococcus epidermidis. J Infect Dis* 157:713–22, 1988
99. Tollefson DF, Bandyk DF, Kaebnick HW, et al: Surface biofilm disruption. Enhanced recovery of microorganisms from vascular prostheses. *Arch Surg* 122:38–43, 1987
100. Tsang TK, Pollack J, Chodash HB: Inhibition of biliary endoprostheses occlusion by ampicillin-sulbactam in an *in vitro* model. *J Lab Clin Med* 130:643–8, 1997
101. van Pelt AW, Weerkamp AH, Uyen MH, et al: Adhesion of *Streptococcus sanguis* CH3 to polymers with different surface free energies. *Appl Environ Microbiol* 49:1270–5, 1985
102. Verheyen CC, Dhert WJ, de Blieck-Hogervorst JM, et al: Adherence to a metal, polymer and composite by *Staphylococcus aureus* and *Staphylococcus epidermidis. Biomaterials* 14:383–91, 1993
103. Wang IW, Anderson JM, Jacobs MR, et al: Adhesion of *Staphylococcus epidermidis* to biomedical polymers: contributions of surface thermodynamics and hemodynamic shear conditions. *J Biomed Mater Res* 29:485–93, 1995
104. Watanabe T: [Observation of *Pseudomonas aeruginosa* biofilm with confocal laser scanning microscope]. *Kansenshogaku Zasshi* 69:114–22, 1995
105. Webb LX, Myers RT, Cordell AR, et al: Inhibition of bacterial adhesion by antibacterial surface pretreatment of vascular prostheses. *J Vasc Surg* 4:16–21, 1986
106. Wu-Yuan CD, Eganhouse KJ, Keller JC, et al: Oral bacterial attachment to titanium surfaces: a scanning electron microscopy study. *J Oral Implantol* 21:207–13, 1995
107. Yu J, Montelius MN, Paulsson M, et al: Adhesion of coagulase-negative staphylococci and adsorption of plasma proteins to heparinized polymer surfaces. *Biomaterials* 15:805–14, 1994
108. Yu JL, Andersson R, Ljungh A: Protein adsorption and bacterial adhesion to biliary stent materials. *J Surg Res* 62:69–73, 1996
109. Zdanowski Z, Ribbe E, Schalen C: Bacterial adherence to synthetic vascular prostheses and influence of human plasma. An *in vitro* study. *Eur J Vasc Surg* 7:277–82, 1993
110. Zdanowski Z, Ribbe E, Schalen C: Influence of some plasma proteins on *in vitro* bacterial adherence to PTFE and Dacron vascular prostheses. *Apmis* 101:926–32, 1993

# 9

# Laboratory Culture
# and Analysis of Microbial Biofilms

**Chris H. Sissons,[1] Lisa Wong,[1] and Yuehuei H. An[2]**

*[1]Dental Research Group, Department of Pathology and Molecular Medicine*
*Wellington School of Medicine, University of Otago, Wellington, New Zealand*
*[2]Department of Orthopaedic Surgery, Medical University of South Carolina,*
*Charleston, SC, USA*

## I. INTRODUCTION

Microbial biofilms develop when bacteria adhere to a substratum and grow inside a secreted extracellular matrix. They can be defined as "matrix-embedded microbial populations adherent to each other and/or to surfaces of interfaces".[31] This is the growth mode for most bacteria. Biofilms are important in human health and disease; for example, the body's normal flora resists pathogen invasion but can itself turn pathogenic. Biofilm infections are a major problem, especially of prosthetic devices, as 1 to 3% of all orthopedic implant patients experience severe infection following surgery as the probable result of biofilm formation.[2] Biofilm formation within a tube can increase frictional resistance over 200%.[23] Antibacterial agents, antibiotics, phagocytic white blood cells, and other biocides are much less effective against the bacteria within a biofilm than against planktonic bacteria.[52]

The protection a biofilm provides to bacteria is a persistent problem in many industrial processes as well as in medicine. A biofilm layer within pipes not only causes increased friction leading to slower flow, but increased heat transfer resistance, corrosion, and can act as an antimicrobial-resistant nidus of re-infection. Biofilm fouling of the hulls of boats and in hydroelectric turbines results in increased frictional resistance and markedly decreased efficiency.[23] However, biofilms can be beneficial industrially through such processes as water quality improvement by pollutant removal, sewage treatment and manufacturing processes in biotechnology.

For these reasons it is important to explain, predict, modify or counter biofilm development and behavior. Studies *in situ* or in vivo are often impracticable. Laboratory culture and modeling is then needed. A major advantage of in vitro over in vivo biofilm studies is the degree of control over the microbes, environment, nutrient supply, and substrata, and increased options for experimental protocols, sampling and analysis. In this review we will analyze issues involved in modeling different types of biofilms and focus on appropriate in vitro methods for their cultivation and analysis. While we will

*Handbook of Bacterial Adhesion: Principles, Methods. and Applications*
Edited by: Y. H. An and R. J. Friedman © Humana Press Inc., Totowa, NJ

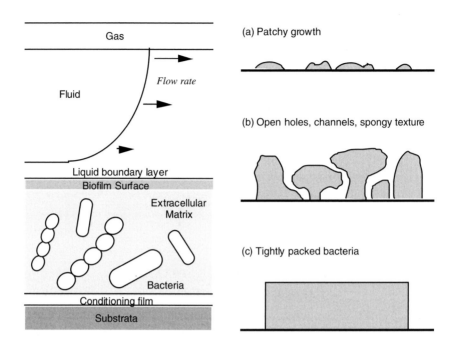

**Figure 1.** Major features of biofilms.          **Figure 2.** Types of biofilm structure.

emphasize a range of systems which are appropriate to biofilms impacting on human health from among the vast range of experimental systems available, these systems and the issues involved are generic and widely applicable. Some major references on the topics include the reviews by McFeters et al.,[100] Characklis,[22] Ladd and Costerton,[76] An et al.,[3] Gilbert and Allison,[50] and Sissons.[131]

## II. BIOFILMS

### A. Types of Biofilms

Bacterial biofilms have a set of common structural features (Fig. 1). These include a substratum, usually solid, but which may be a liquid–liquid or gas–liquid interface, or it may be other bacteria. A "conditioning" film of molecules from the fluid environment is adsorbed onto the substratum. The microbes in their extracellular macromolecule matrix adhere to the conditioning film. The opposite exterior surface of the biofilm, usually a defined extracellular matrix boundary, forms a surface–liquid interface with a semi-viscous laminar-flow 10 to 100 μm liquid boundary layer. This liquid boundary layer merges through a transition region into a mobile, usually turbulent, fluid phase.[25] The liquid phase interacts with a gas phase, possibly at some distance from the biofilm in such structures as tubes and pipes. In modeling natural biofilms using laboratory culture systems, appropriate decisions need to be made regarding all these features.

A range of concepts is encompassed by the term biofilm, different biofilm systems having different structures (Fig. 2). They range from the patchy microcolony formation and subsequent coalescence characteristic of monocultures in a low nutrient environment, to open spongy structures of microcolonies and communities with pores and channels,[30,148]

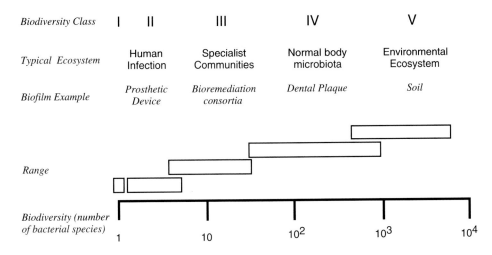

**Figure 3.** Biodiversity class of biofilm.

to dense biofilms of highly packed bacteria such as those seen in mature dental plaque by electron microscopy.[111,159] There are three major general influences on biofilm structure: fluid hydrodynamics, nutrient concentration and biodiversity of the microbiota.

Fluid hydrodynamic and associated shear forces deform the biofilm and provide a desorptive mechanism countering adhesive forces. These shear forces may be modified by substrata shape to give sheltered sites or by moving surfaces impacting on the biofilm. The particular fluid hydrodynamics involved is a key variable in deciding on an appropriate culture system and we will classify biofilm culture technologies on this basis. The second strong influence is the nutrient concentration in the fluid environment, ranging from 0.1 mg/L in some aquatic environments to $10^5$ mg/L in body fluids. Computer modeling using cellular automaton models suggests that open biofilm structures tend to form in low nutrient environments, and more densely-packed structures at high nutrient concentrations.[169] The nutrient composition, concentration and application regime need to be set with regard to the natural system being modeled. The third major factor is the species diversity of the biofilm.

## B. Biofilm Biodiversity: Classification and Implications

The nature of questions needing investigation and, therefore, the appropriate technologies, differ according to the biodiversity of the biofilm. Biofilms can be regarded as falling into five overlapping classes on the basis of their biodiversity (Fig. 3).

Class I biofilms are single species. Class II are those with a few species. Examples include infection of long-term prostheses by pathogens, transdermal devices by staphylococci, and urinary tract and bladder catheter infections by one of a range of gram negative species (*see* Chapter 1). These are the biofilm equivalent of single pathogen infections investigated using Koch's criteria which says a pathogen is always present in clinical disease, can be isolated, purified and causes the disease by transmission. Biofilms contaminating industrial processes are mainly Class I and II. The properties of the individual strains of bacteria present, i.e., rate of extracellular polymer production, is an important influence on these biofilms. Investigation of questions regarding individual bacteria from more complex biofilms is also usually carried out using similar techniques.

Class III biofilms consist of communities comprising a small number of species, perhaps up to 30. They are found, for example, as metabolic consortia formed in situations where a major substrate predominates in the environment. Examples include the xenochemical bioremediation consortia studied by Caldwell and colleagues[19] that emerge as a specialized subset of a more complex and biodiverse environmental ecosystem. Useful laboratory systems which are constructed to model more biodiverse biofilms also fall into this category, such as the 10-species plaque-reconstruction consortia developed by Marsh et al.[10,99,131]

Class IV and V biofilms are heterogeneous, high biodiversity microbial biofilms ecosystems constantly evolving in response to the environment yet showing considerable resistance to perturbation. For humans, the normal body microbiota is an example, especially in the mouth and other regions of the alimentary tract. Dental plaque has at least 500 known cultivable taxa at species level[105] and perhaps an equal number yet to be cultivated.[146,160] The colon has a similar biodiversity.[85] Questions of pathogenicity relate more to shifts in the composition of the biofilm microbiota to increase the proportion of existing "normal" bacteria with pathogenic properties than to acquisition of particularly virulent strains within species, although the balance between these options is unknown and they are not mutually exclusive. Koch's criteria are basically inapplicable.[17,41] The species diversity by DNA renaturation analysis of Class V biofilms, e.g., those from soil ecosystems, is probably several thousand.[154,155] This high level of biodiversity may insure against changes in the environment and enhance ecosystem resistance to adversity.[153] There may be no clear distinction between Class IV and V high biodiversity biofilms, with basic differences relating more to their particular environments.

A fundamental feature of high biodiversity biofilms is that their properties cannot be predicted from studying their individual bacteria in isolation as they are essentially complex communities of highly-interacting organisms and their properties are emergent from these myriad interactions in response to the biofilm environment (nutrients, fluid hydrodynamics, pH, $pO_2$, $pCO_2$, etc.) and formation of intraplaque gradients. These interactions include specific interbacterial adhesion relationships, nutrient cross-feeding, complementary macromolecule breakdown and defense against antibacterial factors, competition for nutrients and space, and direct antagonisms by metabolites, enzymes, and bacteriocins. Such biodiverse biofilms perhaps might best be regarded as equivalent to multicellular organisms. Culture, analytical and conceptual approaches differ from those of simpler biofilms. High biodiversity biofilms can reach thicknesses of several millimeters or more whereas monocultures rarely exceed 50 μm,[27] a major technical consideration.

## III. EXAMPLES OF LOW AND HIGH BIODIVERSITY BIOFILMS RELEVANT TO HUMAN HEALTH

We will focus on culture systems appropriate to Class I biofilms, where *Staphylococcus epidermidis* infection of intravascular prosthetic devices and catheters is a typical example, and Class IV biofilms, with dental plaque as an typical example. Between them, these two cover the range of issues relevant to most biofilm systems. First we will describe distinguishing characteristics of both systems.

### A. *S. epidermidis Biofilms on Prostheses: A Low Biodiversity Class I Biofilm*

The nature and importance of prosthesis infection is described in Chapter 1. There are basically three environmental fluids involved: 1) those exposed to the cardiovascular system which have a fluid phase of pulsating blood and an infection risk from bacteremia, 2) urinary catheters exposed to urine and urinary pathogens, with periodic flushing, and 3) joint prostheses exposed to extracellular fluids but possibly subject to considerable mechanical disturbance.

Intravascular biofilms of slime-producing *S. epidermidis* or other coagulase-negative staphylococci (CNS) are formed when the bacteria adhere to a catheter surface. Following attachment, bacteria continue to produce slime, composed usually of polysaccharides and other extracellular material. As more and more bacteria adhere and grow, a biofilm is formed. They form a working community; the bacteria on the surface of the biofilm help capture organic and inorganic molecules from the bulk liquid. and the bacteria within the biofilm work together to promote removal of waste and toxins from the biofilm.[24,121]

Capsules and slime are both subclasses of extracellular polymeric substances, usually polysaccharides. The polysaccharide coat of slime produced by CNS is a loose hydrogel associated through ionic interactions. These exopolysaccharides are composed of neutral monosaccharides including D-glucose, D-galactose, D-mannose, L-fucose and L-rhamnose; amino sugars; polyols; and uronic acid.[57] Strains not producing slime are less adherent and less pathogenic.[28,29,34,72] CNS strains that are completely slime negative are hard to find. Bacteria washed to remove slime have a similar adhesion to poly(tetrafluorethylene-co-hexafluorpropylene) as nonwashed bacteria[57] and hence slime is not thought to be involved in the primary adhesion of the bacteria to a material surface, but to be important for intercell connection during surface colonization.[65,66] The current concept is that production of slime is especially important for events which include colonization of various surfaces, protection against phagocytosis, interference with cellular immune responses, and reduction of antibiotic effects. CNS which do not quickly adhere to the surfaces are rapidly killed by the immune system. Slime-forming CNS are less susceptible to antibiotics after adhering to biomaterials than bacteria grown in planktonic culture.[106] Such antibiotic resistance may be partly due to the slow growth rate of CNS in biofilms or to the limited transport of nutrients, metabolites and oxygen to and from the biofilm surface. Biofilms also protect bacteria by sequestering them from host defense systems including inhibition of polymorphonuclear chemotaxis, phagocytosis and oxidative metabolism; depression of immunoglobulin synthesis; mononuclear lympho-proliferative response; killer cell cytotoxicity; and antibiotic action. Chronic infection occurs when a critical inoculum size is reached which overcomes the local host defenses.

## B. Dental Plaque: A High Biodiversity Class IV Biofilm

Dental plaque is a highly biodiverse Class IV microbial biofilm on teeth, different sites having a maximum thickness from about 50 μm to 5 mm.[131] Teeth are essentially immortal nonshedding surfaces which penetrate the oral epithelium, hence plaque is constantly threatening invasion of the body. Subgingival, smooth-surface, fissure, and approximal plaques are recognised as distinct[94] but in fact plaque is essentially microheterogeneous. It contains hundreds of species of bacteria in spatially and temporally organized interspecies structures and microcolonies. Its composition changes over time[144] yet it demonstrates considerable homeostasis and varies widely between intraoral sites and people.[91] A major growth mode may be clonal stochastic blooms limited by intra-biofilm interactions with the rest of the microbiota, nutrient supply, and host immunological and

**Table 1. Typical Objectives of Biofilm Studies**

*All biofilms*

To understand, control or prevent biofilm development.
To establish biofilm substructure.
To measure/lower biofilm viability.
To measure biofilm metabolism quantitatively.
To modify biofilm properties.
To model adverse biofilm properties, e.g., pathogenicity, corrosiveness.
To understand the interelationships between biofilm and planktonic cells.
To understand and enhance colonization resistance
To remove biofilm.
To understand and reduce resistance to antimicrobials.
To examine substrata/conditioning film effects on biofilm formation.
To examine effects of the biofilm on the substrata.
To establish the significance of the biofilm as a source of infection.

*Mainly specific to Monoculture/Group I biofilms*

To establish adhesion mechanisms to substrata and other species of bacteria and then interfere with them.
To establish biofilm-specific physiology.
To establish physiological changes during biofilm maturation.
To establish the composition and role of the extracellular matrix.
To prevent biofilm formation and growth and facilitate biofilm removal.
To identify and manipulate biofilm virulence factors in pathogenicity.
To examine the response to superinfection.
To establish mechanisms of strain/mutant variation in important properties.
To understand and prevent adaptation of cells to increased antimicrobial resistance.

*Mainly specific to High Biodiversity Groups III/IV biofilms*

To examine microbiota composition and succession during biofilm development.
To delineate community relationships.
To measure overall emergent properties, e.g., growth rates, pH behavior, calcification, pathogenicity.
To identify and quantify mechanisms underlying emergent properties.
To establish shifts in biofilm composition and behavior and the nature of any homeostasis in response to
    changes in the environment and perturbation.
To understand how biofilm composition, properties and behavior can be modified by environmental controls.
To establish the nature and causes of composition shifts in, e.g., disease states.
To investigate and prevent selection by antimicrobials of resistant organisms.
To establish the degree of relationship between different biofilms.

chemical defenses. Its thickness is also limited by removal through abrasion, oral hygiene procedures and, probably only to a very small extent, by fluid shear.[131] Plaque development, behaviour and pathogenicity is the outcome of a vast number of specific adherence,[83] nutrient,[20] and metabolic interactions between the bacteria present in the biofilm including; micro-environment modification, synergism, competition and antagonism, in a background of fluctuating, interacting physicochemical intra-plaque gradients. Plaque biofilms are sited in constantly changing, locus-specific oral environments dominated by laminar flows of thin films of saliva and gingival crevicular fluid (an inflammatory exudate), with occasional periodic exposure to host meals. Dental disease is correspondingly site-specific.[35] In vivo experimental protocols are limited, variability of results is high and their interpretation difficult.

**Table 2. Considerations Affecting Choice of Protocols and Technology**

Existing knowledge of the natural situation being modeled.
Objectives for study, and biodiversity of biofilm (Table 1).
Origin of bacteria used: laboratory strains, adventitious infections, natural flora.
Growth protocols: physical conditions (temperature, gas phase), nutrient regime
    (continuous or discontinuous), duration of experiment.
Environmental fluid composition: concentration, chemically-defined, undefined media, biological fluids,
    pH, $pCO_2$, etc.
Fluid flow and shear: turbulent, plug, laminar, dropwise "impact/laminar" flows.
Substrata: simple synthetic surface, model of natural surface, natural surface itself
Conditioning film: source and application regime.
Sampling protocol: replication, subsampling; continuous, pre- or post-experimental treatment
    and end point analysis, time series
Analytical techniques: and their integration (Section VIII).
Planktonic bacteria: their relationship to the biofilm, reinfection of biofilm
Experimental treatment protocols, i.e., of antimicrobials

The most realistic laboratory systems for biodiverse, complex Class IV biofilms are microcosms evolved from the natural microflora.[131,152] A microcosm is "a laboratory subset of the natural system from which it originates but from which it also evolves"[165] and features the genetic, temporal and structural heterogeneity of the natural system. It may be of any size. A microcosm after culture and evolution in an in vitro environment is a holistic, controllable model, closely related to the natural system and which embodies much of the original complexity.[165,166] They are structurally heterogeneous and their composition is likely to be irreproducible but emergent properties such as the "resting" pH may be reproducible.[142] Plaque microcosms are cultured from either saliva, plaque-enenriched saliva or plaque and the resulting microcosm biofilms self-organize under the environmental conditions set.

## IV. GENERAL ISSUES
## RELEVANT TO BIOFILM CULTURE TECHNOLOGIES IN VITRO

A wide range of potential biofilm culture technologies is available. A particular system obviously should be chosen to deliver defined objectives in the simplest manner. All have strengths, limitations and difficulties. The objectives amenable to study depend to some extent on the biodiversity of the biofilm (Table 1). Some are relevant to all biofilms. Some require Class I and II biofilms (especially monocultures), such as most detailed genetic and physiological studies. Some relate to high biodiversity biofilms where changing heterogeneous bacterial populations, and the involvement of unknown processes or particular species in the whole ecosystem properties of interest may play a role. The particular technology used determines the experimental protocols and analyses possible.

The structure and nature of biofilms (Fig. 1) dictate some key decisions about the culture technology to be used. Table 2 lists these items and some of the options. These decisions about the bacteria and their culture need to be made taking into consideration the natural situation and how closely it needs to be modeled in order to achieve meaningful and worthwhile results. To obtain worthwhile data, the bacteria and their culture conditions obviously also need to be relevant to the natural situation being modeled. The major initial choice is whether to use defined bacterial species monocultures or consortia, or use the natural flora to develop microcosms. Decisions on modeling of the liquid phase, its composition, mode of application and relation to the gaseous environment, the choice of substrata, conditioning film, and growth configuration (affecting sampling possibilities), will largely determine the culture system of choice. Further considerations include the inoculation procedure and whether to include continuous exposure to planktonic bacterial reinfection with recycle protocols, and how experimental treatments are to be applied. Other key issues include: available expertise for technologies and microbial systems which can be quite sophisticated, whether a number of different simultaneous growth conditions is required, and the amount of focus on biofilm morphology, flexibility of experimental protocols, number of replicates and type of samples, and the available techniques for biofilm and substratum analysis. Just how flexible or sophisticated an experimental system is needed is a significant consideration.

Culture systems that realistically model biofilms are almost all "open" systems, that is, they have a continuous flow of environmental fluids containing nutrients. This involves a significant increase in technical complexity compared to simple "batch" culture generated biofilms. These open systems can be classified as:

1. Preformed, nongrowing pseudo-biofilm models. A particularly powerful example is the cell-agarose laminar plug-flow system developed by Dawes to model quantitatively effects of biofilm thickness and thin-film saliva flow rates on biofilm pH responses.[87,88]
2. Those with internal recycling of liquid and bacteria such as chemostats with submerged surfaces.
3. Derivatives of tube reactors with a series of removable sample areas and turbulent or slow "plug" fluid flow such as the Robbins device,[98] operated with or without recycle protocols.
4. McGlohorn's open channel-based flow chamber, developed to allow a range of substrata samples under potentially thick biofilms to be removed without disturbing the balance and functioning of the system.[101]
5. Parallel plate flow chambers with predominantly laminar flow used frequently to allow direct observation of biofilms by microscopy.
6. Radial, predominantly laminar flow systems with a very thin fluid film such as the "artificial mouth" for dental plaque culture developed by Sissons et al.,[131,134] and radial flow reactors.
7. Systems with novel approaches to fluid application and/or shear forces. These include: (a) the radial flow fermenter with radially-reducing fluid shear, (b) the Rototorque with controlled turbulent fluid-shear,[22] (c) the "constant-depth film fermenter" which applies the fluid with a wiper blade that also removes the top surface of the biofilm as it grows,[118] (d) the "growth-rate controlled biofilm fermenter" where the media is perfused from the base of the biofilm and the rate of nutrient supply controls the biofilm growth rate,[51] and (e) the simple Sorbarod filter system which is an in-line system for growth of biofilms on cellulose fibers.[61]

The thickness of mature biofilms tends to be greater with high biodiversity biofilms.[27] This affects the technology which can be used. All technologies can be used with Class I and II biofilms. Development of thick biofilms restricts the use of narrow bore biofilm reactors and flow cells for Class IV and V biofilms to early stages of biofilm development. For these biofilms, technologies are required which allow essentially unrestricted growth, or constrain it physically or by shear. These issues also affect the duration needed for experiments which will range from a few hours to days to weeks depending on the

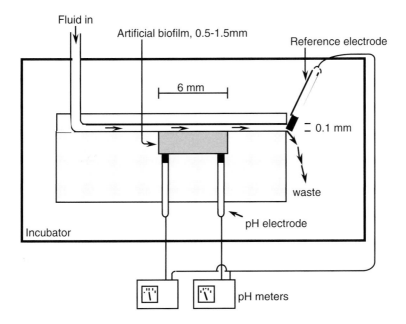

**Figure 4.** Dawes laminar flow pseudo-biofilm device.

experimental objectives, the time frame of biofilm formation in the natural and in the experimental system, and the technology available.

Options which can also be considered include:

1. Very simple closed-system "batch culture" biofilm methods. Despite their major limitations, they have specific uses which may replace the need for more sophisticated systems.
2. In vivo or *in situ* experiments, if feasible will, despite their difficulties, give results in the natural environment which for that reason alone will relate more closely to the natural situation. They may also involve the natural microbiota, thereby increasing their realism.
3. Computer-based modeling based on measurements made of living biofilms, although not involving biofilm culture *per se*, can summarize quantitatively what is known about particular properties of the biofilm, give powerful and useful insights into biofilm behavior, and generate and test hypotheses.[38,169] These are not further considered in this review.

## V. CONTINUOUS-FLOW IN VITRO METHODS FOR BIOFILM CULTIVATION

### A. Preformed, Nongrowing, Pseudo-Biofilm Biochemical Reactors

Biofilms which are constructed instead of cultured are particularly important in establishing metabolic parameters and biofilm-specific behaviour for mathematical modeling in relatively short-term experiments (minutes to hours). They usually involve single species of bacteria. Examples with oral bacteria have included cells suspended in agarose with a gel of defined composition, geometry and fluid flow (e.g., to study pH changes[86,87]), cells packed onto enamel,[84] or cells layered between membranes to study diffusion and metabolism.[37,116]

The Macpherson and Dawes cell-agarose laminar flow model system was designed to model the effect of the flow rate of the thin 100μm laminar-flow saliva film on plaque pH responses over several hours (Fig. 4). Substrate concentration and cell concentration using *Streptococcus oralis* and *Streptococcus vestibularis* (which metabolises urea) were used as microbial models.[87,88] The equipment consists of a rectangular flow chamber in an incubator, which has a rectangular block of low-temperature melting agarose containing the bacteria, and a plate fixed at 100 mm above the surface to set the thickness of the liquid film (Fig. 4). A pump moves the fluid along the chamber at estimated in vivo saliva flow rates, the equivalent of slow plug-flow in larger systems.[22] pH electrodes are positioned in the biofilm proximal and distal to the point of fluid entry and the different "biofilm" pH responses to substrates of the two electrodes are recorded as a function of variation in fluid film flow rate. These experiments have yielded realistic pH curves and allowed a quantitative assessment of the relative importance of the liquid film flow rate and several other key variables to the pH response induced by carbohydrate and urea. This general approach to constructing and analyzing the behaviour of an agarose-bacteria "synthetic biofilm" has considerable potential for deriving metabolic parameters of other biofilm systems in relatively short-term experiments.

## B. Chemostat-Based (Internal Recycle) Systems with Internal Submerged Surfaces

Experimental surfaces can be submerged in chemostats, which are nutrient-limited continuous stirred tank reactors (CSTR). These systems have proved particularly useful in oral microbiology.[1,8,16,60,71,79,80,93] In chemostats, there is a controlled fluid supply determining the nutritionally-limited growth-rate of a planktonic phase which interacts biologically with the biofilm being formed. Biofilm formation usually changes the steady-state conditions in the chemostat, which is then no longer operating under chemostat conditions. Potentially, a large number of replicate samples and a variety of surfaces can be used, but to compare biofilms grown under different conditions requires operation of separate chemostats. Sequential changes in nutrient supply and experimental treatments using fresh surfaces can be applied in long term experiments (over several weeks) although replacement may cause contamination problems. Conditions are likely to vary during the course of the experiment due to the presence of the biofilm, and selection of bacteria with changed properties may occur. Removing the surfaces outside the chemostats to give external biofilm growth reactors, e.g., the Robbins device and flow cells (Sections V.D.1 and V.E), increases options for modeling appropriate sample configurations, fluid hydro-dynamics, biofilm growth and analysis and application of experimental treatments.

## C. Robbins Device Tube Reactors and Derivatives

### 1. Robbins Device

The Robbins device is a simple, widely-used, successful method to establish and analyze surface-associated biofilms in tubes or pipes with a reasonably rapid and turbulent fluid flow.[63,98,108] It is a multiport sampling device with evenly spaced sampling ports, the samples being mounted on removable plugs to fit flush with the inside surface of the pipe. Multiple biofilm samples cultivated for varying lengths of time can be removed and analyzed independently without upsetting the balance of system. The fluid flow normally needs to be sufficiently rapid to ensure the absence of proximal–distal variation in conditions, the opposite situation to that obtained with the Macpherson and

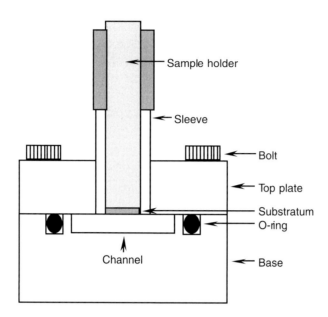

**Figure 5.** Modified rectangular-section Robbins device. (Adapted from ref.[75])

Dawes system (Section V.A.). Slow fluid flow allowing plug-flow conditions would be possible with a small bore reactor. The classical Robbins device has been widely used, for example, to examine colonization of engineered material surfaces in waste-water treatment and industrial situations. In most situations cases it has been operated as an external loop from a CSTR although this can be modified to reflect the situation being modeled. Its major advantages include simplicity of the device itself and the ability to remove samples without disturbing other samples. In common with several other biofilm systems, the classical Robbins device has a significant cost derived from its machined construction, especially if there is a need (not always required) to configure biofilm substrata to match the curvature of the tube so that biofilm growth is smoothly continuous with the wall of the tube. Development of boundary layers occurs, and there can be a problem with tearing the biofilm layer when samples are being removed. However, one further advantage of the system is that various modifications of the basic design can be made to overcome specific difficulties.

*2. Variation A: Use of Whole Catheters*

In this very simple approach, developed by Ladd and colleagues to study biofilm formation in catheters,[77] the entire catheter itself, in its sterile envelope, is attached to the effluent of a CSTR until colonized and then removed, rinsed, cut into sections and analyzed. Ladd et al.[77] analyzed *Pseudomonas aeruginosa* biofilms by microscopy and by radiorespirometry of glutamate to measure their sensitivity to tobramycin. This is a typical method for studying bacterial adhesion to stents.

*3. Variation B: Unmodified Glass Tubes*

Quite simple components can be developed into quite a sophisticated model. An example is the system set up by Byers and Characklis[13] with two cylindrical glass flow

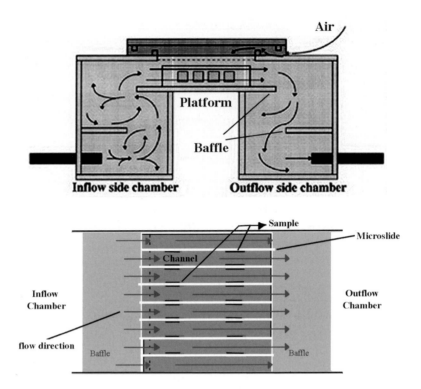

**Figure 6.** McGlohorn channel-based laminar flow device: (Top) Chamber; (Bottom) Top view of channels.

reactors in sequence connected to a CSTR with an internal recycle circuit, planktonic cells being generated within both the mixing vessel and recycle circuit. The first reactor contained fifteen glass cylinders 5.1 cm long and 1.27 cm ID in series, housed in a stainless steel outer sleeve. Individual glass cylinders which were periodically removed to measure biofilm accumulation from much larger surface areas than possible with plug samples in classical Robbins devices. The second flow tube was a cylinder 91 cm long and 1.27 cm ID with pressure taps at each end to measure the pressure drop. The system was mounted on a photomicroscope to allow image analysis of the area covered by the biofilm as it developed.

*4. Rectangular Section Flow Chambers with Sample Plugs*

A popular modification of the Robbins device is to use a rectangular section Perspex flow chamber with a removable top containing sample plugs with experimental substrata glued onto the end, i.e., for studying adhesion and biofilm growth on catheter sections.[5,42] This is either connected to a CSTR as in Part 2 above, or inoculated over several hours by recycle from a batch flow fermenter (which does not allow one to distinguish between ongoing adhesion or biofilm growth), or the plugs can be preinoculated by letting a dense bacterial suspension adhere for 1 h,[5] a variant of the adhered-cell outgrowth method (Section VI.3.).

As well as uses for examining biofilm development, this reactor (and other Robbins reactors) can be inserted into a disinfection rig to examine biofilm removal (clean-in-place) procedures.[5] A further option is to place the removed plugs with their attached biofilms into test procedures. An example of this is exposure of *P. aeruginosa* biofilms on a polyethylene plug to antibiotic (12 micrograms/mL gentamicin sulfate) and ultrasound delivered directly from the opposite face of a chamber at 10 mW/cm$^2$ and different frequencies.[122] By this technique, it could be shown that although ultrasound by itself did not reduce biofilm viability, it increased kill by gentamicin, especially at lower frequencies (e.g., 70 kHz).

## D. McGlohorn's Channel-Based Laminar Flow Chamber[101,102]

The McGlohorn chamber is an eight open channel flow chamber containing removable microscope slides to which samples of various materials as test substrata (e.g., titanium $10 \times 15 \times 1$ mm) can be attached with doubled sided tape to facilitate removal and biofilm analysis (Fig. 6). Special attention was paid to the fluid dynamics of the system to ensure uniform shear stress, flow velocity, and flow pattern environment of selected biomaterials test substrata. The flow chamber itself has identical mixing chambers at either end to allow reversal of fluid flow, each containing a baffle to allow turbulent mixing on fluid entry but reduced turbulence at the entry to the flow chamber. In each channel, the flow is laminar, at the same flow rate and degree of fluid shear, which, however, decreases along the flow cells due to the wall boundary effects. For samples to be comparable, they have to be positioned the same distance along the flow cells. On the first day the media reservoir is inoculated (e.g., with *S. epidermidis*) and fresh media is then supplied each day for 10 d at 37°C. The chamber has a removable cover so that samples can be withdrawn and replaced without disturbing the others, or the balance of the system. Below the lid there is free airspace, and the flow is in channels not tubes, so creation of anaerobic conditions would need further modification. Because the whole slide with sample is removed, the biofilm is not torn before analysis as may be a problem with the Robbins devices. It is simple and cheap to construct, straightforward to operate and allows time course studies, potentially on almost any material. Because of its open construction it could be used togrow thick biofilms, e.g., Class IV microcosms.

## E. Laminar Flow, Parallel Plate Flow Chambers

### 1. Continuous Flow Slide Culture

Development of noninvasive methods for the real-time observation of growing biofilms, based mainly on various forms of microscopy, has led to the development of a range of arrangements where a biofilm is cultured in a chamber between two parallel plates, usually microscope slides or coverslips, for study *in situ*.[18,19] Often multiple channels are present to increase replication. Of particular importance is the use of confocal laser scanning microscopy (CLSM) combined with fluorescent probes[1,9] A technique useful for quantifying early biofilm growth is Attenuated Total Reflectance Fourier-Transform Infra-Red Spectroscopy, requiring growth on germanium prisms but providing a quantitative IR spectrum of the developing biofilm.[60,70]

### 2. Gradient Culture Methods

Caldwell and colleagues have developed a series of gradient flow diffusion systems called microstats. The Wolfaardt microstat was based on diffusion of substances from

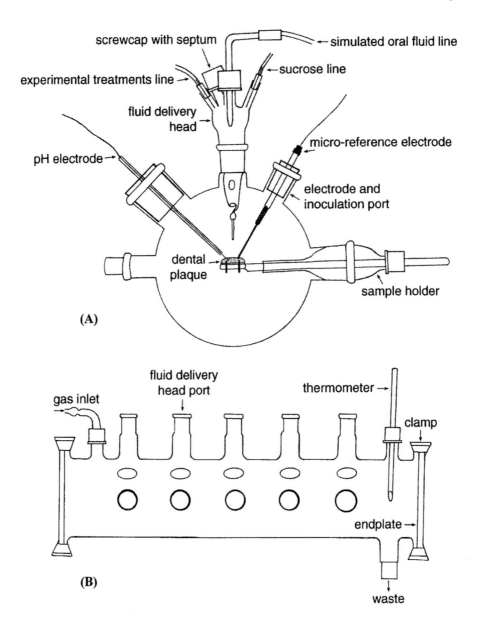

**Figure 7.** Multiplaquechannel Artifical Mouth: (A) Cross section of biofilm growth station; (B) Longitudinal section of culture chamber.

adjacent edges of square gels. Laminar wedge microstats use wedge-shaped channels to develop gradients across the flow cells and, due to plug-flow conditions, a decreasing concentration gradient along the cell. The laminar wedge microstat allows gradients to be switched on and off. The biofilm is cultured on microporous sintered glass. These techniques allow growth optimization in the presence of interacting gradients, and formation of specific consortia or communities of bacteria occupying habitats formed at particular points on the dual gradients.

## F. Laminar Flow, Radial "Artificial Mouth" Systems

The term "artificial mouth" is generally applied to biofilm culture systems modeling dental plaque on teeth, tooth tissues or artificial surfaces. They tend to feature independently-grown plaque biofilms with dropwise application of fluids. After the initial impact of the fluid drop, liquid flow over the biofilm approximates thin film laminar flow, as in the oral cavity. They allow flexible and precisely-timed, discontinuous fluid application regimes and can grow a number of replicate plaques under different controlled nutrient conditions. They essentially lack a planktonic phase and are limited in size to a small degree only by liquid shear. Biofilm size is restrained by the growth geometry as excess falls off, and by the nutrient supply. Because of stochastic structured biofilm growth, exact sample replication within each plaque is limited but subsampling and effluent analysis is usually possible. Growth and properties of plaques over weeks can be studied. Treatments with semisolids such as toothpaste can be applied outside the culture chamber.[175] Their facility for long-term continuous monitoring of biofilm pH behaviour by electrodes is a major advantage.[64,125,133,172]

The "Multi-plaque Artificial Mouth" (MAM) designed and refined by Sissons, Wong and colleagues, is a generic artificial mouth which has developed into a flexible plaque biofilm culture system appropriate to a range of experimental regimes and modifications (Fig. 7).[33,127,128,133,134,139,141,172] It was developed from designs by Russell and Coulter[125] which included sectioned teeth, pH and Eh electrodes and effluent analysis, and by Dibdin and colleagues[39] which had up to six replicated chambers for independent controlled plaque growth and control of nutrients by pumps. Current design features include long-term independent growth of initially five (Fig. 7B), but now eight, replicate plaques from the same inoculum, at the same temperature in the same gas phase.[134] There is a wide choice of possible plaque growth supports and substrata, including tooth tissues,[33,127,133] with a standard plaque support system comprising a 2.5 cm diameter Thermanox tissue-culture cover-slip to give radially symmetrical plaques, positioned on a glass ring with posts to standardize fluid dynamics.[133] There are three or more independent computer-controlled fluid lines to each plaque supplying simulated oral fluid, carbohydrate and experimental treatments.[172] Data acquisition from micro-pH and reference electrodes is also computer-controlled.[133,172]

In standard growth protocols a nutritional analogue of saliva-containing mucin, modified from Glenister et al.[54,134] (basal medium mucin, BMM), is supplied at 3.6 mL per hour per plaque, and every 6 or 8 h, 1.5 mL of 5% or 10% sucrose is supplied for 6 min to mimic meals. All these parameters can be varied. Growth of microcosm plaque biofilms is initiated by inoculation with saliva which has been enriched for plaque bacteria by the donor abstaining from oral hygiene procedures for 24 h. In many experiments the developing biofilms are reinoculated 3 and 5 d later to facilitate acquisition of species requiring an established biofilm plaque environment to colonize and grow.[134] Mixtures of pelleted and resuspended stationary-phase species are used to initiate biofilm consortia. Experiments can last from a few days to several weeks. After about 3 wk growth, plaques are convex, reach up to 5 mm maximum depth and 1 to 2 g weight. During growth, plaque wet-weight, continuous pH measurement and sampling for a range of chemical, biochemical, microbiological and microscopy measurements can be made on plaques still attached to their substratum.[131,137] The MAM enables investigation in many areas of plaque ecology and pathology with a wide range of protocols, and allows realistic testing of antimicrobial agents.[139]

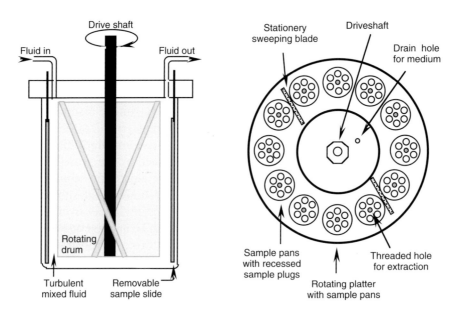

**Figure 8.** Shear-dominant fermenters: (A) Roto-Torque; (B) Constant-depth film fermenter.

Plaque microcosms have been shown to be similar to natural plaque in composition and structure, growth and pH behaviour. Complex fluctuating intraplaque pH gradients have been demonstrated directly.[133,141] The regulation of urease by urea, ammonia and arginine has been studied, enabled by the control of the environment and nutrient supply.[138] The plaque resting pH (no effect of meals) and hence pH range can be controlled metabolically through the continuous supply of urea at levels found in saliva.[142] Among a range of analyses of specific components and whole ecosystem emergent behavior, detailed analysis of growth patterns and rates has shown that they are the same as in vivo growth rates.[137] Plaque mineralizaton can be analysed.[134,171] We have also been successfully pursuing the complementary approach of constructing synthetic plaque-like consortia with major plaque species, using four to six species of putative caries pathogens.[127,128]

A different artificial mouth lineage involves rotation of a platter containing enamel blocks or hydroxyapatite disks under the nutrient fluid supply. In one example, the Orofax, slowly melted frozen saliva which was dripped onto hydroxyapatite disks mounted between slides in a slide projector carousel but it had problems with reproducible growth.[173] The other major system in current use, developed by Noorda et al.,[109] consists of a computer-controlled fluid supply and rotating platter which contains a series of enamel blocks or hydroxyapatite disks enclosed in a fermenter. Inoculation is from batch or chemostat culture.[32,110,129] The constant-depth film fermenter (CDFF-Section V.G.3) is a variant of this approach.

### G. Culture Systems with Distinctive Fluid Flow or Shear Conditions

#### 1. Radial Flow Reactor

In the radial flow biofilm reactor, liquid is applied to flow radially from the center of a circular disk-shaped fermenter with a gap of about 500 μm and a pre-inoculated bottom

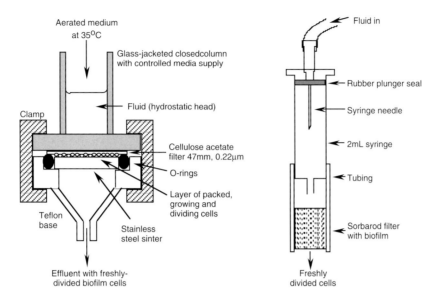

**Figure 9.** Growth-rate controlled devices: (A) Growth-rate biofilm fermenter; (B) Sorborod perfused biofilm model

surface. Fluid shear forces theoretically decrease in proportion to the square of the distance from the center. However, with significant biofilm growth, the inner biofilm affects the outer by deposition of cells, the plug-flow conditions, and increases frictional drag. This system is best used for examining the effects of fluid shear on initial biofilm growth.

### 2. Fluid Shear-Controlled Roto-Torque

The Roto-Torque (rotating annular reactor) is a CSTR reactor, essentially a chemostat having an internal cylinder with angled holes which rotates at controlled speeds creating very turbulent flow and hence generating considerable and uniform shear forces in the annular fluid volume (Fig. 8A).[22] The outside wall has up to 12 removable slides attached which enable analysis of biofilm growth and properties. The effect of shear forces on biofilm growth and properties can be examined, and used to limit biofilm thickness. Fluid drag due to biofilm growth can be measured directly with a torque transducer. External recycle additions can be used to increase mixing, adjust gas concentrations and to apply experimental treatments.

### 3. Mechanical Shear Controlled Constant-Depth Film Fermenter (CDFF)

The CDFF designed by Wimpenny and colleagues[118,167,170] limits biofilm thickness by mechanical shear. Biofilms develop in a set of 15 pans in a revolving platter, each containing 5 or 6 disk depressions of preset depth, usually 300 µm. Biofilm growth above the surface of the platter is removed by a Teflon blade sweeping the surface which also distributes the fluid supply, which had been delivered dropwise to the rotating platter. All this is mounted inside a custom-made fermenter. Until the maximum biofilm thickness is reached, there is a planktonic culture above the biofilm forming a second phase. Insertion of electrodes during growth is not feasible. The CDFF used for dental plaque studies can be regarded as a type of artificial mouth. In addition to providing constant size biofilms this system provides a large number of replicates for analysis. It can be operated as a two-stage system, i.e., with a nine-species inoculum established as a steady-state consortium in an anaerobic chemostat and aerobic CDFF biofilm growth to yield what are claimed to be pseudo-steady-state biofilms.[73] Class IV biofilms such as dental plaque microcosms can also be studied in this system.

### 4. Growth-Rate Controlled Biofilm Fermenter (GRBF)

The GRBF gives media-limited outgrowth of thin filter-deposited biofilms, analogously to the chemostat, but without the intervening planktonic phase present in chemostat systems.[50-52] It is a development of the Helmstetter and Cummings[59] method for selection of freshly divided bacteria. A cell layer (about $5 \times 10^9$ exponential phase cells) is packed under vacuum onto 47 mm diameter 0.22 µm pore size cellulose acetate membrane to form a "constructed" pseudo-biofilm. The filter is incubated with the cells face down over a steel sinter, with the rate of perfused media applied to the upper side controlled by the hydrostatic head. After the initial loosely adhered cells are removed, the cell division rate is proportional to the low rate of media perfusion, i.e., the biofilm is growth-rate controlled. An important feature is fluid flow through the biofilm instead of across its surface. This nutrient and fluid supply is probably common in mucosal surface biofilms. Analysis of both the biofilms during steady-state growth and the effluent containing a cohort of freshly-divided daughter cells enables a distinction to be made between biofilm-specific and growth-rate specific effects on such parameters as antimicrobial resistance. This approach works well with monocultures. Higher biodiversity biofilms might yield interesting results. The main differences between the growth rate-controlled biofilm fermenter and chemostat-related systems is its lack of intervening planktonic phase, fluid flow through the biofilm instead of across its surface, possibilities for cell-cycle analysis with the eluted cells, but limited sample size, replication and choice of substratum. This system has proved unsuitable for long-term experiments on *Staphylococcus aureus* or *P. aeruginosa*.

### 5. The Sorbarod Perfused Biofilm Model (SPBM)[61]

A simplified modification of the GRBF suitable for *S. aureus* and *P. aeruginosa* is based on biofilm growth in a Sorbarod filter of a cellulose fiber (Ilaron, Kent, UK) housed in a PVC tube and perfused from a syringe following inoculation with approx $10^9$ cfu bacteria.[61] The cells grow attached, probably as microcolony-mode biofilms to the cellulose fibers. Biofilm and eluted cells can be separately analyzed. Although the biofilms grow slowly with a reproducible and measurable growth rate, control by elution rate seen with the GRBF does not occur. Likewise, perfusion through the biofilm from its base no longer occurs. The advantages of the SPBM are simplicity, cheapness, ease of use and replication, and each filter yields three orders of magnitude more cells than the GRBF. High biodiversity biofilms have not been studied.

### H. Ancillary Equipment for Culture Chambers and Reactors

All continuous flow systems, except the most simple (e.g., SPBM, Section V.G.5.), require a flow of fluid from a reservoir which is connected by tubing, via a pump, to the reactor or culture chamber. When the fluid flows out of the reactor and chamber it either enters a waste system or a recycling circuit. All this equipment needs to be kept sterile for the duration of the experiment and is usually housed in an incubator. Mechanical timers or computer controls are required if there are timed fluid supplies (e.g., media), carbohydrate or experimental treatments. There may also be a need to control a humidified gas phase in the reservoirs and reactors. Many useful technical details are described by Drew.[40] We describe some which have proved useful for the MAM system.

## 1. Reservoirs

Reservoirs can be constructed from screw-cap laboratory bottles (Schott, Mainz, Germany). For the MAM simulated oral fluid supply (BMM), 2 liter bottles were modified with the addition of a 4 cm length of 6 mm diameter glass tube for attaching a 45 mm diameter vent filter, three GL14 and one GL25 screw cap vertical outlets at the top. One of the GL14 screw caps takes 7 mm glass capillary tubing to the bottom for the fluid exit line, another has 7 mm capillary glass tubing to the base leading to a 16 gauge syringe needle with its point ground down, for sterile sampling of the reservoir, and the third is closed off with a silicone septum but is available to install a gas or fluid addition line. The GL25 has a silicone septum for the injection of supplements or addition of more fluid through a needle. Reservoirs for additions such as carbohydrate or experimental treatments are modified 500 mL laboratory bottles with a vent and two GL14 screw caps, one for a line out and the other with a silicone septum.

## 2. Tubing

The tubing for most of the fluid lines is Silastic silicone (3.8 mm OD, 1.67 mm ID or 2.16 mm OD, 1.02 mm ID) with some connections of butyl rubber. Quick-connects are made from 2.5 cm × 12 gauge and 5 cm × 14 gauge, or from 2 cm × 15 gauge and 4 cm × 18 gauge hypodermic tubing, for the larger and smaller bore tubing respectively.[40] The larger bore tube of the connection acts as a protective sleeve for the smaller bore tube which makes an internal seal on the silicone tubing. For sterility, the reservoirs and lines can be autoclaved fully connected but we have found that if the individual quick-connects are separated, wrapped in aluminum foil and autoclaved, they can be connected aseptically, maintaining sterility even for media lines.

## 3. Pumps

Generally, peristaltic pumps are used. Where there is a need for low flow and critical control over periods of weeks, it is necessary to use very high performance pumps, such as the Watson-Marlow 505 Series pumps with cassette heads (Watson-Marlow Limited, Falmouth, UK) and Marprene pump tubing (Watson-Marlowe Limited, Falmouth, UK). These pumps are used for the MAM to pump the simulated oral fluid. For less critical pumping, such as reagent addition, or for single channels, there are high quality alternative peristaltic pumps. Where timed discontinuous fluid applications are needed, mechanical or electronic timers can be used to directly switch the pumps on and off. Control by computer programs, such as LabVIEW (National Instruments Corporation, Austin TX),[172] is more complex but more satisfactory.

## 4. Gas Supply

For systems where there is a gas phase above the biofilm and the fluid reservoir, a gas supply is required, e.g., 5% carbon dioxide in nitrogen is used for the MAM. To maintain water saturation of the system, the low pressure gas passes through a vent filter, then is pre-humidified by bubbling through sterile water.

*5. Incubators*

Many culture systems can be temperature controlled in a standard incubator or with a water jacket.[40] Construction of a custom-designed incubator of 6 or 10 mm thick Perspex allows unobstructed access to important parts of the reactor and appropriate entry of lines kept neatly together—in the MAM there are 24 tubing lines and potentially 16 electrode (pH and reference) cables.

## VI. CLOSED SYSTEM "BATCH CULTURE" BIOFILM METHODS

These methods are very simple and include: the analysis of colonies on agar plates, wall-growth on test tubes, on beads or wires, in microplates, and on various substrata. Only small amounts of biofilm are produced by most of these methods and they have the major disadvantage of changing culture conditions which usually are far removed from the natural biofilm environment. They are generally suitable for studies of monocultures or Class II biofilms only.

### A. The Bacterial Colony

Colony formation on agar media in a Petri dish is in practice the simplest and most common in vitro analogue of a biofilm.[164] Microbial functions elicited by growth on substrata and or by high-density of cells will be manifested. Gradients of substrates, $pO_2$, metabolites, etc., are generated similarly to biofilms, and complex structures can form, even with *Escherichia coli*[126] or *Bacillus subtilis*.[95] Usually colonies are pure mono-cultures but consortia of interdependent bacteria can also form colonies. Colony morphology is strongly affected by changing cell surface structures and extracellular polymers. A useful example is slime production by CNS which can be detected by Congo red uptake and colony blackening from brain heart infusion-5% sucrose-0.08% Congo red media, a robust, reproducible and sensitive method which yields viable colonies for further analysis.[44]

### B. Wall-Growth Methods

The wall adherence assay in culture tubes reported by Christensen and colleagues is a qualitative macro-method for detecting the existence of bacterial biofilm in vitro.[28] Adherent growth by *S. epidermidis* lining the inner surface of the tube is demonstrated by safranin O or trypan blue staining. Further criteria such as concentrated surface growth at the meniscus has proved useful for oral bacteria.[128] Growth on nichrome wires can also be assessed.[97] A microplate modification of the test for slime production by CNS adapted to a spectrophotometric micro-test assay was reported by Pfaller et al[120] and used to demonstrate reproducible quantitative differences in slime production among different strains and species of CNS. It may be useful in studying the effects of conditions such as antibiotic exposure on slime production. See also Section VIII.C.2.

### C. Adhered-Cell Outgrowth Method

In 1987, Prosser et al.[121] reported a simple method to produce biofilms in vitro. Cells (*E. coli*) were grown overnight on Mueller-Hinton agar, resuspended in buffer and dispensed on 0.5 cm$^2$ catheter disks. The disks were incubated for 1 h at 37°C, washed, and incubated in broth for a further 20 to 22 h, by which time thick biofilms were established.[121]

### D. Growth on Glass Beads

To get a relatively large amount of biofilm, Giridhar et al.[53] placed 10 gm of glass beads in a CNS culture (100 mL of TSB-1% glucose-5% NaCl) to increase the surface area for bacterial adhesion and subsequent biofilm formation. It was grown unstirred at 37°C for 4 d, the beads separated by filtering, the biofilm extracted, and collected by filtering with a 0.45 μm filter. A more sophisticated nonbatch culture variant of this technique involves glass-bead columns with media percolating through them, used successfully for oral microbiota microcosms, a Class IV biofilm.[150] Any particulate substratum can be used. At larger scales, gas-lift fermenters give more efficient mass-transfer and biofilm growth.

### E. Batch Culture Biofilm Growth on "Natural" Surfaces

These involve immersion of materials of interest in batch cultures of bacteria. Colonization and growth of *P. aeruginosa* on the surfaces of longitudinal halves of 1.0 and 1.5 cm catheter sections inoculated in a synthetic urine actively-growing planktonic bacteria has been described by Nickel and colleagues.[108] This technique produces 50 replicates at a time for antibiotic sensitivity testing. In dental caries models, hydroxyapatite beads and blocks of tooth enamel have been incubated in batch cultures of bacteria in standard culture media and $Ca^{2+}$ solubilization measured.[26] These systems, however unrealistic the bacterial growth state, allow comparative studies of selected properties and antimicrobial strategies.

## VII. IN VIVO METHODS FOR BIOFILM CULTIVATION

### A. Specimens Harvested from Human Patients

Biofilms to be studied can be harvested from the surface of infected prosthetic devices (such as joint prostheses, artificial heart valves, or vascular and urinary catheters, urethral stents), infected or dead bone surfaces, and different locations of dental plaque (smooth surfaces, approximal surfaces, fissures and subgingival pockets).[56,123] These biofilms are a sample from the real human condition, but suffer the drawback that often the biofilm structure is disrupted, the sample is pooled from several distinct sites thereby averaging out real differences, and it may be complicated by the presence of proteins, tissue cells or debris, compared to samples from an in vitro controlled experiment. Another type of sample, which can be analyzed to reflect in vivo activities, is typified by saliva or salivary sediment.[74,132] Salivary sediment is the centrifuged mixed salivary bacteria that are pooled from all oral microbial biofilm ecosystems, including dental plaque and biofilms on mucosa. Most of the bacteria are clumped or adherent to epithelial cells, and in the pelleted state form a nongrowing pseudo-biofilm (*see* Section V.A.). The pooling of bacteria from heterogeneous locations in salivary sediment is an argument for using salivary bacteria to overcome source heterogeneity problems when initiating dental plaque microcosms, allowing the environment to "select" the appropriate species.[131]

## B. Human In Vivo, In Situ *and Intraoral Models*

Apart from removal and analysis of infected prosthetic devices, opportunities to carry out in vivo or *in situ* studies in humans are severely limited. Studies are confined to the alimentary tract and mainly to the most accessible region, the oral cavity. Intraoral model systems include plaque growth on plastic strips,[12] under bands,[9] on enamel attached to teeth[124] and to intra-oral appliances in different configurations.[67,89,112,117,176] Compared to the study of plaque in vivo, these models increase access and sampling possibilities, and those involving removable intraoral appliances allow experimental treatments.[92,113,176] One intra-oral model of dental caries involves *Streptococcus mutans* layered between enamel blocks in a palatal appliance.[176] In another model, bacterial monolayers were applied to the enamel followed by growth of the natural microflora.[81] Intraoral models benefit from but are limited by their siting in the natural, site-specific, uncontrolled oral environment.

## C. Animal Models

Most of the in vivo animal models of bacterial biofilms are foreign body infection models. Implants together with adhered bacteria are implanted into a subcutaneous tissue pouch, peritoneal cavity,[14,47] or the medullary canal of bone.[68,96] Bacteria can also be injected into the implant site after implantation. For in vivo study, implants are often left in animal tissue for days, weeks, or even months.[14,47]

## VIII. METHODS FOR ANALYZING OF BIOFILMS

Techniques for biofilm analysis are both numerous and varied. This section could not possibly describe all of them and selectively outlines a few fundamental techniques and briefly only mentions some others. Part III (Chapters 13 to 19) contains further description and the detail of many of these. For biofilms more complex than Class I monocultures, especially Class IV and V biofilms such as the normal human microbiota, techniques are required to assess their taxa composition, population structure, biodiversity and similarity, and these are also outlined.

## A. *Thickness, Weight, Area, Density Measurements*

Biofilm thickness, area, wet-weight, and dry-weight measurements and density estimates are basic parameters in biofilm studies. Thickness measurement by light microscopy is usually effective but may not work with thick biofilms.[4,156] The biofilm is placed on the stage of a microscope which has calibration scales on the fine control and the objective is lowered until the biofilm surface is in focus and the fine adjustment dial setting of the microscope recorded. The visibility of the biofilm surface can be enhanced by blowing fine powder (e.g., fine talcum powder) over the surface.[168] The microscope objective is then focused on the substrata surface, preferably in an area with no biofilm. The difference in fine adjustment settings is compared with a calibration curve constructed to allow for refractive index differences, and the thickness calculated. Several determinations may be needed to establish a thickness profile.[4] A simple manual gauge-needle method[62] and an electronic probe to measure biofilm thickness[90] have been described. Properly prepared SEM samples (e.g., freeze-dried cross-sections of a Foley bladder catheter) enable an estimation of biofilm thickness and also reveal layering of embedded bacterial cells.[48] Biofilm thickness measurement allows volume calculations in conjunction with area measurement and is important in estimating likely mass-transfer diffusion limitations for substrate access and metabolite removal at different levels of the biofilm. Biofilm area is usually known for non-patchy biofilms. For patchy biofilms it can be either measured directly with a calibrated grid (i.e., in a microscope eyepiece) or by image analysis techniques.

Biofilm wet-weight is a useful biomass measure, especially of samples on tared substrata as it is a very simple, quick procedure. In the MAM system, the whole biofilm on its preweighed substratum is rapidly weighed to 0.1 mg with a five-place electronic balance to give values which correlate closely with dry weight and total protein, allowing growth curves as biomass accumulation to be constructed.[137] For dry weight estimation, biofilm samples can be dried at more than 60°C (60°C, 103°C, and 105°C have been reported) for several hours,[63,104,156] or placed in a desiccator over $P_2O_5$ for several days until constant weight is achieved.[137] The substrata can be either weighed before biofilm growth (with the assumption of no substrata solubilization) or cleaned, dried, and weighed again in order to derive the dry biofilm mass.

Biofilm density (r) is usually measured as the dry-mass per unit volume ($kg/m^3$). Measurements of thickness, surface area, and dry weight can be used to calculate the density. If there are both wet and dry-weight measurements on the same sample, an approximate density measurement may be made by assuming that the volume of the biofilm sample is the same as the water volume estimated as the wet-weight minus the dry weight. For comparative purposes, an area film density can be calculated as dry-weight per unit of substratum area.

## B. Morphology and Substructure

Light microscopy techniques are the fundamental set of methods for biofilm observation and measurement, either directly *in situ*[156] or of histologic sections.[21] CLSM is a particularly important biofilm analysis technique but is restricted to 50 mm to 200 mm thick biofilms, depending on instrumentation (*see* Chapter 15). Scanning and transmission electron microscopy have proved invaluable for examining the structure of biofilms.[43,56,82,104] For electron and optical microscopy, care is needed during sample preparation as bacterial biofilms are highly hydrated and thus readily deformed. The Electroscan wet scanning electron microscope (Electroscan Corporation, Wilmington, MA) avoids this problem and visualizes the matrix surface but does not penetrate well into the biofilm.[151] Chapter 15 and reviews by Gristina and Costerton,[55,56] and Ladd and Costerton[76] describe SEM and TEM sample preparation. NMR (nuclear magnetic resonance), and ATR-FTIR (attenuated-total-reflection/Fourier-transform-infrared-spectroscopy)[60,69] are also newly developed valuable methods for morphological observation of biofilms.[6] Availability is often a problem with these newer and expensive technologies.

## C. Measurement of Biofilm Biomass and Extracellular Matrix (ECM)

### 1. Chemical Analyses of Biomass

"Biomass" has a variety of definitions associated with different approaches. These range from wet or dry-weight measures of the whole biofilm, to measurements which focus primarily on the cell content, to those which focus on cellular activities or viable cell biomass (Section VIII.C.3. below). Viability is also a term with a range of meanings. Viable cells have metabolic potential but nonviable cells also may have structural functions or serve as a nutrient source. Most biomass measures focus on cells, in effect defining biomass as living biomass, and aim to distinguish cells and their contents from the ECM, frequently a useful distinction.

Direct measures of cell contents include total protein, DNA, lipids, and enumeration of total or viable cells, if necessary, in a dispersed biofilm sample. Total Folin protein is a reliable measure which we favor as a biomass base. The procedure requires solubilization of a biofilm sample (a wet or dried sample or a 7% trichloroacetic or 0.5 M perchloric acid precipitate, i.e., from mineral extraction or other procedure) in 500 µL of 1.0 M NaOH at 35°C for 18 h. Sodium hydroxide at 0.5 M does not dissolve all the protein and higher solubilization temperatures run the risk of degrading tryptophan. A further modification[136] of the Peterson Folin reaction[119] is then carried out. This method allows duplicate reactions to be carried out using as little as one milligram wet-weight of biofilm.

Total protein measurement is not specific to cells; there are proteins in the ECM but generally at lower levels. Total DNA is more cell-specific and can be measured by a variety of reactions such as the orcinol reaction (insensitive) and a range of fluorimetric reactions. Various lipid fractions show either a close correlation with total biomass (e.g., lipid phosphate or phospholipid-associated fatty acids, PLFA) nutritional status (poly-$\beta$-hydroxyalkanoic acid) or may reflect the population structure by quantitative analysis of different "signature" PLFAs.[163] Total carbohydrate using, for example, the Dubois phenol-$H_2SO_4$ reaction, is commonly used to increase focus on ECM polymers which are predominantly carbohydrates (*see* Section VIII.C.2.). Other compounds analyzed include cell-wall components (lipopolysaccharide, muramic acid) but many of these vary with cell composition (*see* Part III, Chapters 13 and 17). A variant of this approach is to specifically radiolabel various components such as DNA and carbohydrates but care needs to be taken to label for a sufficient period to equilibrate the radiolabel in precursor metabolic pools to avoid underestimation, potentially a particular problem in thick mature complex biofilms (*see* Section VIII.C.2. and Chapter 16).

*2. Extracellular Matrix and CNS Slime*

Variants of wall-growth biofilm systems (Section VI.B) have been used to measure the proportion of ECM in CNS biofilms, a virulence factor. Most methods distinguishing ECM from bacterial cells do not completely discriminate between the two biofilm phases. Tsai and colleagues[157] showed that toluidine blue (and safranin O) are solubilized by 0.2 M NaOH (85° C for 1 h) from the ECM and cells of wall-growth biofilms in a standard batch culture after fixation with Carnoy's fixative. The solubilized dyes can be quantified spectrophotometrically to yield an estimate of activity in slime production. Growth of CNS in a chemically-defined medium containing $^{14}C$-glucose leads to heavy labeling of the extracellular polysaccharide, a major component of the slime.[65,66] Van Pett and colleagues introduced dual radiolabeling of biofilms by $^{3}H$-thymidine of bacterial DNA and $^{14}C$-glucose for ECM, respectively, a simple technique to quantify the extracellular matrix of different strains of *S. epidermidis.*[158] An immunochemical method targeted to a water-soluble ECM component has been developed to analyze the ECM of *S. epidermidis.*[75]

## D. Viability

Viable cells are those with the ability or potential ability to metabolize, grow and replicate. Bacteria can have a spectrum of increasingly slow metabolism and inactive states before permanent cell death. Different techniques discriminate at different points in this spectrum. There are basically three approaches: culture, differential staining and detection of metabolic activity. All have limitations. Total cell enumeration is required to establish percentage viability.

Culture on solid media to establish colony forming units (cfu), or dilution in liquid media to estimate cfu by least probable number procedures requires complete dispersion of the samples with no destruction of bacteria, often impossible with biofilms, and is problematic with chain-forming bacteria, e.g., streptococci. With complex biofilm populations, slow growing bacteria may be overgrown on the plates by faster growing bacteria. Elective plates place a cultural window on a complex flora; those outside the window may greatly exceed those inside. Bacteria such as staphylococci can exist in a low activity metabolic state where they are killed by the "step-up" conditions of the rich culture media normally used.[149] Hence culture gives an underestimate of viability which may be substantial, e.g., cultured plaque biofilms from smooth tooth surfaces (exposed to saliva) show a rapid hundred-fold increase in cfu, due mainly to a viability increase.[103]

With intact biofilms, microscopy following staining by fluorescent dyes is now widely used (*see* Chapter 15). Techniques include staining all cells (e.g., ethidium bromide), staining viable cells by dye exclusion, or by a metabolic activity such as fluorescein diacetate hydrolysis, or using fluorescein (which cannot permeate intact cell membranes).[162] A nonfluorescent alternative is formazan production from 1-iodonitrotetrazolium indicating a functioning electron transport chain.[76] Biochemical methods of viability estimation include measurement of cellular ATP or adenylate charge, an index of the metabolic integrity of energy systems,[170] and short-term or pulse radiolabelling of cell-constituents such as DNA or protein (note caveats concerning metabolite pools—*see* Chapter 16).

## E. Whole Biofilm Activities

Whole biofilms can be regarded as an entities, and their overall functioning measured. Approaches include using microelectrode or biosensor technology to measure appropriate activities, measurement of total activities or metabolic rates per unit of biomass, and use of bioreporters.

Microelectrode and biosensor technologies can be used in two ways. If the sensor is an appropriate size, robust, stable, and difficulties of calibration of the response in the presence of the biofilm can be accommodated, the activity can be measured in the intact functioning biofilm, e.g., the studies of pH responses in microcosm plaques by Sissons and colleagues.[133,135,140-142] Esophageal pH (Microelectrodes, Bedford, NH) and reference (Diamond-General Corporation, Ann Arbor, MI) electrodes provide a sensitive system which can function when installed in thick biofilms for at least 2 wk.[133] The major problem is drift due to the changes in reference electrode function. These studies have demonstrated direct formation of substantial pH dynamic intrabiofilm pH gradients in response to substrates like sucrose[133] or urea.[140] Stable electrodes, for such activities as temperature and redox potential, can be used directly or are increasingly converted into biosensors.[15] The alternative to implanting electrodes long-term is to sample biofilms, and measure activities such as $pO_2$ gradients away from the growth system, or by endpoint analysis of a sectioned, perhaps, fixed sample.[170]

Measurement of quantitative rates of metabolic processes is a fundamental requirement for understanding the functioning of biofilms and computer modeling of their activities.[38] The activity of enzyme systems, and their response to changes in biofilm nutrients and other environmental conditions is of particular importance. An example of this approach is our studies of urease levels in microcosm dental plaques to examine their regulation by environmentally-supplied urea and ammonia.[138] Metabolic fluxes are usually measured by pulse-labeling. Potential enzyme and metabolic systems for analysis are almost endless but a focus on central metabolic process and enzymes is probably most productive. These processes include nutrient acquisition, energy metabolism, DNA, RNA and protein synthesis, biofilm defense, and activities of particular significance such as control of pH, elaboration of virulence factors, etc.

A newer approach, if the bacterial composition of the biofilm can be controlled or at least manipulated (mainly Class I to III biofilms), is to directly monitor the activity of bacteria in the biofilm with natural or more usually, genetically engineered reporter activities,[15] particularly useful if combined with CLSM (*see* Chapter 15). One common reporter includes the *lux* gene of the firefly luciferase, which requires $O_2$ for activity, a problem in thick biofilms. Another reporter is the gene for Green Fluorescent Protein from *Aequorea victoria*.

Overall activities of importance relating to specific biofilms also include pathogenic mechanisms in human biofilms, substrata attack such as corrosion of metals, and effects distant from the biofilm caused by liberated biofilm products, or the biofilm providing a nidus for downstream infections, in fact, any activity which affects the health of a host or functioning system. Because these activities are highly specific to the particular biofilm system being studied, the approaches taken depend on the particular system, and the questions to be asked are usually evident to the investigator. These are outside the scope of this review.

### F. Analysis of Biofilm Population Abundance and Structure

In biodiverse biofilms, issues such as pathogenicity usually relate more to shifts in the proportions of species, bacterial populations or communities already present rather than colonization by an exogenous pathogen.[7,107,143] In highly biodiverse biofilm systems the experimental approaches are essentially the same as those used for whole microbial ecosystems.

#### 1. Low Biodiversity Biofilms

If the biodiversity is low (Class I to III biofilms) techniques aimed at identifying single species of bacteria are directly applicable. These include: 1) Cultural analysis using either selective media (which are seldom completely specific) or elective media followed by isolation of usually 30 to 50 colonies, and identification of isolates using mainly commercial kits. The "Marsh" 10-species consortium dental plaque model system is an example of a powerful experimental approach using cultural analysis to study interrelationships among microbiota and made possible by selection of strains allowing a total cultural analysis of the microbiota.[10,11,99] 2) Identification frequently uses labeled antibodies, either monoclonal antibodies to specific epitopes, potentially varying among strains of

the same species, or polyclonal antibodies from various species which may need their specificity enhanced by pre-adsorption to bacterial species closely related to the target taxon and which might cross-react. Common labels are enzymes giving solid products, (e.g., peroxidase, phosphatase), fluorescent labels (e.g., fluoroscein isothiocyanate, as described in other chapters, especially Chapters 9 and 14) and for electron microscopy, electron-dense labels such as gold. Further details are described in Chapters 9 and 12. A range of molecular biology techniques are also used, principally, 1) polymerase chain reaction (PCR) techniques, usually with ribosomal DNA primers, and 2) DNA hybridization with labeled probes ranging from whole cell DNA to oligonucleotides. These become increasingly important as the biodiversity increases. *In situ* identification using labeled antibodies or DNA probes is particularly useful in delineating intermicrobial relationships in biofilms.[32,130]

## 2. High Biodiversity Biofilms

In Class IV and V biofilms with potentially hundreds to thousands of species present, analysis of population abundance can present considerable difficulty. In these systems, much of the microbiota may be unspeciated.[78,160] The boundaries between bacterial species, genus and higher taxa is uncertain, even defining what a microbial species consists of is uncertain.[147,161] This has led to nontaxonomic analyses using a variety of molecular biology and functional characterization techniques to examine structure, diversity and similarities of ecosystems and high biodiversity biofilms. None of the techniques are without theoretical limitations and practical difficulties.

Conventional techniques usually are targeted against one or a limited selection of the flora present. For example, characterization of 50 colonies cultured on an elective medium from a biofilm such as dental plaque, containing possibly 500 cultivable species and perhaps as many again "not-yet-cultivable" species, identifies only relatively high abundance species (except for excluding high abundance "not-yet-cultivables").[78,146,160] It does not yield a detailed or particularly accurate analysis of the microbiota. However, even simple analysis using selective media or antibodies may be useful as the changing abundance of just a few important species in these biofilms will reflect the overall biofilm response to environmental change.

## 3. Molecular Biology Techniques

There is a vast range of these techniques primarily involving different combinations of gene cloning, PCR amplification, and DNA:DNA hybridization and DNA sequencing. It is outside the scope of this review to describe them. Reviews include those of Stahl,[45,147] Torsvik,[154] and Fuhrman.[46] Here we will outline some of the basic issues involved.

There is still a lack of coherence between systematics and taxonomies based on cultural isolation and phenotypic speciation, and those based on DNA similarity (and RNA similarity) which measure phylogenetic relationship. Phenotypic speciation relates directly to ecological function, DNA-based taxonomic methods as yet do not. The boundaries delineating species, genus and higher orders of relationship are not agreed for both approaches. Hence DNA studies give phylogenetic species relationships distinct from current functional phenotypic assessment of relationships between species and the main reason for classification in these biofilms and ecosystems is to illuminate their functioning and ecology.

There are technical limitations and systematic bias with these techniques which need appropriate controls and interpretation. First is the difficulty of obtaining a representative sample of DNA from biofilms containing a wide variety of bacteria and viability states because differential breakage is likely to occur. There may be difficulties with interference from the substrata or the biofilm environment as contamination by impurities may affect the subsequent analysis. PCR amplification of different gene sequences is unequal, leading to bias which undermines abundance estimates based on amplified products. Hybridization stringency conditions depend on the melting temperature, in turn dependent on the GC ratio, and impurities, so that standardized conditions are needed with careful controls. Despite these limitations, molecular approaches have provided a powerful, widely-used diverse set of tools for the microbiologist studying biodiverse ecosystems.

The most commonly used PCR techniques focus on appropriately conserved or variable regions of 16sRNA species. Specific functional genes are similarly studied and have a closer link to ecology. PCR is often an integral step in hybridization technologies using small probes based on specific genes. Several valuable techniques are based on whole genome DNA probes: analysis of renaturation kinetics to establish genome biodiversity; reciprocal whole ecosystem DNA hybridization to quantify the overall degree of DNA similarity (best for very similar or dissimilar communities); multiple probe analysis of membrane-bound ecosystem DNA (e.g., the "Checkerboard" hybridization procedure) and the converse, reverse sample genome probing (RSGP). "Checkerboard" DNA–DNA hybridization was developed for dental plaque analysis.[145] It hybridizes at high stringency, lanes of forty digoxigenin-labeled probes of bacterial species of interest at right angles across 28 lanes of denatured biofilm DNA plus two composite bacterial standards at $10^5$ and $10^6$ cells per lane. The sensitivity threshold (approx $10^4$ cells) eliminates detection of minor cross-hybridizations although the specificity needs to be established for each probe. The resulting quantification of forty species is relatively precise, well controlled, and after probe preparation comparatively simple, quick and cheap. The RFLP RSGP technique is similar except that the probes are membrane bound and dot-blot hybridization procedure is used.

### 4. Phospholipid-Associated Fatty Acid Analysis (PLFA)

A chemical technique for analyzing population structure is based on PLFA analysis by mass spectrometry, a technique established as a taxonomic tool based on the differences in the identity and proportion of PLFA in different species.[163] A single PLFA analysis of a biofilm community, such as dental plaque, by comparisons with PLFA in known component species allows deconvolution into (currently) 11 groups of species.[114] Changes in nutrient supply to microcosm dental plaques have been analyzed by this procedure which has considerable potential as an alternative method for microbial population analysis independent of DNA analysis.[115]

### 5. Nontaxonomic and Fingerprint Analysis of Community Population Structure

Some DNA and RNA techniques do not relate directly to species composition but provide a characteristic pattern of the ecosystem DNA such as a gel banding pattern. Restriction fragment electrophoresis patterns, useful for monocultures and low biodiversity systems, are too complicated with high biodiversity DNA with current

techniques. For low biodiversity biofilms, the low molecular weight RNA's (5s ribosomal and tRNA's) provide a stable characteristic pattern.

Denaturing gel gradient electrophoresis (DGGE) is based on regional melting point differences in DNA sequences as the DNA migrates into increasingly denaturing regions of the gel. DGGE analysis of PCR-amplified 16s rDNA fragments yields a rapid characterization of major species in the community (within the limitations of bias in amplification and chimera formation) and the products can be isolated and sequenced. Both these techniques or even simple polyacrylamide gel electrophoresis can be combined with taxon-specific probes to estimate the phylogenetic distance of the ecosystem DNA from the probe DNA by detection of heteroduplex formation.[36]

The functional diversity of biofilm ecosystem behavior can be measured as the biofilm's overall metabolic activities.[174] Measurement of microbial ecosystem metabolic profiles using Biolog plates (Biolog Inc., Hayward, CA) also developed originally as a microbial identification system, is a technique developed established in soil microbial ecosystems.[49,58,174] The metabolic capabilities of dispersed microbial communities on special microtitre plates containing 95 different substrates is assessed by the appearance of tetrazolium violet as a result of microbial electron transport activity. Subsequent cluster, principal component and community ordination analysis of overall similarity, allows comparisons of the "functional diversity" of whole microbial ecosystems.

### 6. Community Structure

Microbial communities are often vaguely defined and overlap with concepts of ecosystem, consortia, and populations. For the present purposes, communities are defined as specific assemblages of bacteria with direct structural and functional interrelationships. This definition is more general than the concept of consortia and, for example, would apply to both specific assemblages in single species biofilms or colonies[41,95] and complex communities carrying out macromolecule metabolism. Evidence for communities in biofilms is mainly derived indirectly from identification of structural patterns or demonstration of specific coaggregation relationships between species. Use of CLSM with fluorescent probes or other noninvasive techniques for studying structure within the biofilm is essential for studying spatial relationships. Metabolic interrelationships of community members, such as food chains, also suggest that direct functional links exist. Criteria and techniques developed by Caldwell and colleagues in this difficult area of detection and analysis of community formation include concepts of self-organization (autopoesis), synergy, homeostasis and communality (*see* Chapter 10). Detection of discontinuities along environmental gradients with coordinate changes in species abundance (ecotomes) is major evidence for communality.

## IX. CONCLUDING REMARKS

The field of biofilm research is developing rapidly. Use of laboratory biofilm culture systems with careful consideration of the key features of the natural situation being modeled, precise definition of experimental objectives, and appropriate technologies, yields exciting insights into relationships between humans and bacteria. Although the wide range of biofilm culture and analysis technologies available all have technical constraints which limit the validity of findings, they are a powerful set of tools to address some of the most important questions in microbiology.

# REFERENCES

1. Allison C, Watson GK, Singleton S, et al: Modulation of physiological responses and population structure of mixed culture oral biofilms grown *in vitro*. *Adv Dent Res* 11:191, 1997

2. An YH, Friedman RJ: Prevention of sepsis in total joint arthroplasty. *J Hosp Infect* 33:93–108, 1996

3. An YH, Friedman RJ, Draughn RA, et al: Bacterial adhesion to biomaterial surfaces. In: Wise DE, ed: *Human Biomaterials Applications*. Humana Press, Totowa, NJ, 1996:19–57

4. Bakke R, Olsson PQ: Biofilm thickness measurements by light microscopy. *J Microbiol Meth* 5:93–8, 1986

5. Blanchard AP, Bird MR, Wright SJL: Biofilm disinfection with peroxygens. In: Wimpenny JWT, Handley PS, Gilbert P, et al, eds: *Biofilms: Community Interactions and Control*. Bioline, Cardiff, UK, 1997:235–44

6. Blenkinsopp SA, Costerton JW: Understanding bacterial biofilm. *Tebtech* 9:138–142, 1991

7. Bowden GH: Which bacteria are cariogenic in humans? In: Johnson NW, ed: *Risk Markers for Oral Diseases. Volume 1. Dental Caries*. Cambridge University Press, Cambridge, UK, 1991:266–86

8. Bowden GHW, Li YH: Nutritional influences on biofilm development. *Adv Dent Res* 11:81–99, 1997

9. Boyar RM, Thylstrup A, Kolmen L, et al: The microflora associated with the development of initial enamel decalcification below orthodontic bands *in vivo* in children living in a fluoridated area. *J Dent Res* 68:1734–8, 1989

10. Bradshaw DJ, Marsh PD, Watson GK, et al: Inter-species interactions in microbial communities. In: Wimpenny JWT, Handley PS, Gilbert P, et al., eds: *Biofilms: Community Interactions and Control*. Bioline, Cardiff, UK, 1997:63–71

11. Bradshaw DJ, McKee AS, Marsh PD: Effects of carbohydrate pulses and pH on population shifts within oral microbial communities *in vitro*. *J Dent Res* 68:1298–1302, 1989

12. Brecx M, Ronstrom A, Theilade J, et al: Early formation of dental plaque on plastic films. 2. Electron microscopic observations. *J Periodont Res* 16:213–27, 1981

13. Bryers J, Characklis W: Early fouling biofilm formation in a turbulent flow system: Overall kinetics. *Water Res* 15:483–91, 1981

14. Buret A, Ward KH, Olson ME, et al: An *in vivo* model to study the pathobiology of infectious biofilms on biomaterial surfaces. *J Biomed Mater Res* 25:865–74, 1991

15. Burlage RS: Emerging technologies: bioreporters, biosensors, and microprobes. In: Hurst CJ, Knudsen GR, McInerney MJ, et al, eds: *Manual of Environmental Microbiology*. ASM Press, Washington, DC, 1997:115–23

16. Burne RA, Chen Y-YM, Penders JEC: Analysis of gene expression in *Streptococcus mutans* in biofilms *in vitro*. *Adv Dent Res* 11:100–9, 1997

17. Caldwell DE, Atuku E, Wilkie DC, et al: Germ theory versus community theory in understanding and controlling the proliferation of biofilms. *Adv Dent Res* 11:4–13, 1997

18. Caldwell DE, Lawrence JR: Study of attached cells in continuous-flow slide culture. In: Wimpenny JWT, ed: *CRC Handbook of Laboratory Model Systems for Microbial Ecosystems*. CRC Press Inc., Boca Raton, FL, 1988:117–38

19. Caldwell DE, Wolfaardt GM, Korber DR, et al: Cultivation of microbial consortia and communities. In: Hurst CJ, Knudsen GR, McInerney MJ, et al, eds: *Manual of Environmental Microbiology*. ASM Press, Washington, DC, 1997:79–90

20. Carlsson J: Bacterial metabolism in dental biofilms. *Adv Dent Res* 11:75–80, 1997

21. Chang CC, Merritt K: Microbial adherence on poly(methyl methacrylate) (PMMA) surfaces. *J Biomed Mater Res* 26:197–207, 1992

22. Characklis WG: Laboratory biofilm reactors. In: Characklis WG, ed: *Biofilms*. Wiley, New York, 1990:55–89

23. Characklis WG: Microbial fouling. In: Characklis WG, ed: *Biofilms*. Wiley, New York, 1990:523–84

24. Characklis WG, Mcfeters GA, Marshall KC: Physiological ecology in biofilm systems. In: Characklis WG, ed: *Biofilms*. Wiley, New York, 1990:341–94

25. Characklis WG, Tirakhia MH, Zelver N: Transport and interfacial transfer phenomena. In: Characklis WG, ed: *Biofilms*. Wiley, New York, 1990:265–340

26. Chestnutt IG, Macfarlane TW, Stephen KW: The dissolution of mineral substrates in the determination of the cariogenic potential of *Streptococcus mutans*. *Microb Ecol Health Dis* 7:145–52, 1994

27. Christensen BE, Characklis WG: Physical and chemical properties of biofilms. In: Characklis WG, ed: *Biofilms*. Wiley, New York, 1990:93–130

28. Christensen GD, Simpson WA, Bisno AL, et al: Adherence of slime-producing strains of *Staphylococcus epidermidis* to smooth surfaces. *Infect Immun* 37:318–26, 1982

29. Christensen GD, Simpson WA, Bisno AL, et al: Experimental foreign body infections in mice challenged with slime-producing *Staphylococcus epidermidis*. *Infect Immun* 40:407–10, 1983

30. Costerton J, Lewandowski Z: Overview - The biofilm lifestyle. *Adv Dent Res* 11:192–5, 1997

31. Costerton JW, Lewandowski Z, Caldwell DE, et al: Microbial biofilms. *Ann Rev Microbiol* 49:711–45, 1995

32. Cummins D, Moss MC, Jones CL, et al: Confocal microscopy of dental plaque development. *Binary* 4:86–91, 1992

33. Cutress TW, Sissons CH, Pearce EIF, et al: Effects of fluoride-supplemented sucrose on experimental dental caries and dental plaque pH. *Adv Dent Res* 9:14–20, 1995

34. Davenport DS, Massanari RM, Pfaller MA, et al: Usefulness of a test for slime production as a marker for clinically significant infections with coagulase-negative staphylococci. *J Infect Dis* 153:332–9, 1986

35. Dawes C, Macpherson LMD: The distribution of saliva and sucrose around the mouth during the use of chewing gum and the implications for the site-specificity of caries and calculus formation. *J Dent Res* 72:852–7, 1993

36. Delwart EL, Shpaer EG, Louwagie J, et al: Genetic relationships determined by a DNA heteroduplex mobility assay: analysis of HIV-1 *env* genes. *Science* 262:1257–61, 1993

37. Dibdin GH: Diffusion of sugars and carboxylic acids through human dental plaque *in vitro*. *Arch Oral Biol* 26:515–23, 1981

38. Dibdin GH: Mathematical modeling. *Adv Dent Res* 11:127–32, 1997

39. Dibdin GH, Shellis RP, Wilson CM: An apparatus for the continuous culture of micro-organisms on solid surfaces with special reference to dental plaque. *J Appl Bacteriol* 40:261–8, 1976

40. Drew SW: Liquid culture. In: Gerhardt P, Murray RGE, Costilow RN, et al., eds: *Manual of Methods for General Bacteriology*. American Society for Microbiology, Washington, DC, 1981:151–78

41. Dworkin M: Multiculturalism versus the single microbe. In: Shapiro JA, Dworkin M, eds: *Bacteria as Multicellular Organisms*. Oxford University Press, Oxford, UK, 1997:3–13

42 Evans RC, Holmes CJ: Effect of vancomycin hydrochloride on *Staphylococcus epidermidis* biofilm associated with silicone elastomer. *Antimicrob Agents Chemother* 31:889–94, 1987

43. Fletcher M, Floodgate GD: An electron-microscopic demonstration of an acidic polysaccharide involved in the adhesion of a marine bacterium to solid surfaces. *J Gen Microbiol* 74:325–34, 1973

44. Freeman DJ, Falkiner FR, Keane CT: New method for detecting slime production by coagulase negative staphylococci. *J Clin Pathol* 42:872–4, 1989

45. Fry NK, Raskin L, Sharp R, et al: *In situ* analyses of microbial populations with molecular probes. In: Shapiro JA, Dworkin M, eds: *Bacteria as Multicellular Organisms.* Oxford University Press, Oxford, UK, 1997:292–336

46. Fuhrman JA: Community structure: bacteria and archaea. In: Hurst CJ, Knudsen GR, McInerney MJ, et al., eds: *Manual of Environmental Microbiology.* ASM Press, Washington, DC, 1997:278–83

47. Gallimore B, Gagnon RF, Subang R, et al: Natural history of chronic *Staphylococcus epidermidis* foreign body infection in a mouse model. *J Infect Dis* 164:1220–3, 1991

48. Ganderton L, Chawla J, Winters C, et al: Scanning electron microscopy of bacterial biofilms on indwelling bladder catheters. *Eur J Clin Microbiol Infect Dis* 11:789–96, 1992

49. Garland JL: Analytical approaches to the characterization of samples of microbial communities using patterns of potential C source utilization. *Soil Biol Biochem* 28:213–21, 1996

50. Gilbert P, Allison DG: Laboratory methods for biofilm production. In: Denyer SP, Gorman SP, Sussman M, eds: *Microbial Biofilms: Formation and Control.* Blackwell Scientific, London, 1993:29–49

51. Gilbert P, Allison DG, Evans DJ, et al: Growth rate control of adherent bacterial populations. *Appl Environ Microbiol* 55:1308–11, 1989

52. Gilbert P, Das J, Foley I: Biofilm susceptibility to antimicrobials. *Adv Dent Res* 11:160–7, 1997

53. Giridhar G, Kreger AS, Myrvik QN, et al: Inhibition of *Staphylococcus* adherence to biomaterials by extracellular slime of *S. epidermidis* RP12. *J Biomed Mater Res* 28:1289–94, 1994

54. Glenister DA, Salamon KE, Smith K, et al: Enhanced growth of complex communities of dental plaque bacteria in mucin-limited continuous culture. *Microb Ecol Hlth Dis* 1:31–8, 1988

55. Gristina AG, Costerton JW: Bacterial adherence and the glycocalyx and their role in musculoskeletal infection. *Orthop Clin North Am* 15:517–35, 1984

56. Gristina AG, Costerton JW: Bacterial adherence to biomaterials and tissue. The significance of its role in clinical sepsis. *J Bone Joint Surg* 67:264–73, 1985

57. Gristina AG, Hobgood CD, Barth E: Biomaterial specificity, molecular mechanisms, and clinical relevance of *S. epidermidis* and *S. aureus* infections in surgery. In: Pulverer G, ed: *Pathogenesis and Clinical Significance of Coagulase-negative Staphylococci.* Fisher Verlag, Stuttgart, Germany, 1987:143–57

58. Haack SK, Garchow H, Klug MJ, et al: Analysis of factors affecting the accuracy, reproducibility, and interpretation of microbial community carbon source utilization patterns. *Appl Environ Microbiol* 61:1458–68, 1995

59. Helmstetter CE, Cummings DJ: Bacterial synchronization by selection of cells at division. *Proc Nat Acad Sci* 50:767–74, 1963

60. Herles S, Olsen S, Afflitto J, et al: Chemostat flow cell system: An *in vitro* model for the evaluation of antiplaque agents. *J Dent Res* 73:1748–55, 1994

61. Hodgson AE, Nelson SM, Brown MRW, et al: A simple *in vitro* model for growth control of bacterial biofilms. *J Appl Bacteriol* 79:87–93, 1995

62. Hoehn RC, Ray AD: Effects of thickness on bacterial film. *J Water Poll Control Fed* 45:2302–20, 1973

63. Holmes CJ, Evans RC, Vonesh E: Application of an empirically derived growth curve model to characterize *Staphylococcus epidermidis* biofilm development on silicone elastomer. *Biomaterials* 10:625–9, 1989

64. Hudson DE, Donoghue HD, Perrons CJ: A laboratory microcosm (artificial mouth) for the culture and continuous pH measurement of oral bacteria on surfaces. *J Appl Bacteriol* 60:301–10, 1986

65. Hussain M, Collins C, Hastings JG, et al: Radiochemical assay to measure the biofilm produced by coagulase-negative staphylococci on solid surfaces and its use to quantitate the effects of various antibacterial compounds on the formation of the biofilm. *J Med Microbiol* 34:62–9, 1992

66. Hussain M, Hastings JG, White PJ: Isolation and composition of the extracellular slime made by coagulase-negative staphylococci in a chemically defined medium. *J Infect Dis* 163:534–41, 1991

67. Igarashi K, Lee IK, Schachtele CF: Effect of dental plaque age and bacterial composition on the pH of artificial fissures in human volunteers. *Caries Res* 24:52–9, 1990

68. Isiklar ZU, Darouiche RO, Landon GC, et al: Efficacy of antibiotics alone for orthopaedic device related infections. *Clin Orthop* 332:184–9, 1996

69. Iwaoka T, Griffiths PR, Kitasako JT, et al: Copper-coated cylindrical internal reflection elements for investigating interfacial phenomena. *Appl Spectrosc* 40:1062–5, 1986

70. Jolley JG, Geesey GG, Hankins MR, et al: *In situ*, real-time FT-IR/CIR/ATR study of the biocorrosion of copper by gum arabic, alginic acid, bacterial culture supernatant and *Pseudomonas atlantica* exopolymer. *Appl Spectrosc* 43:1062–7, 1989

71. Keevil CW, Bradshaw DJ, Dowsett AB, et al: Microbial film formation: dental plaque deposition on acrylic tiles using continuous culture techniques. *J Appl Bacteriol* 62:129–38, 1987

72. Khardori N, Rosenbaum B, Bodey GP: Evaluation of *in vitro* markers for clinically significant infections with coagulase-negative staphylococci. *Clin Res* 35:20A, 1987

73. Kinniment SL, Wimpenny JWT, Adams D, et al: Development of a steady-state oral microbial biofilm community using the constant-depth film fermenter. *Microbiol* 142:631–8, 1996

74. Kleinberg I: Biochemistry of the dental plaque. In: Staple PH, ed: *Advances in Oral Biology*. Academic Press, New York, NY, 1970:43–90

75. Kotilainen P, Maki J, Oksman P, et al: Immunochemical analysis of the extracellular slime substance of *Staphylococcus epidermidis*. *Eur J Clin Microbiol Infect Dis* 9:262–70, 1990

76. Ladd TI, Costerton JW: Methods for studying biofilm bacteria. *Meth Microbiol* 22:287–307, 1990

77. Ladd TI, Schmiel D, Nickel JC, et al: The use of a radiorespirometric assay for testing the antibiotic sensitivity of catheter-associated bacteria. *J Urol* 138:1451–6, 1987

78. Levin IM, Lau CN, Socransky SS, et al: Cultivable and uncultivable species on or in gingival epithelial cells. *J Dent Res* 78:453, 1999

79. Li YH, Bowden GH: Characteristics of accumulation of oral Gram-positive bacteria on mucin-conditioned glass surfaces in a model system. *Oral Microbiol Immunol* 9:1–11, 1994

80. Li YH, Bowden GH: The effect of environmental pH and fluoride from the substratum on the development of biofilms of selected oral bacteria. *J Dent Res* 73:1615–26, 1994

81. Liljemark WF, Bloomquist CG, Coulter MC, et al: Utilization of a continuous streptococcal surface to measure interbacterial adherence *in vitro* and *in vivo*. *J Dent Res* 67:1445–60, 1988

82. Locci R, Peters G, Pulverer G: Microbial colonization of prosthetic devices. I. Micro-topographical characteristics of intravenous catheters as detected by scanning electron microscopy. *Zentralbl Bakteriol Mikrobiol Hyg [B]* 173:285–92, 1981

83. London J, Kolenbrander P: Coaggregation: enhancing colonization in a fluctuating environment. In: Fletcher M, ed: *Bacterial Adhesion: Molecular and Ecological Diversity*. Wiley-Liss, New York, 1996:249–79

84. Luoma H, Alakuijala P, Korhonen A, et al: Enamel dissolution in relation to fluoride concentrations in the fluid of dental plaque-like layers of precultured *Streptococcus sobrinus*. *Arch Oral Biol* 39:177–84, 1994

85. Macfarlane S, McBain AJ, Macfarlane GT: Consequences of biofilm and sessile growth in the large intestine. *Adv Dent Res* 11:59–68, 1997

86. Macpherson LMD, Chen WY, Dawes C: Effects of salivary bicarbonate content and film velocity on pH changes in an artificial plaque containing *Streptococcus oralis*, after exposure to sucrose. *J Dent Res* 70:1235–8, 1991

87. Macpherson LMD, Dawes C: Effects of salivary film velocity on pH changes in an artificial plaque containing *Streptococcus oralis*, after exposure to sucrose. *J Dent Res* 70:1230–4, 1991

88. Macpherson LMD, Dawes C: Urea concentration in minor mucous gland secretions and the effect of salivary film velocity on urea metabolism by *Streptoccocus vestibularis* in an artificial plaque. *J Periodont Res* 26:395–401, 1991

89. Macpherson LMD, Macfarlane TW, Stephen KW: An *in situ* microbiological study of the early colonization of human enamel surfaces. *Microb Ecol Hlth Dis* 4:39–46, 1991

90. Main C, Geddes DAM, McNee SG, et al: Instrumentation for measurement of dental plaque thickness *in situ*. *J Biomed Eng* 6:151–4, 1984

91. Marsh PD: The significance of maintaining the stability of the natural microflora of the mouth. *Brit Dent J* 171:174–7, 1991

92. Marsh PD: The role of microbiology in models of dental caries. *Adv Dent Res* 9:244–54, 1995

93. Marsh PD, Bradshaw DJ: Physiological approaches to the control of oral biofilms. *Adv Dent Res* 11:176–85, 1997

94. Marsh PD, Martin MV: *Oral Microbiology*. Chapman and Hall, London, UK, 1992

95. Matsushita M: Formation of colony patterns by a bacterial cell population. In: Shapiro JA, Dworkin M, eds: *Bacteria as Multicellular Organisms*. Oxford University Press, Oxford, UK, 1997:366–93

96. Mayberry-Carson KJ, Tober-Meyer B, Lambe DW, et al: Osteomyelitis experimentally induced with *Bacteroides thetaiotaomicron* and *Staphylococcus epidermidis*. Influence of a foreign-body implant. *Clin Orthop* 280:289–99, 1992

97. McCabe RM, Keyes PH, Howell A: An *in vitro* method for assessing the plaque forming ability of the oral bacteria. *Arch Oral Biol* 12:1653–6, 1967

98. McCoy WF, Bryers JD, Robbins J, et al: Observations of fouling biofilm formation. *Can J Microbiol* 27:910–7, 1981

99. McDermid AS, McKee AS, Ellwood DC, et al: The effect of lowering the pH on the composition and metabolism of a community of nine bacteria grown in a chemostat. *J Gen Microbiol* 132:1205–14, 1986

100. McFeters GA, Bazin MJ, Bryers JD: Biofilm development and its consequences. In: Marshall KC, ed: *Microbial Adhesion and Aggregation*. Springer-Verlag, Berlin, 1984:109–24, 1984

101. McGlohorn J, An YH, Friedman RJ: A simple flow chamber for producing bacterial biofilm on biomaterial surfaces. *Trans Soc Biomater* 21:484, 1998

102. McGlohorn JB, Bednarski BK, An YH, et al: Cultivation of biofilm on titanium surface using a new continuous flow system. *MUSC Orthop J* 2:20–3, 1999

103. Mikkelsen L: Influence of sucrose intake on saliva and number of microorganisms and acidogenic potential in early dental plaque. *Microb Ecol Hlth Dis* 6:253–64, 1993

104. Molin G, Nilsson I: Degradation of phenol by *Pseudomonas putida* ATCC 11172 in continuous culture at different ratios of biofilm surface to culture volume. *Appl Environ Microbiol* 50:946–50, 1985

105. Moore WEC, Moore LVH: The bacteria of periodontal diseases. *Periodontology* 5:66–77, 1994

106. Naylor PT, Myrvik QN, Gristina A: Antibiotic resistance of biomaterial-adherent coagulase-negative and coagulase-positive staphylococci. *Clin Orthop* 261:126–33, 1990

107. Newman HN: Plaque and chronic inflammatory periodontal disease. A question of ecology. *J Clin Periodontol* 17:533–41, 1990

108. Nickel JC, Ruseska I, Wright JB, et al: Tobramycin resistance of *Pseudomonas aeruginosa* cells growing as a biofilm on urinary catheter material. *Antimicrob Agents Chemother* 27:619–24, 1985

109. Noorda WD, Purdell-Lewis DJ, de Koning W, et al: A new apparatus for continuous cultivation of bacterial plaque on solid surfaces and human dental enamel. *J Appl Bacteriol* 58:563–9, 1985

110. Noorda WD, van Montfort AMAP, Purdell-Lewis DJ, et al: Developmental and metabolic aspects of a monobacterial plaque of *Streptococcus mutans* C67-1 grown on human enamel slabs in an artificial mouth model. I. Plaque data. *Caries Res* 20:300–7, 1986
111. Nyvad B, Fejerskov O: Structure of dental plaque and the plaque-enamel interface in human experimental caries. *Caries Res* 23:151–8, 1989
112. Nyvad B, Kilian M: Microbiology of the early clonization of human enamel and root surfaces *in vivo*. *Scand J Dent Res* 95:369–80, 1987
113. Nyvad B, Kilian M: Microflora associated with experimental root surface caries in humans. *Infect Immun* 58:1628–33, 1990
114. Palmer RJ, Almeida JS, Ringelberg DB, et al: Phospholipid-bound fatty-acid profiles reveal community structure of oral biofilms. *Adv Dent Res* 11:187, 1996
115. Palmer RJ, Wong L, Sissons CH: Community structure and enzyme activity in microcosm dental plaques. *J Dent Res* 77:988, 1998
116. Pearce EIF, Dibdin GH: The diffusion and enzymic hydrolysis of monofluorophosphate and dental plaque. *J Dent Res* 74:691–7, 1995
117. Pearce EIF, Wakefield JSJ, Sissons CH: Therapeutic mineral enrichment of dental plaque visualized by transmission electron microscopy. *J Dent Res* 70:90–4, 1991
118. Peters A, Wimpenny JWT: A constant-depth laboratory model film fermenter. In: Wimpenny JWT, ed: *CRC Handbook of Laboratory Model Systems for Microbial Ecosystems*. CRC Press, Boca Raton, FL, 1988:175–195
119. Peterson GL: A simplification of the protein assay method of Lowry et al. which is more generally applicable. *Anal Biochem* 83:346–56, 1977
120. Pfaller M, Davenport D, Bale M, et al: Development of the quantitative micro-test for slime production by coagulase-negative staphylococci. *Eur J Clin Microbiol Infect Dis* 7:30–3, 1988
121. Prosser BL, Taylor D, Dix BA, et al: Method of evaluating effects of antibiotics on bacterial biofilm. *Antimicrob Agents Chemother* 31:1502–6, 1987
122. Qian Z, Sagers RD, Pitt WG: The effect of ultrasonic frequency upon enhanced killing of *P. aeruginosa* biofilms. *Ann Biomed Eng* 25:69–76, 1997
123. Reid G, Denstedt JD, Kang YS, et al: Microbial adhesion and biofilm formation on ureteral stents *in vitro* and *in vivo*. *J Urol* 148:1592–4, 1992
124. Robinson C, Kirkham J, Shore RC, et al: A quantitative site-specific study of plaque biofilms formed *in vivo*. *J Dent Res* 75:232, 1996
125. Russell C, Coulter WA: Continuous monitoring of pH and Eh in bacterial plaque grown on a tooth in an artificial mouth. *Appl Microbiol* 29:141–4, 1975
126. Shapiro JA: Multicellularity: the rule, not the exception: Lessons from *Escherichia coli* colonies. In: Shapiro JA, Dworkin M, eds: *Bacteria as Multicellular Organisms*. Oxford University Press, Oxford, UK, 1997:14–49
127. Shu M, Sissons CH, Miller JH, et al: Cariogenicity of monoculture and consortia caries pathogen plaque biofilms in an artificial mouth. *J Dent Res* 75:35, 1996
128. Shu M, Wong L, Miller JH, et al: Development of multi-species consortia biofilms of oral bacteria as an enamel and root caries model system. *Arch Oral Biol*: Submitted, 1999
129. Simmonds RS, Naidoo J, Jones CL, et al: The streptococcal bacteriocin-like inhibitory substance, Zoocin A, reduces the proportion of *Streptococcus mutans* in an artificial plaque. *Microb Ecol Hlth Dis* 8:281–92, 1995
130. Singleton S, Treloar R, Warren P, et al: Methods of microscopic characterization of oral biofilms: analysis of colonization, microstructure, and molecular transport phenomena. *Adv Dent Res* 11:133–49, 1997
131. Sissons CH: Artifical dental plaque biofilm model systems. *Adv Dent Res* 11:110–26, 1997
132. Sissons CH, Cutress TW: *In vitro* urea-dependent pH-changes by human salivary bacteria and dispersed, artificial-mouth, bacterial plaques. *Arch Oral Biol* 32:181–9, 1987

133. Sissons CH, Cutress TW, Faulds G, et al: pH responses to sucrose and the formation of pH gradients in thick "artificial mouth" microcosm plaques. *Arch Oral Biol* 37:913–22, 1992

134. Sissons CH, Cutress TW, Hoffman MP, et al: A multi-station dental plaque microcosm (artificial mouth) for the study of plaque growth, metabolism, pH, and mineralization. *J Dent Res* 70:1409–16, 1991

135. Sissons CH, Cutress TW, Wong L, et al: Effect of urea on pH in artificial mouth microcosm plaques. *Caries Res* 27: Abstract # 69:226, 1993

136. Sissons CH, Hancock EM, Perinpanayagam HER, et al: A procedure for urease and protein extraction from staphylococci. *J Appl Bacteriol* 67:433–40, 1989

137. Sissons CH, Wong L, Cutress TW: Patterns and rates of growth of microcosm dental plaque biofilms. *Oral Microbiol Immunol* 10:160–7, 1995

138. Sissons CH, Wong L, Cutress TW: Regulation of urease levels in microcosm dental plaque. *Microb Ecol Hlth Dis* 8:219–24, 1995

139. Sissons CH, Wong L, Cutress TW: Inhibition by ethanol of the growth of biofilm and dispersed microcosm dental plaques. *Arch Oral Biol* 41:27–34, 1996

140. Sissons CH, Wong L, Hancock EM, et al: pH gradients induced by urea metabolism in "artificial mouth" microcosm plaques. *Arch Oral Biol* 39:507–11, 1994

141. Sissons CH, Wong L, Hancock EM, et al: The pH response to urea and the effect of liquid flow in "artificial mouth" microcosm plaques. *Arch Oral Biol* 39:497–505, 1994

142. Sissons CH, Wong L, Shu M: Factors affecting the resting pH of *in vitro* human microcosm plaque and *Streptococcus mutans* biofilms. *Arch Oral Biol* 43:93–102, 1998

143. Socransky SS, Haffajee AD, Cugini MA, et al: Microbial complexes in subgingival plaque. *J Clin Periodontol* 25:134–44, 1998

144. Socransky SS, Manganiello AD, Propas D, et al: Bacteriological studies of developing supragingival dental plaque. *J Periodont Res* 12:90–106, 1977

145. Socransky SS, Smith C, Martin L, et al: "Checkerboard" DNA-DNA hybridization. *Biotechniques* 17:788–92, 1994

146. Spratt DA, Weightman AJ, Wade WG: Diversity of oral asaccharolytic *Eubacterium* species in periodontitis – identification of novel phylotypes representing uncultivated taxa. *Oral Microbiol Immunol* 14:56–9, 1999

147. Stahl DA: Molecular approaches for the measurement of density, diversity, and phylogeny. In: Hurst CJ, Knudsen GR, McInerney MJ, et al., eds: *Manual of Environmental Microbiology*. ASM Press, Washington, DC, 1997:102–14

148. Stoodley P, Boyle JD, Dodds I, et al: Consensus model of biofilm structure. In: Wimpenny JWT, Handley PS, Gilbert P, et al., eds: *Biofilms: Community Interactions and Control*. Bioline, Cardiff, UK, 1997:1–9

149. Straskrabova V: The effect of substrate shock on populations of starving aquatic bacteria. *J Appl Bacteriol* 54:217–24, 1983

150. Sudo S, Gutfleisch JR, Schotzko NK, et al: Model system for studying colonization and growth of bacteria on a hydroxyapatite surface. *Infect Immun* 12:576–85, 1975

151. Sutton NA, Hughes N, Handley PS: A comparison of conventional SEM techniques, low temperature SEM and the electroscan wet scanning electron microscope to study the structure of a biofilm of *Streptococcus crista* CR3. *J Appl Bacteriol* 76:448–54, 1994

152. Tatevossian A: Film fermenters in dental research. In: Wimpenny JWT, ed: *CRC Handbook of Laboratory Model Systems for Microbial Ecosystems*. CRC Press Inc., Boca Raton, FL, 1988:197–227

153. Tilman D, Downing JA: Biodiversity and stability in grasslands. *Nature* 367: 363–5, 1994

154. Torsvik V, Daae FL, Sandaa RA, et al: Novel techniques for analysing microbial diversity in natural and perturbed environments. *J Biotechnol* 64:53–62, 1998

155. Torsvik V, Goksoyr J, Daae FL: High diversity in DNA of soil bacteria. *Appl Environ Microbiol* 56:782–7, 1990

156. Trulear MG, Characklis WG: Dynamics of biofilm processes. *J Water Poll Control Fed* 54:1288–1301, 1982

157. Tsai CL, Schurman DJ, Smith RL: Quantitation of glycocalyx production in coagulase-negative *Staphylococcus*. *J Orthop Res* 6:666–70, 1988

158. Van Pett K, Schurman DJ, Smith RL: Quantitation and relative distribution of extracellular matrix in *Staphylococcus epidermidis* biofilm. *J Orthop Res* 8:321–27, 1990

159. Vrahopoulos TP, Barber PM, Newman HN: The apical border plaque in severe periodontitis. An ultrastructural study. *J Periodontol* 66:113–24, 1995

160. Wade WG, Harper-Owen R, Dymock D, et al: Associations between "not-yet-cultivable" bacterial phylotypes with periodontal health and disease. *J Dent Res* 77:918, 1998

161. Watve MG, Gangal RM: Problems in measuring bacterial diversity and a possible solution. *Appl Environ Microbiol* 62:4299–301, 1996

162. Weiger R, von Ohle C, Decker E, et al: Vital microorganisms in early supragingival dental plaque and in stimulated human saliva. *J Periodont Res* 32:233–40, 1997

163. White DC, Pinkart HC, Ringelberg DB: Biomass measurements: biochemical appproaches. In: Hurst CJ, Knudsen GR, McInerney MJ, et al., eds: *Manual of Environmental Microbiology*. ASM Press, Washington, DC, 1997:91–101

164. Wimpenny JWT: The bacterial colony. In: Wimpenny JWT, ed: *CRC Handbook of Laboratory Model Systems for Microbial Ecosystems*. CRC Press, Boca Raton, FL, 1988:109–39

165. Wimpenny JWT: Introduction. In: Wimpenny JWT, ed: *CRC Handbook of Laboratory Model Systems for Microbial Ecosystems*. CRC Press, Boca Raton, FL, 1988:1–17

166. Wimpenny JWT: On the nature and validity of models. In: Wimpenny J, Handley PS, Gilbert P, et al., eds: *The Life and Death of Biofilm*. Bioline, Cardiff, UK, 1995:1–8

167. Wimpenny JWT: The validity of models. *Adv Dent Res* 11:150–9, 1997

168. Wimpenny JWT: Email communication to Bionet Microbiology Biofilms Newsgroup, September 1998

169. Wimpenny JWT, Colasanti R: A unifying hypothesis for the structure of microbial biofilms based on cellular automaton models. *FEMS Microbiol Ecol* 22:1–16, 1997

170. Wimpenny JWT, Kinniment SL, Scourfield MA: The physiology and biochemistry of biofilm. In: Denyer SP, Gorman SP Sussman M, eds: *Microbial Biofilms: Formation and Control*. Blackwell Scientific Publications, London, 1993:51–94

171. Wong L, Sissons CH, Cutress TW: Characterization of calcium phosphate deposited in mineralised microcosm dental plaques. *J Dent Res* 75:1077, 1996

172. Wong L, Sissons CH, Cutress TW: Control of a multiple dental plaque culture system and long-term, continuous, plaque pH measurement using LabVIEW®. *Binary* 6:173–80, 1994

173. Yaari A, Bibby BG: Production of plaques and initiation of caries *in vitro*. *J Dent Res* 55:30–6, 1976

174. Zak JC, Willig MR, Moorhead DL, et al: Functional diversity of microbial communities: a quantitative approach. *Soil Biol Biochem* 26:1101–8, 1994

175. Zampatti O, Roques C, Michel G: An *in vitro* mouth model to test antiplaque agents: Preliminary studies using a toothpaste containing chlorhexidine. *Caries Res* 28:35–42, 1994

176. Zero DT: *In situ* caries models. *Adv Dent Res* 9:214–30, 1995

# 10

# Monitoring the Organization of Microbial Biofilm Communities

Subramanian Karthikeyan,[1] Darren R. Korber,[1]
Gideon M. Wolfaardt,[2] and Douglas E. Caldwell[1]

[1]*Department of Applied Microbiology and Food Science, University of Saskatchewan, Saskatoon, Saskatchewan, Canada*
[2]*Department of Microbiology, University of Stellenbosch, Stellenbosch, South Africa*

## I. INTRODUCTION

Microbial organization within a biofilm community can be thought of as the product of species composition and spatial positioning of individuals within the biofilm matrix. Species composition within a microbial community, also referred to as community structure,[8] determines the community's overall genetic potential for survival and reproductive success under various environmental conditions. Spatial positioning allows individuals to interact physiologically and genetically. It also allows the creation of favorable microbial microenvironments within hostile macroenvironments. When a biofilm community is subjected to an environmental perturbation (e.g., an introduction of a pollutant or antimicrobial compound), continued reproductive success may be facilitated by a process of reorganization consisting of changes in composition and spatial arrangement of individuals within the community. Thus, the structural and spatial organization of a biofilm community, and its functional significance, should be a consideration when attempting to control or enhance the activities of biofilm communities in industrial or environmental settings.

The analysis of biofilm organization has been limited by the inherent complexity and temporal variability found in natural ecosystems. Even in communities cultivated in laboratory model systems under defined culture conditions, the comprehensive *in situ* analysis and monitoring of structural and spatial organization may be hampered by conceptual and methodological limitations. Despite efforts to develop methodology which is suitable for the analysis of microbial communities, relatively few approaches (e.g., fluorescence *in situ* hybridization and fluorescent antibody techniques in conjunction with epifluorescence or scanning confocal laser microscopy) are useful for the direct analysis of both structural and spatial organization within intact, fully-hydrated biofilm communities. New approaches include the insertion of various Green Fluorescent Protein (GFP)—variant genes into specific biofilm members, thus permitting the direct analysis of the abundance and distribution of these organisms within multispecies systems over

*Handbook of Bacterial Adhesion: Principles, Methods, and Applications*
Edited by: Y. H. An and R. J. Friedman © Humana Press Inc., Totowa, NJ

time. Other techniques involving fluorescent molecular probes such as fluor-conjugated lectins and dextrans are useful to elucidate the physicochemical heterogeneity (charge distribution, diffusion characteristics, exopolymer chemistry, etc.) within intact biofilm communities; however, they are limited in their ability to delineate the structural organization within these communities. Application of most other techniques requires that biofilms be removed from their substratum and disrupted, resulting in the destruction of spatial cellular arrangements (architecture) within the community being analyzed. Attempts to understand biofilm organization often fail if the functional significance of the structural and spatial organization within the community is not determined. Thus, in addition to summarizing techniques useful for the analysis of the microbial organization within intact and disrupted biofilm communities, this chapter also discusses techniques useful for the elucidation of microbial interactions and metabolic characteristics that influence biofilm organization.

## II. BIOFILM ORGANIZATION

A very brief summary of those events that influence the organization of biofilm communities is provided in this review. More information concerning the factors and events associated with biofilm formation can be found elsewhere.[12,13,32,37,38,47]

Abiotic events that precede surface colonization include the rapid formation of a preconditioning film composed of proteins, glycoproteins and inorganic nutrients on the attachment surface after immersion.[3,32,46] Microbial adhesion to preconditioned surfaces is then facilitated by van der Waals forces, electrostatic interactions and specific interactions or by a combination of these, depending on the proximity of the organism to the attachment surface.[88] Initial attachment is followed by a consolidation phase during which production of bacterial exopolysaccharides (EPS) results in a more stable attachment by forming organic bridges between the cells and the substratum.[59] Subsequently, growth and multiplication of firmly attached primary colonizing organisms lead to the formation of microcolonies. Cells which are loosely attached may detach and these cells together with offspring of other sessile cells may recolonize previously uncolonized surfaces,[33] thereby extending the spatial boundaries of the biofilm.

Interaction and networking of community members may indeed be the most important determinant of biofilm organization. Interactive behavior complements the functional capabilities of individuals,[27,85,88] thereby allowing microbial colonization of environments otherwise adverse or unfavorable. The sequence of colonization impacts the structure of a biofilm community, as the primary colonizers often predispose the surface environment for subsequent colonization by specific organisms. A good example is the successional events associated with the formation of dental plaque. Specific protein–protein or carbohydrate–protein molecular interactions between primary and secondary colonizers determine the pattern of bacterial colonization and succession within dental plaques,[45] as well as the patterns of coaggregation whereby specific pairs of bacteria closely interact.[5,25,28,57] The events of coaggregation are highly specific. During the interaction between *Prevotella intermedia* and an *Actinomyces* species, coaggregation could occur only between specific strains of these two genera.[57] The enzymatic activities of early colonizers of the teeth surface, such as *Streptococcus oralis,* might also influence ecological succession.[41]

Sometimes, succession is mediated through modifications in the physical characteristics of the attachment surface by early colonizers. For example, development of saucer shaped cavitations by the wearing-away of necrotic dentine on tooth surfaces previously formed by root lesion microflora favors the subsequent colonization by aciduric flora.[73] In the rumen ecosystem, the exopolymeric glycocalyx produced by primary colonizers has been shown to facilitate the attachment of secondary colonizers.[49] The beneficial role of exopolymers has also been demonstrated in laboratory model systems. The colonization of *Listeria monocytogenes* on glass coverslips was enhanced by mixing the *L. monocytogenes* culture with an exopolymer producing strain of *Pseudomonas fragi*.[69] Thus, spatial organization within a biofilm community may be influenced by the specific order in which various bacteria colonize the surface.

As biofilm development proceeds, attachment and detachment of cells continuously alter the biofilm community structure. This process optimizes cell–cell arrangements and interactions in response to changes in the physical or chemical environment, and may be thought of as a process of self-regulation. Notably, pure culture biofilms also often show a high degree of organization,[30,35] suggesting that in addition to these biotic interactions, there are various other abiotic factors (e.g., characteristics of the attachment surface, physicochemical environment, etc.) which control the resultant biofilm architecture.

The characteristics of the attachment surface are important determinants during biofilm formation. In general, materials that have low free surface energies and low negative surface charges favor the formation and stability of biofilms.[82] However, variation in cell surface hydrophobicity and charge among various strains of microorganisms[50] may actually translate into differences in their ability to attach to surfaces, thereby modulating the final organization of the biofilm community. The organization of the biofilm matrix is also influenced by the nature of the physicochemical environment, including the type, concentration and flux of nutrients, metabolites, or antimicrobial substances, as well as by gradients of pH, Eh or other factors.[4,52,76,86]

Overall, microbial organization within biofilm communities may be influenced by a number of factors, including: 1) the numbers and types of colonizing cells, 2) the characteristics of the bacterial cell and colonization surface, 3) the characteristics of the physicochemical environment, and 4) the nature and extent of microbial interactions (e.g., microbial succession, coaggregation). Structural and spatial reorganization in response to these factors enable biofilm communities to continuously optimize their reproductive success under both steady-state and changing environmental conditions. However, the factors which govern the formation of complex biofilm communities are poorly understood, and thus require further study. There is consequently an ongoing need to develop or refine techniques which permit the nondestructive study of these factors in real time.

## III. CULTIVATION OF BIOFILM COMMUNITIES

The inherent complexity and temporal variability of the physicochemical environment found in natural ecosystems poses many challenges to the study of bacterial organization under *in situ* conditions.[8] One solution to this problem is to allow biofilm communities to develop on artificial surfaces such as stainless steel[42] under *in situ* conditions, and study them *ex situ* under laboratory conditions. In general, monitoring and analysis of these systems are less tedious if the biofilm communities are cultivated in the laboratory. Furthermore, interpretation of the organizational relationships is somewhat simplified when these communities are cultivated using model systems under defined environmental conditions. Examples of typical model systems include flow cells,[33,85,86] dual-dilution continuous culture (DDCC),[31] rototorque annular bioreactors[58] and chemostats.[2] Detailed discussion of the principles and application of these and other systems can be found elsewhere.[8,10]

One advantage of flow cell culture is that it permits instantaneous changes in the flux and concentration of growth substrates. Secondly, biofilm formation occurs on a glass coverslip which facilitates the application of various forms of microscopy[33,34] including scanning confocal laser microscopy (CLSM) for the nondestructive analysis of fully-hydrated communities (discussed in the following section).[35,86] DDCC allows independent dilution of attached and planktonic bacteria within a flowing system, and is therefore useful for correlating the behavioral or functional characteristics of a bacterium to surface colonization.[31] Other systems such as rototorque annular bioreactors and chemostats can be fitted with removable coupons for analysis of biofilms. The design of rototorque annular bioreactors generally consists of two concentrically placed cylinders, one fixed and one rotatable, thereby allowing for the study of friction and shear effects on biofilm formation and organization.[62] Angell et al.[2] used a chemostat to study the effect of biofilm formation on the corrosion of removable coupons. The chemostat was operated at a high dilution rate to select for bacteria adapted to sessile growth and was fed by three separate chemostats containing steady-state populations of *Pseudomonas aeruginosa, Thiobacillus ferrooxidans* and *Desulfovibrio vulgaris*. However, systems which involve the physical removal of beads (such as DDCC) or coupons for subsequent *ex situ* analysis may potentially introduce shear effects and hence change biofilm organization.

## IV. MONITORING ORGANIZATION
## WITHIN INTACT BIOFILM COMMUNITIES

The initial stages of biofilm development can be examined by phase contrast microscopy.[9,60] Caldwell and Lawrence[9] used high magnification phase contrast microscopy (×100) to analyze the growth kinetics of *Pseudomonas fluorescens* microcolonies. Cell monolayers can readily be digitized and analyzed using phase contrast microscopy in conjunction with digital image analysis. This approach, however, is not suitable for quantitative study of thicker or complex biofilms.[9] The high contrast images and enhanced depth of field offered by low magnification darkfield microscopy also favor the analysis of bacterial adhesion and subsequent biofilm formation.[33,34] However, a drawback of darkfield microscopy is that since each cell is represented as a point light source, cells appear larger than they are. Palmer and Caldwell[61] used a combination of microscopic techniques, including low-magnification darkfield microscopy, to analyze the regrowth of plaque biofilms developed in flow cells. However, negative staining (a technique where fluorescein is applied to the bulk liquid solution[6]) used in conjunction with CLSM is often the most useful technique for the temporal analysis of the three-dimensional architecture of a biofilm community in its fully-hydrated state, since it provides optical thin sections of high contrast and resolution. A detailed discussion of the hardware, setup and operation of CLSM as well as application of image analysis techniques in conjunction with laser microscopy can be found in a recent review by Lawrence et al.[36]

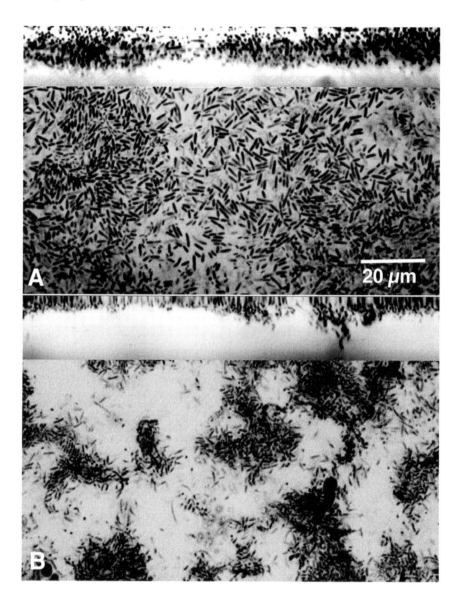

**Figure 1.** Confocal laser micrographs showing the architecture of biofilm communities cultivated using TSB (A) and sodium benzoate (B) as sole sources of carbon. Biofilms were negatively stained with fluorescein (fluorescence exclusion) which permitted visualization of bacteria as dark objects on a bright background. While XY images show the horizontal spatial distribution of cells, vertical (XZ) sections reveal the thickness and surface topography. The nontoxic nature of fluorescein also permits the analysis of temporal changes in spatial organization for the same biofilm community, without having to sacrifice a flow cell channel following each observation.

Negative staining is based on the principle that the brightness of fluorescein is proportional to the pH of the liquid phase, and at an ambient pH greater than that of the intracellular pH, the cells appear as dark objects on a bright background. During most biofilm studies, a higher extracellular fluorescein concentration further contributes to the

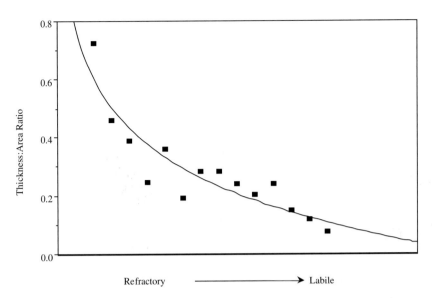

**Figure 2.** Relationship between the recalcitrance of carbon substrate used in biofilm cultivation and the ratio of biofilm thickness to surface coverage. A 2,4,6-T degrading biofilm community was cultivated using 15 substrates of varying degrees of chlorination and complexity. Higher thickness:area ratios obtained during cultivation using refractory substrates suggest that highly-organized spatial arrangements (such as cell clusters) may be required for the utilization of these substrates.

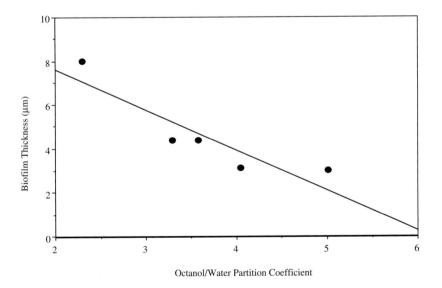

**Figure 3.** Correlation between published $\log_{o/w}$ coefficients for trichloroethylene, dichlorophenol, trichlorophenol, pentachlorophenol and 9-fluorenone, and the extent of biofilm development when the 2,4,6-T community was cultivated using these compounds. A high, negative correlation ($R^2 = 0.82$) suggests that higher partition coefficients may have led to higher intracellular accumulation of the test compounds, and consequent reduction in growth rates due to substrate toxicity.

**Table 1. Responses of a Biofilm Community to Growth
on Substrates of Varying Chemical Structure—
Values Indicated Are Averages of 5 Replicate Measurements**

| Compound | % Biofilm area coverage | Biofilm thickness ($\mu$m) |
|---|---|---|
| 2,4,6-Trichlorobenzoic acid | 50.84 | 6.40 |
| 2,7-Dichloro 9-fluorenone | 29.22 | 6.40 |
| Diphenyl anthracene | 22.31 | 6.80 |
| Benzo(a)pyrene | 10.54 | 5.10 |
| Pentachlorophenol | 2.09 | 3.00 |

bacteria appearing darker than the brighter background. The fully-hydrated nature of EPS commonly results in an almost equal fluorescein concentration within EPS when compared to the bulk fluid phase, thus EPS does not usually interfere with negative staining of bacteria. Negative staining is especially useful for the study of the initial stages of biofilm formation, although this technique can also be used to analyze subsequent attachment, recolonization, emigration, and immigration.[6,86]

The nontoxic nature of fluorescein also permits real time analysis of temporal changes in the spatial organization of the same biofilm community, without having to sacrifice a flow cell channel following each observation. Figure 1 shows confocal laser micrographs of a biofilm community cultivated using a labile (tryptic soy broth) (A) and a refractory (sodium benzoate) (B) carbon source. The XY images are projections of 15 optodigital thin sections collected at 1 $\mu$m increments from the attachment surface. A vertical section (XZ) reveals the thickness and surface topography of the same biofilm community. Other compounds, such as resazurin or fluor-dextran conjugates, may also be useful for negatively staining biofilm communities. Positive staining has also been extensively applied for the study of pure culture biofilms and biofilm communities. Acridine orange, Nile red, Texas red isothiocyanate (TRITC) and fluorescein isothiocyanate (FITC) can be used for this purpose.[7,26,81]

Negative staining in conjunction with CLSM may also be used to provide data on biofilm area, thickness and biomass, all of which are necessary for understanding the effects which the physicochemical environment and biological interactions have on the formation and organization within biofilm communities. For example, digital image analysis of CLSM optodigital thin sections revealed varying responses (in terms of biofilm thickness and area coverage) of a degradative community to sole carbon sources having diverse chemical structure. Some examples are shown in Table 1. Substrates such as pentachlorophenol resulted in less biofilm thickness (~3.0 $\mu$m), whereas more labile substrates such as 2,4,6-trichlorobenzoic acid (2,4,6-T) resulted in thicker biofilms (~6.4 $\mu$m). This data also revealed a positive correlation between substrate recalcitrance and the ratio of biofilm thickness to percent area coverage (Fig. 2). When supplied with a labile carbon source, biofilm cells presumably did not rely on one another in order to metabolize the substrate. Under these circumstances, close positioning of cells would be a disadvantage as the cells would compete for carbon. Therefore, an even distribution of cells in biofilms cultivated on labile carbon sources is typically observed.[52,85] When challenged with a refractory substrate, biofilms typically form tight clusters of cells separated by void spaces.[52,86]

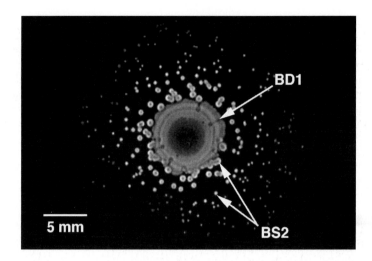

**Figure 4.** Satellite effect showing the formation of satellite colonies of a benzoate-sensitive bacterial strain (*P. fluorescens* strain BS2) around a primary colony of a benzoate-degrader (*P. fluorescens* strain BD1).

The chemical nature of these substrates, especially the octanol-water partition coefficient (o/w), was also found to influence the architecture of the resulting biofilms. In general, high inverse correlations were noted between published[22] $\log_{o/w}$ coefficients such as trichloroethylene (o/w = 2.29), dichlorophenol (3.30), trichlorophenol (4.05), pentachlorophenol (5.01) and 9-fluorenone (3.58), and the thickness ($R^2$ = 0.82) of biofilms formed by the 2,4,6-T community on these substrates (Fig. 3). A high, negative correlation between the $\log_{o/w}$ coefficient and biofilm thickness suggests that higher partition coefficients may have led to higher intracellular accumulations of the test compounds, and consequent reduction in growth rates due to substrate toxicity.

In order to study the effect of various physical, chemical and biological factors on the structural and spatial organization of biofilm communities, it is necessary to determine the relative abundance of single species or groups of bacteria, and also track the fate of specific individuals over time. While negative staining in conjunction with CLSM serves as a nondestructive approach for the analysis of overall biofilm architecture, studies of structural and spatial organization (relative abundance and positioning of various bacteria) within intact biofilm communities can be best achieved by using a combination of either epifluorescence or CLSM and one or more fluorescently labeled oligonucleotide probes or antibodies targeted against specific community members.

Oligonucleotide probes target gene sequences usually within conserved regions of 16S or 23S rRNA of individual species or broader taxonomic groups within the biofilm community.[1,44,51,63,66] Fluorescence *in situ* hybridization (FISH), when applied in conjunction with CLSM, has been shown to provide valuable data on the three-dimensional distribution of specific organisms within biofilms.[53] Furthermore, FISH provides meaningful inferences of the structure-function relationships within the microbial community when combined with microelectrode analyses[14,40] of the physico-chemical environment or with indicators of cell viability, growth or gene expression.[53,54,72]

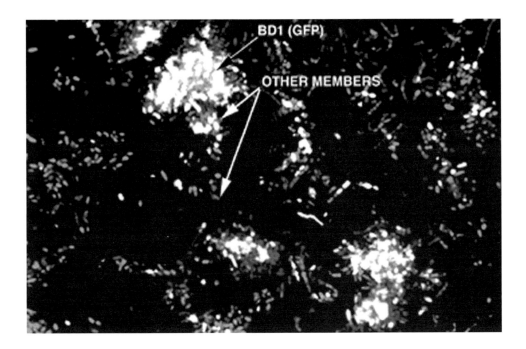

**Figure 5.** Use of GFP to visualize spatial interactions between benzoate-degrading and benzoate-sensitive members of a biofilm community. This biofilm community shown was derived from a pristine soil inoculum and amended with a benzoate-degrading strain of *P. fluorescens* (BD1). Strain BD1 was previously labeled with GFP as a conservative fluorescent marker which permitted visualization of the spatial positioning of BD1 with respect to other members of the biofilm community. Note the formation of organized microcolonies consisting of central BD1 cells surrounded by other members of the biofilm community during growth on sodium benzoate.

Schramm et al.[72] used microelectrodes for $O_2$ and $NO_2/NO_3$ and fluor-conjugated 16S rRNA targeted probes to determine the activity and spatial distribution of nitrifiers within a trickling filter biofilm community. Using these techniques, they showed that closer spatial positioning of *Nitrosomonas* sp. and *Nitrobacter* sp. facilitated the complete conversion of ammonia into nitrate. Moller et al.[53] correlated the rRNA content and size of *Pseudomonas putida* cells within a toluene-degrading biofilter community to rates of toluene degradation. A 16S rRNA targeting probe was used to detect *P. putida* cells. Although the application of FISH does not destroy biofilm architecture, it results in cell death precluding the possibility of repeated analysis of the same biofilm.

Fluorescent antibody probes have also proven valuable for the identification of target organisms from within complex microbial communities. Stewart et al.[78] used fluorescently-labeled monoclonal antibodies to examine the spatial distribution of *Klebsiella pneumoniae* and *P. aeruginosa* within a binary biofilm system. Alternatively, immunogold labeling and fluorescein immunolabeling were used to visualize *Legionella pneumophila* within aquatic biofilms developed on glass and polystyrene surfaces.[65] Interference contrast microscopy was used in an episcopic mode for simultaneous visualization of the immunogold-labeled *L. pneumophila* as well as total biofilm.

More recently, bacteria tagged with the Green Fluorescent Protein (GFP) have seen application in biofilm studies. Organisms genetically tagged in this way may be identified by their green fluorescence (~514 nm) using CLSM or epifluorescence microscopy. This approach has seen usage for the study of structural and spatial organization within biofilm communities.[79] The *gfp* gene, originally derived from the jellyfish *Aequoria victoria*,[11] can easily be introduced into Gram-negative target strains using plasmid vectors.[54] Cloning the *gfp* gene behind a promoter responsible for the expression of a specific functional gene can then be used to monitor *in situ* gene expression. Moller et al.[54] used this approach in conjunction 16S rRNA oligonucleotide probes to examine *tol* gene expression by specific members of a mixed-species biofilm community. Tombolini et al.[83] demonstrated that *gfp* expression was independent of the growth stage of the bacterial strain used, and was detectable even under nutrient-limiting conditions. However, GFP fluorescence is oxygen-dependent[24] and thus may not be suitable for use in oxygen-limited or anaerobic systems. The main advantage of GFP is that it allows direct visualization of specific organisms without involving long staining procedures as required in FISH or fluorescent antibody techniques. It is also noteworthy that a number of GFP variants are now available (www.clontech.com) that have excitation or emission wavelengths different from that of the wild-type gene. Used in conjunction with a multi-line laser excitation source, multiple organisms may simultaneously be tracked within complex biofilm systems.

We recently employed GFP as a cellular marker to examine the spatial organization of a biofilm community grown under different nutrient regimes. When biofilm communities cultivated in flow cells using tryptic soy broth (TSB) were plated on a minimal-salts medium containing 0.15% sodium benzoate as the sole carbon source, satellite colonies of benzoate-sensitive bacteria formed around colonies of benzoate-resistant primary colonies. An example of the satellite effect, involving a benzoate degrading strain of *P. fluorescens* (BD1) and a benzoate-sensitive satellite strain *P. fluorescens* (BS2) is shown in Figure 4. In order to understand the significance of similar protective spatial interactions during attached growth, a microbial inoculum obtained from a pristine soil environment was amended with the benzoate-degrading strain BD1 and cultivated as a biofilm community. Strain BD1 was previously labeled with GFP as a conservative fluorescent marker, permitting localization of BD1 within the biofilm community under different nutrient regimes. When the BD1-amended biofilm community was cultivated using TSB, the cells were randomly distributed. However, a substrate shift to sodium benzoate resulted in more organized microcolonies consisting of central BD1 cells surrounded by other members of the biofilm community (Fig. 5). Notably, the pristine community was unable to grow in flow cells without strain BD1.

Fluorescent Gram stains have been used as group-specific probes to determine the spatial distribution of Gram-negative and Gram-positive organisms within biofilm communities. For example, Wolfaardt et al.[87] applied a fluorescent gram stain to detect regions within a degradative biofilm community associated with a *Bacillus coagulans* strain. However, these and other fluorescently-labeled probes often result in the death of bound cells and hence are not suitable for continuous monitoring of the same biofilm community.

In addition to the species-specific or group-specific probes described above, fluorescent probes are also available for the determination of the physicochemical heterogeneity of biofilms in terms of exopolymer chemistry (fluor-conjugated lectins), charge distribution (fluor-conjugated anionic, cationic and neutral dextrans), diffusion (fluor-conjugated size-fractionated dextrans) and hydrogen ion distribution (carboxyfluorescein).[7,39,52,87] Fluorescent probes are also available for evaluating the metabolic activity or viability of biofilm cells based on the cytoplasmic redox potential (resorufin), electron transport chain activity (cyanoditolyl-tetrazolium chloride) or cell membrane integrity (propidium iodide, ethidium bromide and *Bac*Light™ viability stain). Recently, Korber et al.[29] used the ability of bacteria to undergo plasmolysis as a physical indicator of viability. When subjected to an osmotic shock, only viable cells possessing an intact semipermeable membrane became plasmolyzed. A detailed discussion of the utility and limitations of these and other methods can be found in the recent reviews by Caldwell et al.[8] and Lawrence et al.[36] Overall, the information derived from application of fluorescent molecular probes will facilitate the elucidation of many structure–function relationships within biofilm communities. It should be noted that such approaches are best applied in combination with other analytical methods.

Although scanning and transmission electron microscopy have been frequently applied in biofilm research, specimen preparation often results in shrinkage and dehydration artifacts that alter the biofilm spatial organization. For a more detailed discussion of different microscope systems utilized for biofilm research, including environmental scanning electron microscopy, episcopic differential interference contrast microscopy, Hoffman modulation contrast microscopy and atomic force microscopy, the reader is referred to a recent review by Surman et al.[80]

## V. MONITORING METHODS THAT REQUIRE BIOFILM DISRUPTION

Biofilm communities may also be removed from surfaces, homogenized by sonication, and subjected to various forms of analysis to elucidate the structural organization of the original community using culture-based or DNA-based detection methods. Traditional plating techniques are limited by the unknown species composition of natural systems that precludes the design of a plating medium capable of detecting all or most of the diversity present. Thus, the species composition of a community delineated in this manner represents the "culturable composition" of the community. However, there are several DNA-based techniques available for the determination of structure and diversity of microbial communities obtained from various environments. These techniques can also be applied to biofilm communities when the analysis of spatial organization is not of interest. For example, quantitative fluorescent oligonucleotide hybridization of biofilm cells smeared on glass slides can be performed and subsequently analyzed using epifluorescence or laser microscopy.[53] This allows simultaneous detection and quantification of a specific organism within a biofilm community, but provides no information as to where those specific cells may have been physically located within the biofilm system.

The presence of a specific organism, or a group of bacteria, within an environmental sample can also be detected by fixation of the DNA or RNA extracted from the sample to nylon or nitrocellulose membranes, followed by hybridization to an oligonucleotide probe containing a sequence complementary to a known target sequence. Probes are commonly labeled with [32]P-nucleotide analogs[64,84] as well as biotin- or digoxigenin-containing analogs. Detection of hybridized probes usually involves colorimetric, bioluminescent or chemiluminescent reactions. Fluorescently-labeled probes can directly be detected using epifluorescence microscopy.

Reverse sample genome probing (RSGP) has also been used to detect the presence of specific bacteria in environmental samples. RSGP involves spotting DNA obtained from a target organism onto a membrane filter, followed by the hybridization of labeled environmental bacteria. Shen et al.[75] applied this technique to detect the enrichment of hydrocarbon-degrading bacteria in soil following the addition of easily degradable and refractory hydrocarbons.

The polymerization chain reaction (PCR) can be applied to detect and amplify specific sequences present at a very low level within an environmental sample.[70] Nested PCR entails a preliminary amplification step to amplify a larger gene sequence from the sample, and subsequent PCR reactions on the preliminary PCR products to amplify more specific sequences. For example, Hastings et al.[23] were able to successfully detect *Nitrosospira* 16S rDNA in samples derived from a eutrophic freshwater lake using nested PCR. Most probable number-PCR (MPN-PCR) allows the quantification of specific gene sequences present in an environmental sample by serially diluting the whole-cell DNA or RNA extracts prior to the PCR reaction. Mantynen et al.[43] applied this technique for the detection and enumeration of enterotoxin C-producing *Staphylococcus aureus* NCTC 10655 from fresh cheese. This method was highly sensitive, and *S. aureus* numbers as low as 20 cfu/g could be detected. Degrange et al.[16] similarly applied this method to quantify *Nitrobacter* sp. in coniferous forest soils.

Restriction digestion of products obtained from PCR of repetitive genome sequences (REP-PCR)[17,68] or randomly amplified polymorphic DNA (RAPD) provides organism-specific restriction patterns which can be used to determine of community structure.[55,74] Density-gradient gel electrophoresis (DGGE) of PCR amplified rDNA products can also be applied to resolve differences in community structure.[18,67] Entire rDNA sequences can be amplified using PCR and sequenced to identify organisms within mixed microbial populations. This wide array of PCR techniques is yet to see extensive usage for community-level studies.

Apart from nucleic acid sequence variations, differences in the phospholipid fatty acid (PLFA) content of microorganisms has seen use for the analysis of community structure. For example, Guezennec et al.[19] used signature polar fatty acid profiles of biofilm communities obtained from hydrothermal vent areas to detect the presence of both sulfate-reducing and sulfur-oxidizing bacteria. Neef et al.[56] also applied fatty acid analysis to detect the presence of *Paracoccus* sp. within a denitrifying sand filter used in waste-water treatment. Scholz and Boon[71] used PLFA to analyze the effect of various light regimes on the structure of biofilm communities developing on wooden slides submerged in billabongs. Although fatty acid profiles are useful in comparing microbial communities,[21] the potential of this approach for the elucidation of structural organization within a biofilm community may be limited by culture-dependent differences in the fatty acid content of community members.

## VI. MONITORING ORGANIZATIONAL RELATIONSHIPS OF FUNCTIONAL SIGNIFICANCE

Structural and spatial organization within a biofilm community may best be interpreted when the organizational relationship can be correlated to a specific function of the

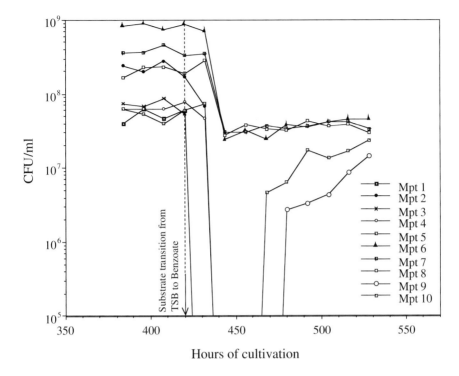

**Figure 6.** Restructuring of a biofilm community in response to a substrate shift from TSB to sodium benzoate. Bacterial adaptation to growth using sodium benzoate as sole carbon source was facilitated by this restructuring process. The number of various colony morphotypes that appeared when effluent collected from the flow cell was plated on tryptic soy agar was used to examine the process of restructuring. Note the appearance, disappearance, or stable maintenance of sets of organisms during this transition.

community. As previously indicated, this can be achieved via a combination of structural gene probes to detect specific members, microelectrode analyses of the ambient environment, or indicators of gene expression, such as GFP or bioluminescence. These methods used in combination often reveal the functional role of specific structural and spatial patterns detected within biofilm communities.[53,54,72]

Bacterial proliferation in various ecosystems almost always involves the association of different organisms for the creation of favorable microenvironments within unfavorable macroenvironments. Traditional plating techniques are sometimes sufficient to detect these functionally significant bacterial interactions. For example, the protective association between benzoate-resistant and benzoate-sensitive strains (Fig. 4) was detected by plating serial dilutions of a biofilm community on a minimal-salts agar containing benzoate as sole carbon source. When the plated cells were positioned sufficiently close together, cell–cell interactions resulted in the formation of satellite colonies of benzoate-sensitive bacteria around primary colonies of benzoate-degrading *Pseudomonas* strains. Thus, it may also be possible to detect similar functionally significant microbial interactions by plating biofilm communities on a medium designed to select for specific functional traits (e.g., pollutant degradation).

Another means of detecting functionally important microbial interactions within a biofilm community involves subjecting the community to environmental perturbations or shifts, and analyzing the changes in community structure (e.g., appearance or disappearance of sets of specific organisms). This approach is based on the assumption that if the organisms within a biofilm community function at the community level, an environmental shift should elicit coordinated responses among community members. The changes in community structure profile should then show sets of responses as opposed to members responding individually. Figure 6 illustrates the changes in the culturable structure (restructuring) of a biofilm community when subjected to a substrate shift from TSB to sodium benzoate. The process of community restructuring was monitored by using the number and types of various colony morphotypes that appeared when emigrants from the flow cell community were plated on tryptic soy agar. When two or more morphotypes appeared or disappeared as part of the culturable and detectable structure of the biofilm community following an environmental transition, it is possible that these organisms were involved in a synergistic or commensalistic association. These organisms may then be isolated and examined by cross streaking against each other on a selective medium to confirm whether the organisms were interacting, or whether they were just a group of independent organisms which happened to immigrate or emigrate simultaneously.

Determination of community-level, sole carbon source (SCS) utilization patterns, in combination with multivariate data analysis, is a relatively new approach to elucidate overall functional capabilities of microbial communities.[15,20,77] Staddon et al.[77] and Haack et al.[20] used principle component analysis (PCA) to resolve sole-carbon utilization differences between soil microbial communities. Degens and Harris[15] similarly used PCA to identify carbon substrates that differentiated five soil communities based on substrate-induced respiration profiles. The extent to which specific substrates were oxidized resulted in SCS utilization patterns allowing for the discrimination of different communities.[20] This approach may also help explain the evolution of functional relationships within biofilms, and thus contribute to our current understanding of the complex interactions which form the basis of co-existence in a communal context.

## VII. CONCLUDING REMARKS

A variety of methods may be applied to determine the structure of microbial communities. Most techniques are limited in their application to analysis of biofilm communities because they often require the biofilm be disrupted to obtain homogenous material for analysis. However, application of fluorescent molecular probes to detect a specific organism or group of organisms can effectively be used to broaden our understanding of the *in situ* spatial organization within biofilm communities. Meaningful inferences on the significance of structural and spatial organization can only be made if the functional value of biofilm organization is determined. Thus techniques that combine the structural and spatial analysis of biofilm communities with their functional roles are preferred for biofilm research.

## REFERENCES

1. Amann RI, Stromley J, Devereux R, et al: Molecular and microscopic identification of sulfate-reducing bacteria in multispecies biofilms. *Appl Environ Microbiol* 58:614–23, 1992

2. Angell P, Machowski WJ, Paul PP, et al: A multiple chemostat system for consortia studies on microbially influenced corrosion. *J Microbiol Meth* 30:173–8, 1997

3. Baier RE: Substrate influence on adhesion of microorganisms and their resultant new surface properties. In: Bitton G, Marshall KC, eds. *Adsorption of Microorganisms to Surfaces.* Wiley, New York, 1980:55–104

4. Brading MG, Jass J, Lappin-Scott HM: Dynamics of bacterial biofilm formation. In: Lappin-Scott HM, Costerton JW, eds. *Microbial Biofilms.* Cambridge University Press, Cambridge, UK, 1995: 46–63

5. Bradshaw DJ, Marsh PD, Watson GK, et al: Role of *Fusobacterium nucleatum* and coaggregation in anaerobe survival in planktonic and biofilm oral microbial communities during aeration. *Infect Immun* 66: 4729–32, 1998

6. Caldwell DE, Korber DR, Lawrence JR: Imaging bacterial cells by fluorescence exclusion using scanning confocal laser microscopy. *J Microbiol Meth* 15:249–61, 1992

7. Caldwell DE, Korber DR, Lawrence JR: Confocal laser microscopy and computer image analysis. *Adv Microb Ecol* 12:1–67, 1992

8. Caldwell DE, Korber DR, Wolfaardt GM, et al: Do bacterial communities transcend Darwinism? *Adv Microb Ecol* 15:105–91, 1997

9. Caldwell DE, Lawrence JR: Growth kinetics of *Pseudomonas fluorescens* microcolonies within the hydrodynamic boundary layers of surface microenvironments. *Microb Ecol* 12:299–312, 1986

10. Caldwell DE, Wolfaardt GM, Korber DR, et al: Cultivation of microbial consortia and communities. In: Hurst CJ, Knudsen GR, McInerney MJ, et al., eds. *Manual of Environmental Microbiology.* American Society for Microbiology Press, Washington, DC, 1997:79–90

11. Chalfie M, Tu Y, Euskirchen G, et al: Green fluorescent protein as a marker for gene expression. *Science* 263:802–5, 1994

12. Characklis WG, McFeters GA, Marshall KC: Physiological ecology in biofilm systems. In: Characklis WG, Marshall KC, eds. *Biofilms.* J. Wiley and Sons, New York, 1990:341–93

13. Cooksey KE, Wigglesworth, Cooksey B: Adhesion of bacteria and diatoms to surfaces in the sea: A review. *Aquatic Microbial Ecology* 9:87–96, 1995

14. DeBeer D, Stoodley P, Roe F, et al: Effects of biofilm structures on oxygen distribution and mass transport. *Biotechnol Bioeng* 43:1131–8, 1994

15. Degens BP, Harris JA: Development of a physiological approach to measuring the catabolic diversity of soil microbial communities. *Soil Biol Biochem* 29:1309–20, 1997

16. Degrange V, Couteaux MM, Anderson JM, et al: Nitrification and occurrence of *Nitrobacter* by MPN-PCR in low and high nitrifying coniferous forest soils. *Plant Soil* 198:201–8, 1998

17. Del-Vecchio VG, Petroziello JM, Gress MJ, et al: Repetitive-sequence PCR. *J Clin Microbiol* 33:2141–4, 1995

18. Gillan DC, Speksnijder AG, Zwart G, et al: Genetic diversity of the biofilm covering *Montacuta ferruginosa* (Mollusca, Bivalvia) as evaluated by denaturing gradient gel electrophoresis analysis and cloning of PCR-amplified gene fragments coding for 16S rRNA. *Appl Environ Microbiol* 64:3464–72, 1998

19. Guezennec J, Ortega MO, Raguenes G, et al: Bacterial colonization of artificial substrate in the vicinity of deep-sea hydrothermal vents. *FEMS Microbiol Ecol* 26:89–99, 1998

20. Haack SK, Garchow H, Klug MJ, et al: Analysis of factors affecting the accuracy, reproducibility, and interpretation of microbial community carbon source utilization patterns. *Appl Environ Microbiol* 61:1458–68, 1995

21. Haack SK, Garchow H, Odelson DA, et al: Accuracy, reproducibility, and interpretation of fatty acid methyl ester profiles of model bacterial communities. *Appl Environ Microbiol* 60:2483–93, 1994

22. Hansch C, Leo A: *Substituent Constants for Correlation Analysis in Chemistry and Biology.* Wiley-Interscience, New York, 1979:339

23. Hastings RC, Saunders JR, Hall GH, et al: Application of molecular biological techniques to a seasonal study of ammonia oxidation in a eutrophic freshwater lake. *Appl Environ Microbiol* 643674–82, 1998

24. Heim R, Prasher DC, Tsien RY: Wavelength mutations and post-translational autoxidation of green fluorescent protein. *Proc Natl Acad Sci USA* 91:12501–4, 1994

25. Holmes AR, Gopal PK, Jenkinson HF: Adherence of *Candida albicans* to a cell surface polysaccharide receptor on *Streptococcus gordonii. Infect Immun* 63:1827–34, 1995

26. Hood SK, Zottola EA: Adherence to stainless steel by foodborne microorganisms during growth in model food systems. *Int J Food Microbiol* 37:145–53, 1997

27. Kinniment SL, Wimpenny JWT, Adams D, et al: Development of a steady-state oral microbial biofilm community using the constant-depth film fermenter. *Microbiol* 142:631–8, 1996

28. Kolenbrander PE, Parrish KD, Andersen RN, et al: Intergeneric coaggregation of oral *Treponema* sp. with *Fusobacterium* sp. and intrageneric coaggregation among *Fusobacterium* sp. *Infect Immun* 63:4584–8, 1995

29. Korber DR, Choi A, Wolfaardt GM, et al: Bacterial plasmolysis as a physical indicator of viability. *Appl Environ Microbiol* 62:3939–47, 1996

30. Korber DR, James GA, Costerton JW: Evaluation of fleroxacin activity against established *Pseudomonas fluorescens* biofilms. *Appl Environ Microbiol* 60:1663–9, 1994

31. Korber DR, Lawrence JR, Caldwell DE: Effect of motility on surface colonization and reproductive success of *Pseudomonas fluorescens* in dual-dilution continuous culture and batch culture systems. *Appl Environ Microbiol* 60:1421–9, 1994

32. Korber DR, Lawrence JR, Lappin-Scott HM, et al: Growth of microorganisms on surfaces, In: Lappin-Scott HM, Costerton JW, eds. *Microbial Biofilms*, Cambridge University Press, Cambridge, UK, 1995:15–45

33. Korber DR, Lawrence JR, Sutton B, et al: Effects of laminar flow velocity on the kinetics of surface colonization by mot+ and mot- *Pseudomonas fluorescens. Microb Ecol* 18:1–19, 1989

34. Lawrence JR, Korber DR, Caldwell DE: Computer-enhanced darkfield microscopy for the quantitative analysis of bacterial growth and behavior on surfaces. *J Microbiol Meth* 10:123–38, 1989

35. Lawrence JR, Korber DR, Hoyle BD, et al: Optical sectioning of microbial biofilms. *J Bacteriol* 173:6558–67, 1991

36. Lawrence JR, Korber DR, Wolfaardt GM, et al: Analytical imaging and microscopy techniques. In: Hurst CJ, Knudsen GR, McInerney MJ, et al., eds. *Manual of Environmental Microbiology.* American Society for Microbiology Press, Washington, DC, 1997:79–90

37. Lawrence JR, Korber DR: Aspects of microbial surface colonization behavior. In: Guerrero R, Pedros-Alio C, eds, *Trends in Microbial Ecology.* Spanish Society for Microbiology, Barcelona, Spain, 1994:113–8

38. Lawrence JR, Wolfaardt GM, Korber DR, et al: Behavioral strategies of surface colonizing bacteria. *Adv Microb Ecol* 14:1–75, 1995

39. Lawrence JR, Wolfaardt GM, Korber DR: Determination of diffusion coefficients in biofilms by confocal laser microscopy. *Appl Environ Microbiol* 60:1166–73, 1994

40. Lewandowski Z, Altobelli SA, Fukushima E: NMR and microelectrode studies of hydrodynamics and kinetics in biofilms. *Biotechnol Prog* 9:40–5, 1993

41. Lo CS , Hughes CV: Identification and characterization of a protease from *Streptococcus oralis* C104. *Oral Microbiol Immunol* 11:181–7, 1996

42. Lutterbach MT, Franca FPD: Biofilm formation monitoring in an industrial open water cooling system. *Revista de Microbiologia* 28:106–9, 1997

43. Mantynen V, Niemela S, Kaijalainen S, et al: MPN-PCR-quantification method for staphylococcal enterotoxin c1 gene from fresh cheese. *Int J Food Microbiol* 36:135–43, 1997

44. Manz W, Amann R, Szewzyk R, et al: *In situ* identification of Legionellaceae using 16S rRNA-targeted oligonucleotide probes and confocal laser scanning microscopy. *Microbiol* 141:29–39, 1995

45. Marsh PD, Bradshaw DJ: Dental plaque as a biofilm. *J Ind Microbiol* 15:169-75, 1995

46. Marshall KC, Stout R, Mitchell R: Mechanisms of the initial events in the sorption of marine bacteria to solid surfaces. *J Gen Microbiol* 68:337–48, 1971

47. Marshall KC: Colonization, adhesion and biofilms. In: Hurst CJ, Knudsen GR, McInerney MJ, et al, eds. *Manual of Environmental Microbiology.* American Society for Microbiology Press, Washington, DC, 1997:358–65

48. Mattila-Sandholm T, Wirtanen G: Biofilm formation in the industry: a review. *Food Rev Int* 8:573–603, 1992

49. McAllister TA, Bae HD, Jones GA, Cheng KJ: Microbial attachment and feed digestion in the rumen. *J Anim Sci* 72:3004–18, 1994

50. Millsap KW, Reid G, Van Der Mei HC, et al: Adhesion of *Lactobacillus* species in urine and phosphate buffer to silicone rubber and glass under flow. *Biomaterials* 18:87–91, 1997

51. Mobarry BK, Wagner M, Urbain V, et al: Phylogenetic probes for analyzing abundance and spatial organization of nitrifying bacteria. *Appl Environ Microbiol* 62:2156–62, 1996

52. Moller S, Korber DR, Wolfaardt GM, et al: Impact of nutrient composition on a degradative biofilm community. *Appl Environ Microbiol* 63:2432–8, 1997

53. Moller S, Pedersen AR, Poulsen LK, et al: Activity and three-dimensional distribution of toluene-degrading *Pseudomonas putida* in a multispecies biofilm assessed by quantitative *in situ* hybridization and scanning confocal laser microscopy. *Appl Environ Microbiol* 62:4632–40, 1996

54. Moller S, Sternberg C, Andersen JB, et al: *In situ* gene expression in mixed-culture biofilms: Evidence of metabolic interactions between community members. *Appl Environ Microbiol* 64:721–32, 1998

55. Morea M, Baruzzi F, Cappa F, et al: Molecular characterization of the *Lactobacillus* community in traditional processing of Mozzarella cheese. *Int J Food Microbiol* 43:53–60, 1998

56. Neef A, Zaglauer A, Meier H, et al: Population analysis in a denitrifying sand filter: Conventional and *in situ* identification of *Paracoccus* sp. in methanol-fed biofilms. *Appl Environ Microbiol* 62:4329–39, 1996

57. Nesbitt WE, Fukushima H, Leung KP, et al: Coaggregation of *Prevotella intermedia* with oral *Actinomyces* species. *Infect Immun* 61:2011–4, 1993

58. Neu TR, Lawrence JR: Development and structure of microbial biofilms in river water studied by confocal laser scanning microscopy. *FEMS Microbiol Ecol* 24:11–25, 1997

59. Notermans S, Doormans JA, Mead GC: Contribution of surface attachment to the establishment of microorganisms in food processing plants: a review. *Biofouling* 5:1–16, 1991

60. O'Toole GA, Kolter R: Flagellar and twitching motility are necessary for *Pseudomonas aeruginosa* biofilm development. *Mol Microbiol* 30:295–304, 1998

61. Palmer RJ Jr, Caldwell DE: A flowcell for the study of plaque removal and regrowth. *J Microbiol Meth* 24:171–82, 1995

62. Peyton BM, Characklis WG: A statistical analysis of the effect of substrate utilization and shear stress on the kinetics of biofilm detachment. *Biotechnol Bioeng* 41:728–35, 1993

63. Poulsen LK, Dalton HM, Angles ML, et al: Simultaneous determination of gene expression and bacterial identity in single cells in defined mixtures of pure cultures. *Appl Environ Microbiol* 63:3698–702, 1997

64. Raskin L, Poulsen LK, Noguera DR, et al: Quantification of methanogenic groups in anaerobic biological reactors by oligonucleotide probe hybridization. *Appl Environ Microbiol* 60:1241–8, 1994

65. Robers J, Keevil CW: Immunogold and fluorescein immunolabelling of *Legionella pneumophila* within an aquatic biofilm visualized by using episcopic differential interference contrast microscopy. *Appl Environ Microbiol* 58:2326–30, 1992

66. Rothemund C, Amann R, Klugbauer S, et al: Microflora of 2,4-dichlorophenoxyacetic acid degrading biofilms on gas permeable membranes. *Syst Appl Microbiol* 19:608–15, 1996

67. Santegoeds CM, Ferdelman TG, Muyzer G, et al: Structural and functional dynamics of sulfate-reducing populations in bacterial biofilms. *Appl Environ Microbiol* 64:3731–9, 1998

68. Sarand I, Timonen S, Nurmiaho LE, et al: Microbial biofilms and catabolic plasmid harbouring degradative fluorescent pseudomonads in Scots pine mycorrhizospheres developed petroleum contaminated soil. *FEMS Microbiol Ecol* 27:115–26, 1998

69. Sasahara KC, Zottola EA: Biofilm formation by *Listeria monocytogenes* utilizes a primary colonizing microorganism in flowing systems. *J Food Prot* 56:1022–8, 1993

70. Schneegurt MA, Kulpa CF Jr: The application of molecular techniques in environmental biotechnology for monitoring microbial systems. *Biotechnol Appl Biochem* 27:73–9, 1998

71. Scholz O, Boon PI: Biofilms on submerged River Red gum (*Eucalyptus camaldulensis* Dhenh. Myrtaceae) wood in billabongs: An analysis of bacterial assemblages using phospholipid profiles. *Hydrobiologia* 259:169–78, 1993

72. Schramm A, Larsen LH, Revsbech NP, et al: Structure and function of a nitrifying biofilm as determined by microelectrodes and fluorescent oligonucleotide probes. *Water Sci Technol* 36:263–70, 1997

73. Schupbach P, Osterwalder V, Guggenheim B: Human root caries: Microbiota of a limited number of root caries lesions. *Caries Research* 30:52–64, 1996

74. Segonds C, Bingen E, Couetdic G, et al: Genotypic analysis of *Burkholderia cepacia* isolates from 13 French cystic fibrosis centers. *J Clin Microbiol* 35:2055–60, 1997

75. Shen Y, Stehmeier, LG, Voordouw G: Identification of hydrocarbon-degrading bacteria in soil by reverse sample genome probing. *Appl Environ Microbiol* 64:637–45, 1998

76. Speers JGS, Gilmour A: The influence of milk and milk components on the attachment of bacteria to farm dairy equipment surfaces. *J Appl Bacteriol* 59:325–32, 1985

77. Staddon WJ, Duchesne LC, Trevors JT: Microbial diversity and community structure of post disturbance forest soils as determined by sole-carbon-source utilization patterns. *Microb Ecol* 34:125–30, 1997

78. Stewart PS, Camper AK, Handran SD, et al: Spatial distribution and coexistence of *Klebsiella pneumoniae* and *Pseudomonas aeruginosa* in biofilms. *Microb Ecol* 33:2–10, 1997

79. Stretton S, Techkarnjanaruk S, McLennan AM, et al: Use of green fluorescent protein to tag and investigate gene expression in marine bacteria. *Appl Environ Microbiol* 64:2554–9, 1998

80. Surman SB, Walker JT, Goddard DT, et al: Comparison of microscope techniques for the examination of biofilms. *J Microbiol Meth* 25:57–70, 1996

81. Tang RJ, Cooney JJ: Effects of marine paints on microbial biofilm development on three materials.*J Industr Microbiol Biotechnol* 20:275–80, 1998

82. Teixeira P, Oliveira R: The importance of surface properties in the selection of supports for nitrification in airlift bioreactors. *Bioprocess Eng* 19:143–7, 1998

83. Tombolini R, Unge A, Davey ME, et al: Flow cytometric and microscopic analysis of GFP-tagged *Pseudomonas fluorescens* bacteria. *FEMS Microbiol Ecol* 22:17–28, 1997

84. Wang GCY, Wang Y: Rapid differentiation of bacterial species with multiple probes of different lengths in a single slot blot hybridization. *Appl Environ Microbiol* 61:4269–73, 1995

85. Wolfaardt GM, Lawrence JR, Robarts RD, et al: The role of interactions, sessile growth, and nutrient amendments on the degradative efficiency of a microbial consortium. *Can J Microbiol* 40:331–40, 1994

86. Wolfaardt GM, Lawrence JR, Robarts RD, et al: Multicellular organization in a degradative biofilm community. *Appl Environ Microbiol* 60:434–46, 1994

87. Wolfaardt GM, Lawrence JR, Roberts RD, et al: In situ characterization of biofilm exopolymers involved in the accumulation of chlorinated organics. *Microb Ecol* 35:213–23, 1998

88. Zottola EA, Sasahara KC: Microbial biofilms in food processing industry—should they be a concern? *Int J Food Microbiol* 23:125–48, 1994

# 11

# Models and Measurement
# of Bacterial Growth Rates on Polymers

**William G. Pitt[1] and Alan J. Barton[2]**

[1]*Chemical Engineering Department, Brigham Young University, Provo, UT, USA*
[2]*Department of Pediatrics and Communicable Diseases, University of Michigan Hospitals, Ann Arbor, MI, USA*

## Nomenclature

| | |
|---|---|
| $C_I(t,x)$ | concentration of planktonic bacteria at the adsorbing interface |
| $g$ | doubling time |
| $k$ | average growth rate constant (adsorbed bacteria) |
| $k_{eff}$ | effective adsorption rate constant |
| $k_{off}$ | average desorption rate constant |
| $k_{on}$ | average adsorption rate constant |
| $k_p$ | average growth rate constant of planktonic bacteria |
| $L$ | length from beginning of adsorbing surface to the viewing area |
| $N(t,x)$ | number of cells per area as a function of time and position |
| $N_o$ | initial number of cells per viewing area at time zero |
| PLA | poly(lactic acid) |
| POE | poly(ortho ester) |
| PSF | polysulfone |
| $R_{on}$ | rate of adsorption |
| $t$ | time |
| $x$ | spatial position |
| $\alpha$ | proportionality constant relating the bacterial concentration at the interface to the total cells desorbed into the volume of liquid in the flow cell |

## I. INTRODUCTION

The historical investigation of the role of bacteria in disease and infection has been to study the growth of planktonic bacteria and its response to various treatments. However, as more medical devices are used inside the human body, there is increasing concern about the pathogenesis of infections associated with the surface of the implant. This includes understanding the source of infection, the development into a biofilm, and the response of the infectious organism to medical treatment. Such surface-associated bacteria are a critical concern in medicine because they appear to be recalcitrant to the conventional antibiotic therapy that has been developed to target planktonic bacteria.[2,15]

*Handbook of Bacterial Adhesion: Principles, Methods, and Applications*
Edited by: Y. H. An and R. J. Friedman © Humana Press Inc., Totowa, NJ

In most situations, the infection can be suppressed by antibiotic therapy, but it usually returns after the therapy is terminated; thus removal of the infected device is required to eliminate the infection. While many reports in the literature are concerned with events or chemistry leading to the initial adhesion event, it is the subsequent growth of the bacteria on the surface that leads to the symptoms of infection. Thus understanding and measuring growth rates on surfaces is an important aspect of understanding and controlling bacterial infections on implanted devices.

Sessile, or surface-adherent bacteria differ greatly from their planktonic counterparts.[10,15,16,19] Various genes are repressed or promoted, leading to differences in protein expression, exopolysaccharide production, and metabolic rate. As in suspension, the metabolic rate of sessile bacteria appears to be a strong function of the growth conditions.[11,18,20,31] Van Loosdrect[31] has reviewed the comparison of planktonic and sessile growth rates and found that some sessile bacteria grow at about the same rate as planktonic bacteria, while other sessile bacteria have a minor increase or decrease in growth rate. Since that publication, there have been other reports of increased[18] or decreased[6,20] growth rates of adherent bacteria. One complication of measuring bacterial growth is that many studies approximate surface growth by quantifying the amount of bacteria on the surface at various times after exposure to a suspension of bacteria.[7,8,17,20] However, they do not always distinguish between the bacteria that adsorbed from the suspension and those that multiplied on the surface after adsorption. In addition, the possibility of, or the measurement of desorption is often neglected.[13]

The purpose of this work is to develop mathematical models that include the adsorption, desorption and growth of bacteria during early stages of colonization on a surface. When combined with good experimentation, these models give insight and allow regression of growth rates or adsorption rates of bacteria, thus allowing experimental studies of sessile growth rates as a function of the substrate and growth conditions. Following our development of the models, an example is presented in which experimental data was used to obtain the growth rates of three bacteria on three different medically important polymers. Hopefully such models and experiments will be valuable in developing methods to understand and eventually resist bacterial colonization and growth on medical implants.

## II. MATHEMATICAL MODELS

This section describes the mathematical formulation of several models of the population of bacteria growing on surfaces, and the population when combined with desorption from the surface and adsorption to the surface from the planktonic bacteria adjacent to the surface. The models are arranged from simple to more complex, and there are some cases of simplification of the more complex models.

### A. Surface Growth Only

This first model describes the surface population of bacteria when there is neither adsorption to nor desorption from the surface. The change in population is only due to cell growth (or death) on the surface. To develop this model, we will start with the following assumptions.

1. All cells are growing at a constant average rate.
2. There is no lag phase in the growth.
3. At an initial time ($t = 0$), the surface has a distribution of bacteria.
4. The is no desorption or adsorption of cells.

In experimental systems, cells on a surface have a distribution of growth and division rates. However if the population is large enough, the growth rate of the collection of cells on the entire surface or the cells within an observation can be represented by an average growth rate that can be determined from experimental measurements. We suggest that at least 30 cells be used to estimate an average growth rate. After an initial lag phase, constant growth rate is a fairly good assumption as long as the environmental conditions are constant (temperature, pH, nutrients, oxygen tension, etc.), and there is sufficient convection to prevent the buildup of inhibitory metabolic byproducts.

Assumption 3 above simply states that there is a distribution of cells on the surface. The cells do not necessarily need to be in a uniform distribution, and in many cases, cell to cell interactions preclude uniform or random distributions.[29] However, these mathematics are valid for any initial distribution as long as the same area of surface is observed and compared throughout the experiment. We assume that within the area of observation at time $t$ there is $N(t)$ number of cells per area, and this number changes with time. After a time increment $\Delta t$, the new population of cells is

$$N(t+\Delta t) = N(t)+kN(t)\Delta t \tag{1}$$

where $k$ is an average growth rate constant. Rearranging this equation and taking the limit as $\Delta t$ goes to zero gives

$$\lim_{\Delta t \to 0} \left( \frac{N(t + \Delta t) - N(t)}{\Delta t} \right) = kN(t) \tag{2}$$

or

$$\frac{dN(t)}{dt} = kN(t). \tag{3}$$

Thus the rate of growth is proportional to the surface population. Integration of this equation from an initial time ($t = 0$) with $N_o$ bacteria present, to any observation at time $t$, gives the population

$$N(t) = N_0 \exp(kt). \tag{4}$$

A common term used in microbiology is $g$, the amount of time required for a population to double in size. Using this definition of $g$, the population as a function of time is given by

$$N(t) = N_0\, 2^{t/g} \tag{5}$$

If we equate Eqs. (4) and (5), we arrive at the relationship between the doubling time and the average growth rate constant:

$$g = \ln2/k. \tag{6}$$

Equations (4) and (5) give continuous models for the population of bacteria as a function of time. However, most lab measurements are discrete. If the interval between discrete observations is uniform, the population at the $i^{th}$ measurement after time zero is given by

$$N_i = N_0\, 2^{i\Delta t/g} \tag{7}$$

or

$$N_i = N_0 \exp(ik\Delta t) \tag{8}$$

where $N_i$ is the population after the $i^{th}$ time interval. Either of these equations models exponential growth on a surface when $k > 0$. These equations can also be used to model killing of bacteria on a surface, such as when antimicrobial agents are applied. In such a case, $k < 0$, and $N$ would represent the population of living bacteria (i.e., dead bacteria remaining on the surface are not included in the population).

### B. Surface Growth and Desorption

In this model, the growing cells are allowed to leave the surface, but no adsorption is allowed. Assumptions 1–3 in the previous model are applied, and the fourth assumption is replaced with the assumption that all bacteria have an equal probability of desorption. This latter assumption may not always be valid, since we have observed that following division, the newly formed cells have a higher probability of desorption.[6] Meinders et al. have proposed that there is a time dependence to desorption such that the desorption probability decreases as the surface residence time increases.[25] In such a case, the $k_{off}$ would not be a time-invariant constant for each individual cell. However, since the population would have a distribution of individual values of $k_{off}$, an average $k_{off}$ would remain constant because the cells are dividing at a constant average rate, producing a fairly stable distribution of individual values of $k_{off}$. The variability in $k_{off}$ due to average versus individual $k_{off}$ can be reduced by selecting a viewing area with a large bacterial population.

A consequence of our fourth assumption is that the rate of desorption is proportional to the surface population of bacteria. Thus after a discrete time interval, the new surface population is

$$N(t+\Delta t) = N(t)+kN(t)\Delta t\text{-}k_{off}N(t)\Delta t \tag{9}$$

where $k_{off}$ is the average desorption rate constant. Now Eqs. (2) – (4) become

$$\lim_{\Delta t \to 0} \left( \frac{N(t + \Delta t) - N(t)}{\Delta t} \right) = kN(t) - k_{off}N(t) = (k - k_{off})N(t) \tag{10}$$

$$\frac{dN(t)}{dt} = (k - k_{off})N(t) \tag{11}$$

$$N(t) = N_0 \exp[(k\text{-}k_{off})t]. \tag{12}$$

The discretized form of the latter equation becomes

$$\frac{N_i}{N_o} = \exp[(k - k_{off})i\Delta t] = 2^{[(1/g - k_{off}/\ln 2)i\Delta t]}. \tag{13}$$

There exists a problem in estimation of $k_{off}$ from experimental data. Since $k$ and $k_{off}$ always appear together, a measurement of $N$ as a function of time cannot give $k$ and $k_{off}$ independently; thus one of these parameters will have to be measured independently. In our lab, we have measured $k_{off}$ independently by video microscopy coupled with image analysis.[6] In a large field of bacteria, the number of desorbing bacteria was averaged over time to get an estimate of $k_{off}$.

Equations (12) and (13) show that, although the growth rate may be positive, if the desorption rate is greater than the growth rate, the population on the surface will decrease with time. Conversely, if $(k - k_{off}) > 0$, the surface population will grow exponentially with time. However, in the lab we do not observe this runaway growth at long times for several reasons. Foremost, when the bacteria grow to the concentration of a multilayer biofilm, cells on the bottom of the biofilm eventually have less access to nutrients and oxygen, and their growth rate decreases, or even stops. In addition, these mature biofilms usually develop a nonuniform structure containing pores, channels, and clumps or mushroom-shaped structures of bacterial aggregates held together by an exopolysaccharide matrix.[15,23,30,33] Because the convoluted surface presents an increased area available for desorption, and the desorption can occur as cell aggregates in addition to single cells, the desorption rate usually increases. Eventually a balance is reached between growth and desorption, and a quasi-steady state biofilm thickness is obtained. Modeling of the non-uniform growth rates and structures of mature biofilms is beyond the scope of this present analysis.

## C. Surface Growth with Adsorption

The addition of an adsorption term greatly complicates the growth model because adsorption itself is such a complex phenomenon. For example, adsorption may occur from a planktonic suspension in which the population changes with position or time, or in which the flux or convection of bacteria to the surface is governed by geometry and fluid flow. Once a bacterium approaches within about a micron of the surface, then the probability of adsorption may be governed by fluid shear stresses, chemistry of the surface and bacterium, electrical charge of the surface and bacterium, and many other parameters.[1,3,12,27] If we assume that for a given surface–bacterium pair, the probability of adsorption (or the rate of adsorption) is only a function of the concentration of the planktonic bacteria adjacent to the surface, the adsorption will be governed by a rate constant $(k_{on})$ and the flux of bacteria to the surface, or the concentration of bacteria presented to the surface. The bacterial flux has been modeled as a function of time and position for some flow geometries including rotating disks,[21,28,32] radial flow chambers,[7,17] and other geometries.[25]

The growth rate of the bacterial surface population is the sum of the growth of adherent bacteria and the addition of adsorbing bacteria, and is given by

$$\frac{dN(t,x)}{dt} = kN(t,x) + k_{on}C_I(t,x) \tag{14}$$

where $C_I(t,x)$ is the concentration of planktonic bacteria at the interface and is a function of time and position. The solution to this equation is an integral equation that requires knowledge of $C_I(t,x)$ to solve for $N(t,x)$:

$$N(t,x) = N_o e^{kt} + k_{on} e^{kt} \int_0^t C_I(t,x) e^{-kt} dt. \tag{15}$$

If $C_I(t,x)$ is a complex function, this equation may be difficult to solve analytically; however, it may be solved numerically if the time and position dependence of $C_I(t,x)$ are known. Additional complications arise when the bacteria are suspended in a growth medium because the planktonic population has a growth rate constant $k_p$, and thus $C_I(t,x)$ will increase exponentially with time. This may be one of the reasons that many researchers resuspend their bacterial culture in buffer when studying the rate of adsorption.[17,21,26,29] However, resuspension in a non-nutrient solution may reduce the metabolic activity and profoundly impact the growth rate under investigation. One study found that the percentage of metabolically active cells on a surface was only 2% in a buffer, compared to 67% in a minimal growth media.[20]

There are some simplifying cases that give insight into the behavior of Eq. (15). First, if either $k_{on}$ or $C_I(t,x)$ is zero (no adsorption or no planktonic bacteria), then Eq. (15) reduces to Eq. (4). If $C_I$ is a constant (in time and position), then Eq. (15) reduces to

$$N(t) = N_o e^{kt} + \frac{k_{on} C_I e^{kt}}{k} (1 - e^{-kt}). \tag{16}$$

A constant value of $C_I$ is usually a good assumption when the planktonic population is constant (as in a chemostat), the planktonic bacteria are not growing (as when resuspended in buffer), the planktonic concentration at the interface is very high compared to the adsorption (or desorption) rate, or when turbulent flow reduces any concentration boundary layer at the interface. If these conditions are not met, one should assume that $C_I$ is a function of time and position.

If one assumes that $N_o = 0$, Eq. (16) can be further reduced to the equation developed by Caldwell et al. to model adsorption and growth on surfaces.[9,13,22,24]

### D. Surface Growth with Adsorption and Desorption

The combination of growth, adsorption and desorption is not much more difficult to formulate mathematically; in Eq. (14) – (16), the growth constant $k$ is replaced by $k - k_{off}$. For example, Eq. (15) becomes

$$N(t,x) = N_o e^{(k-k_{off})t} + k_{on} e^{(k-k_{off})t} \int_0^t C_I(t,x) e^{-(k-k_{off})t} dt. \tag{17}$$

In most experimental systems, growth, adsorption and desorption all occur simultaneously. This complicates the calculation of the value of $C_I$ because desorbing bacteria increase the bacterial concentration at the interface, and thus $C_I$ becomes a function of $t$, $x$, $N(t,x)$, $k_p$ and $k_{off}$. Even if the surface is flushed with buffer or saline to remove planktonic bacteria, any bacterial desorption occurring will create a concentration of planktonic bacteria at the interface that can subsequently adsorb.

We have developed an experimental system that allows us to simplify Eq. (17) such that we can extract the average growth rate constant $k$, which is the primary objective of many of our studies of bacteria growth on medically important polymers. In these experiments, the surface is seeded with a partial layer of bacteria by flowing a bacterial culture over the surface for 1 h. Then a buffer is introduced to purge the flow system of planktonic bacteria. Finally a growth medium is passed over the surface so the adherent bacteria can grow. This eliminates any planktonic bacteria flowing into the system. Although the incoming media has no bacteria, cells on the surface upstream from the viewing area desorb and flow downstream over the viewing area where they may adsorb. We assume that the desorption rate upstream is proportional to the surface concentration upstream, and that the desorption from all the area between the entrance and viewing area continually adds to and increases the local $C_I$ in the particular viewing area. Thus the local $C_I$ is proportional to $N(t)$, $k_{off}$, and the length $L$ between the beginning of the adsorbing surface and the local viewing area. Since the surface population (and thus the desorption rate) may vary along the length of the surface we must integrate along the length of the surface:

$$C_I(t,x) = \alpha \int_0^L k_{off} N(t,x) dx \qquad (18)$$

where $\alpha$ is a proportionality constant relating the concentration at the interface to the total cells desorbed into the volume of liquid in the flow cell. We assume that the initial distribution of adsorbed cells is uniform, and that the growth, desorption and adsorption are fairly uniform along the length $L$, so that $N$ is only a function of time. This assumption allows Eq. (18) to be integrated to

$$C_I(t,x) = \alpha K_{off} N(t,x) L. \qquad (19)$$

In this equation we retain the possibility that $N$ may be a function of position as described in the next paragraph.

Of the many assumptions in this model, the one subject to the most scrutiny is that the initial distribution of adsorbed cells is uniform. Obviously if the initial deposition of cells is similar throughout the flow cell, this assumption is valid. However, models of initial adsorption show that the flux of bacteria to a surface in a laminar flow chamber is not uniform, and increased deposition would be expected near the entrance to the flow chamber.[17,25] Even in this or another situation of nonuniform initial deposition, there is an initial proportionality between the surface concentration upstream and the surface concentration in the viewing area; and if populations in both areas grow exponentially with the same rate constant, the ratio or proportionality of the two populations remains constant. Such a case would simply introduce another proportionality constant in Eq. (19) that could be combined with the constant $\alpha$; and thus the form of Eq. (19) would still be valid.

The adsorption rate, $R_{on}$, is the product of the adsorption rate constant and $C_I$ and can be related to an effective adsorption rate constant, $k_{eff}$, and the surface population by

$$R_{on} = k_{on} C_I(t,x) = \alpha k_{on} k_{off} N(t,x) L = k_{off} N(t,x). \qquad (20)$$

The total bacterial accumulation in one viewing field is the sum of the growth, desorption and adsorption,

$$\frac{\partial N(t,x)}{\partial t} = kN(t,x) - k_{off} N(t,x) + k_{eff} N(t,x) \qquad (21)$$

and the integrated form of this equation is

$$\frac{N(t,x)}{N_o} = e^{(k-k_{off}+k_{eff})t}. \tag{22}$$

Since $k$, $k_{off}$, and $k_{eff}$ all appear together in this equation, one cannot regress all three parameters simultaneously from the same data set. In our lab, we make measurements of the adsorption and desorption rates to obtain the latter two parameters, and then regress the value of $k$. Similarly, independent measurement of any two parameters will allow the regression of the third from a set of data. Eq. (22) was used to estimate the growth rate of bacteria on medically relevant polymers as described below.

## III. METHODS AND MATERIALS

### A. Bacteria Preparation

The day prior to the experiment, tryptic soy broth (TSB) without dextrose was inoculated with a colony of *P. aeruginosa* (GNRNF-Ps-1), *S. epidermidis* (ATCC 12228) or *E. coli* (ATCC 10798) and incubated for 16 h at 37°C. One mL of the overnight culture was pipetted into 100 mL of TSB, and the new culture was incubated 6 h at 37°C.

### B. Polymer Preparation and Characterization

Polyorthoester (POE, mol wt of 82,000) was obtained from the lab of Dr. A. U. Daniels at the University of Utah. Poly(L-lactic acid) (PLA) was obtained from Boehringer-Ingelheim. The polysulfone (PSF) was obtained from Amoco (Udel™, Naperville, IL). Glass slides were cleaned and primed by dipping in a 1% solution of aminopropyl-triethoxysilane as described previously.[14] These slides were then dipped in a 1% solution of POE, PLA, or PSF in methylene chloride and dried for 48 h in a vacuum oven at 70°C and 21 kPa absolute pressure. Microscopic analysis of the polymer coatings verified that they had smooth and homogeneous surfaces. The surface composition was determined using X-ray photoelectron spectroscopy (XPS) on a Hewlett Packard 5950 XPS and is described elsewhere.[5] The surface energies of the polymers and bacteria, as well as the free energies of bacterial adhesion were presented previously.[5]

### C. Flow Cell with Video Microscope

Bacterial adhesion was monitored in a flow cell mounted to an Olympus BH-2 microscope equipped with a Sony video camera. The flow cell consisted of a Lexan base plate that was milled out to form a shallow flow chamber and had an inlet and exit made from stainless steel tubing. A polymer-coated cover slip was clamped on top of the base plate to seal the flow cell without the use of a gasket. In these experiments the flow rate was 0.81 mL/min which produced a wall shear rate of 1.91 s$^{-1}$. In these creeping flow conditions, shear induced stripping of the cells from the surface is minimal, if not non-existent.[17]

The microscope was mounted on its side with the barrel horizontal, and the flow cell was positioned with the flow direction vertical. The polymer surface was viewed at a magnification of ×1000 which provided a 100 μm × 80 μm viewing area. The video camera was connected to an 8 mm video recorder and a color monitor, allowing continuous monitoring and recording of the positions of the bacteria on the cover slip surface. The bacterial suspension, valves, flow cell, and microscope were maintained at 37°C in a thermostatic chamber.

**Table 1. Theoretical and Experimental Polymer Composition**

| Polymer | Theoretical composition (%) | | | XPS analysis (%) | | | |
|---------|--------|--------|--------|--------|--------|--------|---------|
|         | Carbon | Oxygen | Sulfur | Carbon | Oxygen | Sulfur | Silicon |
| PLA | 60 | 40 | 0 | 63 | 36 | 0 | 1 |
| PSF | 84 | 13 | 3 | 90 | 9 | 1 | 1 |
| POE | 75 | 25 | 0 | 76 | 23 | 0 | 1 |

## D. Growth Procedure

The following experimental procedure was designed to meet the conditions and assumptions that make Eq. (22) valid. At the beginning of the adhesion experiment, a prerinse of 3 mL of phosphate-buffered saline (PBS) was pumped into the flow cell. Then the 6-h culture was pumped continuously through the flow cell for 1 h. Next, the flow cell was rinsed with 3 mL of PBS to clear out any planktonic bacteria. Then fresh TSB flowed continuously through the flow cell for 2 h. During this growth phase of the experiment, the video microscope system continuously recorded the development of the bacterial colonies on the substrates.

The video tapes were analyzed, and the total number of bacteria adhering to the polymer substrate was counted at 15 min time increments. In addition, the numbers of bacteria leaving the view area, as well as the number of newly adhering bacteria, were recorded during each time period. The average desorption rate and adsorption rate were determined in each time period, and the $k_{off}$ and $k_{eff}$ were calculated for each time period by diving the rates by the total number of cells at the beginning of the time period (*see* Eq. [20]). The growth rate constant was regressed from Eq. (22) by a nonlinear least-squares fit of the data.[4]

The growth rate of the bacteria in planktonic suspension in TSB was measured by making a 1:1000 dilution of an overnight culture into TSB in test tubes, and then measuring the planktonic concentration at 1 h intervals. For all species, a log-linear growth rate was preceded by a lag phase and followed by a stationary phase (data not shown).[4] Equation (4) was applied to the planktonic concentration during the log-linear growth phase from which the planktonic growth rate constant was regressed.

## IV. RESULTS

### A. Polymer Characterization

Observation of the polymer films under a microscope showed that the coatings were smooth and continuous. The XPS analysis of the polymers closely matched the theoretical composition as shown in Table 1, indicating that the polymers were clean. Because a silane coupling agent was used to anchor the polymers to the glass, the presence of a small amount of silicon was not unexpected.

### B. Bacterial Growth

Figure 1 presents the averages (*n* = 4) and standard deviations of the surface populations on POE. The solid lines are the best fit of Eq. (22) to the data. The growth

**Table 2. Rate Constants**

| Bacteria/polymer | $k_{off}$ (min$^{-1}$) ×1000 | $k_{eff}$ (min$^{-1}$) ×1000 | $k$ (min$^{-1}$) ×1000 | g (min) |
|---|---|---|---|---|
| *P. aeruginosa* | | | | |
| PLA | $7.5 \pm 2.0^a$ | $3.8 \pm 1.0$ | $13.6 \pm 2.1$ | $51.1 \pm 7.9$ |
| PSF | $12 \pm 2.8$ | $4.3 \pm 1.2$ | $15.5 \pm 2.5$ | $44.6 \pm 7.0$ |
| POE | $14 \pm 3.4$ | $6.6 \pm 2.7$ | $18.9 \pm 3.3$ | $36.7 \pm 6.3$ |
| Planktonic | N/A | N/A | $22.4 \pm 3.0$ | $30.9 \pm 4.1$ |
| *E. coli* | | | | |
| PLA | $9.2 \pm 9.3$ | $1.9 \pm 3.1$ | $18.2 \pm 8.9$ | $38.1 \pm 15.6$ |
| PSF | $11 \pm 12$ | $0.7 \pm 1.6$ | $20.8 \pm 11.2$ | $33.4 \pm 14.6$ |
| POE | $5.6 \pm 4.7^b$ | $2.4 \pm 5.1$ | $19.7 \pm 5.5$ | $35.2 \pm 9.2$ |
| Planktonic | N/A | N/A | $31.7 \pm 5.7$ | $21.9 \pm 3.8$ |
| *S. epidermidis* | | | | |
| PLA | $2.9 \pm 1.8$ | $1.1 \pm 1.4$ | $13.9 \pm 1.9$ | $49.9 \pm 6.8$ |
| PSF | $2.4 \pm 0.8^b$ | $1.1 \pm 0.4$ | $11.8 \pm 0.8$ | $58.9 \pm 3.9$ |
| POE | $1.8 \pm 0.5^b$ | $5.1 \pm 1.5$ | $9.8 \pm 1.4$ | $70.5 \pm 9.8$ |
| Planktonic | N/A | N/A | $18.2 \pm 3.3$ | $38.1 \pm 6.7$ |

[a] Mean and 95% confidence interval for 4 replicates ($n = 4$)
[b] $n = 8$ in these experiments

kinetics on the other polymers were similar, all showing an exponential increase with time. Values of $k_{off}$ and $k_{eff}$ were determined directly from the video observations and are presented in Table 2. Using these values and the surface population data, the average surface growth rate constants were regressed. The doubling time ($g$) was calculated from $k$ using Eq. (6). The 95% confidence intervals of $k_{off}$ and $k_{eff}$ were determined using Student's t test for small sample sizes. The 95% confidence intervals for $k$ and $g$ were determined by the partial differential method of propagation of error using the average adsorption and desorption rate constants and their variances.

There were no significant differences ($p > 0.05$) in the growth rate constants for *E. coli* on the various polymers; nor were they significantly different from the planktonic growth rate constants. However, for *P. aeruginosa*, the $k$ for growth on PLA was significantly smaller ($p < 0.05$) than for growth on POE or the growth in suspension. For *S. epidermidis*, the $k$ for growth on POE was significantly smaller than for growth on PLA or the growth in suspension. Four additional repeats of this experiment verified such a small growth rate.

Figure 2 compares the growth rate constants of the adhering bacteria to the values of $k$ of planktonic bacteria. Although the constants are less than those of the planktonic bacteria, the differences are statistically significant only as mentioned above and shown in the figure.

## V. DISCUSSION

The mathematical formulation of the increase in the surface population of bacteria can be very complex, particularly if the adsorption of bacteria occurs from an interfacial concentration that changes with time and position, which is usually the case in laminar

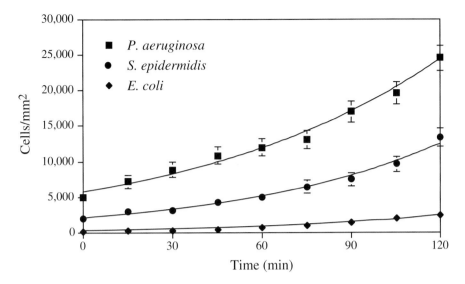

**Figure 1.** The number density of adherent bacteria on POE for the three species. The solid line is the best fit of Eq. (22) to the data. Error bars represent the standard deviation. Absent error bars indicate the standard deviation is smaller than the symbol size.

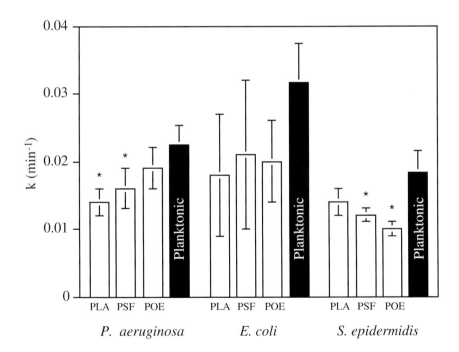

**Figure 2.** The average surface growth rate constants for the three bacteria on PLA, PSF, and POE (open bars). The solid bars indicate the growth rate constant for the log phase planktonic bacteria in TSB. Error bars represent the 95% confidence intervals. The asterisks indicate which growth rate constants are statistically less than those of the planktonic bacteria.

flow, or if the bacteria are growing in a nutrient solution. Determination of a surface growth rate can be a difficult task in such a case, and a nutrient solution is required if one wants to obtain surface growth rates representative of clinical or industrial situations where nutrients are present. We have shown that flushing the flow system of planktonic bacteria, followed by perfusing a nutrient medium (initially without bacteria), leads to surface populations that follow very closely an exponential increase with time (Fig. 1), as predicted by Eq. (22). Furthermore, by an independent measurement of the desorption rate constant, $k_{off}$, and the effective adsorption rate constant, $k_{eff}$, one can unambiguously regress an average growth rate constant. Using this technique, the surface growth rate can be studied as a function of temperature, pH, limiting nutrients, substrate composition, adsorbed proteins, and many other parameters pertaining to the growth of bacteria on surfaces. The experiments are fairly simple to perform, requiring a simple flow cell, video microscopy equipment, and image analysis software.

The data presented above show that the average growth rate constants have relatively narrow 95% confidence intervals, despite the fairly large confidence intervals for the $k_{off}$ and $k_{eff}$. This occurs because in the TSB solution, the increase in surface population due to growth is much greater than the adsorption or desorption rates. If the experiments were done using less nutrient, or employing a limiting nutrient, the growth rate would be less, and the confidence intervals of $k$ would probably be larger.

It is interesting that none of the growth rate constants on the surface are larger than the growth rate constants for the planktonic bacteria. All of the surface growth rate constants are of the same order of magnitude as the planktonic constants, indicating that even though adsorption reduces the growth rate of *P. aeruginosa* and *S. epidermidis* on some polymers, this is not a drastic reduction, at least on these experiments which are not nutrient limited or in which thick biofilms have not yet formed. Habash et al. reported that *P. aeruginosa* growing on silicone rubber under minimal growth conditions had a growth rate constant about one third of the values reported herein for *P. aeruginosa*.[20] As nutrients become limited, or as the bacteria become more crowded, such a reduction in growth rate would be expected. Malone and Caldwell also reported a growth rate constant of *P. aeruginosa* on glass and in suspension that was about one half of the values reported herein.[24]

The experimental model presented in this example is much simpler than an in vivo situation in which the colonizing bacteria would be bathed in blood or interstitial fluid, instead of nutrient-rich TSB; thus the measured growth rates are different than one would expect for growth in a clinical infection. Although we only studied monocultures in this work, clinical infections are often polymicrobial. This technique could be extended to study the interactions of multiple species as long as each could be identified by the video microscopy and image analysis system.

## VI. CONCLUSIONS

These results demonstrate that the polymer surface does have a significant impact on bacterial growth, and these differences can be quantified under varying growth conditions. Thus this technique can be used to study the growth rates of bacteria on surfaces of medical or industrial significance during the initial stages of biofilm development.

The mathematical models presented above show that the bacterial population on a surface is determined by the concentration of planktonic bacteria at the interface (or flux of bacteria to the surface), and the rate constants for growth, desorption and adsorption. If these parameters are known, the surface population can be predicted as a function of time. The equations modeling surface population are fairly complex, but can be simplified to match experiments done in which the surface is seeded with bacteria, and then the planktonic bacteria are flushed out of the system. Even in such a case it is difficult to regress the rate constants from surface population alone, but if the desorption rate constant and the effective adsorption rate constant are measured independently, the growth rate constant can be obtained.

The mathematical model from which the growth rates are calculated is based upon several assumptions, some of which may fail under certain circumstances. However, the shortcomings of the model are minimized by selecting a video observation area away from the flow cell entrance, and by making the observation area large enough to view at least 30 bacteria.

**Acknowledgments:** The authors acknowledge funding from NIH grants HL59923 and CA76562, from the Center for Biopolymers at Interfaces, and from Tau Beta Pi who provided AJB with a graduate fellowship. We also thank Jared Nelson for his careful and thorough review of this paper.

## REFERENCES

1. An YH, Friedman RJ: Concise review of mechanisms of bacterial adhesion to biomaterial surfaces. *J Biomed Mater Res* 43:338–48, 1998.
2. Anwar H, Strap JL, Costerton JW: Establishment of aging biofilms: possible mechanism of bacterial resistance to antimicrobial therapy. *Antimicrob Agents Chemother* 36:1347–51, 1992.
3. Bar-Or Y: The effect of adhesion on survival and growth of microorganisms. *Experientia* 46:823–6, 1990
4. Barton AJ: *The Adhesion and Growth of Bacteria on Orthopedic Implant Polymers.* Brigham Young University, M.S. Thesis, 1994
5. Barton AJ, Sagers RD, Pitt WG: Bacterial adhesion to orthopedic implant polymers. *J Biomed Mater Res* 30:403–10, 1996
6. Barton AJ, Sagers RD, Pitt WG: Measurement of bacterial growth rates on polymers. *J Biomed Mater Res* 32:271–8, 1996
7. Baumgartner JN, Cooper SL: Influence of thrombus components in mediating *Staphylococcus aureus* adhesion to polyurethane surfaces. *J Biomed Mater Res* 40:660–70, 1998
8. Benson DE, Burns GL, Mohammad SF: Effects of plasma on adhesion of biofilm forming *Pseudomonas aeruginosa* and *Staphylococcus epidermidis* to fibrin substrate. *ASAIO J* 42:M655–60, 1996
9. Brannan DK, Caldwell DE: Evaluation of a proposed surface colonization equation using *Thermothrix thiopara* as a model organism. *Microb Ecol* 8:15–21, 1982
10. Brown MRW, Allison DG, Gilbert P: Resistance of bacterial biofilms to antibiotics: a growth-rate related effect? *J Antimicrob Chemother* 22:777–83, 1988
11. Brown MRW, Collier PJ, Gibert P: Influence of growth rate on susceptibility to antimicrobial agents: modification of the cell envelope in batch and continuous culture studies. *Antimicrob Agents Chemother* 34:1623–8, 1990
12. Busscher HJ, Weerkamp AH: Specific and non-specific interactions in bacterial adhesion to solid substrata. *FEMS Microbiol Rev* 46:165–73, 1987
13. Caldwell DE, Malone Ma, Kieft TL: Derivation of a growth rate equation describing microbial surface colonization. *Microb Ecol* 9:1–6, 1983

14. Cook AD, Sagers RD, Pitt WG: Bacterial adhesion to poly(HEMA)-based hydrogels. *J Biomed Mater Res* 27:119–26, 1993
15. Costerton JW, Lewandowski A, Caldwell DE, et al: Microbial biofilms. *Ann Rev Microbiol* 49:711–45, 1995
16. Davies DG, Parsek MR, Pearson JP, et al: The involvement of cell-to-cell signals in the development of a bacterial biofilm. *Science* 280:295–8, 1998
17. Dickinson RB, Cooper SL: Analysis of shear-dependent bacterial adhesion kinetics to biomaterial surfaces. *AIChE J* 41:2160–74, 1995
18. Geesey GG, White DC: Determination of bacterial growth and activity at solid-liquid interfaces. *Ann Rev Microbiol* 44:579–602, 1990
19. Gristina AG: Biomaterial-centered infection: Microbial adhesion versus tissue integration. *Science* 237:1588–95, 1987
20. Habash MB, van der Mei HC, Reid G, et al: Adhesion of *Pseudomonas aeruginosa* to silicone rubber in a parallel plate flow chamber in the absence and presence of nutrient broth. *Microbiology* 143:2569–74, 1997
21. Higashi JM, Wang I, Shlaes DM, et al: Adhesion of *Staphylococcus epidermidis* and transposon mutant strains to hydrophobic polyethylene. *J Biomed Mater Res* 39:341–50, 1998
22. Kieft TL, Caldwell DE: A computer simulation of surface microcolony formation during microbial colonization. *Microb Ecol* 9:7–13, 1983
23. Lawrence JR, Korber DR, Hoyle BD, et al: Optical sectioning of microbial biofilms. *J Bacteriology* 173:6558–67, 1991
24. Malone JA, Caldwell DE: Evaluation of surface colonization kinetics in continuous culture. *Microb Ecol* 9:299–305, 1983
25. Meinders JM, Mei HCvd, Busscher HJ: Physicochemical aspects of deposition of *Streptococcus thermophilus* B to hydrophobic and hydrophilic substrata in a parallel plate flow chamber. *J Colloid Interface Sci* 164:355–63, 1994
26. Nomura S, Lundberg F, Stollenwerk M, et al: Adhesion of staphylococci to polymers with and without immobilized heparin in cerebrospinal fluid. *J Biomed Mater Res* 38:35–42, 1997
27. Reid G, Beg HS, Preston CA, et al: Effect of bacterial, urine and substratum surface tension properties on bacterial adhesion to biomaterials. *Biofouling* 4:171–6, 1991
28. Sapatnekar S, Kao WJ, Anderson JM: Leukocyte-biomaterial interactions in the presence of *Staphylococcus epidermidis*: Flow cytometric evaluation of leukocyte activation. *J Biomed Mater Res* 35:409–20, 1997
29. Sjollema J, Mei HCvd, Uyen HM, et al: Direct observations of cooperative effects in oral streptococcal adhesion to glass by analysis of the spatial arrangement of adhering bacteria. *FEMS Microbiol Let* 69:263–79, 1990
30. Stoodley P, de Beer D, Lewandowski Z: Liquid flow in biofilm systems. *Appl Env Micro* 60:2711–6, 1994
31. Van Loosdrecht MC, Lyklema J, Norde W, et al: Influence of interfaces on microbial activity. *Microb Rev* 54:75–87, 1990
32. Wang I, Anderson JM, Marchant RE: *Staphylococcus epidermidis* adhesion to hydrophobic biomedical polymer is mediated by platelets. *J Inf Dis* 167:329–36, 1993
33. Yu FP, Callis GM, Stewart PS, et al: Cryosectioning of biofilms for microscopic examination. *Biofouling* 8:85–91, 1994

# 12

# Analysis of Gene Expression in Biofilm Bacteria

### Robert A. Burne and Yi-Ywan M. Chen

*Department of Microbiology and Immunology and Center for Oral Biology
University of Rochester School of Medicine, Rochester, NY, USA*

## I. INTRODUCTION

### A. Rationale for Exploring Gene Expression in Sessile Populations

Microbial processes that are of economic, medical, dental or environmental consequence occur with the organisms immobilized on a surface. Early studies with in vitro grown biofilms demonstrated that the phenotypic properties of adherent organisms differed from their planktonic counterparts.[7] These results are now being reinforced by new studies that are using a wide variety of bacterial species and novel methods. It is also becoming apparent that the phenotypic changes seen with adherent organisms, as with suspended populations, are due to sensing by the organisms of environmental stimuli which signal changes in the patterns of expression of particular genes. In some cases, the organisms appear to sense binding to the surface, as in the case of *Escherichia coli* or *Pseudomonas aeruginosa*.[8,14] More commonly, local environmental conditions arising as a result of mass transport limitations, synergistic and antagonistic activities of biofilm bacteria, intercellular signaling and so forth, may induce bacteria to modulate expression of genes differently than suspended populations growing under very similar conditions.[2] Clearly, if a comprehensive view of the characteristics of adherent populations is to be developed, it will be essential to incorporate a thorough understanding of 1) how gene expression differs between suspended and sessile organisms, 2) what, if any, genetic control circuits are peculiar to biofilm growth, and 3) how global and/or biofilm specific genetic circuits crosstalk with one another in adherent bacteria.

Direct biochemical measurements of the activity or abundance of gene products is often sufficient to monitor gene expression in biofilm populations. However, not all gene products are readily assayable and frequently such measurements are not adequately sensitive or readily adaptable to kinetic studies in adherent populations. Difficulties can also arise in accurately comparing gene expression in adherent cells with suspended samples because of potentially significant differences in half-lives of components in biofilms versus planktonic cells. This is especially relevant if the gene product is secreted to, or beyond, the cell surface. Quantitative data about gene expression has been obtained by hybridization with fluorescently labeled oligonucleotides,[10] but an appropriately equipped microscope is required and some method to permeabilize the biofilm cells to the

*Handbook of Bacterial Adhesion: Principles, Methods, and Applications*
Edited by: Y. H. An and R. J. Friedman © Humana Press Inc., Totowa, NJ

probes is needed. A number of investigators have circumvented these and other short-comings of direct measurements with the use of reporter gene fusions to explore gene regulation in biofilm bacteria. A major advantage of this technique is that real time, continuous monitoring of gene expression can be achieved in biofilm chambers which can be mounted for microscopy. Some examples of the application of this technique include the use of operon fusions to a chloramphenicol acetyltransferase (*cat*) gene to show that the expression of the major polysaccharide synthase enzymes of the pathogenic bacterium, *Streptococcus mutans*, were different in attached bacteria compared with cells in suspension.[4] Davies et al.[8] used a LacZ gene fusion to the promoter of a gene (*algC*) involved in alginate production in *P. aeruginosa* to show that contact with surfaces stimulated transcription of the alginate genes, consistent with the finding that alginate production increases in adherent *P. aeruginosa*. Recently, random mutagenesis has been used to disclose genes necessary for stable biofilm formation in *Pseudomonas fluorescence* and such strategies may be useful for getting strains of other species which are similarly deficient in surface growth.[11] Logical experiments to follow the identification of "biofilm-specific" genes will undoubtedly involve exploring the differences in expression of these genes when cells are planktonic versus sessile. The ability to monitor gene expression in adherent populations will be vital component of efforts to understand the molecular basis for the so-called biofilm phenotype. This chapter details some methods and considerations for examining gene expression in adherent bacteria.

## II. METHODS

### A. Closed vs Open Systems

A primary concern in designing and implementing studies of gene expression in adherent organisms is the selection of the model system. In many cases, the system to be used is dictated by placing priority on attempting to mimic the natural environment of the organisms, with nutrient concentrations and flow rates adjusted to that which is typical of the situation being evaluated. However, approaches to contrasting sessile and planktonic gene expression have also been driven by a desire to introduce the maximum level of control over the environment of the organisms and to create, in so far as one can, conditions in which the only major difference is that one population is adherent. In both cases, the use of open systems, such as chemostats, continuous flow bioreactors, or flow cells, should be favored over closed systems. While closed systems, including microtiter plates and batch grown biofilms, have many valid applications, such as antimicrobial testing and mutant screening, the inability to control factors such as growth rates, nutrient availability, growth domain, and pH results in the introduction of many confounding variables into the experimental system. While one cannot exercise complete control over biofilm populations in open systems, one can minimize variation from run to run while having more flexibility than can be achieved in closed systems.

### B. Control of Biofilm Physiology

Batch cultivation systems, which are closed systems, have many advantages over continuous flow systems. They provide higher throughput, and they generally do not require substantial capital investments, as do some continuous flow systems. There are also batch systems, particularly those in which the organisms grow over a narrow period of time, that are very useful in analysis of gene expression.

Continuous chemostat culture affords a very high level of control over the physiologic state of cells. At steady-state, the specific growth rate is directly related to the dilution rate, so control of growth rate and domain are easily achieved. Modulation of pH, oxygen tension, and limiting nutrient is also straightforward. Biofilm model systems based on chemostats into which a suitable substratum is inserted for bacterial growth are probably the systems which allow for the closest approximation of the control that can be achieved in a single stage chemostat. Functionally similar reactors, like the Rototorque,[5] while not quite as flexible as a chemostat system with respect to the insertion of pH and $O_2$ probes, are also a fairly close approximation of the chemostat. I have discussed considerations related to growth parameters in continuous flow reactors elsewhere, so details and equations will not be reviewed here.[3] The main difference between chemostats and continuous flow bioreactors, in terms of monitoring gene expression, is that the populations on the surface are not homogeneous, as they are at steady state in a chemostat. Consequently, interpretations of the data need to account for the fact that subpopulations that differ dramatically from the bulk of the population may influence the data to a great extent.

## C. Use of Reporter Genes: Considerations

There are a handful of reporter genes available for the study of bacterial gene expression. The one most commonly used is the *E. coli lacZ* gene, which is simple to assay with a variety of relatively inexpensive colorigenic and fluorescent substrates available. Consequently, LacZ activity can be measured by traditional biochemical methods in dispersed biofilms or *in situ* using fluorescent or confocal laser scanning microscopy (CLSM). Based on our experiences and the experience of others, there are two major drawbacks with using LacZ. The first is that the enzyme loses activity rapidly as the pH falls below neutrality. In the streptococci, which allow their intracellular pH to fluctuate, use of LacZ is largely prohibited at pH values around 5.5 and below. In bacteria which maintain a cytoplasmic pH near neutrality, low pH inactivation of the enzyme may not be as serious a concern, but caution is warranted in studies where cells may be exposed to acid stresses. The second potential problem with LacZ is that the enzyme is fairly stable, so down-regulation of the promoter to which *lacZ* is fused may not be readily detectable. CAT (chloramphenicol transferase) has also been used effectively to monitor transcriptional activity in a wide range of organisms and in biofilm bacteria. Although assay of CAT is not as easy or inexpensive as LacZ, and the fluorescent substrates for CAT are not able to penetrate cells, we have empirically determined that CAT may have advantages for biofilm study. First, CAT seems fairly stable in cells growing at pH values between 7.5 and 4.0. Also, we found that CAT, at least in a number of streptococci, has a short half-life, so measuring CAT activity yields a very consistent picture of transcription and steady state mRNA levels.

Three other reporter genes which are readily adaptable to biofilm study include green fluorescent protein (GFP), luciferase, and β-glucuronidase. GFP has the advantage that no substrate need be provided, but it is not yet clear how reliable GFP would be in the oxygen restricted conditions that may develop in deep biofilms. The enzyme β-glucuronidase should in theory be similar to LacZ in terms of its applications, but there is not much data available on its utility in biofilm studies. Luciferase may also have advantages and specific applications, but the coupling of fluorescence to energy expenditure by cells may limit its utility.

### D. Collection of Cells

Other chapters in this volume deal in greater detail with the specifics of methods to remove biofilms from different substrata, and the most appropriate method will need to be determined largely based on the type of biofilms and the sensitivity of the organisms to disruption during the removal process. We have worked mostly with oral streptococci, which are resistant to prolonged sonication and osmotic shock, so rather harsh methods are suitable. Conversely, organisms which are more readily lysed may need to be removed with gentle means if leakage of reporter enzyme is to be prevented during the cell collection phase.

### E. Sample Preparation

Intercellular interactions seem to significantly contribute to the stability of biofilms, but the tendency of bacteria to coadhere and coaggregate can interfere with accurate measurement of reporter activity, particularly if complete disruption of the cells is not part of the assay procedure. For example, measurement of LacZ activity in biofilm cells can be achieved by dispersing the cells with mild sonication or vortexing in the presence of glass beads (1/3–1/2 volume of glass beads, 0.1 mm diameter) followed by toluene or toluene:acetone (1:9) permeabilization. If aggregates are not fully dispersed, diffusion of substrate to cells can be restricted, yielding apparently lower activity. Dispersal of aggregates should be monitored by phase contrast microscopy and can be accompanied by spectrophotometric measurement of the supernatant fluid at $A_{280}$ to ensure that dispersal is not causing substantial cell disruption.

### F. Disruption of Cells

Heating or creation of significant gas–liquid interfacial areas during the disruption process can cause rapid loss of enzymatic activity. Great care should be exercised not only in reducing the possibility that enzymes will be denatured and inactivated during the disruption process, but also in monitoring the empirical determination of the stability after disruption. The latter is recommended because a significant portion of the material that is collected as a biofilm is extracellular material, which may be rich in proteolytic or inhibitory activities. Also noteworthy is that if there is little information about the potential for inhibition of the enzyme to be measured by biofilm components, it is worthwhile constructing a system in which the biofilms are grown, usually of the parent organism without the gene fusion, and processed as desired. Then, the preparations can be "spiked" with known quantities of enzymatic activities to determine if inhibition or proteolysis is occurring. If rapid loss of activity is observed, processes such as treatment of the extracts using protease inhibitors or dialysis can be tried to remove the inhibitory activity.

The preferred method of disruption of bacteria in our laboratory is homogenization in the presence of glass beads using a Bead Beater™ (Biospec Products). This process is rapid, little air is introduced during the procedure, and intermittent cooling insures the preservation of enzyme activity. Below is the method we have used for oral streptococci in the preparation of cell free lysates for assay of CAT.

1. Biofilms are removed by scraping into 35–40 mL of 10 mM Tris-HCl, pH 7.8, and collected by centrifugation at 5000 rpm in a Sorvall centrifuge equipped with an SS34 rotor, at 4°C, for 15 min.
2. Cells are washed once with an equal volume of 10 mM Tris-HCl, pH 7.8. We generally concentrate the cells some 25–50-fold. Obviously yield will depend on biofilm age and growth conditions, so as a reference point, the biofilms we have studied contain about $10^8$ cfu per $mm^2$. We have found that this amount of cells is adequate for populations expressing CAT at levels as low as 5–10 nmoles/min/mg.
3. The concentrated cell suspension is transferred to a 2 mL screw-cap microfuge tube containing 0.75 g (roughly 1/3 vol) of glass beads (0.1 mm avg. diam., Sigma, St. Louis, MO). The tube should be filled to the top to minimize introduction of bubbles during the homogenization process. The screw cap tubes we use are equipped with O-rings to prevent leakage and can be obtained from numerous manufacturers, including Sarstedt (#72.694.006).
4. Cell suspensions are homogenized in the Bead Beater for 30 s, at 4°C, followed by chilling the tubes on ice for 2 min. This step is then repeated.
5. Cleared lysates are generated by centrifugation at 10,000 rpm for 10 min at 4°C in an Eppendorf Microcentrifuge, and the protein content of each lysate is determined by using the Bio-Rad Protein Assay, based on the method of Bradford.[1] Bovine serum albumin serves as the standard. Generally, we use the cell extracts immediately.

Again, the oral streptococci are extremely refractile to lysis. For example, thorough lysis by sonication requires continuous sonication for 7 to 8 min at 350W in the presence of glass beads, whereas *E. coli* is completely disrupted in 30 s under the same conditions. Empirical determination of the minimum time needed to homogenize a given suspension of bacteria can be achieved by monitoring $A_{280}$ until no further increases are observed. A variety of beads are also available through the supplier of the Bead Beater, and some beads may be better for homogenization of a given bacterium compared with others. The supplier will provide this information, but since the data was not generated using biofilm cells, testing of a few different beads may be necessary to optimize disruption.

### G. Measurement of Reporter Gene Activity-Chloramphenicol Acetyltransferase

There are a number of methods for measuring CAT activity. We use the kinetic assay as detailed by Shaw.[13] The assay should be carried out with a recording spectrophotometer equipped with a temperature-controlled cuvette chamber. Small variations in reaction temperature will impact the results. We employ a Beckman DU640 equipped with a Peltier effect temperature-control system with a 6 place cuvette holder. With the 6 cuvette holder, we load three identical test samples and one blank, which contains everything but chloramphenicol (Cm). Using the kinetics software on the instrument, plots of rates and subtraction of any background are done automatically. For assay of CAT in cell free lysates, the following method is used:

1. Prepare the reaction mixture freshly by dissolving 4 mg 5,5'-Dithiobis-2-nitrobenzoic acid (DTNB) in 9.8 mL 100 mM Tris-HCl, pH 7.8. To each 9.8 mL mixture, 0.2 mL of 5 mM acetyl-CoA solution is added. Equilibrate the mixture to 37°C prior to use. The solution should be prepared immediately prior to use as some spontaneous reduction of DTNB occurs over time and the solution turns yellow, raising the background.
2. Add prewarmed (37°C) reaction mixture to each 1 cm light path cuvette and then add prewarmed cell lysate to a final volume of 955 μL. All assays are carried out at 37°C, in triplicate, with one blank (no Cm added).
3. Initiate the reactions by adding Cm to a final concentration of 0.1 mM.
4. The rate of increase in absorption at 412 nm is monitored and differences between the samples and the blank are recorded. The rates of Cm acetylation of each protein lysate are calculated

as described elsewhere.[13] The net change in extinction per minute is divided by 13.6 to obtain µmoles per min of Cm-dependent DTNB reduction, and the specific activities are expressed as µmoles of Cm acetylated/min/mg protein.

Other methods for measuring CAT include the use of radioactive Cm, ELISAs with anti-CAT antibodies, and CAT assays using fluorescent CAT substrates. The use of fluorescent CAT substrates may be adaptable to biofilms which have been gently fixed, but at this stage no publications have reported using such a method.

### H. Normalization of Data

The formation of large quantities of exopolymeric material, accretion of material from the liquid phase, and the retention of proteins and other components which normally diffuse away from planktonic cells results in a much different overall composition of biofilms compared to planktonic cells. In some cases, these differences can be fairly extreme and exopolysaccharide can actually compose in excess of 90% of the total dry weight recovered from the surface. If the experimental goal is to relate the absolute levels of expression of a reporter gene construct in biofilm cells to that of planktonically growing bacteria, it is essential that the data be normalized correctly. Use of total wet weight or dry weight is certainly not appropriate for comparing sessile and planktonic populations, and is questionable for making comparisons between two biofilm populations growing under different conditions unless it can be verified analytically that large differences in composition of the films were not induced by changing the growth conditions. Based on our experience, normalization to total protein is the easiest, fastest and most reliable method. Specifically, we have grown strains of oral streptococci under a wide variety of conditions, e.g., carbohydrate excess and limiting growth, different pH values, and different carbohydrate sources. We have tried several different protein assays and have supported our data with the anthrone method for total carbohydrate and use of amino acid analysis, which will also give amino sugars. By taking a known dry weight of biofilm and subjecting it to these analyses, we have validated that the protein assay method of Bradford is a very accurate measure of total protein. Depending on the organism and if agents which could interfere with the Bradford assay are going to be used to treat the biofilms, alternative protein measurement techniques could be used with the caveat that they should be supported by analytical methods.

### I. Normalizing Data When Cells Are Not Homogenized

We have had occasions where it was desirable to measure enzyme activity in biofilm cells using intact or permeablized cells, in the case of the urease of *Streptococcus salivarius* (Li and Burne, manuscript in preparation) or LacZ expressed in *S. mutans*[4] respectively. The urease of *S. salivarius* rapidly loses activity after release from the cell,[6] whereas it is highly stable in intact cells. In the case of LacZ, use of permeablized cells is warranted, although we have not investigated in great detail whether cell lysis affects measurements. At any rate, using urease as an example, we have normalized that activity to dry cell weight in samples obtained from continuous culture and batch grown organisms. However, our recent measurements of *S. salivarius* biofilms indicate that they are composed of greater than 60% carbohydrate, unlike cells grown in batch or continuous culture (Li and Burne, manuscript in preparation). Although there are multiple ways to circumvent this problem, a simple method involves determining total protein for the entire sample and then calculating the amount of protein input per unit of intact cells. Specifically:

1. Biofilms are collected as above, washed and resuspended in 10 mM potassium phosphate buffer, pH 7.0. Cells are dispersed by brief sonication.
2. The sample is split into two to give sufficient material for urease assays and protein determination.
3. A known volume of cell suspension is used for urease assays.
4. A known volume of the other portion of the cell suspension is concentrated, resuspended in a known volume, and homogenized as above.
5. Protein concentration of the cell-free lysate is determined by the Bradford method.
6. Knowing the protein concentration per unit volume, one can extrapolate to total protein per unit of cell suspension used in the urease assays.

## J. Microscopy for Studying Gene Regulation In Situ

Detailed methods for microscopic examination of biofilms can be found elsewhere[9] and in earlier chapters of this volume. The intent of this section is not to provide detailed methods for microscopy, but instead to highlight a few points about the use of microscopy to study gene expression. First, the equipment required for techniques such as scanning confocal laser microscopy (CLSM) is expensive and enlisting the services of someone with expertise in microscopy is strongly recommended. Still, fluorescence microscopy and CLSM have proven to be powerful methods for the study of structural aspects of bacterial biofilms, and use of microscopy is really the only way to study gene expression as a function of such variables as biofilm depth and association with morphotypes. Using these techniques, quantitative information can be gathered about the physiologic state of populations, and if appropriate reporter gene systems are used, about the relative levels of expression of particular genes in single and mixed species biofilms. We have utilized strains of *S. mutans* expressing *lacZ* which were stained with either 4-methylumbelliferyl-β-D-galactopyranoside (4-MUG) or fluorescein di-galactoside (Molecular Probes) to follow expression of a gene encoding a sucrose transporter in monospecies biofilms. In this case, biofilms were removed from a continuous flow reactor and the substrates were applied at the recommended concentrations. The biofilms were incubated in a sealed polypropylene container in the dark at 37°C for 30 min and the biofilms were rinsed in $dH_2O$. During the staining period, water saturated paper towels were placed in the same container to maintain hydration of the films.

A preferred approach to using fluorescent substrates is to perfuse them into a flow cell in which biofilms are forming or established. If the right type of flow cell is used, continuous monitoring of the development of fluorescence can be achieved. This method is preferred to that described above since it is not necessary to disturb the biofilms.

## K. Alternative Methods

The use of reporter gene technology is a relatively inexpensive and straightforward way to explore gene expression and regulation in biofilm populations. In addition, the method is largely nondisruptive and no fixation of the cells is required. Other methods have been utilized which directly measure mRNA in biofilm cells using a quantitative hybridization with fluorescent DNA oligonucleotides.[10,12] Gentle methods for fixation and permeabilization of the biofilm bacteria to allow penetration of the probes are available, and coupled with use of a flow cell, the technique seems to preserve the overall structure and spatial organization of the bacteria. Although detailed methods will not be presented here, use of labeled probes has a number of advantages. First, it is not necessary to construct specific strains of bacteria carrying a gene fusion. This may be a major consideration for organisms which lack a well developed genetic system. Secondly, multiple genes can be probed in a single population by labeling the probes with different fluors. Finally, pertinent to mixed populations, speciation can be done concurrently with quantitative gene expression studies,[10] for example, to confirm that the morphotype expressing the activity is the desired strain or whether high or low expressors exist in close association with a bacterial species.

### L. Cautions

The heterogeneity of bacterial biofilms is well established. In mature films, cells exist in all phases of growth and quiescence, microenvironments likely differ widely, and diffusion into the biofilms of substrates is probably not uniform. Consequently, as our experience and that of others indicates, unless very young, thin biofilms are used for reporter gene studies, the biofilms appear very heterogeneous. Interpretations of data collected by microscopy need to be subjected to close scrutiny and rigorous evaluation of the significance of differences in gene expression should include statistical methods of a large number of samples before conclusions can be reached.

## III. CLOSING REMARKS

The amount of fundamental and practical knowledge gathered using suspended bacteria is staggering. By far, most of what we know and understand about the genetics and physiology of bacteria have arisen from studies of suspended populations growing in batch culture. Nevertheless, it is equally clear that sessile populations of bacteria are dramatically different than their planktonic counterparts. Given the importance of biofilm bacteria in diseases and in the environment, dissecting the molecular bases for the observed differences in biofilm phenotypes is essential. It may eventually be found that biofilm bacteria are responding to a stimulus in essentially the same way as planktonic cells would, and that regulatory circuits in biofilms cells are fundamentally the same as in suspended organisms. Alternatively, new pathways, unique to biofilm bacteria, may be found which, for example, can be exploited to develop novel therapies for infectious diseases or to reap economic and social benefit by using adherent cells to carry out beneficial processes more efficiently. Either way, we must expand our studies of bacterial gene regulation to include what is certainly the most relevant state of bacterial growth, as adherent populations.

**Acknowledgment:** This work was supported by PHS Grant DE12236 from the National Institute of Dental and Craniofacial Research.

## REFERENCES

1. Bradford MM: A rapid and sensitive method for the quantitation of microgram quantities of protein utilizing the principle of protein-dye binding. *Anal Biochem* 72:248–54, 1976.

2. Burne RA: Regulation of gene expression in adherent populations of oral streptococci. Conference Proceedings. In: LeBlanc DJ, Lantz MS, Switalski LM, eds. *Microbial Pathogenesis: Current and Emerging Issues.* Indiana University Press, Indianapolis, IN, 1998:55–70

3. Burne RA, Chen YM: The use of continuous flow bioreactors to explore gene expression and physiology of suspended and adherent populations of oral streptococci. *Meth Cell Sci,* In Press

4. Burne RA, Chen YM, Penders JEC: Analysis of gene expression in *Streptococcus mutans* in biofilms *in vitro. Adv Dent Res* 11:100–9, 1997

5. Characklis WG: Laboratory biofilm reactors. In: Characklis WG, Marshall KC, eds. *Biofilms.* John Wiley and Sons, New York, NY, 1990:55–89

6. Chen YM, Clancy KA, Burne RA: *Streptococcus salivarius* urease: Genetic and biochemical characterization and expression in a dental plaque streptococcus. *Infect Immun* 64:585–92, 1996

7. Costerton JW, Cheng K-J, Geesey GG, et al: Bacterial biofilms in nature and disease. *Ann Rev Microbiol* 41:435–64, 1987

8. Davies DG, Chakrabarty AM, Geesey GG: Exopolysaccharide production in biofilms: substratum activation of alginate gene expression by *Pseudomonas aeruginosa. Appl Environ Microbiol* 59:1181–6, 1993

9. Lawrence JR, Korber DR, Hoyle BD, et al: Optical sectioning of microbial biofilms. *J Bacteriol* 173:6558–67, 1991

10. Møller S, Sternberg C, Andersen JB: *In situ* gene expression in mixed-culture biofilms: Evidence of metabolic interactions between community members. *Appl Environ Microbiol* 64:721–32, 1998

11. O'Toole GA, Kolter R: Initiation of biofilm formation in *Pseudomonas fluorescens* WCS365 proceeds via multiple, convergent signalling pathways: a genetic analysis. *Mol Microbiol* 28:449–61, 1998

12. Poulsen LK, Ballard G, Stahl DA: Use of rRNA fluorescence *in situ* hybridization for measuring the activity of single cells in young and established biofilms. *Appl Environ Microbiol* 59:1354–60, 1993

13. Shaw WV: Chloramphenicol acetyltransferase activity from chloramphenicol-resistant bacteria. *Meth Enzymol* 43:737–55, 1979

14. Zhang JP, Normark S: Induction of gene expression in *Escherichia coli* after pilus-mediated adherence. *Science* 273:1234–6, 1996

# PART III

## Techniques for Studying Microbial Adhesion and Biofilm

# 13

# Methods for Evaluating Attached Bacteria and Biofilms

*An Overview*

**Gordon D. Christensen,[1] W. Andrew Simpson,[1,2]
Jeffrey O. Anglen,[3] and Barry J. Gainor[3]**

[1]*Departments of Internal Medicine and Molecular Microbiology and Immunology,
University of Missouri–Columbia, Columbia, MO, USA
[2]Orthopaedic and Infectious Diseases Research Lab.,
Harry S. Truman Memorial Veterans' Hospital, Columbia, MO, USA;
[3]Department of Orthopaedic Surgery, University of Missouri–Columbia, Columbia, MO, USA*

## I. INTRODUCTION

If you consider this text a cookbook and yourself—the investigator—the cook, how do you choose which recipe—or experimental procedure—to use? Or should you cook up something new? Like any good chef, the recipe you choose will reflect the meal you wish to prepare, the equipment in your kitchen, the supplies in your pantry, and your own past experience as a cook. But suppose you are not bound by past experience, materials, and equipment, then how do you choose an experimental approach? The answer lies in the experimental question. What hypothesis do you want to test? What meal do you wish to prepare?

This chapter should help you answer this question. We begin with a general approach to the study of microbial colonization of biomaterials, including a discussion of critical controls and potential pitfalls. Our discussion then continues with an overview of the most commonly used methods; this overview should help guide the reader to the other chapters in this textbook that present the details of these methods. Finally, we close with a detailed presentation of the specific methods used to estimate microbial colonization by optical density, since this technique is not covered elsewhere and represents the authors' contribution to this field.

## II. EXPERIMENTAL DESIGN

As the reader knows, scientific investigation proceeds in a cycle that begins with an exploratory study into the unknown where we observe a phenomenon and propose an explanation ("hunch," "theory," or "hypothesis") to explain our observation. We then test this explanation *a priori* by an experiment to determine how well the explanation can

*Handbook of Bacterial Adhesion: Principles, Methods, and Applications*
Edited by: Y. H. An and R. J. Friedman © Humana Press Inc., Totowa, NJ

predict the experimental outcome. Confirmation of our predictions increases our confidence in the validity and applicability of our scientific explanation. The final stage of investigation requires other scientists to confirm our explanation, either by replicating our experiments or by making logical deductions based on our explanation and testing those deductions. Each stage of this scientific investigation dictates its own experimental design.

### A. Exploratory Studies

While the initial exploratory stage allows the investigator the most leeway in experimental design, one must bear in mind that the scientific audience has the least confidence in an explanation made after the observation or *"post hoc."* Scientific confirmation—both by oneself and by others—requires testing *a priori.* For this reason, even though the research is young, investigators should still make every effort to ensure the reproducibility of their work by including proper controls (see below), by preserving unique microbes, materials and reagents, and by using widely available methods. While this call for reproducibility is true for all of science, it is particularly important in the study of microbial colonization because the elements of this investigation—the microbe, the surface, and the experimental methods—can be difficult if not impossible to replicate if the investigator fails to take these precautions.

### B. Hypothesis Testing

Testing the hypothesis, the next stage of scientific investigation, begins to lock the investigator into a particular experimental design. Experiments regarding microbial colonization usually address one of the following four testable predictions:

1. Predictions regarding the capacity of a standard strain of microorganisms, under standard experimental conditions, to attach to or colonize different substrata or substrata treated in different manners. These predictions concern the capacity of certain physico-chemical parameters of the material or the presence of certain adsorbed molecules on the surface of the material to promote or repel colonization of the surface. If confirmed, these predictions demonstrate an understanding of the precise surface characteristics that promote or inhibit attachment and colonization.
2. Predictions regarding the capacity of a standard strain of microorganisms to attach to or colonize a standard substratum under varying experimental conditions, such as pH, buffers, temperature, fluid flow rate, etc. These predictions concern the capacity of the bathing solution to modulate microbial attachment and colonization of a surface. If confirmed, these predictions demonstrate an understanding of the precise fluid conditions that promote or discourage an organism to attach to and colonize a surface.
3. Under standard experimental conditions, predictions regarding the capacity of a standard strain of microorganisms (modified in various manners) OR different microorganisms (with different attributes) to attach to or colonize a standard substratum. If confirmed, these predictions demonstrate an understanding of the specific microbial factors that promote or inhibit attachment and colonization of a surface or the microbial factors that control the expression of microbial attachment and inhibition factors. Such information sets the stage for investigations of the role of such factors to function as virulence factors.
4. Predictions regarding the survival and proliferation of organisms on a surface, particularly when the organisms are exposed to noxious agents, such as antimicrocrobial agents or

phagocytes. These predictions concern the manner in which microbes live and die on a surface. If confirmed, these predictions demonstrate an understanding of how the biology of a surface (sessile) organism differs from the biology of organisms in suspension (planktonic organisms). This information is particularly important in understanding how sessile microorganisms have enhanced or restrained capacities to resist the antimicrobial action of therapeutic agents and host defenses.

Hypotheses 1–3 share a common feature, even though the predictions concern widely different phenomena, fundamentally these predictions require the investigator to compare the number of microorganisms on a surface under different experimental conditions. This determination may be qualitative (e.g., the microscopic presence or absence of bacteria on a surface), semiquantitative (e.g., visual estimates of the density of bacteria on a surface), or quantitative (e.g., the actual number of colony forming units of bacteria recovered from a surface). The determination may be a direct count of bacteria (such as the counts made by a scanning electron microscope of the number of bacteria in a predetermined area) or indirect (such as the measurement of the optical density of the biomass of bacteria attached to a surface). Either way, in order to validate the predictions made by these hypotheses, the investigator must experimentally determine the number of organisms on the surface. Most, but not all, predictions concerning the fourth hypothesis also require counting procedures. Some studies regarding the fourth hypothesis, however, concern the morphology, fine structure, physiology, and microbiology of surface organisms.

## C. Confirmatory Studies

The final stage of scientific investigation sets the most rigorous experimental design parameters. The investigator must either duplicate, as closely as possible, the experimental conditions of the original report or, more commonly, the investigator will test specific predictions emanating from the original hypothesis. Either way, when confirming the work of others, the scientist should first critique the earlier work (that is, the investigator must be aware of the potential pitfalls in the earlier work as well as any critical controls that may have been overlooked) and then overcome these criticisms. This requires an appreciation for potential pitfalls and critical controls.

## III. PITFALLS

Microbial colonization of a surface proceeds in stages. It begins with the exposure of a surface to suspended microbes, followed by the almost instantaneous "reversible" association (through physicochemical interactions) of the microbe with the surface, followed by permanent attachment (through a variety of molecular mechanisms) to the surface, followed by proliferation over the surface (by multiplication), and concluded by the release of daughter cells that leave the surface to colonize more distant sites. It is not possible to study the entire colonization process *in toto;* investigators must instead divide the process into constituent stages and study one or two stages in isolation. A frequent source of confusion in describing the resulting experiments can be the imprecise terminology used in describing the different stages of colonization. Notable confusion arises in this regard from the use of the term "adherence."[15]

Different authors use the word "adherence" to refer to different colonization stages. For instance, some authors use the term when referring exclusively to the immediate attachment of an organism to a surface, as for example, in experiments with incubation periods of 2 to 180 min.[2,3,9,20,29,31,33,38,39,42,51,59,64,67] Other authors use "adherence" to refer to both the immediate and permanent attachment of a microbe to a surface, as for example, in experiments with an incubation period of 5 h.[46] Still other investigators use "adherence" to refer to the stages of immediate and permanent attachment as well as the later stage of microbial proliferation, as for example, in experiments with incubation periods of 18 to 96 h.[12,13,17,25,74] Confusion arises when different authors use the same word to describe data regarding different phenomena. For this reason the term should probably be avoided.[15,49]

We prefer the terms "adhesion," "binding," or "attachment" to describe experiments covering the early stages of reversible binding and permanent attachment. These experiments require incubation or exposure periods measured in minutes to 1 to 2 h. If the experiment can distinguish between reversible binding and permanent attachment, then that distinction should be noted. When the experiments use a longer time scale, like 6 to 24 h, the observations can include the initial adhesion as well as the subsequent multiplication of the microorganisms on the surface; a better term for this data would be "colonization." Alternative terms that focus primarily on this last colonization stage include "microcolony formation," "biofilm formation," "microbial mat formation," and "slime production."

Another common pitfall is to overlook the contribution of the unstudied phases of colonization toward the complete phenomenon; this oversight can lead to erroneous, even absurd, conclusions. Investigators should bear in mind that colonization is cyclical with each phase dependent upon the preceding phase. In other words, you can not have biofilm formation without microbial adhesion, adhesion without exposure to microorganisms, and exposure to microorganisms without release of microorganisms from some more distant site, etc.

## IV. CRITICAL CONTROLS

The four previously stated testable predictions regarding microbial colonization reflect the four basic elements of microbial colonization: the substratum, bathing fluid, microorganism, and time. The systematic testing of any one of these elements requires the other three elements to be closely controlled. Failure to adequately recognize and control for these elements is a frequent criticism of scientific reports.

### A. Substratum

Important substratum variables include the architecture of the object, surface microtopography, chemical composition, inclusion of leachable chemicals, surface contamination with foreign substances, surface modifications, and the physicochemical characteristics of the surface (such as electrostatic charge and hydrophobicity).

Microbial colonization is a function of the surface area available for colonization; surface area in turn is a function of the dimensions of the device, its geometric shape and its microtopography (meaning "smoothness" or "roughness" on a microscopic scale).

Comparable objects should have comparable surface areas. For example, some protocols examine microbial adhesion to small beads.[68,73] A minor difference in the diameter of the beads used in two different preparations could lead to significant differences in the surface area presented by the two materials. Assuming an equal number of beads identical in all other respects, there will be greater adhesion to the larger beads simply because the larger beads have the greater surface area.

Some material differences may not be apparent to the investigator; similar appearing objects may have deceivingly different surfaces at the microscopic level. If one surface is rougher than the other is, microbial colonization can be greater for the rougher surface by virtue of the larger surface area[74] as well as by the entrapment of microbes in the microscopic pits, grooves, and interstices of the rougher material.[43] Another problem arises when investigators modify the shape of an object by cutting and trimming so that the object can be placed in the test chamber or compared to another similarly shaped object. By cutting the material and exposing bacteria to fresh, rough material surfaces, the investigator may significantly change the capacity of the object to be colonized by microbes.[50]

Materials can be simply but deceivingly labeled, like calling plastic by the polymeric name, polystyrene, polyethylene, polypropylene, etc. These terms can mislead the investigator into assuming that plastics with the same polymeric name have the same chemical composition. Yet the chemical composition of plastics can vary from manufacturer to manufacturer and even from lot to lot.[62] Likewise plastics include plasticizers which make the material durable and pliable, but the precise composition and amount of the plasticizers in the material often varies from manufacturer to manufacturer. Details regarding the precise chemical composition of a particular plastic material are usually unavailable to the investigator and may be a trade secret or simply unknown. These ccompositional details can have profound effects upon the attachment of organisms and substances, such as proteins,[9,62] to the plastic surface by perturbing material physicochemical properties such as surface charge and hydrophobicity.

Materials are also subject to postmanufacture modification of the surface. In some situations the manufacturer may purposefully treat the surface to modify the adhesion characteristics; a notable example of this is the difference between 96 well *microtiter* plates (used for ELISA assays) and 96 well *tissue culture* plates. Although both plates have similar appearances and are made out of similar plastic, the manufacturer irradiates tissue culture plates to increase surface charge and reduce hydrophobicity thereby increasing the attachment of tissue culture cells to the plate. When compared to non-irradiated microtiter plates, this irradiation also promotes the attachment of bacterial cells to the plates.[13,52]

Inapparent but significant accidental soiling and abrasion of the surface can also influence microbial colonization of plastic materials. Furthermore as the plastic ages, the surface may oxidize and plasticizers will evaporate; heat sterilization, chemical cleaning, and exposure to fluids may cause additional surface changes. All of these changes have the potential for changing microbial adhesion and colonization.

These comments should alert the reader to the multiple material variables that require controlling when comparing the susceptibility of different objects and materials to colonization by a microorganism. The investigator can minimize these variables by attention to detail regarding the manufacturer and lot of the materials, by determining the physicochemical characteristics of the surface, by taking care in the handling, storage, and cleaning of objects, and by making allowances for the design and size of the test objects. When material composition is a critical element in the study, the National Institutes of Health provides investigators with standard reference materials with known composition that can be used as performance standards for use when testing materials of uncertain composition.

### B. Bathing Fluid

The bathing fluid carries the suspended microbes and covers the test object. Two aspects of this fluid require control: the composition of the fluid and the flow of the fluid over the test object.

Important compositional variables include the concentration of dissolved and suspended materials, particularly ions[52] (including pH),[21,22,30] chelating agents,[21,31,51] proteins,[9,23] detergents,[52] and atmospheric gasses.[19] Electrolytes, particularly multivalent cations, like $Ca^{2+}$[19,51] and $Mg^{2+}$,[21] can promote adhesion by crosslinking anionic groups on bacterial cells and surfaces.[52] The concentration of electrolytes can also influence hydrophobic interactions[52] and bacterial adhesion may exhibit pH optimums and minimums.[21,31,52] The presence of detergents in water can diminish surface tension and interrupt attractive van der Waals forces to create a net repulsive force between particle and substratum leading to detachment or desorption.[52,70]

Dissolved and suspended materials not only affect the fluid, they also affect the substratum. With exposure to a fluid, a proportion of the suspended and dissolved materials will bind to the submerged surfaces and modify the surface chemistry; these modifications can influence microbial attachment. For example, the adsorption of serum proteins onto plastic surfaces[9,23] repels some organisms. Paradoxically, these same proteins promote colonization for other bacterial strains by providing unique sites for the targeted attachment of the organisms via adhesin-to-protein binding.[23,72] Likewise, the choice of culture media[12] for providing important nutrients and the concentration of substances like glucose,[12,17] iron,[18] oxygen and carbon dioxide,[5,19] can greatly influence the biofilm formation.

Common sources of error arising from the bathing fluid include:

- Failure to recognize the presence of ions and detergents in the test chamber.
- Failure to account for the effect of oxygenation and media on microbial growth.
- Failure to recognize the influence on colonization of organic substances, particularly proteins, that are carried over from the culture media or are present in the test chamber fluid.

Appropriate controls to prevent these errors include the use of reagent grade materials and the preparation of buffers and reagents in pure water. Washing bacterial cells and the avoidance of complex culture media and protein enriched solutions will minimize the carry-over of organic substances; when these effects can not be eliminated, appropriate allowances should be made in the experimental design.

Investigators should also consider the dynamic aspects of the bathing fluid. Most assays are performed when the substratum is submerged under static (motionless) conditions; such systems are dominated by gravity and may not be a desirable model of a natural dynamic situation. On the other hand, as noted by Busscher and van der Mei,[8] the dipping and rinsing procedures that are often used in these assays can artifactually reduce the number of attached microorganisms. They emphasize that passing a biofilm through a liquid–air interface will produce significant sheer forces that can potentially remove organisms from the material surface that are actually attached to the surface. When using dynamic models, however, the investigator should ensure that the flow of fluids over the test objects is equivalent. Bacterial adhesion under high, and particularly turbulent, flow rates requires higher energy bonds than under low flow rates, laminar flow, or static conditions.[8] Likewise the temperature of the fluid will influence the thermodynamic stability of the surface–microbe bond as well as the subsequent microbial proliferation. These conditions can be controlled by attention to detail regarding the positioning and design of the test objects and by maintaining standard conditions of temperature and fluid flow.

## C. Microorganisms

Like the substratum, the choice and preparation of the test organism appears to be a simple matter but is really a complex set of variables that can be virtually impossible to control. For understandable convenience, protocols generally treat a culture of the test microorganism as a single immutable object. This convention can cause one to overlook the fact that microbial cultures are actually asynchronous transient populations of short lived life-forms; individual members of this population are likely to exhibit variations in phenotype.[27] Variability is particularly likely for organisms that colonize solid surfaces; since sessile life forms invariably include a planktonic stage in their life cycles, cultures of surface growing organisms are likely to include a mixture of both adhesive and non-adhesive phenotypes.[14,27] The precise proportions of the adhesive and nonadhesive microbes will be subject to genotypic drift, natural variation, and the purposeful or accidental enrichment of one form or the other by environmental factors and storage and propagation conditions. Since time always separates individual experiments, different experiments will always be different in terms of the number of microbial generations separating the different test cultures as well as the culture propagation and suspension conditions. For these reasons, it is virtually impossible to duplicate any two preparations of microorganisms, yet these minute differences can have important effects on the capacity of the microbial culture to colonize a surface. Furthermore, the cumulative impact of these effects only increases over a series of experiments and between different laboratories.[14] Fortunately, colonization capacities of different strains tend to be fairly predictable given similar test conditions. Nevertheless, the investigator interested in colonization of surfaces must be willing to tolerate a wide variability in test results and be vigilant for sources of experimental variation.

Investigators can limit the inherent variation of microbial preparations by attention to details regarding the propagation, harvesting, and storage of cultures so that equivalent experiments have similarly prepared test organisms. One approach is to prepare a large batch of organisms, divide the batch into aliquots, and store the aliquots at –70°C. When thawed, the different aliquots are likely to have equivalent colonization properties; however, the population of organisms that survive this treatment may not accurately reflect the original test population. Furthermore, depending upon survival under frozen conditions, aliquots of different age may have different mixtures of adhesive and nonadhesive phenotypes, and of course once the aliquots are exhausted the experiment can never be truly duplicated. An alternate approach is to use a minute starter inoculum to propagate a microbial strain under set culture conditions and then harvest the culture so that the preponderance of the organisms are products of known culture conditions and are at a similar stage in the life cycle (like harvesting in midlogarithmic phase or early stationary phase). This approach has the appeal that the culture conditions are readily duplicated, but for the previously stated reasons the products of the culture may not be the same from experiment to experiment.

Another common problem is for investigators to select one or more clinical strains that are convenient or interesting and proceed with detailed studies of the capacity of these organisms to colonize a material. The problem with this approach is that without access to the study organisms other investigators cannot replicate the experiments. To avoid this problem one does not have to limit studies to a few laboratory strains, but it is advisable to include a well-studied, widely available laboratory strain in the experiments as a performance standard and as a positive or negative control. Table 1 lists some organisms, available from international culture collections, whose capacity to attach to plastic surfaces has been well studied. If an appropriate reference organism does not exist, then many investigators will include the taxonomic type strain or an antimicrobial reference strain in their studies as the performance standard. Published organisms with interesting or well established colonization mechanisms should be submitted to an international culture collection for later reference.

### D. Time

We have already pointed out the distinction between early attachment, i.e., adhesion, and attachment followed by accumulation, i.e., colony formation (or slime production), and the necessity for precision in referring to the two phenomena. Experiments that focus upon adhesion usually use concentrated microbial suspensions ($10^{6-8}$) and short incubation periods (5 min–2 h) under non-nutritive conditions (buffer and 0–4°C). Experiments that focus upon microbial proliferation and survival usually use low inocula ($10^{1-6}$) and long incubation periods (6–24 h) under nutritive conditions (liquid media and 20–37°C).

## V. OVERVIEW OF EXPERIMENTAL METHODS

The following section presents an overview of the most notable experimental methods for studying bacterial attachment and colonization, grouped into direct and indirect observation methods, and stressing the advantages and disadvantages of each approach. The reader can find specific protocols for these techniques in later chapters.

### A. Direct Observation

We can directly observe microbial colonization with the aid of an optical microscope, a scanning electron microscope, a transmission electron microscope, and by laser-scanning confocal microscopy.

### 1. Light Microscope

Relatively inexpensive, simple to use, and readily available, the light microscope has proven the oldest and most versatile instrument for studying microbial attachment to surfaces. It has, for example, been applied to counting *Candida albicans* on acrylic sheets,[46] *E. coli* on polymethacrylate films,[32] *Pseudomonas* on polystyrene petri dishes,[24] and *E. coli* and *S. epidermidis* on glass cover slips.[28] The primary limitation is the requirement for an optically clear, planar material for the substratum. The procedure also destroys the organisms and does not provide information on microbial viability or the three dimensional structure of a biofilm. Staining and fixing a slide can also introduce artifacts.

Under the proper circumstances light microscopy is the quickest way to characterize the interaction between microbe and surface. In addition to estimating the amount of microbial attachment and colonization, light microscopy also enables the microscopist to study the nature of the attachment. By examination of the specimen, the microscopist may find microbes bound to the surface in individual cells, in clumps of cells, in mats, or with the expression of extracellular materials or associated with special surface structures. By making a smear or collecting free floating organisms with a centrifuge, the microscopist can compare the morphology of the sessile form to the planktonic form. Unfortunately, most of the submicrobial structures of interest to the microscopist studying microbial attachment, such as fimbriae and adhesive surface proteins, are below the resolving power of the light microscope and cannot be examined.

The light microscopist can readily transform qualitative optical observations into semiquantitative observations such as "present," "absent," "rare," "abundant," etc. and use these observations to compare organisms, substrata, and bathing fluids. Quantitative observations, however, require the light microscopist to overcome the tedium of counting by hand (or photographing and counting) individual microbes on a large number of microscopic fields (of known size and shape). If the microbial cells colonize the surface in a dense consortia, the inability to count individual cells will compromise the visual counts. At the other extreme, when microbial attachment and colonization takes place at a low level, the light microscopist may not be able to detect low, but nevertheless important, concentrations of organisms on a surface. Under these circumstances the investigator can use a microscope equipped for epifluorescence and fluorescent dyes like acridine orange, to increase the sensitivity by five-to-tenfold. Finally, we should note that bacterial counts made by light microscopy often vary widely from sample to sample and experiment to experiment, an observation that can frustrate the inexperienced investigator.

## 2. Laser-Scanning Confocal Microscope

The laser-scanning confocal microscope overcomes some of the important limitations of the simple light microscope. The equipment combines laser illumination, confocal imaging, plan-apochromatic objectives, and computer based image processing to generate high-resolution, three-dimensional images of the specimen. The instrument works by constructing a precise two-dimensional image of the specimen at a precise focal level. By varying the focus level and by obtaining multiple images, the computer component constructs a three-dimensional image of the specimen, from top-to-bottom, which includes both internal and external structures. By using fluorescent dyes, the microscopist can label and visualize specific microscopic elements. Because this instrument can see through transparent specimens while constructing a three-dimensional images, the laser-scanning confocal microscope can be invaluable for understanding the *in situ* morphology of complex microbe-surface interactions, such as the structure of microbial biofilms and the interaction of microbes with eukaryotic cells.[10]

The work of Sanford and colleagues[60] beautifully illustrates the application of this technique to biofilms. They used the laser-scanning confocal microscope to examine slime layers produced by the RP62A strain of *S. epidermidis*. By examining the preparation at a wavelength that excited green autofluorescence the investigators visualized the bacterial cells; by re-examining the same preparations treated with a Texas Red-labeled lectin that specifically bound the extracellular slime and was excited by another wavelength, these investigators could separately visualize the extracellular matrix as a red fluorescence. Combining these observations with a three-dimensional computer reconstruction of the image, the investigators determined the living architecture of the slime layer. Rather than a uniform biofilm, the organism grew in conical multicellular structures separated by channels that presumably allowed the deepest layers of the biofilm to receive nutrients and release wastes.[60]

Like the optical microscope, the wavelength of light limits the resolving power of the laser-scanning confocal microscope. Although image processing enhances this resolution over light microscopy, most submicrobial structures can not be visualized with the laser scanning confocal microscope. The laser-scanning confocal microscope is an expensive instrument, not available to most investigators. Fortunately, the study of simple microbe–surface interactions does not usually require such sophisticated equipment.[48]

### 3. Transmission Electron Microscope

The transmission electron microscope shows the greatest utility for visualization and characterization of internal and external microbial adherence structures.[40] By combining this method with gold-labeled antibodies, the microscopist can locate specific antigenic structures inside and outside of the microbe. The technique, however, has several limitations, foremost of which is the restriction to a soft substratum that can be sectioned. When sparse attachment occurs, the microscopist may have trouble locating cells in contact with the substratum. Artifacts caused by fixation and sectioning plague the technique, and the technique does not easily lend itself to comparative counts of colonizing microorganisms. Like light microscopy, the procedure destroys the organisms and does not provide information on microbial viability.

### 4. Scanning Electron Microscope

The advent of the scanning electron microscope has allowed investigators to observe in fine detail the attachment of microorganisms to surfaces.[40] The instrument has the greatest utility for exploring the attachment of microbes to a wide variety of complex opaque materials, like intravascular catheters,[55] Dacron grafts,[61] pacemaker leads,[56] plant leaves,[50] and biliary stents.[76] These observations allow the microscopist to not only determine where microorganisms preferentially attach to an object but also the nature of this attachment and the three dimensional appearance of microbial biofilms.

Like the optical microscope, the method can be converted from qualitative observations to quantitative observations by simply counting the number of organisms over a given surface area. The procedure has the unique advantage of allowing direct counts of the number of organisms attached to opaque or highly textured surfaces which is impossible by any other method. It has, for example, been used to count bacteria attached to smooth and sandblasted materials,[74] metals,[54,74] plastics,[54,74] ceramics,[54,74] intravascular catheters,[64] surgical biomaterials,[54] urethral catheters,[48] polymethylmethacrylate beads,[45] corneas,[34] and mucosal surfaces.[40] Obviously the procedure requires access to a scanning electron micro-scope and because the field of view is so much smaller, this procedure can be more tedious (consuming large amounts of instrument time) and less sensitive than similar observa-tions made by an optical microscope. Because the instrument visualizes the specimen at an angle from the side, maintaining the same field size between different fields and deter-mining the field dimensions (observed area) can be difficult for objects with a uniform surface and nearly impossible for objects with a convoluted surface. The scanning electron microscope has the same limitations as the optical microscope in that individual members of a consortium can not be reliably counted. The resolving power of the scanning electron microscope greatly exceeds the power of the light microscope, allowing the visualization of some attachment structures. The scanning electron microscope, however, does not have the resolving power of the transmission electron microscope. Furthermore, the prepara-tion and fixation process frequently introduces artifacts,[50] such as the condensation of polysaccharide films that obscure or distort surface structures.[10] Like the light micro-scope, bacterial counts tend to fluctuate widely from sample to sample and experiment to experiment.

### B. Indirect Observation

We have divided the indirect methods into two groups, procedures that first detach the microorganisms from the surface and then count the detached organisms, and procedures that estimate the number of attached microorganisms *in situ* by measuring some attribute of the attached organism.

### 1. Counts of Living Detached Organisms

#### A. PLATE COUNTS

One of the more gratifying approaches to enumerating surface microorganisms calls for the investigator to detach the organisms and count the number of colony forming units (cfu) recovered from the detachment procedure. The resulting data are quantitative, rapidly generated, and demonstrate the number of living organisms on the surface. The procedures are straightforward and generally do not involve special technology or materials. Although gratifying this approach is inherently paradoxical, in that we study attached organisms by first detaching the organisms. There are also drawbacks as the detachment procedure may not be complete and it may be harmful to the cells. If the detached cells are present in particles or packets rather than individual cells, results based on cfu may not accurately reflect the true numbers of attached cells. Despite these limitations there are two basic approaches which have wide applicability to medical devices.

#### B. ROLL TECHNIQUE (ALSO TOUCH AND *IN SITU* CULTURES)

This approach is a variation on the clinical method Maki[47] developed to assess the degree of bacterial colonization of intravascular catheters as an index of colonization and possible infection. Sheth and Franson applied the technique to determine the capacity of catheters of different materials[25,64] to accept surface colonization. The technique can be applied to any small gauge cylindrical objects (like catheters or rods); similar touch and *in situ* cultures have also been applied to other materials like vascular grafts[75] and plastic petri dishes.[46] The method is easily modified to address particular experimental questions.

For example, Greenfield and coworkers recently applied the technique to study the capacity of antiseptic coated catheters to inhibit biofilm formation in an experimental pig model.[30] These investigators modified the roll technique by doing the procedure twice on each catheter segment in order to distinguish between moderately bound organisms (removed by the first roll) and tightly bound organisms (removed by the second roll).[30] The roll method is simple, expedient, and requires very little in supplies or equipment, but it does have limitations. As the method is limited to external surfaces, it does not measure the intraluminal colonization of catheters. Because in order to count individual organisms macroscopic distances must separate the attached bacteria, this is at best a semi-quantitative method. It can not distinguish between moderate and heavy colonization and it may not detect polymicrobial colonization. Perhaps the major problem is that the technique depends upon dislodging organisms by touching the object to the tacky surface of the agar, tightly held organisms may not be displaced by this procedure.

C. SONICATION

Investigators have used a variety of means, like vortexing and scraping, to actively remove microorganisms from the surface of an object. Cleaning with sonic energy is the culmination of this approach. Unlike vortexing,[75] this approach is efficient and easily standardized. Silverhus et al.[65] used this procedure to study bacterial colonization of two forms of Dacron vascular graft material, Sherertz et al.[63] refined the technique for assessing colonization of vascular catheters, and Wengrovitz et al.[75] have demonstrated the superiority of this method over *in situ* cultures and vortexing. The approach has wide versatility, since it can be performed on a variety of objects with complex shapes,[75] it is quantitative and provides information on living organisms. While specialized, sonication equipment is not terribly expensive, but operating the equipment can damage the operator's hearing if the operator does not wear protective headgear. The technique may not uniformly strip bacteria from the surface of the object, as certain places on the object may be protected from the sonic energy; this problem is particularly true of objects with complex shapes. The major limitation is the tedium of performing serial dilutions to quantify released bacteria. If available, a Coulter counter adjusted to counting particles the size of microbes can relieve this tedium. Counting by this automated method may not work if the organism attaches to the glass surfaces and plastic tubing of the counter.

*2. Estimation of Attached Microorganisms* In Situ

A. RADIOLABELED BACTERIA

Perhaps the most sensitive and versatile method for an investigator to study microbial adhesion to surfaces is to radiolabel the organism. The utility of this approach is illustrated by the wide variety of radionuclides, microorganisms, and substrata used in radiolabeling experiments. Most investigators label the organisms by propagation in liquid media that includes a radiolabeled essential nutrient; for example, [$^3$H]thymidine has been used to label various Gram positive and Gram negative bacteria for adhesion to intravascular catheters,[72] suture materials,[67] and glass[73] and polystyrene[68] beads; likewise [$^{14}$C]glucose has been used to measure the adhesion of *Candida* to intravascular catheters[59] and Gram positive and Gram negative bacteria to intravascular catheters,[3] needles,[3] and suture material;[39] [$^{111}$]indium-oxine has been used to measure the attachment of *S. epidermidis* and *P. aeruginosa* to fibrin coated glass cover slips;[7] and both [$^{35}$S]methionine-labeled amino acids and [$^{14}$C]-labeled amino acids have been used to measure the adhesion of bacteria to plant leaves.[50] Radiolabeled bacteria have also been harvested from agar containing [$^{3}$H]glucose for studies of adhesion to crushed silicone rubber,[6] and passively labeled with [$^{51}$Cr] for studies of adhesion to silicone neurosurgical prostheses.[29] Prakobohl and coworkers[57] have recently introduced a novel variation using bacteria radiolabeled with [$^{35}$S]methionine and a centrifuge to quantify the strength of the microbial bond to proteins adsorbed onto microtiter plates. They begin the assay by allowing the bacteria to bind to the plate, then they flip the plate, spin off the weakly bound organisms, and count the radioactivity of the residual organisms that are tightly bound to the floor of the plate.[57]

The primary limitation to the use of radionuclides is that the ratio of counts-per-minute(cpm)-to-microbe is unstable. Microbial replication dilutes the label and metabolic processes destroy the radionuclide-microbe link. Under most experimental conditions, the investigator can follow experiments using radiolabeled bacteria for only a few hours, limiting this technique to experiments concerning the microbial adhesion phase of colonization. Radionuclides and scintillation fluid are hazardous materials; purchase and disposal are expensive and counting requires specialized instruments. Radiolabeled organisms should be well washed prior to use and discarded supernatant from the washing procedure should be included in each experiment as a quality control. A standard curve comparing radioactive counts by cfu-to-cpm of the radiolabeled culture should be constructed for each organism. Likewise a blank consisting of the sterile object should be run in parallel with the test samples and investigators should report the specific activity of the radionuclide when describing their methods.[15]

B. ENZYME-LINKED IMMUNOSORBENT ASSAY (ELISA)

This technique is similar in many ways to radiolabeling organisms, except the detection system uses an enzyme-linked sandwich antibody instead of a radiolabel. The ELISA avoids the radiation hazard of radiolabels, but retains the utility of application to a wide variety of substrata and microorganisms. The method requires the investigator to first use an antibody to a bacterial antigen exposed on the surfaces of attached bacteria. The investigator measures the amount of bound antibody by using a second enzyme-linked (usually either horseradish peroxidase or alkaline phosphatase) antibody directed at the first antibody. The investigator exposes the bound enzyme to a chromogenic substrate and measures the color change with a spectrophotometer. Other similar methods include the enzyme-linked lectinosorbent assay (ELLA), which substitutes an enzyme-linked lectin for the antibody sandwich, and the enzyme-linked biotin-avidin assay (ELBA), which measures the attachment of biotinylated bacteria to a surface.[53]

In all of these assays, the amount of color change is proportional to the number of attached organisms, but the proportion is not strictly stoichiometric. For the ELISA the number of first antibodies binding to antigenic sites on the organism, the number of second antibodies binding to first antibodies as well as the reaction kinetics of the enzyme can interfere with this proportion. Obviously the ELISA requires the investigator to have access to an antibody that is specific for the microorganism and does not bind, specifically or nonspecifically, in an appreciable amount to the substratum. Less obvious, but equally important, is for the target antigen to be both expressed by attached bacteria and not to be obscured by overlying extracellular materials like slime or capsule. Unlike radiolabel assays, the ELISA will not count bound organisms embedded in a biofilm. For these reasons the assay shows greatest utility for detecting the attachment of small numbers of uncoated bacteria to a surface,[53] but loses utility in studying the formation of dense biofilms covered by extracellular materials.

Like biologic assays, the ELISA must first be standardized by another counting technique. Most investigators use quantitative cultures of serial dilutions of unattached, planktonic, organisms. Like biologic assays (see below), this standardization step assumes that the expression of the target antigen in planktonic cells will be the same as sessile cells, which for geometric reasons alone is incorrect.

## C. Biologic Assays

Every method introduced in this chapter has a major disadvantage. Direct microscopic counts are tedious and insensitive. Radiation assays are hazardous and limited by a short observation window. Assays using sonication to detach organisms do not study the organisms *in situ* and the procedure can harm the organisms. Touch preparations – like the roll technique – are insensitive, have limited applicability, and also do not study the organisms *in situ*. Enzyme-linked assays are not strictly stoichiometric and may not count all attached organisms in a biofilm. Likewise, the procedure that we will present at the end of the chapter, the optical density of bacterial films, is limited to specialized optically clear materials. Biologic assays—assays that measure the production of a microbial product as an indirect assay for the number of microorganisms on a surface—avoid most of the disadvantages presented by these techniques.

Biologic assays are sensitive linear measurements that are easily performed, generally do not require hazardous materials, and allow the investigator to follow all phases of microbial colonization *in situ* over a wide variety of complex objects under a variety of conditions. A major disadvantage to this approach is that the methods often require specialized equipment and reagents. Another disadvantage is the requirement for the investigator to standardize the assay by correlating the amount of biologic product to the number of microbes generating the product. Because experimental conditions usually rrequire the investigator to perform the standardization with planktonic organisms, standardization can introduce a serious theoretical flaw into the observations by making the assumption that the generation of biologic products by planktonic organisms is similar to the generation of biologic products by sessile organisms.[27,66] Biologic assays may also require the investigator to rerun the standardization curve for each test organism.[66] Nevertheless, monitoring colonization by monitoring the production of a biologic product has great utility, particularly for following the activities of well-established microbial communities. To perform these studies investigators have monitored ATP production via light release (bioluminescence) from a solution of firefly luciferin and luciferase,[31,37,42,44] the number of attached bacterial cells via the production of cell associated urease,[20,21] and electron transport via formazan production.[26,43]

## D. Stained Bacterial Films

Firmly attached microbial colonies are easily stained and visualized. The appearance of these deposits can form the basis for qualitative and semiquantitative visual estimates of microbial colonization and for quantitative assays of microbial colonization by spectrophotometric determination of the optical density of the colony. Two approaches have evolved. The first, known as the "test tube" or "tube" method, is a simple estimate of the macroscopic presence or absence of a stained bacterial film on test tubes that had contained broth cultures of bacteria. The second approach, known as the "microtiter plate" method, uses an automatic spectrophotometer to determine the optical density (OD) of a stained bacterial film on the floor of a multiwell tissue culture plate.

Christensen et al.[12,13] introduced the test tube method as both a demonstration of slime production and as an assay for slime production among strains of coagulase negative staphylococci (CNS) associated with intravascular catheter infection. When introduced the procedure called for using either trypan blue or safranin as the staining agent; however, most investigators have used safranin.[1,16,35] Since both reagents primarily stain the microbial cells and not the slimy matrix material, the OD of the stained residue is a rough approximation of the bacterial density on the test tube surface, not the extracellular slime. Other investigators,[36,41] have used alcian blue, which in addition to staining the bacterial cells also stains the extracellular matrix.[9,12,13] This dual staining appears to be a more reliable index of extracellular matrix production. Since the test tube procedure is simple, inexpensive, and expedient, it has been used by a number of investigators, primarily in the context of clinical isolates of CNS, although it has also been extended to *S. aureus*.[35]

The test tube procedure is deceivingly simple; minor factors, such as the inoculum and the surface-to-volume ratio of the culture vessel, can have a major effect on the appearance of slime. For example, because slime production depends on the multiplication of bacteria on a surface, small inocula—leading to many bacterial generations and a thicker biofilm—will more likely produce positive results than large inocula that lead to relatively few bacterial generations and a thinner biofilm. Likewise, the low surface-to-volume ratio of a deep narrow test tube under static conditions can produce low concentrations of oxygen at the bottom of the tube; for some slime producing strains of CNS, low oxygen concentration discourages slime production and can lead to a false negative test.[5] The tube test is qualitative; results are interpreted as either (strongly or weakly) positive or negative. Significant observer-to-observer and laboratory-to-laboratory variation can occur.[13] The basic procedure has been converted to a quantitative assay by Alexander and Rimland[1] who used a spectrophotometer (Junior Spectrophotometer, Coleman Instruments, Maywood, IL) at 550 nm to directly read the optical density (OD) of the safranin stained bacterial film and by Tsai et al.[69] who measured the OD at 590 nm of bacterial films stained with toluidine blue and solubilized with base (0.2 $M$ NaOH at 85°C for 1 h). The assay is obviously limited to culture tubes and is not easily expanded to other materials.

Christensen et al.[13] introduced the microtiter plate method to correct many of the deficiencies of the tube test. The approach is a modification of the methodology introduced by Fletcher who used the OD of stained bacterial films on plastic petri dishes as a model for the colonization of marine surfaces by aquatic microorganisms.[23] After exposing plastic petri dishes to bacterial solutions, Fletcher rinsed the dishes four times with buffer, then fixed the residual attached bacteria with Bouin's fixative and stained the cells with ammonium oxalate-crystal violet. The plates were then air dried and the OD read at

590 nm with a spectrophotometer. Christensen et al.[13] adapted this approach to 96-well tissue culture plates, which increased data acquisition by the use of an automatic spectrophotometer. The resulting method is inexpensive, easily performed, and produces reliable quantitative data.

The primary drawback to the procedure is the confinement to multiwell (usually 96-well) plates; most investigators will use this approach for experiments that focus on microbial processes rather than on the substratum. Nevertheless, if the substratum material is available in optically clear sheets, it is possible to adapt this procedure to studying material variables by cutting the material into sheets or discs and placing the specimens on the floor of the test chamber. Because 96-well plates have a higher surface-to-volume ratio than test tubes, the results of colonization experiments with this method can be at variance with the test tube method, primarily by the identification of additional slime producing strains which are dependent upon higher oxygenation conditions.[5] Since the microtiter plate assay depends on the optical density of attached cells, the method will not detect low levels of surface colonization.

In setting up experiments using the microtiter plate, the investigator must consider the choices of spectrophotometer, 96-well plate, and fixative procedure. For example the MicroELISA Auto Reader (Dynatech Laboratories, Chantilly, VA) automatic spectrophotometer has an OD ceiling of 1.500 while the BioRad EIA reader (BioRad Laboratories, Richmond, CA) records optical densities that approach 7.5.[4] Although the significance of optical densities at the high end of the BioRad scale is uncertain, important data in the OD range of 1.5 to 3.0 may be lost with the Dynatech instrument but recovered by the BioRad instrument. Different 96-well plates exhibit different colonization characteristics. As previously noted, the surfaces of microtiter plates tend to be hydrophobic whereas the manufacturer modifies the surfaces of tissue culture plates to reduce hydrophobicity.[13,52] Bacteria generally attach to a greater extent to tissue culture plates, which are the substratum of choice. The choice of fixative is the final decision. As the microtiter plate procedure has become more popular, different investigators have introduced a variety of fixatives,[4] probably because the original procedure as developed by Fletcher called for Bouin's fixative. Bouin's fixative includes picric acid which is potentially explosive and discarding excess reagent can be both difficult and expensive. Alternative fixatives, like methanol, formalin, Carnoy's, etc., however, do not work as well as Bouin's fixative and can lead to widely variable results, even with the same strain of bacteria.[4] Simple air-drying, however, appears to work nearly as well as Bouin's fixative.[4]

### E. Microtiter Plate Procedure

For determination of slime production, bacteria are first propagated overnight in Trypticase soy broth (TSB) and then diluted 1:100 in fresh TSB. Individual wells of a flat bottomed 96-well tissue culture plate are then filled with 0.2 mL of the inoculated broth and the plates are incubated in a stationary position at 37°C for 18–24 h. For adhesion experiments, bacteria are suspended in buffer to a concentration of $10^7$–$10^8$ cfu/mL, following which 0.2 mL portions of the suspension are dispensed into individual wells and the plate incubated for 5–120 min, usually in a stationary position at 37°C. At the conclusion of the incubation period the liquid contents of the wells are gently aspirated (usually with an eight or twelve prong aspirator connected to low vacuum) by tipping the plate forward, placing the tip of the aspirator on the lowest surface of the well wall (at the air–fluid meniscus), and as the fluid is aspirated slowly following the retreating meniscus into the corner created by the sides and floor of the well. (It is important to not disturb the film in the center of the well which is used to measure the OD.) The wells are then refilled with 0.2 mL of PBS and reaspirated; the procedure is repeated for a total of four changes of PBS. The residual adhesive bacterial film is fixed by drying at 60°C for 1 h and stained by flooding the wells with Hucker crystal violet. The stained plates are rinsed under running tap water, emptied by shaking the plate upside down, and set aside to dry. The dried plates are read in a BioRad EIA reader at a wavelength of 570 nm. Appropriate positive controls include ATCC 35984 (RP62A) or ATCC 35983 (RP12) in TSB; appropriate negative controls include ATCC 35983 in TSB without glucose and ATCC 35982 (SP2) in TSB.

F. INTERPRETATION

Each plate should contain a media blank, the OD of which is subtracted from the sample OD. The absolute contribution of the blank reading to the data, however, tends to be negligible and in many experiments can be safely ignored. Samples are usually run in multiples (usually 4–8 wells) and assays are repeated at least once. In experiments following the colonization of plates by one (or a few) strains (relative values), investigators have analyzed the data by using Student's t test of the averaged sample OD. For studies that compare the slime producing capacity of different strains, we have previously proposed an arcane procedure which combines the OD of bacterial films produced in TSB with the OD produced in TSB without glucose.[13] Currently, however, most investigators simply average the OD of the bacterial films produced in TSB. For studies looking at group data the average OD readings are compared by Student's t test; for determinations of the slime producing capacity of a particular strain (absolute values), the organism is labeled either negative, or weakly positive, or strongly positive based on the averaged OD. The accepted convention for these categories[13,17] is to use as the ceiling OD for negative strains three standard deviations above the mean value of a series of blank wells (usually an OD of 0.12 to 0.3). Strains with an OD that exceeds this ceiling value but is less than double this value can be considered weakly positive (usually an OD of 0.12 to 0.6), and strains with an OD greater than double (usually 0.24 to 0.6) the ceiling value for negative strains can be considered strongly positive. The problem with this approach is that many strains have OD readings of 1.5 or greater and this potentially important data is not captured by such a categorizing system.

**Acknowledgments:** The authors' research is supported by research funds from the Department of Veterans Affairs; the Missouri Foundation for Medical Research; Department of Orthopedics, University of Missouri–Columbia; and the Orthopedic Trauma Association.

## REFERENCES

1. Alexander W, Rimland D: Lack of correlation of slime production with pathogenicity in continuous ambulatory peritoneal dialysis peritonitis caused by coagulase negative staphylococci. *Diagn Microbiol Infect Dis* 8:215–20, 1987

2.  Appelgren P, Ransjo U, Bindslev L, et al: Surface heparinization of central venous catheters reduces microbial colonization *in vitro* and *in vivo*: Results from a prospective, randomized trial. *Crit Care Med* 24:1482–9, 1996

3.  Ashkenazi S, Weiss E, Drucker M: Bacterial adherence to intravenous catheters and needles and its influence by cannula type and bacterial surface hydrophobicity. *J Lab Clin Med* 107:136–40, 1986

4.  Baldassarri L, Simpson WA, Donelli G, et al: Variable fixation of staphylococcal slime by different histochemical fixatives. *Eur J Microbiol Infect Dis* 12:34–37, 1993

5.  Barker LP, Simpson WA, Christensen GD: Differential production of slime under aerobic and anaerobic conditions. *J Clin Microbiol* 28:2578–9, 1990

6.  Barrett S: Staphylococcal infection of plastic inserts; a method to measure staphylococcal adhesion. *Brit J Clin Pract* (Suppl 25) 37:81–5, 1983

7.  Benson DE, Burns GL, Mohammad SF: Effects of plasma on adhesion of biofilm forming *Pseudomonas aeruginosa* and *Staphylococcus epidermidis* to fibrin substrate. *ASAIO J* 42: M655–60, 1996

8.  Busscher HJ, van der Mei HC: Use of flow chamber devices and image analysis methods to study microbial adhesion. *Meth Enzymol* 253:455–76, 1995

9.  Carballo J, Ferreiros CM, Criado MT: Importance of experimental design in the evaluation of the influence of proteins in bacterial adherence to polymers. *Med Microbiol Immunol* 180:149–55, 1991

10. Costerton JW, Lewandowski Z, Caldwell DE, et al: Microbial biofilms. *Ann Rev Microbiol* 49:711–45, 1995

11. Cree RG, Phillips I, Noble WC: Adherence characteristics of coagulase-negative staphylococci isolated from patients with infective endocarditis. *Microbial Pathogen* 43:161–8, 1995

12. Christensen GD, Simpson WA, Bisno AL, et al: Adherence of slime-producing strains of *Staphylococcus epidermidis* to smooth surfaces. *Infect Immun* 37:318–26, 1982

13. Christensen GD, Simpson WA, Younger JJ, et al: Adherence of coagulase-negative staphylococci to plastic tissue culture plates, a quantitative model for the adherence of staphylococci to medical devices. *J Clin Microbiol* 22:996–1006, 1985

14. Christensen G, Baddour LM, Madison BM, et al: Colonial morphology of staphylococci on Memphis agar: phase variation of slime production, resistance to beta-lactam antibiotics and virulence. *J Infect Dis* 161:1153–69, 1990

15. Christensen GD, Baldassarri L, Simpson WA: Methods for studying microbial colonization of plastics. *Meth Enzymol* 253:477–500, 1995

16. Davenport DS, Massanari RM, Pfaller MA, et al: Usefulness of a test for slime production as a marker for clinically significant infections with coagulase-negative staphylococci. *J Infect Dis* 153:332–9, 1986

17. Deighton MA, Balkau B: Adherence measured by microtiter assay as a virulence marker for *Staphylococcus epidermidis* infections. *J Clin Microbiol* 28:2442–7, 1990

18. Deighton M, Borland R: Regulation of slime production in *Staphylococcus epidermidis* by iron limitation. *Infect Immun* 61:4473–9, 1993

19. Denyer SP, Davies MC, Evans JA, et al: Influence of carbon dioxide on the surface characteristics and adherence potential of coagulase-negative staphylococci. *J Clin Microbiol* 28:1813–7, 1990

20. Dunne WM, Burd EM: *In vitro* measurement of the adherence of *Staphylococcus epidermidis* to plastic by using cellular urease as a marker. *Appl Environ Microbiol* 57:863–6, 1991

21. Dunne WM, Burd EM: The effects of magnesium, calcium, EDTA, and pH on the *in vitro* adhesion of *Staphylococcus epidermidis* to plastic. *Microbiol Immunol* 36:1019–27, 1992

22. Fletcher M, Floodgate GD: An electron-microscopic demonstration of an acidic polysaccharide involved in the adhesion of a marine bacterium to solid surfaces. *J Gen Microbiol* 74:325–34, 1973

23. Fletcher M: The effects of proteins on bacterial attachment to polystyrene. *J Gen Microbiol* 94:400–4, 1976

24. Fletcher M: The effects of culture concentration and age, time, and temperature on bacterial attachment to polystyrene. *Can J Microbiol* 23:1–6, 1977

25. Franson TR, Sheth NK, Menon L, et al: Persistent *in vitro* survival of coagulase-negative staphylococci adherent to intravascular catheters in the absence of conventional nutrients. *Clin Microbiol* 24:559–61, 1986

26. Gagnon RF, Harris AD, Prentis J, et al: The effects of heparin on rifampin activity against *Staphylococcus epidermidis* in biofilms. *Adv Periton Dial* 5:138–42, 1989

27. Gilbert P, Collier PJ, Brown MR: Influence of growth rate on susceptibility to antimicrobial agents: biofilms, cell cycle, dormancy, and stringent response *Antimicrob Agents Chemother* 34:1865–8, 1990

28. Gilbert P, Evans DJ, Evans E, et al: Surface characteristics and adhesion of *Escherichia coli* and *Staphylococcus epidermidis*. *J Appl Bacteriol* 71:72–7, 1991

29. Gower, DJ, Gower VC, Richardson SH, et al: Reduced bacterial adherence to silicone plastic neurosurgical prosthesis. *Ped Neurosci* 12:127–33, 1986

30. Greenfeld JI, Sampath L, Popilskis SJ, et al: Decreased bacterial adherence and biofilm formation on chlorhexidine and silver sulfadiazine-impregnated central venous catheters implanted in swine. *Crit Care Med* 23:894–900, 1995

31. Harber MJ, Mackenzie R, Asscher AW: A rapid bioluminescence method for quantifying bacterial adhesion to polystyrene. *J Gen Microbiol* 129:621–32, 1983

32. Harkes G, Feijen J, Dankert J: Adhesion of *Escherichia coli* on to a series of poly(methacrylates) differing in charge and hydrophobicity. *Biomaterials* 12:853–60, 1991

33. Harris JM, Martin LF: An *in vitro* study of the properties influencing *Staphylococcus epidermidis* adhesion to prosthetic vascular graft materials. *Ann Surg* 206:612–20, 1987

34. Hazlett LD: Analysis of ocular adhesion. *Meth Enzymol* 253:53–66, 1995

35. Hebert GA, Crowder CG, Hancock GA, et al: Characteristics of coagulase-negative staphylococci that help differentiate these species and other members of the family *Micrococcaceae*. *J Clin Microbiol* 26:1939–49, 1988

36. Hogt AH, Dankert J, Feijen J: Encapsulation, slime production and surface hydrophobicity of coagulase-negative staphylococci. *FEMS Microbiol Lett* 18:211–5, 1983

37. Hussain M, Hastings JG, White PJ: Comparison of cell-wall teichoic acid with high-molecular-weight extracellular slime material from *Staphylococcus epidermidis*. *J Med Microbiol* 37:143–75, 1991

38. Ishak MA, Groschel DM, Mandel GL, et al: Association of slime with pathogenicity of coagulase-negative staphylococci causing nosocomial septicemia. *J Clin Microbiol* 22:1025–9, 1985

39. Katz S, Izhar M, Mirelman D: Bacterial adherence to surgical sutures. A possible factor in suture induced infection. *Ann Surg* 194:35–41, 1981

40. Knutton S: Electron microscopical methods in adhesion. *Meth Enzymol* 253:145–58, 1995

41. Kotilainen P, Nikoskelainen J, Huovinen P: Antibiotic susceptibility of coagulase-negative staphylococcal blood isolates with special reference to adherent, slime-prouducing *Staphylococcus epidermidis* strains. *Scand J Infect Dis* 23:325–32, 1991

42. Kristinsson KG: Adherence of staphylococci to intravascular catheters. *J Med Microbiol* 28:249–57, 1989

43. Ladd TI, Costerton JW: Methods for studying biofilm bacteria. *Meth Microbiol* 22:285–307, 1990

44. Ludwicka A, Jansen B, Wadstrom T: Attachment of staphylococci to various synthetic polymers. *Zbl Bakt Mikrobiol Hyg* A 256:479–89, 1984

45. Lyons VO, Henry SL, Faghiri M, et al: Bacterial adherence to plain and tobramycin-laden polymethylmethacrylate beads. *Clin Orthop* 278:260–4, 1992

46. Mackenzie AM, Rivera-Calderon RL: Agar overlay to measure adherence of *Staphylococcus epidermidis* to four plastic surfaces. *Appl Env Microbiol* 50:1322–4, 1985

47. Maki DG, Weise CE, Sarafin HW: A semiquantitative culture method for identifying intravenous-catheter-related infection. *New Eng J Med* 296:1305–9, 197

48. Manning PA: Use of confocal microscopy in studying bacterial adhesion and invasion. *Meth Enzymol* 253:159–66, 1995

49. Marshall KC: Introduction. In: Marshall KC, ed. *Microbial Adhesion and Aggregation.* Dahlem Konferenzen, Springer-Verlag, Berlin, 1984:1–3.

50. Matthysse AG: Observation and measurement of bacterial adhesion to plants. *Meth Enzymol* 253:189–206, 1995

51. McCourtie J, Douglas J: Relationship between cell surface composition of *Candida albicans* and adherence to acrylic after growth on different carbon sources. *Infect Immun* 32:1234–41, 1981

52. McEldowney S, Fletcher M: Variability of the influence of physicochemical factors affecting bacterial adhesion to polystyrene substrata. *Appl Environ Microbiol* 52:460–5, 1986

53. Ofek I: Enzyme-linked immunosorbent-based adhesion assays. *Meth Enzymol* 253:528–36, 1995

54. Oga M, Sugioka Y, Hobgood CD, et al: Surgical biomaterials and differential colonization by *Staphylococcus epidermidis. Biomateials* 9:285–9, 1988

55. Peters G, Romano L, Pulverer G: Adherence and growth of coagulase-negative staphylococci on surfaces of intravenous catheters. *J Infect Dis* 146:479–82, 1982

56. Peters G, Sabrowski F, Romano L, et al: Investigations on staphylococcal infection of transvenous endocardial pacemaker electrodes. *Am Heart J* 108:359–65, 1984

57. Prakobohl A, Leffler H, Fisher SJ: Identifying bacterial receptor proteins and quantifying strength of interactions they mediate. *Meth Enzymol* 253:132–42, 1995

58. Roberts JA, Fussell EN, Kaack MB: Bacterial adherence to urethral catheters. *J Urol* 144:264–9, 1990

59. Rotrosen D, Gibson TR, Edwards JE: Adherence of candida species to intravenous catheters. *Infect Dis* 147:594, 1983

60. Sanford BA, de Feijter AW, Wade MH, et al: A dual fluorescence technique for visualization of *Staphylococcus epidermidis* biofilm using scanning confocal laser microscopy. *J Indust Microbiol* 16:48–56, 1996

61. Schmitt DD, Bandyk DF, Pequet AJ, et al: Mucin production by *Staphylococcus epidermidis Arch Surg* 121:89–95, 1986

62. Shekarchi IC, Sever JL, Lee YJ, et al: Evaluation of various plastic microtiter plates with measles, toxoplasma, and gamma globulin antigens in enzyme-linked immunosorbent assays. *J Clin Microbiol* 19:89–96, 1984

63. Sherertz RJ, Raad II, Belani A, et al: Three-year experience with sonicated vascular catheter cultures in a clinical microbiology laboratory. *J Clin Microbiol* 28:76–82, 1990

64. Sheth NK, Rose HD, Franson TR, et al: *In vitro* quantitative adherence of bacteria to intravascular catheters. *J Surg Res* 34:213–8, 1983

65. Silverhus DJ, Schmitt DD, Edmiston CE, et al: Adherence of mucin and non-mucin-producing staphylococci to preclotted and albumin-coated velour knitted vascular grafts. *Surg* 107:613–9, 1990

66. Stollenwerk M, Fallgren C, Lundberg F, et al: Quantitation of bacterial adhesion to polymer surfaces by bioluminescence. *Zbl Bakteriol* 287:7–18, 1998

67. Sugarman B, Musher D: Adherence of bacteria to suture materials. *Proc Soc Exp Biol Med* 167:156–60, 1981

68. Timmerman CP, Fleer A, Besnier JM, et al: Characterization of a proteinaceous adhesin of *Staphylococcus epidermidis* which mediates attachment to polystyrene. *Infect Immun* 59:4187–92, 1991

69. Tsai CL, Schurman DJ, Smith RL: Quantitation of glycocalyx production in coagulase-negative *Staphylococcus*. *J Orthop Res* 6:666–70, 1988

70. van Oss CJ, Charny CK, Absolom DR, et al: Detachment of cultured cells from microcarrier particles and other surfaces by repulsive van der Waals forces. *Biotechniques* 1983:194

71. Vaudaux P, Pittet D, Haeberli A, et al: Host factors selectively increase staphylococcal adherence to inserted catheters: a role for fibronectin and fibrinogen or fibrin. *J Infect Dis* 160:865–75, 1989

72. Vaudaux P, Pittet D, Haeberli A, et al: Fibronectin is more active than fibrin or fibrinogen in promoting *Staphylococcus aureus* adherence to inserted intravascular catheters. *J Infect Dis* 167:633–41, 1993

73. Vergeres P, Blaser J: Amikacin, ceftazidime, and flucloxacillin against suspended and adherent *Pseudomonas aeruginosa* and *Staphylococcus epidermidis* in an *in vitro* model of infection. *J Infect Dis* 165:281–9, 1992

74. Verheyen CC, Dhert WJ, de Blieck-Hogervorst JM, et al: Adherence to a metal, polymer and composite by *Staphylococcus aureus* and *Staphylococcus epidermidis*. *Biomaterials* 14:383–91, 1993

75. Wengrovitz M, Spangler S, Martin LF: Sonication provides maximal recovery of *Staphylococcus epidermidis* from slime-coated vascular prosthetics. *Am Surgeon* 57:161–4, 1991

76. Yu J-L, Andersson R, Parsson H, et al: A bacteriologic and scanning electron microscope study after implantation of foreign bodies in the biliary tract in rats. *Scand J Gastroenterol* 31:175–81, 1996

# 14

# Evaluating Adherent Bacteria and Biofilm Using Electron Microscopy

**Theresa A. Fassel[1] and Charles E. Edmiston[2]**

[1]*Core Electron Microscope Unit, The Scripps Research Institute, LaJolla, CA, USA*
[2]*Department of Surgery, Medical College of Wisconsin, Milwaukee, WI, USA*

## I. INTRODUCTION

### A. Bacterial Glycocalyx and Electron Microscopy

The secreted exopolysaccharide that constitutes the bacterial glycocalyx allows bacterial proliferation in a well-protected environment.[2,19] Bacterial adherence, proliferation and biofilm production on a biomedical device can have catastrophic consequences for the patient, resulting in infection or failure of the device. Therefore, laboratory studies of the pathogenesis of these latent device-related infections is critical.[3] However, attempts to study the bacterial biofilm by electron microscopy employing only conventional fixation techniques suffer from inadequately preserved or stained glycocalyx. This is largely due to the characteristics of the glycocalyx as a highly hydrated, polymerized anionic matrix of variably substituted polysaccharides. Polysaccharides are not well stabilized by the conventional fixatives, namely the aldehydes, glutaraldehyde and paraformaldehyde, or osmium tetroxide. Graded dehydration with alcohol to gradually remove water from this highly hydrated structure may further distort delicate features resulting in observation of condensed or collapsed structure.[2,15,16,18] Not naturally electron-dense, the conventional means for adding contrast with the poststains uranyl acetate and lead citrate is usually not sufficient to enable exopolysaccharide constituents to appear dark or electron-dense on the phosphorescent screen of the transmission electron microscope. Too electron-translucent on its own, and unable to gain contrast from conventional poststains, the glycocalyx is often indistinguishable from the embedding resin background.[16]

### 1. Ruthenium Red, Alcian Blue, and Lysine
*in Electron Microscopy Procedures for Visualization of the Bacterial Glycocalyx*

The use of the cationic reagents, such as ruthenium red and alcian blue, provides added stability to the glycocalyx throughout the rigorous stages of processing for electron microscopy. Ruthenium red, a 1.1 nm sphere of +6 charge,[11,13] was extensively utilized by Luft[13,14] to improve preservation of highly polymerized anionic mammalian

*Handbook of Bacterial Adhesion: Principles, Methods, and Applications*
Edited by: Y. H. An and R. J. Friedman © Humana Press Inc., Totowa, NJ

**Table 1. Sample Loss and Sample Survival by Different Forms of Lysine**

| Fixation method | n | Sample loss by 1 h | Additional sample loss | | | Sample 24 h |
|---|---|---|---|---|---|---|
| | | | 2 h | 4 h | 24 h | |
| Free amino lysine | | | | | | |
| GA-lysine | 9 | 4 | 3 | 2 | 0 | 0 |
| GA-RR-lysine | 9 | 3 | 1 | 1 | 2 | 2 |
| GA-AB-lysine | 9 | 1 | 3 | 2 | 1 | 2 |
| PF-GA-lysine | 9 | 0 | 0 | 0 | 3 | 6 |
| PF-GA-RR-lysine | 9 | 0 | 1 | 2 | 4 | 2 |
| PF-GA-AB-lysine | 9 | 0 | 0 | 0 | 3 | 6 |
| | | | | | | |
| Monohydrochloride lysine | | | | | | |
| PF-GA-lysine | 9 | 0 | 0 | 0 | 0 | 9 |
| PF-GA-RR-lysine | 9 | 0 | 0 | 0 | 0 | 9 |
| PF-GA-AB-lysine | 9 | 0 | 0 | 0 | 0 | 9 |
| | | | | | | |
| Acetate lysine | | | | | | |
| PF-GA-lysine | 9 | 0 | 0 | 0 | 0 | 9 |
| PF-GA-RR-lysine | 9 | 0 | 0 | 0 | 0 | 9 |
| PF-GA-AB-lysine | 9 | 0 | 0 | 0 | 0 | 9 |

Abbreviations:
  GA = glutaraldehyde
  PF = paraformaldehyde
  RR = ruthenium red
  AB = alcian blue

glycocalyx as well as several unidentified bacterial species.[14] However, the Luft approach was more successful for some bacterial species than others.[5,6] Species of clinically important staphylococci require a platform for enhancing structural stabilization especially where a direct attachment surfaces or sufficient protein sites were lacking.[2,6] Alcian blue, a planar copper phthalocyanin with a +4 charge distributed on quaternary amino side chains,[17] while providing stability for the anionic glycocalyx of some bacterial species,[5] has shown limited success when used on its own in chemical fixation procedures for the staphylococci.[6]

When the positively charged diamine lysine, previously effective in the preservation of the mammalian glycocalyx,[1] was added to glutaraldehyde and ruthenium red, the effectiveness of this fixative was notably increased, in particular for species where ruthenium red alone was inadequate.[12] Elaborate bacterial glycocalyces, including those of the staphylococci, were observed by scanning electron microscopy (SEM)[4] and transmission electron microscopy (TEM).[6] When lysine was included with alcian blue, abundant glycocalyx was often observed surrounding and extending between staphylococci cells with an elimination of condensed artifactual features previously described.[7]

The major limitation with employing lysine as a fixative is due to the inflexibility of the 20 min prefixation step. As prefixation extends beyond 20 min, the sample often solidifies into a gelled mass.[8,9] For critical samples obtained in the clinical realm, an improvement in the time effectiveness of this method was greatly needed. This improvement was partially attained by inclusion of paraformaldehyde in the fixative with

**Table 2. Key Chemical Fixation Components**

| Component | Concentration | Source |
|---|---|---|
| Sodium cacodylate buffer | 0.1 M pH 7.2 | Fisher Scientific |
| Glutaraldehyde | 2.5% | Electron Microscopy Sciences, Fort Washington, PA |
| Paraformaldehyde | 2% | Electron Microscopy Sciences, Fort Washington, PA |
| Ruthenium red | 0.075% | Sigma Chemical Co.; R-2751 40% dye content |
| Alcian blue | 0.075% | Aldrich, Inc., Milwaukee, WI #19,978-8 Lot 009112ML 25% dye content |
| Lysine free amino | 75 mM | Sigma Chemical Co.; L-5501 (L-2,6 diaminohexanoic) |
| Lysine monohydrochloride | 75 mM | Sigma Chemical Co.; L-5626 |
| Lysine acetate | 75 mM | Sigma Chemical Co., L-1884 |

glutaraldehyde-lysine and ruthenium red,[8] and this time restriction of 20 min could be extended to several hours and even up to 24 h, as summarized in Table 1. However, solidification and loss of some samples still would occur in a variable and unpredictable manner with increased fixation time. A further improvement was attained by replacement of the lysine-free amino form previously used,[6,7,8] with the lysine-monohydrochloride or lysine-acetate form.[9] In these studies, sample loss at 24 h was eliminated, as shown in Table 1. Ultrastructural observation of extended fibrous glycocalyces was maintained. Alcian blue could also be used with the inclusion of paraformaldehyde and/or lysine-monohydrochloride or lysine-acetate forms; although condensed features were occasionally seen in fixations less than 2 h.[10] In these studies, three species of Gram-positive coagulase-negative staphylococci were utilized; a laboratory reference strain, *S. aureus* ATCC 25923, and two clinical isolates frequently studied in biomedical device infection models, *Staphylococcus hominis* SP2 and *S. epidermidis* RP62.[3]

## II. MATERIALS AND METHODS

### A. Cell Culture and Handling

Cells were recovered from −70°C frozen storage and plated on blood agar plates for 24 h to test viability. Inoculation of five colonies to trypticase soy broth (TSB) was followed by incubation for an additional 18 h at 35°C.

For SEM studies, 1 cm² segments of test biomaterial substrates of polyurethane foam or silicone rubber (Surgitek, Racine, WI) were sterilized and added to the TSB prior to inoculation.

For TEM studies, aliquots of lysine in 1.5 mL eppendorf tubes were vortexed with freshly prepared chemical prefixatives. This new mixture was immediately added to pellets of cells (ultracentrifuged at 4°C, 4,000 rcf, 10 min). Resuspension with a pasteur

## Control Samples, Short and Extended Time Fixation by Composition of Fixative and Form of Lysine

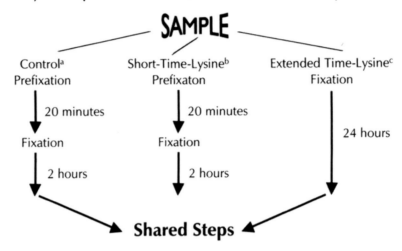

**Figure 1.** Procedures from sample to shared steps for SEM or TEM are shown for sample controls, short time and extended time fixation with prefixation (20 min), fixation (2 h) versus extended fixation (24 h) respectively. Note: a = control samples detailed in section II.C.2.a.; b = short-time-lysine methods detailed in sections II.C.2.b. and II.C.2.c.; c = extended time methods detailed in section II.C.2.d.

pipette was gentle but thorough. After each successive fluid exchange and incubation, cells were microcentrifuged at 12,000-16,000 rcf for 2 min and carefully resuspended in the successive reagents. Before the first dehydration stage, the cells were enrobed in 4% agar and handled as 1 $mm^3$ blocks thereafter.

### B. Chemical Fixative Components

A listing of key chemical components is indicated in Table 2. Sodium cacodylate buffer at 0.1 M, pH 7.2, is used for all aqueous fixative solutions and washes. Glutaraldehyde (Electron Microscopy Sciences, Fort Washington, PA) is used at 2.5% and paraformaldehyde (Electron Microscopy Sciences, Fort Washington, PA) at 2%. The cationic reagents ruthenium red (Sigma Chemical Co., St. Louis, MO, #R-2751; 40% dye content) and alcian blue (Aldrich, Inc. Milwaukee, WI, #19,978-8; 25% dye content) were used at identical 0.075% solutions throughout. All forms of lysine (L-2,6 diaminohexanoic) were used at 75 mm and were obtained from Sigma Chemical Co.; free amino (L-5501), monohydrochloride (L-5626) and acetate (L-1884).

### C. Procedures for Electron Microscopy

#### 1. Shared Aspects of Procedures

Procedures are shown in the following flowcharts. Figure 1 illustrates the chemical fixation methods by time for controls, short time and extended time procedures. The short time lysine methods have a 20 min prefixation and 2 h fixation. The extended time lysine methods have 24 h fixation.

## Shared Steps in Scanning
## Electron Microscopy (SEM)

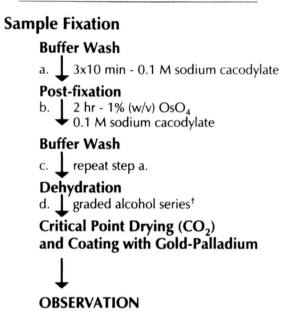

**Sample Fixation**

**Buffer Wash**

a. ↓ 3x10 min - 0.1 M sodium cacodylate

**Post-fixation**

b. ↓ 2 hr - 1% (w/v) $OsO_4$
↓ 0.1 M sodium cacodylate

**Buffer Wash**

c. ↓ repeat step a.

**Dehydration**

d. ↓ graded alcohol series[†]

**Critical Point Drying ($CO_2$)**
**and Coating with Gold-Palladium**

↓

**OBSERVATION**

**Figure 2.** Processing for SEM has several shared steps for samples irrespective of preceding prefixation/fixation stages which may have different chemical components. The graded-ethanol series consists of 10%, 25%, 50% for 10 min each, 70% overnight and 95%, 100%, 100% 1 h each.

Samples for SEM share several steps of processing in common as shown in Figure 2. These include postfixation in osmium tetroxide, ethanol dehydration (10%, 25%, 50%, 75%, 95%, 100%, 100%), critical point drying and metal coating to improve conduction of the electron beam in the scanning electron microscope.

Samples for TEM share several steps of processing in common as shown in Figure 3. After postfixation in osmium tetroxide and ethanol dehydration (10%, 25%, 50%, 75%, 95%, 100%, 100%), samples are infiltrated with the embedding resin, LR White (Electron Microscopy Sciences, Fort Washington, PA) for about 24 h. Samples are then embedded in gelatin capsules and placed in an oven for heat polymerization, preferably under a weak vacuum to assist consistent LR White polymerization, overnight. Thin sections are post-stained with uranyl acetate and lead citrate before observation in the transmission electron microscope.

## 2. Different Aspects of Procedures

A. Control Methods

The control fixation methods exclude lysine. All solutions are prepared in 0.1 M sodium cacodylate buffer.

Control 1: 2.5% glutaraldehyde.
Control 2: 2% paraformaldehyde, 2.5% glutaraldehyde.

# Shared Steps in Transmission Electron Microscopy (TEM)

## Sample Fixation

### Buffer Wash
a. ↓ 3x10 min - 0.1M sodium cacodylate

### Post-fixation
b. ↓ 2 hr - 1% (w/v) OsO$_4$
   ↓ 0.1M sodium cacodylate

### Buffer Wash
c. ↓ repeat step a.

### Dehydration
d. ↓ graded alcohol series[†]

### Infiltration
e. ↓ overnight - LR white resin
   ↓ 2 x 3 hr - LR white resin

### Embedment (gelatin capsules) and Polymerization
f. ↓ 60°C overnight under weak vacuum

### Thin Section and Staining
g. ↓ 10 min - 2% aq. uranyl acetate
   ↓ 5 min - lead citrate

## OBSERVATION

**Figure 3.** Processing for TEM has several shared steps for samples irrespective of preceding prefixation/fixation stages which may have different chemical components. Note: a weak vacuum draw (24 h) will assist in consistent LR White polymerization. The graded ethanol series consists of 10%, 25%, 50% for 10 min each, 70% overnight and 95%, 100%, 100% 1 h each.

Control 3:  0.075% ruthenium red, 2.5% glutaraldehyde.

Control 4:  0.075% ruthenium red, 2% paraformaldehyde, 2.5% glutaraldehyde.

Control 5:  0.075% alcian blue, 2.5% glutaraldehyde.

Control 6:  0.075% alcian blue, 2% paraformaldehyde, 2.5% glutaraldehyde.

B. Free Amino Lysine Methods

These short time methods do not use paraformaldehyde nor alternative forms of lysine. All solutions are prepared in 0.1 M sodium cacodylate buffer.

Method A: prefixation in 75 mM lysine, 2.5% glutaraldehyde; fixation in 2.5% glutaraldehyde.

Method B: prefixation in 75 mM lysine, 0.075% ruthenium red, 2.5% glutaraldehyde; fixation in 0.075% ruthenium red, 2.5% glutaraldehyde.

Method C: prefixation in 75 mM lysine, 0.075% alcian blue, 2.5% glutaraldehyde; fixation in 0.075% alcian blue, 2.5% glutaraldehyde.

## C. PARAFORMALDEHYDE AND LYSINE METHODS

These short time methods use paraformaldehyde and any of the three forms of lysine; free amino, monohydrochloride or acetate. All solutions are prepared in 0.1 M sodium cacodylate buffer.

Method 1: prefixation in 75 mM lysine, 2% paraformaldehyde, 2.5% glutaraldehyde; fixation in 2% paraformaldehyde, 2.5% glutaraldehyde.

Method 2: prefixation in 75 mM lysine, 0.075% ruthenium red, 2% paraformaldehyde, 2.5% glutaraldehyde; fixation in 0.075% ruthenium red, 2% paraformaldehyde, 2.5% glutaraldehyde.

Method 3A: prefixation for 20 min in 75 mM free amino lysine, 0.075% alcian blue, 2% paraformaldehyde, 2.5% glutaraldehyde; fixation in 0.075% alcian blue, 2% paraformaldehyde, 2.5% glutaraldehyde.

Method 3B: fixation for 2 h in 75 mM lysine (monohydrochloride or acetate), 0.075% alcian blue, 2% paraformaldehyde, 2.5% glutaraldehyde.

## D. EXTENDED TIME METHODS

For the extended time methods, paraformaldehyde and any of the three forms of lysine are included; however, some samples processed with the free amino form may solidify by 24 h. This does not occur with the monohydrochloride or acetate forms. All solutions are prepared in 0.1 M sodium cacodylate buffer.

Method 4: fixation in 75 mM lysine, 2% paraformaldehyde, 2.5% glutaraldehyde.

Method 5: fixation in 75 mM lysine, 0.075% ruthenium red, 2% paraformaldehyde, 2.5% glutaraldehyde.

Method 6: fixation in 75 mM lysine, 0.075% alcian blue, 2% paraformaldehyde, 2.5% glutaraldehyde.

## III. OBSERVATIONS AND DISCUSSION

### A. SEM Observation of Staphylococci on Biomaterials

The advantage of SEM is its three-dimensional image of the biomaterial and the biofilm that has grown upon its surface. Silicone rubber provides a flat relatively uniform surface (Fig. 4A). In contrast, the honeycomb of polyurethane foam (Fig. 4B) provides numerous opportunities for bacterial adhesion and growth in three dimensions. With a conventional fixation of glutaraldehyde (Control 1), the smooth round *S. epidermidis* RP62 cocci (arrow) can be seen building upward from the single dimension of silicone rubber (Fig. 4C). The cocci of *S. hominis* SP2 proliferate into a mass projecting well away from the foam surface taking advantage of its interstitial space (Fig. 4D). A glutaraldehyde-only fixative approach (Control 1) reveals only smooth appearing cells with only a few strains of secreted exopolysaccharide material (arrows) rarely seen. With ruthenium red added to the glutaraldehyde (Control 3), a few more strands of exopolysaccharide (arrows) are observed between smooth cells of *S. hominis* SP2 grown upon the flat silicone rubber (Fig. 4E). For the copious glycocalyx producer *S. epidermidis* RP62, some sparse fibrous material is seen on the surfaces of some cells (Fig. 4F, arrows) with this fixation (Control 3).

Clusters of cells of *S. epidermidis* RP62 (arrows) proliferate along the curving edge of polyurethane foam (Fig. 5A). The extensive fibrous material covering these cells (arrow-

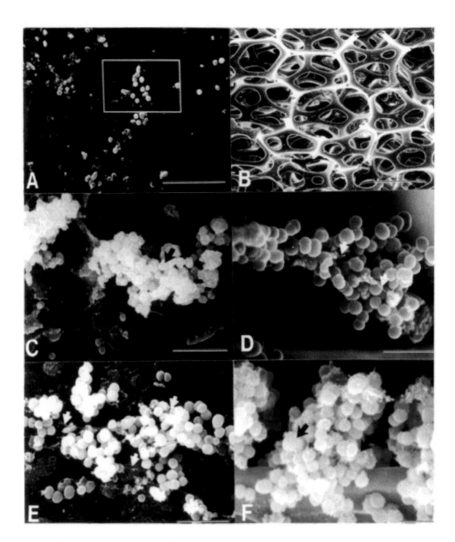

**Figure 4.** (A) Staphylococci grow outward from the flat one-dimensional silicone rubber surface. Bar = 10 μm. (B) Polyurethane foam provides opportunity for growth throughout a three-dimensional honeycomb-like network. Bar = 500 μm. (C) With glutaraldehyde fixation (Control 1), *S. epidermidis* RP62 cocci appear smooth as cells build outward from the silicone rubber surface. Bar = 5 μm. (D) Smooth *S. hominis* SP2 cocci proliferate as a mass into the interstitial space of the polyurethane foam. A few strands of fibrous material (arrows) rarely occurs (Control 1). Bar = 5 μm. (E) By glutaraldehyde-ruthenium red (Control 3) processing, more exopolysaccharide strands (arrows) between smooth cells are illustrated for *S. hominis* SP2 grown upon silicone rubber. Bar = 5 μm. (F) By glutaraldehyde-ruthenium red (Control 3) processing, the copious glycocalyx producer of *S. epidermidis* RP62 exhibits some sparse fibrous material on cell surfaces (arrows) as shown on polyurethane foam. Bar = 2.5 μm.

heads) as well as strands of material between cell masses and polyurethane foam substrate (arrow) are seen (Fig. 5B) when Method B (glutaraldehyde-ruthenium red-lysine-free amino) fixative is used. For *S. aureus* ATCC 25923 grown on polyurethane foam (Fig. 5C), considerable fibrous glycocalyx is also seen on cells by Method B. When grown on silicone rubber, a biomaterial preferred by *S. aureus* ATCC 25923,[4] a biofilm with

**Figure 5.** The successful use of glutaraldehyde-ruthenium red-lysine-free amino (Method B) is demonstrated for staphylococci grown on either silicone rubber or polyurethane foam. (A) Clusters (arrows) of *S. epidermidis* RP62 proliferate along the curving edge of polyurethane foam. Bar = 100 μm. (B) Extensive fibrous material (arrowheads) covers *S. epidermidis* RP62. Plentiful strands of material are seen between cell masses and the polyurethane foam substrate (arrow). Bar = 5 μm. (C) For *S. aureus* ATCC 25923 grown on polyurethane foam, considerable fibrous glycocalyx is visualized. Bar = 5 μm. (D) and (F) When *S. aureus* ATCC 25923 is grown on its preferred biomaterial silicone rubber, a biofilm with thickened strands is seen (long arrows). Bars = 50 and 5 μm, respectively. (E) and (F) Additionally, fibrous glycocalyx is elaborate upon these cells. Bars = 5 μm. (G) For *S. hominis* SP2 grown on silicone rubber, elaborate extracellular material is seen between cells (arrows) and upon cells (arrowhead). Bar = 10 μm.

thickened strands builds (Fig. 5D, 5F, long arrows). Fibrous material is also seen on cells (Fig. 5E, 5F, arrows). For *S. hominis* SP2 (Fig. 5G) by this fixation (Method B), extensive extracellular material is seen between cells (arrows) and upon cells (arrowhead) grown on silicone rubber.

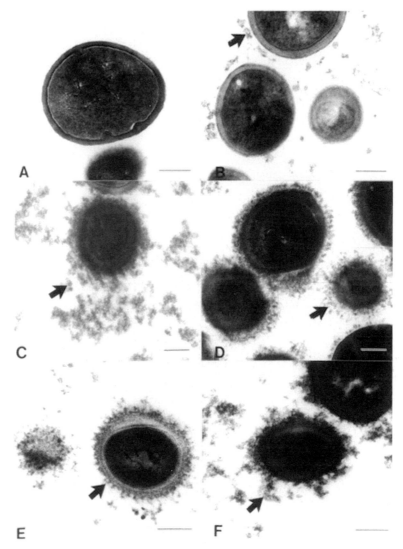

**Figure 6.** (A) For paraformaldehyde–glutaraldehyde fixation (Control 2), glycocalyx preservation is inadequate as shown for *S. hominis* SP2. (B) With ruthenium red added (Control 4), any material (arrow) is sparse as shown for *S. epidermidis* RP62 cocci. (C) and (D) For glutaraldehyde-lysine-free amino (Method A), fibrous glycocalyx material (arrows) surrounds cocci of *S. aureus* ATCC 25923 and *S. hominis* SP2, respectively. (E) and (F) For glutaraldehyde-ruthenium red-lysine-free amino (Method B), abundant glycocalyx (arrows) is visualized as shown for *S. hominis* SP2 and *S. epidermidis* RP62, respectively. Bars = 0.25 µm.

## B. TEM Observations of Staphylococci

The inadequate preservation of the staphylococci glycocalyx is seen in TEM in Figure 6A for *S. hominis* SP2 by paraformaldehyde-glutaraldehyde fixation (Control 2). With ruthenium red added for times corresponding to either short or extended fixation times (Control 4), only sparse material (arrow) is seen in thin section as shown for *S. epidermidis* RP62 cocci (Fig. 6B). With lysine-free amino added to glutaraldehyde (Method A),

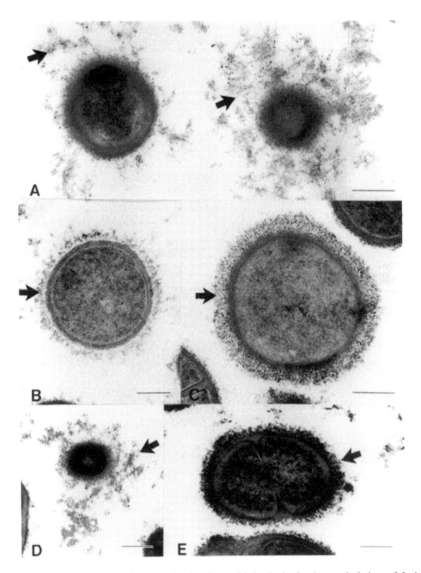

**Figure 7.** (A) and (B) For paraformaldehyde-glutaraldehyde-lysine/extended time (Method 4), fibrous glycocalyces (arrows) surrounds cells of *S. epidermidis* RP62 and *S. hominis* SP2, respectively. (C) For paraformaldehyde-glutaraldehyde-ruthenium red-lysine-free amino/short time (Method 1), an abundantly preserved uniform glycocalyx (arrow) surrounds a cocci of *S. hominis* SP2. (D) For paraformaldehyde-glutaraldehyde fixative with lysine-monohydrochloride (Method 4), a fibrous glycocalyx (arrow) is illustrated for *S. aureus* ATCC 25923. (E) For paraformaldehyde-glutaraldehyde-ruthenium red-lysine-monohydrochloride at short fixation time (Method 2), cells of *S. hominis* SP2 are surrounded by a fibrous glycocalyx (arrow). Bars = 0.25 μm.

a considerable increase in fibrous glycocalyx material (arrows) surrounds cocci of *S. aureus* ATCC 25923 (Fig. 6C) and *S. hominis* SP2 (Fig. 6D). When ruthenium red was included with lysine-free amino and glutaraldehyde (Method B), elaborate and abundant glycocalyx (arrows) is seen as shown for *S. hominis* SP2 (Fig. 6E) and *S. epidermidis* RP62 (Fig. 6F). Note the increase in abundant fibrous glycocalyx material (Fig. 6B) gained by glutaraldehyde-ruthenium red-lysine-free amino fixation (Method B) versus the

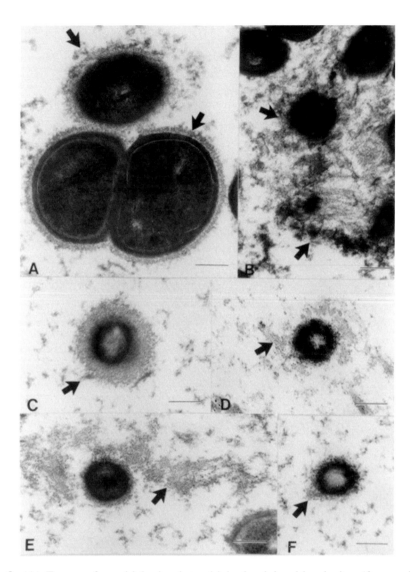

**Figure 8.** (A) For paraformaldehyde-glutaraldehyde-alcian blue-lysine (free amino) at extended fixation time (Method 6), an abundant glycocalyx (arrows) completely surrounds cells of *S. hominis* SP2. (B) By the same method (Method 6*), S. epidermidis* RP62 cocci have elaborate and extensive glycocalyx material (arrows). (C) and (D) For paraformaldehyde-glutaraldehyde-alcian blue-lysine-monohydrochloride at extended fixation time (Method 6), abundant glycocalyx (arrows) is seen for *S. hominis* SP2 and *S. epidermidis* RP62. (E) and (F) For paraformaldehyde-glutaraldehyde-alcian blue-lysine-acetate at extended fixation time (Method 6), fibrous material (arrows) is elaborate for cells of *S. hominis* SP2 and *S. aureus* ATCC 25923. Bars = 0.25 μm.

limited material (Fig. 6E, 6F) seen by the addition of ruthenium red only to the aldehydes (Control 4). The limit of a strict time of short prefixation hampers application of this improved methodological approach.

With the addition of paraformaldehyde, the time of fixation can be increased (Table 1) with the gain of improved glycocalyx preservation maintained. The views of fibrous matrices surrounding cells (arrows) is seen for *S. epidermidis* RP62 (Fig. 7A) and *S. hominis* SP2 (Fig. 7B) by Method 4 (paraformaldehyde-glutaraldehyde-lysine/extended time). A very well-preserved uniform glycocalyx (arrow) surrounds a cocci of *S. hominis* SP2 (Fig. 7C) after ruthenium red is added to paraformaldehyde-glutaraldehyde-lysine-free amino, shown at short fixation time (Method 1). The time of fixation can be extended to 24 h without sample loss by use of an alternative lysine, monohydrochloride or acetate (Table 1). The observation of the ultrastructural appearance of the glycocalyx is also maintained by use of these alternative lysines. A fibrous glycocalyx (arrows) is illustrated for *S. aureus* ATCC 25923 (Fig. 7D) by lysine-monohydrochloride in paraformaldehyde-glutaraldehyde fixative at extended time (Method 4) and for *S. hominis* SP2 (Fig. 7E) by ruthenium red added to lysine-monohydrochloride in paraformaldehyde-glutaraldehyde at short fixation time (Method 2). Alcian blue can successfully be employed in the same procedures.

By paraformaldehyde-glutaraldehyde-alcian blue-lysine (free amino) at extended fixation time (Method 6), a uniform glycocalyx (arrows) completely surrounds a dividing cell and an adjacent cell of *S. hominis* SP2 (Fig. 8A). By the same method (Method 6), glycocalyx material is elaborate for cocci of *S. epidermidis* RP62 (Fig. 8B, arrows). For the alternative lysine-monohydrochloride in paraformaldehyde-glutaraldehyde-alcian blue at extended fixation time (Method 6), abundant glycocalyx is seen for *S. hominis* SP2 (Fig. 8C, arrows) and *S. epidermidis* RP62 (Fig. 8D). For the other alternative, lysine-acetate in paraformaldehyde-glutaraldehyde-alcian blue at extended fixation time (Method 6), fibrous material (arrows) is elaborate for cells of *S. hominis* SP2 (Fig. 8E) and *S. aureus* ATCC 25923 (Fig. 8F).

Thus, a variety of fixation methods based upon the same principal components can be used to enhance visualization of the bacterial glycocalyx by electron microscopy.

## REFERENCES

1. Boyles JK: The use of primary amines to improve glutaraldehyde fixation. In: Johari O, ed. *Science of Biological Specimen Preparation for Microscopy and Microanalysis*. SEM, Inc., Chicago, IL, 1984:7–21
2. Costerton JW, Irvin RT, Cheng KJ: The bacterial glycocalyx in nature and disease. *Ann Rev Microbiol* 35:299–324, 1981
3. Edmiston CE, Schmitt DD, Seabrook GR: Coagulase-negative staphylococcal infections in vascular surgery: Epidemiology and pathogenesis. *Infect Control Hosp Epidemiol* 10:111–7, 1989
4. Fassel TA, Van Over JE, Hauser CC, et al: Adhesion of staphylococci to breast prosthesis biomaterials: an electron microscopic evaluation. *Cells and Materials* 1:199–208, 1991
5. Fassel TA, Schaller MJ, Remsen CC: Comparison of alcian blue and ruthenium red effects on preservation of outer envelope ultrastructure in methanotrophic bacteria. *Microscopy Research and Technique* 20:87–94, 1992
6. Fassel TA, Van Over JE, Hauser CC, et al: Evaluation of bacterial glycocalyx preservation and staining by ruthenium red, ruthenium red-lysine and alcian blue for several methanotroph and staphylococcal species. *Cells and Materials* 2:37–48, 1992
7. Fassel TA, Sanger JR, Edmiston CE: Lysine effect on ruthenium red and alcian blue preservation and staining of the staphylococci glycocalyx. *Cells and Materials* 3: 327–36, 1993
8. Fassel TA, Mozdziak PE, Sanger JR, et al: Paraformaldehyde effect on ruthenium red and lysine preservation and staining of the staphylococcal glycocalyx. *Microscopy Research and Technique* 36:422–7, 1997

9. Fassel TA, Mozdziak PE, Sanger JR, et al: Superior preservation of the staphylococcal glycocalyx with aldehyde-ruthenium red and select lysine salts using extended fixation times. *Microscopy Research and Technique* 41:291–7, 1998

10. Fassel TA, Mozdziak PE, Sanger JR, et al: Alcian blue effect on aldehyde-lysine preservation and staining of the staphylococcal glycocalyx. Submitted, 1999

11. Hanke DE, Northcote DH: Molecular visualization of pectin and DNA by ruthenium red. *Biopolymers* 14:1–17, 1975

12. Jacques M, Graham L: Improved preservation of bacterial capsule for electron microscopy. *J Electron Microsc Tech* 11:167–9, 1989

13. Luft JH: Ruthenium red and violet. I. Chemistry, purification, methods of use for electron microscopy and mechanism of action. *Anat Rec* 171:347–68, 1971

14. Luft JH: Ruthenium red and violet. II. Fine structural localization in animal tissue. *Anat Rec* 171:369–415, 1971

15. Progulske A, Holt SC: Transmission-scanning electron microscopic observations of selected *Eikenella corodens* strains. *J Bact* 143: 1003–18, 1980

16. Roth IL: Physical structure of surface carbohydrates. In: Sutherland I, ed. *Surface Carbohydrates of the Prokaryotic Cell.* Academic Press, New York, 1977:5–26

17. Scott JE: Histochemistry of alcian blue: II. The structure of alcian blue 8GX. *Histochemie* 30: 215–4, 1972

18. Sutherland IW: Bacterial exopolysaccharides. *Adv Microbial Physiol* 8:143–213, 1972

19. van Iterson W: Coverings of the outer cell wall surface. In: van Iterson W, ed. *Outer Structures of Bacteria.* Van Nostrand Reinhold Company, New York, 1984:155–200

# 15

# Confocal Laser Scanning Microscopy for Examination of Microbial Biofilms

Colin G. Adair,[1] Sean P. Gorman,[1]
Lisa B. Byers,[1] David S. Jones,[1] and Thomas A. Gardiner[2]

[1]*Medical Devices Group, The School of Pharmacy,*
*and* [2]*Department of Ophthalmology, The Queen's University of Belfast, UK*

## I. INTRODUCTION

The principle behind confocal laser scanning microscopy (CLSM) was first described in 1961 by Minsky,[27] but it has been the advent of faster scanning laser systems and increased computational capabilities that has led to the recent rapid development of CLSM as a major biological tool.[1,2,27]

With conventional light microscopy, the whole area of interest is imaged onto a screen or viewed directly by the eye. While suitable in many cases, this does not permit subsequent electronic processing or easy adapting to take advantage of resolution enhancement apparatus. Electron microscopy, on the other hand, requires extensive specimen preparation in which dehydration and slicing may lead to disruptive shrinkage and artefacts. The cryosectioning associated with transmission electron microscopy (TEM) further compounds sample distortion. The development of immunofluorescence using highly specific stains and antibodies has contributed to fluorescence microscopy becoming one of the most widely techniques in biological research. However, this method can produce fluorescence emissions throughout the entire depth of the specimen causing out-of-focus blurring.[6,26,34] While this may be partially corrected by restricting sample thickness to about 10 µm or using flattened cell cultures, these practices can distort the views obtained, while the cutting of sections complicates the interpretation of three dimensional structures.

Confocal laser scanning microscopy (CLSM) forms a bridge between light, fluorescence and electron microscopy. It has found wide application, including the study of infections associated with the formation of microbial biofilm on implanted medical devices, and the challenges posed by its eradication. CLSM affords penetrative, noninvasive views of specimens in which the biofilm is not dehydrated, background "flare" is virtually absent and 3-dimensional (3D) and quantitative imaging are possible.[11,18]

*Handbook of Bacterial Adhesion: Principles, Methods, and Applications*
Edited by: Y. H. An and R. J. Friedman © Humana Press Inc., Totowa, NJ

**Table 1. Stains Commonly Used in CLSM**

| Stain | Absorption max (nm) | Emission max (nm) | Site of stain |
|---|---|---|---|
| Fluorescein | 490 | 520 | General |
| Rhodamine | 570 | 590 | General |
| Texas red | 596 | 620 | General |
| Acridine orange | 490 | 590 | Nucleic acid |
| Ethidium bromide | 510 | 595 | Nucleic acid |
| Propidium iodide | 536 | 623 | Nucleic acid |
| Di-I | 546 | 565 | Lipophilic |

## II. CONFOCAL LASER SCANNING MICROSCOPY

In CLSM the basic optical arrangement is similar to that used in scanning optical and scanning electron microscopy (SEM), except for the source of illumination, which employs a scanning laser beam, usually a low-power air-cooled argon laser. This can emit a variety of wavelengths of which the two strongest lines are at 488 nm (the excitation maximum of fluorescein) and at 514 nm (capable of causing emissions from rhodamine and Texas red). Argon–krypton laser confocal microscopes operate at longer excitation wavelengths (488 nm, 568 nm, 647 nm) and are suited to fluorochromes excited by green light.[6,31] However, this is less stable, has a shorter lifetime and is more susceptible to chromatic aberration than the argon laser. Nevertheless, both laser systems provide high intensity illumination giving good sensitivity and fluorescence resolution.[6]

CLSM is particularly useful in biomedical studies in that it overcomes the inherent limitation of conventional light microscopy—the out-of-focus blurring or fluorescence flare. This is brought about because any part of the image which is outside the very narrow depth of focus of the CLSM appears black and therefore is not visible.[15] By removing this background out-of-focus haze, the confocal system allows imaging of these structures. It has the added advantage that preparations normally considered overstained or deemed to have unacceptable levels of background staining can be successfully viewed. However this narrow depth of field has its disadvantages in that the filtering of the confocal system can reduce the signal from weak fluorescent probes to an unacceptably low level, while small vertical displacements, such as that caused by play in the microscope stage, can have a large effect on the image. Problems may also occur in locating specific areas within a fluorescent specimen as objects outside the depth of field appear black and are therefore not visible.[15]

Bleaching of the specimen by the very intense laser beam can be problematic. This limits the time a specimen can be exposed to the laser and may partially be solved in living specimens by use of dyes that tend to be less photosensitive, such as Texas red or rhodamine.[31] Applications of some stains commonly used in CLSM are shown in Table 1. In fixed cells, combinations of antibleaching agents may be effective.[10,16] Photobleaching, however, may not always be detrimental to specimen investigation. It has been suggested that the measurement of the recovery time after photobleaching may be utilized to indicate properties of that tissue such as permeability or the determination of diffusion coefficients.[19]

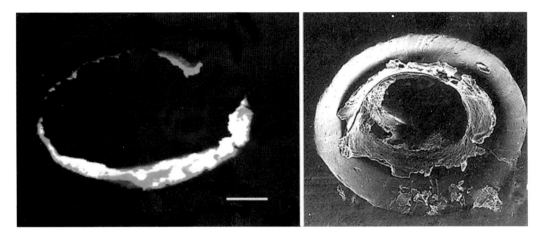

**Figure 1.** Low-power, color-enhanced confocal laser scanning micrograph of microbial biofilm attached to a pore of a long-dwell CAPD catheter. The lighter areas represent regions of higher bacterial concentration (bar = 100 µm). For comparison, a catheter pore with biofilm, viewed by SEM, is shown on the right hand frame.

By using point scanning in the CLSM, rather than full field illumination, it is possible to scan in the $x,z$ plane instead of the usual $x,y$ plane. Images of optical sections parallel to the optical axis of the microscope are thus generated. Current systems permit optical sectioning down to 0.1 µm and these sections may then be digitally enhanced to provide a three-dimensional non-invasive image of subsurface organelles. Hence, cell interiors within living tissue can be investigated without the artifacts introduced by preparing specimens for scanning electron microscopy, and cellular functional architecture (the structural relationship between various cell organelles) may be elucidated. Much of this architecture occurs in the range 0.1–1.0 µm which cannot adequately be resolved by conventional light microscopes. Colloidal gold immunocytochemical labeling has revolutionized electron microscopic localization of cellular organelles and has found successful application in confocal imagery.[29,36] As the gold particles are not subject to photobleaching, four-dimensional investigations are possible in which a series of three-dimensional confocal images of living cells may be viewed over specified time intervals.

### III. CLSM TO DEFINE THE STRUCTURE OF BIOFILM

Unlike electron microscopy, CLSM is well suited to investigation of the structure of microbial biofilm because of its ability to scan below the biofilm surface ($x,z$ plane) enabling a 3-D image of the biofilm to be constructed.[28] Patients on continuous ambulatory peritoneal dialysis (CAPD) frequently suffer from recurrent episodes of peritonitis and may require their catheters to be surgically removed.[22] Examination by scanning electron microscopy (SEM) of such catheters often reveals the presence of microbial biofilm, considered to be the cause of recurring peritonitis.[7,22] In the majority of catheter sections examined, however, only erythrocytes and inflammatory cells are observed on the catheter surface, although bacterial presence may be confirmed by microbiological procedures and TEM.[21] CLSM of the catheter surface was employed to enable

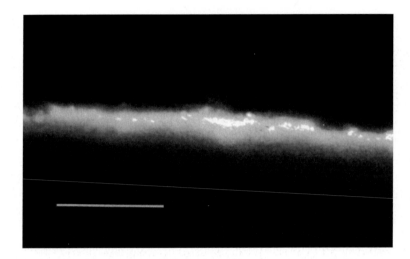

**Figure 2.** Color-enhanced confocal laser scanning micrograph of *S. epidermidis* biofilm attached to a CAPD catheter. Clusters of spherical entities 1–2 μm in diameter are in the size range expected for bacterial cocci (bar = 25 μm).

direct observation of fluorescent dye stained microbial biofilm underlying the occluding erythrocyte and inflammatory cell layer.[23]

Peritoneal catheters were obtained from CAPD patients with recurrent peritonitis and 1 cm sections cut from the catheter cuff. These were split longitudinally, with one portion placed in 5% (w/v) glutaraldehyde in cacodylate buffer (0.1 M, pH 7.4) for examination by CLSM and the other portion retained for microbiological identification of adherent bacteria. Catheter portions were attached to glass microscope slides and stained with acridine orange (0.001%) for 5 min. Samples were examined at a wavelength of 488 nm by CLSM (Biorad Lasersharp MRC 500) with long working distance lenses.[15]

Figure 1 shows a low power enhanced confocal fluorescent micrograph of a CAPD catheter pore with entrained biofilm in which areas of high microbial density are seen as white regions. A penetrative view of the biofilm specimen below the superficial inflammatory cells and erythrocytes and demonstrate discrete entities, 1–2 μm in diameter, clearly visible within the centre of the glycocalyx matrix (Fig. 2).

The distribution of bacteria within the hydrated matrix of the biofilm viewed by CLSM is more dilute than is observed in TEM observations, reflecting the effects of the dehydration process in the preparation of specimens for TEM. This observation is supported by other CLSM studies indicating that living biofilms are highly hydrated, having between 50–90% void space.[3] Recently, the algal, bacterial and exopolymeric components of natural biofilms were quantified using a simple triple fluorescent approach. Measurement of each of these components in biofilm was undertaken by staining for bacterial mass (a fluorescent nucleic acid stain), exopolymeric substances (fluor-conjugated lectins) and algal biomass (autofluorescence). Results from their investigation showed that biofilm was composed of 85% exopolymeric substance, 4.5% bacteria and 10.6% algae.[20]

By using optical sectioning it has been demonstrated that *Staphylococcus epidermidis* biofilms on CAPD catheters from patients with peritonitis were in the order of 30 μm in

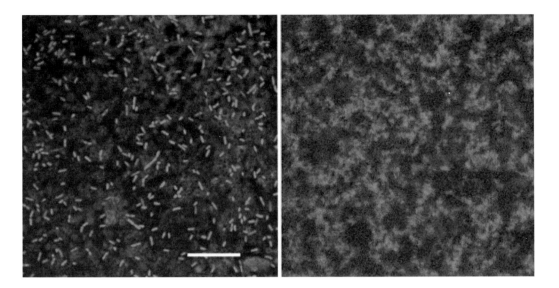

**Figure 3.** CLSM of *P. aeruginosa* biofilm formed on PVC before (a) and after exposure to 10 mg/L ciprofloxacin (b). Specimens were fixed in buffered formalin and stained with propidium iodide (for 60 min), before mounting in a fluorescent medium with and antifade agent (Vectashield). Biofilm thickness was 18 μm and scan depth was 10 μm (bar = 10 μm).

thickness.[21] These data are similar to those from *Pseudomonas aeruginosa* biofilms (medium 33 μm and range 13–60 μm).[34] The technique of optical sectioning can also be used to determine the distribution of microbial cells within a biofilm and how this is affected by the microorganism forming the biofilm. For example, *P. aeruginosa* biofilms are characterized by a dense cell mass near the biofilm base (Fig. 3), whereas those of *Vibrio parahaemolyticus* adopt an inverted structure with the majority of the biomass near the surface.[24] The coverage of a given area with biofilm may be quantified by CLSM in which the confocal image is converted into a series of black and white pixels using novel image analysis software. The percentage of biofilm (black pixels) can then be calculated as a percentage of the white pixels (the viewed area without biofilm).[32]

## IV. PENETRATION INTO BIOFILMS

The prevention of biofilm attachment to implanted medical devices and its eradication has posed a challenging problem. Attempts at preventing the attachment of pathogens to implanted medical devices, by incorporating antibiotics into biomaterials, have met with little success. Once established, microorganisms within the glycocalyx are relatively insensitive to the action of antimicrobial agents at clinical doses.[12]

In providing insight into possible solutions to this challenge it is necessary to determine how materials enter and move within the biofilm. Traditional approaches to the measurement of diffusion in biofilm have included the use of dialysis membranes and agar gels as models from which the results could be extrapolated to microbial systems.[17] The technique of fluorescence recovery after photobleaching (FRAP) has been used to determine mobility of molecules through biological media.[18,19] The principle involves the irreversible bleaching of the fluorescent–labeled molecules in biofilm by brief exposure to intense light and monitoring the replacement of the bleached molecules with new fluorescent molecules. This technique has been used to measure the diffusion coefficients of fluorescein-conjugated dextrans (size range 4–2000 K) through biofilms.[19] The method could be easily adapted to determine the interaction between antimicrobial agents and exopolymeric matrix.

Microinjection of fluorescent dyes highlighted how small, nonbinding molecules were able to move freely within the voids, while the movement of larger molecules was impeded by the matrix, suggesting the exopolymeric substance pore diameter was in the order of 80 nm. The growth patterns and transport processes of biofilm have been studied by tracing the movement of inert fluorescent microparticles (1 μm diameter) through a developing biofilm using CLSM. When added to a bacterial culture, these microparticles rapidly penetrated into a thick biofilm via water channels and pores. As the biofilm developed, the void space was filled with growing biomass and the beads were gradually displaced to the surface. Beads were shown to persist in biofilm for up to 20 d.[25] In addition to visualizing pores and quantifying biofilm void space and biomass, CLSM may be used to describe the movement of fluids within biofilm. Using CLSM particle tracking, voids have been demonstrated to be connected to the bulk liquid and the flow velocity inside the biofilm was proportional to the bulk flow velocity.[4] Flow velocities may also been determined using microelectrodes with CLSM.[9]

The effects of ultrasonic eradiation and gentamicin on biofilm disruption were viewed by coupling a flow cell to the CLSM and staining biofilm with ethidium bromide (nucleic acid stain). Visualizing the 3-dimensional structure of a 24 h biofilm showed that it appeared to grow under sonication, rather than undergoing disruption. The authors concluded that sonication, rather than disrupting the biofilm, may facilitate antibiotic transport into the biofilm causing cell death.[37]

## V. SURFACE TOPOGRAPHY

Surface topography, including the degree of surface roughness (microrugosity), plays an important role in initial bacterial adherence.[35] Implants with porous surfaces are much more susceptible to infection after implantation as microorganisms tend to sequester themselves in these cavities and avoid host defences. Over time, implanted medical devices, which were originally smooth, can develop roughened or cracked surfaces which enable bacterial adhesion which encouraged biofilm formation.[5,8,24,30]

CLSM offers resolution of fine surface detail and such data can be electronically manipulated to create three-dimensional and topographical images. The software allows simulated fluorescence processing (SFP, shadow-imaging), topographical imaging with 3-D plot (T3D) and topographical imaging with measurement line (TML).[15]

Figure 4 shows topographical imaging by CLSM of an unused CAPD catheter and one which had been *in situ* for 18 mo. The long-dwell catheter displays a greater degree of surface roughness than unused catheter. For this study, the ratio values obtained by computer 3-D measurement were typically in the range 4.5 to 5.0 for long-dwell catheters and 3.5 to 4.0 for control catheters. The ratio is a direct measure of the roughness of the product surface. Adherence of bacteria to long dwell catheters was significantly higher than control catheters because of their rougher surface. While commercially available catheters may have similar surface microrugosities, CLSM may prove useful in

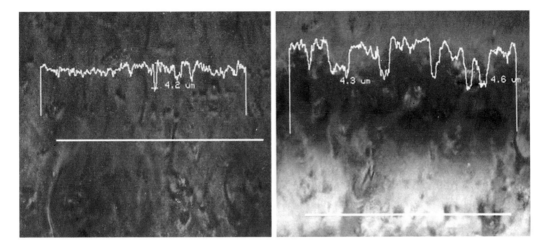

**Figure 4.** CLSM topographical imaging showing an unused catheter (a) in comparison to the roughened surface of a long-dwell catheter (b). Biofilm was removed from catheters by vortex mixing and gentle sonication. One cm sections from the tip region were fixed in glutaraldehyde and stained with acridine orange (0.05% w/v in 0.1 M sodium citrate buffer) for 3 min, before examination by CLSM.

determining their durability and changing surface characteristics as a result of prolonged dwell-time in the patient.[13]

## VI. CONCLUSION

CLSM is a useful, noninvasive method for providing penetrative views of microbial biofilms, together with quantitative data on biofilm structure and functioning. By enabling observation below the surface, the extensive sample preparation and sectioning associated with electron microscopy, is minimized, allowing examination of biofilms in their hydrated form. CLSM can also facilitate the tracing of diffusion markers through biofilm and, in eradicating biofilm from, for example, medical devices, the technique of photobleaching will be useful in monitoring biocide penetration into biofilm. Further developments in computational capabilities and the advent of faster scanning laser systems will undoubtedly enhance CLSM's role as a major biological tool.

## REFERENCES

1. Agard DA, Sedat JW: Three dimensional architecture of a polyene nucleus. *Nature* 302:676–81, 1983
2. Baak JP, Thunnissen FB, Oudejans CB, et al: Potential clinical uses of laser scanning microscopy. *Appl Opt* 26:3413–6, 1987
3. Costerton JW, Lewandowski Z, Caldwell DE, et al: Microbial biofilms. *Ann Rev Microbiol* 49:711–45, 1995
4. de Beer D, Stoodley P: Relationship between the structure of an aerobic biofilm and transport phenomena. *Water Sci Technol* 32:11–8, 1995
5. Dinnen P: The effect of suture material in the development of vascular infection. *Surgery* 91:61–3, 1977

6. Dixon AJ, Benham GS: Applications of confocal scanning fluorescence microscope in biomedical research. *Int Lab* April 1998

7. Evans RC, Holmes CJ: Effect of vancomycin hydrochloride on *Staphylococcus epidermidis* biofilm associated with silicone elastomer. *Antimicrob Agents Chemother* 31:889–94, 1987

8. Fessia SL, Amirana O, Carr KL: Biofilm formation on commercially available plastic tubings. *Adv Peritoneal Dialysis* 4:253–6, 1988

9. Fuhu X, Beyenal H, Lewandowski Z: An electrochemical technique to measure local flow velocity in biofilms. *Water Sci Technol* 32:3631–6, 1995

10. Giloh H, Sedat JW: Fluorescence microscopy: Reduced photobleaching of rhodamine and fluorescence protein conjugates by n-propygallate. *Science* 217:1252–5, 1982

11. Goldner M, Coquis-Rondon M, Carlier JP: Demonstration by confocal laser scanning microscopy of invasive potential with *Bacteroides fragilis. Microbiologica* 14:71–5, 1991

12. Gorman SP: Microbial adherence and biofilm production. In: Denyer SP, Hugo WB, eds. *Mechanisms of Action of Chemical Biocides.* Blackwell Scientific Publications Technical Series No. 27, Oxford, UK, 1991:271–95.

13. Gorman SP, Mawhinney WM, Adair CG: The influence of catheter surface microrugosity on the development and persistence of recurrent peritonitis in CAPD. *Pharmacotherapy* 11:267, 1991

14. Gorman SP, Adair CG, Mawhinney WM: Microscopy of biofilms of adherent micro-organisms. In: Gorman SP, Denyer SP, eds. *Microbial Adherence, Biofilms and Their Control.* Blackwell Scientific, Oxford, 1993

15. Gorman SP, Mawhinney WM, Adair CG, et al: Confocal laser scanning microscopy of peritoneal catheter surfaces. *J Med Microbiol* 38:411–7, 1993

16. Johnson GD, Davidson KC, McNamee K, et al: Fading of immunofluorescence during microscopy: a study of the phenomenon and its remedy. *J Immunol Methods* 55:231–42, 1982

17. Korber DR, Lawrence JR, Hendry MJ, et al: Analysis of Spatial variability within MOT+ and MOT- *Pseudomonas fluorescens* biofilms using representative elements. *Biofouling* 7:339–58, 1993

18. Lawrence JR, Korber DR, Hoyle BD, et al: Optical sectioning of microbial biofilms. *J Bacteriol* 173:6558–67, 1991

19. Lawrence JR, Wolfraadt GM, Korber DR: Determination of diffusion coefficients in biofilms by confocal laser microscopy. *Appl Environ Microbiol* 60:1166–73, 1994

20. Lawrence JR, Neu TR, Swerhone GD: Application of a multiple parameter imaging for the quantification of algal, bacterial and exopolymeric components of microbial biofilms. *J Microbiol Methods* 32:253–61, 1998

21. Mawhinney WM, Adair CG, Gorman SP: Confocal laser scanning microscopic investigation of microbial biofilm associated with CAPD catheters. *Pharmacotherapy* 10:241, 1990

22. Mawhinney WM, Adair CG, Gorman SP: Development and treatment of peritonitis in continuous ambulatory peritoneal dialysis. *Int J Pharm Pract* 1:10–8, 1991

23. Mawhinney WM, Adair CG, Gorman SP: Examination of microbial biofilm on peritoneal catheters by electron and confocal laser scanning microscopy. *Proceedings of the 10th Pharmaceutical Technology Conference*, Bologna, Italy, 2:652–60, 1991

24. Merritt K, Schafer JW, Brown SA: Implant site infection rates with porous and dense materials. *J Biomed Mater Res* 13:101–8, 1979

25. Okabe S, Yasuda T, Watanabe Y: Uptake and release of inert fluorescence particles by mixed population biofilms. *Biotechnol Bioeng* 53:459–69, 1997

26. Paradiso AM, Tsien RY, Machen TE: Digital image processing of intracellular pH in gastric oxyntic and chief cells. *Nature* 325:447–50, 1987

27. Petran M, Hadravsky M, Benes J, et al: The tandem scanning reflected light microscope: Part 1: the principle, and its design. *Proc R Microsc Soc* 20:125–9, 1985

28. Richards SR, Turner RJ: A comparative study of techniques for the examination of biofilms by scanning electron microscopy. *Water Res* 18:767–73, 1984

29. Shotton D, White N: Confocal scanning microscopy: three dimensional biological imaging. *Trends Biochem Sci* 14:435–9, 1989

30. Schmitt D, Bandyk DF, Pequet AJ, et al: Mucin production by *Staphyloccocus epidermidis*. A virulence factor promoting adherence to vascular grafts. *Arch Surg* 121:89–95, 1986

31. Shuman H, Murray JM, DiLullo C: Confocal microscopy: An overview. *Biotechniques* 7:154–159, 1989

32. Silyn R, Lewis G: A technique in confocal laser microscopy for establishing biofilm coverage and thickness. *Water Sci Technol* 36:117–24, 1997

33. Stewart PS, Peyton BN, Drury WJ, Murga R: Quantitative observation of heterogeneities in *Pseudomonas aeruginosa* biofilms. *Appl Environ Microbiol* 59:327–9, 1993

34. Tsien RY, Rink TJ, Poenie M: Measurement of cytosolic free Ca2+ in individual small cells using fluorescence microscopy with dual excitation wavelengths. *Cell Calcium* 6:145–57, 1985

35. Wilkins KM, Martin GP, Hanlon GW, et al: The influence of critical surface tension and microrugosity on the adhesion of bacteria to polymer monofilaments. *Int J Pharmaceutics* 57:1–7, 1989

36. White JG, Amos WB, Fordham M: An evaluation of confocal versus conventional imaging of biological structures by fluoresence light microscopy. *J Cell Biol* 105:41–8, 1987

37. Zhen Q, Stoodley P, Pitt WG: Effect of low intensity ultrasound upon biofilm structure from confocal scanning laser microscopy observation. *Biomaterials* 17:1975–80, 1996

# 16

# Quantitation of Bacterial Adhesion to Biomaterials Using Radiolabeling Techniques

## Syed F. Mohammad and Reza Ardehali

*Department of Pathology, University of Utah, and Utah Artificial Heart Institute*
*Salt Lake City, UT, USA*

## I. INTRODUCTION

Infection is recognized as a major concern with temporary as well as permanent medical implants.[6,13,24] Clinical experience suggests that bacterial adhesion followed by colonization and biofilm formation may be critical in the sequence of events that lead to device infection. Biofilm may form a barrier and prevent effective penetration of antibiotics, and bacteria sequestered in the biofilm may evade phagocytic cells thus dodging the hostile immune system. Device-associated infections are also difficult to resolve with antimicrobial therapy, and, in many instances, force the removal or replacement of the infected device.

## II. QUANTIFICATION OF BACTERIAL ADHESION AND COLONIZATION

Bacterial colonization of a prosthetic device is considered to be a multiphasic process. Colonization begins with either a direct or random encounter of the microbe with the biomaterial surface. Depending upon the nature and composition of the surface, a pathogen may find the local conditions conducive to rapid adhesion. Interfacial energy, surface texture, hydrophobicity/hydrophilicity, and the composition of the fluid that coats the surface at the time encounter occurs, all affect bacterial interaction with devices. Once the adhesion has been established, the adherent cells reorganize to consolidate a relatively firm attachment with the surface.[33] It is well recognized that interaction of microorganisms with animate tissues involves specific interaction of microbial structures (i.e., adhesins) with reciprocal substratum structures (i.e., receptors), usually through a lectin–saccharide interaction, or through surface integrin receptors. Such a targeted attachment may take place for inanimate structures (i.e., device surface) as well if the substratum has adsorbed specific ligands that make bacterial adhesion possible. Successful colonization of a surface may set the stage for production of an adhesive extracellular polymer coating known as biofilm or slime.[33] Sustained growth of microorganisms on the device surface may lead to local or systemic infection.

*Handbook of Bacterial Adhesion: Principles, Methods, and Applications*
Edited by: Y. H. An and R. J. Friedman © Humana Press Inc., Totowa, NJ

Since the nature of the material and the composition of the biological milieu may dictate whether or not bacteria would adhere at an interface, efforts have been made to modify the interfacial properties to prevent bacterial adhesion. These modifications include imparting hydrophilic properties[17] or incorporation of antibacterial agents.[2,10,21] Such efforts have triggered the need for quantitative assessment of adhesion and colonization of bacteria on artificial surfaces. While a number of methods have been developed for quantitation of bacterial adhesion to surfaces, no one technique has proven to be satisfactory for studying bacterial adhesion under all conditions.

The methods most often used to enumerate adherent bacteria involve determination of absolute numbers of bacteria by direct counting when use of a microscope is feasible. Morphological observations of adherent bacteria have become possible by using scanning and/or transmission electron microscopy. Furthermore, when specific antibodies are available, it is possible to estimate the relative numbers of bacteria using enzyme-linked immunosorbent assay (ELISA), or quantitation of a biochemical metabolite, such as ATP, by bioluminescence. Bacteria can also be labeled metabolically using radioactive tracers and absolute numbers of bound bacteria determined by liquid scintillation or gamma counting.[1,5] Another approach involves labeling bacterial surfaces with biotin, followed by use of avidin-peroxidase to obtain an estimate of relative number of adherent bacteria.[12] Other researchers have implemented existing techniques in studying bacterial adhesion under the influence of fluid flow, including mass transport and wall shear rate.[11,32] These measurements utilize parallel plate chamber, rotating disc, and stagnation point flow techniques. However, each of these procedures has certain drawbacks. Microscopic assessment is tedious and possible only when cells are adherent as a monolayer. Specific antibodies are not readily available for various immunochemical assays and the use of radioactive labels invokes safety issues. Labeling of bacterial surfaces with biotin or other compounds could have deleterious effects that are difficult to account for and systems involving dynamic fluid flow limit consideration of other relevant parameters that may have an impact on bacterial adhesion.

Among all the techniques employed to evaluate bacterial adhesion, the use of radioactive tracers, particularly the use of gamma emitters, appears most promising for bacterial adhesion studies in vitro. Similarly, bacteria with genetic markers appear most promising for evaluation of bacterial interaction with devices in vivo.

## III. METHODS FOR EVALUATION
## OF BACTERIAL ADHESION TO SOLID SURFACES

### A. Microscopic Evaluation of Adherent Bacteria

A number of methods have been utilized to determine bacterial adhesion to biomaterials. It is outside the scope of this chapter to describe each of these in detail; however, a brief mention of a few techniques is warranted.

Light microscopy is the most common approach used to evaluate bacterial adhesion.[3,25] Since this method does not require special staining or processing of samples, viable bacteria can be viewed readily at different time points during an experiment. Light microscopes, equipped with a CCD camera to generate computerized images, allow the observation of adhesion and colonization of bacteria at a solid–liquid interface in real time. However, for quantification purposes, adherent cells can be fixed with a crosslinking agent (glutaraldehyde), and several random fields of small areas (i.e., 0.0252 mm²) selected for manual counting of adherent bacteria. The obvious limitation of this approach is the requirement for optically clear, planar material as the substratum. Manual counting is also tedious and time-consuming, and individual cells in dense consortia are difficult to identify. Furthermore, light microscopic evaluation of adherent bacteria is problematic when dealing with a large number of samples.

Scanning electron microscopy (SEM) makes it possible to observe, in fine detail, the attachment of microorganisms to surfaces. Bacteria can be enumerated on transparent as well as opaque surfaces with ease.[8,23] However, SEM also suffers from many of the same limitations indicated for light microscopy. As with optical microscopy, the method can be converted from qualitative to quantitative observations by simply counting the number of organisms over a given surface area as long as they are adherent as a monolayer. Bacteria embedded in the biofilm or adherent in multiple layers cannot be resolved, and the small field of view precludes assessment of a large surface area.

Transmission electron microscopy (TEM) has been used in specific instances to assess bacterial adhesion. While TEM is a powerful tool for identification and characterization of adherent components, technical limitations preclude its use for quantitation of bacteria adherent to surfaces.

Confocal microscopy provides a means by which discrete thin sections can be viewed individually, in a desired sequence, or subjected to image analysis to provide high resolution, three dimensional images of specimens. With the use of computer generated optical images, qualitative analysis of bacterial interaction with the substrate can be studied. However, confocal microscopy, like other microscopic methods, can be useful for qualitative but not quantitative assessment of bacterial adhesion.

## B. Biochemical and Immunochemical Methods

Many investigators have monitored the production of a microbial product as an indirect assay for the number of microorganisms on a surface.[26,29,30] Biological assays have the advantage of being sensitive, and allow investigators to follow microbial colonization on a wide variety of materials under many conditions. The disadvantage of these approaches is that the assays provide an indirect assessment and require a standardization step to correlate the amount of product to the number of microbes generating the product. Since this standardization is usually done with planktonic organisms, it is not necessarily correct to assume that it can be applied to adherent microorganisms as well. One of the most common methods used involves monitoring ATP production by luciferin and luciferase reaction.

Immunochemical staining methods have also been used for evaluation of bacterial adhesion. An ELISA-based system has been developed as a convenient and sensitive means to enumerate adherent microorganisms. These methods rely on the detection of adherent bacteria by treatment with an antibody against bacteria (primary antibody) and then a secondary antibody derivatized with horseradish peroxidase or alkaline phosphatase. ELISA has been used extensively in the investigation of bacterial adhesion to animal cells immobilized on microtiter plates.[27] While ELISA offers a highly sensitive approach for detection, the major drawback is that it requires specially treated plates which limits its use for the assessment of bacterial adhesion on various biomaterials.

### C. Introduction of Reporter Genes

Bacterial interaction with implanted devices can be evaluated by employing micro-organisms with stable marker systems with an easily detectable phenotype.[7,14,20,31] The most popular reporter enzyme in bacteria is β-galactosidase. Transduction of bacteria with LacZ gene results in the expression of the β-galactosidase enzyme, which can be easily detected by biochemical methods. The adherence of such genetically engineered microorganisms to a substrate can be evaluated by the degree of β-gal expression. A standard curve correlating the actual number of bacteria to the strength of the blue staining of β-gal could be used to quantitatively assess bacterial adhesion. Such methods can be used for in vivo systems to monitor the fate, survival, and colonization of genetically altered bacteria. However, the use of β-galactosidase as a reporter enzyme has been limited because it is present in a number of bacteria, such as *Escherichia coli*, and deletions of the LacZ gene must be constructed before its use. Recently, Kalabat et al. have reported the development of N,N'-diacetylchitobiase (chitobiase) (β-N-acetyl-β-D-glucosaminidase).[18] This enzyme hydrolyzes the disaccharide chitobiase to N-acetyl glucosamine. The advantages of the reporter gene encoding chitobiase are that chitobiase and N-acetyl-β-D-glucosaminidase activities are missing in *E. coli* strains (one of the widely used species for bacterial adhesion studies) and that bacterial chitobiase activity can be measured quantitatively and monitored using blue/white colony indicator plates. Furthermore, convenient substrates for this enzyme (such as p-nitrophenyl-N-acetyl-β-glucosaminide [PNAG] and 5-bromo-4-chloro-3-indolyl-N-acetyl-β-D-glucosaminide [X-Gluc]) are commercially available. The use of chitobiase as a reporter enzyme is generally applicable to the study of bacterial adhesion and its assessment in those bacteria that do not contain N-acetyl-β-D-glucosaminidase.

Along with the biochemical analyses to monitor the expression of exogenous genes, light emission can be used to evaluate the expression of bioluminescent proteins. The green fluorescent protein (GFP), a 27-kDa protein from the marine bioluminescent jellyfish *Aequorea victoria*, is a unique marker that can be identified by noninvasive methods. Neither substrates, complex media, nor expensive equipment are required for detection of GFP.[14,20] Green fluorescent protein absorbs light with an excitation maximum of 395 nm and fluoresces with an emission maximum at 510 nm, so it can be simply detected by shining a hand-held UV lamp on GFP-containing colonies and observing green light emission. GFP has been introduced into a number of bacterial species, and Tresse et al have reported the detection of individual green fluorescent colonies of transformants (*Moraxella* sp.) for up to 2 weeks after inoculation.[22] The relatively long expression of the exogenous reporter gene makes this system an attractive method for long-term experiments in both in vitro and in vivo models. This is a stable and useful marker that allows for easy, rapid, and inexpensive detection of adherent bacteria.

### D. Detachment of Adherent Bacteria

A simpler approach often employed for quantitation of adherent microorganisms is to detach the organisms and count the number of colony-forming units (cfu) recovered from the stripping procedure.[9] The procedure itself is straightforward and generally does not involve special tools or materials. Nevertheless, the approach is inherently paradoxical: it relies on studying attached organisms by first detaching them from the surface. Moreover, the detachment procedure, if too aggressive, may injure the cells, and if not too aggressive, may fail to remove all the adherent cells. Further, if the detached cells are present in clusters rather than individual cells, results based on cfu may not accurately reflect the true numbers of adherent cells.

Despite the limitations, the following approaches have been used to remove adherent microorganisms to detect and quantify infection in medical devices:

The roll technique was developed to assess the degree of bacterial colonization of intravascular catheters. It has also been used to determine the capacity of catheters of different materials for bacterial adhesion and colonization. The technique is simple, expedient, and requires little in supplies or equipment. Although simple, the approach has its limitations as it does not measure intraluminal colonization. This technique has also been used for measuring bacterial adhesion to solid cylinders of different material. The major problem with the roll technique is that it depends on detachment of adherent organisms when a cylindrical object is rolled over the agar surface; tightly held organisms may not be displaced readily from all materials.

Investigators have used a variety of means, such as vortexing and scraping, to remove adherent microorganisms from the surface. Another commonly employed technique relies on ultrasonication to recover adherent microorganisms and quantitation of dislodged cells by serial dilution.[4] Ultrasonication is particularly suited when dealing with devices with complex shapes, inaccessible lumens, seams or crevices. However, careful optimization of the technique is necessary since the heat generated during ultrasonication may adversely affect adherent as well as recovered cells.

## IV. USE OF RADIOACTIVE TRACERS FOR LABELING OF BACTERIA

Perhaps the most sensitive and versatile approach to evaluating microbial adhesion to surfaces involves the use of radiolabeled organisms.[24] The utility of this approach is illustrated by the wide variety of radionuclides, microorganisms, and substrata used in radiolabeling experiments.

### A. Use of β-emitters

Most investigators label the organisms by propagation in liquid medium that includes a radiolabeled essential nutrient. The use of $^3H$, $^{14}C$, $^{32}P$, $^{35}S$, $^{51}Cr$, and $^{75}Se$ as radioactive tracers has been employed for labeling microorganisms.[2,29,30] $^3H$-thymidine labeled Gram-positive and Gram-negative bacteria have been utilized for adhesion to intravascular catheters, suture materials, and glass and polystyrene beads. Likewise $^{14}C$-glucose and $^{35}S$-methionine have been used to study the adhesion of a variety of bacterial strains to intravascular catheters, vascular grafts, needles, and suture materials. Radiolabeled bacteria have also been harvested from agar containing $^3H$-glucose for studies of adhesion to a variety of biomaterials,[15,22] and passively labeled with $^{51}Cr$ for studies of adhesion to neurosurgical prosthesis. However, the use of a metabolic label entails prolonged incubation of cells with nuclide, and β-emitters require further processing of samples for quantitation of adherent bacteria.

### B. Use of γ-emitters

To facilitate the investigation of bacterial adhesion to biopolymers, bacteria labeled with γ-emitters are more desirable because surfaces with labeled cells can be assessed quickly without the need for solubilizing the cells or relying on the use of scintillation fluids. Moreover, increased sensitivity with the use of $\gamma$-emitters is particularly useful when dealing with a low number of adherent cells. Microorganisms labeled with [111]Indium are particularly suited for assessment of bacterial adhesion on a wide variety of materials or devices.[24] [111]Indium-oxine has been used for two decades to label platelets for localization of deep vein thrombosis and leukocytes for localization of infection by gamma imaging.[28,34] The technique used for labeling platelets and leukocytes was adapted by the authors for labeling of microorganisms with [111]Indium-oxine. The experience with this procedure has suggested that most microorganisms can be labeled easily with [111]Indium, and labeled cells exhibit no adverse effects thus facilitating the quantitation of bacteria adherent to materials.

## V. RADIOLABELING OF MICROORGANISMS

A brief outline of methods employed for [111]In or [35]S labeling of bacteria is provided for the benefit of readers who may have interest in using cells labeled with $\gamma$- or $\beta$-emitters to study bacterial interaction with materials or devices.

### A. Preparation of Bacteria

Cultures of *Staphylococcus epidermidis, Staphylococcus auras,* and *Pseudomonas aeruginosa* (ATCC 12228, 25923, and 27853, respectively) were incubated for 36 h in tryptic soy broth (TSB; Remel, Lenexa, KS) at 37°C. The cell suspension was centrifuged at 1200 $g$ for 15 min at 4°C and cells were washed three times with phosphate-buffered saline (0.014 M $Na_2HPO_4$, 0.003 M $NaH_2PO_4$, 0.15 M NaCl, pH 7.4; PBS). After the final washing, the sedimented bacteria were resuspended in 10 mL of TSB (approx 1 to $5 \times 10^8$ cells/mL).

### B. Radiolabeling Technique

Cultures of bacteria at a concentration of 1 to $5 \times 10^8$ cfu/mL in TSB were labeled with trans [35]S-methionine (ICN Radiochemicals, Irvine, CA) or [111]indium-oxine (Amersham, Arlington Heights, IL) under identical conditions. Approximately 100 µCi (1103 Ci/mmol) of trans [35]S-methionine was added dropwise to a 4 mL bacterial suspension in TSB. The cells were incubated with continuous mixing for 48 h at 37°C, to obtain a desirable level of [35]S uptake. In the experiments with [111]indium, the cells were incubated with approximately 30 µCi of [111]indium-oxine for 2 h, under identical conditions. Following the incubation, which ended at the same time for both experiments, the cells were washed three times with PBS to remove the unbound labels. The washed, labeled bacteria were suspended in 4 mL of TSB; porous caps were used to allow aerobic growth of the bacteria where necessary. Due to the high density of bacteria, doubling approximately every 20 min, cells were subcultured in fresh TSB every 48 h.

### C. Rate of Incorporation of Radioactivity

Aliquots of cell suspensions were taken at various time intervals during incubation to determine the rate of [111]indium-oxine or [35]S-methionine incorporation into the bacteria. At predetermined times, 500 µL of bacterial suspension was centrifuged, and after washing the cells twice with PBS, the amount of radioactivity associated with the sedimented cells was determined either by gamma counting ([111]In) or scintillation counting ($^{35}$S). The incorporation of radioactivity was closely monitored over a 2-h period in the case of $^{111}$indium-oxine, and 48 h in the case of $^{35}$S-methionine-labeled bacteria.

### D. Test for Bacterial Viability

The number of bacteria in a given suspension medium was determined every 48 h by measuring the optical density (OD) at 600 nm in TSB solution. Suspensions of *S. aureus, S. epidermidis*, or *P. aeruginosa* in TSB were adjusted to an OD of 1.0 to obtain $10.7 \times 10^8$, $9.3 \times 10^8$, or $11.0 \times 10^8$ cfu/mL, respectively.

### E. Bacterial Adhesion Studies

To examine whether the radiolabeling interferes with bacterial adhesion, experiments were conducted using solid glass beads (size 3 mm, Scientific Products, McGaw Park, IL). $^{111}$Indium-oxine or $^{35}$S-methionine-labeled bacteria at a concentration of approximately $10^9$ cfu/mL in fresh platelet-poor plasma (PPP) were incubated with the glass beads at 37°C. At desired time intervals, nonadherent bacteria were removed by washing the beads twice with PBS. The surfaces were then prepared for gamma or scintillation counting, or fixed in buffered 2.5% glutaraldehyde for scanning electron microscopy. Adherent bacteria (percent of total) were quantified by the amount of radio-activity associated with the glass beads.

### F. Gamma and Scintillation Counting

Every 48 h, the bacteria were centrifuged at 1200 $g$ for 15 min at 4°C, and washed with 4 mL of PBS. Half the population of cells was subcultured in fresh TSB, the other half prepared for scintillation or gamma counting to measure the amount of radioactivity associated with the cells. $^{35}$S-labeled bacteria were solubilized in Protosol (New England Nuclear, Boston, MA) overnight and then mixed with 6 mL of scintillation cocktail and counted in a scintillation counter. For $^{111}$In-labeled bacteria, aliquots of cells were transferred into a vial for quantification of radioactivity in a gamma counter.

### G. Scanning and Transmission Electron Microscopy

Samples of each strain of bacteria, non-labeled (control), labeled with $^{35}$S-methionine or labeled with $^{111}$In-oxine, were fixed in 2.5% buffered glutaraldehyde for 2 h. The speci-mens were then dehydrated in graded alcohol, critical-point-dried, coated with gold-palladium, and examined under a JEOL SM-35 scanning electron microscope to evaluate the effect of radiolabeling on the morphology of bacteria. Additional samples were post-fixed in 1.5% osmium tetroxide for 1 h, and processed for transmission electron microscopy.

## VI. RESULTS

### A. Rate of $^{111}$In or $^{35}$S Incorporation

It was noted that $^{111}$In-oxine incorporation into the cells was significantly more rapid than $^{35}$S incorporation. Although the bacteria were incubated with $^{111}$In-oxine for 2 h, results summarized in Figure 1A demonstrate that more than 90% of the radioactivity was incorporated by the cells within the first 10 min. The rate of $^{35}$S-methionine uptake by bacteria under identical conditions was found to be slow and time-dependent (Fig. 1B).

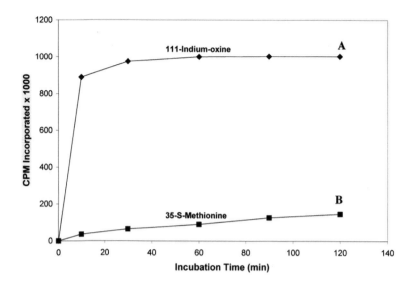

**Figure 1.** Relative uptake of [111]Indium oxine and [35]S-methionine by *S. aureus*. Note that approx 90% of [111]Indium was taken up in 10 min.

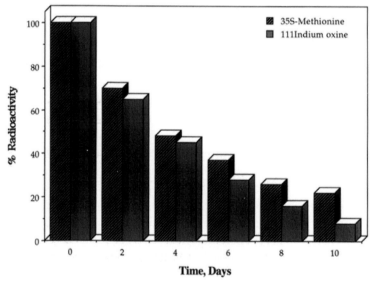

**Figure 2.** Release of radioactive label over a 10-d period. As cells replicate, a portion of the cellular contents is released into the suspension medium.

## B. Radiolabeling Efficiency

Figure 2 shows the amount of radioactivity associated with bacteria at 2-d intervals. The results show no significant difference when bacteria were labeled with either [35]S-methionine or [111]In-oxine. Bacteria labeled with [35]S-methionine released approximately 15±2.5% of their radioactivity every 24 h, while [111]In-labeled bacteria released 20±2.5% of radioactivity in the same time period. The rate of displacement of the

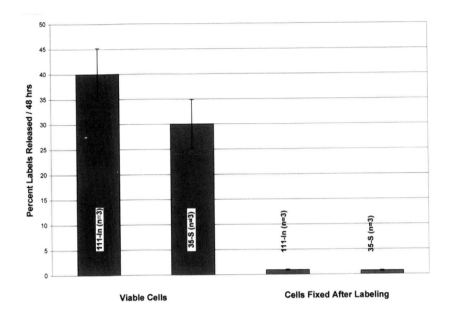

**Figure 3.** This figure demonstrates that radioactive labels are released by viable cells only; cells which were fixed after labeling retained their radioactivity.

label was not dependent on the amount of radioactivity. Since certain intracellular components are presumably released in the surrounding medium by the dividing cells, bacteria were labeled with [111]In-oxine or [35]S-methionine and then immediately fixed with 2.5% buffered glutaraldehyde to prevent cell division and then incubated for several days to investigate spontaneous release of labels. The rate of release of incorporated isotopes by fixed cells was less than 1.0% for both isotopes in 48 h (Fig. 3).

## C. Bacterial Viability

[111]In-oxine- and [35]S-methionine-labeled bacteria showed normal growth patterns. When compared with non-labeled bacteria under identical conditions, the growth kinetics were indistinguishable (Table 1). These results show that a viable bacterial population can be labeled efficiently with either of the two methods. Scanning electron microscopic observations showed no morphological changes in nonlabeled (control) or radiolabeled bacteria. However, ultrastructural assessments utilizing scanning and transmission electron microscopy showed no detectable differences between the non-labeled and labeled cells. A few damaged cells were found among control as well as labeled cells with equal frequency, suggesting that neither the [35]S-methionine nor [111]In-oxine caused detectable cellular damage.[24] Moreover, many cells were found to be undergoing cell division confirming that the labeling of the cells did not adversely affect cell growth or viability.

## D. Bacterial Adhesion

Incubation of [35]S-methionine or [111]In-oxine bacteria with glass beads resulted in identical adhesion patterns for bacteria labeled by either technique (Table 2). This suggests that [111]In-oxine labeling does not interfere with bacterial adherence, when compared with bacteria labeled with [35]S-methionine.

**Table 1. Quantification of Bacterial Growth in the Presence of A: No Labeling Agent (Control), B: $^{111}$In-Oxine, and C: $^{35}$S-Methionine[a]**

| Days | S. aureus | | | S. epidermidis | | | P. aeruginosa | | |
|---|---|---|---|---|---|---|---|---|---|
| | A | B | C | A | B | C | A | B | C |
| 0 | 14.50 | 14.89 | 14.12 | 9.10 | 8.77 | 9.12 | 12.76 | 12.83 | 12.71 |
| 2 | 24.29 | 23.71 | 23.50 | 17.21 | 17.69 | 18.01 | 27.00 | 26.14 | 26.43 |
| 4 | 52.05 | 53.85 | 53.26 | 43.77 | 43.05 | 44.47 | 53.00 | 52.78 | 54.58 |
| 6 | 115.43 | 112.18 | 114.88 | 91.92 | 90.58 | 93.04 | 109.20 | 119.48 | 111.88 |
| 8 | 206.83 | 210.97 | 207.86 | 176.64 | 170.08 | 179.52 | 225.12 | 218.72 | 226.40 |
| 10 | 387.33 | 383.36 | 391.99 | 349.00 | 336.64 | 339.52 | 461.44 | 453.76 | 451.60 |

Note: A: $n = 7$, B: $n = 7$, C: $n = 3$,
[a] expressed as total number of bacteria in the suspension medium $\times 10^8$.

**Table 2. Comparison of the Adhesion
of $^{111}$In-Oxine- and $^{35}$S-Methionine-Labeled *S. aureus* on Glass Beads[a]**

| Time (hours) | $^{111}$In-Oxine labeling (% adhesion) | $^{35}$S-Methionine labeling (% adhesion) |
|---|---|---|
| 1 | $12.2 \pm 1.1$ | $10.2 \pm 0.9$ |
| 12 | $7.6 \pm 0.4$ | $8.7 \pm 1.1$ |
| 24 | $4.2 \pm 0.4$ | $6.0 \pm 0.8$ |

[a] $n = 3$.

## E. Conclusions

The results summarized in the preceding paragraphs show that: 1) bacteria can be effectively labeled with β- or γ-emitters; and 2) there is no indication that labeling significantly alters cell behavior. The advantages of labeling bacteria with $^{111}$indium include short incubation time, enhanced sensitivity and the ease with which cells adherent on surfaces can be quantitated.

## VII. AREAS OF CONCERN AND IMPROVEMENT

The risk of infection with implanted devices, and emerging literature suggesting that bacteria may exhibit enhanced affinity for certain biomaterials, has triggered the need for careful assessment of bacterial interaction with materials. Since the use of prosthetic devices is rapidly increasing, the need for an enhanced pace of efforts to prevent device infection has also increased. Consequently, efforts are in progress to modify the interfaces by a variety of methods including the incorporation of antibacterial agents. These efforts have also brought to focus the need for better methods to investigate pathogen–material interaction. This includes development of methods to closely monitor and quantify bacterial adhesion and colonization. Labeling with $^{111}$In appears to be well suited for quantitative assessment of bacterial adhesion. The advantages of $^{111}$In relate to its labeling efficiency, γ-emitting properties, short half-life, and low toxicity. Unlike β-emitters, where labeled microorganisms adherent to surfaces must be removed quantitatively for scintillation counting, $^{111}$In-labeled bacteria can be readily assessed sequentially *in situ* during the course of an experiment. Other γ-emitting ligands have not shown the same promise as $^{111}$In. Studies conducted with $^{75}$Se-selenomethionine, also a γ-emitting radioisotope, have shown that $^{75}$Se has low labeling efficiency (5-21% for *P. aeruginosa* and *S. aureus*), and therefore is not as suitable an isotope for labeling bacteria as $^{111}$In.[19]

Concerns relating to the detrimental effects of $^{111}$In on cells have been raised in the published literature. Therefore, to employ the $^{111}$In labeling method for studying bacterial interaction with biomaterials in vitro or in vivo, it is important that labeled cells remain viable and exhibit normal functions. The cellular damage could occur in part due to either the radiation emitted by the radioactive tracer, or the toxicity of the oxine.[24] $^{111}$In forms a saturated (3:1) complex with oxine similar to that of oxine-iron complex. The extent of cell damage caused by nominally carrier-free $^{111}$In-oxine has been investigated, and other $^{111}$In-chelates have been developed. However, it has been documented that oxine, when used under appropriate conditions, does not adversely affect the cells. Based on growth kinetics and ultrastructural studies, neither the toxicity of oxine nor the $^{111}$In radiodecay showed any detectable effect on the labeled cells. Nonetheless, further studies are warranted to investigate the possible mutagenic effects of the labeling agents.

The mechanism by which [111]In-oxine complex is incorporated into the cells is not well understood.[2] However, the lipophilic nature of the oxine chelating agent allows the complex to penetrate the cell membrane. [111]In-labeling studies performed on platelets have revealed that once within the cell, [111]In-oxine complexes dissociate, resulting in the binding of [111]In ions to cytoplasmic components with mol wt of approx 25,000–46,000. This has led to the suggestion that [111]In released from the cells will not be re-utilized.[2] The uniform distribution of the labels within the cell and the precise localization of radioactivity in the cell have not been established and further research in this field is warranted.[2,34]

## VIII. CONCLUDING REMARKS

Considering that all implanted devices carry the risk of infection with associated increase in morbidity and mortality, the need for a better understanding of the mechanism(s) by which pathogens adhere and colonize a device is not only a scientific curiosity but also an economic necessity. Various approaches to studying bacterial interaction with biomaterials have been enumerated in this chapter. While each method has advantages and limitations, and bacteria labeled with [111]In offer a simple and direct approach to assess bacterial adhesion to materials, a need exists for a universal method that would allow investigators to track adhesion followed by colonization in short term as well as long term implants. Bacteria labeled with radioactive tracers cannot be used to investigate long term interactions, and labeled bacteria have limited utility for studies in vivo. The use of molecular biology tools and incorporation of reporter genes hold significant promise particularly for studies involving interaction of pathogen with devices in vivo.

## REFERENCES

1. Ardehali R, Mohammad SF: [111]Indium labeling of microorganisms to facilitate investigation of bacterial adhesion. *J Biomed Mater Res* 27:269–75, 1993
2. Baumbauer R, Mestres P, Schiel R, et al: Surface treated large bore catheters with silver based coatings versus untreated catheters for extracorporeal detoxification methods. *ASAIO J* 44:303–8, 1998
3. Baumbauer R, Schiel R, Mestres P, et al: Scanning electron microscopic investigation of catheters for blood access. *Blood Purif* 14:249–56, 1996
4. Benson DE, Burns GL, Mohammad, SF: Effect of plasma on adhesion of biofilm forming *Pseudomonas aeroginosa* and *Staphylococcus epidermidis* to fibrin substrate. *ASAIO J* 42:M655–60, 1996
5. Brewer AR, Stromberg BV: *In vitro* adherence of bacteria to prosthetic grafting materials. *Ann Plast Surg* 24:134–8, 1990
6. Burns GL: Infections associated with implanted blood pumps. *Int J Artif Org* 16:771–6, 1993
7. Chalfie M, Tu Y, Euskirchem G, et al: Green fluorescent protein as a marker for gene expression. *Science* 263:802–5, 1994
8. Chang CC, Merrit K: Effect of *Staphylococcus epidermidis* on adherence of *Pseudomonas aeruginosa* and *Proteus micabilis* to polymethyl methacrylate (PMMA) and gentamicin-containing PMMA. *J Orthop Res* 9:284–8, 1991.

9. Chiang BY, Burns GL, Pantalos GM, et al: Microbially infected thrombus in animals with total artificial hearts. *Trans Am Soc Artif Intern Organs* 37:M256–7, 1991

10. Colburn MD, Moore WS, Chvapil M, et al: Use of an antibiotic-bonded graft for *in situ* reconstruction after prosthetic graft infection. *J Vasc Surg* 16:651–8, 1992

11. Dickinson RB, Nagel JA, Proctor RA, et al: Quantitative comparison of shear dependent *S. aureus* adhesion to three polyurethane ionomer analogs with distinct surface properties. *J Biomed Mater Res* 36:152–62, 1997

12. Fogarasi M, Pullman J, Winnard P, et al: Pretargeting of bacterial endocarditis in rats with streptavidin and [111]In-labeled biotin. *J Nucl Med* 40:484–90, 1999

13. Gristina AG: Biomaterial centered infection: microbial adhesion versus tissue integration. *Science* 237:1588–95, 1987

14. Handfield M, Schweizer HP, Mahan MJ, et al: ASD-GFP vectors for *in vivo* expression technology in *Pseudomonas aeruginosa* and other gram-negative bacteria. *Biotechniques* 24:261–4, 1998

15. Harris JM, Martin LF: An *in vitro* study of the properties influencing *S. epidermidis* adhesion to prosthetic vascular graft materials. *Ann Surg* 206:612–20, 1987

16. Hudson EM, Ramsey RB, Evatt BL: Subcellular localization of [111]Indium in [111]Indium labeled platelets. *J Lab Clin Med* 97:577–82, 1981

17. John SF, Derrick MR, Jacob AE, et al: The combined effects of plasma and hydrogel coating on adhesion of *S. epidermidis* and *S. aureus* to polyurethane catheters. *FEMS Microbiol Lett* 144:241–7, 1996

18. Kalabat DY, Froelich JM, Phuong TK, et al: Chitobiase, a new reporter enzyme. *Biotechniques* 25:1030–5, 1998

19. Kishore R, Baltch AL, Hammer M, et al: Bacterial labeling with gamma emitting radionuclide [75]Se and its use in phagocytosis. *Radiochem Radioanal Lett* 43:55–62, 1980

20. Leff LG, Leff AA: Use of green fluorescent protein to monitor survival of genetically engineered bacteria in aquatic environment. *Appl Environ Microbiol* 62:3486–8, 1996

21. Li H, Fairfax MR, Dubocq F, et al: Antibacterial activity of antibiotic coated silicone grafts. *J Urol* 160:1910–3, 1998

22. Lopez-Lopez G, Pascual A, Perea EJ: Effect of plastic catheter material on bacterial adherence and viability. *J Med Microbiol* 34:349–53, 1991

23. Maki DG, Weise CE, Sarafin HW: A semiquantitative method for identifying intravenous catheter-related infection. *New Eng J Med* 296:1305–9, 1977.

24. Mohammad SF: Association between thrombosis and infection. *Trans Amer Soc Artif Intern Org* 40:226–30, 1994

25. Narendran V, Gupta G, Todd DA, et al: Bacterial colonization of indwelling vascular catheters in newborn infants. *J Pediatr Child Health* 32:391–6, 1996

26. Olmsted SB, Norcross N: Effect of specific antibody on adherence of *Staphylococcus aureus* to bovine mammary epithelial cells. *Infect Immun* 60:249–56 1991

27. Osseewaarde JM, Rieffe M, van Doornam GJ, et al: Detection of amplified *Chlamydia trachomatis* DNA using microtiter plate-based enzyme immunoassay. *Eur J Clin Microbiol Infect Dis* 13:732–40, 1994

28. Schauwecker DS, Carlson KA, Miller GA, et al: Comparison of [111]In-leukocytes in a canine osteomyelitis model. *J Nucl Med* 32:1394–8, 1991

29. Sexton M, Reen D: Characterization of antibody-mediated inhibition of *Pseudomonas aeruginosa* adhesion to epithelial cells. *Infect Immun* 60: 3332–8, 1991

30. Sloan AR, Pistole TG: A quantitative method for measuring the adherence of group B streptococci to murine peritoneal exudate macrophages. *J Immunol Methods* 154:217–23, 1992.

31. Tresse O, Errampalli D, Kostrzynska M, et al: Green fluorescent protein as a visual marker in a *p*-nitrophenol degrading *Moraxella* sp. *FEMS Microbiol Lett* 164:187–93, 1998

32. Vacheethasanee K, Temenoff JS, Higashi JM, et al: Bacterial surface properties of clinically isolated *S. epidermidis* strains determine adhesion on polyethylene. *J Biomed Mater Res* 42:425–32, 1998

33. van Pett K, Schurman DJ, Smith RL: Quantitation and relative distribution of extracellular matrix in *S. epidermidis* biofilm. *J Orthop Res* 8:321–7, 1990

34. Wessels P, Heyns AP, Pieters H, et al: An improved method for the qualification of the *in vivo* kinetics of a representative population of [111]Indium labeled human platelets. *Eur J Nucl Med* 10:522–7, 1985

# 17

# Evaluating Adherent Bacteria and Biofilm Using Biochemical and Immunochemical Methods

**W. Michael Dunne, Jr.**

*Division of Microbiology, Henry Ford Hospital, Detroit, MI, USA*

## I. INTRODUCTION

In its infancy, the study of microbial adhesion relied heavily upon direct measurements of cell number or cell mass as a means of quantitation. This was generally accomplished by microscopic enumeration of surface-bound microorganisms, spectrophotometric measurements of stained or unstained biofilm layers, scintillation counting of radiolabeled microorganisms, or colony count determinations of organisms dislodged from a colonized surface. These assays were often time consuming and labor intensive, required handling and disposal of radionuclides or provided only crude measurements of biofilm density. As the study of microbial adhesion grew in complexity over the past decade to include an examination of interactions with eukaryotic cell surfaces and synthetic or biological polymers, the need to develop assays with increased versatility, sensitivity, and economy of scale became paramount. In contrast to earlier methods, the techniques used to evaluate microbial adhesion that will be reviewed in this chapter rely upon the quantitation of indirect markers as a proxy for microbial cell count or biofilm density. These procedures can be further subdivided into two basic categories: immunologic methods that target cell surface antigens as an indication of microbial adhesion and biological assays that measure organism-specific enzyme activity as evidence of surface colonization. While there are numerous descriptions of each type of assay reported for a wide variety of microorganisms, only a select number of reports will be discussed as examples of each.

## II. ENZYME IMMUNOASSAY

The enzyme-linked immunoassay (EIA) was initially developed to supplant the radio-immunoassay as a comparably sensitive diagnostic workhorse in the diagnostic clinical laboratory without the problems associated with the use of radioactive materials. EIAs have been used for a multitude of purposes in that arena but most often to establish a serologic response to a variety of infectious agents or as a means of detecting the presence of endogenous or exogenous antigens characteristic of a specific disease process. In one iteration of the EIA, an antibody conjugated with an enzyme is used to detect a target

*Handbook of Bacterial Adhesion: Principles, Methods, and Applications*
Edited by: Y. H. An and R. J. Friedman © Humana Press Inc., Totowa, NJ

antigen that has been adsorbed to a solid phase. Upon addition of the appropriate substrate, a colored, fluorescent, or chemiluminescent reactant is produced that can be quantitatively measured using a spectrophotometer or luminometer.[4] It wasn't long before researchers in the field of bacterial adhesion recognized the potential for this technique in the study of microbial binding to both animate and inanimate substrata. Not only do EIAs preclude the use of radiolabeled substances, but they are adaptable to automation, allow a large number of analyses to be run simultaneously, are far less labor intensive and more objective than microscopic quantitation and eliminate the need for transparent substrata. In addition, the EIA is particularly suited for studies involving the adhesion of microorganisms to eukaryotic cell surfaces.

In one of the first published reports to exploit an EIA specifically for the purpose of measuring bacterial adhesion, Stanislawski et al.[32] examined the attachment of type 1 fimbriated *Escherichia coli* (mannose-sensitive) and *Streptococcus pyogenes* to human skin and lung fibroblast cell lines and human buccal epithelial cells as a function of endogenous and exogenous fibronectin concentrations. For this assay, a standardized bacterial inoculum ($\sim 5 \times 10^8$ cfu/mL) was prepared in PBS from broth cultures. All cell lines were cultured to confluence in 96-well polystyrene tissue culture plates, washed, and fixed with 0.25% glutaraldehyde for 10 min at 4°C followed by a 30 min incubation with 0.2 M glycine and 0.1% bovine serum albumin (BSA) at room temperature. The authors had previously determined that fixation caused no obvious changes in bacterial adhesion when compared to nonfixed monolayers and prevented the significant loss of cells in the monolayer during the numerous wash steps required by the EIA. To perform the assay, the cell monolayers were washed with PBS and exposed to 100 μL of the bacterial inoculum containing 10 mg/mL of hemoglobin to prevent nonspecific adhesion. The wells were incubated for 30 min at 37°C, washed five times with PBS, and overlaid with 100 μL of appropriately diluted rabbit anti-*S. pyogenes* or *E. coli* antisera. The plates were incubated at room temperature for 30 min, and washed five times with 0.9% NaCl containing 0.05% Tween 20 and 0.02% $NaN_3$. One-hundred μL of diluted, affinity-purified goat anti-rabbit IgG conjugated to alkaline phosphatase was added to each well and the plates were incubated for 40 min at room temperature. After five additional washes, 100 μL of substrate (*p*-nitrophenyl phosphate) was added to each well and the plates were incubated for 20 min after which the absorbance of each well was measured at 405 nm ($A_{450}$) using a microEIA reader. Bacterial adhesion was expressed in relative EIA Units based upon $A_{405}$ readings or as a percentage of a control when binding in the presence of increasing exogenous fibronectin was examined. No attempt was made in this study to correlate EIA Units with the actual number of bacteria bound to the cellular substratum.

In 1986, Ofek et al.[26] modified this basic procedure and also introduced the concept of replacing the primary antibody/antibody-conjugate detection system with biotin-labeled bacteria. Once again, the study involved the adhesion of *S. pyogenes* and various strains of *E. coli* to human buccal epithelial cells and to porcine enterocytes. For this study cells were collected by scraping, washed and suspended in PBS to the desired concentration. The wells of flat-bottomed microtiter plates were treated with 1 M lysine, 1.25% glutaraldehyde, and PBS, respectively, prior to the addition of cells. The cells were allowed to settle for 10 min after which the plates were centrifuged, the supernatant removed, and the plates dried overnight at 37°C. To perform the adherence assay with antibodies, all wells were incubated with 100 μL of blocking agent (see below) for 1 h at

37°C and washed four times with PBS prior to the addition of 100 μL of the bacterial inoculum. The inoculated plates were rotated horizontally for 1 h at room temperature and washed five times with PBS. The plates were fixed by heating at 65°C for 10 min. From this point forward, the procedure was essentially identical to the method described previously[32] except for the use of a peroxidase-conjugated detection antibody and a 5-aminosalicylic acid substrate. The reaction in this case was monitored at 450 nm.

Several unique developments resulted from this investigation that have laid the foundation for continued use of EIA as a means of assessing microbial adhesion. First, by comparing methodologies, the authors were able to conclude that there was good correlation between EIA and microscopic determinations of cell-bound *S. pyogenes* ranging from 10 to 200 bacteria per cell. Secondly, the investigators observed that the choice of blocking agent is critical to prevent nonspecific bacterial adhesion to the activated plastic wells and yet must be optimized so as not to interfere with specific bacterial attachment to cells. In this study, a 5% solution of bovine serum albumin (BSA) in PBS was used for assays involving *E. coli* while a 20 mg/mL solution of hemoglobin in PBS was selected for use with *S. pyogenes*. Finally, one of the most significant contributions to the development of EIA for studies of bacterial adhesion involves the use of biotinylated bacteria. For this, a strain of *E. coli* was suspended in 0.1 M carbonate buffer (pH 8.2) to an absorbance of 0.7 $A_{450}$. Biotin-*n*-hydroxysuccinimide (10 mg/mL in dimethyl sulfoxide) was added to the bacterial suspension (12.5 μL/1 mL of bacteria) and incubated for 2 h at room temperature. The cells were then washed and resuspended in PBS to the desired concentration. Fifty μL of the biotin-labeled bacteria were added to washed and blocked wells and incubated for 40 min at 37°C. The plates were washed and heat fixed prior to the addition of an avidin-peroxidase conjugate for 30 min at 37°C. After washing, 100 μL of the *O*-phenylene-diamine substrate was added to each well and the color reaction was measured at $A_{450}$ as described above. This process circumvents the need for primary and conjugated antibodies and eliminates one step in the assay. The authors concluded that biotinylation did not alter the binding characteristics of *E. coli* as the results for experiments using the EIA or biotin-labeled assays were nearly identical. Both assays were effectively employed to evaluate promoters and inhibitors of bacterial adhesion to cellular targets including BSA and lipoteichoic acid for *S. pyogenes*, and mannose analogs and fimbriae for *E. coli*.

Over the next several years, the innovative use of EIA that emerged from the study of Ofek et al.[26] was used repeatedly to explore a variety of bacterial–cellular interactions including the adhesion of fimbriated *E. coli* to renal tubular cell lines,[21] the binding of group B streptococci to peritoneal macrophages,[31] and investigations involving antibody-mediated inhibition of: 1) *Hemophilus influenzae* type b binding to buccal epithelial cells[12] 2) *Pseudomonas aeruginosa* adhesion to buccal epithelial cells[29] and, 3) *S. aureus* attachment to bovine mammary epithelial cells.[27] Athamna and Ofek[2] and Athamna et al.[3] later modified the EIA to examine time-dependent attachment and ingestion of *Klebsiella pneumoniae* by macrophages of alveolar, peritoneal, and peripheral blood origin. Two of these investigations took advantage of the direct biotinylation EIA procedure[12,27] while the remaining studies used the standard antibody sandwich EIA format. Sexton and Reen[29] determined that a linear relationship existed between EIA absorbance values and the cell density of the bacterial inoculum used to coat individual wells which ranged from $10^6$ to $10^8$ cfu/mL for uncoated plastic wells and from $10^6$ to $10^9$ bacteria/well for wells coated with buccal epithelium. Others were also able to demonstrate a linear relationship between absorbance value and the actual number of organisms per well ranging from $2\times10^5$ and $5\times10^6$ cfu/well for *K. pneumoniae*[2] and from $6\times10^4$ to $6\times10^6$ cfu/well for group B streptococci[31] by constructing a standard curve using a known number of organisms immobilized to microtiter wells after correcting for the loss of bacteria caused by washing. An EIA protocol similar to the assay of Ofek et al.[26] was also developed by Filler and colleagues to explore the binding of *Candida albicans* to human umbilical vein endothelial cells.[10] These authors found a linear relationship between the $\log_{10}$ absorbance value and the $\log_{10}$ number of attached cells/well.

While the majority of studies in which EIA was used to assess microbial adhesion have been limited to interactions between organisms and animate (cellular) substrata, Fish et al.[11] adapted the assay to evaluate the hydrophobicity of *Bordetella pertussis* and *B. bronchiseptica* by examining the binding of individual strains to polystyrene microtiter wells. Rather than developing standard curves, absorbance values were normalized for the binding affinity of each strain for the conjugate antibody by comparison with a reference strain. This value was then used to correct adhesion values and to determine the percent adhesion (also by comparison to the reference strain).

In 1994, Skurnik et al.[30] developed an innovative "on-slide" adaptation of the EIA that allowed quantitative evaluation of bacterial adhesion to tissue sections. For this assay, 8 µm frozen sections of tissue (in this case, human distal ileum or proximal colon) were mounted on sterile microscope slides and fixed in methanol containing 2% $H_2O_2$ for 30 min. The slides were then washed, blocked with 3% BSA in PBS, washed, blocked again with 10% sheep serum in 3% BSA-PBS (to decrease background absorbance) and washed once more. The sections were overlaid with 100 µL of a bacterial suspension (in this report, *Yersinia enterocolitica* or *E. coli*), incubated in a moist chamber with gentle rotation at 4°C for 15 min and washed to remove unbound bacteria. The sections were then overlaid with organism-specific murine monoclonal antibody, incubated for 15 min at room temperature, washed, and overlaid with 100 µL of diluted peroxidase-conjugated rabbit anti-mouse antibody for 15 min at room temperature. The sections were then washed twice and overlaid with 100 µL of substrate (3 mg 1,2-phenylendiamine/mL of citrate buffer with 10 µL of 30% $H_2O_2$/15 mL of buffer) for 10 min at room temperature. From each section, 75 µL of the substrate was transferred to a microtiter plate well and 125 µL of 1 M HCL was added to stop the color reaction. The absorbance of the well was then read at $A_{492}$. The assay showed a linear relationship between absorbance values and the bacterial concentration from $5\times10^6$ to $5\times10^8$ organisms/section. The real beauty of this assay is that the same procedure can be used to examine bacterial adhesion by three different methods; gram-stain, immunoperoxidase staining, and EIA. Recently, the on-slide EIA assay was used not only to evaluate bacterial attachment to tissue sections (rat kidney) but also to quantitate bacterial binding to glass slides coated with extracellular matrix molecules cluding collagen, laminin, and fibronectin.[9]

The final variation on the EIA theme to be reviewed in this section was described by Nilsson et al.[25] and was used to examine the binding of strains of *E. coli* (P-fimbriated and nonfimbriated) to immobilized carbohydrate receptor molecules. The nuances of this protocol are far too detailed to describe here but can be summarized as follows. Microbeads (latex or Dynosphere, 0.2-5) were coupled with [Gal($\alpha$1-4)Gal($\beta$)]-BSA and

## Table 1.  Reported Configurations of EIA-Based Assays
### for the Investigation of Microbial Adhesion

| Substratum | Organism(s) | Inhibitors | Ref. |
|---|---|---|---|
| Murine macrophages | *K. pneumoniae* | mannose analogs | 2 |
| Guinea pig and human macrophages | *K. pneumoniae* | Capsular polysaccharide, mannan, mono-, di-, and oligosaccharides | 3 |
| Human colon and Lewis rat kidney sections; collagen, fibronectin and BSA-coated glass slides. | *Y. enterocolitica* *E. coli* | | 9 |
| Human umbilical vein endothelium | *C. albicans* | | 10 |
| Polystyrene | *B. pertussis* *B. bronchiseptica* | | 11 |
| Human buccal epithelium | *H. influenzae* type b | Pili antibodies | 12 |
| Porcine renal tubular cells | *E. coli* | Fimbriae inhibitors and mutations | 21 |
| Ligand-coated polystyrene film | *E. coli* | | 25 |
| Human buccal epithelium Porcine enterocytes | *S. pyogenes* *E. coli* | Lipoteichoic acid, albumin, mannose analogs, purified fimbriae | 26 |
| Bovine mammary epithelium | *S. aureus* | Immune sera and milk | 27 |
| Human buccal epithelium | *P. aeruginosa* | | 29 |
| Human distal ileum and proximal colon sections | *Y. enterocolitica* *E. coli* | Collagen, laminin, and YadA | 30 |
| Human skin and lung fibroblasts, buccal epithelium | *S. pyogenes* *E. coli* | Fibronectin | 32 |
| Murine macrophages | Group B streptococci | | 31 |

with alkaline phosphatase. Polystyrene strips were coated with [Gal($\alpha$1-4)Gal($\beta$)]-BSA alone. The strips were blocked, washed, and incubated for 30 min in suspensions of the strains of *E. coli*. After washing again, the strip was then incubated for 30 min. with the microbead suspension. The strips were then washed, incubation with *p*-nitrophenyl phosphate substrate, and the color intensity was measured spectrophotometrically at $A_{405}$. The sensitivity of the assay ranged from $10^6$ to $10^7$ cfu/mL of inoculum and thus provided as sensitive and flexible means of evaluating ligand-receptor interactions among microorganisms.

A summary of the uses of EIA discussed in this section for the quantitation of microbial adhesion is presented in Table 1.

## III. FLUORESCENT AND IMMUNOFLUORESCENT ASSAYS

Fluoroprobe-based assays have traditionally been developed for the same purposes as enzyme immunoassay, i.e., for the detection of antibodies and antigens in biological tissues or fluids without the need for radiolabeled materials. Fluorescent immunoassays have been used in both qualitative and quantitative formats with the latter demonstrating sensitivities comparable to radioimmunoassay,[23] so it is not surprising that direct fluorescent and immunofluorescent methods have found use in the field of bioadhesion. Of the examples chosen to review here, the study of Sveum et al.[34] was the first to describe a quantitative fluorescent method to investigate the binding of *Streptococcus pneumoniae* to human peripheral blood monocytes (PBMC). For this assay, pneumococci ($1.5×10^9$ heat-killed organisms) were labeled with Lucifer Yellow (1 mg/mL in 0.1 $M$ NaHCO$_3$, pH 9.5) for 2 h at room temperature and washed. Labeled pneumococci were incubated with PBMC (50 organisms/cell) for 30 min at 0°C to permit attachment but reduce ingestion. Unattached bacteria were removed by centrifugation and attached organisms were stained with anti-Lucifer Yellow antibodies conjugated with biotin, washed, and reacted with a streptavidin-Texas Red conjugate and washed again. The monocytes were then fixed in 1% paraformaldehyde and analyzed by dual laser flow cytometry. The total number of bound bacteria was determined by subtracting the mean fluorescent value of the monocytes alone from monocytes with attached organisms and then dividing by the mean fluorescent value of the pneumococci alone. The authors noted good correlation between the flow cytometry results and direct quantitation by phase contrast fluorescent microscopy. Dunn and colleagues[5] used a classic indirect immunofluorescent assay in conjunction with flow cytometry to examine the attachment of unlabeled *Helicobacter pylori* to various gastric carcinoma cell lines. The basic protocol was nearly identical to that described by Sveum et al.[34] except that cell-bound organisms were detected by reacting cells first with anti-*H. pylori* rabbit serum, followed by washing, incubating with fluorescein isothiocyanate-conjugated goat anti-rabbit antisera, washing, fixing, and flow cytometry analysis. A similar strategy was employed by Li and Walker[18] to evaluate the binding of *Rickettsia conorii* and *R. rickettsii* to a mouse fibroblast cell line. In contrast, Almeida et al[1] avoided the use of antibody and conjugates by developing a direct fluorescent labeling procedure for their investigation of the adhesion of *Streptococcus uberis* to polystyrene microwells coated with fibronectin, collagen, and laminin. To accomplish this, bacteria ($5×10^6$ cfu/200 µL) were mixed with 350 µL of 2',7'-bis-(2-carboxyethyl)-5-carboxy fluorescein cetomethyl ester (45 g/mL) for 45 min at 37°C.

An inoculum containing $1×10^7$ labeled bacteria is added to each microwell, incubated for 60 min and washed. Bacteria bound to the substratum were measured fluorometrically and the results were expressed as a percentage of the initial fluorescence of the inoculum.

## IV. BIOLUMINESCENT ASSAYS

Unlike the previous section, the assays to be presented under this heading represent a diverse collection of procedures that have been designed to study the interactions between a wide variety of microorganisms and substrata. One of the most frequently cited and universally applied methods within this category is the ATP bioluminescence assay. In 1983, Harber et al.[13] reported on the use of a rapid bioluminescence assay to examine the adhesion of strains of *E. coli* to polystyrene tubes. For this procedure, 300 µL of a bacterial inoculum containing ~$1 \times 10^8$ cfu/mL was placed in triplicate polystyrene tubes which were incubated for 10 min at 37°C. The inocula was aspirated and the tubes were washed twice with PBS. ATP was extracted from surface-bound organisms using a nucleoti de-releasing reagent (Lumac, Sterilin, UK). The extracted ATP was measured using a luminometer after adding 100 µL of ATP-monitoring reagent (firefly luciferin/luciferase, LKB) to 200 µL of sample extract. The adhesion ratio per 1000 inoculum cells was calculated as:

$$\frac{\text{Mean mV of extract} \times 1000}{\text{mV reading of extract of bacterial inoculum} \times 15 \text{ (volume correction factor)}}$$

The intra-assay precision of this method in terms of coefficient of variation for 12 replicate samples was 8.5% and the limit of detection was 0.2 pmol ATP in a 200 µL sample: approx 5 ($10^5$ organisms/mL of inoculum) or about 1 ($10^4$ attached bacteria/tube). Later, Ludwicka et al.[19,20] modified the protocol to examine the adhesion of staphylococcal species to synthetic polymers. Polymer pieces were cut into 6 mm disks, sterilized, and incubated with 250 µL of a bacterial inoculum ($10^8$ cells/mL) for 1 h at 20°C. Unattached bacteria were removed by washing and bacterial ATP was extracted from the surface using 50 µL of trichloracetic acid (TCA). ATP monitoring reagent (1 mL) was then added to the ATP extract and the light emitted (I) was measured in a luminometer. Immediately after measurement, 20 µL of an ATP standard was added to the cuvette and the increase in light emission ($I_{std}$) was recorded. The bacterial ATP content was calculated as $I/I_{Std} \times$ ATP concentration of the standard. The number of attached bacteria/cm² was calculated from standard curves prepared with known numbers of organisms in suspension. However, the authors cautioned that the standard curve must be prepared at the same time that polymer disks are extracted because the ATP content of bacteria in suspension decreases with time. Mean values of 3 samples of each polymer were taken as the final result. The ATP content of the extract was found to be linear with respect to bacterial density between $10^3$ and $10^8$ cfu/mL.

The same procedure with minor modifications was used over the past decade to examine the adhesion of *S. aureus* and *S. epidermidis* to catheter segments,[16] strains of *S. epidermidis* to polystyrene microtiter wells,[15] and staphylococcal species to films of poly(ether urethane) and polyethylene in the presence and absence of serum and plasma proteins.[28,33] The correlation coefficient of the assay in terms of viable bacterial count versus ATP concentration (in nmoles) ranged from 0.62[19] to 0.98.[28] However, Stollenwerk et al.[33] noted that a standard curve had to be generated for each strain tested because of the wide divergences of bacterial ATP content. Furthermore, growth conditions also influenced the correlation coefficient of the curve for a single strain such that separate standard curves would be required for different media. These authors did observe a linear relationship between the log of the bacterial ATP content and the log of the bacterial count.

Hibma et al.[14] used a more creative approach to the bioluminescent assay. In their study, the authors used a strain of *Listeria monocytogenes* containing a plasmid for the *lux* AB gene which allowed for the expression of endogenous luciferase. By using this strain, and an L-form of the strain selected by growth in ampicillin, the authors were able to directly evaluate bacterial adhesion to pieces of stainless steel and intravenous tubing by adding a luciferase substrate to the substratum after planktonic bacteria had been removed. As before, the light produced was measured in a luminometer. Regression analysis of viable bacterial counts and bioluminescent readings for classical and L-form cultures produced correlation coefficients of 0.96 and 0.79 respectively.

## V. ENDOGENOUS MICROBIAL ENZYME ACTIVITY

The activity of surface-bound microorganisms has also been estimated indirectly by measuring the rate of substrate utilization by endogenous microbial enzymes. In the assay described by Ladd and Costerton,[17] the bacterial glucose oxidase activity of a bacterial biofilm is measured using the method of Trinder[35] where D-glucose is oxidized to D-gluconic acid by glucose oxidase which is then converted to a quinoneimine dye by peroxidase. For this procedure, bacterial-coated substrata are placed in vials containing a 1 mg/mL glucose substrate. The vials are incubated and the glucose concentration of the solution is monitored over time and compared to a standard curve. Samples (25 μL) are transferred from the vials to cuvettes containing 2.5 mL of glucose oxidase/peroxidase reagent. The mixture is incubated at 37°C for 10 min. and read spectrophotometrically at 505 nm. However, because the glucose solution can stimulate growth, the bacterial density of the test vials must be continuously compared to a control vial containing no glucose by determination of colony counts scraped from the surface of the substratum over time.[24]

In 1987, Minami and colleagues[22] devised an assay using endogenous bacterial galactosidase activity as a means of quantitating the adhesion of enteropathogenic *E. coli* to Hep-2 monolayers. To carry out this assay, strains of *E. coli* were grown in 1% tryptone broth with 0.2 mM isopropul-thio-D-galactose (α-galactosidase inducer) while Hep-2 cells were cultured in 96-well tissue culture plates. Prior to use, the monolayers were washed with PBS and inoculated with 10–20 μL of a bacterial suspension to give $1\times10^7$ organisms per well. The wells were supplemented with 200 μL of cell culture medium containing 0.5% D-mannose to inhibit adhesion by type 1 pili. The wells were then incubated for 30 min at 37°C, and washed three times to remove planktonic bacteria. Eighty μL of assay buffer (0.1 M sodium phosphate, pH 7.0 with 1 mM $MgSO_4$ and 0.1 M 2 mercaptoethanol saturated with toluene) was added to each well. The plates were incubated for 30°C for 10 min followed by the addition of 80 μL of assay buffer containing 4 mg/mL of orthonitropheyl-galactoside. The volume of each well was adjusted to 210 μL with assay buffer and the absorbance of each well was read at 405 nm after incubation for 2 or 4 h at 30°C. By constructing a standard curve of $A_{405}$ versus viable attached bacteria, the authors were able to demonstrate that the assay was linear from $5\times10^4$ to $2\times10^6$ organisms per well. The sensitivity of the assay could be improved by prolonging the incubation period after the addition of substrate from 2-4 h. However, the authors also noted that α-galactosidase activity varies from strain to strain, possibly necessitating the construction of standard curves for each individual strain.

Dunne and Burd[6,7,8] took advantage of endogenous urease production by *Staphylococcus epidermidis* to measure the effects of divalent cations, EDTA, pH, albumin, fibronectin, and fragments of fibronectin on the adhesion of this species to polystyrene microwells. To perform the assay, bacteria from an overnight broth culture were harvested by centrifugation, washed in PBS, and diluted to the desired bacterial density. Eight wells of a sterile 96-well flat-bottomed tissue culture plate were inoculated with 100 µL of the bacterial suspension and incubated for 1 h at 37°C. Studies had shown that extended incubations did not result in a significant increase in adhesion. The wells were washed four times with distilled water to remove planktonic bacteria and 100 µL of a commercial EIA-grade urease reagent was added to each well. Eight uninoculated wells served as a reagent control. The plates were again incubated at 37°C and the color intensity was recorded at 10 min intervals at 570 nm using a microEIA reader which had been set to zero using the reagent control wells. Surface-bound bacteria produced a linear increase in absorbance with time until the substrate was depleted. On the basis of rate kinetics, a standardized inoculum containing $2\times10^8$ cells/mL was selected. The sensitivity of the assay did not extend below $3\times10^4$ bacteria/well and above $3.5\times10^8$ organisms, substrate depletion became rate limiting. The linearity of the assay extended from $3\times10^4$ to $3.5\times10^7$ organisms per well with a correlation coefficient of 0.974.

## VI. MISCELLANEOUS ASSAYS

A number of protocols have been established using radiospirometry to estimate the heterotrophic activity of microbial biofilms. These methods have been thoroughly reviewed by Ladd and Costerton[17] and will not be examined in detail here. Briefly, the premise behind radiospirometric techniques relies upon the evolution of $^{14}CO_2$ from $^{14}C$-labeled substrates such as glucose and glutamic acid by actively metabolizing, surface-associated microorganisms. The $^{14}CO_2$ product is trapped, measured, and compared to acid-killed controls in either a single, or kinetic readings. For the most part, radiospirometric determinations have been used qualitatively and, due to the inherent complexity of the measurements, have been replaced by technically less demanding, nonisotopic methods.

Similar to spirometric methods, the formazan reduction assay targets actively respiring organisms within a biofilm. The assay is based upon the ability of the electron transport system of aerobic organisms to use 2-(*p*-iodophenyl)-3-(*p*-nitrophenyl)-5-phenyl tetrazolium chloride (INT) as a terminal electron acceptor. During the process of oxidative phosphorylation, the straw-colored dye is reduced to a dark red formazan product which is deposited intracellularly. The assay was originally designed as a microscopic determination[36] but has since been formatted for use as a semi-quantitative spectrophotometric method.[17] The assay is performed by placing a sample of the colonized surface into a sterile vial containing 5 mL of sterile PBS (pH 6.8) and adding 1 mL of a 0.2% aqueous solution of INT. The vial is incubated for 20 to 30 min in the dark at room temperature after which the sample is removed, rinsed with PBS, and placed in a separate vial. The intra-cellular formazan is extracted with 4 mL of absolute ethanol at 37°C for 1 h. The extracted formazan is then measured spectrophotometrically at 495 nm. A number of problems are inherent with this assay including incomplete extraction of the formazan dye and interference with copper and iron ions or erythrocytes. Although not reported, it is apparent that the formazan assay could be easily adapted for use with 96-well polystyrene microtiter plates.

## VI. SUMMARY

The diverse spectrum of methods reviewed in this chapter has allowed researchers to examine the process of microbial adhesion and biofilm formation in greater detail and with greater efficiency. Immunological and biochemical-based procedures have improved the sensitivity and objectivity of adhesion assays and at the same time permitted the use of the 96-well microtiter plate format for large-scale investigations For the most part, these methods are highly versatile in that they can be used for a wide variety of organism-substrate combinations including both animate and synthetic surfaces. Even though the full potential of these methodologies has yet to be realized, it is possible that many of them will eventually be replaced by far more sensitive quantitative genomic amplification assays.

## REFERENCES

1. Almeida RA, Luther DA, Kumar SJ, et al: Adherence of *Streptococcus uberis* to bovine mammary epithelial cells and to extracellular matrix proteins. *Zentralbl Veterinarmed [B]* 43:385–92, 1996
2. Athamna A, Ofek I: Enzyme-linked immunosorbent assay for quantitation of attachment and ingestion stages of bacterial phagocytosis. *J Clin Microbiol* 26:62–6, 1988
3. Athamna A, Ofek I, Keisari Y, et al: Lectinophagocytosis of encapsulated *Klebsiella pneumoniae* mediated by surface lectins of guinea pig alveolar macrophages and human monocyte-derived macrophages. *Infect Immun* 59:1673–82, 1991
4. Carpenter AB: Enzyme-linked immunoassay. In: Rose NR, de Macario EC, Folds JD, et al, eds. *Manual of Clinical Laboratory Immunology.* ASM Press, Washington, DC, 1997:20–9
5. Dunn BE, Altmann M, Campbell GP: Adherence of *Helicobacter pylori* to gastric carcinoma cells: analysis by flow cytometry. *Rev Infect Dis* 13(Suppl 8):S657–64, 1991
6. Dunne WM Jr, Burd E: *In vitro* measurement of the adherence of *Staphylococcus epidermidis* to plastic by using cellular urease as a marker. *Applied Environ Microbiol* 57:863–6, 1991
7. Dunne WM Jr, Burd E: The effects of magnesium, calcium, EDTA, and pH on the *in vitro* adhesion of *Staphylococcus epidermidis* to plastic. *Microbiol Immunol* 36:1019–27, 1992
8. Dunne WM Jr, Burd E: Fibronectin and proteolytic fragments of fibronectin interfere with the adhesion of *Staphylococcus epidermidis* to plastic. *J Applied Bacteriol* 74:411–6, 1993
9. El Tahir Y, Toivanen P, Skurnik M: Application of an enzyme immunoassay to monitor bacterial binding and to measure inhibition of binding to different types of solid surfaces. *J Immunoassay* 18:165–83, 1997
10. Filler SG, Der LC, Mayer CL, et al: An enzyme-linked immunosorbent assay for quantifying adherence of *Candida* to human vascular endothelium. *J Infect Dis* 156:561–6, 1987
11. Fish F, Navon Y, Goldman S: Hydrophobic adherence and phase variation in *Bordetella pertussis. Med Microbiol Immunol* 176:37–46, 1987
12. Forney LJ, Gilsdorf JR, Wong DCL: Effect of pili-specific antibodies on the adherence of *Hemophilus influeanzae* type b to human buccal cells. *J Infect Dis* 165:464–70, 1992
13. Harber MJ, Mackenzie R, Asscher AW: A rapid bioluminescence method for quantifying bacterial adhesion to polystyrene. *J Gen Microbiol* 129:621–32, 1983
14. Hibma AM, Jassim SAA, Griffiths MW: *In vivo* bioluminescence to detect the attachment of L-forms of *Listeria monocytogenes* to food and clinical contact surfaces. *Int J Food Microbiol* 33:157–67, 1996
15. Hussain M, Hastings JGM, White PJ: A chemically defined medium for slime production by coagulase-negative staphylococci. *J Med Microbiol* 34:143–7, 1991
16. Kristinsson KG: Adherence of staphylococci to intravascular catheters. *J Med Microbiol* 28:249–57, 1989
17. Ladd TI, Costerton JW: Methods for studying biofilm bacteria. *Methods Microbiol* 22:285–308, 1990
18. Li H, Walker DH: Characterization of rickettsial attachment to host cells by flow cytometry. *Infect Immun* 60:2030–5, 1992

19. Ludwicka A, Jansen B, Wadström T, Pulverer G: Attachment of staphylococci to various synthetic polymers. *Zbl Bact Hyg A* 256:479–89, 1984
20. Ludwicka A, Switalski LM, Lundin A, et al: Bioluminescent assay for measurement of bacterial attachment to polyethylene. *J Microbiol Methods* 4:169–77, 1985
21. Marre R, Kreft B, Hacker J: Genetically engineered S and F1C fimbriae differ in their contribution to adherence of *Escherichia coli* to cultured renal tubular cells. *Infect Immun* 58:3434–7, 1990.
22. Minami J, Okabe A, Hayashi H: Enzymic detection of adhesion of enteropathogenic *Escherichia coli* to Hep-2 cells. *Microbiol Immunol* 31:851–8, 1987
23. Nakamura RM, Bylund DJ: Fluorescence immunoassay. In: Rose NR, de Macario EC, Folds JD, et al, eds. *Manual of Clinical Laboratory Immunology.* ASM Press, Washington, DC, 1997:39–48.
24. Nickel JC, Ruseska I, Wright JB, et al: Tobramycin resistance of *Pseudomonas aeruginosa* cells growing as a biofilm on urinary catheter material. *Antimicrob Agents Chemother* 27:619–24, 1985
25. Nilsson B, Larsson A-C, Pålsson K, et al: An amplification technology for improving sensitivity when measuring components in biological samples. *J Immunol Methods* 108:237–44, 1988
26. Ofek I, Courtney HS, Schifferli DM, et al: Enzyme-linked immunosorbent assay for adherence of bacteria to animal cells. *J Clin Microbiol* 24:512–6, 1986
27. Olmsted SB, Norcross NL: Effect of specific antibody on adherence of *Staphylococcus aureus* to bovine mammary epithelial cells. *Infect Immun* 60:249–56, 1992
28. Paulsson M, Kober M, Freij-Larsson C, et al: Adhesion of staphylococci to chemically modified and native polymers, and the influence of preadsorbed fibronectin, vitronectin and fibrinogen. *Biomaterials* 14:845–53, 1993
29. Sexton M, Reen DJ: Characterization of antibody-mediated inhibition of *Pseudomonas aeruginosa* adhesion to epithelial cells. *Infect Immun* 60:3332–8, 1992
30. Skurnik M, El Tahir Y, Saarinen M, et al: YadA mediates specific binding of enteropathogenic *Yersinia enterocolitica* to human intestinal submucosa. *Infect Immun* 62:1252–61, 1994
31. Sloan AR, Pistole TG: A quantitative method for measuring the adherence of group B streptococci to murine peritoneal exudate macrophages. *J Immunol Methods* 154:217–23
32. Stanislawski L, Simpson WA, Hasty D, et al: Role of fibronectin in attachment of *Streptococcus pyogenes* and *Escherichia coli* to human cell lines and isolated oral epithelial cells. *Infect Immun* 48:257–9, 1985
33. Stollenwerk M, Fallgren C, Lundberg F, et al: Quantitation of bacterial adhesion to polymer surfaces by bioluminescence. *Zent bl Bakteriol* 287:7–18, 1998
34. Sveum RJ, Chused TM, Frank MM, et al: A quantitative fluorescent method for measurement of bacterial adherence and phagocytosis. *J Immunol Methods* 90:257–64, 1986
35. Trinder P: Determination of glucose in blood using glucose oxidase with an alternative oxygen acceptor. *Ann Clin Biochem* 6:24–7, 1969
36. Zimmermann R, Iturriaga R, Becker-Birck J: Simultaneous determination of the total number of aquatic bacteria and the number thereof involved in respiration. *Appl Environ Microbiol* 36:926–35, 1978

# 18

# Evaluating Bacterial Adhesion Using Atomic Force Microscopy

**Anneta P. Razatos[1] and George Georgiou[2]**

[1]*Department of Chemical Engineering, Arizona State University, Tempe, AZ, USA*
[2]*Department of Chemical Engineering, University of Texas at Austin, Austin, TX, USA*

## I. INTRODUCTION TO ATOMIC FORCE MICROSCOPY

The Atomic Force Microscope (AFM) is a scanning probe microscope invented in 1986 by Binnig, Quate, and Gerber in order to image surface features of planar substrates with atomic resolution.[3,22,26] The AFM consists of a cantilever that scans a planar substrate resulting in a topographic map of the surface features of the substrate. Standard AFM cantilevers have sharp, pyramidal, silicon nitride tips at their ends that contact the surface during imaging (Fig. 1).[26] During imaging the AFM cantilever experiences a repulsive force that originates from the overlap of electron orbitals between atoms of the tip and the substrate.[9] As the cantilever scans the substrate, features on the surface cause the cantilever to deflect. Cantilever deflection is detected using a laser which is reflected off of the cantilever into a photodiode, in accordance with the optical lever rule (Fig. 1).[22,51]

In addition to imaging, the AFM can also be used to measure the forces of interaction between cantilevers and planar surfaces. In force mode, the substrate is first moved towards, and then retracted from the cantilever. Because this chapter focuses on the initial interaction of bacteria as they approach a surface, only the approach portion of AFM force curves will be considered. As the surface approaches the cantilever, at some critical distance of separation, the cantilever tip will either be attracted to and deflected down towards the surface or repelled away from and deflected up away from the surface. Therefore, tip deflection reflects the attractive or repulsive interaction of the cantilever with the substrate as a function of distance of separation.

Silicon nitride tips, known to assume a net negative charge in neutral solutions,[10,48] are limited in the types of interactions they can represent for AFM studies. A number of groups, however, have used the AFM to analyze interactions between planar surfaces and materials other than silicon nitride by attaching particles onto AFM cantilever tips.[1,2,10,14,15,17,52] For example, AFM cantilevers have been modified by attaching microspheres or even metallic shards to their tips in order to measure the interactions of colloid particles with surfaces.[1,2,10,14,15,17,52]

*Handbook of Bacterial Adhesion: Principles, Methods, and Applications*
Edited by: Y. H. An and R. J. Friedman © Humana Press Inc., Totowa, NJ

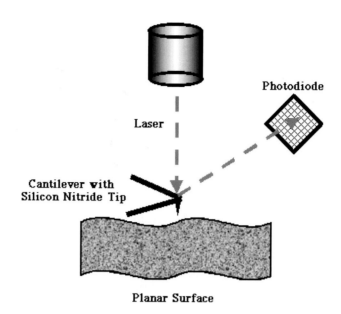

**Figure 1.** Schematic of atomic force microscope (AFM)

## II. BIOLOGICAL APPLICATIONS OF AFM

Atomic force microscopy is ideally suited for biological studies because it does not require a conductive surface, and as a result: 1) the pretreatment of the biological samples for AFM experiments is minimal in comparison to manipulations required by other microscopic techniques (e.g., SEM), and 2) experiments can be executed under physiological conditions. Milder treatment of biological samples minimizes artifacts in the ensuing images. Moreover, images and force measurements can be executed in buffer solutions representative of natural biological environments. The AFM has been used to image microscopic biological particles ranging from DNA to membrane proteins to mammalian cell surfaces and intracellular structures.[24,25,28,46,49,51] In addition to imaging, the AFM has also been used to measure the forces involved in various biological interactions. The AFM, which is sensitive enough to detect forces in the picoNewton range,[11] has been employed to quantitatively measure the forces between complementary DNA base pairs,[4] DNA strands,[32] biotin-avidin,[11,19] and antibody-antigen pairs.[7,13,36,44] Recently, the AFM has also been adapted to cell adhesion studies. Bowen et al.[6] measured the pull-off force between single yeast cells glued to AFM cantilevers and planar substrates. The force required to remove cervical carcinoma cells attached to protein-coated hydrophilic and hydrophobic polystyrene substrates was measured by using the AFM cantilever to displace or push a cell off of the substrates.[45] Sagvolden et al.[45] employed an inclined AFM cantilever and a laser beam deflection system to measure the force applied to displace individual cells. Carcinoma cells were found to attach to protein-coated hydrophilic surfaces stronger and faster at 37°C in comparison to 23°C.[45] Nevertheless, these studies do not address the more relevant problem in the context of surface-associated infections, namely what are forces exerted on cells as they initially approach a synthetic surface.

The first and least understood step in the adhesion of bacteria to inanimate surfaces is the initial attraction or repulsion of bacteria by the substrate. This initial interaction between the bacteria and a surface is a reversible, nonspecific process governed by the physiochemical properties of the bacterial cell and substratum surfaces (i.e. the respective surface free energies and surface charge densities), as well as the properties of the interstitial fluid.[8,12,23,31,33-35,38,40] Whether a cell is attracted to or repelled away from a surface depends on the balance between van der Waals forces (which for bacteria interacting with planar surfaces scale approximately with $1/D^2$, the distance of separation[29]), electrostatic interactions, short-range hydration and/or hydrophobic interactions, and steric effects due to the overlap of surface bound polymers.[39] If the approach of bacteria to a surface is unfavorable, cells must overcome an energy barrier in order to establish direct contact with the surface. Only when the bacteria are in close proximity to the surface do short-range interactions become significant. Short-range interactions include protein–ligand binding events mediated by a plethora of microbial adhesins, and, in some cases, the production of extracellular polymers, rendering the binding process practically irreversible.[12,23,38]

In earlier studies, bacterial adhesion was evaluated by enumerating the cells remaining attached to a surface following periods of incubation and rinsing.[18,38] There are several problems with this approach. First, the number of bacteria that remain associated with the surface depend on both long-range attractive/repulsive interactions as well as short range biospecific interactions. Second, the data obtained from such studies depend strongly on the experimental protocol such as incubation and rinsing conditions, and are difficult to reproduce. Third, results provide no quantitative information on the magnitude of the forces between the cells and surfaces.

The AFM was adapted by Razatos et al.[43] to measure the reversible, long-range interactions between bacteria and various substrates as the bacteria initially approach the solid-liquid interface. This AFM-based methodology was found to be a highly sensitive, reproducible and versatile means for evaluating the bacterial cell and substrate surface properties involved in bacterial adhesion.[39,42,43] The AFM can be employed to investigate the interactions between practically any bacterium and substrate under physiological conditions. For example, atomic force microscopy could detect the effect of deleting three sugar residues in the lipopolysaccharide layer coating the *E. coli* cell surface on the interaction of the bacteria with various materials.[43]

## III. PROTOCOL

The critical step in the use of the AFM for biological specimens is immobilization of biological samples onto a rigid support surface such that the samples resist removal under the force of the AFM probe.[51] In order to measure the force of interaction between bacteria and substrates, two possible configurations were explored: 1) bacteria were irreversibly immobilized onto a planar surface and probed by an AFM cantilever, and 2) bacteria were irreversibly immobilized onto AFM cantilever tips and used to probe planar substrates (Fig. 2). Razatos et al.[43] used a simple cell immobilization protocol to form confluent cell layers on both planar glass substrates and the silicon nitride tips of AFM cantilevers.

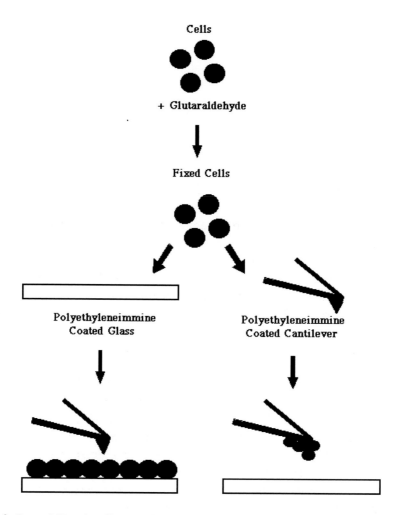

**Figure 2.** Immobilization Protocol. Bacterial cells are fixed with glutaraldehyde and then deposited onto planar glass or AFM cantilevers that have been coated with poly-ethyleneimmine.

A uniform, confluent bacterial lawn coating planar substrates or AFM cantilevers is necessary to ensure that a measured interaction is in fact between bacteria and the surface of interest. For example, in the case of the AFM cantilever probing bacteria immobilized on a planar substrate, in the absence of a confluent bacterial layer, it would be necessary to first image the surface in order to position the AFM cantilever tip over a bacterium. During imaging, the AFM cantilever tip contacts the immobilized bacteria such that biological macromolecules could adsorb onto the tip. Adsorption of biological molecules could then change the surface properties of the cantilever tip, and hence alter subsequent force measurements. We circumvented the need for prior imaging of bacteria by forming a confluent layer of bacteria on the solid support surface. Since the entire support surface is covered by bacteria, the approaching tip is certain to interact with cells and not the underlying support. Moreover, after every force measurement, AFM or SEM images of the bacteria immobilized on the planar substrates or the AFM cantilevers respectively are taken to verify that the measured interactions are indeed between bacteria and the substrate.

## A. *Immobilization Technique*

The *Escherichia coli* (*E. coli*) K-12 strains, D21 and D21e7, are Gram negative bacteria whose cell surface lipopolysaccharide compositions have been well characterized (*E. coli* Genetic Stock Center, Dept. of Biology, Yale University, New Haven, CT, USA).[5,16] Bacterial cell cultures are grown aerobically in Luria Broth at 37°C and harvested in mid-exponential phase by centrifugation at 8,000 rpm for 10 min. The cells are washed in phosphate-buffered saline (PBS, pH 7.2) and then stirred into a 2.5% v/v glutaraldehyde solution for two hours at 4°C to a final concentration of 0.6–0.8 mg dry cell weight/ml.[20] The glutaraldehyde solution is prepared from a 25% v/v stock solution diluted to 2.5% v/v in PBS and purified by stirring with 50 mg/mL charcoal at 4°C for 24 h.[20] After treatment with glutaraldehyde, the bacterial cells are rinsed and resuspended in 1 mM Tris buffer (tris{hydroxymethyl}aminomethane, pH 7.5). The cell suspension is incubated at 4°C overnight. Cells are rinsed repeatedly and resuspended in 1 mM Tris (pH 7.5).

Prior to immobilization of bacteria, planar glass substrates and AFM cantilevers are first coated with polyethyleneimine (PEI). 100% PEI stock solution ($M_r$ 1200) is diluted to 1% v/v in distilled, deionized water (ddH$_2$O); pH is adjusted to 8. Glass to be coated with PEI is cleaned by soaking in 1 M HNO$_3$ overnight, rinsing with ddH$_2$O followed by methanol, and finally dried with sterile air. A drop of 1% v/v PEI is placed on one side of the glass and allowed to adsorb for 3 h. The PEI solution is decanted and the glass slides are rinsed in dH$_2$O and stored at 4°C. Standard AFM cantilevers (Digital Instruments, Santa Barbara, CA) are immersed in 1% v/v PEI for 3 h, rinsed in ddH$_2$O and stored at 4°C.

In order to immobilize bacteria onto planar glass substrates, a drop of the glutaraldehyde-treated cell suspension is placed on PEI-coated glass slips which are then placed in a vacuum dessicator at room temperature until excess water has evaporated (2–3 h); the cells themselves must not be dessicated. To immobilize bacteria onto AFM cantilevers, a pellet of glutaraldehyde-treated cells is manually transferred onto PEI coated tips. The pellet is further treated with a drop of 2.5% v/v glutaraldehyde and incubated at 4°C for 1 to 2 h. The cantilevers are rinsed in dH$_2$O and excess water is allowed to evaporate at room temperature.

## B. *AFM Operation*

AFM measurements were performed using a Nanoscope III Contact Mode AFM and Nanoprobe cantilevers with silicon nitride tips (Digital Instruments, Santa Barbara, CA). New, fresh cantilevers were used for every force measurement—cantilevers and substrates were used only once. All force measurements were performed in an AFM fluid cell (Digital Instruments, Santa Barbara, CA) filled with buffer solutions.

Force measurements are carried out by engaging the AFM without touching the tip to the surface thus preventing tip or substrate contamination from the bacteria. The substrate, mounted on a piezo motor, is approached towards the AFM cantilever in 100 nm increments with a specified Z scan size of 300 nm at a frequency of 1Hz. Bacteria immobilized onto planar surfaces are imaged by AFM after every force measurement to confirm the presence of a confluent bacterial lawn. Similarly, bacteria-coated tips are

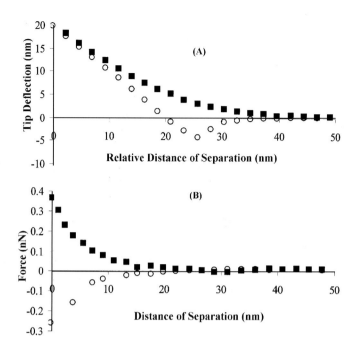

**Figure 3.** Typical AFM results of *E. coli* D21 bacteria interacting with glass (O) or polysty-rene (■) in 1 mM Tris. (A) Results presented as tip deflection (nm) vs relative distance of separation (nm). (B) Results presented as force (nN) vs distance of separation (nm).

imaged by SEM after every force measurements to confirm the presence of a confluent bacterial cell layer on the tip.

### C. Data Presentation and Analysis

Data are acquired in terms of tip deflection (nm) versus relative distance of separation (nm). All deflection curves are normalized so that tip deflection is zero where there is no interaction, and the slope of the constant compliance region (portion of curve where cantilever moves with the surface) is equal to the rate of piezo displacement.[15] Representative curves are plotted together by aligning the zero deflection and constant compliance regions of the force curves to produce tip deflection (nm) versus relative distance of separation (nm). Figure 3A depicts a typical tip deflection versus relative distance of separation curve where D21 bacteria immobilized on the AFM cantilever are attracted to glass and repelled away from a polystyrene surface.

Zero distance of separation, the point of contact between the surface and the tip, is defined as the onset of the constant compliance region. Absolute distance of separation (nm) is calculated as the sum of tip deflection and piezo position relative to zero distance of separation.[15] Force is calculated by treating the cantilever as a spring with a characteristic spring constant (k in nN/nm) according to: $F = k \cdot \Delta Y$ where $\Delta Y$ is the tip deflection.[15] Values for the spring constants of the AFM cantilever used to convert the data are provided by the manufacturer (Digital Instruments, Santa Barbara, CA). Finally, force (nN) is plotted versus absolute distance of separation (nm) according to the method described by Ducker et al. (Fig. 3B).[15]

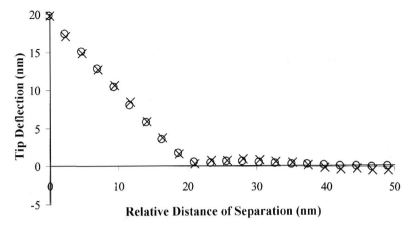

**Figure 4.** Bacterial lawns immobilized on glass are not elastic, and hence, do not deform under the force of the AFM cantilever. In 1 mM Tris, *E. coli* D21e7 (×) immobilized on glass neither attract nor repel the cantilever tip. The resulting force curve is identical to that obtained for a cantilever tip and mica (O) in 1 mM Tris.

## IV. CONTROLS

### A. Immobilization Protocol

Control experiments were performed to ensure that the cell immobilization protocol does not introduce artifacts into AFM force measurements. First, in order to confirm that PEI does not desorb off the glass substrate and contaminate the AFM cantilever tip, standard AFM cantilevers were incubated with PEI coated glass in the AFM fluid cell filled with buffer. Force measurements performed on freshly cleaved mica (a molecularly smooth surface) using a clean tip versus a tip that had been pre-incubated in the presence of PEI coated glass were identical (unpublished data).

Glutaraldehyde treatment is a necessary step in the establishment of stable, confluent, uniform bacterial lawns on glass or AFM cantilevers coated with PEI. Glutaraldehyde crosslinks cellular proteins, but does not react with the lipopolysaccharide and exopolysaccharide molecules on the surface of Gram-negative bacteria. Our studies, as well as previous reports, indicate that contributions by these polysaccharide molecules dominate the long-range interactions between bacteria and substrates.[30,50]

A variety of control experiments were performed to ensure that glutaraldehyde treatment does not alter the physiochemical properties of the bacterial cell surface such as surface free energy and surface charge density. Contact angles which reflect the surface free energy of the bacterial cell surface[8,40] can be measured by sessile drop method on lawns of bacteria filtered onto cellulose acetate membranes.[43] Contact angles measured with polar and apolar liquids were found to be identical for *E. coli* with and without glutaraldehyde treatment.[43] For example, contact angles measured with water for *E. coli* D21 with and without glutaraldehyde treatment are 27±7° and 26±3° respectively. Similarly, zeta potential measurements which reflect the surface charge density of particles do not change with glutaraldehyde treatment.[43] For instance, zeta potentials for *E. coli* D21 are –28.8±0.7 mV before and –28.9±1.7 mV after glutaraldehyde treatment. Finally, varying glutaraldehyde concentration from 2.5 to 5.0% v/v does not alter the ensuing force measurements (unpublished data). Concentrations of glutaraldehyde lower than 2.5 v/v% do not produce stable bacterial lawns, and as a result bacteria detach from the PEI coated surfaces under the force of the AFM probe. Finally, standard bacterial adhesion studies performed in a flow cell confirm that the adhesive behavior of *E. coli* does not change following glutaraldehyde treatment. No differences are observed in the number of cells adhering onto either hydrophilic or hydrophobic surfaces for two strains of *E. coli* with and without glutaraldehyde treatment (unpublished data).

The experiments described above suggest strongly that glutaraldehyde treatment affects neither the tendency of *E. coli* to adhere onto hydrophilic or hydrophobic surfaces nor the surface energy or surface charge density of the bacteria.

### B. Elastic Deformation of Biological Samples

The AFM has been used extensively to measure the elasticity or extent of deformation of biological samples under the force of the AFM probe.[21,28,41,47] With respect to bacterial adhesion studies, however, bacterial lawns proved to be rigid, therefore ensuring that AFM force curves reflect long range interactions between bacteria and substrates and not elastic deformation of the bacterial cells. The petidoglycan layer of Gram-negative and Gram-positive bacteria renders bacterial cells rigid and resistant to deformation.[37]

The elastic deformation of soft biological samples (i.e., proteins and mammalian cells) under the force of the AFM probe is determined by the deviation of the constant compliance region (region where the tip and sample are in contact and move at the same rate) from the expected value.[21,28,41,47] We found the constant compliance region for *E. coli* lawns to be the same as that for mica (Fig. 4). Therefore there is no apparent elastic deformation of bacterial lawns immobilized onto glass substrates due to the increasing load force applied when the sample and tip are in contact. Moreover, in the absence of repulsive electrostatic interactions, the resulting force curves between silicon nitride tips and bacterial lawns resemble those measured between the silicon nitride tip and mica (Fig. 4). This result illustrates that the portion of the force curves prior to the constant compliance region depicts long range attractive or repulsive interactions and not artifacts due to elastic deformation.

## V. INTERPRETATION OF RESULTS

### A. Reproducibility

The AFM-based methodology described herein was found to be highly reproducible. Figure 5A presents tip deflection versus relative distance of separation curves for six different AFM cantilevers coated with D21 bacteria probing clean glass in 1 mM Tris buffer. Clearly the shape and the magnitude of the interaction between bacterial-coated cantilevers and the substrate are readily reproducible. The average and standard deviations of the six curves depicted in Figure 5A are presented in Figure 5B. The standard deviation for interactions measured using the AFM is less than 30%.

### B. Agreement of Two Configurations

Due to similarities in surface characteristics of glass and silicon nitride,[48] the force of interaction between a silicon nitride tip and a lawn of bacteria was expected to agree with the force of interaction between an AFM tip coated with bacteria and planar glass

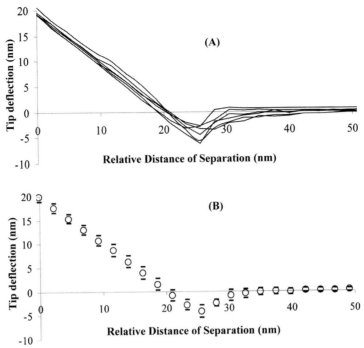

**Figure 5.** Reproducibility of AFM measurements. (A) Multiple AFM force curves from six independent experiments of D21-coated cantilevers interacting with planar glass in 1 m*M* Tris. (B) Average (O) plus/minus one standard deviation (–) for the six curves presented in Fig. 5A.

substrate. Indeed, the force curves for each of the cases were experimentally indistinguishable for *E. coli.*[43]

## VI. LIMITATIONS

The following considerations need to be kept in mind when determining the force of interaction between bacteria and surfaces by atomic force microscopy: 1) In order to analyze AFM force curves, the point of contact between the cantilever and the substrate is assumed, therefore the actual distance of separation is not absolute.[29] Moreover, the exact tip geometry and radius of curvature of the contact area are unknown for not only AFM tips modified with bacteria, but for AFM cantilevers in general.[29] These unknowns make interpretation and quantitative comparison of AFM results to theoretical predictions difficult.[29] 2) A limitation specific to the AFM-based methodology described for bacterial adhesion studies is that surfaces and cantilevers can only be used once, therefore, requiring the preparation of many samples. 3) We were not successful in our efforts to measure reproducible pull-off forces following contact of the bacteria with the surface. Analysis of pull-off forces was difficult due to inconstancies observed in the retraction curves. Hysteresis in retraction curves was used only as an indicator that contact between the cells and the substrates had been achieved during a force measurement. 4) Frequently, in AFM experiments, hydration forces become significant if a shell of ordered water molecules organizes itself at the interface between colloid particles and planar substrates, resulting in monotonic, short-range repulsive interactions.[29] In order to measure the true interaction between surfaces, this hydration shell must be displaced.[27] Repulsive hydration forces were observed for AFM experiments where bacterial lawns immobilized on planar substrates were probed with 30 μm glass or polystyrene beads attached to AFM cantilevers (unpublished data). Direct contact between the bacteria and the beads was never established as determined by the lack of hysterisis in the retraction curves. Hydration forces, however, are negligible when using smaller beads or particles attached to AFM cantilevers, because the magnitude of the hydration force decreases linearly with decreasing the radius of curvature of the apex.[27]

## REFERENCES

1. Basu S, Sharma MM: Measurement of critical disjoining pressure for dewetting solid surfaces. *J Colloid Interface Sci* 181:443–55, 1996
2. Biggs S: Steric and bridging forces between surfaces bearing adsorbed polymer: an atomic force microscopy study. *Langmuir* 11:156–62, 1995
3. Binnig G, Quate CF, Gerber C: Atomic force microscope. *Phys Rev Lett* 56:930–3, 1986
4. Boland T, Ratner BD: Direct measurements of hydrogen bonding in DNA nucleotide bases by atomic force microscopy. *Proc Natl Acad Sci USA* 92:5297–301, 1995
5. Boman HG, Monner DA: Characterization of lipopolysaccharides from *Escherichia coli* K-12 mutants. *J Bacteriol*121:455–64, 1975
6. Bowen WR, Hilal N, Lovitt RW, Wright CJ: Direct measurement of the force of adhesion of a single biological cell using an atomic force microscope. *Colloids Surf A* 136:231–34, 1998
7. Browning-Kelly ME, Wadu-Mesthrige K, Hari V, et al: Atomic force microscopic study of specific antigen/antibody binding. *Langmuir* 13:343–50, 1997
8. Busscher HJ, Weerkamp AH, Mei HC, et al: Measurement of the surface free energy of bacterial cell surfaces and its relevance for adhesion. *Appl Environ Microbiol* 48:980–83, 1984
9. Butt HJ: Electrostatic interactions in atomic force microscopy. *Biophys J* 60:777–85, 1991
10. Butt HJ, Jaschke M, Ducker W: Measuring surface forces in aqueous elelctrolyte solution with the atomic force microscope. *Bioelectrochem Bioenergetics* 38:191–201, 1995
11. Chilkoti A, Boland T, Ratner BD, et al: The relationship between ligand-binding thermo-dynamics and protein-ligand interaction forces measured by atomic force microscopy. *Biophysical J* 69:2125–30, 1995
12. Christensen GD, Baldassarri L, Simpson WA: Methods for studying microbial colonization of plastics. *Methods Ezymol* 253:477–500, 1995
13. Dammer U, Henger M, Anselmetti D, et al: Specific antigen/antibody interactions measured by force microscopy. *Biophysical J* 70:2437–41, 1996
14. Ducker WA, Senden TJ, Pashley RM: Direct measurement of colloidal forces using an atomic force microscope. *Nature* 353:239–41, 1991
15. Ducker WA, Senden TJ, Pashley RM: Measurement of forces in liquids using a force micro-scope. *Langmuir* 8:1831-36, 1992
16. Eriksson-Grennberg KG, Nordström K, Englund P: Resistance of *Escherichia coli* to penicillins. *J Bacteriol* 108:1210–23, 1971
17. Feldman K, Tervoort T, Smith P, Spencer ND: Toward a force spectroscopy of polymer surfaces. *Langmuir* 14:372–78, 1998
18. Fletcher M, Pringle JH: The effect of surface free energy and medium surface tension on bacterial attachment to solid surfaces. *J Colloid Interface Sci* 140:5–13, 1985
19. Florin E-L, Moy VT, Gaub HE: Adhesion forces between individual ligand-receptor pairs. *Science* 264:415–17, 1994
20. Freeman A, Abramov S, Georgiou G: Fixation and stabilization of *Escherichia coli* cells displaying genetically engineered cell surface proteins. *Biotechnol Bioeng* 52:625–30, 1996

21. Goldmann WH, Ezzell RM: Visoelasticity in wild-type and vinculin-deficient mouse embryonic carcinoma cells examined by atomic force microscopy and rheology. *Exp Cell Res* 226:226–37, 1996

22. Gould SAC, Drake B, Prater CB, et al: The atomic force micrscope: a tool for science and inducstry. *Ultramicroscopy* 33:93–8, 1990

23. Gristina AG: Biomaterial-centered infection: microbial adhesion versus tissue integration. *Science* 237:1588–95, 1987

24. Hansma HG, Hoh JH: Biomolecular imaging with the atomic force microscope. *Annu Rev Biophys Biomol Struct* 25:115–39, 1994

25. Hansma HG, Revenko I, Kim K, et al: Atomic force microscopy of long and short double-stranded, single-stranded and triple-stranded nucleic acids. *Nucleic Acids Res* 24:713–20, 1996

26. Hansma PK, Elings VB, Marti O, et al: Scanning tunneling microscopy and atomic force microscopy: applications to biology and technology. *Science* 242:209–16, 1988

27. Ho R, Yuan JY, Shao Z: Hydration force in the atomic force microscope: a computation study. *Biophysical J* 75: 1076–83, 1998

28. Hoh JH, Schoenenberger CA: Surface morphology and mechanical properties of MDCK monolayers by atomic force microscopy. *J Cell Sci* 107:1105–14, 1994

29. Israelachvili J: *Intermolecular and Surface Forces.* 2nd ed. Academic Press Limited, San Diego, CA, 1992

30. Klein NJ, Ison CA, Peakman M, et al: The influence of capsulation and lipooligosaccharide structure on neutrophil adhesion molecule expression and endothelial injury by *Neisseria meningitidis. J Infect Dis* 173:172–9, 1996

31. Klotz SA: Role of hydrophobic interactions in microbial adhesion to plastics used in medical devices. In: Doyle RJ, Rosenberg M, eds. *Microbial Cell Surface Hydrophobicity.* American Society for Microbiology, Washington DC, 1990:107–36

32. Lee GU, Chrisey LA, Colton RJ: Direct measurements of the forces between complementary strands for DNA. *Science* 266:771–3, 1994

33. Loosdrecht MCM, Lyklema J, Norde W, et al: Electrophoretic mobility and hydrophobicity as a measure to predict the initial steps of bacterial adhesion. *Appl Environ Microbiol* 53:1898–901, 1987

34. Loosdrecht MC, Lyklema J, Norde W, et al: The role of bacteria cell wall hydrophobicity in adhesion. *Appl Environ Microbiol* 53: 1893–7, 1987

35. Mozes N, Marchal F, Hermesse MP, et al: Immobilization of microorganisms by adhesion: interplay of electrostatic and nonelectrostatic interactions. *Biotechnol Bioeng* 30:439–50, 1987

36. Mulhern PJ, Blackford BL, Jericho MH, et al: AFM and STM studies of the interaction of antibodies with the S-layer sheath of the archaeobacterium *Methanospirillum hungatei. Ultramicroscopy* 42-44:1214–21, 1992

37. Neidhardt FC, Ingraham JL, Lin EC, et al, eds. *Escherichia coli and Salmonella typherium: Cellular and Molecular Biology.* 2nd ed. ASM Press, Washington DC, 1996

38. Ofek I, Doyle RJ: *Bacterial Adhesion to Cells and Tissues.* Chapman Hall, NY, 1994

39. Ong YL, Razatos A, Georgiou G, Sharma MM: Adhesion forces between *E. coli* bacteria and biomaterial surfaces. *Langmuir* in press, 1999

40. Oss CJ: The forces involved in bioadhesion to flat surfaces and particles – their determination and relative roles. *Biofouling* 4:25–35, 1991

41. Radmacher M, Fritz M, Cleveland JP, et al: Imaged adhesion forces and elasticity of lysozyme adsorbed on mica with atomic force microscopy. *Langmuir* 10:3809–14, 1994

42. Razatos A, Ong YL, Sharma MM, et al: Evaluating the interaction of bacteria with biomaterials using atomic force microscopy. *J Biomater Sci Polymer Educ* 9:1361–73 1998

43. Razatos A, Ong YL, Sharma MM, et al: Molecular determinants of bacterial adhesion monitored by atomic force microscopy. *Proc Natl Acad Sci USA* 95:11059–64, 1998

44. Ros R, Schwesinger F, Anselmetti D, et al: Antigen binding forces of individually addressed single-chain Fv antibody molecules. *Proc Natl Acad Sci USA* 95:7402–5, 1998

45. Sagvolden G, Giaever I, Pettersen EO, et al: Cell adhesion force microscopy. *Proc Natl Acad Sci USA* 196:471–6, 1999

46. Schabert FA, Engel A: Reproducible acquisition of *Escherichia coli* porin surface topographs by atomic force microscopy. *Biophysical J* 67:2394–403, 1994

47. Schoenenberger CA, Hoh JH: Slow cellular dynamics in MDCK and R5 cells monitored by time-lapse atomic force microscopy. *Biophysical J* 67:929–36, 1994

48. Senden TJ, Drummond CR: Surface chemistry and tip-sample interactions in atomic force microscopy. *Colloids Surf A* 94:29–51, 1995

49. Shao Z, Yang J, Somlyo A: Biological atomic force microscopy: from microns to nanometers and beyond. *Annu Rev Cell Dev Biol* 11:241–65, 1995

50. Valkonen KH, Wadström T, Moran AP: Interaction of lipopolysaccharides of *Helicobacter pylori* with basement membrane protein laminin. *Infect Immun* 62:3640–8, 1994

51. Yang J, Tamm LK, Somlyo AP, et al: Promises and problems of biological atomic force microscopy. *J Microscopy* 171:183–98, 1993

52. Yoon RH, Flinn DH, Rabinovich YI: Hydrophobic interactions between dissimilar surfaces. *J Colloid Interface Sci* 185:363–70, 1997

# 19

# Direct Measurement of Long-Range Interaction Forces Between a Single Bacterium and a Substrate Using an Optical Trap

**Richard B. Dickinson, Aaron R. Clapp, and Stephen E. Truesdail**

*Department of Chemical Engineering and NSF Engineering Research Center for Particle Science and Technology, University of Florida, Gainesville, FL, USA*

## I. INTRODUCTION

As discussed in Chapter 2, stable adhesion of a bacterium on a surface requires the following events: transport to the surface, attachment to the surface, and subsequent resistance to detachment. In general, long-range forces (including macromolecular bridging) govern the probability of cell attachment to a surface, whereas short-range forces govern the strength of the ultimate adhesion. Brownian motion, convection or cell motility must transport the cell over any energy barrier created by repulsive forces. The magnitude of these forces need not be large to repel a Brownian particle from a surface. We can roughly estimate the force acting over a certain range that would create an insurmountable potential energy barrier for a Brownian particle as follows: The probability of crossing the energy barrier through energy fluctuations scales with the factor, $\exp(-\phi_{max}/kT)$, where $k$ is Boltzmann's constant ($1.38 \times 10^{-23}$ J/K), and $T$ is absolute temperature (~300 K);[19] therefore, an energy barrier of $10\,kT$ would be sufficient to prevent attachment of a Brownian particle. Assuming the repulsive forces act over a range of around 50 nanometers, then only about one *picoNewton* ($10^{-12}$ Newtons) of force is required to create the $10\,kT$ of work.

Because of the complex macromolecular structure of the bacterial cell surface, quantitative prediction of the magnitude of the long-range interaction forces and their range of interaction is an extremely difficult problem. One approach is to rely on colloid theory such as DLVO theory[8,28] to predict these forces. However, as discussed in Chapter 2, this approach has several pitfalls. The primary reason for the lack of development of more suitable theories is that, until recently, there has been no way to directly measure these long-range forces on bacteria. Razatos et al.[21,20] recently applied atomic force microscopy (AFM) to measure force versus separation distance on bacterial lawns, which is described in the following chapter. AFM is an excellent tool for studying forces involved in bacterial adhesion because of the fine spatial resolution and sensitivity of the

*Handbook of Bacterial Adhesion: Principles, Methods, and Applications*
Edited by: Y. H. An and R. J. Friedman © Humana Press Inc., Totowa, NJ

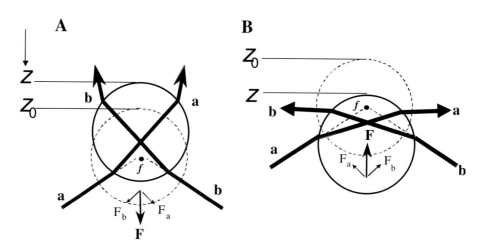

**Figure 1.** Principle of the gradient trapping force from the ray optics perspective. The refraction of rays **a** and **b** of the focused trapping beam result in forces $F_a$ and $F_b$, which sum to the restoring force, $F$. The net force of all rays over the surface of the particle, $F_{trap}(z)$, is always restoring back to the equilibrium position, $z_0$, shown by the dotted line. For small displacements, the trapping force is proportional to the displacement, i.e., $F_{trap}(z) = -\gamma(z-z_0)$, where $\gamma$ is the trap stiffness. (Adapted from ref.[2])

instrument. However, the force sensitivity atomic force microscopy is still limited to, at best, tens of picoNewtons, which is substantially larger than the attractive or repulsive forces on a single bacterium that may be relevant during attachment.

In this chapter, we describe a novel technique for directly measuring the long-range interaction force as a function of separation distance between a single colloidal particle, such as a bacterium, and a test surface.[7] In this approach a single-beam gradient optical trap (three-dimensional optical trap—3DOT or "optical tweezers") is used as a force transducer and evanescent wave light scattering (EWLS) is used to precisely measure the separation distance between the particle and the surface. Like AFM, nanometer spatial resolution is possible with the 3DOT-EWLS technique, but the sensitivity of the optical trap force transducer is significantly improved, down to tens of *femtoNewtons* (1 fN $=10^{-15}$ N). This technique can therefore be used to measure the weaker long-range forces that are relevant to deposition of the particle to the surface.

## II. OPTICAL TRAP FORCE TRANSDUCER

As first demonstrated by Ashkin,[1,3,4] small dielectric particles can be trapped by radiation pressure that is created by focusing a laser beam through a high numerical aperture objective lens. The focused laser creates a three-dimensional intensity gradient around the focal point of the objective. When the particle is displaced from the trap center, a net force is directed back toward the trap center. This force results from a net momentum transfer due to anisotropic scattering of the incident laser radiation and is linearly proportional to the displacement of the particle from the trap center for small displacements, analogous to a spring displaced from equilibrium. From the ray optics perspective as illustrated in Figure 1, the net trapping force is created by the change in

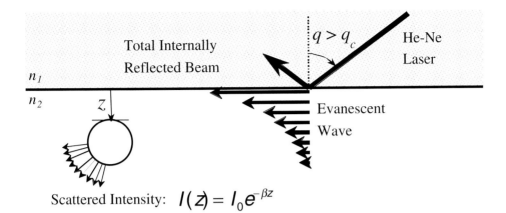

Scattered Intensity: $I(z) = I_0 e^{-\beta z}$

**Figure 2.** Principle of evanescent wave light scattering. An internally reflected laser beam at the solid-fluid interface is totally internally reflected if the angle of incidence is greater than the critical angle of reflection. The reflection creates and evanescent wave that penetrates into the rarer fluid medium. Both the intensity of the evanescent wave and the intensity of scattered light from a particle in the vicinity of the surface decay exponentially with separation distance from the surface.

momentum of light rays as they are refracted over the surface upon crossing the surface of the particle.[2]

Although the trap has both axial (along the axis of propagation of light) and transverse components, only the axial component is used for the force measurement technique described here. One should note that there is also a net force (*scattering force*) in the direction of the propagation of light, but for laser light sufficiently focused through a high numerical aperture objective lens, this force is small compared to trapping force (*gradient force*). The net effect of this scattering force is to shift the equilibrium position slightly in the direction of the light propagation. It has been shown theoretically that the force on a spherical colloidal particle in an optical trap is expected to be linear for deflection distances up to about one particle radius.[2,27] This has led to the application of optical traps as force transducers in several studies.[13,14,26,24] In this linear regime, the axial trapping force is given by $F_{trap}(z) = -\gamma(z-z_0)$, where $\gamma$ is the axial trap stiffness (i.e., spring constant) and $z_0$ is the position of the trap center. The trap potential energy is therefore $\phi_{trap}(z) = \gamma(z-z_0)^2/2$.

### III. EVANESCENT WAVE LIGHT SCATTERING

As illustrated in Figure 2, total internal reflection of light on an interface between a denser and a rarer medium creates an evanescent wave that propagates parallel to the surface with intensity that decays exponentially with distance into the rarer medium. As shown both theoretically[6] and experimentally,[16] the intensity of scattered light, $I(z)$, from a particle in an evanescent wave decays exponentially with the separation distance, $z$, from the reflecting surface; i.e.,

$$I(z) = I_0 exp(-\beta z) \tag{1}$$

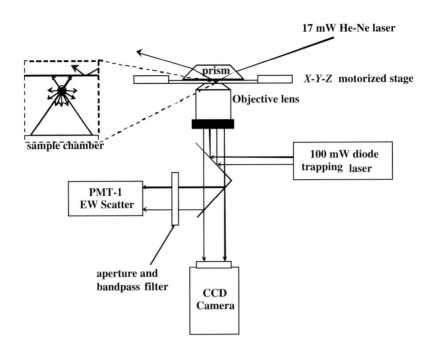

**Figure 3.** Schematic of the apparatus. A single particle is trapped near the test surface with 100 mW diode laser focused through the objective lens of an inverted microscope. The objective lens collects the scattered light from the evanescent wave that is created at the surface from the totally internally reflected 17 mW Helium–Neon laser beam. The separation distance is determined from the measured intensity of scattered light from the evanescent wave. A CCD camera images the experiment. Stepper motors control the X-Y-Z position of the optical trap.

where $I_0$ is the intensity from a particle in contact with the surface ($z = 0$). The evanescent wave decay constant (inverse penetration depth) is calculated by

$$\beta = (4\pi/\lambda_0)[n_1^2 sin^2\theta_i - n_2^2]^{1/2} \tag{2}$$

where $\theta_i$ is the angle of incidence, and $n_1$ and $n_2$ are the refractive indices of the flat plate and the fluid medium, respectively.

A related technique known as total internal reflection microscopy (TIRM) developed by Prieve et al.[17] exploits this known functional relationship between scattered light intensity and separation distance in order to measure interaction potential energy, $\phi(z)$, of conservative forces between a single spherical particle and a surface. The interaction potential is calculated based on the measured stationary distribution of particle positions, $p(z)$, as the particle fluctuates near the surface. Based on Boltzmann's distribution law,

$$p(z) \propto exp(-\phi(z)/kT) \tag{3}$$

where $k$ is Boltzmann's constant, and $T$ is the absolute temperature. As discussed below, this same principle is exploited to determine the potential energy of the optical trap force transducer, thereby calibrating the trap. By exploiting the precise measurement of particle position by evanescent wave light scattering, TIRM has been successfully used to mea-sure interaction potential energies up to ~6 $kT$ for particles larger than 5 μm in diameter.[5,9,10,15,18,22,23]

The novel technique described here combines direct force measurement capability of the three-dimensional optical trap with the precise position measurement of EWLS. In contrast to TIRM, force is measured directly based on the displacement of the most probable position of the particle from the center of the trap. Therefore, the trap serves an analogous function to the cantilever in AFM force measurements. A force–distance profile is generated by scanning the trap position toward the surface.

## IV. APPARATUS

The three-dimensional optical trap/evanescent wave light scattering (3DOT-EWLS) apparatus is illustrated in Figure 3. The sample chamber consists of fluid in a thin gap (generally 10 μm thick) spaced by appropriately sized polystyrene microspheres. The upper surface of the chamber is the test surface and consists of an optically smooth transparent plate (typically a glass microscope slide), which can be surface-modified as desired. The lower surface of the cell is a glass coverslip.

A 100 mW diode trapping laser ($\lambda_0$ = 830 nm; Cell Robotics, Albuquerque, NM) is mounted beneath the objective turret of a Nikon Diaphot 200 inverted microscope and focused through a high numerical aperture (NA) objective lens (Plan Fluor ×100, 1.3 NA, oil immersion; Nikon, Melville, MY) to form the single-beam gradient optical trap. The trapped particle can be manipulated laterally by a motorized stage and axially with a high-resolution stepping motor (Ludl Electronic Products, Hawthorne, NY) to allow the object lens to be moved in 10 nm increments (which corresponds to trap movements of 8.8 nm due the difference in refractive index of the aqueous medium).

The particle position is determined by measuring the scatter of the particle in an evanescent wave at the solid–liquid interface, which is created by total internal reflection of a 17 mW He-Ne laser ($\lambda_0$ = 632.8 nm; Melles Griot, Irvine, CA). A pair of mirrors mounted on an optical post directs this laser beam to the hypotenuse face of a dove prism optically coupled with index-matching oil to the upper plate of the sample chamber. The angle of the top-most mirror is precisely manipulated with a micrometer-driven rotation stage (Melles Griot, Irvine, CA) that adjusts the vertical pitch of the mirror with a resolution of 0.1°. The dove prism will direct the He–Ne laser beam to the appropriate angle for total internal reflection at the test surface. A CCD camera mounted to the eye-port of the microscope is used for visualization of the particles and the image is viewed on a dedicated monitor. The objective lens the collects light scattered by the particle from the evanescent wave. The scattered intensity is measured by a photomultiplier tube (PMT; Oriel Instruments, Stratford, CT) mounted in the side camera port of the inverted micro-scope. The PMT is fitted with a 1 mm diameter circular aperture and interference filters (Oriel Instruments, Stratford, CT) to measure only the scattered light from the particle at the appropriate wavelength. The PMT delivers a continuous voltage signal to a PC data acquisition board (DAS 1801ST, Keithley, Cleveland, OH) for data collection. This provides a real-time measurement of the particle position near the surface.

## V. MEASUREMENT OF FORCE–DISTANCE PROFILE

The optically trapped Brownian particle is subjected to both the trapping force and the surface force. The trap is calibrated in the course of each force measurement by measuring the intensity, $I(t)$, at multiple trap positions that are far from the surface where surface forces are negligible. However, the particle must also be near enough to the flat surface to scatter light from the penetrating evanescent wave to give an accurate measurement of the position.

In the absence of external forces, the particle fluctuates via Brownian motion around the trap center. Boltzmann's distribution law implies that the stationary distribution of particle positions, $p(z)$, is proportional to $exp(-\phi_{tot}(z)/kT)$, where $\phi_{tot}(z)$ is the total potential energy of the particle, which is generally due to both the trap and the surface (gravity is typically negligible but is nonetheless automatically factored into the trap calibration). However, far from the surface where surface forces are negligible,

$$\phi_{tot}(z) = \phi_{trap}(z) = g(z-z_0)^2/2 \tag{4}$$

Hence, $p(z)$ is Gaussian with a stationary mean position, $<z>$, that is equal to the position of the trap center, $z_0$, and the stationary variance, $s_z^2$, is equal to $kT/g$. Because $I(z)=I_0exp(-bz)$, the stationary distribution of $I(t)$ is lognormal with mean, $m \int <I>$, and variance, $s_I^2$, from which $<z>$ and $s_z^2$ (hence $g$ and $z_0$) can easily be calculated based on the properties of the lognormal distribution. Optimal estimates of $g$ and $z_0$ are calculated using a maximum likelihood optimization based on measurements of $m$ and $s_I^2$ over ten to fifty consecutive trap positions as the trap is stepped toward the surface, as described by Clapp et al.[7]

As the position of the optical trap approaches the surface, the surface force eventually becomes significant, and the total potential energy of the particle is the sum of the trap potential plus the interaction potential energy, i.e., $\phi_{tot}(z) = \phi(z) + \phi_{trap}(z)$. Using statistical techniques detailed in Clapp et al.[7] the maximum in the distribution $p(z)$ (i.e., the 'mode', $z_p$) is determined at each trap position. Because of Boltzmann's distribution law, the mode, $z_p$, also corresponds to the minimum in $f_{tot}(z)$, which is the separation distance where the net surface force exactly balances the trapping force. This can be shown by

$$\phi'_{tot}(z_p) = \phi'(z_p) + \phi'_{trap}(z_p) = -F(z_p) - F_{trap}(z_p) = 0 \tag{5}$$

The surface force at $z_p$ is then calculated from the force balance, $F(z_p) = \gamma(z_p-z_0)$. The calculation is performed at each new trap position, which ultimately yields a force-distance profile. The potential energy profile can then be estimated by numerical integrating the force data.

## VI. CORRECTION FOR NOISE

Two significant sources of signal noise must be addressed: background noise (so-called "dark noise") and electronic shot noise. Over the range of measurement, these noise contributions combined are generally less than 5% of the total voltage signal. Nonetheless, a correction for background and shot noise is made to improve the accuracy of the calibration and force measurements. The mean and variance of the dark noise are

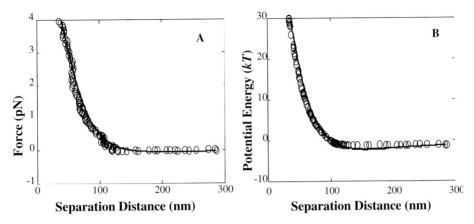

**Figure 4.** The measured force and potential energy between a 1.5 μm silica microsphere and a flat glass surface in 0.1 mM NaCl aqueous solution are plotted as a function of (relative) separation distance. The solid lines represent predictions from DLVO theory. The potential energy curve was obtained by numerically integrating the force data. The solid lines represent fit of theoretical curves using DLVO theory (Hogg-Healy-Fuerstenau approximation).[12] Because $I_0$ (hence absolute distance) was not known for these measurements, the data points were shifted on the abscissa to overlay the theoretical curve.

estimated from the intensity measurements with the particle far from the surface where $I(t) \sim 0$. In the trap calibration, the background values are simply subtracted from mean and variance of the total measured intensity. The variance in the PMT voltage due to shot noise has been found proportional to the mean signal (minus mean background signal),[7] as expected theoretically.[11] Because the signal decreases exponentially with separation distance, this effect becomes more significant at larger separation distances. The proportionality constant, $c$, can be determined by measuring the signal variance at various EW intensities using fully attached particles which have no intensity fluctuations due to Brownian motion.[7] Because $c$ is only a function of the PMT power and pre-amplifier settings, it need not be re-estimated for each sample. The signal variance due to particle fluctuations alone was obtained by subtracting the variance due to shot noise.

## VII. RESULTS AND DISCUSSION

Figure 4 shows example data of force measurement between a 1.5 mm diameter silica particle probed against glass in 0.1 mM electrolyte solution. Each force data point corresponded to 16384 intensity measurements at the trap position taken at 0.5 ms intervals. As shown by the solid line representing predictions from DLVO theory,[8,28] the technique accurately measures the expected exponential decay due to the overlapping counter-ion clouds of the electrical double layers. Although the force is measured directly, the potential energy can be approximated by numerically integrating the force data. In the data shown here, $\phi(z)$ was obtained by numerically integrating a spline interpolation of the data points. In low electrolyte solution, the repulsive force due to electrical double layer decays exponentially with the Debye length as the characteristic decay length. The Debye length can be precisely calculated from the measured conductivity of the solution and does not depend on the surface properties of the particle

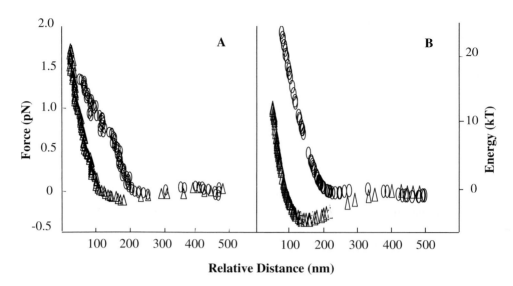

**Figure 5.** The measured force and potential energy between a *Staphylococcus aureus* bacterium and a glass surface is plotted as a function of relative separation distance in aqueous electrolyte solutions of 0.15 mM KCl (circles) and 0.92 mM KCl (triangles).

or flat plate; therefore, the measured decay length of the repulsive part of the force distance curve is a robust measure of accuracy. The experimentally measured repulsive electrostatic force consistently matched closely with the predicted force, as in the example shown in Figure 4. Furthermore, the exponential force profile continues for particle displacements from the trap center up to about a particle radius, suggesting trap linearity over this range.

The measurement precision depends strongly on the number of intensity samples, the total time of measurement at any trap position, and the total curvature of the potential energy well (due to both trap force and surface force). For the prototype system, 16384 intensity samples at 0.5 ms intervals was sufficient to provide ~3 nm precision in particle positions for polystyrene and silica particles.

One challenge in using evanescent wave light scattering to determine absolute distance is to estimate the intensity at zero separation, $I_0$ (*see* Eq. [1]). Without $I_0$, only relative separation distances can be measured. For systems where the particle can contact the flat plate, $I_0$ can be measured by addition of sufficient electrolyte to eliminate repulsive forces, allowing the particle to fully contact the surface. However, for many systems of interest, such as those with macromolecules in the interface between the particle and surface, this approach is not possible because full contact between the particle and the surface may not be possible for direct measurement of $I_0$ (i.e., the particle cannot be "salted out" by adding electrolyte). One approach is to assume that scattering intensities of the particles are sufficiently homogeneous such that $I_0$ can be calibrated under conditions where attachment is possible; the calibrated value $I_0$ would then be used for subsequent measurements. However, for bacteria, this is not possible because the macromolecular surface structure is an inherent part of the system and "zero separation distance" lacks a clear reference point. Therefore, one must be satisfied with measurements of forces versus relative separation distance.

Although most of the development of work of the 3DOT-EWLS technique has been with relatively ideal silica or polystyrene microspheres, we are currently exploring the range of applicability for real bacteria. The theoretical basis of linear trapping force and the exponential decay of the scattered intensity require a spherical particle; therefore, measurements to date have focused on spherical bacteria such as *Staphylococcus aureus*. Figure 5 shows example data for the measured force between a *S. aureus* bacterium and a glass surface at two different electrolyte concentrations. Because the maximum force is quite weak, we do not expect significant cell deformation as the cell probes against the surface, although the compression of the glycocalyx or other surface appendages may be a significant factor in the force measurements. This may be the explanation for the observation that measured repulsive force generally decays more slowly with separation distance as compared to DLVO theoretical predictions. The stiffness of the trap is smaller for bacteria relative to silica and polystyrene; therefore, the magnitude of measurable forces is somewhat diminished. However, this could be remedied in part by a higher power laser for the optical trap, which would enhance the trap stiffness, $\gamma$.

A common concern in optical trapping experiments, especially for biological particles, is the effect of laser heating of the particle. However, the growing literature on the effect of heating due to absorption on the trapping force on biological particles suggests that heating is minimal at near-infrared wavelengths and moderate laser intensity. [25]

We are currently extending the technique to use for measuring viscous forces and macromolecular bridging interactions, using both real bacteria and model cells. The versatility and capability of this sensitive force-measurement technique remains to be fully explored. Because it is uniquely capable of measuring the weaker long-range interaction forces involved in the initial attachment of a bacterium to a surface, we anticipate that 3DOT-EWLS will prove to be a useful tool in investigating bacterial adhesion mechanisms.

**Acknowledgments:** This work was generously supported by a National Science Foundation CAREER Award, #BES-9704236, and the NSF Engineering Research Center for Particle Science and Technology at the University of Florida.

## REFERENCES

1. Ashkin A: Acceleration and trapping of particles by radiation pressure. *Phys Rev Lett* 19:283–5, 1970
2. Ashkin A: Forces of a single-beam gradient laser trap on a dielectric sphere in the ray optics regime. *Biophys J* 61:569–82, 1992
3. Ashkin A, Dziedzic JM, Bjorkholm JE, et al: Observation of a single-beam gradient force optical trap for dielectric particles. *Optics Letters* 11:288–90, 1986
4. Ashkin A, Dziedzic JM, Yamane T: Optical trapping and manipulation of single cells using infrared laser beams. *Nature* 330:769–71, 1987
5. Bike SG, Prieve DC: Measurement of double-layer repulsion for slightly overlapping counterion clouds. *Int J Multiphase Flow* 16:727–40, 1990
6. Chew H, Want DS, Kerker M: Elastic scattering of evanescent electromagnetic waves. *Appl Optics* 18:2679–87, 1979
7. Clapp AR, Ruta AG, Dickinson RB: Three-dimensional optical trapping and evanescent wave light scattering for direct measurement of long range forces between a colloidal particle and a surface. *Rev Sci Inst* 70(6):2627–36, 1999

8. Derjaguin BV, Landau LD: Theory of the stability of strongly charged lyophobic sols and of the adhesion of strongly charged particles in solutions of electolytes. *Acta Phys Chim* (USSR) 14:633, 1941

9. Flicker SG, Bike SG: Measuring double layer repulsion using total internal reflection microscopy. *Langmuir* 8:257–62, 1993

10. Frej NA, Prieve DC: Hindered diffusion of a single sphere very near a wall in a nonuniform force field. *J Chem Phys* 98:7552–64, 1993

11. Gardiner CW: *Handbook of Stochastic Methods*. Springer Series in Synergetics, ed. H. Haken. Vol. 13, Springer-Verlag, Berlin Heidelberg, 1983

12. Hogg R, Healy TW, Fuerstenau DW: Mutual coagulation of colloidal dispersions. *Trans Faraday Soc* 62:1638–51, 1960

13. Kuo SC, Sheetz MP: Force of single kinesin molecules measured with optical tweezers. *Science* 260:232–4, 1993

14. Liang H, Wright WH, Rieder CL, et al: Directed movement of chromosome arms and fragments in mitotic newt lung cells using optical scissors and optical tweezers. *Exp Cell Res* 213:308–12, 1994

15. Liebert RB, Prieve DC: Species-specific long range interactions between receptor/ligand pairs. *Biophys J* 69:66–73, 1995

16. Meixner AJ, Bopp MA, Tarrach G: Direct measurement of standing evanescent waves with a photon tunneling microscope. *Appl Optics* 33:7995–8000, 1994

17. Prieve DC, Frej NA: Total internal reflection microscopy: A quantitative tool for the measurement of colloidal forces. *Langmuir* 6:396–403,1990

18. Prieve DC, Luo F, Lanni F: Brownian motion of a hydrosol particle in a colloidal force field. *Faraday Discuss Chem Soc* 83:297–307, 1987

19. Prieve DC, Ruckenstein E : Effect of London forces upon the rate of deposition of Brownian particles. *AI Ch E J* 20:1178–87, 1974

20. Razatos A, Ong YL, Sharma MM, et al: Evaluating the interaction of bacteria with biomaterials using atomic force microscopy. *J Biomater Sci Polym Ed* 9:1361–73, 1998

21. Razatos A, Ong YL, Sharma MM, et al: Molecular determinants of bacterial adhesion monitored by atomic force microscopy. *Proc Natl Acad Sci U S A* 95:11059–64, 1998

22. Robertson SK, Bike SG: Quantifying cell-surface interactions using model cells and total internal reflection microscopy. *Langmuir* 14:938–4, 1998

23. Sharma A, Walz JY: Direct measurement of depletion interaction in a charged colloidal dispersion. *J Chem Soc Faraday Trans* 92:4997–5004, 1996

24. Simmons RM, Finner JT, Chu S, et al: Quantitative measurements of force and displacement using an optical trap. *Biophys J* 70:1813–22, 1996

25. Svoboda K, Block SM: Biological applications of optical forces. *Ann Rev Biophys Biomol Struct* 23:247–85, 1994

26. Svoboda K, Block SM: Force and velocity measured for single kinesin molecules. *Cell* 77:773–84, 1994

27. Tlusty T, Meller A, Bar-Ziv R: Optical gradient forces of strongly localized fields. *Phys Rev Letters* 81:1738–41, 1998

28. Verway EJW, Overbeek JTG: *Theory of Stability of Lyophobic Colloids*. Elsevier, Amsterdam, The Netherlands, 1948

# PART IV

## STUDYING MICROBIAL ADHESION TO BIOMATERIALS

# 20

# Staphylococcal Factors Involved in Adhesion and Biofilm Formation on Biomaterials

**Dietrich Mack, Katrin Bartscht, Sabine Dobinsky, Matthias A. Horstkotte, Kathrin Kiel, Johannes K.-M. Knobloch and Peter Schäfer**

*Institut for Medizinische Mikrobiologie und Immunologie, Universitäts-Krankenhaus Eppendorf, Hamburg, Federal Republic of Germany*

## I. INTRODUCTION

Today *Staphylococcus aureus* and coagulase-negative staphylococci both rank among the five most frequent causative organisms of nosocomial infections.[144,145] There has been a tremendous increase in the incidence of nosocomial sepsis, beginning in the 1980s. This has resulted mainly from the increase of cases caused by coagulase-negative staphylococci, mostly *Staphylococcus epidermidis*, and to a more modest degree by *S. aureus*.[29] The reasons for these changes are probably multifactorial and include treatment of more severely ill patients and their longer survival, increasingly invasive procedures used in the management of these patients, and a tremendous increase in use of indwelling medical devices and prostheses in modern medicine.[139]

*S. aureus* is a frequent pathogen causing a variety of severe, often life-threatening infections such as sepsis, infectious endocarditis, pneumonia, osteomyelitis, surgical wound infections, and staphylococcal toxic shock syndrome.[144] *S. aureus* is a common causative organism of infections related to intravascular catheters and other implanted medical devices, ranking second only to *S. epidermidis*.[8]

*S. epidermidis*, the most prominent coagulase-negative staphylococcal species, colonizes human skin and mucous membranes and exhibits a rather low pathogenic potential in the normal human host.[69,106] Most infections caused by this organism are related to implanted medical devices like intravascular and peritoneal dialysis catheters, cerebrospinal fluid shunts, prosthetic heart valves, prosthetic joints, vascular grafts, cardiac pacemakers and intraocular lenses.[29,70,123,127] These infections are nearly always nosocomial in origin, cause a substantial additional morbidity and mortality, and lead to increased cost of hospital stay.[58,90,127] The infecting organisms normally belong to the endogenous microflora of the patient's skin or mucous membranes.[127]

Therapy of prosthetic device-related infections is problematic due to the increasing incidence of multiple antibiotic resistance of staphylococcal nosocomial isolates and the

*Handbook of Bacterial Adhesion: Principles, Methods, and Applications*
Edited by: Y. H. An and R. J. Friedman © Humana Press Inc., Totowa, NJ

frequent clinical ineffectiveness of antibiotics tested as sensitive in vitro, which regularly necessitates removal of the implanted prosthetic device for successful therapy.[1,13,20,21,25,125,157]

The cause of the pronounced virulence of *S. epidermidis* in the setting of prosthetic device-related infections has therefore attracted considerable attention in the last decade. In parallel, mechanisms leading to colonization of host and polymer surfaces by *S. aureus* were intensively explored. Considerable progress has been made in deciphering the molecular mechanisms leading to surface colonization by staphylococci, which will be discussed in the following.

## II. SURFACE COLONIZATION BY STAPHYLOCOCCI

Using scanning electron microscopy, *S. epidermidis* was shown to colonize intravascular catheters in large adherent biofilms composed of multilayered cell clusters embedded in an amorphous extracellular material, which consists of exopolysaccharides referred to as slime or glycocalyx.[16,23,36,89,120,121] Most strikingly, the majority of staphylococcal cells in these biofilms have no direct contact with the polymer surface, indicating that the cells have to express intercellular adhesion to reside in the adherent biofilm. Similarly, multilayered biofilms containing staphylococcal cells encased into an amorphous matrix were observed on the surface of catheters, pacemaker leads and peritoneal dialysis tubing infected by *S. aureus*.[87-89,124]

After implantation, native polymer surfaces become rapidly modified by adsorption of host derived plasma proteins, extracellular matrix proteins, and coagulation products, i.e., platelets and thrombi, followed often by even more intense integration of the foreign-body implant into host tissue.[38,44] It is believed that attachment of *S. aureus* to implanted biomaterials proceeds primarily through the specific recognition of host protein factors deposited on the surface of the polymer by specific receptors in the bacterial cell surface.[151] However, a significant contribution of attachment to native polymer surfaces cannot be completely excluded. Biofilm accumulation as seen in scanning electron micrographs of attached *S. aureus* cells may proceed later by aggregation of proliferating cells mediated by host derived protein factors. Certain *S. aureus* strains were reported to produce biofilm in vitro using similar conditions as for *S. epidermidis*.[5]

In vitro a proportion of *S. epidermidis* strains are able to produce a macroscopically visible, adherent biofilm on test tubes or tissue culture plates with a morphology in scanning electron micrographs very similar to that of infected intravascular catheters.[6,16,17,61,134] This phenotype is often referred to as slime production.[16] However, as the term slime defines exopolysaccharides of bacteria noncovalently attached to the cell wall and is used for different phenomena by different authors[7,16,62,122] this phenotype is now often referred to as biofilm formation. In this article, these strains are differentiated by the term biofilm-producing as opposed to biofilm-negative strains.

It was proposed that biofilm formation is essential for the pathogenicity of *S. epidermidis*. This hypothesis was supported by several clinical studies which revealed that coagulase-negative staphylococci, which were biofilm-producing as detected with the tube-test[16] or its semi-quantitative modifications using 96-well tissue culture plates,[17] were significantly more often related to significant infections than biofilm-negative strains.[19,158] It has therefore been believed that the specific virulence of *S. epidermidis* in device-related infections is linked to an unusual ability to colonize polymer surfaces.

# Phases of biofilm formation

**Phase 1:**
primary attachment by
nonspecific and specific
adhesion mechanisms

**Phase 2:**
bacterial accumulation
glycocalyx formation
intercellular adhesion

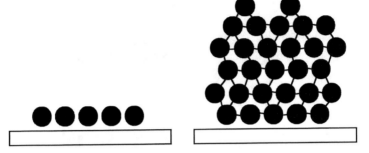

**Figure 1.** Phases of biofilm formation of staphylococci.

The formation of biofilm on a polymer surface by staphylococci can be separated into two main phases: 1) rapid primary attachment to the polymer surface followed by 2) biofilm accumulation in multilayered cell clusters on the polymer surface, which requires the expression of intercellular adhesion.[80] The latter phase is often paralleled by production of a matrix or glycocalyx, which encases the bacterial cells (Fig. 1).

## III. COLONIZATION OF BIOMATERIALS BY *S. AUREUS*

### A. Attachment of S. aureus *to Biomaterial Surfaces*

#### 1. Adhesion to Host Protein Factors by S. aureus

Attachment mechanisms vary with respect to modification of the polymer surface with host protein factors and native surfaces. Most studies investigating staphylococcal adhesion to polymer surfaces indicate that *S. aureus* and coagulase-negative staphylococci attach much better to native, unmodified polymer surfaces as compared to the same surface modified with serum, plasma, or albumin.[31,51,71,91,119,147,148] Despite this fact, adhesion of *S. aureus* to polymer surfaces modified with host proteins is thought to be most important in the pathogenesis of medical device-related infections.[151] The investigation of adhesion of *S. aureus* to polymer surfaces has therefore mainly concentrated on the specific interaction of this organism with host protein factors of the extracellular matrix, plasma, and other origins, potentially involved in surface colonization.

After the initial observation by Kuusela[72] that fibronectin is specifically bound by *S. aureus* in solution, similar specific binding of several other host matrix proteins to *S. aureus* in solution was demonstrated including collagen,[137] laminin,[77] bone sialoprotein,[132] thrombospondin,[52] von Willebrand factor,[54] vitronectin,[15,75] and elastin.[110] Specific binding of *S. aureus* to fibrinogen leading to clumping of the cells in plasma is used in the clinical microbiology laboratory for routine identification of this organism. For most of these protein factors it was shown that they promote bacterial attachment in vitro to polymer surfaces coated with purified fibronectin,[51,73,131,146-150] fibrinogen,[14,27,51,53,73,149,150] collagen,[98,141] vitronectin,[39] laminin,[39,51] thrombospondin,[52] and von Willebrand factor.[54] The quantity of attached bacteria was linearly dependent on the concentration of the respective matrix protein used for coating the respective polymer surface in the range of one to 100 µg/ml of fibronectin,[39,91,148,150] fibrinogen,[92,150] thrombospondin,[52] vitronectin,[39] and von Willebrand factor.[54] When attachment of *S. aureus* Cowan I to polymethylmethacrylate (PMMA) coverslips coated with equal concentrations of different matrix proteins was compared, about tenfold more bacteria adhered to fibronectin- and fibrinogen-coated coverslips as compared to cover slips coated with laminin and thrombospondin.[52] Attachment to vitronectin coated surfaces was fivefold less as compared to thrombospondin.[52] Similar relative adherence promotion was observed with several clinical *S. aureus* isolates from catheter related infections for fibronectin, fibrinogen, and laminin.[51] Apparently, significant concentration-dependent attachment of *S. aureus* to polymer surfaces is promoted by the different matrix proteins, which could therefore be of relevance in the pathogenesis of *S. aureus* foreign body related infections.

To ascertain the clinical relevance of attachment mediated by certain matrix proteins, attachment of *S. aureus* strains was evaluated on inserted catheter segments recovered from patients. Attachment of *S. aureus* was significantly promoted on inserted catheters compared to albumin-blocked control catheters.[149-151] Attachment to inserted catheters varied with the respective biomaterial and the amount of immunologically detected fibronectin, but not fibrinogen.[26,150] In contrast, attachment of *S. aureus* to polyvinyl chloride or polyethylene catheter tubing exposed to canine blood for a short time of up to 6 h depended significantly on adsorbed fibrinogen and to a much lesser extent on fibronectin.[152] These results correlate favorably with the observation that fibronectin and also thrombospondin adsorb poorly to polymethylmethacrylate in the presence of serum, but that adsorption of fibronectin under these conditions is greatly increased by prior adsorption of collagen to the polymer surface.[52,148] These results indicate that early after implantation fibrinogen specific adhesins promote attachment of *S. aureus* to biomaterials. With increasing time of contact with host factors adsorption of fibronectin is promoted by the changing conditioning film on the polymer surface, whereas the attachment promoting activity of fibrinogen is inactivated by proteolytic breakdown.[151] From these studies it appears that fibronectin, fibrinogen, collagen, thrombospondin, and activated platelets are most important for attachment of *S. aureus* to polymer surfaces in a clinical setting. However, the contribution of the other matrix proteins remains to be determined.

## 2. Specific S. aureus *Adhesins for Host Protein Factors*

Several factors relevant for the adhesion of *S. aureus* to various matrix proteins have been characterized at a molecular level (Table 1). Several of these factors are protein adhesins of the MSCRAMM (microbial surface components recognizing adhesive matrix molecules) family, which in most cases are covalently anchored to the cell wall peptidoglycan.[35,102,117]

### A. FIBRONECTIN BINDING PROTEINS OF S. AUREUS

Using affinity chromatography, a 210-kDa protein was isolated from a lysate of *S. aureus* strain Newman, which apparently represented a fibronectin binding protein.[37] The

**Table 1. Adhesion Factors of *Staphylococcus aureus***
**Mediating Attachment to Polymer Surfaces**

| Host factor | S. aureus binding protein | Gene identified | Binding in fluid phase | Attachment to solid phase |
|---|---|---|---|---|
| Fibronectin | FnBPA | *fnbA* | Yes | Yes |
|  | FnBPB | *fnbB* | Yes | Yes |
| Fibrinogen/fibrin | ClfA | *clfA* | Yes | Yes |
|  | ClfB | *clfB* | Yes | Yes |
| Collagen | Cna | *cna* | Yes | Yes |
| Elastin | EbpS | *ebpS* | Yes | n.t. |
| Vitronectin | 60 kDa | n.r. | Yes | Yes |
| Bone sialoprotein | 97 kDa | n.r. | Yes | n.t. |
| Multiple matrix proteins | Map 60–72 kDa | *map* | Yes | n.t. |
|  | Eap 60 kDa | *eap* | Yes | Yes |
| von Willebrand factor | n.r. | n.r. | Yes | Yes |
| Laminin | n.r. | n.r. | Yes | Yes |
| Thrombospondin | n.r. | n.r. | Yes | Yes |

n.r.: identification of specific protein or gene not reported
n.t.: not tested

high molecular weight of this receptor correlated favorably with an earlier report on isolation of a 198-kDa fibronectin binding protein.[30] These observations led to the molecular cloning of a gene *fnbA* of a fibronectin binding protein of *S. aureus* 8325-4.[33] FnbA codes for a protein (FnBPA) containing 1018 amino acids with a predicted mol wt of 108 kDa, which exhibits fibronectin binding activity when expressed in *Escherichia coli*.[33,136] A gene (*fnbB*) for a second fibronectin binding protein (FnBPB) was discovered 682 nucleotides downstream the *fnbA* gene.[64] The sequences of both genes resembled the structure of typical Gram-positive cell surface proteins.[64,102,136] An amino-terminal signal sequence is followed by the large A domains, which display only 45% homology between the two fibronectin binding proteins. Two B repeats are present in *fnbA,* but are not detected in the *fnbB* gene. In contrast, the C and D domains and the wall-spanning regions required for anchoring the receptor in the cell wall are very similar in *fnbA* and *fnbB*.[64] Binding to fibronectin occurs via the D region containing 3 to four 35-amino acid repeats to the amino terminal 29 kDa fragment of the fibronectin molecule.[96] In the heparin binding peptide near the C-terminus of fibronectin there is a second binding site for the staphylococcal fibronectin binding protein, which is functional only in fibronectin attached to a solid support.[11] FnBPB has a second region binding to fibronectin in addition to the binding sites located in the D repeats.[64] Insertional inactivation of only one of the two fibronectin binding proteins did not alter the fibronectin binding phenotype of the respective *S. aureus* strains. However, inactivation of both genes led to an adhesion negative phenotype.[42] Not all clinical *S. aureus* isolates possess both fibronectin binding proteins, which is exemplified by the observation that a single Tn918 insertion in the promoter region of *fnbA* significantly inactivated fibronectin binding of *S. aureus* 879R4S.[74,152] In this strain only a single gene of a fibronectin binding protein homologous to *fnbA* could be detected.[43] From these studies it is clear that *fnbA* and *fnbB* are able to mediate attachment of *S. aureus* to biomaterials coated with fibronectin. There is debate whether anti-FnBP-antibodies could be useful for blocking attachment of *S. aureus* to fibronectin in an approach to prevent infection. This is due to the observation that sera of patients after *S. aureus* infection contained antibodies against the ligand binding domains of FnBPs; however, these did not inhibit fibronectin binding.[12,34,35]

## B. Fibrinogen Binding Proteins of *S. aureus*

In a search for a specific receptor mediating adhesion of *S. aureus* to fibrinogen and clumping in plasma, three fibrinogen-binding proteins with mol wt 87000, 60000, and 19000 were identified.[9] The 87-kDa protein represented coagulase of *S. aureus* and the 60-kDa protein also had coagulase activity. All three proteins were primarily secreted into the culture medium. However, it was postulated that the clumping factor was a cell bound form of coagulase. McDevitt et al.[92,93] presented evidence that coagulase and clumping factor are different molecular entities, as an isogenic Tn917 transposon mutant unable to clump with fibrinogen still produced coagulase, whereas a mutant with an inactivated coagulase gene constructed by allelic replacement mutagenesis was still able to clump. Expression of the cloned clumping factor *clfA* revealed a protein with apparent mol wt 130000. From the nucleotide sequence a protein with 896 amino acids and a deduced mol wt 92000 was predicted.[93] Recently, a second gene *clfB* homologous to *clfA* was identified and, in contrast to ClfA, its gene product was only detectable in the early exponential phase.[104] Both proteins have a N-terminal signal sequence followed by a 540 amino acid residue A domain, which is rather divergent between *clfA* and *clfB* with only 26% identical amino acids. The A domain, which contains the active binding domain for fibrinogen,[94] is linked to the wall and membrane spanning domains of the protein through a unique Ser-Asp-dipeptide domain of 272 amino acid residues, which is essential for the functional presentation of the A domain on the bacterial surface.[45] The clumping factor is covalently linked to the cell wall via the common LPXTG-motif.[102] A *clfA*-negative mutant did not bind to fibrinogen-coated polymethylmethacrylate coverslips.[93,152] When expressed, ClfB also mediated adherence to immobilized fibrinogen.[104] As no differences were noted regarding binding to immobilized fibrinogen between wild-type *S. aureus* and a mutant with a genetically inactivated coagulase gene, it seems clear that binding of *S. aureus* to biomaterials mediated by fibrinogen occurs essentially by ClfA. However, when ClfB is expressed in vivo this receptor may also contribute to attachment. Using a probe for the unique Ser-Asp dipeptide repeat region, another gene locus *sdrCDE* was identified which contained three genes homologous to *clfA* and *clfB*.[66] The derived proteins could represent specific receptors for which the respective ligands have not yet been identified. Another gene *fib*, now referred to as *efb*, encoding the 19-kDa fibrinogen-binding protein of *S. aureus* Newman revealed homology to the fibrinogen-binding domain of coagulase.[10,154] Efb, similar to coagulase, probably does not contribute significantly to attachment of *S. aureus* to fibrinogen modified surfaces.

## C. Specific Adhesins for Other Host-Derived Proteins

A 135-kDa collagen receptor was purified from *S. aureus* Cowan I.[140] Using antibodies raised against this protein a gene *cna* encoding the receptor was cloned and sequenced.[114] *Cna* encodes a protein containing 1185 amino acids, which is composed of a N-terminal signal sequence followed by a large A domain. This A domain is linked via one to four B repeats to the cell wall and membrane spanning domains containing the LPXTG motif for covalent cell wall linkage.[102] The collagen receptor mediates binding of receptor-positive *S. aureus* strains to cartilage and collagen-coated microtiter wells, which is inhibited by specific antibodies against the collagen receptor.[141] Attachment of *S. aureus* cells to cartilage surfaces was completely inhibited in isogenic mutants in which the collagen receptor was inactivated by allelic replacement mutagenesis.[116] Attachment to collagen-coated surfaces was stable even under conditions of sheer stress.[98] The binding site for collagen is localized in the A domain of the receptor protein.[115,118] In almost all clinical *S. aureus* isolates collagen binding is associated with the presence of the single *cna* gene.[41] At present, it is not known whether specific adhesion to collagen contributes to the pathogenesis of foreign body infections early after implantation. However, it seems reasonable to speculate that collagen may contribute significantly after prolonged insertion of biomaterials in the host by either firmly attaching fibronectin in the presence of serum proteins[148,150] or by directly promoting adhesion in collagen receptor positive *S. aureus* strains.

A 25-kDa elastin-binding protein was purified from *S. aureus* by affinity chromatography, which mediated specific binding of elastin to *S. aureus* in solution.[110] Cloning of the respective gene *ebpS* revealed an acidic protein composed of 202 amino acids with a predicted mol wt 23000.[111] Using a synthetic peptide approach, the binding site for elastin was mapped to amino acid residues 14 to 34 of EbpS.[112] However, as specific attachment of *S. aureus* to elastin coated surfaces has not yet been reported, the possible significance of elastin binding of *S. aureus* in biomaterial related infections remains to be determined.

60- and 72-kDa proteins of *S. aureus* strains have been described which exhibit broad binding specificity leading to interaction with several matrix proteins including fibrinogen, fibronectin, thrombospondin, vitronectin, and collagen in ligand blotting experiments.[95] Molecular cloning of the gene of this protein called Map (MHC class II analogous protein) from *S. aureus* FDA 574 revealed a protein of 689 amino acids with a predicted molecular mass of 77 kDa, which is dominated by six repeats of a 110 amino acid long domain.[65] The typical C-terminal sequence representing a cell wall anchor region was not detected, indicating that the protein might be secreted from the cell. Recently, the secreted 60 kDa fibrinogen binding protein of *S. aureus* Newman described earlier,[9] now referred to as Eap (extracellular adherence protein), was found to bind to fibrinogen, fibronectin, prothrombin, and other unidentified plasma proteins.[108,109] The amino-terminal amino acid sequence of Eap had homology with Map. Eap agglutinated *S. aureus* cells and actually enhanced binding of *S. aureus* to fibroblasts and epithelial cells. Eap coated to a plastic surface led to surface attachment of *S. aureus*. The purification of a 60 kDa vitronectin binding protein from *S. aureus* strain V8 was reported.[76] Due to its amino acid composition this vitronectin binding protein could be differentiated from Map. The significance of this adhesin class including Map and Eap for biomaterial related infections remains to be determined.

A 97-kDa protein was purified from *S. aureus*, which specifically binds to bone sialoprotein.[156] A synthetic peptide containing a Leu-Lys-Arg sequence, which comprises residues 56 to 65 of bone sialoprotein apparently included the staphylococcal binding site.[133] Although it can be anticipated that specific binding of *S. aureus* to bone sialoprotein might be of relevance in infections involving prosthetic joints no data have been reported to date, which indicates that *S. aureus* indeed adheres to surface attached bone sialoprotein.

### B. Accumulation of S. aureus *into a Biofilm*

Electron micrographs of biomaterials infected by *S. aureus* clearly show accumulation of *S. aureus* cells in multiple layers.[87-89,124] Although aggregation of *S. aureus* cells by matrix proteins found in plasma could easily lead to cell accumulation in these *S. aureus* biofilms in vivo, it is very interesting that a proportion of *S. aureus* strains were reported to produce biofilm on polystyrene in vitro using similar conditions as for *S. epidermidis*.[5] This observation was supported by the fact that almost all *S. aureus* strains examined were reported to contain a gene locus homologous to *icaADBC* of *S. epidermidis*, which is essential for biofilm accumulation and synthesis of the polysaccharide intercellular adhesin (PIA) of *S. epidermidis*.[24,46,82,85] Sequence comparison revealed an amino acid identity of 56–76% between different *S. aureus* strains and *icaADBC* of *S. epidermidis*.[24] Apparently, mechanisms similar to that of *S. epidermidis* may be operative in a significant proportion of clinical *S. aureus* strains leading to biofilm accumulation.

## IV. BIOFILM FORMATION BY *S. EPIDERMIDIS*

### A. Surface Attachment of S. epidermidis

#### 1. Primary Attachment to Polymer Surfaces

Several studies have indicated that primary attachment of *S. epidermidis* to polymer surfaces is a complex process. The ability to attach to polymer surfaces is widespread among *S. epidermidis* strains. Differences are observed primarily in the quantitative degree of attachment between strains.[31,57,100,113,143] Quantitative differences in attachment between individual strains are related to cell surface hydrophobicity and to the respective polymer used.[56,78] Proteolytic digestion of the bacterial cell surfaces inhibits attachment of *S. epidermidis*, indicating functional involvement of surface proteins.[57,113] Similarly to *S. aureus*, attachment to native polymer surfaces is markedly inhibited in the presence of serum, plasma, and albumin.[31,51,56,99,113,149] Compared with polymer surfaces blocked with albumin, attachment of *S. epidermidis* is markedly increased in the presence of matrix proteins like fibronectin and to a lesser extent by fibrinogen, whereas laminin did not significantly promote attachment.[51,105,149] However, quantitative binding to fibronectin or fibrinogen coated surfaces is still rather low compared with binding of bacterial cells to unmodified native polymers.[99] Therefore, there is debate whether native polymer surfaces or surfaces conditioned by host matrix proteins are more relevant for attachment of *S. epidermidis* in a clinical setting. Even under shear stress in a rotating disc system attachment of *S. epidermidis* RP62A was significantly inhibited by plasma proteins compared to an unmodified surface.[155] However, in the presence of activated platelets adhering to the surface, attachment of *S. epidermidis* RP62A was primarily mediated by platelets as compared with the plasma-modified surface.[155] In addition, lipoteichoic acids were described as relevant factors in binding of *S. epidermidis* to fibrin–platelet clots.[22]

**Table 2. Factors Functionally Involved
in *Staphylococcus epidermidis* Biofilm Formation**

| Primary attachment | Bacterial accumulation |
| --- | --- |
| Hydrophobic interactions | Polysaccharide intercellular adhesin (PIA) |
| Capsular polysaccharide/Adhesin (PS/A) | Hemagglutinin |
| Autolysin AtlE | Accumulation associated protein (AAP) |
| | |
| Staphylococcal surface protein (Ssp1) | |
| Matrix protein binding: | |
|     Fibrinogen binding protein (Fbe) | |

*2. Specific Primary Attachment Factors of* S. epidermidis

Several different specific molecular entities involved in attachment of *S. epidermidis* to polymer surfaces have been described recently (Table 2).

A capsular polysaccharide adhesin (PS/A) mediating primary attachment to silastic catheter surfaces was purified from *S. epidermidis* RP62A.[143] PS/A was detected in bacterial extracts as an activity, which inhibited attachment of *S. epidermidis* strains to silastic catheter tubing. Staphylococcal attachment was also inhibited by antiserum raised against purified PS/A. As almost half of all PS/A producing *S. epidermidis* strains were biofilm-negative, PS/A expression alone apparently is not sufficient for biofilm-production of *S. epidermidis*.[100] Isogenic biofilm-negative transposon mutants of the PS/A-positive *S. epidermidis* M187 displayed decreased attachment to silastic and expressed significantly reduced amounts of PS/A.[101] Attachment to polyethylene under shear stress was not different between the wild-type M187 and the PS/A-negative, attachment-impaired mutant *S. epidermidis* M187-sn3, which is in contrast to the above findings.[55] The pathogenic role of *S. epidermidis* PS/A in a low-inoculum rabbit model of prosthetic valve endocarditis has also been reported recently.[138] Direct contamination of an intraventricular foreign body by low levels of PS/A-positive *S. epidermidis* results in endocarditis in rabbits, but at suitably high doses PS/A-negative strains have sufficient virulence to infect cardiac vegetations.

A monoclonal antibody was described which significantly inhibited attachment of *S. epidermidis* 354 to polystyrene spheres in a concentration-dependent manner.[142] This mAb reacted with a 220-kDa cell wall associated protein of *S. epidermidis* 354.[142] The protein identified by the mAb referred to as Ssp1 is organized in a fimbria-like structure on the surface of *S. epidermidis* 354. Proteolytic processing of Ssp1 apparently occurs, rendering a variant of lower molecular weight (Ssp2), which appears to have lower activity regarding attachment of bacterial cells.[153]

Recently, Heilmann et al[48] isolated a Tn917 insertion mutant Mut1 of the biofilm-producing *S. epidermidis* O-47, which was severely impaired in attachment to poly-styrene. The cell surface of Mut1 was less hydrophobic and lacked 5 prominent proteins of 120, 60, 52, 45, and 38 kDa. The minimal cloned wild-type DNA fragment complementing the phenotype of Mut1 encoded synthesis of the 60-kDa protein. Genetic analysis revealed that the gene of the major autolysin AtlE of *S. epidermidis* was inactivated in Mut1 by a transposon induced deletion of about 8 kb of DNA. *AtlE* encodes a deduced protein of 1335 amino acids with a predicted molecular mass of 148 kDa.[48] AtlE contains two bacteriolytically active domains, a 60-kDa amidase and a 52-kDa glucosaminidase, and is expressed on the cell surface. It is presently unsettled, whether AtlE mediates attachment directly or by exposing or presenting the actual adhesin, as antiserum against the 60 kDa fragment of AtlE did not inhibit attachment. In addition to mediating attachment of *S. epidermidis* to unmodified polystyrene it also has affinity for binding to vitronectin.[48] Another member of the family of autolysins of staphylococci homologous to *atlE*, the autolysin/hemagglutinin (*aas*) of *Staphylococcus saprophyticus*, has binding affinity for fibronectin.[50] No data have been reported whether the other homologous autolysins of *S. aureus* and *S. saprophyticus*, Atl and Aas, respectively, are involved in attachment of cells of the respective species to native polymer surfaces similar to AtlE.[48,50,107]

A gene (*fbe*) encoding a fibrinogen-binding protein was cloned recently from a phage display library of *S. epidermidis* HB, which corresponds favorably with the property of some *S. epidermidis* strains to bind fibrinogen-coated polymer surfaces.[105] Fbe is a protein with a deduced molecular mass of 119 kDa, which displays homology with the *sdrCDE* family of cell surface receptors of *S. aureus*.[66] The A domain of *fbe* displays the highest degree of homology with *sdrE* of *S. aureus*.[66] PCR analysis indicated that a large proportion of clinical *S. epidermidis* isolates possessed the gene for Fbe.[105] Despite the presence of the *fbe*-gene a very heterogeneous binding activity to fibrinogen was observed in most *S. epidermidis* strains analyzed. This may reflect different expression levels of Fbe in different strains or it may even indicate that Fbe binds to a ligand different from fibrinogen when expressed on the cell surface.

## B. Factors Involved in Biofilm Accumulation of S. epidermidis

### 1. Characterization of a Polysaccharide Intercellular Adhesin (PIA)

An essential prerequisite for the accumulation of multilayered biofilms is the ability of the bacterial cells to display intercellular adhesion. In search for such an intercellular adhesin Mack et al[81] made use of a rabbit antiserum raised against the biofilm-producing *S. epidermidis* 1457 grown as a biofilm, which was absorbed with several biofilm-negative *S. epidermidis* strains. Using the absorbed antiserum reactivity with an antigen exclusively expressed by biofilm-producing *S. epidermidis* strains was detected.[81,83] Expression of the antigen varied in parallel with formation of biofilm under different physiologic growth conditions.[81] The immunologic reactivity of the antigen was completely abolished by periodate oxidation, indicating its carbohydrate nature. A significant proportion of cells of biofilm-producing as compared to biofilm-negative *S. epidermidis* strains grown in trypticase soy broth were located in large cell clusters exceeding 50 cells. Periodate oxidation of the cell preparations of biofilm-producing *S. epidermidis* strains led to complete disintegration of these cell clusters, indicating functional activity of the antigen in intercellular adhesion.[81,83]

Two completely biofilm-negative isogenic mutants, M10 and M11, were isolated using Tn917 for transposon mutagenesis of the biofilm-producing *S. epidermidis* 13-1.[82] Linkage of the transposon insertions of mutants M10 and M11 to the altered phenotype was demonstrated using phage transduction.[82,103] Transfer of the transposon insertions of mutants M10 and M11 into heterologous biofilm-producing *S. epidermidis* strains led to an identical biofilm-negative phenotype.[86,103] Primary attachment of the mutants M10 and M11 to polystyrene spheres was not significantly different as compared to the wild-type. Cell clustering as an indication of intercellular adhesion was not detected with any mutant. These results demonstrate that the mutants were impaired in the accumulative phase of biofilm production. Mutants M10 and M11 and all transductants did not produce detectable amounts of the specific polysaccharide antigen of biofilm-producing *S. epidermidis*[82,86] providing direct genetic evidence for an essential functional role of this antigen in intercellular adhesion.

To differentiate the polysaccharide from other *S. epidermidis* polysaccharides already described it was purified to homogeneity by gel filtration and anion-exchange chromatography.[85] The polysaccharide was separated by Q-Sepharose chromatography into a major cationic polysaccharide I (>80%) and a minor polysaccharide II (<20%), which was moderately anionic.[85] As shown by chemical analyses and NMR-spectroscopy polysaccharide I is a linear homoglycan of at least 130 β-1,6-linked 2-deoxy-2-amino-D-glucopyranosyl residues. On average 15-20% of them are not N-acetylated and positively charged. Cation exchange chromatography separated molecular species whose content of non-N-acetylated glucosaminyl residues varied between 2% and 26%. Polysaccharide II is structurally related to polysaccharide I but has a lower content of non-N-acetylated D-glucosaminyl residues and contains phosphate and ester-linked succinate, rendering it anionic.[85] The structure of the polysaccharide is so far unique and according to its function in *S. epidermidis* is referred to as polysaccharide intercellular adhesin (PIA).

In a population of 179 *S. epidermidis* strains there was a significant positive association between biofilm-production and expression of PIA.[84] There was a linear association between the amount of PIA produced as detected by ELISA-inhibition and the amount of biofilm produced in 49 *S. epidermidis* strains representing a continuum from biofilm-negative to strongly biofilm-producing.[84] Apparently, PIA is essential for biofilm accumulation of the majority of clinical *S. epidermidis* isolates.

### 2. Genetic Basis of PIA Synthesis

Heilmann et al.[47] independently isolated transposon mutants Mut2 and Mut2a, which have a biofilm-negative phenotype similar to mutants M10 and M11.[82] Transformation of Mut2 and Mut2a with plasmid pCN27, which contained the *ica* (intercellular adhesion) gene cluster cloned from *S. epidermidis* RP62A, restored the biofilm producing phenotype of the mutants.[46] Expression of pCN27 in the heterologous host *Staphylococcus carnosus* led to cell cluster formation and production of PIA as detected by the PIA-specific absorbed antiserum.[46] Apparently, genes encoded on pCN27 are sufficient for PIA synthesis in a heterologous staphylococcal background. The recombinant *S. carnosus* did not detectably produce biofilm on polystyrene due to insufficient attachment, although a biofilm was formed on a glass surface.[46] Other workers recently reported that the recombinant *S. carnosus* strain produced biofilm on polystyrene microtiter plates.[97] These results are best explained by a different degree of primary attachment of *S. carnosus* to differently modified polystyrene surfaces prepared for bacteriologic use or for use in tissue culture experiments, respectively.

*Ica* encodes four genes *icaADBC* which are organized in an operon-like structure.[40,46] IcaA is a transmembrane protein containing predicted 412 amino acids and has homology to N-acetylglucosaminyltransferases. IcaB with predicted 289 amino acids is probably a secreted protein, and IcaC is a hydrophobic integral membrane protein with predicted 355 amino acids. Recently, a fourth gene referred to as *icaD* was described, which partially overlaps *icaA* and *icaB* and is integrated in the order *icaADBC*.[40] At the 5'-end of the *icaADBC* locus an additional gene is located, which is referred to as *icaR*. IcaR is proposed to be transcribed in the opposite direction as *icaADBC*.[159] No detailed analysis of this genetic element has yet been reported, which may function in the regulation of *icaADBC* expression. Frame shift mutation within either of the three genes *icaA, icaB*, or *icaC* of pCN27 abolished the intercellular adhesion-positive phenotype of the recombinant *S. carnosus*.[46]

Mutant Mut2 has a Tn917 insertion 72 basepairs proximal of the ATG start codon of the *icaA*-gene, whereas the insertion site of Mut2a is nearly 1 kb proximal to the *icaA* translational start site,[46,47] which probably explains the residual PIA synthesis of Mut2a.[49] In contrast, Tn917 insertions of mutants M10 and M11 are located within the *icaA* coding sequence at nucleotides 931 and 87.[86] From these data it is apparent that the *ica* locus contains synthetic genes required for PIA synthesis. Recently, it was reported that in several biofilm-negative phase variants of *S. epidermidis* RP62A the reversible biofilm-negative phenotype resulted from inactivation of *icaADBC* by insertion and excision of insertion sequence element IS256 into *icaA* and *icaB* and into several different locations of *icaC*.[159] No insertions of IS256 into *icaD* of the *icaADBC* locus were observed.

PIA synthesis has been reconstituted in vitro using extracts prepared from recombinant *S. carnosus* expressing various genes of the *ica* operon.[40] IcaA alone exhibits a low N-acetyglucosaminyltransferase activity. Coexpression of both IcaA and IcaD largely increased the glycosyltransferase activity of the extract, resulting in N-acetylglucosamine oligomers with a chain length of up to 20 sugar residues. Synthesis of oligosaccharide chains reactive with the PIA-specific antiserum required expression of IcaC in parallel with IcaA and IcaD. It has not been differentiated whether these immunoreactive oli gomers represent full length PIA carbohydrate chains containing approximately 130 sugar residues[85] or whether these oligomers are protein-bound intermediates of PIA synthesis. Probably additional enzymatic activities are required to generate non-N-acetylated glucosamine residues characteristic for PIA,[85] as these were not detected in in vitro synthesized oligomers. This putative enzymatic reactivity may be located in a different cellular compartment not contained in the in vitro reaction. Surprisingly, *S. carnosus* expressing recombinant *icaADC* are capable of forming intercellular cell aggregates indicating that functional PIA molecules are synthesized by only these three genes.[40] This is in contrast to previous results, where mutations in *icaB* lead to an aggregation negative phenotype in *S. carnosus*[46] and a biofilm-negative phenotype in *S. epidermidis*.[159]

### 3. Hemagglutination

*S. epidermidis* strains have the ability to hemagglutinate erythrocytes of several species.[126,128] There was a linear relation of hemagglutination titers and the amount of biofilm accumulating in the adherence assay in several collections of *S. epidermidis* strains.[128] Hemagglutination was not associated with hydrophobicity. Periodate oxidation and glycosidase digestion abolished the hemagglutinating activity.[128] The hemagglutinin could be extracted into a cell-free supernatant and appeared to be a polysaccharide.[128] Direct evidence for a functional role of PIA in hemagglutination has been obtained recently: 1) transfer of the transposon insertions of the biofilm-negative mutants M10 and

M11, which lead to impaired PIA synthesis by inactivation of *icaA*, into three independent hemagglutination-positive *S. epidermidis* wild-type strains led to a hemagglutination-negative phenotype, 2) specific anti-PIA antibodies, and 3) purified PIA inhibited hemagglutination in several different *S. epidermidis* strains.[86] These results are supported by an epidemiological association of the presence of the *icaADBC* locus and a hemagglutination-positive phenotype in 39 clinical *S. epidermidis* isolates and several well characterized mutant pairs.[32] These results define PIA as the hemagglutinin of *S. epidermidis* or at least as its major functional component.[32,86] The biofilm-negative mutant Mut1, whose phenotype results from impaired primary attachment due to inactivation of *atlE*, is still hemagglutination positive, which is consistent with unimpaired PIA synthesis in this mutant.[32,49, Mack et al., unpublished]

### 4. PIA Leading to Biofilm Accumulation in the Pathogenesis of Biomaterial-Related Infection

Recently, the pathogenic potential of the isogenic biofilm-negative mutant 1457-M10 and its wild-type *S. epidermidis* 1457 was compared in two different animal infection models.[129,130] In a subcutaneous catheter infection model in mice the wild-type strain caused abscesses significantly more often than the isogenic mutant 1457-M10.[130] In addition, the wild-type strain was significantly less commonly eliminated from the site of infection and higher cell concentrations were detected on catheters infected with the wild-type *S. epidermidis* 1457.[130] These differences were due to altered biofilm accumulation resulting from abolished PIA synthesis and not to secondary effects of altered PIA synthesis on fibronectin binding of the biofilm-negative mutant,[130] as proposed recently.[4] In a central venous catheter infection model in rats the wild-type *S. epidermidis* 1457 was significantly more likely to induce a catheter-related infection (71% vs 14%) resulting in bacteremia and metastatic disease than its isogenic biofilm-negative mutant 1457-M10.[129] These results confirm for the first time the importance of biofilm accumulation mediated by PIA and the activity of *icaADBC* for the pathogenesis of *S. epidermidis* biomaterial-related infections using a genetically and phenotypically well characterized pair of strains.

### 5. Accumulation Associated Protein (AAP)

A biofilm-negative mutant M7 of biofilm-producing *S. epidermidis* RP62A generated by mitomycin mutagenesis is still competent for initial attachment to glass and polystyrene but did not accumulate on these surfaces.[135] No differences were noted between wild-type and M7 regarding growth rate, cell wall composition, surface characteristics, DNA restriction and antibiotic resistance profiles. However, a 140-kDa extracellular protein of M7 was no longer detectable.[63,135] Biofilm accumulation of several biofilm-producing *S. epidermidis* strains was inhibited in a concentration-dependent manner by an antiserum raised against the purified protein, whereas pre-immune serum had no effect.[63] Thirty two out of 58 *S. epidermidis* strains produced the 140-kDa protein. However, 7 of 26 140-kDa protein-negative strains produced substantial amounts of biofilm.[63] As M7 is still capable of producing PIA[63, Mack et al., unpublished] the functional role of the 140-kDa protein is presently unknown.

## C. Polysaccharide Components of S. epidermidis *Glycocalyx*

The composition of the *S. epidermidis* glycocalyx or slime has attracted major interest and numerous attempts were started to define the respective polysaccharide components

**Table 3. Polysaccharides of the *Staphylococcus epidermidis* Glycocalyx**

| Polysaccharide | Main components | Authors year (Ref) |
|---|---|---|
| Polysaccharide intercellular adhesin (PIA) | | Mack et al. 1996[85] |
| Polysaccharide I | GlcNAc, GlcN | |
| Polysaccharide II | GlcNAc, GlcN, Succinate, Phosphate | |
| Major slime component/wall teichoic acid | Glc, Glyc, Ala, GlcNAc, Phosphate | Hussain et al. 1992[60] |
| Slime-associated antigen (SAA) | Glc, GlcN, GlcA, GalA | Christensen et al. 1990[18] |
| | GlcNAc | Baldassarri et al. 1996[3] |
| Capsular Polysaccharide/Adhesin (PS/A) | Gal, GalN, GlcN, Urs | Tojo et al. 1988[143] |
| | GlcN, Succinate, Acetate | McKenney et al. 1998[97] |
| Extracellular slime substance (ESS) | Gal, Man, Glc, GlcNAc | Ludwicka et al. 1984[79] |
| Sulfated slime polysaccharide | GlcN, Glc, Fuc, Xyl, Sulfate | Arvaniti et al. 1994[2] |

Abbreviations: Glc: Glucose; Gal: Galactose; Man: Mannose; Fuc: Fucose; Xyl: Xylose; GlcN: Glucosamine; GlcNAc: N-Acetylglucosamine; GalN: Galactosamine; Urs: Uronic acids; GlcA: Glucuronic acid; GlaA: Galacturonic acid; Ala: D-Alanine; Glyc: Glycerole

(Table 3). Only in 1990 it became apparent by the work of Drewry et al.[28] and Hussain et al.[59] that slime preparations isolated from *S. epidermidis* strains grown in complex medium and/or on agar were regularly contaminated by carbohydrate compounds rich in galactose largely derived from agar. Contamination with phosphate, glucose, galactose and galactosamine derived from tryptic soy broth was also possible. To obtain meaningful analytical results chemically defined medium and solidification of medium with silica gel instead of agar have to be used. Alternatively analysis of polysaccharide preparations obtained in parallel from isogenic strain pairs, which do or do not produce the respective polysaccharides, leads to identification of the specific components of the respective polysaccharides.

Several polysaccharide components of *S. epidermidis* have been described, which are major constituents of *S. epidermidis* slime or glycocalyx, or have been proposed to be functionally significant for biofilm production.

Our group recently elucidated the structure of the polysaccharide intercellular adhesin (PIA) of *S. epidermidis* 1457 and RP62A.[85] Several measures were taken to ensure the specificity of the material purified from bacterial cells grown in trypticase soy broth. First, the isogenic biofilm-negative, PIA-negative transposon mutant 1457-M11 was analyzed in parallel.[85] In comparison with *S. epidermidis* 1457 only hexosamine was determined as a specific component of PIA. In addition, several independent methods such as chemical analysis, immunochemical analysis, and NMR-spectroscopy were used in the structural analysis of the PIA. PIA prepared from *S. epidermidis* 1457 and 1457-M11 grown in the chemically defined medium HHW[60] had a composition very similar to PIA prepared from TSB-grown cells.[Mack and Krokotsch, unpublished material] In addition, it was demonstrated that the PIA-specific absorbed antiserum indeed was reactive with β-anomeric N-acetylglucosamine residues, as ELISA inhibition with various monosaccharides revealed the β-anomeric form and the acetylated amino group of the D-glucosaminyl residues as important for the reactivity with the specific antiserum.[85]

Hussain et al.[60] compared the major high molecular weight slime component of biofilm-producing *S. epidermidis* RP62A grown in a chemically defined medium and of purified cell wall teichoic acid and found a similar composition containing primarily glucose, glycerol, D-alanine, *N*-acetylglucosamine, and phosphate. Apparently, teichoic acids have no significant functional role in biofilm formation, as similar amounts of teichoic acid were produced by biofilm-producing and biofilm-negative *S. epidermidis* strains.[59,60] Nevertheless, cell wall teichoic acids may stabilize the established biofilm representing major components of the *S. epidermidis* glycocalyx.

A slime associated antigen (SAA) was proposed to be associated with biofilm-production of *S. epidermidis* strains.[18] Semi-purified SAA prepared by Sephadex G 200 chromatography contained 64 % hexose and only 0.91 % hexosamine as determined by colorimetric assay.[18] Gas–liquid chromatography revealed as constituents 59 % glucose and about 7 % N-acetylglucosamine and N-acetylgalactosamine. As the composition of SAA was not compared to a similar extract of a biofilm-negative phase variant or a SAA-negative mutant it remained unclear, which of the determined sugars were specific components of SAA. Recently, Baldassarri et al.[3] reevaluated the preparation and purification of SAA of *S. epidermidis* RP62A and noted that in a preparative procedure very similar to the preparation of PIA, SAA was almost exclusively composed of N-acetylglucosamine. In a control preparation of the biofilm-negative variant HAM892, which did not produce SAA, no hexosamines were detected, indicating that N-acetylglucosamine is a specific component of SAA. These results led these authors to speculate that SAA and PIA might be related. As this may easily be so in the latter case, the relation of the former glucose-rich SAA preparation remains unclear.

Analysis of purified PS/A revealed a composition of 54 % hexoses, 20 % amino sugars and 10% uronic acids.[143] As specific sugars 22% galactose, 15% glucosamine, and 5% galactosamine were detected. PS/A was prepared from the spent medium of *S. epidermidis* RP62A grown in trypticase soy broth in a fermentor. However, comparison of the composition of a similar polysaccharide preparation prepared from a PS/A-negative control strain was not reported. Recently, the composition of PS/A from *S. epidermidis* RP62A and M187 grown in a chemically defined medium was reevaluated.[97] There was an almost identical reactivity of anti-PS/A and anti-PIA antisera with several well-characterized *S. epidermidis* strains including the PS/A-producing *S. epidermidis* M187 and the PS/A-negative transposon mutant M187-sn3, the PIA-producing *S. epidermidis* 1457 and the isogenic PIA-negative mutant 1457-M11, and the recombinant *S. carnosus* containing the cloned *icaADBC* locus.[97] These data suggest that PS/A and PIA are structurally related or even identical. PS/A had peculiar properties and readily precipitated unless solubilized in buffer containing desoxycholate or at low pH and displayed an apparent molecular mass in gel filtration >250 kDa.[97] The authors postulated that PS/A is a polysaccharide of high molecular mass composed of β-1,6-linked glucosamine residues with a high degree of substitution with succinate and acetate.[97] In addition, these authors proposed that PS/A is synthesized by the gene products of *icaADBC.* [97] As detailed data on the structural analysis of PS/A have not been reported it remains to be determined if PS/A is identical with polysaccharide II of PIA or represents an additional variant of PIA.

A high molecular weight polysaccharide extracellular slime substance (ESS) of *S. epidermidis* consisting mainly of galactose and mannose was proposed to be functionally involved in attachment or accumulation of *S. epidermidis* on polymer surfaces.[79,122] As agar was used consistently in preparation of ESS, it is reasonable to assume that the material was heavily contaminated. Recently, analytical data of purified polysaccharide compounds in slime preparations derived from several different *S. epidermidis* strains were reported.[2,67,68] As no measures were taken to control for contamination with medium components and as no reference material was analyzed in parallel from strains lacking a specific function or differing in production of a specific polysaccharide, these data are very difficult to interpret with regard to the significance of the individual carbohydrate compounds for the composition of the *S. epidermidis* glycocalyx.

**Acknowledgment:** Work in the laboratory of the authors was supported in part by grants of the Deutsche Forschungsgemeinschaft to D.M.

## REFERENCES

1. Archer GL, Climo MW: Antimicrobial susceptibility of coagulase-negative staphylococci. *Antimicrob Agents Chemother* 38:2231–7, 1994
2. Arvaniti A, Karamanos NK, Dimitracopoulos G, et al: Isolation and characterization of a novel 20-kDa sulfated polysaccharide from the extracellular slime layer of *Staphylococcus epidermidis. Arch Biochem Biophys* 308:432–8, 1994
3. Baldassarri L, Donnelli G, Gelosia A, et al: Purification and characterization of the staphylococcal slime-associated antigen and its occurrence among *Staphylococcus* epidermis clinical isolates. *Infect Immun* 64:3410–5, 1996
4. Baldassarri L, Donelli G, Gelosia A, et al: Expression of slime interferes with *in vitro* detection of host protein receptors of *Staphylococcus epidermidis. Infect Immun* 65:1522–6, 1997
5. Baselga R, Albizu I, De La Cruz M, et al: Phase variation of slime production in *Staphylococcus aureus*: implications in colonization and virulence. *Infect Immun* 61:4857–62, 1993
6. Bayston R, Penny SR: Excessive production of mucoid substance in *Staphylococcus* SIIA: a possible factor in colonisation of Holter shunts. *Dev Med Child Neurol* 14 (Suppl 27): 25–8, 1989
7. Bayston R, Rodgers J: Production of extra-cellular slime by *Staphylococcus epidermidis* during stationary phase of growth: its association with adherence to implantable devices. *J Clin Pathol* 43:866–70, 1990
8. Bisno AL, Waldvogel FA: *Infections Associated With Indwelling Medical Devices.* 2nd ed. American Society of Microbiology, Washington, DC: 1994
9. Boden MK, Flock JI: Evidence for three different fibrinogen-binding proteins with unique properties from *Staphylococcus aureus* strain Newman. *Microb Pathog* 12:289–98, 1992
10. Boden MK, Flock JI: Cloning and characterization of a gene for a 19 kDa fibrinogen-binding protein from *Staphylococcus aureus. Mol Microbiol* 12:599–606, 1994
11. Bozzini S, Visai L, Pignatti P, et al: Multiple binding sites in fibronectin and the staphylococcal fibronectin receptor. *Eur J Biochem* 207:327–33, 1992
12. Casolini F, Visai L, Joh D, et al: Antibody response to fibronectin-binding adhesin FnbpA in patients with *Staphylococcus aureus* infections. *Infect Immun* 66:5433–42, 1998

13. Chambers HF: Methicillin resistance in staphylococci: molecular and biochemical basis and clinical implications. *Clin Microbiol Rev* 10:781–91, 1997
14. Cheung AL, Fischetti VA: The role of fibrinogen in staphylococcal adherence to catheters *in vitro*. *J Infect Dis* 161:1177–86, 1990
15. Chhatwal GS, Preissner KT, Muller-Berghaus G, et al: Specific binding of the human S protein (vitronectin) to streptococci, *Staphylococcus aureus*, and *Escherichia coli*. *Infect Immun* 55:1878–83, 1987
16. Christensen GD, Simpson WA, Bisno AL, et al: Adherence of slime-producing strains of *Staphylococcus epidermidis* to smooth surfaces. *Infect Immun* 37:318–26, 1982
17. Christensen GD, Simpson WA, Younger JJ, et al: Adherence of coagulase-negative staphylococci to plastic tissue culture plates: a quantitative model for the adherence of staphylococci to medical devices. *J Clin Microbiol* 22:996–1006, 1985
18. Christensen GD, Barker LP, Mawhinney TP, et al: Identification of an antigenic marker of slime production for *Staphylococcus epidermidis*. *Infect Immun* 58:2906–11, 1990
19. Christensen GD, Baldassarri L, Simpson WA: Colonization of medical devices by coagulase-negative staphylococci. In: Bisno AL, Waldvogel FA, eds. *Infections Associated With Indwelling Medical Devices*. 2nd ed. ASM, Washington, DC: 1994, 45–78
20. Chuard C, Vaudaux P, Waldvogel FA, et al: Susceptibility of *Staphylococcus aureus* growing on fibronectin-coated surfaces to bactericidal antibiotics. *Antimicrob Agents Chemother* 37:625–32, 1993
21. Chuard C, Vaudaux PE, Proctor RA, et al: Decreased susceptibility to antibiotic killing of a stable small colony variant of *Staphylococcus aureus* in fluid phase and on fibronectin-coated surfaces. *J Antimicrob Chemother* 39:603–8, 1997
22. Chugh TD, Burns GJ, Shuhaiber HJ, et al: Adherence of *Staphylococcus epidermidis* to fibrin-platelet clots in vitro mediated by lipoteichoic acid. *Infect Immun* 58:315–9, 1990
23. Costerton JW, Cheng KJ, Geesey GG, et al: Bacterial biofilms in nature and disease. *Ann Rev Microbiol* 41: 435–64, 1987
24. Cramton SE, Gerke C, Schnell N, et al:: The intercellular adhesion (*ica*) locus is present in *Staphylococcus aureus* and is required for biofilm formation. *Infect Immun* 67(10), 1999, in press
25. Diaz-Mitoma F, Harding GKM, Hoban DJ, et al: Clinical significance of a test for slime production in ventriculoperitoneal shunt infections caused by coagulase-negative staphylococci. *J Infect Dis* 156: 555–60, 1987
26. Delmi M, Vaudaux P, Lew DP, et al: Role of fibronectin in staphylococcal adhesion to metallic surfaces used as models of orthopaedic devices. *J Orthop Res* 12:432–8, 1994
27. Dickinson RB, Nagel JA, McDevitt D, et al: Quantitative comparison of clumping factor- and coagulase-mediated *Staphylococcus aureus* adhesion to surface-bound fibrinogen under flow. *Infect Immun* 63:3143–50, 1995
28. Drewry DT, Galbraith L, Wilkinson BJ, et al: Staphylococcal slime: a cautionary tale. *J Clin Microbiol* 28:1292–6, 1990
29. Emori TG, Gaines RP: An overview of nosocomial infections, including the role of the microbiology laboratory. *Clin Microbiol Rev* 6: 428–42, 1993
30. Espersen F, Clemmensen I: Isolation of a fibronectin-binding protein from *Staphylococcus aureus*. *Infect Immun* 37:526–31, 1982
31. Espersen F, Wilkinson BJ, Gahrn-Hansen B, et al: Attachment of staphylococci to silicone catheters *in vitro*. *APMIS* 98:471–8, 1990
32. Fey PD, Ulphani JS, Gotz F, et al: Characterization of the relationship between polysaccharide intercellular adhesin and hemagglutination in *Staphylococcus epidermidis*. *J Infect Dis* 179:1561–4, 1999
33. Flock JI, Froman G, Jonsson K, et al: Cloning and expression of the gene for a fibronectin-binding protein from *Staphylococcus aureus*. *EMBO J* 6:2351–2357, 1987

34. Flock JI, Brennan F: Antibodies that block adherence of *Staphylococcus aureus* to fibronectin. *Trends Microbiol* 7:140–141, 1999
35. Foster TJ, Hook M: Surface protein adhesins of *Staphylococcus aureus*. *Trends Microbiol* 6:484–488, 1998
36. Franson TR, Sheth NK, Rose HD, et al: Scanning electron microscopy of bacteria adherent to intravascular catheters. *J Clin Microbiol* 20: 500–5, 1984
37. Froman G, Switalski LM, Speziale P, et al: Isolation and characterization of a fibronectin receptor from *Staphylococcus aureus*. *J Biol Chem* 262:6564–71, 1987
38. Fuller RA, Rosen JJ: Materials for medicine. *Sci Am* 255:118–125, 1986
39. Fuquay JI, Loo DT, Barnes DW: Binding of *Staphylococcus aureus* by human serum spreading factor in an in vitro assay. *Infect Immun* 52:714–7, 1986
40. Gerke C, Kraft A, Sussmuth R, et al: Characterization of the N-acetylglucosaminyltransferase activity involved in the biosynthesis of the *Staphylococcus epidermidis* polysaccharide intercellular adhesin. *J Biol Chem* 273:18586–93, 1998
41. Gillaspy AF, Lee CY, Sau S, et al: Factors affecting the collagen binding capacity of *Staphylococcus aureus*. *Infect Immun* 66:3170–8, 1998
42. Greene C, McDevitt D, Francois P, et al: Adhesion properties of mutants of *Staphylococcus aureus* defective in fibronectin-binding proteins and studies on the expression of fnb genes. *Mol Microbiol* 17:1143–52, 1995
43. Greene C, Vaudaux PE, Francois P, et al: A low-fibronectin-binding mutant of *Staphylococcus aureus* 879R4S has Tn918 inserted into its single fnb gene. *Microbiology* 142:2153–60, 1996
44. Gristina AG: Biomaterial-centered infection: microbial adhesion versus tissue integration. *Science* 237:1588–95, 1987
45. Hartford O, Francois P, Vaudaux P, et al: The dipeptide repeat region of the fibrinogen-binding protein (clumping factor) is required for functional expression of the fibrinogen-binding domain on the *Staphylococcus aureus* cell surface. *Mol Microbiol* 25:1065–76, 1997
46. Heilmann C, Schweitzer O, Gerke C, et al: Molecular basis of intercellular adhesion in the biofilm-forming *Staphylococcus epidermidis*. *Mol Microbiol* 20:1083–91, 1996
47. Heilmann C, Gerke C, Perdreau-Remington F, et al: Characterization of Tn917 insertion mutants of *Staphylococcus epidermidis* affected in biofilm formation. *Infect Immun* 64:277–82, 1996
48. Heilmann C, Hussain M, Peters G, et al: Evidence for autolysin-mediated primary attachment of *Staphylococcus epidermidis* to a polystyrene surface. *Mol Microbiol* 24:1013–24, 1997
49. Heilmann C, Gotz F: Further characterization of *Staphylococcus epidermidis* transposon mutants deficient in primary attachment or intercellular adhesion. *Zentralbl Bakteriol* 287:69–83, 1998
50. Hell W, Meyer HG, Gatermann SG: Cloning of aas, a gene encoding a *Staphylococcus saprophyticus* surface protein with adhesive and autolytic properties. *Mol Microbiol* 29:871–81, 1998
51. Herrmann M, Vaudaux PE, Pittet D, et al: Fibronectin, fibrinogen, and laminin act as mediators of adherence of clinical staphylococcal isolates to foreign material. *J Infect Dis* 158:693–701, 1988
52. Herrmann M, Suchard SJ, Boxer LA, et al: Thrombospondin binds to *Staphylococcus aureus* and promotes staphylococcal adherence to surfaces. *Infect Immun* 59:279–88, 1991
53. Herrmann M, Lai QJ, Albrecht RM, et al: Adhesion of *Staphylococcus aureus* to surface-bound platelets: role of fibrinogen/fibrin and platelet integrins. *J Infect Dis* 167:312–22, 1993
54. Herrmann M, Hartleib J, Kehrel B, et al: Interaction of von Willebrand factor with *Staphylococcus aureus*. *J Infect Dis* 176:984–91, 1997

55. Higashi JM, Wang IW, Shlaes DM, et al: Adhesion of *Staphylococcus epidermidis* and transposon mutant strains to hydrophobic polyethylene. *J Biomed Mater Res* 39:341–50, 1998

56. Hogt AH, Dankert J, Feijen J: Adhesion of *Staphylococcus epidermidis* and *Staphylococcus saprophyticus* to a hydrophobic biomaterial. *J Gen Microbiol* 131:2485–91, 1985

57. Hogt AH, Dankert J, Hulstaert CE, et al: Cell surface characteristics of coagulase-negative staphylococci and their adherence to fluorinated poly(ethylenepropylene). *Infect Immun* 51:294–301, 1986

58. Huebner J, Goldmann DA: Coagulase-negative staphylococci: role as pathogens. *Annu Rev Med* 50:223–36, 1999

59. Hussain M, Hastings JG, White PJ: Isolation and composition of the extracellular slime made by coagulase-negative staphylococci in a chemically defined medium. *J Infect Dis* 163:534–41, 1991

60. Hussain M, Hastings JG, White PJ: Comparison of cell-wall teichoic acid with high-molecular-weight extracellular slime material from *Staphylococcus epidermidis*. *J Med Microbiol* 37:368–75, 1992

61. Hussain M, Wilcox MH, White PJ, et al: Importance of medium and atmosphere type to both slime production and adherence by coagulase-negative staphylococci. *J Hosp Infect* 20:173–84, 1992

62. Hussain M, Wilcox MH, White PJ: The slime of coagulase-negative staphylococci: biochemistry and relation to adherence. *FEMS Microbiol Rev* 10:191–207, 1993

63. Hussain M, Herrmann M, von Eiff C, et al: A 140-kilodalton extracellular protein is essential for the accumulation of *Staphylococcus epidermidis* strains on surfaces. *Infect Immun* 65:519–24, 1997

64. Jonsson K, Signäs C, Muller HP, et al: Two different genes encode fibronectin binding proteins in *Staphylococcus aureus*. The complete nucleotide sequence and characterization of the second gene. *Eur J Biochem* 202:1041–8, 1991

65. Jonsson K, McDevitt D, McGavin MH, et al: *Staphylococcus aureus* expresses a major histocompatibility complex class II analog. *J Biol Chem* 270:21457–60, 1995

66. Josefsson E, McCrea KW, Ni Eidhin D, et al: Three new members of the serine-aspartate repeat protein multigene family of *Staphylococcus aureus*. *Microbiology* 144:3387–95, 1998

67. Karamanos NK, Panagiotopoulou HS, Syrokou A, et al: Identity of macromolecules present in the extracellular slime layer of *Staphylococcus epidermidis*. *Biochimie* 77:217–24, 1995

68. Karamanos NK, Syrokou A, Panagiotopoulou HS, et al: The major 20-kDa polysaccharide of *Staphylococcus epidermidis* extracellular slime and its antibodies as powerful agents for detecting antibodies in blood serum and differentiating among slime-positive and -negative *S. epidermidis* and other staphylococci species. *Arch Biochem Biophys* 342:389–95, 1997

69. Kloos WE: Taxonomy and systematics of staphylococci indigenous to humans. In: Crossley KB, Archer GL: *The Staphylococci In Human Disease*. Churchill Livingstone, New York. 1997:113–7

70. Kloos WE, Bannerman TL: Update on clinical significance of coagulase-negative staphylococci. *Clin Microbiol Rev* 7:117–40, 1994

71. Kristinsson KG: Adherence of staphylococci to intravascular catheters. *J Med Microbiol* 28:249–57, 1989

72. Kuusela P: Fibronectin binds to *Staphylococcus aureus*. *Nature* 276:718–20, 1978

73. Kuusela P, Vartio T, Vuento M, et al: Attachment of staphylococci and streptococci on fibronectin, fibronectin fragments, and fibrinogen bound to a solid phase. *Infect Immun* 50:77–81, 1985

74. Kuypers JM, Proctor RA: Reduced adherence to traumatized rat heart valves by a low-fibronectin-binding mutant of *Staphylococcus aureus*. *Infect Immun* 57:2306–12, 1989

75. Liang OD, Maccarana M, Flock JI, et al: Multiple interactions between human vitronectin and *Staphylococcus aureus*. *Biochim Biophys Acta* 1225:57–63, 1993

76. Liang OD, Flock JI, Wadstrom T: Isolation and characterization of a vitronectin-binding surface protein from *Staphylococcus aureus*. *Biochim Biophys Acta* 1250:110–6, 1995

77. Lopes JD, dos Reis M, Brentani RR: Presence of laminin receptors in *Staphylococcus aureus*. *Science* 229:275–7, 1985

78. Ludwicka A, Jansen B, Waldström T, et al: Attachment of staphylococci to various synthetic polymers. *Zbl Bakt Hyg*, I Abt Orig A 256: 479–89, 1984

79. Ludwicka A, Uhlenbruck G, Peters G, et al: Investigation on extracellular slime substance produced by *Staphylococcus epidermidis*. *Zentralbl Bakteriol Mikrobiol Hyg* [A] 258:256-67, 1984

80. Mack, D: Molecular mechanisms of *Staphylococcus epidermidis* biofilm formation. *J Hosp Infect* 43 (Suppl A) 1999; in press.

81. Mack D, Siemssen N, Laufs R: Parallel induction by glucose of adherence and a polysaccharide antigen specific for plastic-adherent *Staphylococcus epidermidis*: evidence for functional relation to intercellular adhesion. *Infect Immun* 60:2048–57, 1992

82. Mack D, Nedelmann M, Krokotsch A, et al: Characterization of transposon mutants of biofilm-producing *Staphylococcus epidermidis* impaired in the accumulative phase of biofilm production: genetic identification of a hexosamine-containing polysaccharide intercellular adhesin. *Infect Immun* 62:3244–53, 1994

83. Mack D, Siemssen N, Laufs R: Identification of a cell cluster associated antigen specific for plastic-adherent *Staphylococcus epidermidis* which is functional related to intercellular adhesion. *Zentralbl Bakteriol* Suppl 26: 411–3, 1994

84. Mack D, Haeder M, Siemssen N, et al: Association of biofilm production of coagulase-negative staphylococci with expression of a specific polysaccharide intercellular adhesin. *J Infect Dis* 174:881–4, 1996

85. Mack D, Fischer W, Krokotsch A, et al: The intercellular adhesin involved in biofilm accumulation of *Staphylococcus epidermidis* is a linear beta-1,6-linked glucosaminoglycan: purification and structural analysis. *J Bacteriol* 178:175–83, 1996

86. Mack D, Riedewald J, Rohde H, et al: Essential functional role of the polysaccharide intercellular adhesin of *Staphylococcus epidermidis* in hemagglutination. *Infect Immun* 67:1004–8, 1999

87. Marrie TJ, Nelligan J, Costerton JW: A scanning and transmission electron microscopic study of an infected endocardial pacemaker lead. *Circulation* 66:1339–41, 1982

88. Marrie TJ, Noble MA, Costerton JW: Examination of the morphology of bacteria adhering to peritoneal dialysis catheters by scanning and transmission electron microscopy. *J Clin Microbiol* 18:1388–98, 1983

89. Marrie TJ, Costerton JW: Scanning and transmission electron microscopy of in situ bacterial colonization of intravenous and intraarterial catheters. *J Clin Microbiol* 19:687–93, 1984

90. Martin MA, Pfaller MA, Wenzel RP: Coagulase-negative staphylococcal bacteremia Mortality and hospital stay. *Ann Intern Med* 110:9–16, 1989

91. Maxe I, Ryden C, Wadstrom T, et al: Specific attachment of *Staphylococcus aureus* to immobilized fibronectin. *Infect Immun* 54:695-704, 1986

92. McDevitt D, Vaudaux P, Foster TJ: Genetic evidence that bound coagulase of *Staphylococcus aureus* is not clumping factor. *Infect Immun* 60:1514–23, 1992

93. McDevitt D, Francois P, Vaudaux P et al: Molecular characterization of the clumping factor (fibrinogen receptor) of *Staphylococcus aureus*. *Mol Microbiol* 11:237–48, 1994

94. McDevitt D, Francois P, Vaudaux P, et al: Identification of the ligand-binding domain of the surface-located fibrinogen receptor (clumping factor) of *Staphylococcus aureus*. *Mol Microbiol* 16:895–907, 1995

95. McGavin MH, Krajewska-Pietrasik D, Ryden C, et al: Identification of a *Staphylococcus aureus* extracellular matrix-binding protein with broad specificity. *Infect Immun* 61:2479–85, 1993

96. McGavin MJ, Raucci G, Gurusiddappa S, et al: Fibronectin binding determinants of the *Staphylococcus aureus* fibronectin receptor. *J Biol Chem* 266:8343–7, 1991

97. McKenney D, Hubner J, Muller E, et al: The ica locus of *Staphylococcus epidermidis* encodes production of the capsular polysaccharide/adhesin. *Infect Immun* 66:4711–20, 1998

98. Mohamed N, Teeters MA, Patti JM, et al: Inhibition of *Staphylococcus aureus* adherence to collagen under dynamic conditions. *Infect Immun* 67:589–94, 1999

99. Muller E, Takeda S, Goldmann DA, et at: Blood proteins do not promote adherence of coagulase-negative staphylococci to biomaterials. *Infect Immun* 59:3323–6, 1991

100. Muller E, Takeda S, Shiro H, et al: Occurrence of capsular polysaccharide/adhesin among clinical isolates of coagulase-negative staphylococci. *J Infect Dis* 168:1211–8, 1993

101. Muller E, Hubner J, Gutierrez N, et al: Isolation and characterization of transposon mutants of *Staphylococcus epidermidis* deficient in capsular polysaccharide/adhesin and slime. *Infect Immun* 61:551–8, 1993

102. Navarre WW, Schneewind O: Surface proteins of gram-positive bacteria and mechanisms of their targeting to the cell wall envelope. *Microbiol Mol Biol Rev* 63:174–229, 1999

103. Nedelmann M, Sabottke A, Laufs R, et al: Generalized transduction for genetic linkage analysis and transfer of transposon insertions in different *Staphylococcus epidermidis* strains. *Zentralbl Bakteriol* 287:85–92, 1998

104. Ni Eidhin D, Perkins S, Francois P, et al: Clumping factor B (ClfB), a new surface-located fibrinogen-binding adhesin of *Staphylococcus aureus*. *Mol Microbiol* 30:245–57, 1998

105. Nilsson M, Frykberg L, Flock JI, et al: A fibrinogen-binding protein of *Staphylococcus epidermidis*. *Infect Immun* 66:2666–73, 1998

106. Noble WC: Staphylococcal carriage and skin and soft tissue infection. In: Crossley KB, Archer GL. *The Staphylococci in Human Disease*. Churchill Livingstone, New York, 1997:401–12

107. Oshida T, Sugai M, Komatsuzawa H, et al: A *Staphylococcus aureus* autolysin that has an N-acetylmuramoyl-L-alanine amidase domain and an endo-beta-N-acetylglucosaminidase domain: cloning, sequence analysis, and characterization. *Proc Natl Acad Sci U S A* 92:285–9, 1995

108. Palma M, Wade D, Flock M, et al: Multiple binding sites in the interaction between an extracellular fibrinogen-binding protein from *Staphylococcus aureus* and fibrinogen. *J Biol Chem* 273:13177–81, 1998

109. Palma M, Haggar A, Flock JI: Adherence of *Staphylococcus aureus* is enhanced by an endogenous secreted protein with broad binding activity. *J Bacteriol* 181:2840–5, 1999

110. Park PW, Roberts DD, Grosso LE, et al: Binding of elastin to *Staphylococcus aureus*. *J Biol Chem* 266:23399–406, 1991

111. Park PW, Rosenbloom J, Abrams WR, et al: Molecular cloning and expression of the gene for elastin-binding protein (*ebpS*) in *Staphylococcus aureus*. *J Biol Chem* 271:15803–9, 1996

112. Park PW, Broekelmann TJ, Mecham BR, et al: Characterization of the elastin binding domain in the cell-surface 25-kDa elastin-binding protein of *Staphylococcus aureus* (EbpS). *J Biol Chem* 274:2845–50, 1999

113. Pascual A, Fleer A, Westerdaal NAC, et al: Modulation of adherence of coagulase-negative staphylococci to teflon catheters *in vitro*. *Eur J Clin Microbiol* 5:518–22, 1986

114. Patti JM, J^nsson H, Guss B, et al: Molecular characterization and expression of a gene encoding a *Staphylococcus aureus* collagen adhesin. *J Biol Chem* 267:4766–72, 1992

115. Patti JM, Boles JO, H^^k M: Identification and biochemical characterization of the ligand binding domain of the collagen adhesin from *Staphylococcus aureus*. *Biochemistry* 32:11428–35, 1993

116. Patti JM, Bremell T, Krajewska-Pietrasik D, et al: The *Staphylococcus aureus* collagen adhesin is a virulence determinant in experimental septic arthritis. *Infect Immun* 62:152–61, 1994

117. Patti JM, Allen BL, McGavin MJ, et al: MSCRAMM-mediated adherence of micro-organisms to host tissues. *Ann Rev Microbiol* 48:585–617, 1994

118. Patti JM, House-Pompeo K, Boles JO, et al: Critical residues in the ligand-binding site of the *Staphylococcus aureus* collagen-binding adhesin (MSCRAMM). *J Biol Chem* 270:12005–11, 1995

119. Paulsson M, Kober M, Freij-Larsson C, et al: Adhesion of staphylococci to chemically modi-fied and native polymers, and the influence of preadsorbed fibronectin, vitronectin and fibrinogen. *Biomaterials* 14:845–53, 1993

120. Peters G, Locci R, Pulverer G: Microbial colonization of prosthetic devices: II. Scanning electron microscopy of naturally infected intravenous catheters. *Zbl Bakt Hyg*, I Abt Orig B 173: 293–9, 1981

121. Peters G, Locci R, Pulverer G: Adherence and growth of coagulase-negative staphylococci on surfaces of intravenous catheters. *J Infect Dis* 146:479–82, 1982

122. Peters G, Schumacher-Perdreau F, Jansen B, et al: Biology of *S. epidermidis* extracellular slime. *Zbl Bakt* Suppl 16:15–31, 1987

123. Pfaller MA, Herwaldt LA: Laboratory, clinical, and epidemiological aspects of coagulase-negative staphylococci. *Clin Microbiol Rev* 1:281–99, 1988

124. Raad II, Bodey GP: Infectious complications of indwelling vascular catheters. *Clin Infect Dis* 15:197–210, 1992

125. Raad I, Alrahwan A, Rolston K: *Staphylococcus epidermidis*: emerging resistance and need for alternative agents. *Clin Infect Dis* 26:1182–7, 1998

126. Rupp ME, Archer GL: Hemagglutination and adherence to plastic by *Staphylococcus epidermidis*. *Infect Immun* 60:4322–47, 1992

127. Rupp ME, Archer GL: Coagulase-negative staphylococci: pathogens associated with medical progress. *Clin Infect Dis* 19:231–43, 1994

128. Rupp ME, Sloot N, Meyer HG, et al: Characterization of the hemagglutinin of *Staphylococcus epidermidis*. *J Infect Dis* 172:1509–18, 1995

129. Rupp ME, Ulphani JS, Fey PD, et al: Characterization of *Staphylococcus epidermidis* polysaccharide intercellular adhesin/hemagglutinin in the pathogenesis of intravascular catheter-associated infection in a rat model. *Infect Immun* 67:2656–9, 1999

130. Rupp ME, Ulphani JS, Fey PD, et al: Characterization of the importance of polysaccharide intercellular adhesin/hemagglutinin of *Staphylococcus epidermidis* in the pathogenesis of biomaterial-based infection in a mouse foreign body infection model. *Infect Immun* 67:2627–32, 1999

131. Russell PB, Kline J, Yoder MC, et al: Staphylococcal adherence to polyvinyl chloride and heparin-bonded polyurethane catheters is species dependent and enhanced by fibronectin. *J Clin Microbiol* 25:1083–7, 1987

132. Ryden C, Yacoub AI, Maxe I, et al: Specific binding of bone sialoprotein to *Staphylococcus aureus* isolated from patients with osteomyelitis. *Eur J Biochem* 184:331–6, 1989

133. Ryden C, Tung HS, Nikolaev V, et al: *Staphylococcus aureus* causing osteomyelitis binds to a nonapeptide sequence in bone sialoprotein. *Biochem J* 327:825–9, 1997

134. Schmidt DD, Bandyk DF, Pequet AJ, et al: Mucin production by *Staphylococcus epidermidis*. A virulence factor promoting adherence to vascular grafts. *Arch Surg* 121:89–95, 1986

135. Schumacher-Perdreau F, Heilmann C, Peters G, et al: Comparative analysis of a biofilm-forming *Staphylococcus epidermidis* strain and its adhesion-positive, accumulation-negative mutant M7. *FEMS Microbiol Lett* 117:71–8, 1994

136. Signäs C, Raucci G, Jönsson K, et al: Nucleotide sequence of the gene for a fibronectin-binding protein from *Staphylococcus aureus*: Use of this peptide sequence in the synthesis of biologically active peptides. *Proc Natl Acad Sci U S A* 86:699–703, 1989

137. Speziale P, Raucci G, Visai L, et al: Binding of collagen to *Staphylococcus aureus* Cowan 1. *J Bacteriol* 167:77–81, 1986

138. Shiro H, Meluleni G, Groll A, et al: The pathogenic role of *Staphylococcus epidermidis* capsular polysaccharide/adhesin in a low-inoculum rabbit model of prosthetic valve endocarditis. *Circulation* 92: 2715–22, 1995

139. Swartz MN: Hospital-aquired infections: diseases with increasingly limited therapies. *Proc Natl Acad Sci USA* 91:2420–7, 1994

140. Switalski LM, Speziale P, Höök M: Isolation and characterization of a putative collagen receptor from *Staphylococcus aureus* strain Cowan 1. *J Biol Chem* 264: 21080–6, 1989

141. Switalski LM, Patti JM, Butcher W, et al: A collagen receptor on *Staphylococcus aureus* strains isolated from patients with septic arthritis mediates adhesion to cartilage. *Mol Microbiol* 7:99–107, 1993

142. Timmerman CP, Fleer A, Besnier JM, et al: Characterization of a proteinaceous adhesin of *Staphylococcus epidermidis* which mediates attachment to polystyrene. *Infect Immun* 59:4187–92, 1991

143. Tojo M, Yamashita N, Goldmann DA, et al: Isolation and characterization of a capsular polysaccharide adhesin from *Staphylococcus epidermidis*. *J Infect Dis* 157:713–22, 1988

144. US Department of Health and Human Services, Public Health Service: National nosocomial infectious surveillance (NNIS) report. Data summary from October 1986-April 1996, issued May 1996. *Am J Infect Control* 24:380–8, 1996

145. US Department of Health and Human Services, Public Health Service: National nosocomial infectious surveillance (NNIS) report. Data summary from October 1986-April 1997, issued May 1997. *Am J Infect Control* 24:477–87, 1997

146. Valentin-Weigand P, Timmis KN, Chhatwal GS: Role of fibronectin in staphylococcal colonisation of fibrin thrombi and plastic surfaces. *J Med Microbiol* 38:90–95, 1993

147. Vaudaux P, Suzuki R, Waldvogel FA, et al: Foreign body infection: role of fibronectin as a ligand for the adherence of *Staphylococcus aureus*. *J Infect Dis* 150:546–53, 1984

148. Vaudaux P, Waldvogel FA, Morgenthaler JJ, et al: Adsorption of fibronectin onto polymethylmethacrylate and promotion of *Staphylococcus aureus* adherence. *Infect Immun* 45:768–74, 1984

149. Vaudaux P, Pittet D, Haeberli A, et al: Host factors selectively increase staphylococcal adherence on inserted catheters: a role for fibronectin and fibrinogen or fibrin. *J Infect Dis* 160:865–75, 1989

150. Vaudaux P, Pittet D, Haeberli A, et al: Fibronectin is more active than fibrin or fibrinogen in promoting *Staphylococcus aureus* adherence to inserted intravascular catheters. *J Infect Dis* 167:633–41, 1993

151. Vaudaux PE, Lew DP, Waldvogel FA: Host factors predeposing to and influencing therapy of foreign body infections. In: Bisno AL, Waldvogel FA, eds. *Infections Associated With Indwelling Medical Devices*. 2nd ed. American Society of Microbiology, Washington, DC, 1994:1–29

152. Vaudaux PE, Francois P, Proctor RA, et al: Use of adhesion-defective mutants of *Staphylococcus aureus* to define the role of specific plasma proteins in promoting bacterial adhesion to canine arteriovenous shunts. *Infect Immun* 63:585–90, 1995

153. Veenstra GJ, Cremers FF, van Dijk H, et al: Ultrastructural organization and regulation of a biomaterial adhesin of *Staphylococcus epidermidis*. *J Bacteriol* 178:537–41, 1996

154. Wade D, Palma M, Lofving-Arvholm I, et al: Identification of functional domains in Efb, a fibrinogen binding protein of *Staphylococcus aureus*. *Biochem Biophys Res Commun* 248:690–5, 1998

155. Wang IW, Anderson JM, Marchant RE: *Staphylococcus epidermidis* adhesion to hydrophobic biomedical polymer is mediated by platelets. *J Infect Dis* 167:329–36, 1993

156. Yacoub A, Lindahl P, Rubin K, et al: Purification of a bone sialoprotein-binding protein from *Staphylococcus aureus*. *Eur J Biochem* 222:919–25, 1994

157. Younger JJ, Christensen GD, Bartley DL, et al: Coagulase-negative staphylococci isolated from cerebrospinal fluid shunts: importance of slime production, species identification, and shunt removal to clinical outcome. *J Infect Dis* 156:548–54, 1987
158. Ziebuhr W, Heilmann C, Gˆtz F, et al: Detection of the intercellular adhesion gene cluster (ica) and phase variation in *Staphylococcus epidermidis* blood culture strains and mucosal isolates. *Infect Immun* 65:890–6, 1997
159. Ziebuhr W, Krimmer V, Rachid S, et al: A novel mechanism of phase variation of virulence in *Staphylococcus epidermidis*: evidence for control of the polysaccharide intercellular adhesin synthesis by alternating insertion and excision of the insertion sequence element IS256. *Mol Microbiol* 32:345–56, 1999

# 21

# Studying Bacterial Adhesion to Irregular or Porous Surfaces

## Lucio Montanaro and Carla R. Arciola

*Research Laboratory on Biocompatibility of Implant Materials,
Rizzoli Orthopaedic Institute, Bologna, Italy*

## I. INTRODUCTION

### A. Bacterial Adhesion and Peri-Implant Infection

Bacterial adhesion to surfaces of artificial materials is considered the basal pathogenetic mechanism of the prosthesis-associated infections.[13,27,31,35] The relationships between a bacterium and a biomaterial, namely the interactions contracted, in periprosthesis tissues, between the bacterial external structure and the material surface, are the subject of more and more detailed studies, as the infection still represents today a very serious problem in implant surgery. Peri-implant infections are not resolved by antibiotics or surgical cleaning, only by prosthesis removal, and represent the main reason for failure of transdermal or implanted medical devices.

The severity of these infections must be essentially ascribed to the expression of the adhesive aptitude of the microorganisms, that in turn is conditioned by the characteristics of the material by which the medical device is constituted, is favored by the weakening of the periprosthesis tissues and is obviously triggered by an initial contamination of the implant.

Over the last few years several studies have been carried out in order to elucidate what structures and what mechanisms are implied with the capacities of some bacterial species – usually regarded as opportunistic because of their poor pathogenic power – to cause severe and irreducible infections associated to biomaterials.[25] Therefore many studies have been addressed to clear up the chemical and physical properties of the materials relevant in favoring the infection, in order to develop infection-resistant materials, namely new materials and new coatings with an anti-adhesive surface.

The physicochemical properties of the material (i.e., surface charge, hydrophobicity, surface roughness, and surface configuration) and the structural modification of the prosthesis during the application may modulate the adhesion.[35] Irregularities or porosity of the biomaterial surface seem to promote adhesion and offer a physical protection to the microorganisms against antibiotics and host's immune defenses. Bacteria can also provoke breaks of the implant surface by changing the microenvironment (pH,

*Handbook of Bacterial Adhesion: Principles, Methods, and Applications*
Edited by: Y. H. An and R. J. Friedman © Humana Press Inc., Totowa, NJ

oxygenation level, and enzyme activities). The micromovements of the implanted material or the release of wear debris can damage the tissues around the prosthesis, thus enhancing the chance of microorganism adhesion. Moreover, substances released by corrosion and wear can be used by bacteria for their metabolism so as to change their response to antibodies.

Thus in tissue surrounding prostheses a *"locus minoris resistentiae"* occurs which helps bacteria attachment.[49] The surface of the material provides the microorganism with the opportunity to colonize and this opportunity is preferentially taken by microorganisms provided with adhesion mechanisms.

### B. Aspecific and Specific Factors Promoting Bacterial Adhesion and Infection

Many aspects have to be evaluated when studying peri-implant infection and bacterial adhesion mechanisms. The factors promoting a high infection incidence can be both aspecific (i.e., common to all postsurgical infections) or specific (i.e., typical of peri-implant infections). The pathogenesis of peri-implant infections differs from post-surgical infections for the phenomena strictly related to the biomaterial.

Postsurgical infections, along with prosthesis-associated infections, may be based on several mechanisms, related to the patient, the microorganisms, the surgical tissue damage and the hospital environment. Patients can show an increased sensitivity to peri-implant infection due to depression of the immune defenses. Such deficiency can be determined by intercurrent diseases or by immune-suppressive treatments, including antiblastic chemotherapy, radiotherapy or corticosteroid administration.

Microorganisms can be endogenous or exogenous. In the first case, the organisms normally growing on the skin and in the nose and oral cavity (e.g., *Streptococcus* sp. and *Staphylococcus* sp.) or in the gastrointestinal tract (e.g., *Bacteroides* and *Escherichia coli*) become virulent and cause an infection. In the exogenous transmission, the patient is infected either by other patients or by the hospital personnel, directly or by air. As mentioned above, microorganisms responsible for hospital infections take advantage of the weakening of the body defenses to establish an infection. Among prosthesis-associated infection, more than 50% are caused by Gram-positive bacteria, with *Staphylococcus epidermidis* and *Staphylococcus aureus* as the most frequent agents. *Enterobacteriaceae* have been shown to be the most commonly encountered Gram-negative pathogens (25–30%). Streptococci and *Pseudomonas* sp. also are often involved.

The widespread use of antibiotics has been associated with the development of resistant bacteria following suppression of the host's endogenous flora.

The extent of surgical damage to tissues is related to the complexity of the surgical procedures (e.g., use of cauterizers or retractors). The duration of surgery, with prolonged exposure of the wound, the hematoma formation, and thrombosis occurrence are additional factors promoting the entrapment of microorganisms.

The hospital plays an important role as far as the chances to acquire postsurgical infections and the nature of such infections are concerned. The operating rooms are focal sites for the development of infections, both because of the exposure of the host's deep tissues and because of the presence of many infection sources, mainly people. Prosthesis-associated infections are enhanced by specific factors, including the biomaterial and its structure.

Virulence properties of bacteria are brought about by bacterial envelopes, including the cell membrane, wall and capsule. Some bacterial species with glycocalyx, such as *S. epidermidis* have been shown to adhere onto biomaterials and to form a slime.[12,39,44] This consists of a mudlike glycidic material that protects bacteria. It is thought to be important for intercellular connection during surface colonization. The current concept is that the production of slime will be particularly important for events after the initial phase of adhesion, which include colonization of various surfaces, protection against phagocytosis, interference with the cellular immune response and reduction of antibiotic effects.[46] The penetration of antibiotics is hampered by slime, whereas the capture of nutrients and the resistance against the host's defenses are enhanced.[34,53] Studies on *S. epidermidis* have shown that only some strains are able to form slime.[16] Bacterial strains that do not produce slime are less adherent and less pathogenic. Seemingly, not all bacteria have a glycocalyx, but those lacking this envelope have also been shown to adhere to the artificial surfaces. Actually, the bacterial cell has often more than one structure for the adhesion, also, plasma membrane proteins are involved.[67] Other external components of the bacterial cell, such as the phospholipids of the plasma membrane or the lipoteicoic acid of the *S. aureus* wall are involved in the adhesion mechanisms.

Biomaterials can favor infections also by increasing the bacterial replication or inducing resistance to antibiotics.[32,33] Physicochemical characteristics of the material and the biomaterial surface organization is fundamental for promoting bacterial adhesion, development of protected colonies, and therefore infection. Biomaterials can damage tissues both during surgery, such as polymethylmethacrylate (PMMA), which develops a high polymerization temperature, or after some time, due to prosthesis micromovements and release of wear particles. Moreover, the host's immune system may be depressed by biomaterials.

## II. POROUS MATERIALS AND SURFACE CHARACTERISTICS INFLUENCING ADHESION

### A. Bacterial Factors

Adhesion is a close relationship—often a true receptorial binding—between the "living" biological surface of the bacterium and an artificial surface, that of the material (which in vivo is coated, moistened, modified by biological molecules and reactive fluids of the host). The characteristics both of the adherent bacterium and of the colonized material are therefore essential in bringing about the modalities of such a relationship.

Among the bacterium characteristics, besides the specific adhesive structures (slime and/or adhesins) two more properties deserve to be mentioned: hydrophobicity and the surface charge.

Bacterial hydrophobicity of surface is an important physical factor for adhesion, especially when the substrata surfaces are either hydrophilic or hydrophobic. The hydrophobicity of bacteria varies according to bacterial species and is influenced by growth medium, bacteria age and bacterial surface structure. Generally, bacteria with hydrophobic properties prefer hydrophobic material surfaces, the ones with hydrophilic characteristics prefer hydrophilic surfaces.

Bacterial surface charge may be another important physical factor for bacterial adhesion. Most particles acquire an electric charge in aqueous suspensions due to the ionization of their surface groups. The surface charge attracts ions of opposite charge in the medium and results in the formation of an electric double layer. Isoelectric point, zeta potential and electrophoretic mobility characterize the surface charge. Bacteria in aqueous suspensions are generally negatively charged. Bacterial species, growth medium, bacterial age and bacterial surface structure affect surface charge.[18,41]

As far as the material is concerned, both the chemical and the physical properties of its surface affect the adhesion. According to Gristina and Costerton,[31] *S. epidermidis* preferentially adheres to polymers and *S. aureus* to metals. The implant surface properties may influence the adherence of staphylococci to prosthetic materials.[4,13] Finishing the material surface with a repellent coat to reduce bacterial adherence can reduce the risk of developing an infection. Many studies were focused on the search for coatings able to make the material repellent to bacterial adherence. Surfaces of biomaterials soaked with antibiotics, or coated with polyacrylamide films bound to antiseptic substances, or with quaternary amines containing organosilicon salts have shown antibacterial properties.[19,30,37,65] *Staphylococcus* adherence on polystyrene was limited by modifying artificial surface with surfactants.[15] A significant reduction in bacterial adherence was noted on polyethylene therephthalate surface modified with polyethylene oxide.[21] Biologic molecules, such as heparin, were also successfully used for this purpose.[3,23]

Most of the above listed treatments were performed on polymers, but metal surfaces can also be modified. Dunkirk et al.,[23] for example, have proposed photochemical coupling process both on polymers and metals for the prevention of bacterial colonization. Opalchenova et al. have demonstrated a significant antimicrobial effect of calcium phosphate ceramics.[55] Among the most important chemical characteristics of the material surface affecting bacterial adhesion are hydrophobicity and hydrophilicity. Depending on the hydrophobicity of both bacteria and material surfaces, bacteria adhere differently to materials. Besides chemical characteristics, equally important are physical properties, particularly porosity, roughness and surface configuration.

### B. Porosity

Porous materials are used in a variety of biomedical applications including implants and filters for extracorporeal devices. In other applications, porosity may be an undesirable characteristic since pores concentrate stress and decrease mechanical strength. Perhaps the most important physical quantity associated with porous materials is the solid volume fraction, $V_s$. The porosity, often expressed as a percent figure, is given by:

$$Porosity = 1 - V_s. \tag{1}$$

It is also noted that there are three measurements of volume, i.e., true, apparent, and total (bulk) volume: true volume = total volume – total pore volume; apparent volume = total volume – open pore volume; total pore volume = open pore volume + closed pore volume:

The pore size is important in situations in which tissue ingrowth is to be encouraged, or if the permeability of the porous material is of interest. Porous materials may be characterized by a single pore size, or may exhibit a distribution of pore sizes. Porosity and pore size can be measured in a variety of ways. If the density of the parent solid is known, a measurement of apparent density of a block of material suffices to determine the porosity. Mercury intrusion porosimetry is a more precise method that measures the porosity and the pore size distribution. In this method, mercury is forced into the pores under a known pressure and the relationship between pressure and mercury volume is determined. Since the mercury has a high surface tension and does not wet most materials, higher pressures are required to force the mercury into progressively smaller pores. If a single pore size predominates, it can be measured by optical or electron microscope.[56]

### C. Surface Roughness

Surface roughness is a two-dimensional parameter, which can be measured by "roughness measuring systems" such as the stylus system and is commonly described as the arithmetic average roughness. It is a distance measurement between the peak and the valley part on a material surface and does not represent the morphological configuration of the surface.[1,2]

### D. Surface Configuration

Closely related to surface roughness is the surface configuration. This is a three-dimensional parameter much more complex than the simple roughness. It gives a morphological analytical description of the pattern of a material surface, or a gridlike surface. Physical configuration is routinely evaluated by scanning electron microscopy.

Examined by electron microscopy, microorganisms seem to have a preference for the irregular and porous surfaces. These actually offer a wider exposed surface and provide hospitable niches in which bacteria can lurk and establish protected colonies. All kinds of material (metals, polymers, ceramics, composites, etc) can have surface irregularities, particularly when built as complex prostheses or made by assembling different pieces. The same applies to porosity, which, however, is generally more noticeable with some materials, such as, for example, resorbable ceramics like hydroxyapatite. McAllister et al.[45] found that the irregularities of polymeric surfaces promote bacterial adhesion and slime formation. Baker and Greenham[11] found that roughening the surface of either glass or polystyrene with a grindstone greatly increased the rate of bacterial colonization. Surfaces of implant materials with high porosity which are to be located in the oral cavity were found to be 25-fold more adhesive (and thus more prone to bacterial plaque formation) than smoother surfaces. Typical examples of the importance of the surface configuration are suture threads. They are both differently rough according to the constituting material (a higher roughness associated with natural threads, as silk and cotton, smoother with synthetic threads, like nylon) and to their weave, depending on whether they are made by a single or miltiple filaments. Merritt et al.[47] found that implant site infection rates are obviously different between porous and dense dental materials, and porous materials have a much higher rate of infection. This finding is consistent with the knowledge that bacteria preferentially colonize porous surfaces.

### III. METHODS

### A. Microscopic Observations

The optical microscope is probably the simplest instrument for the observation and counting of adherent microorganisms The obvious limitation is the requirement for optically clear, planar material for the substratum. The procedure also involves cell fixation, destroying the organisms, and does not provide information on viability. Moreover, if bacteria gather in clusters, chains or bunches, or if bacteria are covered by slime both the staining and the counting may present difficulties. With some thin and transparent material, the microscopic observation of adherent bacteria is easier by means of phase contrast microscopy, or with an inverted microscope.[10] After staining with fluorescent dyes specific for nucleic acids, bacteria adherent to opaque materials can also be observed. By this technique bacteria adherent to individual filaments of a braided suture can be localized and counted (unpublished data).

The scanning electron microscope (SEM) has a particular advantage in counting the number of organisms on opaque or highly textured surfaces; for example, it has been used to count bacteria on metals, plastics, ceramics, and catheters. The scanning electron microscope has the same limitations of the optical microscope in that individual members of aggregates cannot be reliably counted. Transmission electron microscopy (TEM) allows to estimate the clustering of bacteria and in some cases to locate the preferential sites of bacterial colonization onto porous materials.[4]

### B. Detaching and Counting Adherent Bacteria

The European Standards EN 1174-1 and EN 1174-2 (Sterilization of medical devices - Estimation of the population of microorganisms on product) describe the techniques for detaching of microorganisms from medical devices. Detached microorganisms are then collected onto filters and the filters are placed onto agar plates for culturing and counting of the colonies. In the stomaching technique the material and a known volume of eluent are enclosed in a sterile stomacher bag. This method is particularly suitable for soft, fibrous materials.

*Ultrasonication* is particularly suitable for solid impermeable samples and for products with complex shapes. The sonication energy and time of sonication should not be so great as to cause disruption and death of microorganisms. In our laboratory, in order to detach bacteria adherent to the biomaterial surfaces, the tubes containing the tested materials and 5 mL of saline were placed in the ultrasonic bath cleaner (Branson DTH 2210) operating at 47 kHz, 234 W, and sonicated for 6 min.

Other techniques for the removal of microorganisms include *shaking* with or without glass beads with a mechanical shaker; *vortex mixing* (only suitable for small materials with regular surfaces); *flushing* (the eluent is passed through the internal lumen of the product); *blending* or disintegration in a known volume of eluent.

### C. In Vitro Quantification of Bacterial Adhesion onto Porous Materials

Bacteria adherent to biomaterials are quantified using microbiological techniques (e.g., turbidimetric technique, radioactive labeling, dye elution, agar overlaying and contact plate). The quantitative evaluation of bacterial adhesion has to be assessed on sample with the same area. The adhesion has to be assessed on samples with the same surface area. The device in the final configuration is to be preferred to the raw material when testing bacterial adhesion, as the adhesive phenomena strongly depend on the surface geometry. Nylon is usually chosen as a negative control, since it is refractory to bacterial adhesion.

In our study, a very high adhesion rate was observed onto natural sutures, whereas it was lower onto synthetic resorbable sutures and at a minimum onto unresorbable synthetic ones. Braided sutures promote adhesion much more than monofilaments. Sugarman and Musher examined the effects of physical configuration of suture materials on bacterial adhesion: adhesion of bacteria to gut suture was up to 100 times greater than to nylon, and adhesion to polyglycolic acid or silk was intermediate.[66] When silicones are tested, adhesion to irregular surface of soft silicones is higher than that measured for hard silicones.[4]

Bacterial adhesion can be measured by the method of dye elution described by Merritt et al.,[48] consisting of dyeing adherent bacteria with crystal violet, eluting the dye with ethanol and measuring the eluted dye with a spectrophotometer. Materials are cut in 6 mm diameter disks and sterilized by dipping in 95% ethanol for 3 min and rinsed three times with sterile saline solution. Disks are incubated in 12-well plates with 2 mL of trypticase soy broth for 48 h at 37°C in the absence and in the presence of *S. epidermidis* RP62A, a standard strain for adherent organisms. Bacteria are seeded in each well at a final concentration of $10^6$ colony forming units/mL. At the end of incubation, disks are transferred to a new 12-well plate, washed five times with 4 mL of saline and fixed for 5 min with 2 mL of formalin. The fixative is removed by washing 5 times with 4 mL of distilled water and disks are dyed for 3 min with Hucker crystal violet solution. Excess dye is cleared by washing disks 5 times with 4 mL of saline following which the dye is eluted with 0.5 mL of ethanol for 5 min with gentle agitation. At the end of the elution 100 μL aliquots of each eluate are transferred to a 96-well plate and the absorbance measured with a spectrophotometric microplate reader at 540 nm.

Radiolabeling is a sensitive method for counting the number of microorganisms on the material surface. $^3$H, $^{14}$C, $^{32}$P, $^{35}$S, $^{111}$In, $^{75}$Se were employed as radioactive tracers for labeling microorganisms.[6,24]

The agar overlaying consists of coating the surfaces of the material with a molten agar culture medium (at a maximum temperature of 45°C). After the incubation period colonies are visible. Contact plates or slides are means by which solidified culture medium can be applied to a surface with the intention that viable microorganisms will adhere to the surface of the medium. The plate or slide can then incubated to produce colonies, which can then be counted. Results are directly related to the area in contact with the solidified culture medium.

### D. Capability of Modifying Bacterial Structure Following Adherence to Prostheses

Bacterial adherence is not restricted to a simple physical contact between the microorganism and the biomaterial surface; a biochemical interaction is needed. Therefore, the changes in bacterial surface arrangement following adhesion to biomaterials have to be studied. A possibility is represented by the study of the structural variations in microorganisms, which can be quantified after the adhesion step. In our laboratory, the phospholipids of the bacterial cell membrane have been identified by capillary column gas chromatography. By this procedure, the long chain fatty acids of the bacterial cell wall can be identified both before and after adherence to materials. This is accomplished by calculating the retention time and the relative area underlying the peaks of the chromatograph. Following adhesion of *S. aureus* onto PMMA and heparin-surface-modified PMMA (HSM-PMMA), the bacterial fatty acids were found significantly altered upon adhesion to the modified surface.[3]

### E. Resistance to Antibiotics and Bacterial Adherence

Enlightenment regarding the pathogenetic mechanisms of the biomaterial-associated infections is the necessary premise for the adoption by clinicians of efficacious prophylactic and therapeutic measures. The available antibiotic therapies are usually ineffective, because of the antibiotic resistance of the bacterial strains isolated from peri-prosthesis tissues. The onset of antibiotic resistance in staphylococci responsible for nosocomial infections has been ascribed to the positive selective pressure carried on by the large clinical usage of broad-spectrum antibiotics.

With regard to the antibiotic resistance of bacteria isolated from peri-prosthesis infections, it can be conceived that the biomaterial, acting as a substrate for the bacterial adherence, leads to a selection, among the whole contaminant bacterial population, of variants endowed with more marked adhesive properties as well as with increased resistance towards antibiotics. However, the relationship between adherence and antibiotic resistance has not yet been elucidated, nor is it clear whether the correlation between the two bacterial features is always positive.

The resistance to antimicrobial agents can also be acquired through transfer of genetic material from a bacterium to another by transduction, transformation or conjugation. Therefore a population of bacteria susceptible to antibiotics, provided it is large enough, is likely to contain some antibiotic-resistant mutants which can preferentially multiply under the selective pressure of the drug. Moreover mutational changes conferring antibiotic resistance may simultaneously alter virulence factors and thus modify the bacterial pathogenicity.[36,60,64]

The role of biomaterials in the selection of more adhesive and more antibiotic-resistant mutants could be investigated by employing a strain slightly resistant to antibiotics before the exposure to biomaterials.

Antimicrobial susceptibility testing after adhesion to devices has been carried out with antibiotics commonly used in surgery. The area of inhibited growth around each antibiotic disk was measured using the image analyzer.[51] All materials intended for implantation should be assayed, as the possibility that an increased resistance to antibiotics is induced cannot be excluded.

### F. Inhibition of Bacterial Replication by Biomaterials

In vivo bacterial adhesion is associated with bacterial growth in protected colonies. The possibility that biomaterials affect bacterial replication cannot be disregarded. An increase in bacterial growth could be one of the ways through which the device enhances infectability. Bacterial replication in vitro in the presence of biomaterials is compared with controls without materials.

### G. Phenotypic Characterization of Slime Production

The production of slime by the strains detached from infected biomaterials can be studied by means of two different methods: the Christensen method (this method involves incubation of the microorganisms on polystyrene plates, dyeing with crystal violet and reading the optical density),[17] and by the culture of the strains on Congo Red agar.[28] Plates are incubated for 24 h at 37°C and subsequently overnight at room temperature. On these plates the nonslime-forming colonies appear red and black the slime-forming ones appear black. The *Staphylococcus* strains are classified into slime-forming and nonslime-forming.

## H. Research on Bacterial Genes for Adhesion and Antibiotic Resistance

*S. epidermidis* and *S. aureus* are opportunistic pathogens involved in biomaterial-centered infection. Genes encoding antibiotic-binding proteins (mecA) and genes required for specific adhesion mechanisms (adhesin genes, i.e. the genes encoding surface receptors for collagen, fibronectin, fibrinogen) are searched in resistant/adhering bacteria of periprosthetic infections by means of polymerase chain reaction (PCR) techniques.[38,40,52]

The adhesins that mediate the binding of host proteins, termed MSCRAMMs to denote their role as "microbial surface components recognizing adhesive matrix molecules," could mediate adhesion on prostheses by binding to host protein that in vivo cover the implant surface.[29,43,57-59] What is the role of MSCRAMMs? We suggest that a study of the presence and expression of genes for adhesion molecules may help in clarifying the relevance of the different adherence mechanisms in the pathogenesis of prostheses associated infections.[50] Genes encoding MSCRAMMs of *Staphylococcus aureus* that bind fibronectin, fibrinogen, elastin, osteopontin and collagen have been identified.

## I. Bacterial Adhesion to Porous Materials

The material's surface and the biological environment's features affect the adhesive behavior of microorganisms. In order to reduce the incidence of prosthesis-associated infections, several biomaterial surface treatments have been proposed. It has been, moreover, inferred that infections should be controlled not only by direct inhibition of bacterial adherence, but also by enhancement of tissue compatibility or integration. Over the last few years, with the progress in knowledge, the concept itself of biocompatibility has changed. If once the ideal biomaterial was considered that maximally inert, today the search is for a bioactive material, able to positively interact with the host tissues.

A metal prosthesis surface treatment with hydroxyapatite (HA) was recently devised in order to promote integration between bone tissue and prosthesis.[63] Clinical studies performed on implants, up to now, highlight how HA-coating improves the bone-biomaterial interface, conferring therefore on the prostheses good osteointegration capabilities. Some studies suggest that use of HA-coated prostheses is related with a lower infection incidence. In prostheses more exposed to the risk of contamination, such as screws for external fixation, this behavior could result extremely advantageous. Hydroxyapatite is a highly bioactive material, which rather quickly undergoes surface modifications after implantation.

In our laboratory the adherence of a staphylococal strain to HA-coated stainless steel screws has been evaluated in vitro as compared with the adherence measured on uncoated stainless steel screws.[7] Adherence was also evaluated after prolonged immersion of the screws in saline, in order to modify the surface properties as it happens in vivo. Kummer et al.[42] demonstrated increased instability of HA-coatings when tested in physiological solutions. Radin et al.[61] studied the dissolution of HA in simulated interstitial fluid. Surface reactivity, i.e., capability to interact in vivo with surrounding tissues and fluids as related to bone formation and bone tissue bonding, was evaluated by Ducheyne et al.[22] by means of zeta-potential measurement in several physiological solutions; these measurements act as a suitable method to assess the actual state of the solid solution interface *in situ*. In our study, HA-coating modifications were evaluated by determining Ca and P concentrations in the medium in relation with immersion time. The adherence tests were performed:

1. on coated and uncoated screws exposed to air;
2. on coated and uncoated screws kept for 72 h in 0.9% NaCl, then removed from the solution, washed with saline and reincubated in fresh medium;
3. on coated and uncoated screws kept for 168 h in 0.9% NaCl, then removed from the solution, washed with saline and reincubated in fresh medium.

Bacterial adherence on HA-coated screws resulted significantly lower than on uncoated screws. In HA-coated screws adherence decreases significantly after immersion in solution; concurrent release of calcium and phosphorus from screws has been observed. These observations point out that both surface properties and their changes after immersion in saline solution influence bacterial adherence. Surface properties can therefore favor in vivo the material–tissue integration, and at the same time can limit materials infectability. In the specific case of the HA-coated screws examined in our study, if it is assumed that the Ca and P release is representative of the material's tendency to integrate with bone tissue, and that such a tendency matches with the tendency to hamper bacterial adherence, it can be inferred that the most refractory surface to bacterial adherence is the one which allows the maximum prosthesis integration. This approach allowed the evaluation of, in conditions close to the real in vivo situation, the anti-adhesive properties of the HA coating and suggested that HA-coating limits infectability both by favoring the material–tissue integration and reducing bacterial adherence.

### J. New Technology: Atomic Force Microscopy

Atomic force microscopy (AFM) has been shown to be an exquisitely sensitive tool for analyzing whether interactions between bacteria and biomaterial surfaces are attractive or repulsive and for understanding the nature of the underlying forces.[62] To perform AFM measurements a confluent bacterial lawn on the surface of a suitable support is established. In the absence of a confluent cell layer AFM is used in imaging mode prior to force measurements, in order to locate bacteria. Then the tip of the cantilever, made of silicon nitride, is approached towards the surface in 100 nm increments and the force curves recorded to determine the initial interactions between surfaces and bacteria. Measurements of forces between bacteria and different biomaterials are made by attaching microspheres of the material under investigation onto the cantilever and probing bacterial lawns. This approach is limited by the scarcity of biomaterials that can be shaped as microspheres 10–30 μm in diameter. Alternatively the cantilever tip can be coated with a confluent layer of bacteria and force curves between the modified tip and a planar surface of the material to be investigated can be obtained. In this mode AFM can be readily employed to measure the interactions between virtually any bacterium and surface of interest.

### REFERENCES

1. An YH, Friedman RJ: Concise review of mechanisms of bacterial adhesion to biomaterial surfaces. *J Biomed Mater Res* 43:338–48, 1998

2. An YH, Friedman RJ, Draughn RA, et al: Rapid quantification of staphylococci adhered to titanium surfaces using image analysis epifluorescence microscopy. *J Microbiol Methods* 24:29–40, 1995
3. Arciola CR, Caramazza R, Pizzoferrato A: *In vitro* adhesion of *Staphylococcus epidermidis* on heparin-surface-modified-intraocular lenses. *J Cataract Refract Surg* 20:195–202, 1994
4. Arciola CR, Cenni E, Caramazza R, et al: *In vitro* seven surgical silicones retain differently *Staphylococcus aureus*. *Biomaterials* 16:681–4, 1995
5. Arciola CR, Maltarello Mc, Cenni E, et al: Disposable contact lenses and bacterial adhesion. *In vitro* comparison between high water content/ionic and low water content/non-ionic lenses. *Biomaterials* 16:685–90, 1995
6. Arciola CR, Montanaro L, Caramazza R, et al: Inhibition of bacterial adherence to high-water content polymer by a water-soluble, nonsteroidal anti-inflammatory drug. *J Biomed Mater Res* 42:1–5, 1998
7. Arciola CR, Montanaro L, Moroni A, et al: Hydroxyapatite-coated orthopaedic screws as infection resistant material: *in vitro* study. *Biomaterials* 20: 323–7, 1999
8. Arciola CR, Pizzoferrato A, Bertoluzza A, et al: Soft contact lens interactions with bacteria. In: Caramazza R, ed. *Biomaterials in Ophthalmology*. Studio ERC, Bologna, Italy, 1990:71–4
9. Arciola CR, Radin L, Alvergna P, et al: Heparin surface treatment of PMMA alters adhesion of a *Staphylococcus aureus* strain: utility of bacterial fatty acid analysis. *Biomaterials* 14:1161–4, 1993
10. Arciola CR, Monti P, Caramazza R, et al: Evaluation of adhesion of conjunctival *Staphylococcus epidermidis* on polymeric materials. In: De Putter C, et al, eds. *Implant Materials In Biofunction*. Elsevier, Amsterdam, 1988:349–54
11. Baker AS, Greenham LW: Release of gentamicin from acrylic bone cement: elution and diffusion studies. *J Bone Joint Surg* 70:1551–7, 1988
12. Baldassarri L, Christensen GD, Donelli G: Is slime the virulence factor in staphylococcal biomaterial-associated infections? *Microecol Therapy* 25:103–7, 1995
13. Barth E, Myrivik QN, Wagner W, et al: *In vitro* and *in vivo* comparative colonization of *Staphylococcus aureus* and *Staphylococcus epidermidis* on orthopaedic implant materials. *Biomaterials* 10:325–8, 1989
14. Bertoluzza A, Arciola CR, Simoni R, et al: Structural studies of the factors affecting bacterial adhesion on soft contact lenses. *Transactions of the 4th World Biomaterials Congress*, Berlin, 1992, 8:20
15. Bridgett MJ, Davies MC: Control of staphylococcal adhesion to polystyrene surfaces by polymer surface modification with surfactants. *Biomaterials* 13:411–6, 1992
16. Christensen GD: Adherence of slime-producing strains of *Staphylococcus epidermidis* to surfaces. *Infect Immun* 37:318–26, 1982
17. Christensen GD, Simpson WA, Younger JJ, et al: Adherence of coagulase-negative staphylococci to plastic tissue culture plates: a quantitative model for the adherence of styaphylococci to medical devices. *J Clin Microbiol* 22:996–1006, 1985
18. Dankert J, Hogt AH, Feijen J: Biomedical polymer: bacterial adhesion, colonization and infection. *CRC Crit Rev Biocompat* 2:219–301, 1986
19. Deitch EA, Marino AA, Gillespie TE, et al: Silver-Nylon: a new antimicrobial agent. *Antimicrob Agents Chemother* 23:356–9, 1983
20. Delmi M, Vadaux P, Lew DP, et al: Role of fibronectin in staphylococcal adhesion to metallic surfaces used as models of orthopaedic devices. *J Orthop Res* 12:432–8, 1994
21. Desai NP, Hossainy SF, Hubbel JA: Surface-immobilized polyethylene oxide for bacterial repellence. *Biomaterials* 13:417–20, 1992
22. Ducheyne P, Kim CS, Pollak SR: The effect of phase differences on time-dependent variation of the zeta-potential of hydroxyapatite. *J Biomed Mater Res* 26: 147–68, 1992
23. Dunkirk, SG, Gregg SL, Duran LW, et al: Photochemical coatings for the prevention of bacterial colonization. *J Biomater Appl* 6:131–56, 1991

24. Farber BF, Wolff AG: Salycilic acid prevents the adherence of bacteria and yeast to silastic catheters. *J Biomed Mater Res* 27:599–602, 1993

25. Foster TJ, Mcdevitt D: Molecular basis of adherence of staphylococci to biomaterials In: Bisno AL, Waldvogel FA, eds. *Infections Associated with Indwelling Medical Devices.* 2nd ed., American Society for Microbiology, Washington DC, 1994

26. Francois P, Vaudaux P, Foster TJ, et al: Host-bacteria interactions in foreign body infections. *Infect Control Hosp Epidemiol* 17:514–20, 1996

27. Francois P, Vaudaux P, Nurdin N, et al: Physical and biological effects of a surface coating procedure on polyurethane catheters. *Biomaterials* 17:667–78, 1996

28. Freeman DJ, Falkiner FR, Keane CT: New method for detecting slime production by coagulase-negative staphylococci. *J Clin Pathol* 42:872–4, 1989

29. Gillaspy AF, Patti JM, Pratt FP Jr, et al: The *Staphylococcus aureus* collagen adhesin-encoding gene (*cna*) is within a discrete genetic element. *Gene* 196:239–248, 1997

30. Greco RS, Harvey RA, Srnilow PC, et al: Prevention of vascular prosthetic infections by benzalkonium-oxacillin bonded polytetrafluoroethylene graft. *Surg Gynecol Obstetr* 155:28–32, 1982

31. Gristina AG, Costerton JW: Bacterial adherence to biomaterials and tissue. *J Bone Joint Surg* 67:264–73, 1985

32. Gristina AG, Hobgood CD, Webb LX: Adhesive colonization and antibiotic resistance. *Biomaterials* 8:423–6, 1987

33. Gristina AG, Jennings RA, Naylor PT, et al: Comparative *in vitro* antibiotic resistance of surface-colonizing coagulase-negative staphylococci. *Antimicrob Agents Chemother* 33:813–6, 1989

34. Hamory BH, Parisi JT, Hutton P: *Staphylococcus epidermidis*: a significant nosocomial pathogen. *Am J Infect Control* 15:59–74, 1987

35. Harches G, Feijen J, Dankert J: Adhesion of *Escherichia coli* to a series of poly(methacrylates) differing in charge and hydrophobicity. *Biomaterials* 12: 853–60, 1991

36. Hartman BJ, Thomas A: Low affinity binding protein associated with beta-lactam resistance in *Staphylococcus aureus*. *J Bacteriol* 158:513–6, 1984

37. Ikeda T, Yamaguchi H, Tazuke S: Self sterilizing materials 2: Evaluation of surface antibacterial activity. *J Bioact Compat Pol* 1:301–8, 1986

38. Inman RD, Gallegos KV, Brause BD, et al: Clinical and microbial features of prosthetic joint infection. *Am J Med* 77:47–53, 1984

39. Ishak MA, Groshel DH, Mandell GL, et al: Association of slime with pathogenicity of coagulase-negative staphylococci causing nosocomial septicemia. *J Clin Microbiol* 22:1025–9, 1985

40. Kreiswirth B, Kornblum J, Arbeit RD, et al: Evidence for a clonal origin of methicillin resistance in *Staphylococcus aureus*. *Science* 259:227–30, 1993

41. Krekeler C, Ziehr H, Klein J: Physical methods for characterization of microbial cell surfaces. *Experientia* 45:1047–54, 1989

42. Kummer JK, Jaffe WL: Stability of a cyclically loaded hydroxyapatite coating: Effect of substrate material, surface preparation, and testing environment. *J Appl Biomater* 3:211–5, 1992

43. Lech MS, Speziale P, Hook M: Isolation and characterization of a putative collagen receptor from *Staphylococcus aureus* strain Cowan 1. *J Biol Chem* 264:280–6, 1998

44. Martin MA, Pfaller MA, Massanari MN, et al: Use of cellular hydrophobicity, slime production and species identification markers for the clinical significance of coagulase-negative staphylococcal isolates. *Am J Infect Control* 17:130–5, 1989

45. McAllister EW, Carey LC, Brady PG, et al: The role of polymeric surface smoothness of biliary stents in bacterial adhesion, biofilm deposition and stent occlusion. *Gastrointest Endosc* 39:422–5, 1993

46. Mempel M, Feucht H, Ziebuhr W, et al: Lack of mecA transcription in slime-negative phase variants of methicillin-resistant *Staphylococcus epidermidis*. *Antimicrob Agents Chemother* 38:1251–5, 1994

47. Merritt K, Shafer JW, Brown SA: Implant site infection rates with porous and dense materials. *J Biomed Mater Res* 13:101-8, 1979

48. Merritt K, Gaind A, Anderson JM: Detection of bacterial adherence on biomedical polymers. *J Biomed Mater Res* 39:415–22, 1998

49. Montanaro L: Infezione. *Enciclopedia Medica Italiana* 7:1859–63, 1979

50. Montanaro L, Arciola CR, Borsetti E, et al: A polymerase chain reaction (PCR) method for the identification of collagen adhesin gene (*cna*) in *Staphylococcus*-induced prosthesis infections. *Microbiologica* 21:359–63, 1998

51. Montanaro L, Arciola CR, Cenni E: *In vitro* response to methicillin of a *Staphylococcus aureus* susceptible strain (ATCC 25923) after adhesion on both PMMA and silicon. Research of *mec*A gene by means of polymerase chain reaction (PCR). Submitted, 1999

52. Murakami K, Minamide W, Wada K, et al: Identification of methicillin-resistant strains of staphylococci by polymerase chain reaction. *J Clin Microbiol* 29:2240–4, 1991

53. Needham CA, Stempsey W: Incidence, adherence and antibiotic resistance of coagulase-negative *Staphyloccus* species causing human disease. *Diagn Microbiol Infect Dis* 2:293–9, 1984

54. Nomura S, Lundberg F, Stollenwerk M, et al: Adhesion of staphylococci to polymers with and without immobilized heparin in cerebrospinal fluid. *J Biomed Mater Res* 38:35–42, 1997

55. Opalchenova G, Dyulgerova E, Petrov O: Effect of calcium phosphate ceramics on Gram-negative bacterial resistant to antibiotics. *J Biomed Mater Res* 32:473–9, 1996

56. Park JB, Lakes RS: *Biomaterials An Introduction*. 2nd ed., Plenum Press, New York. 4:71–3, 1992

57. Patti JM, Alle BL, Mcgavin MJ, et al: MSCRAMM-mediated adherence of microorganisms to host tissue. *Ann Rev Microbiol* 48:585–617, 1994

58. Patti JM, Jonsson H, Guss B, et al: Molecular characterization and expression of a gene encoding a *Staphylococcus aureus* collagen adhesin. *J Biol Chem* 267:4766–72, 1992

59. Paulsson MA, Kober C, Frei-Larsson C, et al: Adhesion of staphylococci to chemically modified and native polymers, and the influence of preadsorbed fibronectin, vitronectin and fibrinogen. *Biomaterials* 14:845–53, 1993

60. Pratt WB: Drug resistance. In: Pratt WB, Taylor P, eds. *Principles of Drug Action, The Basis of Pharmacology*. 3rd ed., Churchill Livingstone, New York, 1990:565–637

61. Radin SR, Ducheyne P: The effect of varying the calcium phosphate composition and ture on dissolution and reprecipitation reactions in simulated interstitial fluid. In: Davies JE, ed. *The Bone-Biomaterial Interface*. University of Toronto Press, Toronto, 1991

62. Razatos A, Ong YL, Sharma MM: Molecular determinants of bacterial adhesion monitored by atomic force microscopy. *Proc Natl Acad Sci USA* 95:11059–64, 1998

63. Rosen HM: The response of HA to contiguous time infection. *Plast Reconstr Surg* 88:1076–80, 1991

64. Sanderson PJ: Infection in orthopaedic implants. *J Hosp Infect* 18 (Suppl A):367–75, 1991

65. Speir JL, Malek JR: Destruction of microorganisms by contact with solid surfaces. *J Colloid Interface Sci* 89:68–76, 1982

66. Sugarman B, Musher D: Adherence of bacteria to suture materials. *Proc Soc Exp Biol Med* 167:156–60, 1981

67. Timmermann CP, Verhoef J: Characterization of a proteinaceous adhesin of *Staphylococcus epidermidis* which mediates attachment to polystyrene. *Infect Immun* 59:4187–92, 1991

68. Vaudaux P, Yasuda H, Velazco MI, et al: Role of host and bacterial factors in modulating staphylococcal adhesion to implanted polymer surfaces. *J Biomater Appl* 5:134–53, 1990

# Studying Bacterial Colonization of Tubular Medical Devices

## Gregor Reid

*Lawson Research Institute, London, Ontario, Canada.*
*Department of Microbiology and Immunology, The University of Western Ontario, Canada*

## I. INTRODUCTION

In medical practice, in both developed and developing countries, use of tubular devices such as drainage systems to remove urine, blood or wound fluids, or tubes to keep respiratory airways clear, is essential for sustaining the life of many millions of patients. These devices can be made by using a patient's own tissues, such as blood vessels transferred from one site to another, or intestinal tissues used to make a urinary conduit. In addition, because of advances made in polymer chemistry many artificial options such as urinary, peritoneal and central venous catheters, stents, shunts, tracheal and drainage tubes are now available.[8,11,17] Indeed, the majority of patients entering an intensive care setting or undergoing invasive surgery will require temporary use of one or more of such tubular devices.

While biocompatibility of a prosthesis is important, tubular devices tend not to be in place for long periods, tend not to be integrated into host tissues, and most are removed through open access (for example, via skin or urethra) once their function has been completed. It is the infection of these devices which constitutes the main clinical problem and the most significant barrier to their use.[25] For example, many hospitalized patients who are catheterized in the urinary tract will develop infections.[14,31] The primary stages of infection usually involve microbial adhesion to the conditioning film covered biomaterial, and this process will form the focus for the following sections with respect to how such adhesion can be studied in vitro and in vivo.

The formation of microbial biofilms has distinct stages, depending to some extent upon the location and accessibility of microbes. This process has recently been described in full elsewhere.[10,16,17] Some of the tools used to study biofilms also provide insight into adhesion and colonization.

## II. DIFFERENT TYPES OF SURFACES

Many different materials are employed in tubular form for human usage. These can be man-made or naturally derived. The most commonly used substances are polyethylene,

*Handbook of Bacterial Adhesion: Principles, Methods, and Applications*
Edited by: Y. H. An and R. J. Friedman © Humana Press Inc., Totowa, NJ

polytetrafluoroethylene, polyurethane, polyvinyl chloride, silicone, styrenic thermo-plastic elastomer and latex rubber as well as metals.[23] Other materials, including metals such as titanium, biodegradable meshes coated with collagen, and living tissues, are the subject of experimentation and appear to have potential for human use.[22]

Prior to testing any materials for bacterial adhesion, it is wise to have comprehensive knowledge of their chemical and physical properties. A series of techniques have been used for this purpose including scanning electron microscopy (SEM), X-ray photo-electron spectroscopy (XPS), energy dispersive X-ray analysis (EDX), secondary ion mass spectrometry (SIMS), polar and nonpolar fluid contact angle measurements, Fourier Transform infrared (FTIR) spectroscopy, atomic force microscopy (AFM), and high performance liquid chromatography (HPLC).[5,18,26,28] These types of analyses indicate the hydrophobicity, charge, chemical composition, roughness and topography of the materials; all these factors can influence microbial adhesion.

### A. *Urinary and Peritoneal Catheters*

Various types of catheters are available commercially. It is important to find out from the manufacturer or distributor whether or not the material has been surface treated. For example, some hydrophilic catheters are prepared by dipping the end section into the coating. This means that if this device was used for testing in vitro, the exact position of the coat would need to be determined, otherwise sections which were coated could be tested with ones which were uncoated. Also, coatings have different properties and some will come off the material within a short timeframe, while others will expand and alter the dimensions of the material. Not all coatings cover the inside and outside surfaces, and this should be determined prior to analysis. Some materials have antimicrobials used as coatings or integrated into the material, and these will influence microbial adhesion. Coloration should be consistent between batches. The internal dimensions of the catheter should be indicated on the packaging: this is important for reproducibility to make sure that the surface area of each section to which organisms are being exposed is identical. Lastly, many catheters have holes punched along the device: care must be taken to analyze sections with the same hole patterns or without any holes, otherwise adhesion results will be variable.

The nature of catheter infections, especially in the peritoneal cavity means that host proteins are often densely covering the inner parts of the device. Care must be taken not to disrupt this film when processing devices collected from patients.

### B. *Stents*

Ureteral, prostatic and biliary stents are composed of similar materials to those used in catheters. Stents were invented in 1970 and are defined as endoluminal mechanical support.[24] They tend to be much smaller in their internal diameter. If analyzing stents removed from patients, the ends must be marked (for example kidney versus bladder). Often there are loops at the end to enhance placement. These represent the parts closest to the organs. The nature of the bending make them more difficult to examine microscopically.

Urinary and biliary stents tend to become encrusted with calcium, magnesium and phosphate compounds which are extremely brittle. Thus, care must be taken when dissecting in vivo sections for analysis. In the near future, bioabsorbable and biodegradable stents will become commercially available.[3,30] These will have to be handled with particular care because by design they will degrade under certain conditions.

## C. Vascular Tubes

Medical devices and host blood vessels are used as vascular grafts. While the study of human vessels themselves is important, to best mimic the in vitro situation, the junctions of the stitching where the new blood vessel (or material) meets the old one should be a focus for investigation. It is at this interface where organisms are most likely to adhere and infect. If analyzing host tissues, it is vital to maintain their viability for the duration of the experimentation. Flow cell studies will provide the most useful data, as they can better simulate tubular flow found in vasculature. Selection of material for study could include polyurethane with heparin covalently attached, as this has been found to reduce microbial adhesion on central venous catheters.[2]

In general terms, for any given tube, sections between 1 and 5 cm should be adequate for study, with triplicate samples being tested.

## III. SELECTION OF ORGANISMS AND SUSPENDING FLUIDS

Invariably the organisms chosen for experimentation represent the pathogens which cause the most severe diseases. In rare instances, researchers may use bacteria because of their special, defined properties or because they represent normal flora which are believed to reduce the risk of infection. The difficulty comes when selecting the actual strains and their condition at the time of experimentation. There has been some debate in this area. One line of thinking is that the bacteria should be grown to express certain key virulence factor(s) or to optimize their ability to adhere. Others believe that the organism must originate from and also mimic the environment in which they infect the host, and not be subjected to repeated subculturing. A further consideration is that whatever culture conditions are chosen, it is preferred that the surface properties of the strains be known. For example, growth phase can alter hydrophobicity and expression of adhesins such as fimbriae.

Once grown, the organisms are usually washed or filtered to remove remnants of the growth media (unless say urine is used for growth and for adhesion testing). The washing often exposes the organisms to centrifugal forces which can affect and potentially remove surface adhesins. Reincubation of the final suspension at 37°C for 30 min may somewhat stabilize the organisms. Selection of concentration is important. Many bacterial adhesion studies have used 1000 times more bacteria than cells[29] or concentrations of $1\times10^8$ per mL[15] for 1cm long biomaterial sections. Lower concentrations are useful if the experiment is over a longer duration (as growth may occur and be important to the end point) or to simulate a lower inoculum at a particular infection site.

The fluid in which microbes and biomaterials are suspended can influence greatly the ability for adhesion to occur.[1] Where possible, the suspending fluid found in vivo should be used. However, as this might be difficult to sterilize without damaging its make-up, or it may increase or decrease growth of the organisms, physiological buffers can be used. Phosphate buffered saline has tended to be the fluid of choice, but other options such as artificial urine[12] can be used, but they may not contain hormones, intermediate

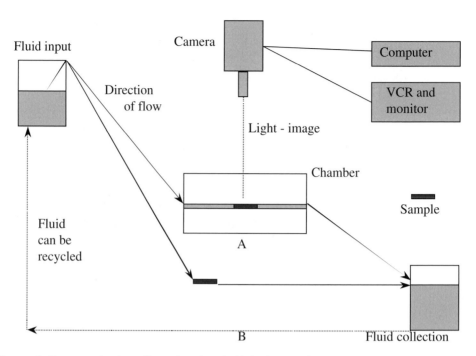

**Figure 1.** In scenario A, a flow chamber is linked to an image analysis system and the material is flat. In scenario B, a tubular piece of a device is examined. The bacterial suspension flows over the material and is dispensed or recycled.

metabolites (e.g., phenols, lactic acid), proteins (e.g., Tamm Horsfall glycoprotein) and cellular components (e.g., antibodies, red and white blood cells). The pH, electrolyte content and osmolarity should be checked for consistency prior to use.

## IV. METHODS OF DETERMINING MICROBIAL ADHESION AND DATA ANALYSIS

Experiments can be performed in static or flow conditions. The former mimics the contact made between a material and organisms on a tissue, such as the urethra or skin, while flow conditions simulate blood and urinary flow through tubes. If the primary goal of the experiment is to determine microbial adhesion to the inner portion of tubes, it is best to have a flow system in place (Fig. 1). Ideally, this should take into account Eddy currents, friction and air bubbles which can cause backflow and disruption of biofilms, respectively. Two devices have proved useful: the Robbins device[6] which has several exit points for inserting materials and the flow cell chamber.[4] Flow times can vary from a few minutes to hours, and even days. Most adhesion experiments are run for a few hours, while those for biofilms are run for two to five days. The fluids can be recycled and reused or replaced at a constant rate. For lengthy biofilm studies, replacement of fluids being flushed into the system is essential. The flow rate can mimic the slow dripping from the urethra into a catheter or the more rapid flow of micturition upon intermittent catheter insertion or peritoneal dialysis exchange.

A disadvantage of these two methods is that the surfaces should be flat when tested. In order to maintain a cylindrical shape during testing, the material can be attached at both

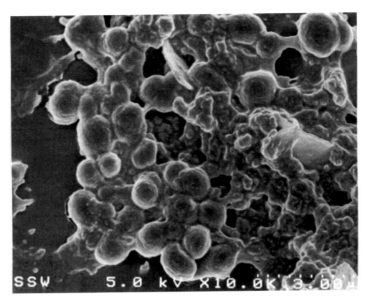

**Figure 2.** SEM image of Gram-positive coccal biofilm and encrustations deposited onto a ureteral stent recovered from a patient.

ends to tubing which provides an input and outlet for flowing the microbial suspension (Fig. 1). Controls should use the suspending fluids without microbes.

The flow cell chamber allows for real time analysis and quantification via a camera and video system if the material is opaque or used as a flat surface. The material can be further analyzed upon removal from the chamber. In general, enumeration of viable microbial counts requires removal of the organisms from the surface by water bath sonication, followed by dilution and inoculation onto an agar plate. Additional scraping and vortexing have been found to supplement the sonication.[13] Simply rolling the device over an agar plate is inadequate for accurate enumeration as the organisms do not easily come off the devices. Duplicate and triplicate plate counts should be collected and analyzed statistically.

Many studies of microbial adhesion have utilized SEM.[7,13,21,27,32] This can provide a useful visualization of the process which has taken place, such as bacterial biofilm formation (Fig. 2) and encrustation deposition (Fig. 3). While samples can be fixed prior to gold coating, it is possible to air dry them and process without fixation, thereby minimizing the hydration effects of the fixatives. Of course, to view the inner surface of tubes, the material must be dissected, and this itself can disrupt adherent organisms. The difficulty with SEM is to obtain a reliable count of adherent bacteria especially when multilayered biofilms and host proteins and encrustations are present. Also, if there is a mixed culture of two rod shaped species, SEM will not easily differentiate (except if labeling and back scatter mode is used).

## V. CONCLUSIONS

A large number of tubular devices are used each year in clinical and surgical practice to drain wounds, provide new blood vessels and conduits, and manage excretory output. Many types of microorganisms adhere to and colonize these structures, which are host

**Figure 3.** SEM image of ureteral stent recovered from a patient. The stent was opened logitudinally to show the extent of encrustation which was almost sufficient to block the flow of urine.

originated or made from polymers and metals. Therefore, there is much interest in the mechanisms whereby such adhesion processes occur and how infectious agents can be either prevented from adhering or removed once attached. The practical methods used to analyze material from in vitro investigations and materials removed from the host are the subject of this review. It is hoped that the reader will be provided with insight into the complexity of adhesion and be stimulated to contribute further knowledge to this important field.

There are many aspects to be considered when studying microbial adhesion to tubes. It is essential to have clear and feasible objectives from the outset and to utilize surfaces, organisms, fluids and test systems best suited to providing reliable answers (Fig. 4). Bacterial adhesion studies very often have quite large sample variations and so repetition is essential. Variations are also noted in vivo making it somewhat difficult to interpret adhesion counts. For example, a count of 5 bacterial per cell can be found on patients with symptomatic urinary tract infection, as can a count of 100 per cell. So is 100 significantly higher than 5 — yes, but what is the clinical significance? Just so, if 10,000 bacterial adhere to 1 cm of a polyurethane material and 100,000 to a polystyrene surface does this translate into clinical significance? The situation is made more complex by the finding of adherent bacteria, without signs and symptoms of infection, in certain patient populations — spinal cord injured, peritoneal dialysis and ureteric stent patients.[9,19,21] Therefore, the study of adhesion must, I believe, coincide with practical relevance, and the interpretation of data with respect to clinical significance must be done with a degree of caution.

**Acknowledgment**: The support of Cook Urology, Bayer Canada and Microvasive is appreciated. I acknowledge Ms. Christina Tieszer for the help with the electron micrographs.

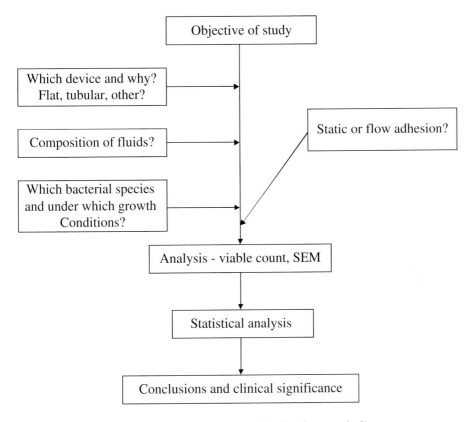

**Figure 4.** Stages in the study of microbial adhesion to tubular structures

## REFERENCES

1. Absolom DR, Lamberti FV, Policova Z, et al: Surface thermodynamics of bacterial adhesion. *Appl Environ Microbiol* 46:90–7, 1983
2. Appelgren P, Ransjo U, Bindsley L, et al: Surface heparinization of central venous catheters reduces microbial colonization *in vitro* and *in vivo*: results from a prospective randomized trial. *Crit Care Med* 24:1482–9, 1996.
3. Brauers A, Thissen H, Pfannschmidt O, et al: Development of a biodegradable ureteric stent: surface modification and *in vitro* assessment. *J Endourol* 11:399–403, 1997
4. Busscher HJ, Doornbusch GI, van der Mei HC: Adhesion of mutans streptococci to glass with and without a salivary coating as studied in a parallel-plate flow chamber. *J Dent Res* 71:491–500, 1992
5. Colliver TL, Brummel CL, Pacholski ML, et al: Atomic and molecular imaging at the single cell level with TOF-SIMS. *Anal Chem* 69:2225–31, 1997
6. Costerton JW, Lappin-Scott HM: Introduction to microbial biofilms. In: Lappin-Scott HM, Costerton JW, eds. *Microbial Biofilms*. University Press, Cambridge, UK, 1995:1–11
7. Cox AJ, Hukins DWL, Sutton TM: Infection of catheterized patients: bacterial colonization of encrusted Foley catheters shown by scanning electron microscopy. *Urol Res* 17:349–52
8. Dankert J, Hogt AH, Feijen J: Biomedical polymers: bacterial adhesion, colonization, and infection. *CRC Crit Rev Biocompatability* 2:219–301, 1986
9. Dasgupta MK, Bettcher KB, Ulan RA, et al: Relationship of adherent bacterial biofilms to peritonitis in chronic ambulatory peritoneal dialysis. *Perit Dial Bull* 7:168–73, 1987

10. Denstedt J, Wollin T, Reid G: Biomaterials used in urology: current issues of biocompatibility, infection and encrustation. *J Endourol* 12:493–500, 1998

11. Elliott TSJ, Faroqui MH: Infections and intravascular devices. *Brit J Hosp Med* 48:496–503, 1992

12. Geshnizgani AM, Onderdonk AB: Defined medium simulating genital tract secretions for growth of vaginal microflora. *J Clin Microbiol* 30:1323–6, 1992

13. Keane PF, Bonner MC, Johnston SR, et al: Characterization of biofilm and encrustation on ureteric stents *in vivo. Brit J Urol* 73:687–91, 1994

14. Nickel JC, Feero P, Costerton JW: Incidence and importance of bacteriuria in postoperative, short-term urinary catheterization. *Can J Surg* 32:131–2, 1989

15. Reid G: Adhesion of urogenital organisms to polymers and prosthetic devices. *Meth Enzymol* 253:514–9, 1995

16. Reid G: Bacterial colonization of prosthetic devices and measures to prevent infection. *New Horizons: The Science and Practice of Acute Medicine* 6:S58–63, 1998

17. Reid G: The development of microbial biofilms of medical prostheses. In: Tannock GW, ed. *Biofilms*. Chapman and Hall, London, 1998:477–86

18. Reid G, Busscher HJ, Sharma S, et al: Surface properties of catheters, stents and bacteria associated with urinary tract infections. *Surface Sci Rep* 7:251–73, 1995

19. Reid G, Charbonneau-Smith R, Lam D, et al: Bacterial biofilm formation in the urinary bladder of spinal cord injured patients. *Paraplegia* 30:711–7, 1992

20. Reid G, Denstedt J, Tieszer C: Encrustation and microbial adhesion on stents. In: Yachia D, ed. *Stenting the Urological System*. ISIS Medical Media, Oxford, UK, 1998:111–8.

21. Reid G, Denstedt JD, Kang YS,et al: Microbial adhesion and biofilm formation on ureteral stents *in vitro* and *in vivo. J Urol* 148:1592-4, 1992

22. Reid G, Millsap K, Denstedt J: Biomaterials in Urology II: future usage and management. *J Endourol* 13:1–7, 1999

23. Slepian MJ, Hossainy SFA: Characteristics of materials used in implants: polymers. In: Yachia D, ed. *Stenting the Urological System*. ISIS Medical Media, Oxford, UK, 1998:45–52

24. Slepian MJ, Yachia D: Urological stents: material, mechanical and functional classification. In: Yachia D, ed. *Stenting the Urological System*. ISIS Medical Media, Oxford, UK, 1998: 3–10

25. Stamm WE: Catheter-associated urinary tract infections: epidemiology, pathogenesis and prevention. *Am J Med* 91 (Supp 3B):65S–71S, 1991

26. Steele A, Goddard DT, Beech IB: An atomic force microscopy study of the biodeterioration of stainless steel in the presence of bacterial biofilms. *Int Biodeter Biodegrad* 34: 35–46, 1995

27. Stickler D, King J, Nettleton J, et al: The structure of urinary catheter encrusting bacterial biofilms. *Cells Mater* 3:315–20, 1993

28. Suci PA, Siedlecki KJ, Palmer RJ, et al: Combined light microscopy and attenuated total reflection Fourier transform infrared spectroscopy for integration of biofilm structure, distribution, and chemistry at solid-liquid interfaces. *Appl Environ Microbiol* 63:4600–3, 1997

29. Svanborg Eden C, Hanson LA, Jodal U, et al: Variable adherence to normal human urinary-tract epithelial cells of *Escherichia coli* strains associated with various forms of urinary-tract infection. *Lancet* 1:490–2, 1976

30. Talja M, Valimaa T, Tammela T, et al: Bioabsorbable and biodegradable stents in urology. *J Endocrinol* 11:391–7, 1997

31. Trexler Hessen H, Kaye D: Infections associated with foreign bodies in the urinary tract. In: Bisno A, Waldvogel F, eds. *Infections Associated with Indwelling Medical Devices*. American Society for Microbiology, Washington, DC, 1989:199–213

32. Wollin TA, Tieszer C, Riddell JV, et al: Bacterial biofilm formation, encrustations and anti-biotic adsorption to ureteral stents indwelling in humans. *J Endourol* 12:101–11, 1998

# 23

# Studying Plaque Biofilms on Various Dental Surfaces

**Doron Steinberg**

*Department of Oral Biology, Hebrew University–Hadassah, Jerusalem, Israel*

## I. INTRODUCTION

Oral biofilms harboring pathogenic bacteria, are the major contributing virulent factors associated with diseases of the oral cavity, such as tooth decay, gingivitis and periodontitis. In addition to the virulent properties of the biofilm, adhesion to oral surfaces is a key factor in the survival of bacteria in the oral cavity. Bacteria that are not part of a biofilm are ordinarily flushed out of the oral cavity. The dental biofilm is not only a site harboring oral bacteria, but may serve as a protective environment for the embedded microorganisms. Within the biofilm ecosystem, microorganisms are less susceptible to antibacterial agents, and are better protected from the host defense system. It is also conceivable that bacteria in the biofilm exhibit different phenotypic and genotypic characteristics than do planktonic bacteria. In addition, cell-free proteins and enzymes may differ in their characteristics and activity, probably due to conformational changes upon being immobilized on the surface, and due to local conditions in the biofilm environment.

Biofilms in the oral cavity are diverse microbial communities embedded in a matrix of bacterial and host-origin constituents. The formation and maturation of the oral biofilm follow a series of dynamic biological events. The preliminary stage is the formation of an acquired pellicle (conditioning film) comprised of cell-free host constituents, mainly salivary components, and bacteria-free constituents. In the second stage, early colonizers bacteria adhere to the pellicle. This is followed by adhesion of late bacteria colonizers and co-adhesion occurs. Next, bacteria propagate within the biofilm, after which a steady state is achieved with the surrounding environment.

Diversity is the key feature of the oral cavity. Numerous types of hard and soft surfaces are part of the oral cavity. Among the hard, nonshedding surfaces are enamel, restorative materials, implants, prosthetic and orthodontic appliances. All of these differ in their chemical and physical surface properties and in their surface topography. Some of these surfaces are temporarily placed in the oral cavity, as orthodontic appliances, removable prosthetic appliances or temporary restorations. Yet, even the permanent hard surfaces may undergo modification in course of time due to changes of the oral environment. For

*Handbook of Bacterial Adhesion: Principles, Methods, and Applications*
Edited by: Y. H. An and R. J. Friedman © Humana Press Inc., Totowa, NJ

example, enamel surfaces undergo surface fluoridation following exposure to fluoride. Sedimentation of minerals may occur on the enamel surface, thus changing the surface properties of the tooth. Changes in surface topography may also occur when restorative material and enamel surfaces undergo abrasion, erosion or microcracking processes.

The different locations of the various surfaces within oral cavity may generate further variations in biofilm formation, although these hard surfaces are regarded as part of the same physiological organ and are bathed in the same physiological medium. Patterns of salivary flow in the oral cavity, proximity to the opening of the salivary gland ducts, airflow, differences in exposure to food and beverage intake, may be accounted for additional variability within types of biofilms.

Taken together, it is likely that a wide variety of biofilms are present in the oral cavity. Indeed, the available data suggest that the oral biofilms are unique in their constituents and properties. Such diversity makes it imperative to investigating each type of biofilm individually while examining its effect on the pathogenicity of dental diseases.

In this chapter we shall review the formation and composition of biofilms on several hard surfaces in the oral cavity and in models. By far, most of the research on oral biofilms has focused on enamel plaque, and information on oral biofilms of other dental surface stems largely from the extensive research conducted on enamel biofilm.

## II. DENTITION

Dental plaque biofilm, related to tooth surfaces, has been studied extensively over the last fifty years both in vitro and in vivo. The tooth biofilm plays a major role in the pathogenesis of tooth decay by harboring cariogenic bacteria that generate acid due to metabolism. The resulting organic acids cause decalcification of the enamel, which leads to cavitation of the tooth.

The formation of dental biofilm on the tooth surface is a dynamic process involving several phases.[15,61] The initial adsorption of cell-free host and bacterial constituents onto enamel surfaces is a rapid process.[16,58,93] The interactions of the adsorbents with the enamel surface are of high affinity and are highly dependent on the nature of the outer enamel surface. The composition of this primary conditioning film (acquired pellicle) is of utmost importance in the transformation of the cell-free film to bacteria-containing biofilm. This stage involves specific and nonspecific interactions between the acquired pellicle and the adsorbed bacteria. The biofilm maturation process involves further adhesion, co-adhesion and bacterial growth, during which a shift in the bacteria population occurs, until a dynamic steady state is achieved between the biofilm and the oral environment.

The bare enamel surface is the primary layer of the tooth that interacts with the developing biofilm.[33,61] The properties of the outer enamel surface can influence this initial adsorption of constituents of the pellicle, and most probably influences the subsequent processes of pellicle maturation and dental plaque biofilm formation. The enamel surface is a heterogeneous layer exposing various chemical functional groups. This heterogeneity is further induced due to exposure to food, chemical agents, and drugs. For example, phosphate groups on the outer enamel surface attract cations, as calcium, to the Stern layer (liquid adhesion surface adjacent to the solid) which can further affect the formation of the pellicle.[7] Treatment of enamel with fluoride applications results in the formation of $CaF_2$ deposits on the surface.[8,92] The $CaF_2$ deposits, in turn, influence the adsorption of proteins demonstrating a reduction in the total adsorbed proteins.[88] However, the affinity of albumin to fluoridated hydroxyapatite (HA) is increased while the affinity of lysozyme was without significant effect compared with HA.[32]

Surface free energy, contact angle, hydrophobicity and wettability are additional important properties of the enamel surface which affect the initial adsorption. The physical parameters are a consequence of the enamel's interactions with its immediate environment. These parameters, as wettability and contact angle may even change during the course of the day time.[68] During the first five minutes of pellicle formation, the free-energy on the surface is increased, after which it remains constant for at least 2 h.[24] Although this may seem to be a violation of the second law of thermodynamics, the authors explain that the driving force for this process is most likely the free-energy of the proteins, which is decreased upon adsorption. Chemical agents may also alter the surface free energy of the enamel. For example; aminofluoride and sodium fluoride decrease the surface free energy of bare enamel.[25,47]

Tooth surfaces are covered with the acquired pellicle film. While the overall protein pattern of the acquired pellicle shows characteristics typical of saliva, the salivary pellicles formed in vitro and in vivo may show major differences in the appearance of several constituents as compared to saliva, which may prevail in different parts of the mouth.[17] Those differences in the salivary composition of pellicle may be responsible for the differences in the nature of the developing biofilm.

Mucins are among the major salivary components of a 2 h in vivo acquired pellicle.[2,35] According to Al-Hashimi and Levine[2], the predominant mucins found in the acquired pellicle are the high molecular weight submandibular-sublingual mucins (MG1). These mucins have higher affinity for HA than do the low molecular weight submandibular-sublingual mucins (MG2).[118] Glycoproteins inhibit the adsorption of MG1 to HA with no effect on MG2. Cystein-containing salivary phosphoproteins (CCP) also inhibit the adsorption of MG2 to HA. These effects may result from competition for the same binding sites on the pellicle by proteins that have different affinity to these sites on the surface.

Proline-rich proteins (PRP) are other important salivary constituents found in the pellicle proteins.[9,12,13,67,86] This family of proteins undergoes conformational changes upon their adsorption to surfaces, which affect their biological activity.

Amylases are a group of enzymes which are present in saliva, and as in the cell free form in dental pellicle.[95] Amylases were found in an in vitro 2 h pellicle[74,75,81] and also in in vivo pellicles.[2] According to Stiefel[112] amylase is adsorbed to enamel from saliva, but its affinity to the surface is low. Interestingly, early work on amylase indicated that amylase has enhanced activity on the surface compared with its activity in solution.[59]

Glucosyltransferase (GTF) and fructosyltransferase (FTF) are among the bacteria-derived constituents of dental pellicle. GTF are extracellular enzymes expressed by several cariogenic bacterial strains of oral bacteria, such as *Streptococcus sobrinus, Streptococcus mutans* serotypes a, b, c, d.[21] GTF catalyzes the synthesis of glucan type polysaccharides from sucrose substrate by polymerizing the glucosyl moiety of the sucrose into glucans. Cell-free GTF is found in saliva collected from human subjects,[97,98] in plaque,[122] and in in vitro pellicle originating from human saliva.[81,82,126] There is evidence indicating that cell-free GTF is readily adsorbed to surfaces while still retaining its ability to synthesize glucans.[82,97,100,105,106,124,125] Furthermore, due to the immobilization of GTF on HA surfaces, the enzyme undergoes changes in its kinetic properties compared to the enzyme in solution.[107] A burst effect in GTF activity is recorded on the surface, which rapidly levels off, most likely due to the production of polysaccharides that act as diffusion barriers. The Km values for the synthesis of glucans from sucrose by GTF were lower when the enzyme was adsorbed to a surface compared to solution.[128] Changes in activity resulted also in structural different glucans synthesized by GTF on surfaces compared to those in solution.[57] Susceptibility of GTF to antienzymatic drugs also differs on the surface compared to solution.[105,123,135] In addition, optimal pH, thermostability and of GTF adsorbed onto experimental pellicle differ from the same parameters of GTF in solution.[100]

FTF is an extracellular enzyme which synthesizes fructan polymers from sucrose. FTF originates from oral bacteria such as *Streptococcus salivarius* or *Actinomyces* sp. Cell-free FTF can be adsorbed onto the tooth surface from human collected saliva while retaining its activity.[82] Collectively, the data regarding GTF and FTF indicate that these enzymes on the surface have distinct properties which differ from those in solution. Glucan and fructans synthesized by these enzymes play a major role in the pathogenesis of the dental plaque. Clearly, the activity of these enzymes on the tooth surface can significantly affect the virulence of the dental biofilm.

Adherence of oral bacteria to acquired pellicle leads to the development of the dental plaque biofilm. The mechanisms of bacterial adhesion to the pellicle are complex. Biofilm formation starts with random movements of bacteria towards the tooth surface. Once the bacteria are within a range of physical interaction with the surface, the adhesion process is initiated. Several studies have indicated that hydrophobic interactions appear to play a role in the initial adhesion of bacteria to surfaces. Rosenberg et al.[83] have shown that most microorganisms cultivated from human plaque are hydrophobic. A positive correlation was found between the adhesion of *S. sanguis* to biomaterials and the hydrophobic properties of the bacteria.[91] Changing the hydrophobicity of the surfaces can also influence adhesion of bacteria.[73] Hydrophobicity may play a significant role in the adherence of bacteria and biofilm formation to surfaces, however, those kinds of interactions are prone to environmental changes.

In addition to the physical attraction forces, the salivary components, expressed on the pellicle, are also potential binding sites for the initial bacterial adhesion. These domains may serve as specific binding sites for bacteria.[72,103] On the other hand, salivary components may aggregate bacteria in saliva,[18] thereby reducing their adhesion potential to the dental plaque. It is conceivable that bacteria which can interact with salivary pellicle in a highly selective fashion are the early colonizer-type bacteria. *S. sanguis* and *Actinomyces viscosus* were found to adhere more strongly than *S. mutans* to saliva-coated surfaces.[18] Further support for the assumption that *Actinomyces* sp. are predominately early colonizers comes from the fact that specific salivary binding sites promote adhesion of those bacteria. It has been found *A. viscosus* can adhere to the acquired pellicle through specific constituents of the pellicle. PRP adsorbed onto the surface promotes the adhesion several *Actinomyces* sp. to pellicle.[40] This interaction occurs only when salivary PRP is adsorbed onto the surface. The cryptitopes, molecular domains which mediate this adhesion, are exposed only upon PRP adsorption to the surface. This specific adhesion is mediated by Type-1 fimbriae on the surface of *Actinomyces*. Statherin, another salivary constituent of the pellicle, promotes adhesion of *A. viscosus* to pellicle, although the amounts required are higher than that of PRP.[40]

It has been suggested that the adhesion of *S. sanguis* to pellicle is mediated by a biphasic binding process. At first, the initial adhesion is reversible and desorption of the bacteria from the surface occurs rapidly. Next, high affinity bonds immobilize the bacteria onto the surface. This adhesion involves multiple binding sites on the pellicle, of at least two types, with different affinity towards the bacteria,[41] which interact with several distinct functional adhesion epitopes on the surface of *S. sanguis*.[43]

*S. sanguis* in the biofilm, similar to other oral bacteria, originates from the salivary planktonic phase of the mouth. These oral planktonic bacteria are bathed in saliva, thus salivary coatings of bacterial surfaces should not be overlooked in the bacteria adhesion process. Salivary pre-treatment of *S. sanguis* cells altered the manner in which these bacteria adhered to HA, saliva-coated HA, or bovine serum albumin-coated HA. The different adsorption patterns seem to be due to changes of affinity and the maximal number of binding sites on the HA beads to the pre-coated bacteria.[119]

Salivary α-amylase is one of the most abundant enzymes in human saliva, which is found also on tooth surfaces.[95] Salivary amylases are known to bind specifically to several species of oral streptococci resembling *S. mitis*, *S. gordonii* and *S. salivarius*.[94] Indeed, it was found that α-amylase promoted the adhesion of *S. gordonii* to HA.[96] Incubation of *S. gordonii* in the presence of starch and maltotriose increased the binding ability of this strain to amylase-coated HA; however, the adhesion of *S. sanguis* to amylase-coated HA was not affected by these saccharides. Animals with salivary amylase activity were found to have more of the amylase-binding bacteria in their oral cavity than did animals which did not exhibit this activity.[94] These results suggest that amylase plays a role in the formation of the dental biofilm by serving as a specific binding site for bacteria on the pellicle.

Several other constituents of the saliva have been shown to promote adhesion of oral bacteria.[51,72] A number of other studies have indicated that mutans streptococci might also have affinity to salivary constituents. It was found that *S. mutans* adhered better to pellicles from parotid saliva than from whole saliva, while *S. sanguis* adhered better to whole saliva than parotid saliva pellicles.[18] The adhesion of *S. mutans* to salivary pellicle-coated tooth surfaces involves a major cell surface protein termed antigen (Ag) I/II. The high affinity of Ag I/II adhesin for salivary constituents of saliva-coated HA is mediated by the N-terminal part of the molecule.[45] Two major salivary proteins of apparent relative molecular mass of 28,000 and 38,000 respectively, containing high proportions of pro-line, glycine, and glutamic acid, and overall compositions similar to basic proline-rich salivary proteins, were found to have high affinity to Ag I/II.[87]

Dietary sucrose has been shown to have a profound effect on the colonization of bacteria to tooth surfaces. Sucrose-dependent adhesions of oral streptococci facilitate ac-cumulation of several serotypes of oral bacteria. This type of adhesion is mediated by the synthesis of extracellular glucans by cell-free or cell-associated GTF. Sucrose-dependent adhesion of oral streptococci to saliva-coated HA is comparable to those bacteria whose other binding mechanisms are impaired[129] which may indicate that this mechanism prevails over other types of adhesion mechanisms.

The observation that sucrose, promotes accumulation and adhesion of *S. mutans* to teeth and to solid surfaces strongly suggests that GTF and glucans are involved in this type of bacterial adhesion.[38,56] Several studies have emphasized the importance of *in situ* glucan synthesis by cell-free GTF on the adherence of *S. mutans* to experimental pellicle.[99,101] These glucans, synthesized on the surface of saliva-coated HA, served as binding sites for *S. mutans*. It was further found that the α,1-6 type glucan is important in this type of adhesion process. *S. gordonii* can also adhere tightly to HA surfaces through *de novo* glucan synthesis by the mutans streptococci glucosyltransferase that are present in the experimental salivary pellicles.[48] Conversely, glucans synthesized on salivary pellicle can mask binding sites for *Actinomyces* strains, thereby decreasing the adhesion of several strains of *Actinomyces* to glucan-coated salivary pellicle as compared to salivary pellicle.[109] This latter example indicates the functional diversity of glucans in controlling dental biofilm formation by either promoting or decreasing specific bacterial adhesion to the pellicle.

The *in situ* synthesis of glucans by cell-free or cell-bound GTF is one of the pathways which has a significant impact on bacterial adhesion to acquired pellicle and dental plaque. However, important as it is, sucrose-mediated adhesion is only one of several routes enabling bacteria to adhere to the tooth surface. The elimination of sucrose from the diet did not diminish the presence of *S. mutans* from the oral cavity.[26,132] The ability of *S. mutans* to persist in the oral cavity in the absence of sucrose argues against the obligatory sucrose-dependent adhesion concept and supports the notion that other mechanisms may facilitate bacterial adhesion in the oral cavity.

In addition to the adhesion effect, these polysaccharides synthesized by GTF and FTF increase the total mass and volume of the developing biofilm, thus affecting diffusion rates across the dental plaque. Clearly, the diffusion of acids, anti-caries agents and bacterial metabolites affect the virulence of the biofilm. As expected, and according to Fick's first law, an increase in coating thickness decreases the diffusion coefficient across the barrier. For example, the thickness of the biofilm affects the pH gradient across the plaque. With thick plaque, a low pH was often not achieved at the inner surface but at some intermediate depth.[23] Tortuosity of the biofilm is another aspect influencing diffusion; increasing the plaque tortuosity reduced the diffusion rate of sucrose across dental biofilm. Increasing the concentrations of water-insoluble glucans in the plaque effectively reduced the diffusion rate of sucrose.[120] It may be argued that water soluble polysaccharides, as dextrans or water soluble fructans, synthesized in the biofilm may increase the available channels in the biofilm, thereby facilitating a faster diffusion rate across the biofilm. Interactions between the biofilm and the molecules which diffuse, may also affect the diffusion rate. NaF, a relatively small molecule, diffused only 38% slower in cell-free glucan sediment than in water, suggesting that glucans *per se* do not form a diffusion barrier to NaF. However, it should be noted that the total diffusion time through plaque may be increased if the presence of extracellular polysaccharides results in thicker layers of plaque.[64]

The above-mentioned studies indicate that salivary pellicles are selectively formed on enamel surfaces, which most likely control and mediate the formation of the biofilm on teeth surfaces and consequently affecting the progression of dental caries.

## III. RESTORATIVE AND PROSTHETIC MATERIALS

Tooth restoration is a widely accepted clinical procedure in restorative and prosthetic dentistry. A major surface area of the tooth may be covered with restorative material, while other materials replace the natural crown of the tooth. Today, numerous types of restorative materials are being used in the dental office. Among the most popular types are the classic amalgam fillings, which are mercury alloys. Cements are another type of restorative materials used as a filling material and are employed also to retain restorations or appliances in the mouth. A unique type of cements are the glass ionomer cements, consisting of alumino silicate glass with a high fluoride content. These unique materials are capable of releasing fluoride. Composites are another type of restorative materials, consisting of two or more essentially insoluble phases. Porcelain, a mixture of feldspar, silica, and kaolin is widely used as an artificial tooth but also can be used as a filling restorative materials.

Biofilms on restorative materials can serve as a reservoir of bacteria in the oral cavity by adding to the critical mass of cariogenic bacteria in the mouth. In addition, the accumulation of bacteria on filling restorative materials can lead to secondary caries, resulting in decay under the restorative material. The breakdown of the marginal areas between enamel and restorative material can provide potential pathways for bacterial reinfection and reoccurrence of caries. Plaque accumulation should be minimized on restorative materials adjacent to the gingival tissues in order to avoid tissue irritations that may lead to periodontal diseases.

Restorative materials, like other surfaces in the oral cavity, are covered with dental biofilm consisting mainly of host and bacterial constituents. However, the different chemical properties and different topography of the various restorative material surfaces may lead to the formation of biofilms that differ from one another in their components and properties.

Amalgam fillings are still the most frequently applied dental materials, although other alternatives as composites and cements are currently being used more frequently. In vitro assessment of microbial plaque formation on five different types of amalgams has demonstrated variations in both the amount and viability of the microbial counts between the different amalgam samples.[29] Significantly less plaque was accumulated on amalgam as compared with on composite cement.[28] Differences in plaque composition were also found between amalgam and glass ionomer cements, demonstrating a significantly lower percentage of viable mutans streptococci counts in samples taken from glass ionomer restorations than from amalgam.[116,117] Benderli et al.[11] have compared plaque formation by *S. mutans* on five types of glass-ionomer cements and two composite materials. After five days of incubation, the amount of bacteria on the restorative surfaces was evaluated. Differences in sucrose-dependent adhesion patterns were found between the tested materials. Differences in plaque accumulation were found within the various types of glass ionomer cements. The most amounts of plaque were found on lining cement and base cement. Two types of glass ionomer cements and one composite demonstrated similar capability of bacterial adsorption. A broader study, using several types of oral bacteria, was conducted on various types of restorative materials.[77] Glass ionomer cements were found to reduce sucrose-dependent accumulations of *A. viscosus, S. mitis, S. mutans, S. sanguis* and *Lactobacillus casei* by over 80% compared to enamel surfaces. It seems that the local release of fluoride from the glass ionomer cements affected bacterial growth, since elevations in short-term fluoride release levels were positively correlated with bacterial growth inhibition. The inhibiting effect was not observed when the samples were prewashed before biofilm formation. This points to a burst effect in fluoride release, after which the release is sharply reduced, thereby minimizing the effect of fluoride on biofilm formation.[63]

Saliva seems to play an important role in bacterial adsorption to restorative surfaces, similar to described above for teeth surfaces. A salivary coating of the dental material may alter its physical properties, including hydrophobicity, contact angle, wetting values and zeta potential. Clearly, this conditioning film on the surface can affect bacterial adhesion. Salivary coating reduces the adsorption of oral bacteria onto restorative resin composites as compared to uncoated resins.[90,91,113] These assays were conducted in the absence of sucrose, indicating that hydrophobic and electrostatic interactions between the bacteria and the substratum probably played an important role in this type of adhesion. A correlation with surface hydrophobicity was found in the adhesion of *S. sanguis,* while adhesion of *S. mutans,* a less hydrophobic bacteria, was positively correlated to the zeta potential.[90,91] It appears that a salivary coating on restorative materials has a regulatory effect on the surface properties. For example, while the contact angles of various uncoated restorative materials differ, the contact angle values between these materials after salivary coating are similar.[65] Shahal et al.[102] have found similar adsorption profiles of human saliva on different types of restorative materials, further indicating the regulatory effect of salivary coating. This in vitro result corresponds with the in vivo results of Hannig[46] who showed that pellicle formed on different restorative materials is similar with regard to ultrastructural appearance.

Shahal et al.[102] modified the biofilm model for testing plaque accumulation on restorative materials. The biofilm coating of restorative materials in their model included human saliva, cell-free FTF, cell-free GTF, and *in situ* production of polysaccharides by these enzymes. Next, *S. mutans* was adsorbed onto the conditioning film via a sucrose-dependent mechanism. Similar protein profiles of the initial layer of adsorbed saliva onto the surfaces were found on different types of tested restorative materials. Consequently, the sucrose-dependent adsorption profile of *S. mutans* on these restorative materials was also comparable in all tested dental materials.

The effects of plaque biofilm formation on restorative materials have been studied also in vivo. Leonhardt et al.[60] conducted a study on human subjects to evaluate qualitative and quantitative differences in bacterial colonization on amalgam surfaces. Two pieces of amalgam were placed in the oral cavity for 10 min, 1, 3, 6, 24, and 72 h. Total viable bacterial counts increased on all surfaces during the experiment time period. Similar colonization patterns of *Streptococcus* sp., *Neisseria* sp., *Fusobacterium* sp., *Prevotella* sp., *Hemophilus parainfluenzae,* and *Actinomyces naeslundii* were recorded on amalgam surfaces as compared to titanium (Ti) or HA surfaces. It appears, then, that Ti surfaces, similar to amalgam and HA surfaces, do not seem to have a marked influence on the early bacterial colonization pattern in vivo. Yamamoto et al.[136] have found early streptococci colonizers such as *S. mitis* and *S. sanguis* on composite resins placed in the mouth for 2 h. They did not find *S. mutans* nor *S. salivarius* on composite resins of human subjects after the short exposure in the oral cavity. In a long-term evaluation, lasting for 2 mo, no statistical differences were found in the amount of *S. mutans* accumulation on different types of amalgams.[131] This study is in agreement with another study which found that early plaque formation on solid surfaces such as amalgam, composites, glass ionomer cements and enamel is influenced predominantly by the oral environment rather than by material-dependent parameters.[46] These in vivo findings correlate with aforementioned in vitro studies, and may be ascribed to the presence of a primary pellicle layer, which apparently masks surface differences among restorative materials, thus supporting the notion of the regulatory effect of salivary coating on dental surfaces.

Levels of mutans streptococci in plaque samples taken from the margins of amalgam composite and glass-ionomer restorations were compared in vivo.[117] The percentage of mutans streptococci from the total bacterial count in plaque was three times higher on composite as compared to amalgam, and ten times higher than on glass-ionomer. This study is in agreement with another in vivo study[36] which showed a marked reduction in levels of mutans streptococci near glass ionomer cements. Similar to the in vitro findings, these differences may be attributed to the elevated fluoride levels found in plaque adjacent to glass ionomer cements, resulting from the release of fluoride from these types of dental restorative materials. Different observations were found in a study[37] collecting plaque from sites adjacent to the glass ionomer cement fillings and from the contralateral teeth, after application of a 1.2% fluoride gel. No significant differences in the proportion of mutans, streptococci and lactobacilli in plaque from glass ionomer cements and the contralateral teeth were found. The assays suggest that the fluoride concentration of plaque growing on old glass ionomer fillings is low, resulting in limited effect on the cariogenic microflora. In another study, the quantity of *S. mutans*, total streptococci, and lactobacilli on sound enamel surfaces and one-year-old glass ionomer cement and composite resin placed subgingivally was compared intra-individually.[127] The number of lactobacilli and *S. mutans* recovered from glass ionomer cement and composite resin surfaces was the same as for the enamel surfaces. Fluoride levels in plaque adjacent to glass ionomer cement were not high enough to inhibit accumulation of the investigated bacteria, again suggesting the short-term effect of the fluoride released from glass ionomer cements.

Tooth-bleaching is a commonly used procedure in dentistry today. Whitening teeth can be performed at the dental office or at home using various methods and active agents. Whitening the dentition also exposes the restorative materials to the bleaching agents, which in turn can induce surface changes on the exposed restorative surfaces.[84,85] In a recent study, the effect of bleaching agents on bacterial adherence to polished surfaces of composite resin restorations was assessed in vitro.[66] A three-day pre-treatment of composite resin with a 10% solution of carbamide peroxide and hydrogen peroxide caused a significant increase in surface adherence of *S. mutans* and *S. sobrinus*. In a follow-up study Steinberg et al.[110] examined the effect of in vitro salivary film on the sucrose-dependent adherence of oral bacteria to bleached and unbleached restorative composite material. Salivary film coating the restorative material surface significantly decreased sucrose-dependent adhesion of *S. sobrinus* and *S. mutans* to the bleached and unbleached surfaces, compared to uncoated specimens. Saliva had a minor effect on adhesion of *A. viscosus*.

Ceramic and porcelain are other popular restorative materials used in the prosthetic dental field. Plaque accumulation on artificial crowns and surfaces adjacent to the gingival tissues should be minimized in order to maintain good gingival health.[104,130] No significant differences were recorded between the plaque-retaining capacities of metal ceramic porcelain and Dicor ceramic surfaces after 12 and 24 h. However, there was less plaque accumulation on glazed surfaces than on nonglazed surfaces.[19] Differences in plaque accumulation were found between two types of ceramics. It was found that single-crystal aluminum ceramic retards less plaque than polycrystal aluminum.[76] Adamczyk et al.[1] have found differences in plaque accumulation between ceramic crowns and enamel surfaces. The study by Hahn et al.[44] also substantiated the finding that ceramics accumulate less plaque than tooth enamel, although no significant differences in plaque accumulation on ceramics compared to amalgam or glass ionomer cements was found.[46] In another comparative study, it was shown that cerestore full-ceramic crowns have little soft debris retention as compared to ceramometal crowns, natural teeth or cast gold restorations or acrylic resin veneer crowns.[20] Acrylic resins seem to have a selective pattern of saliva adsorption.[30] The identification of these molecules on the acquired denture pellicle may elucidate the mechanism of the specific fungal cell colonization on these type of dental surfaces.

Surface topography is yet another important factor that can foster the accumulation of dental plaque to dental restorative materials.[121,138] The effect of surface roughness on bacterial adherence is complex. It is generally accepted that bacteria accumulate to a greater degree on rough abraded surfaces than on a highly polished surface. Yamamoto et al.[137] found no differences in adhesion of *S. oralis* to composites of different degrees of roughness. In contrast, Yamaguchi et al.[138] found that *S. mutans* adhered better to smooth surfaces, while Tullberg[121] reported that polishing increased the adhesion capacity of gold and polymethyl metacrylate.

Adhesion of oral bacteria to restorative materials plays an important role in the pathogenesis of oral diseases. New restorative materials are currently being introduced to the dental market. These innovative materials differ in their chemical composition as well as in their surface topography. Restorative materials which can influence biofilm formation have a great advantage in maintaining good oral hygiene.

## IV. IMPLANT MATERIALS

The use of implants in dentistry today is an well-accepted prosthetic technique. The implant material is surgically placed in the alveolar bone, after which an artificial tooth is affixed on the implant. Titanium (Ti), Ti alloys and Ti based materials are considered to be the dental implant materials of choice, due to their highly biocompatible properties.[27,49] Ti, once exposed to air atmosphere, forms a highly stable oxide surface ($TiO_2$). Due to its high dielectric constant, Ti oxide undergoes further modifications through the binding of various ions, and proteins.[114,115]

While considerable information is available regarding oral bacterial adhesion to enamel, less is known about the mechanism of biofilm formation on implant materials in the oral cavity. Several studies have investigated the presence of bacteria around implants in edentulous patients,[69,70,71] partially edentulous patients,[6,53,89] and patients with unsuccessful and successful implants.[10,71,78,80] It was found that the presence of Gram positive cocci is significantly greater than other bacteria around implants in edentulous mouths. No significant differences could be found between the subgingival flora of teeth and Ti implants in a periodontally healthy mouth. However, the occurrence of *Actinobacillus actinomicetemcomitans* and *A. viscosus* in the supragingival plaque was higher for teeth as compared to the implants.

Salivary and serum constituents can adsorb onto Ti surfaces. The composition of the primary Ti-acquired pellicle plays a pivotal role in determining the type and amount of bacteria that will adsorb onto the surface. This process may later determine the rate of success or failure of the implant procedure.

The primary conditioning film formed on Ti surfaces is mediated by the presence of calcium ions. Pretreatment of Ti with calcium increased the amount of salivary proteins adsorbed onto Ti, while EDTA solution reversed this process.[108] Salivary proteins adsorb onto untreated Ti, thus indicating that pretreatment of Ti with calcium is apparently not a prerequisite for protein adsorption and that other forces may facilitate the initial adsorption of salivary constituents onto Ti surfaces. It has been suggested that the adsorption of human serum albumin to biomedical polymers can be mediated also via hydrophobic interactions, ligand-specific and electrostatic interaction.[42,50] The non-specific adhesion of salivary proteins to the surface of Ti is probably secondary to the specific calcium-mediated process, since calcium ions are abundant in saliva. Monovalent cations such as potassium do not mediate the adsorption of albumin to Ti, but the presence of magnesium ions had a similar effect on albumin adhesion as did calcium.[52] Several other studies highlighted the role of calcium on adsorption of biomolecules to Ti. Ellingsen[34] showed that calcium is incorporated into the Ti oxide surface and that Ti pre-treated with calcium adsorbs mainly human serum albumin and IgG from the serum. Collis and Embery[22] demonstrated that connective tissue components such as glycosaminoglycan chondroitin-4-sulfate, adsorb to Ti only in the presence of calcium.

One of the major whole saliva protein adsorbed onto in vitro calcium-treated Ti was identified as albumin.[108] The mechanism of albumin adsorption to Ti in the presence of calcium followed a Langmiur adsorption isotherm.[52] Upon replotting to a Scatchard plot, two peaks were revealed, possibly indicating the presence of dual binding sites of albumin to the Ti. Support for this finding is provided from a clinical study which analyzed salivary components adsorbed onto Ti after exposure of the abutments to the oral environment for a short period of 2 h[55] and a prolonged period of 2 to 6 wk.[54] The proteins found on the Ti in both studies were mainly serum albumin and alpha-amylase.

The presence of a primary coat on Ti has a substantial influence on bacterial adhesion and biofilm formation. In an in vitro study, Wolinsky et al.[133] found that *A. viscosus* adhered to saliva-treated Ti surfaces in lower values compared to saliva-treated enamel. Other types of bacteria, such as fresh isolates and reference stains of *S. oralis*, and *S. salivarius,* were also found to adhere less to saliva-coated Ti as compared to Ti without saliva coating.[31]

The mechanism by which salivary or serum film on Ti inhibits bacteria accumulation is not fully understood. As mentioned above, salivary albumin seems to be one of the proteins found on Ti implants. It is believed that the presence of albumin on dental implants determines the ability of bacteria to adhere and accumulate on those hard surfaces. Indeed, treatment of various types of biomaterials and dental materials with albumin led to the reduced adsorption of bacteria to the respective surface[39,111] probably by interferes with hydrophobic-type adhesion of bacteria to surfaces. Furthermore, albumin may mask binding sites on the pellicle by means of steric hindrance, similar to the mechanism of *in situ* glucans formation on pellicle[109] thereby reducing bacterial adhesion. Albumin in the oral cavity may undergo degradation. Crosslinked albumin adsorbed onto Ti was found to be far less susceptible to degradation, while maintaining its capacity to inhibit bacterial adhesion and reduce infections around implants.[3,5,62]

Few clinical studies have been performed to examine biofilm formation on Ti dental implants. A short-term study was conducted in vivo to evaluate qualitative and quantitative differences in in vivo bacterial colonization on Ti, HA, and amalgam surfaces.[60] Microbiological samples were taken during a three day period. Various streptococci species predominated the sample sites, usually constituting over 50% of total viable counts. During the study period, no significant differences were found among the tested materials regarding colonization of the investigated bacteria. Therefore, it seems that Ti, HA, and amalgam surfaces do not have a marked influence on the early colonization pattern in vivo. Support for the notion that early plaque formation on Ti is similar to that of other hard surfaces as amalgams, casting alloys, ceramics, glass cements, composite resins, unfilled resins, and bovine enamel was obtained using scanning electron microscopy[46] demonstrating no significant changes in the ultra-structural appearance of the early plaque formed on these different material surfaces. These findings may be ascribed to the presence of the pellicle layer, which apparently masks any differences among materials with regard to surface properties and biocompatibility.

Similar to other materials reviewed in this chapter, the surface topography of Ti implants may affect biofilm formation. Testing the adsorption properties of oral bacteria to Ti discs with different surface morphology revealed that smooth surfaces promoted poor attachment for *S. sanguis* and *A. viscosus*. However, *Porphyromonas gingivalis* attached equally well to smooth and grooved Ti surfaces.[134] Four Ti abutments with different degrees of surface roughness were randomly placed in partially edentulous patients. Only the two roughest abutments harbored spirochetes after a period of 1 mo. After 3 mo, the composition of the subgingival flora showed little variation between the different types of abutment. Anaerobic bacterial culturing demonstrated comparable amounts of colony-forming units for all abutment types, both supragingivally and subgingivally.[79] Extending the exposure time to twelve months has not revealed significant differences in the microbiota on the various types of Ti surfaces being tested.[14] The above results indicate that a reduction in surface roughness below a threshold of R(a) = 0.2 had no further effect on the quantitative and qualitative nature of microbiological adhesion or colonization, neither supragingivally nor subgingivally. An et al.[4] have also demonstrated that adsorption of *S. epidermidis* to pure Ti pre-treated with 120-1200 grit sandpaper had little effect on adhesion.

## V. CONCLUSION

Biofilm formation around the various dental related surfaces has a marked effect on tooth decay, periodontal diseases, osseointegration and biocompatibility processes. Understanding the formation and composition of the various types of biofilms will enhance our ability control the progression of the major diseases of the oral cavity.

## REFERENCES

1. Adamczyk E, Spiechowicz E: Plaque accumulation on crowns made of various materials. *Int J Prosthodont* 3:285–91, 1990
2. Al-Hashimi I, Levine MJ: Characterization of *in vivo* salivary-derived enamel pellicle. *Arch Oral Biol* 34:289–95, 1989
3. An YH, Bradley J, Powers DL, et al: The prevention of prosthetic infection using a cross-linked albumin coating in a rabbit model. *J Bone Joint Surg Br* 79:816–9, 1997

4. An YH, Friedman RJ, Draughn RA, et al: Rapid quantification of staphylococci adhered to titanium surfaces using image analyzed epifluorescence microscopy. *J Microbiol Meth* 24:29–40, 1995

5. An YH, Stuart GW, McDowell SJ, et al: Prevention of bacterial adherence to implant surfaces with a crosslinked albumin coating *in vitro*. *J Orthop Res* 14:846–9, 1996

6. Apse P, Ellen RP, Overall CM, et al: Microbiota and crevicular fluid collagenase activity in the osseointegrated dental implant sulcus: a comparison of sites in edentulous and partially edentulous patients. *J Periodont Res* 24:96–105, 1989

7. Arends J, Jongebloed WL: The enamel substrate-characteristics of the enamel surface. *Swed Dent J* 1:215–24, 1977

8. Arends J, Reintsema H, Dijkman TG: "Calcium fluoride-like" material formed in partially demineralized human enamel *in vivo* owing to the action of fluoridated toothpastes. *Acta Odontol Scand* 46:347–53, 1988

9. Armstrong WG: Characterisation studies on the specific human salivary adsorbed *in vitro* by hydroxyapatite. *Caries Res* 5:215–27, 1971

10. Becker W, Becker BE, Newman MG, et al: Clinical and microcrobiologic findings that may contribute to dental implant failure. *Int J Oral Maxillofac Implants* 5:31–38, 1990

11. Benderli Y, Ulukapi H, Balkanli O, et al: *In vitro* plaque formation on some dental filling materials. *J Oral Rehabil* 24:80–3, 1997

12. Bennick A, Cannon M: Quantitative study of the interaction of salivary acidic proline-rich proteins with hydroxyapatite. *Caries Res* 12:159–69, 1978

13. Bennick A, Chau G, Goodlin R, et al: The role of human salivary acidic proline-rich proteins in the formation of acquired dental pellicle *in vivo* and their fate after adsorption to the human enamel surface. *Arch Oral Biol* 28:19–27, 1983

14. Bollen CM, Papaioanno W, Van Eldere J, et al: The influence of abutment surface roughness on plaque accumulation and peri-implant mucositis. *Clin Oral Implants Res* 7:201–11, 1996

15. Bowden GH, Hamilton IR: Survival of oral bacteria. *Crit Rev Oral Biol Med* 9:54–85, 1998

16. Busscher HJ, Uyen HM, Stokroos I, et al: A transmission electron microscopy study of the adsorption patterns of early developing artificial pellicles on human enamel. *Arch Oral Biol* 34:803–10, 1989

17. Carlen A, Borjesson AC, Nikdel K, et al: Composition of pellicles formed *in vivo* on tooth surfaces in different parts of the dentition, and *in vitro* on hydroxyapatite. *Caries Res* 32:447–55, 1998

18. Carlen A, Olsson J, Ramberg P: Saliva mediated adherence, aggregation and prevalence in dental plaque of *Streptococcus mutans*, *Streptococcus sanguis* and *Actinomyces* spp. in young and elderly humans. *Arch Oral Biol* 41:1133–40, 1996

19. Castellani D, Bechelli C, Tiscione E, et al: *In vivo* plaque formation on cast ceramic (Dicor) and conventional ceramic. *Int J Prosthodont* 9:459–65, 1996

20. Chan C, Weber H: Plaque retention on teeth restored with full-ceramic crowns: a comparative study. *J Prosthet Dent* 56:666–71, 1986

21. Ciardi JE: Purification and properties of glucosyltransferase from *Streptococcus mutans*: a review. In: Doyle RJ, Ciardi JE, eds. *Glucosyltransferases, Glucans, Sucrose And Dental Caries. Sp. Supp. Chemical Senses*. IRL Press, Washington, 1983:51–64

22. Collis JJ, Embery G: Adsorption of glycosaminoglycans to commercially pure titanium. *Biomaterials* 13:548–52, 1992

23. Dawes C, Dibdin GH: A theoretical analysis of the effects of plaque thickness and initial salivary sucrose concentration on diffusion of sucrose into dental plaque and its conversion to acid during salivary clearance. *J Dent Res* 65:89–94, 1986

24. de Jong HP, de Boer P, Busscher HJ, et al: Surface free energy changes of human enamel during pellicle formation. An *in vivo* study. *Caries Res* 18:408–15, 1984

25. de Jong HP, de Boer P, van Pelt AW, et al: Effect of topically applied fluoride solutions on the surface free energy of pellicle-covered human enamel. *Caries Res* 18:505–8, 1984

26. de Stoppelaar JD, van Houte J, Backer DIRKS O: The effect of carbohydrate restriction on the presence of *Streptococcus mutans, Streptococcus sanguis* and iodophilic polysaccharide-producing bacteria in human dental plaque. *Caries Res* 4:114–23, 1970

27. Dion I, Baquey C, Monties JR, et al: Haemocompatibility of $Ti_6Al_4V$ alloy. *Biomaterials* 14:122–6, 1993

28. Dummer PM, Harrison KA: *In vitro* plaque formation on commonly used dental materials. *J Oral Rehabil* 9:413–7, 1982

29. Dummer PM, Wills-Wood M: *In vitro* plaque formation on dental amalgam. *J Oral Rehabil* 11:539–45, 1984

30. Edgerton M, Levine MJ: Characterization of acquired denture pellicle from healthy and stomatitis patients. *J Prosthet Dent* 68:683–91, 1992

31. Edgerton M, Lo SE, Scannapieco FA: Experimental salivary pellicles formed on titanium surfaces mediate adhesion of streptococci. *Int J Oral Maxillofac Implants* 11:443–9, 1996

32. Eggen KH, Rolla G: Surface properties of fluoride treated hydroxyapatite as judged by interactions with albumin and lysozyme. *Scand J Dent Res* 91:347–50, 1983

33. Eggen KH, Rolla G: Further studies on the composition of the acquired enamel pellicle. *Scand J Dent Res* 91:439–46, 1983

34. Ellingsen JE: A study on the mechanism of protein adsorption to $TiO_2$. *Biomaterials* 12:593–6, 1991

35. Fisher SJ, Prakobphol A, Kajisa L, et al: External radiolabeling of components of pellicle on human enamel and cementum. *Arch Oral Biol* 32:509–17, 1987

36. Forss H, Jokinen J, Spets-Happonen S, et al: Fluoride and mutans streptococci in plaque grown on glass ionomer and composite. *Caries Res* 25:454–8, 1991

37. Forss H, Nase L, Seppa L: Fluoride concentration, mutans streptococci and lactobacilli in plaque from old glass ionomer fillings. *Caries Res* 29:50–3, 1995

38. Fujiwara T, Tamesada M, Bian Z, et al: Deletion and reintroduction of glucosyltransferase genes of *Streptococcus mutans* and role of their gene products in sucrose dependent cellular adherence. *Microb Pathog* 20:225–33, 1996

39. Gibbons, RJ, Etherden I: Albumin as a blocking agent in studies of streptococcal adsorption to experimental salivary pellicles. *Infect Immun* 50:592–4, 1985

40. Gibbons RJ, Hay DI: Human salivary acidic proline-rich proteins and statherin promote the attachment of *Actinomyces viscosus* LY7 to apatitic surfaces. *Infect Immun* 56:439–45, 1988

41. Gibbons RJ, Moreno EC, Etherden I: Concentration-dependent multiple binding sites on saliva-treated hydroxyapatite for *Streptococcus sanguis*. *Infect Immun* 39:280–9, 1983

42. Gombotz WR, Wang GH, Horbett TA, et al: Protein adsorption to poly(ethylene oxide) surfaces. *J Biomed Mater Res* 25:1547–62, 1991

43. Gong K, Herzberg MC: *Streptococcus sanguis* expresses a 150-kilodalton two-domain adhesin: characterization of several independent adhesin epitopes. *Infect Immun* 65:3815–21, 1997

44. Hahn R, Weiger R, Netuschil L, et al: Microbial accumulation and vitality on different restorative materials. *Dent Mater* 9:312–6, 1993

45. Hajishengallis G, Koga T, Russell MW: Affinity and specificity of the interactions between *Streptococcus mutans* antigen I/II and salivary components. *J Dent Res* 73:1493–502, 1994

46. Hannig M: Transmission electron microscopy of early plaque formation on dental materials *in vivo*. *Eur J Oral Sci* 107:55–64, 1999

47. Hay DI, Moreno EC: Differential adsorption and chemical affinities of proteins for apatitic surfaces. *J Dent Res* 58:930–42, 1979

48. Hiroi T, Fukushima K, Kantake I, et al: *De novo* glucan synthesis by mutants streptococcal glucosyltransferases present in pellicle promotes firm binding of *Streptococcus gordonii* to tooth surfaces. *FEMS Microbiol Lett* 15:193–8, 1992

49. Kasemo, B: Biocompatibility of titanium implants: surface science aspects. *J Prosthet Dent* 49:832–7, 1983

50. Keogh JR, Velander FF, Eaton JW: Albumin-binding surfaces for implantable devices. *J Biomed Mater Res* 26: 441–56, 1992

51. Kishimoto E, Hay DI, Gibbons RJ: A human salivary protein which promotes adhesion of *Streptococcus mutans* serotype c strains to hydroxyapatite. *Infect Immun* 57:3702-7, 1989

52. Klinger A, Steinberg D, Kohavi D, et al: Mechanism of adsorption of human albumin to titanium *in vitro*. *J Biomed Mater Res* 36:387–92, 1997

53. Kohavi D, Greenberg R, Raviv E, et al: Subgingival and supragingival microbial flora around healthy osseointegrated implants in partially edentulous patients. *Int J Oral Maxillofac Implants* 9:673–8, 1994

54. Kohavi D, Klinger A, Steinberg D, et al: Adsorption of salivary proteins onto prosthetic titanium components. *J Prosthet Dent* 74:531–4, 1995

55. Kohavi D, Klinger A, Steinberg D, et al: alpha-Amylase and salivary albumin adsorption onto titanium, enamel and dentin: an *in vivo* study. *Biomaterials* 18:903–6, 1997

56. Koga T, Asakawa H, Okahashi N, et al: Sucrose-dependent cell adherence and cariogenicity of serotype c *Streptococcus mutans*. *J Gen Microbiol* 28:73–83, 1986

57. Kopec LK, Vacca-Smith AM, Bowen WH: Structural aspects of glucans formed in solution and on the surface of hydroxyapatite. *Glycobiology* 7:929–34, 1997

58. Kuboki Y, Teraoka K, Okada S: X-ray photoelectron spectroscopic studies of the adsorption of salivary constituents on enamel. *J Dent Res* 66:1016–9, 1987

59. Ledingham WM, Hornby WE: The action pattern of water-insoluble α-amylases. *FEBS Lett* 5:118-20, 1969

60. Leonhardt A, Olsson J, Dahlen G: Bacterial colonization on titanium, hydroxyapatite, and amalgam surfaces *in vivo*. *J Dent Res* 74:1607–12, 1995

61. Marsh P, Martin M: Acquisition, distribution and adherence of oral microorganisms. In: Marsh P, Martin M, eds. *Aspects of Microbiology Oral Microbiology: Oral Microbiology*. Van Nostrand Reinhold Co, London, UK, 1984:26–47

62. McDowell SG, An YH, Draughn RA, et al: Application of a fluorescent redox dye for enumeration of metabolically active bacteria on albumin-coated titanium surfaces. *Lett Appl Microbiol* 21:1–4, 1995

63. McKnight-Hanes C, Whitford GM: Fluoride release from three glass ionomer materials and the effects of varnishing with or without finishing. *Caries Res* 26:345–50, 1992

64. McNee SG, Geddes DA, Weetman DA, et al: Effect of extracellular polysaccharides on diffusion of NaF and [14C]-sucrose in human dental plaque and in sediments of the bacterium *Streptococcus sanguis* 804 (NCTC 10904). *Arch Oral Biol* 27:981–6, 1982

65. Milosevic A: The influence of surface finish and in-vitro pellicle on contact-angle measurement and surface morphology of three commercially available composite restoratives. *J Oral Rehabil* 19:85–97, 1992

66. Mor C, Steinberg D, Dogan H, et al: Bacterial adherence to bleached surfaces of composite resin *in vitro*. *Oral Surg Oral Med Oral Pathol Oral Radiol Endod* 86:582–6, 1998

67. Moreno EC, Kresak M, Hay DI: Adsorption of molecules of biological interest onto hydroxy-apatite. *Calcif Tissue Int* 36:48–59, 1984

68. Morge S, Adamczak E, Linden LA: Variation in human salivary pellicle formation on biomaterials during the day. *Arch Oral Biol* 34:669–74, 1989

69. Mombelli A, Buser D, Lang NP: Colonization of osseointegrated titanium implants in edentulous patients. Early results. *Oral Microbiol Immunol* 3:113–20, 1988

70. Mombelli A, Mericske-Stern R: Microbiological features of stable osseointegrated implants used as abutment for overdentures. *Clin Oral Impl Res* 1:1–7, 1990

71. Mombelli A, van Oosten MA, Schurch E, et al: The microbiota associated with successful or failing osseointegrated titanium implants. *Oral Microbiol Immunol* 2:145–51, 1987

72. Newman F, Beeley JA, MacFarlane TW: Adherence of oral microorganisms to human parotid salivary proteins. *Electrophoresis* 17:266–70, 1996

73. Olsson J, Carlen A, Holmberg K: Inhibition of *Streptococcus mutans* adherence by means of surface hydrophilization. *J Dent Res* 69:1586–91, 1990

74. Orstavik D, Kraus FW: The acquired pellicle: immunofluorescent demonstration of specific proteins. *J Oral Pathol* 2:68–76, 1973

75. Orstavik D, Kraus FW: The acquired pellicle: enzyme and antibody activities. *Scand J Dent Res* 82:202–5, 1974

76. Otogoto J, Ebashi S, Tanaka K, et al: Subgingival plaque formation on single and poly-crystal aluminum ceramics. *J Nihon Univ Sch Dent* 36:209–15, 1994

77. Palenik CJ, Behnen MJ, Setcos JC, et al: Inhibition of microbial adherence and growth by various glass ionomers *in vitro*. *Dent Mater* 8:16–20, 1992

78. Palmisano DA, Mayo JA, Block MS, et al: Subgingival bacteria associated with hydroxy-apatite-coated dental implants: morphotypes and trypsin-like enzyme activity. *Int J Oral Maxillofac Implants* 6:313–8, 1991

79. Quirynen M, Bollen CM, Papaioannou W: The influence of titanium abutment surface rough-ness on plaque accumulation and gingivitis: short-term observations. *Int J Oral Maxillofac Implants* 11:169–78, 1996

80. Quirynen M, Listgarten MA: Distribution of bacterial morphotypes around natural teeth and titanium implants *ad modum* Branemark. *Clin Oral Implants Res* 1:8–12, 1990

81. Rolla G, Ciardi JE, Bowen WH: Identification of IgA, IgG, lysozyme, albumin, alpha-amylase and glucosyltransferase in the protein layer adsorbed to hydroxyapatite from whole saliva. *Scand J Dent Res* 91:186–90, 1983

82. Rolla G, Ciardi JE, Eggen KH, et al: Free glucosyl- and fructosyltransferase in human saliva and adsorption of these enzymes to teeth *in vivo*. In: Doyle RJ, Ciardi JE, eds. *Glucosyltransferases, Glucans, Sucrose And Dental Caries. Sp. Supp. Chemical Senses.* IRL Press, Washington, 1983:21–30

83. Rosenberg M, Judes H, Weiss E: Cell surface hydrophobicity of dental plaque micro-organisms *in situ*. *Infect Immun* 42:831–4, 1983

84. Rotstein I, Cohenca N, Mor C, et al: Effect of carbamide peroxide and hydrogen peroxide on the surface morphology and zinc oxide levels of IRM fillings. *Endod Dent Traumatol* 11:279–83, 1995

85. Rotstein I, Mor C, Arwaz JR: Changes in surface levels of mercury, silver, tin, and copper of dental amalgam treated with carbamide peroxide and hydrogen peroxide *in vitro*. *Oral Surg Oral Med Oral Pathol Oral Radiol Endod* 83:506–9, 1997

86. Ruan MS, Di Paola C, Mandel ID: Quantitative immunochemistry of salivary proteins adsorbed *in vitro* to enamel and cementum from caries-resistant and caries-susceptible human adults. *Arch Oral Biol* 31:597–601, 1986

87. Russell MW, Mansson-Rahemtulla B: Interaction between surface protein antigens of *Streptococcus mutans* and human salivary components. *Oral Microbiol Immunol* 4:106–11, 1989

88. Rykke M, Sonju T, Skjorland K, et al: Protein adsorption to hydroxyapatite and to calcium fluoride *in vitro* and amino acid analyses of pellicle formed on normal enamel and on calcium-fluoride-covered enamel *in vivo*. *Acta Odontol Scand* 47:245–51, 1989

89. Sanz M, Newman MG, Nachnani S, et al: Characterization of the subgingival microbial flora around endosteal sapphire dental implants in partially edentulous patients. *Int J Oral Maxillofac Implants* 5:247–253, 1990

90. Satou J, Fukunaga A, Morikawa A, et al: Streptococcal adherence to uncoated and saliva-coated restoratives. *J Oral Rehabil* 18:421–9, 1991

91. Satou J, Fukunaga A, Satou N, et al: Streptococcal adherence on various restorative materials. *J Dent Res* 67:588–91, 1988

92. Saxegaard E, Rolla G: Kinetics of acquisition and loss of calcium fluoride by enamel *in vivo*. *Caries Res* 23:406–11, 1989
93. Saxton CA: Scanning electron microscope study of the formation of dental plaque. *Caries Res* 7:102–19, 1973
94. Scannapieco FA, Solomon L, Wadenya RO: Emergence in human dental plaque and host distribution of amylase-binding streptococci. *J Dent Res* 73:1627–35, 1994
95. Scannapieco FA, Torres G, Levine MJ: Salivary alpha-amylase: role in dental plaque and caries formation. *Crit Rev Oral Biol Med* 4:301–7, 1993
96. Scannapieco FA, Torres GI, Levine MJ: Salivary amylase promotes adhesion of oral streptococci to hydroxyapatite. *J Dent Res* 74:1360–6, 1995
97. Scheie AA, Eggen KH, Rolla G: Glucosyltransferase activity in human *in vivo* formed enamel pellicle and in whole saliva. *Scand J Dent Res* 95:212–5, 1987
98. Scheie AA, Rolla G: Cell-free glucosyltransferase in saliva. *Caries Res* 20:344–8, 1986
99. Schilling KM, Blitzer MH, Bowen WH: Adherence of *Streptococcus mutans* to glucans formed in situ in salivary pellicle. *J Dent Res* 68:1678–80, 1989
100. Schilling KM, Bowen WH: The activity of glucosyltransferase adsorbed onto saliva-coated hydroxyapatite. *J Dent Res* 67:2–8, 1988
101. Schilling KM, Bowen WH: Glucans synthesized in situ in experimental salivary pellicle function as specific binding sites for *Streptococcus mutans*. *Infect Immun* 60:284–95, 1992
102. Shahal Y, Steinberg D, Hirschfeld Z, et al: *In vitro* bacterial adherence onto pellicle-coated aesthetic restorative materials. *J Oral Rehabil* 25:52–8, 1998
103. Skopek RJ, Liljemark WF: The influence of saliva on interbacterial adherence. *Oral Microbiol Immunol* 9:19–24, 1994
104. Sorensen JA: A rationale for comparison of plaque-retaining properties of crown systems. *J Prosthet Dent* 62:264–9, 1989
105. Steinberg D, Beeman D, Bowen WH: The effect of delmopinol on glucosyltransferase adsorbed on to saliva-coated hydroxyapatite. *Arch Oral Biol* 37:33–8, 1992
106. Steinberg D, Beeman D, Bowen WH: Interactions of delmopinol with constituents of experimental pellicle. *J Dent Res* 71:1797–802, 1992
107. Steinberg D, Beeman D, Bowen WH: Kinetic properties of glucosyltransferase adsorbed onto saliva-coated hydroxyapatite. *Artif Cells Blood Substit Immobil Biotechnol* 24:553–66, 1996
108. Steinberg D, Klinger A, Kohavi D, et al: Adsorption of human salivary proteins to titanium powder. I. Adsorption of human salivary albumin. *Biomaterials* 16:1339–43, 1995
109. Steinberg D, Kopec LK, Bowen WH: Adhesion of actinomyces isolates to experimental pellicles. *J Dent Res* 72:1015–20, 1993
110. Steinberg D, Mor C, Dogan H, et al: Effect of salivary biofilm on the adherence of oral bacteria to bleached and non-bleached restorative dental material. *Dent Mater* 15:14–20, 1999
111. Steinberg D, Sela MN, Klinger A, et al: Adhesion of periodontal bacteria to titanium, and titanium alloy powders. *Clin Oral Implants Res* 9:67-72, 1998
112. Stiefel DJ: Characteristics of an *in vitro* dental pellicle. *J Dent Res* 55:66–73, 1976
113. Suljak JP, Reid G, Wood SM, et al: Bacterial adhesion to dental amalgam and three resin composites. *J Dent* 23:171–6, 1995
114. Sunny MC, Sharma CP: Titanium-protein interaction: changes with oxide layer thickness. *J Biomater Appl* 6:89–98, 1991
115. Sutherland DS, Forshaw PD, Allen GC, et al: Surface analysis of titanium implants. *Biomaterials* 14:893–9, 1993
116. Svanberg M, Krasse B, Ornerfeldt HO: Mutans streptococci in interproximal plaque from amalgam and glass ionomer restorations. *Caries Res* 24:133–6, 1990
117. Svanberg M, Mjor IA, Orstavik D. Mutans streptococci in plaque from margins of amalgam, composite, and glass-ionomer restorations. *J Dent Res* 69:861–4, 1990

118. Tabak LA, Levine MJ, Jain NK, et al: Adsorption of human salivary mucins to hydroxy-apatite. *Arch Oral Biol* 30:423–7, 1985

119. Tanaka H, Ebara S, Otsuka K, et al: Adsorption of saliva-coated and plain streptococcal cells to the surfaces of hydroxyapatite beads. *Arch Oral Biol* 41:505–8, 1996

120. Tatevossian A: The effects of heat inactivation, tortuosity, extracellular polyglucan and ion-exchange sites on the diffusion of [14C]-sucrose in human dental plaque residue *in vitro*. *Arch Oral Biol* 30:365–71, 1985

121. Tullberg A: An experimental study of the adhesion of bacterial layers to some restorative dental materials. *Scand J Dent Res* 94:164–73, 1986

122. Ugarte MA, Rodriguez P: Presence of an extracellular glycosyltransferase in human dental plaque. *Int J Biochem* 23:719–26, 1991

123. Vacca-Smith AM, Bowen WH: Effect of some antiplaque agents on the activity of glucosyltransferase of *Streptococcus mutans* adsorbed onto saliva-coated hydroxyapatite and in solution. *Biofilm* 1: paper#2, 1996

124. Vacca-Smith AM, Bowen WH: Binding properties of streptococcal glucosyltransferases for hydroxyapatite, saliva-coated hydroxyapatite, and bacterial surfaces. *Arch Oral Biol* 43: 103–10, 1998

125. Vacca-Smith AM, Venkitaraman AR, Quivey RG Jr, et al: Interactions of streptococcal glucosyltransferases with alpha-amylase and starch on the surface of saliva-coated hydroxy-apatite. *Arch Oral Biol* 41:291–8, 1996

126. Vacca-Smith AM, Venkitaraman AR, Schillimg KM, et al: Characterization of glucosyltransferase of human saliva adsorbed onto hydroxyapatite surfaces. *Caries Res* 30:354–60, 1996

127. van Dijken J, Persson S, Sjostrom S: Presence of *Streptococcus mutans* and lactobacilli in saliva and on enamel, glass ionomer cement, and composite resin surfaces. *Scand J Dent Res* 99:13–19, 1991

128. Venkitaraman AR, Vacca-Smith AM, Kopec LK, et al: Characterization of glucosyltransferase B, GtfC, and GtfD in solution and on the surface of hydroxyapatite. *J Dent Res* 74:1695–701, 1995

129. Vickerman MM, Jones GW: Sucrose-dependent accumulation of oral streptococci and their adhesion-defective mutants on saliva-coated hydroxyapatite. *Oral Microbiol Immunol* 10:175–82, 1995

130. Waerhaug J: Effect of rough surface upon gingival tissue. *J Dent Res* 35:323, 1956

131. Wallman-Bjorklund C, Svanberg M, Emilson CG: *Streptococcus mutans* in plaque from conventional and from non-gamma-2 amalgam restorations. *Scand J Dent Res* 95:266–9, 1987

132. Wennerholm K, Birkhed D, Emilson CG: Effects of sugar restriction *on Streptococcus mutans* and *Streptococcus sobrinus* in saliva and dental plaque. *Caries Res* 29:54–61, 1995

133. Wolinsky LE, de Camargo PM, Erard JC, et al: A study of *in vivo* attachment of *Streptococcus sanguis* and *Actinomyces viscosus* to saliva-treated titanium. *Int J Oral Maxillofac Implants* 4:27–31, 1989

134. Wu-Yuan CD, Eganhouse KJ, Keller JC, et al: Oral bacterial attachment to titanium surfaces: a scanning electron microscopy study. *J Oral Implantol* 21:207–13, 1995

135. Wunder D, Bowen WH: Action of agents on glucosyltransferases from *Streptococcus mutans* in solution and adsorbed to experimental pellicle. *Arch Oral Biol* 44: 203–14, 1999

136. Yamamoto K, Noda H, Kimura K: Adherence of oral streptococci to composite resin restorative materials. *J Dent* 17:225–9, 1989

137. Yamamoto K, Ohashi S, Taki E, et al: adherence of oral streptococci to composite resin of varying surface roughness. *Dent Mater J* 15:201–4, 1996

138. Yamauchi M, Yamamoto K, Wakabayashi M, et al: *In vitro* adherence of microorganisms to denture base resin with different surface texture. *Dent Mater J* 9:19–24, 1990

# 24

# Studying Bacterial Adhesion to Biliary Stents

**Jian-Lin Yu[1] and Roland Andersson[2]**

[1]*Infectious Disease Division, Massachusetts General Hospital,
Harvard Medical School, Boston, MA, USA*
[2]*Department of Surgery, Lund University Hospital, Lund, Sweden*

## I. INTRODUCTION

Endoscopic biliary stenting has become an effective treatment for obstructive jaundice since its introduction by Soehendra and Reijinders-Fredrix.[66] Randomized trials have shown it to be the preferred approach for the palliation of malignant biliary obstruction when surgical process is not possible.[1,62,65,69] Stenting provides relief of jaundice with low morbidity, and it significantly improves patients' quality of life.[3] In benign conditions, endoscopic stenting offers effective short term treatment.[24,30] The major limitation to long term biliary stenting is the problem of late stent occlusion.[20] Once a stent is placed in the bile duct, an encrustation of amorphous material and bacteria starts to accumulate on its surfaces.[27] Given sufficient time, the lumen becomes occluded, bile flow ceases, and the patient develops symptoms of recurrent biliary obstruction, complicated by cholangitis and sepsis. Late clogging is clearly the most important complication of long term treatment with stents. Numerous studies of bacterial adhesion to bile stents have been conducted to explore mechanisms behind stent clogging since the late 1980s, a decade after endoscopic biliary stenting was introduced. In this chapter, bacteriology and defense mechanisms in the biliary tract are briefly reviewed and the methodology for the study of bacterial adhesion to biliary stents described.

## II. BACTERIOLOGY AND DEFENSE MECHANISMS OF THE BILIARY TRACT

### A. Bacteriology of the Biliary Tract

While bacteria may get access to the biliary tree through the lymphatic route, portal vein,[1] or ascend through the ampulla of Vater, the biliary tract does not harbor bacteria.[14,21,45,60] Under pathophysiological conditions, however, the reported incidence of bacteria in bile varies in different reports, ranging from 12 to 75%.[45] It is well recognized that bacteria are more commonly found in bile if the patient is jaundiced, particularly if the biliary obstruction is due to stones or a benign bile duct stricture.[84] Bacteria isolated from the gallbladder and the common bile duct in those patients with biliary tract diseases belong to the intestinal microflora with *Escherichia coli*, *Klebsiella* sp. and

*Handbook of Bacterial Adhesion: Principles, Methods, and Applications*
Edited by: Y. H. An and R. J. Friedman © Humana Press Inc., Totowa, NJ

enterococci as the predominant species.[45] Brook[6] found that a higher rate of anaerobes was present in patients with chronic infections as compared to acute infections, independent of presence of gallstones or the use of prophylactic antibiotics.

## B. Defense Mechanisms Against Bacterial Infection in the Biliary Tract

### 1. Sphincter of Oddi

The sphincter of Oddi, which separates the uncolonized biliary tract from the colonized duodenum, acts as another mechanical barrier against microbial colonization. The importance of an intact sphincter of Oddi, preventing duodenal bacteria from invading the biliary tract, is supported by clinical observations.[22,26,77] Experimental studies have shown that disruption of the barrier by biliary stenting across the sphincter into the duodenum allows bacteria to ascend into the biliary tract,[25,75] leading to the formation of bacterial biofilm, i.e., a complex association of microorganisms and microbial products attaching to the surface, that rapidly develops on these stents and causes occlusion of the lumen.

### 2. Bile Flow

Canalicular bile flow results from the active secretion of solute, followed osmotically by obligatory water flow. Bile salts, the most abundant organic anions in the bile, are considered the major driving force in bile formation named "bile salt-dependent flow" (BSDF). A linear correlation has been reported between bile flow and bile salt secretion in the liver.[5,59] Extrapolation of the linear relation between bile flow and bile salt secretion defines a component of canalicular flow that theoretically is present if no bile salts are secreted. This bile salt-independent flow (BSIF) represents about 50% of total canalicular bile flow in humans. An average of 800 to 1000 mL bile per day in humans effectively flushes the bile ducts. The physical movement of bile hinders bacteria from colonizing the biliary mucosa. In biliary obstruction, bile flow and bile salt secretion decline in parallel,[73] indicating that a raised biliary pressure suppresses the BSDF. It has been shown in patients that bacterial colonization in the biliary tract inhibits the secretion of bile acids and bilirubin into bile.[53] The decline in bile production due to suppression of BSDF and BSIF reduces the clearance effects of bile and hence might predispose to biliary infections.

### 3. Bile Salts

Bile salts possess inhibitory effects on the proliferation of enteric microorganisms and might thus aid in preventing infections of the biliary tree[70] and control the bacterial flora of the gastrointestinal tract.[28,83] The bacteriostatic activity of bile salts is related to their hydrophobicity and detergent properties. Therefore, bile salts with fewer hydroxyl groups and only α-hydroxylation are more potent than those molecules with more hydroxyl groups and with β-hydroxylation concerning suppression of bacterial growth.[74]

### 4. Mucus

The gastrointestinal mucus plays an important role for the exclusion of pathogenic microorganisms by forming a barrier against pathogens in the lumen and retaining secretory immunoglobulin A (sIgA) within the mucus to permit specific immune protection.[52] The extrahepatic bile duct is lined by a tall columnar mucus-secreting epithelium. The physiological role of mucus in the biliary tract is largely unexplored, in contrast to that of the intestine. Bile mucin is known to be involved in the formation of cholesterol gallstones,[10,40] but its possible antimicrobial activity remains to be investigated. Mucin might also play an important role in the pathogenesis of brown pigmented stones in the biliary tract.[71]

### 5. Immune Proteins

Humoral factors associated with host defense, known to be present in bile, include complement,[51,85] immunoglobulin,[7,12,16,32] and lysosomal hydrolase.[36] A number of acute phase reactants have been detected in bile, including α-1-antitrypsin,[36] Fn,[33,49,54,85] and C-reactive proteins (CRP).[51] Among the detected immune proteins in bile, the complement system seems more important than others for the local defense mechanisms against bacterial infection in the biliary tract.[85]

The complement system is comprised of a series of glycoproteins that circulate in the extracellular fluid compartment.[50] These molecules interact in a precise sequence of reactions leading to the production of biologically active cleavage fragments capable of interacting with particles, microorganisms and cells, promoting phagocytosis and direct cell damage. Two major pathways of complement activation are recognized. The classical pathway (CP) is generally activated by the interaction of antibodies of the appropriate class and subclasses with an antigenic surface. In humans, IgG 1, 2, and 3 and IgM are known to be capable of activating the CP. The second major pathway of complement activation has been termed the alternative pathway (AP). The activation of AP is initiated by binding of C3 or C3b to a surface.

## III. METHODS FOR STUDY OF BACTERIAL ADHESION TO BILIARY STENTS

### A. Labeling Bacteria with Tritium($^3$H)

This method was based on the principle that when bacteria divide, a thymidine molecule bearing a tritium atom in the source (broth) is incorporated into DNA in the cells. The cells therefore carry β-emission that can be detected in a liquid scintillation counter in the presence of scintillation cocktail. From the counts of total incorporated isotope and the number of bacterial cells, one can easily calculate the ratio of bacterial cells to disintegration per minute (dpm) and the number of adherent bacterial cells (see later in this section).

Bacterial cells are grown in Mueller-Hinton broth[80-82] or colonization factor antigen broth[87,88,90,92] overnight. The overnight culture is inoculated to fresh broth such that the fresh culture contains 1/10 vol of [methyl-$^3$H]-thymidine (Pharmacia Biotech) and the overnight culture is diluted 1:50. Cell growth is continued at 37°C for 3 to 5 h with agitation. The cells are harvested by centrifugation at 3000 $g$ for 10 min, washed twice and resuspended in 3 mL phosphate-buffered saline (PBS, pH 7.2)/mL culture. The bacterial cells in 0.1 mL suspension are counted by serial 10-fold dilution and plating, then total disintegrations per minute of 0.1 mL suspension are counted. Check cfu (colony forming units) from the plating, calculate cfu in 0.1 mL and the ratio of cfu to dpm (cfu/dpm).

### B. Scanning Electron Microscopy (SEM)

SEM has been widely used in the study of bacterial adhesion to biliary stents (Fig. 1).[19,37,68,74,78,89] One of the advantages of SEM is that it offers direct, visual evidence

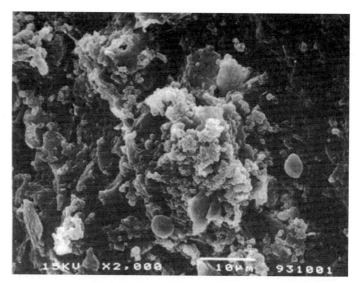

**Figure 1.** SEM micrograph showing biliary sludge and cocci on the surface of an implant that had been kept *in situ* in the rat biliary tract for 8 wk. Reproduced with permission from ref.[90]

of bacterial adhesion. Typical SEM specimen preparation involves sample fixation, treatment with osmium tetroxide and dehydration.[78,89] The following procedure was used in our studies.[89]

Tissue samples and implants were fixed in Millonig's phosphate buffer (MPB, pH 7.2) containing 2.5% glutaraldehyde (GTA) at 4°C for 12 h. The specimens were extensively rinsed in MPB and postfixed in 2% osmium tetroxide in MPB for 2 h. They were then treated with a thiocarbohydrazide and osmium tetroxide series to make them conductive, followed by dehydration through ethanol and Freon 113 series. The specimens were critical-point dried, mounted on stubs, coated in a Polaron E5400 high resolution sputter-coating apparatus, and examined in a JEOL JSM T-330 SEM (Tokyo, Japan) at an accelerated voltage of 15 kV.

### C. Protein Labeling

A modification of Markwell's[47] method has been used for iodination of proteins. Materials and facilities required include Na$^{125}$I (from Amersham Pharmacia Biotech, specific activity 640 MBq or 17.3 mCi $^{125}$I/mg), Iodobeads® or N-chlorobenzenesulfonamide (sodium salt)-derived uniform non-porous polystyrene beads (reducer; Pierce Chemicals, Rockford, IL USA), a gamma counter and desalting column (Sephadex G25; Pharmacia Biotech). In addition, working in a ventilation hood with a shield is strongly recommended. Follow the regulations for radioactive material handling. Carefully clean the area where you have been working. Deliver the radioactive waste to the designated location.

Up to 50 μg protein can be iodinated per labeling using this method. Two microliters (μL) of Na$^{125}$I is pre-incubated with one Iodobead in 180 μL PBS for 5 min. Twenty to twenty-five μg protein in a volume of 18 μL are added to the mixture and the incubation continued for another 15–20 min. 300 μL PBS ia added to the tube and the mixture loaded to the desalting column pre-equilibrated with 1% bovine serum albumin (BSA). The column is washed with PBS and fractions of 0.5 mL are collected. The radioactivity of each fraction is measured. Fractions with highest readings (normally fractions 6–8) are pooled. Unbound isotope should retain in the column. The total radioactivity and specific radioactivity of labeled protein is measured. The specific radioactivity of the labeled proteins should be 1,000,000 to 1,500,000 cpm/μg labeled protein solution.

If the concentration of protein to be labeled is too low, incubate 2 μL Na$^{125}$I with 198 μL protein solution for 20 min to have maximal amount of protein labeled. Add 300 μL PBS to the tube before loading to column as described above.

### D. Detection of Adsorbed Bile Proteins on Biliary Stents

It has been demonstrated that when a device (recognized as a foreign body by the host) is implanted or inserted, host-derived proteins, such as Fn, Fg and immunoglobulins adsorb onto the surfaces of the foreign body and that the adsorbed proteins can be used by microorganisms as adhesion mediators.[81,82] The situation in the biliary tract may be different, as bile salts are thought to be detergents that can "wash" away the adsorbed proteins. However, in a series of studies, we have found that Fn, Vn, and immunoglobulins could adsorb onto the materials placed in the biliary tract and that Fn was found to enhance bacterial adhesion.[89] In an independent immunohistochemical study with clogged stents removed from patients, Chan and colleagues found that immunoglobulins were involved in the process of stent blockage.[12] If, like in other sites, protein adsorption in biliary stenting is the initial event to occur before bacteria attach to the surface, it is maybe the time to reconsider the strategy in choosing materials for stent manufacturing and design in order to minimize protein adsorption.

Biliary stents, removed from patients with malignant obstructive jaundice, from animals, or removed from in vitro perfusion with bile, were washed in PBS (pH 7.2), cut into 1 cm long pieces and probed with antibodies against human Fn, Vn, Fg, or albumin diluted 1:100 in 1% BSA(bovine serum albumin)-PBS. The stent pieces are incubated with $^{125}$I-labeled secondary antibodies diluted 1:100 in 1%BSA-PBS at 20°C for 120 min. The pieces were washed three times between incubations with antibodies. The pieces were then transferred to polystyrene tubes, the radioactivity read in a gamma counter and the amount of adsorbed proteins given as radioactivity of bound secondary antibodies (cpm/cm$^2$).

### E. SDS-Polyacrylamide Gel Electrophoresis (SDS-PAGE) and Western Blot

SDS-PAGE is a useful technique in the biliary stent-related studies, particularly when there is a need to elucidate the pattern of possibly adsorbed bile proteins.[27,88,90] Western blot may also be used when necessary.

To prepare samples for SDS-PAGE, we rinsed biliary stents removed from patients, animal models, in vitro flow models, and rinsed adsorbed materials off the inner surface were eluted with 1 M lithium chloride at 20°C for 60 min. The eluate was then dialyzed against PBS at 4°C overnight and concentrated with PEG 20 M to a volume around 1 mL.

SDS-PAGE is run using a Bio-Rad mini Protean cell (Bio-Rad Laboratories, Richmond, CA, USA) using discontinuous buffer system according to Laemmli.[35] Separated proteins on the gels may be visualized by staining with 0.1% Coomassie brilliant blue in 40% methanol/10% acetic acid for 30 min and destaining in a solution containing 40% methanol and 10% acetic acid.

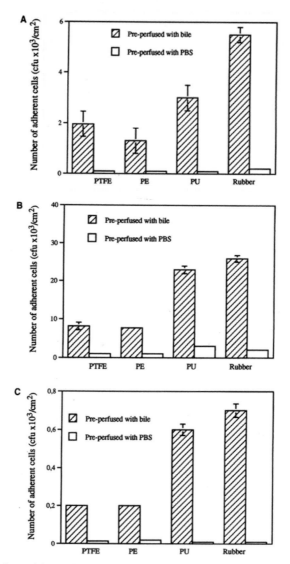

**Figure 2.** Adhesion of bacterial cells onto the surfaces of various stents by three *E.coli* strains after perfusion of these materials with bile. Solid bars indicate that the material had been perfused with bile whereas open bars indicate perfusion with PBS. (A) *E. coli* NG7C; (B) *E. coli* 123 and (C) *E. coli* 4236. Reproduced with permission.[89]

For Western blot, the separated proteins were electrically transferred to a nitrocellulose membrane (NC, pore size 0.45 µm, Schleicher & Schuell GmbH, Dassel, Germany) using a semi-dry electroblotter at a constant current of 190 mA for 90 min. The membrane was quenched with 1% BSA in 0.05 M Tris-HCl buffered saline (TBS, pH 10.3), incubated with rabbit antibodies against human Fn, Vn, Fg, or albumin and, after washing three times, incubated with peroxidase-conjugated swine anti-rabbit immunoglobulin. The reaction was developed in the dark in a mixture of 20 mg 3-amino-9-ethyl carbazole, dissolved in 2.5 mL acetone, 50 mL acetate buffer (pH 5.0) and 25 mL 30% hydrogen peroxide.

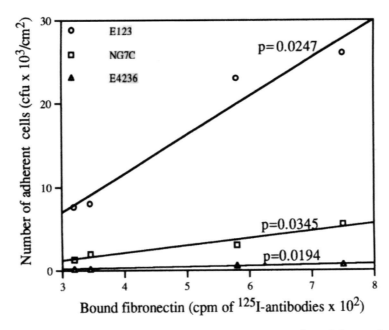

**Figure 3.** Correlation between the amount of adsorbed fibronectin and the number of adherent bacterial cells. Reproduced with permission.[89]

Alternatively, if the membrane is incubated with secondary antibodies labeled with [125]I, after the membrane is washed and dried, the membrane can be placed into an X-ray cassette and an X-ray film exposed against the membrane. This technique is called autoradiography.

There is another technique available to develop the reaction, called ECL (enhanced chemiluminescence) system, developed by Amersham (now known as Amersham Pharmacia Biotech). ECL Western blotting system uses horseradish peroxidase conjugated antimouse and antirabbit antibodies for luminol-based detection for Western blots. Blots probed with mouse or rabbit primary antibodies are incubated with HRP-conjugated secondary antibodies. Addition of ECL detection reagents results in a chemiluminescent signal that can be captured on the film.

This system provides a rapid, sensitive, nonisotopic and quantifiable method for detection of protein immobilized on membranes. As in autoradiography, an X-ray cassette and autoradiography films are required. In addition, a dark room (or a film safe room) is required to install the film and the membrane in the cassette.

## F. Detection of Bacterial Adhesion to Bile Stent in Stationary Phase[87,88,90]

Biliary stents, clean or removed from the perfusion model or from patients, can be cut to size, split open, and incubated in test tubes with radiolabeled bacterial cells at 37°C for 60 min. The materials are washed three times and transferred to a scintillation vial containing 2.5 mL scintillation cocktail and the radioactivity of each piece of material is counted in a scintillation counter (Beckman LS 1800). The number of bacterial cells or cfu can be calculated by multiplying dpm read by cfu/dpm after subtracting background reading, i.e.,

$$N = (R-bkg)xk \tag{1}$$

In Eq. (1), $N$ represents the number of adherent bacterial cells (cfu), $R$ represents the radioactivity counted from the individual sample (dpm), *bkg* represents background reading, $k$ represents the ratio of cfu/cpm (or cfu/dpm), calculated from cfu and radioactivity per 100 µL bacterial cell suspension of each individual strain. Figure 2 shows detectable bacterial adhesion to biliary stent materials which were pre-perfused with bile while Figure 3 shows the correlation of adhered bacterial cells with adsorbed bile protein.

### G. Flow Models

#### 1. Flow Cell [86, 88]

This system consists of a glass tube, a reservoir, a peristaltic pump and silicone tubing which connects the glass tube and the reservoir to form a close circulation system. Material to be tested was placed in the glass tube with support of a steel wire such that the material will be kept in the center of fluid flow in the glass tube. Biliary stents were placed in the flow cell, bile with 0.05% sodium azide or PBS was placed in the reservoir which was placed in 37°C water bath, and the pump was run at a speed of 0.5 mL/min for 24 h. The biliary drains were removed from the flow cell and subject to detection of adsorbed biliary proteins or to perfusion with bacterial suspension.

After pre-perfusion with bile or PBS, $^3$H-labeled bacteria at a density of $1 \times 10^6$ cfu/mL in BSA-PBS were added into the circulation. The experiment continued to run for an additional 120 min at a flow rate of 0.5 mL/min. The materials were then removed, cut and rinsed with PBS and their radioactivity counted in a Beckman LS scintillation counter. Number of adherent bacteria was calculated using Eq. (1) described in section F.

#### 2. Use of Whole Catheter

This method was developed by Ladd[34] and is widely used for studying bacterial adhesion to stents.[18,39] In this model the entire stent or catheter is connected to a closed system which is run by a peristaltic pump at a flow rate of 0.5 mL/min. The stent or catheter can be removed, rinsed, cut into pieces and analyzed.

#### 3. Modified Robbins Device

The modified Robbins device (MRD) is an acrylic sampling device, 42 cm long with a 2×10 mm chamber.[37] It has 25 evenly spaced sampling ports and is designed in such a way that the sampling surfaces lie flush with the inner surface of the flow chamber. It allows the study of various materials one time, or the study of one kind of material for various periods of time. Segments of plastic biliary stents can be attached to rubber discs using fishing line and mounted individually onto the sampling ports. Individual samples can be removed at various times without disturbing other samples or the system.

#### 4. Tsang Perfusion Model of Stent Occlusion with Porcine Bile

Tsang et al. recently developed an in vitro model for studying stent occlusion with porcine bile.[78] The experiment was run for 8 wk and bile in the reservoir was replaced with fresh bile every 7 d. Bile samples were taken at the beginning and end of the week to monitor the concentrations of total calcium, cholesterol, bile salts, phospholipids, total bilirubin and pH. The bacteria concentration was maintained at a consistent level by changing the bile weekly and inoculating with fresh bacteria.

**Table 1. Changes in Plasma Bilirubin, Liver Functions,
and Body Weight During the Study**

| | Before operation | Day 5 | Day 12 |
|---|---|---|---|
| Bilirubin (mmol/L) | 2.14±0.23 | 63.39±13.13** | 2.66±0.49 |
| Aspartate aminotransferase (mkat/L) | 1.61±0.18 | 3.82±0.42** | 1.58±0.25 |
| Alanine aminotransferase (mkat/L) | 1.07±0.06 | 1.80±0.17** | 0.99±0.08 |
| Alkaline phosphatase (mkat/L) | 10.05±0.46 | 12.27±0.93** | 10.05±0.94 |
| Body weight (g) | 342±4 | 325±4 | 352±6* |

* = $p < 0.05$ and ** = $p < 0.01$ compared with values obtained before operation. Reproduced with permission from ref.[94]

## H. Animal Models for the Study of Bacterial Adhesion to Biliary Stents

Animals, including the cat,[44,74,75] the pig,[79] the dog,[9] and the rat[89,91,92] have been used for the biliary related-studies. Because of anatomic similarities to the biliary tract in humans, large animals have an advantage in the terms of placing a stent into to the common bile duct of the animal. The main problems with large animals include the availability of an experienced veterinarian, the availability of facilities and personnel for animal care, and cost. It is clear that using "small animals" will be an option in these circumstances. It is not possible to insert a full size or even a segment of a stent into the biliary tract of a small animal in the normal situation, but pieces of stent material (for example, a piece with a square of 0.5 cm$^2$) can be inserted into the biliary tract in small animals after an appropriate manipulation is made to the biliary tract. Whether you choose to use large or small animals, you will have to apply for animal work permission from your institution's animal ethics committee.

### 1. Feline Model

In this model adult cats were used. The common bile duct strictures were first created surgically, then the animals were stented with 5 French gauge polyethylene tube. The stents were placed with distal ends open to the duodenum.[44,74,75]

### 2. Porcine Model

This model was used to evaluate the histological response to placement and technical aspects of self-expanding spiral nitinol stents.[79]

### 3. Canine Model

This model was initially used for evaluation of a second generation of tantalum biliary stent, but can certainly be applied to the study of bacterial adhesion to biliary stents.[9] The animals were anesthetized with intravenous thiopental, intubated, and maintained with 2-2.5% halothane delivered in 100% $O_2$ during surgical procedure. While the expandable metallic stents were placed to the midcommon bile duct through the cystic duct, a similar procedure may be applied to the insertion of plastic stents.

### 4. Rat Model

Yu et al.[92] developed a novel animal model in the small animal for the study of bacterial adhesion to biliary stents. In this model, using a special device called a minioccluder, the

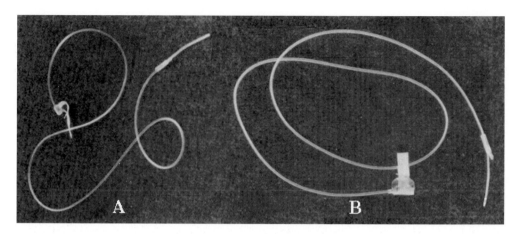

**Figure 4.** (A) A minioccluder consisting of tubing, balloon and two sleeves. (B) A minioccluder with the balloon inflated. Reproduced with permission.[94]

bile flow was first obstructed and the common bile duct dilated. When the minioccluder was removed on day 5, the common bile duct remained dilated, and patency of bile flow was restored. The model offers the possibility of implanting biliary stent pieces into the common bile duct without interfering with the bile flow. The device consists of a miniballoon (2 mm in diameter), a silicone tube connected with the balloon and two leaves around the balloon (Fig. 4). The animals were anesthetized with light ether, the minioccluder was placed around the common bile duct and the balloon was inflated with saline (about 0.5 mL). A pressure against the common bile duct formed and the common duct was dilated. The minioccluder was removed five days after the initial surgery and the common bile duct remains dilated and patent as evidenced from the cholangiogram (Fig. 4) and liver function tests (Table 1).

*5. Working with the Rat Model* [87-92]

A. SAMPLING AND TREATMENT OF SAMPLES

Samples were taken under aseptic conditions under during light ether anesthesia. Blood samples were obtained from the femoral vein by puncture and bile samples were obtained by puncture of the common bile duct. For assaying opsonic activity, blood samples were allowed to clot at 37°C for 60 min and sera were separated. Heparinized plasma was used for the measurements of liver function tests and plasma bilirubin concentrations. Serum, plasma and bile samples should be kept at −70°C until analyzed.

B. STANDARD LIVER FUNCTION TESTS AND PLASMA BILIRUBIN CONCENTRATION

Aspartate aminotransferase, alanine aminotransferase and alkaline phosphatase activities, as well as plasma bilirubin concentration were measured according to the recommendations of the Committee on Enzymes of the Scandinavian Society for Clinical Chemistry and Clinical Physiology.[58]

C. CHOLANGIOGRAPHY

Under light ether anesthesia, cholangiography was made using contrast medium (Fig. 5).[92]

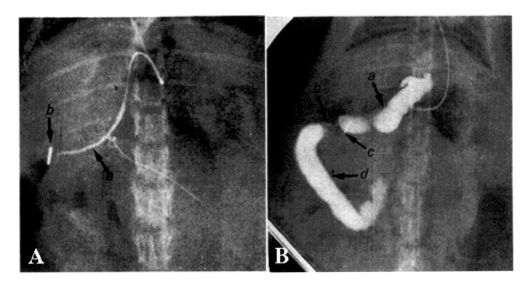

**Figure 5.** Cholangiograms of rat with miniooccluder. (A) Cholangiogram after the miniooccluder was inflated, showing complete obstruction of the common duct. a = common bile duct, b = metal frame of the mini-occluder. (B) Cholangiogram on day 12 (the seventh day after deflation of the mini-occluder), showing that the part of the common duct proximal to the miniooccluder was dilated, and the distal part, including the part surrounded by the miniooccluder, remained normal in diameter and patent. a = dilated part of the common duct, b = the distal part of common duct, c = metal frame of the mini-occluder, and d = duodenum. Reproduced with permission.[94]

## D. SURGERY

Implantation of drain pieces was performed under aseptic conditions. Under ether anesthesia, following a 3 to 3.5 cm incision and gentle access to the common bile duct, choledochotomy with a 0.4 cm incision was made. A piece was implanted into the common bile duct and the incision in the bile duct was closed with 7 to 0 synthetic polyglactin sutures under a microscope. The implanted pieces were kept *in situ* for 1 to 14 wk until removed. The abdominal cavity was closed by continuous suture with 3-0 silk. Animals with implants in the biliary tract were anaesthetized and underwent exsanguination prior to removal of the implant.

## E. BACTERIAL CHALLENGE

An *E. coli* strain (O21:H25) isolated from the bile of a patient with cholangitis was used to challenge the animals in each group from one to four weeks after drain piece implantation in the biliary tract. Cells from overnight cultures in CFA broth at 37°C with agitation were washed in phosphate buffered saline (PBS; pH 7.2) and suspended in sterile 0.9% NaCl. One hundred mL of an appropriate dilution was injected into the common bile duct. The number of viable bacteria injected was checked by plating 100 µL of a $10^{-6}$ dilution of the washed overnight culture containing $2–3 \times 10^8$ cfu/mL. Bile from each animal was obtained by aspiration from the common bile duct after inoculation and serial dilutions were plated out on blood agar at 37°C, which allowed the quantitative determination of cfu/100 µL bile from each animal. The bile was defined as infected when twice the number of initially inoculated cfu was cultured. Spontaneous bacteriobilia was determined by culturing undiluted bile sampled before inoculation. Animals with spontaneous bacteriobilia were excluded from the study.

### F. PREPARATION OF LEUKOCYTES

Polymorphonuclear leukocytes (PMNLs) were obtained from the peritoneal exudate of rats receiving an intraperitoneal injection of 10 mL 12% (wt/vol) sterile sodium caseinate in 0.9% NaCl according to Stossel et al.[72] The animals were sacrificed by exsanguination sixteen hours after the injection, after which laparotomy was made. PMNLs were harvested by abdominal lavage, filtered into ice-cold siliconized glass tubes through eight layers of gauze and washed twice in cold PBS by centrifugation at 250 $g$ for 10 min and suspended in 4 mL of HBSS. The total yield of PMNLs harvested by this technique varied between $5 \times 10^6$ and $10^7$ cells per rat. Cell viability was 90-95%, as revealed by exclusion of 0.1% Trypan blue.

### G. BACTERIOLOGICAL STUDY

Implants as well as specimens taken from the liver and biliary tract mucosa were immersed in 2 mL TSB and vortexed for 30 s. One hundred µL of the suspension and 20 µL of bile were inoculated on blood agar and MacConkey agar and both cultured under aerobic conditions at 37°C. One blood agar plate from each sample was incubated anaerobically at 37°C for up to 4 d before regarded as negative. Identification was done using standard technique.[4]

### H. ASSAYS OF OPSONIC ACTIVITY

Washed bacterial cells from an overnight culture of the *E. coli* strain O21:H25 were suspended in 0.9% NaCl at a concentration of $3.5–4 \times 10^5$ cfu/mL and the phagocytic bactericidal assay described by Lew et al.[41] was used to evaluate opsonic activity of bile and sera. Rat PMNLs, bacteria, and the opsonic sources were mixed in PBS in siliconized glass tubes at a final volume of 1 mL. Each mL of the incubation mixture contained $4 \times 10^5$ PMNLs, $3.5-4 \times 10^4$ cfu of *E. coli*, and 0.5 mL of a serially diluted opsonic source (bile or sera). The mixtures were incubated at 37°C in a shaking water bath for 30 min. Samples were taken at zero time and 30 min after incubation. Serial dilutions were made and incubated on blood agar containing 5% horse erythrocytes at 37°C for 48 h. The reciprocal value of the dilution of bile or sera that resulted in 50% killing of *E. coli* was defined as the opsonic titer. Controls included: 1) heat-inactivated (at 56°C for 30 min) pooled rat bile or sera with PMNLs; 2) pooled rat bile or sera without PMNLs; and 3) PMNLs without bile or sera.

## IV. CURRENT STRATEGIES FOR PREVENTING STENT OCCLUSION

### A. *Stent Diameter*

A larger stent does not prevent bacterial adhesion or biofilm growth but does provide more room for sludge to accumulate before the lumen becomes occluded. Previous studies have demonstrated a clear advantage for stents of 10 French gauge over stents of 8 French gauge. Siegel and coworkers reported a significantly longer patency with 12 French stents when compared with their experience using 10 French stents,[63] but other studies have not been able to confirm this.[23,31] Regardless of whether further increases in stent diameter improve patency, plastic stents larger than 12 French gauge cannot be placed by conventional methods because of the restricted diameter of the instrument channel of the duodenoscopes.

### B. Metal Stents

The expandable metal stents offer significantly longer patency than conventional plastic stents.[11,15] Because a metal stent expands after deployment it attains a much larger final diameter, thus minimizing the risk of occlusion due to sludge accumulation. The mesh-like design provides much less surface for bacteria to adhere. Unfortunately, tumor ingrowth between the interstices of the mesh is a frequent occurrence that limits the advantage of these stents. Another potential problem with expandable stents is their permanence. Even in benign disease, the wire mesh has been shown to penetrate into the submucosa of the bile duct. Subsequent mucosal hyperplasia and tissue ingrowth may eventually compromise the stent lumen.[46] Once deployed, a metal stent is essentially irretrievable by endoscopic means, and when the stent becomes embedded in the submucosa, it is difficult to remove even surgically.[79]

### C. Antibiotics

As the underlying cause of sludge deposition, the biofilm bacteria are obvious targets for interrupting the clogging process. However, inhibition of bacterial growth on foreign bodies is not an easy task. The glycocalyx matrix of biofilm presents a physical barrier to antibiotic penetration, thereby protecting the enclosed bacteria. Once established within a biofilm, bacteria can withstand antibiotic concentrations 100-fold greater than those tolerated by free-floating bacteria.[2]

Long term administration of prophylactic antibiotic is one potential approach to prevent bacterial growth. In theory, free-floating bacteria are more vulnerable to killing before they attach to a foreign body and develop into a sessile colony with protective biofilm. In vitro studies suggest that the administration of low dose antibiotics can retard bacterial adhesion and growth on stents.[17,44] In a recent in vitro study by Tsang and coworkers, continuous use of ampicillin-sulbactam was proven to prevent biofilm formation over 8 wk.[78] However, the benefit of antibiotic treatment has been more difficult to demonstrate in clinical trials.

### D. Surface Coating or Impregnation

In a study comparing cefoxicin-coated biliary stents with conventional ones, Browne and colleagues found that antibiotic-coating did not prolong stent patency.[8] In a recent in vitro study, Rees and colleagues found an antimicrobial benzalkonium chloride (BZC) impregnated polymer to significantly reduce bacterial adhesion.[55] In a series of studies, we tested the effects of phosphatidylcholine (PC) and phosphatidylinositol (PI) on bacterial adhesion in vitro. The study consisted of two parts. The first part tested the effects in vitro. The PC- and PI-coated pieces were incubated with radiolabeled bacterial cells. In the second part the coated pieces were implanted into the common bile duct in our rat model for one and two weeks, after which the materials were removed and incubated with radiolabeled bacterial cells. While preventive effects of phospholipids-coating on bacterial adhesion were noticed in both parts of the study, the effects decreased with time after the material implantation.[87] Silver coating of plastic stents has been tested and suggested as an alternative to traditional antibiotics because emergence of bacterial resistance to silver has not been observed,[38] but the effects in animals or humans have yet to be tested.

## E. Stent Design

Various stent plastics have been investigated for their physicochemical properties. It is a common belief that smoother stent surfaces should be associated with a lower incidence of blockage.[57] It was demonstrated that biliary sludge on the surface was dependent on the material properties, i.e., the smoother the surface, the lesser amount of sludge was found.[18] McAllister and coworkers[48] reported that, when perfused with infected bile in an in vitro model, a polymer with ultrasmooth surface was almost free from bacterial cells on the surface, whereas a surface with defects bound a much greater number of bacterial cells. In another study, it was shown that hydrophilic polymer-coated polyurethane dramatically reduced bacterial adhesion in vitro, as the coating hydrophilic polymer provided an extremely smooth surface.[29]

It has been noticed that side holes may enhance the process of deposition of biliary sludge onto the inner surface, leading to occlusion of the stent lumen. The side holes were initially designed to facilitate biliary drainage, both in the hypothetical situation where the end orifice abuts against the ductal wall and when the terminal holes are occluded by cellular debris, blood clots or mucus plugs. Unfortunately, they were virtually found to accelerate the process of drain occlusion, as biliary sludge accumulated much more on the inner surface of stents with side holes than on those without side holes.[13,18] It was presumed that side holes of stents generate surface irregularity, creating an ideal site for bacterial attachment.[18,64] Furthermore, the side holes create a turbulent flow, which in turn increases the resistance of adherent microorganisms to bile flow.[13,18] To overcome the problem, a newly designed stent without side holes has been introduced and a nonrandomized comparison with pigtail stents has suggested that patency is prolonged.[61] A randomized trial comparison of this new design with straight stents was underway.[43]

## V. CONCLUDING REMARKS

Up to date, there is no optimal stent available. The field of biliary stent-related research, and especially the study of bacterial adhesion to biliary stents, is still under development. The above-described methods provide a powerful set of techniques necessary for these studies.

## REFERENCES

1. Andersen JR, Sorensen SM, Kruse A, et al: Randomized trial of endoprosthesis versus operative bypass in malignant obstructive jaundice. *Gut* 30:1132–5, 1989
2. Anwar H, van Biesen T, Dasgupta MK, et al: Interaction of biofilm bacteria with antibiotics in novel *in vitro* chemostat system. *Antimicrob Agents Chemother* 33:1924–4, 1990
3. Ballinger AB, McHugh M, Catnach SM, et al: Symptom relief and quality of life after stenting for malignant bile duct obstruction. *Gut* 35:467–70, 1994
4. Balows A, Hausler WJ Jr, Herrmann KL, et al, eds: *Manual of Clinical Microbiology 5th Edition*. American Society of Microbiology Press, Washington DC, 1991
5. Boyer JL, Bloomer JP: Canalicular bile secretion in man: Studies utilizing the biliary clearance of $^{14}$C-mannitol. *J Clin Invest* 54:773–881, 1974

6. Brook I: Aerobic and anaerobic microbiology of biliary disease. *J Clin Microbiol* 27:2373–5, 1989

7. Brown WR, Kloppel TM: The liver and IgA: Immunological, cell biological and clinical implications. *Hepatology* 9:763–84, 1989

8. Browne S, Schmalz M, Geenen J, et al: A comparison of biliary and pancreatic stent occlusion in antibiotic-coated vs. conventional stents. *Gastrointest Endosc* 36:206, 1990

9. Cardella JF, Wilson RP, Fox PS, et al: Evaluation of a second-generation tantalum biliary stent in a canine model. *J Vasc Interv Radiol* 6:397–403, 1995

10. Carey MC, Cahalane MJ: Whither biliary sludge? *Gastroenterology* 95:508–23, 1988

11. Carr-Locke DL, Ball TJ, Connors PJ, et al: Multicenter randomized trial of Wallstent biliary endoprosthesis versus plastic stents. *Gastrointest Endosc* 39:310, 1993

12. Chan FK, Suen M, Li JY, et al: Bile immunoglobulins and blockage of biliary endoprosthesis: an immunohistochemical study. *Biomed Pharmacother* 52:403–7, 1998

13. Coene PPLO, Groen AK, Cheng J, et al: Clogging of biliary endoprosthesis: an new perspective. *Gut* 31:913–7, 1990

14. Csendes A, Fernandez M, Uribe P: Bacteriology of the gallbladder bile in normal subjects. *Am J Surg* 129:629–31, 1975

15. Davids PH, Groen AK, Rauws EAJ, et al: Randomized trial of self-expanding metal stents versus polyethylent stents for distal malignant biliary obstruction. *Lancet* 340:1488–92, 1992

16. de Bruijn MA, Mok KS, Out T, et al: Immunoglobulins and α-1-acid glycoprotein do not contribute to the cholesterol crystallization-promoting effect of concanavalin A-binding biliary protein. *Hepatology* 20:626–32, 1994

17. Desta T, Libby E, Liu YL, et al: Prophylactic antibiotic therapy in the prevention of stent blockage. *Gastroenterology* 108:A413, 1995

18. Dowidar N, Kolmos HJ, Matzen P: Experimental clogging of biliary endoprostheses: role of bacteria, endoprosthesis material, and design. *Scand J Gastroenterol* 27:77–80, 1992

19. Dowidar N, Kolmos HJ, Lyon H, et al: Clogging of biliary endoprostheses. A morphologic and bacteriologic study. *Scand J Gastroenterol* 26:1137–44, 1991

20. Dowidar N, Moegaard F, Matzen P: Clogging and other complications of biliary endoprostheses. *Scand J Gastroenterol* 26:1132–6, 1991

21. Edlund YA, Mollstedt BD, Ouchterlony O: Bacteriological investigation of the biliary system and liver in biliary tract disease correlated to clinical data and microstructure of the gall bladder and liver. *Acta Chir Scand* 116:461–76, 1959

22. Feretis CB, Contou CT, Manoruas AJ, et al: Long term consequences of bacterial colonization of the biliary tract after choledochotomy. *Surg Gynecol Obstet* 159:363–6, 1984

23. Finnie IA, O'Toole PA, Rhodes JM, et al: A prospective randomized trial of 10 and 11.5 FG endoprostheses. *Gut* 35(suppl):S50, 1994

24. Foutch PG, Harlan JR, Hoefer M: Endoscopic therapy for patients with a post-operative biliary leak. *Gastrointest Endosc* 39:416–21, 1993

25. Geoghegan JG, Branch MS, Costerton JW, et al: Biliary stent occlude earlier if the distal tip is in the duodenum in dogs. *Gastrointest Endosc* 37:257–61, 1991

26. Gregg JA, Girolami PD, Carr-Lock DL: Effects of sphincteroplasty and endoscopic sphincterotomy on the bacteriologic characteristics of the common bile duct. *Am J Surg* 149:668–71, 1985

27. Groen AK, Out T, Huibregtse K, et al: Characterization of the content of occluded biliary endoprostheses. *Endoscopy* 19:57–9, 1987

28. Hill MJ: Factors controlling the microflora of the healthy upper gastrointestinal tract. In: Hill MJ, Marsh PD, eds. *Human Microbial Ecology.* CRC Press, Boca Raton, FL, 1989:57–86

29. Jansen B, Goodman LP, Ruiten D: Bacterial adherence to hydrophilic polymer-coated polyurethane stents. *Gastrointest Endosc* 39:670–3, 1993

30. Johnson GK, Geenen JE, Venu RP, et al: Treatment of non-extractable common bile duct stones with combination ursodeoxyycholic acid plus endoprostheses. *Gastrointest Endosc* 39:528–31, 1993

31. Kadakia SC, Starnes E: Comparison of 10 French gauge stent with 11.5 French gauge stent in patients with biliary tract diseases. *Gastrointest Endosc* 38:494–9, 1992

32. Kagnoff MF: Immunology of the digestive system. In: Johnson LR, ed. *Physiology of the Gastrointestinal Tract.* Raven Press, New York, 1987:1699–728

33. Korner T, Kropf J, Hackler R, et al: Fibronectin in human bile fluid for diagnosis of malignant biliary diseases. *Hepatology* 23:423–8, 1996

34. Ladd TI, Schmiel D, Nickel JC, et al: The use of radiorespirometric assay for testing the antibiotic sensitivity of catheter-associated bacteria. *J Urol* 183:1451–6, 1987

35. Laemmli UK: Cleavage of structural proteins during the assembly of the head of bacteriophage T4. *Nature* 227:680–5, 1970

36. LaRusso NF: Proteins in bile: how they get there and what they do. *Am J Physiol* 247: G199–205, 1984

37. Leung JW, Liu YL, Desta T, et al: Is there a synergistic effect between mixed bacterial infection in biofilm formation on biliary stents. *Gastrointest Endosc* 48:250–7, 1998

38. Leung JWC, Lau GTC, Sung JY, et al: Decreased bacterial adherence to silver-coated stent material: an *in vitro* study. *Gastrointest Endosc* 38:338–40, 1992

39. Leung JW, Ling TW, Kung JL, et al: The role of bacteria in the blockage of biliary stent. *Gastrointest Endosc* 34:19–22, 1988

40. Levy PF, Smith BF, Mamont JF: Human gallbladder mucin accelerates nucleation of cholesterol in artificial bile. *Gastroenterology* 87:270–5, 1984

41. Lew PD, Zubler R, Vaudaux P, et al: Decreased heat-labile opsonic activity and complement levels associated with evidence of C3 breakdown products in infected pleural effusions. *J Clin Invest* 63:326–34, 1979

42. Libby ED, Coimbre A, Leung JW: Early treatment with ciprofloxacin prevents adherence of biofilm. *Gastrointest Endosc* 40:409 (abstract), 1994

43. Libby ED, Leung JW: Prevention of biliary stent clogging: a clinical review. *Am J Gastroenterol* 91:1301–8, 1996

44. Libby ED, Morck D, McKay S, et al: Ciprofloxacin prevents stent blockage in an animal model. *Gastroenterology* 106:A346 (abstract), 1994

45. Løtveit T: Bacterial infections of the liver and biliary tract. *Scand J Gastroenterol [Suppl]* 85:33–6, 1983

46. Maccioni F, Rossi M, Salvatori FM, et al: Metallic stents in benign biliary strictures: three-year follow up. *Cardiovasc Interven Radiol* 15:360–6, 1992

47. Markwell MAK: A solid reagent for protein iodination. I. Conditions for the efficient labeling of antiserum. *Anal Biochem* 125:427–32, 1982

48. McAllister EW, Carey LC, Brady PG, et al: The role of polymeric surface smoothness of biliary stents in bacterial adherence, biofilm deposition, and stent occlusion. *Gastrointest Endosc* 39:422–5, 1993

49. Miquel JF, Von Ritter C, Del Pozo R, et al: Fibronectin in human gallbladder bile: cholesterol pronucleating and/or mucin "link" protein? *Am J Physiol* 267:G393–400, 1994

50. Morgan BP: Physiology and pathophysiology of complement: Progress and trends. *Crit Rev Clin Lab Sci* 32:265–98, 1995

51. Morrison L, Blamey S, Veitch J, et al: Complement levels in serum and bile in patients with extra-hepatic biliary tract obstruction. *J Clin Lab Immunol* 13:71–74, 1984

52. Neutra MR, Forstner JF: Gastrointestinal mucus: Synthesis, secretion and function. In: Johnson LR, ed. *Physiology of the Gastrointestinal Tract.* Raven Press, New York, 1987: 975–1002, 1987

53. Nishida T, Nakahara M, Nakao K, et al: Biliary bacterial infection decreased the secretion of bile acids and bilirubin into bile. *Am J Surg* 177:38–41, 1999

54. O'Connor MJ, Allen JI, Vennes JA, et al: Protective effects of fibronectin in bile against biliary infection. *Surg Forum* 35:122–4, 1984
55. Rees EN, Tebbs SE, Elliott TS: Role of antimicrobial-impregnated polymer and Teflon in the prevention of biliary stent blockage. *J Hosp Infect* 39:323–9,1998
56. Reuben A: Biliary proteins. *Hepatology* 4:46S–50S, 1984
57. Safrany L, Schrameyer B, Wosiewitz U: Analysis of causes of the occlusion of biliary stents. *Gastrointest Endosc* 32:183A, 1986
58. Scandinavian Committee on Enzymes: Recommended methods for the determination of four elements in blood. *Scand J Clin Lab Invest* 33:291–306, 1974
59. Scharschmidt BR, van Dyke RW: Mechanisms of hepatic electrolyte transport. *Gastroenterology* 85:1199–214, 1983
60. Scott AJ: Bacteria and disease of the biliary tract. *Gut* 12:487–92, 1971
61. Seitz U, Vadeyar H, Soehndra N: Prolonged patency with a new design Teflon biliary prosthesis. *Endoscopy* 26:478–82, 1994
62. Shepherd HA Royle G, Ross AP, et al: Endoprosthesis in the palliation of malignant obstruction of the distal common bile duct: A randomized trial. *Br J Surg* 75:1166–8, 1988
63. Siegel JH, Pullano W, Kodsi JM, et al: Optimal palliation of malignant bile duct obstruction: Experience with endoscopic 12 French prostheses. *Endoscopy* 20:137–41, 1988
64. Smit JM, Out MMJ, Greon AK, et al: A placebo-controlled study on the efficacy of aspirin and doxycycline in preventing clogging of biliary endoprostheses. *Gastrointest Endosc* 35:485–9, 1989
65. Smith AD, Dowsett JF, Russell RC, et al: Randomized trial of endoscopic stenting versus surgical bypass surgery in malignant low bile duct obstruction. *Lancet* 344:1655–60, 1994
66. Soehendra N, Reynders-Frederix V: Palliative bile duct drainage—a new endoscopic method of introducing a transpapillary drain. *Endoscopy* 12:8–11, 1980
67. Speer AG, Cotton P, MacRea KD: Endoscopic management of malignant biliary obstruction: Stents of 10 French gauge are preferable to stents of 8 French gauge. *Gastrointest Endosc* 34:412–7, 1988
68. Speer AG, Cotton PB, Rode J, et al: Biliary stent blockage with bacterial biofilm: a light and electron microscopy study. *Ann Inter Med* 108:546–53, 1988
69. Speer AG, Cotton PB, Russell RC, et al: Randomized trial of endoscopic versus percutanous stent insertion in malignant obstructive jaundice. *Lancet* 2:57–62, 1987
70. Stewart L, Pellergrini CA, Way LW: Antibacterial activity of bile acids against common biliary tract organisms. *Surg Forum* 37:157-9, 1986
71. Stewart L, Smith AL, Pellergrini CA, et al: Pigment gallstone form as a composite of bacterial microcolonies and pigment solids. *Ann Surg* 206:242–50, 1987
72. Stossel TP, Murad F, Mason RJ, et al: Regulation of glycogen metabolism in polymorphonuclear leukocytes. *J Biol Chem* 245:6228–34, 1970
73. Straberg SM, Dorn BC, Small DM, et al: The effect of biliary tract pressure on bile flow, bile salt secretion and bile salt synthesis in the primate. *Surgery* 70:140–6, 1971
74. Sung JY, Costerton JW, Shaffer EA: Bacteriostatic activities of bile salts of different hydrophobicity and in the presence of phospholipid. *Hepatology* 14:262A, 1991
75. Sung JY, Leung JWC, Olson ME, et al: Demonstration of transient bacterobilia by foreign body implantation in feline biliary tract. *Dig Dis Sci* 36:943–8, 1991
76. Sung JY, Leung JW, Shaffer EA, et al: Ascending infection of the biliary tract after surgical sphincterotomy and biliary stenting. *J Gastroenterol Hepatol* 7:240–5, 1992
77. Sung JY, Shaffer EA, Olson ME, et al: Bacterial invasion of the biliary system by way of the portal-venous system. *Hepatology* 14:313–7, 1991
78. Suzuki Y, Kobayashi A, Ohto M, et al: Bacteriological study of transhepatically aspirated bile: Relation to cholangiographic findings in 295 patients. *Dig Dis Sci* 29:109–15, 1984
79. Tsang TK, Pollack J, Chodash HB: Inhibition of biliary endoprostheses occlusion by ampicillin-sulbactam in an *in vitro* model. *J Lab Clin Med* 130:643–8, 1997

80. Van Os EC, Petersen BT, Batts KP: Spiral nitinol biliary stents in a porcine model: evaluation of the potential for use in benign strictures. *Endoscopy* 31:253–9, 1999

81. Vaudaux P, Pittet D, Haeberli A, et al: Host factors selectively increase staphylococcal adherence on inserted catheters: a role for fibronectin and fibrinogen or fibrin. *J Infect Dis* 160:865–875, 1989

82. Vaudaux P, Pittet D, Haeberli A, et al: Fibronectin is more active than fibrin or fibrinogen in promoting *Staphylococcus aureus* adherence to inserted intravascular catheters. *J Infect Dis* 167:633–41, 1993

83. Vaudaux P, Suziki R, Waldvogel FA, et al: Foreign body infection: a role of fibronectin as a ligand for the adherence of *Staphylococcus aureus. J Infect Dis* 150:546–53, 1984

84. Wells CL, Jechorek RP, Erlandsen SL: Inhibitory effect of bile on bacterial invasion of enterocytes: possible mechanism for increased translocation associated with obstructive jaundice. *Crit Care Med* 23:301–7, 1995

85. Wells GR, Taylor EW, Lindsay G, et al: Relationship between bile colonization, high-risk factors and postoperative sepsis in patients undergoing biliary tract operations while receiving a prophylactic antibiotic. *Br J Surg* 76:374–377, 1989

86. Wilton PB, Dalmasso AP, Allen MO: Complement in local biliary tract defense: Dissociation between bile complement and acute phase reactants in cholecystitis. *J Surg Res* 42:434–9, 1987

87. Yu J, Nordman Montelius M, Paulsson M, et al: Adhesion of coagulase-negative staphylococci and adsorption of plasma proteins to heparinized polymer surfaces. *Biomaterials* 15:805–14, 1994

88. Yu JL, Andersson R, Ljungh Å, et al: Reduction of *E. coli* adherence to rubber slices by phospholipid treatment. *APMIS* 101:182–6, 1993

89. Yu JL, Andersson R, Ljungh Å: Protein adsorption and bacterial adhesion to biliary stent materials. *J Surg Res* 62:69–73, 1996

90. Yu JL, Andersson R, Pärsson H, et al: A bacteriology and electron microscope study following implantation in the biliary tract in rats. *Scand J Gastroenterol* 31:175–81, 1996

91. Yu JL, Andersson R, Wang LQ, et al: Fibronectin on the surfaces of implanted intra-biliary material—a role in bacterial adherence. *J Surg Res* 59:595–600, 1995

92. Yu JL, Andersson R, Wang LQ, et al: Experimental foreign body infection in the biliary tract in rats. *Scand J Gastroenterol* 30:478–83, 1995.

93. Yu JL, Ljungh Å, Andersson R, et al: Promotion of *Escherichia coli* adherence to rubber slices by adsorbed fibronectin *in vitro. J Med Microbiol* 41:133–8, 1994

94. Yu JL, Wang LQ, Andersson R, et al: New model of reversible obstructive jaundice model in rats. *Eur J Surg* 159:163–166, 1993

95. Zardi L, Siri A, Carnemolla B, et al: A simplified procedure for the preparation of antibodies to serum fibronectin. *J Immunol Methods* 34:155–65, 1980

# 25

# Studying Bacterial Adhesion to Hydrogel Contact Lenses

**Manal M. Gabriel and Donald G. Ahearn**

*Biology Department, Georgia State University, Atlanta, GA, USA*

## I. INTRODUCTION

More than 25 million individuals in the United States wear hydrogel lenses for therapeutic or cosmetic purposes and hydrogel coated urinary catheters and stents are used daily in hospitals. The initial adherence (or adhesion) of microorganisms to these hydrogels during insertion or the rate of microbial attachment in vivo may significantly affect the overall incidence or time of onset of infections. Microorganisms typically adhere poorly to hydrogels as compared with hydrophobic polymers. Hydrogels are three dimensional matrices that adsorb and entrap water. Typically, the higher the water content the lower the initial adherence of microorganisms, but certain polymers are exceptions.

Contact lenses are chemical gels, i.e., crosslinked water-insoluble polymers of various compositions. The Food and Drug Administration has categorized hydrogel lenses into four major groups on the basis of ionic charge and water content (Table 1). Typically hydrogel lenses are homopolymers and copolymers of methylmethacrylate (MMA) and crosslinked homopolymers and copolymers of hydroxyethylmethacrylate (HEMA). The most common functional groups in the back bone chain for the polymers are hydroxyl (-OH), carboxylic (-COOH), esteric ($-COOCH_3$), and etheric ($-COCH_3$) groups. Hydroxyl or polar groups are important to the hydrogel because they provide the hydrophilicity required for the increased water swelling activity.[7,9] The type of polymer and the organization of ionic groups at the hydrogel surface interacts with water content and affects the deposition of proteins, lipids, and microorganisms on the surface.[1,6] For example, the group IV material (ionic, high water content) attracts protein because of the negative charge that methacrylic acid gives to the material. This favors the deposition of positively charged biomolecules like lysozyme, which in turn could influence the adherence of the microorganisms to the lens.

The hydrogel lenses with nonionic polymer (groups I and particularly II) deposit more lipid than hydrogel lenses with ionic radicals (groups III and IV). Jones et al.[6] indicated that high lipid accumulation is associated with group II lenses because of their N-vinyl pyrrolidone (NVP) content. They demonstrated that a group II lens made of polyvinyl alcohol under identical conditions showed less deposition than a NVP based polymer.

*Handbook of Bacterial Adhesion: Principles, Methods, and Applications*
Edited by: Y. H. An and R. J. Friedman © Humana Press Inc., Totowa, NJ

Miller and Ahearn[8] described a strain of *Pseudomonas aeruginosa* that demonstrated differential degrees of primary adherence to hard and soft hydrogel lenses. A primary characteristic of this strain of *P. aeruginosa* was its ability to form a rapid and firm bond on hydrogels. *P. aeruginosa* is the most important pathogen involved in contact lens associated keratitis. *P. aeruginosa* is ubiquitous in the environment and proliferates readily in aqueous systems, including distilled water, inhalation aerosols, irrigating solutions, eye cosmetics and various contact lens solutions. *P. aeruginosa* produces a glycocalyx (extracellular polysaccharide slime) and adhesins that are associated with adhesion, and survival on hydrogel lenses.

Our laboratory has evaluated this strain further for its differential adherence to hydrogel lenses, intraocular lenses, and urinary catheters with and without lubricious coatings.[3,4,5]

## II. MATERIALS AND SURFACE SELECTIONS

### A. Organisms and Culture Conditions

*Pseudomonas aeruginosa* GSU # 3 was selected after screening numerous strains for their differential adherence to hydrogels.[8] Stock cultures are stored under lyophilization. Working cultures are maintained on Tryptic Soy agar slants (BBL Microbiology Systems, Cockeysville, MD), transferred every 2 mo, and stored at 4°C. All isolates, initially and when taken from lyophylized stock, are characterized by their Gram stain reaction and by biochemical reactions on the API 20E system (Analytab Products, Plainview, NY).

### B. Preparation of Hydrogel Lenses

Sterile, hydrogel contact lenses were obtained from commercially available sources. Different types of lenses are shown in Table 1. All lenses were supplied hydrated in buffered saline solutions. All lenses were removed from their original containers and rinsed three times in sterile saline (100 mL each rinse), and placed in new vials with 3 mL of sterile saline. Lenses may be initially packaged in buffered borate solutions (sometimes with traces of surfactants) that are antimicrobial. Such solutions may interfere with the primary adherence assay. If lenses are received in a dry state, the lenses need to be hydrated in sterile saline for 4 to 6 h prior to testing.

### C. Materials

- Tryptic Soy Broth (TSB) (Difco, Franklin Lakes, NJ)
- Tryptic Soy Agar (TSA)
- Minimal Broth (Difco) (1.0 g D-glucose, 7.0 g $K_2HPO_4$, 2.0 g $KH_2PO_4$, 0.5 g sodium citrate, 1.0 g $(NH_4)_2 SO_4$, and 0.1 g $MgSO_4$ in 1 L distilled water).
- Dissolve by stirring all ingredients except $MgSO_4$ and glucose in 1 L of deionized water (base solution). Prepare separate solutions of 1% (w/v) $MgSO_4$ and 10 % (w/v) glucose in deionized water (supplement stock solutions). Sterilize by autoclaving for 15 min at 121°C/15 psi. Cool the base solution and supplement stock solutions to room temperature. Aseptically transfer 10 mL of each supplement solution per liter of base solution and mix thoroughly. All solutions must be stored under refrigeration.
- Sterile PBS (composition in grams/liter: 8.0 NaCl, 0.2 KCl, and 1.44 $Na_2HPO_4$)
- L-[3,4,5-H] leucine (1–3 µCi/mL; NEN Research Products, DuPont, Wilmington, Delaware).

**Table 1. Lens Types Used in the Adherence of *P. aeruginosa* to Hydrogels**

| Polymer group[a] | Code/Generic name | Water content (%) | Chemical compositions |
|---|---|---|---|
| Low water, nonionic (Group I) | T/Tefilcon A | 37.5 | Poly(HEMA) crosslinked with EGDMA |
| | P/Polymacon A | 38.6 | Poly(HEMA) crosslinked with EGDMA |
| | TA/Tetrafilcon A | 42.5 | Poly(HEMA + NVP + MMA) |
| High water, nonionic (Group II) | V/Vifilcon A | 55 | Poly(HEMA + MAA-γ-povidone) |
| | L/Lidofilcon A | 70 | Poly(NVP + MMA) with allyl methacrylate and ethylene dimethacrylate |
| | S/Surfilcon A | 74 | Poly (MMA +NVP + other methacrylates) |
| | O/Omafilcon A | 59 | HEMA + PC (phosphatidyl choline) |
| Low water, ionic (Group III) | PA/Phemfilcon A | 38 | Poly (HEMA + EOEMA+ MMA) crosslinked with EGDMA |
| | E/Etafilcon A | 43 | Poly(HEMA+sodium methacrylate+ 1,1,1 tri-methylol propane trimethacrylate) |
| | B/Bufilcon A | 45 | Poly[HEMA + N(1,1 dimethyl 3-oxybutyl)-acrylamide + MMA] |
| High water, ionic (Group IV) | PA/Phemfilcon A | 55 | Poly(HEMA + EOEMA + MMA) crosslinked with EGDMA |
| | E/Etafilcon A | 58 | Poly(HEMA+sodium methacrylate + 1,1,1 tri-methylol propane trimethacrylate) |
| | B/Bufilcon A | 55 | Poly[HEMA + N (1,1dimethyl 3-oxybutyl)-acrylamide + MMA] |

[a] adapted from Miller and Ahearn.[5]
Abbreviations:
  HEMA, 2-hydroxyethyl methacrylate
  NVP, n-vinyl pyrrolidone
  MAA, methacrylic acid
  MMA, methylmethacrylate
  EOEMA, 2-ethyoxy ethyl methacrylate
  EGDMA, ethylene glycol dimethacrylate.

- Opti-Fluor scintillation cocktail (Packard Instruments, Downers Grove, IL).
- Liquid scintillation counter (LS-7500, Beckman Instruments, Fullerton, CA).
- 24 well polystyrene culture plate (Corning Glass Works, Corning, NY).
- 20 mL scintillation vials.

## D. Methods of Testing

### 1. Measurement of Adherence to Contact Lenses with Radiolabeled Cells

The relative adherence of representative strains may be determined with modifications of the procedure of Sawant et al.[11] and Gabriel et al.[6] according to the following protocol:

A. RADIOLABELING PROCEDURE

- Inoculate a plate of Tryptic Soy Agar from Tryptic Soy Broth stock culture (usually kept on slants).

Log phase cells of bacterium from TSB

O.D. adjusted to $10^8$ in PBS, serial dilutions, plate counts

Radiolabeled L-[3,4,5-3H]-leucine, 20 min

Hydrogels added (include hydrophobic material as a control example, silicone)

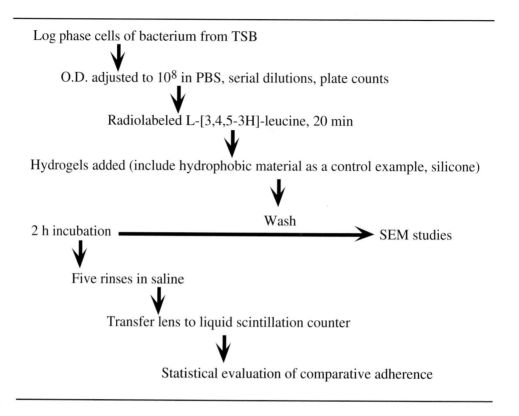

2 h incubation ————————————— Wash ————————————→ SEM studies

Five rinses in saline

Transfer lens to liquid scintillation counter

Statistical evaluation of comparative adherence

**Figure 1.** A typical flow chart for evaluating the adhesion of prelabeled bacteria

- Incubate overnight at 37°C.
- Inoculate 150 mL of TSB with a colony selected from the overnight culture.
- Incubate to midexponential growth phase (about 14–18 h at 37°C) on a rotary shaker.
- Harvest cells by centrifugation at (5000 *g* for 5 min), wash twice with 0.9% saline and resuspend cells in minimal broth.
- Incubate cell suspensions with shaking at 37°C for 1h.
- Add L-[3,4,5-$^3$H] leucine to the cell suspension.
- Incubate the cell suspension for an additional 20 min.
- Harvest the labeled cells, wash 4 times in saline and suspend in PBS to $10^8$ cells per mL.
- Under sterile conditions remove lenses from the vials.
- Incubate the lenses with 3 mL of the radiolabeled cell suspension at 37°C for 2 h on a rotary shaker.
- Remove the lenses aseptically with forceps and rinse by immersing each lens five times in each of three successive changes of PBS (250 mL each).
- Shake the lenses of excess PBS and transfer individual lenses to 20 mL glass scintillation vials containing 10 mL of Opti-Fluor scintillation cocktail.
- Agitate the vials using a vortex-type mixer, and count in a liquid scintillation counter.
- Dispense 100 μL of serially diluted radiolabeled cell suspensions of known concentration into scintillation vials and count as described above. The scintillation counts

are converted to actual cell numbers with a calibration curve relating disintegration per minute (dpm) to viable cell counts.

## B. VIABILITY OF ADHERED CELLS

The radioactivity associated with adhered viable cells (only viable cells actively take up leucine) is distinguished from that of nonspecific absorption.

- Grow the bacteria, wash, suspend in PBS or minimal broth (minimal broth is used in experiments for establish a mature biofilm).
- Incubate the bacterial suspension for 24 h with individual lenses as described above.
- Remove the lenses with a forceps and immerse each lens five times in each of three successive (250 mL) changes of PBS.
- Transfer the lenses to individual 20-mL glass scintillation vials that contain sterile minimal broth with L-[3,4,5-$^3$H] leucine and incubate for 30 min.
- Remove the lenses from the radioactive and immerse it five times in each of five successive changes of PBS. Quantitate the radioactivity associated with the lenses as described above. Include lenses without adhered cells as background controls.

## C. FLOW CHART

*See* Figure 1 for flow chart.

## 2. Bioluminesence (ATP Extraction)

- Incubate the bacteria overnight at 37°C
- Harvest the cells, wash and rinse in PBS. Adjust the O.D. (optical density) to $10^8$ cells/mL.
- Place the hydrogel lenses into the wells of a new polystyrene microtiter plate containing 0.5 mL sterile PBS.
- Dispense 2 mL of cell suspension in each well. Incubate for 2 h.
- Rinse the lenses with adhered bacteria and transfer to wells of a 24-well polystyrene culture plate containing 0.5 mL of PBS.
- Add a nonionic surfactant (0.5 mL, Extractant-XM, Biotrace, Plainsboro, NJ) (with mixing) to each well.
- Agitate the plates gently for 5 min.
- Dispense 0.2 mL from each well into luminometer cuvettes.
- Add 0.1 mL volume of reconstituted luciferase-luciferin reagent (Enzyme-MLX, Biotrace) to each cuvette and analyse the reaction mixture in a luminometer (Uni-Lite X-cel, Biotrace). Light output is recorded in relative light units (RLUs).
- Dispense 100 µL of dilutions of the original inoculum onto spread agar plates, and incubate for 24 h at 37°C. The calibration curve is generated from the plate counts.
- RLUs are converted to actual cell numbers with a calibration curve relating light output to viable cell counts. RLUs obtained from the processing and extraction of enzyme treatment of the uninoculated suspending medium serve as a control and are adjusted for any nonspecific luminescence generated by uninoculated substrata.

## 3. Sonication/Release

- Remove lenses with adhered bacteria from the vials after 0, 6, 16, and 24 h and rinse in PBS.

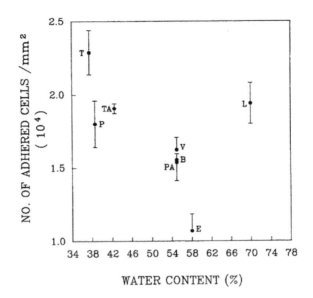

**Figure 2.** Adherence of *Pseudomonas aeruginosa* #3 to different hydrogels. Codes are interpreted in Table 1. Vertical bars represent the standard error (*n* = 3). Adapted from Miller and Ahearn.[8]


- Place each sample in a separate sterile vial with 1 mL of PBS and sonicate for 1 min at 75 kHz in an ultrasonic bath (Branson Ultrasonics, Danbury, CT) filled to a 1 cm depth with deionized water.
- Vortex the vials and then dispense 100 µL of adhered bacteria released from test samples on TSA spread plates.
- Incubate the plates for 24 h at 37°C. Data are expressed as cfu/mm$^{2.}$

*4. Scanning Electron Microscopy*

- Incubate the lenses in suspension of $10^8$ cells/mL in saline at 37°C for varying time periods up to 24 h.
- Place the lenses in glass vials and rinse for 5 min in 3 mL of Sorensen's phosphate buffer pH 7.4.
- Fix the lenses with 2% glutaraldehyde in 0.1 M phosphate buffer for 1 h at room temperature.
- Rinse three times for 10 min in 0.1 M phosphate buffer.
- Postfix in 1% $OsO_4$ in 0.1 M phosphate buffer for 1 h at room temperature.
- Rinse the lenses three times in 0.1 M phosphate buffer for 10 min, dehydrate in a graded ethanol series, and critical point dry in the presence of $CO_2$ in a critical point dryer (Balzer CPD 020).
- Sputter coat the lenses with 7 to 9 nm Au/Pd and examine in a JEOL-35C scanning electron microscope operating at 15 kV.[4]

*5. Statistical Analyses*

For the 2 h or 18 to 24 h assays, repeat experiments are conducted with five or more samples for each type of lens. The adherence data for the 2 h assay are converted from dpm to cfu based on the calibration curve, and expressed as cfu/mm$^2$ of lens area. For the

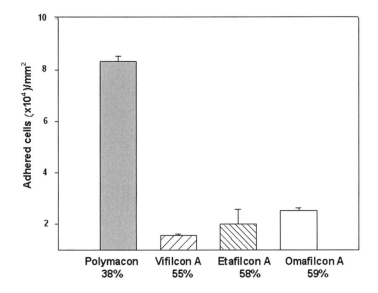

**Figure 3.** Adherence of *Pseudomonas aeruginosa* #3 to hydrogels of varying water content and polymer composition in 2 h. Vertical bars represent the standard error ($n = 5$).

18 to 24 h assays, the data are expressed usually as dpm/mm$^2$. Adherence to various hydrogels is compared statistically using an unpaired student 't-test' with (Sigma-Plot 4.01, Jandel Scientific, Sausalito, CA) or with one way analysis of variance (ANOVA) (results are considered significant at $p < 0.05$). Quality assurance is based on inclusion of a standard hydrophobic silicone polymer control in all adherence studies.

### III. RESULTS AND DISCUSSION

The relative degrees of adherence of *P. aeruginosa* # 3 to various hydrogels as determined by Miller and Ahearn[8] are presented in Figure 2. These data are essentially identical for certain lenses with current data obtained with the strain (Fig. 3). Slight differences were observed for the lenses, but varied lots for a given lens ranged from about 1.5 to 3.0 $\times 10^4$ cells/mm$^2$. Contrary to general findings with increased water-content gels, adherence tended to increase with increased water-content of the gels among Group IV lenses. Analysis of adherence after 2 h via ATP determinations provided similar data for lenses of similar water content but more often this was a trend rather than a statistically significant value (data not shown). Uptake of radiolabeled leucine by cells adhered for 18 h to hydrogels was greater per cell compared to leucine content of labeled cells from the plankton adhered for 2 h (data not shown). These data for 18 h may indicate a difference between alteration of leucine uptake and overall metabolic activity by a substratum rather than a significant reduction in adherence. The degrees of adherence for postlabeled cells are determined only after confirmation with the other test procedures.

Our test procedures have been developed to facilitate the selection of hydrogels that are least susceptible to initial adhesion or firm adherence by microorganisms. We employ a relatively vigorous rinsing procedure which essentially removes all but the firmly adhered cells. Eventually (18–48 h), *P. aeruginosa* GSU #3 will form mature biofilms on most of the hydrogels we have studied, but the rate and extent of formation varies with nutrient conditions and inocula densities.[11] The densities and rate of accumulation of Gram-negative bacteria also have been related to the involvement of cell-to-cell signals.[2] Whether single compounds control the densities of firmly adhered cells on a surface has not been determined. The incorporation of antimicrobials in hydrogels (e.g. antibiotics or silver compounds) also may significantly affect primary adherence. When antimicrobials are present new calibration curves, particularly for ATP analysis, are necessary.

We have addressed test parameters for a selected strain of *P. aeruginosa* that has been studied extensively.[11] Our experience with other species and strains indicates that optimal inocula concentrations, interaction of substratum or inhibitors with reactants for ATP analyses, and rates of leucine uptake must be established for each strain for valid interpretation of the adherence tests.

## REFERENCES

1. Cook AD, Sagers RD, Pitt WG: Bacterial adhesion to protein-coated hydrogels. *J Biomat Application* 2:72–89, 1993
2. Davies DG, Parsek MR, Pearson JP, et al: The involvement of cell-to-cell signals in the development of a bacterial biofilm. *Science* 280:295–8, 1998
3. Gabriel MM, Ahearn DG, Chan K, et al: Comparison of *in vitro* adherence of *Pseudomonas aeruginosa* to intraocular lenses. *J Cataract Refrac Surg* 24:124–9, 1998
4. Gabriel MM, Sawant AD, Simmons, RB, et al: Effects of silver on adherence of bacteria to urinary catheters: *In vivo* studies. *Curr Microbiol* 30:17–22, 1995
5. Gabriel MM, Schultz CL, Wilson LA, et al: Effect of *Staphylococcus epidermidis* on hydrogel contact lens retention on the rabbit eye. *Curr Microbiol* 32:176–8, 1996
6. Jones L, Evans K, Sariri R, et al: Lipid and protein deposition of N-vinyl pyrrolidone-containing group II and group IV frequent replacement contact lenses. *CLAO J* 23:122–6, 1997
7. LaPorte RJ: *Hydrophylic Polymer Coatings for Medical Devices, Structures/Properties, Development, Manufacture and Applications.* Technomic, Lancaster, PA, 1997:57–80
8. Miller MJ, Ahearn DG: Adherence of *Pseudomonas aeruginosa* to hydrophilic contact lenses and other substrata. *J Clin Microbiol* 25:1392–7, 1987
9. Peppas NA: Contact lenses as biomedical polymers: polymer physics and structure-properties evaluation as related to the development of new materials. In: Hartstein J, ed. *Extended Wear Contact Lenses for Aphakia and Myopia.* Mosby, St. Louis, MO, 1982:6–43.
10. Sawant AD, Gabriel MM, Mayo MS, et al: Radiopacity additives in silicone stent materials reduce *in vitro* bacterial adherence. *Curr Microbiol* 22:285–92, 1991
11. Stone JH, Gabriel MM, Ahearn DG: Adherence of *Pseudomonas auruginosa* to various biomaterials. *J Indust Microbiol* Submitted, 1999

# 26

# In Vivo Models for Studying Staphylococcal Adhesion to Biomaterials

**Pierre E. Vaudaux**

*Division of Infectious Diseases, Department of Medicine,*
*University Hospital, Geneva, Switzerland*

## I. INTRODUCTION

Any implanted biomaterial, in particular those used for orthopedic surgery, shows an increased susceptibility to microbial infections, frequently due to staphylococci.[2,27,28,33,50] Although the incidence rate of bacterial infections associated with orthopedic devices has been considerably reduced in many centers, each infection is very detrimental to the patient and very difficult to treat.[2,14,27,28,33,50] *Staphylococcus aureus* and *Staphylococcus epidermidis* not only represent a significant proportion of all pathogens responsible for orthopedic implant infections, but they may lead to dramatic metastatic, life-threatening complications.[37,50]

Although the use of antibiotic prophylaxis to reduce the risk of infection is an established routine,[14,38] the recent emergence of methicillin- and cephalosporin-resistant strains of staphylococci raises important questions about current antibiotic prophylaxis.[28]

The susceptibility of any biomaterial implant to bacterial infections may be influenced by interfacial reactions with host defense mechanisms,[12] and by specifically elaborated microbial factors promoting bacterial colonization.[18] Bacterial adhesion is the first step leading to colonization and infection of orthopedic materials.[18] Alterations in the host defense mechanisms in the vicinity of biomaterial implants are also suggested, although they have rarely been confirmed by experimental data. Since the presence of a foreign body markedly increases the pathogenic potential of organisms of low virulence such as *S. epidermidis*,[36] patients infected with such organisms in connection with biomaterial implants may be considered in some way as locally immunocompromised patients. Arguments in favor of local defects in the host defense against staphylococcal foreign body infections are also provided by experimental studies in animals, indicating that a bacterial inoculum considered as "subinfective" in a particular model of experimental wound infection can become "infective" in the presence of foreign materials such as sutures, hemostats, soil, devitalized and crushed muscle tissue, gelatin or oxidized cellulose, as reviewed by Georgiade et al.[13] For example, the minimal infective dose for *S. aureus* can be as low as 100 cfu, whereas $10^7$ organisms are noninfective in the absence of foreign material.[5]

*Handbook of Bacterial Adhesion: Principles, Methods, and Applications*
Edited by: Y. H. An and R. J. Friedman © Humana Press Inc., Totowa, NJ

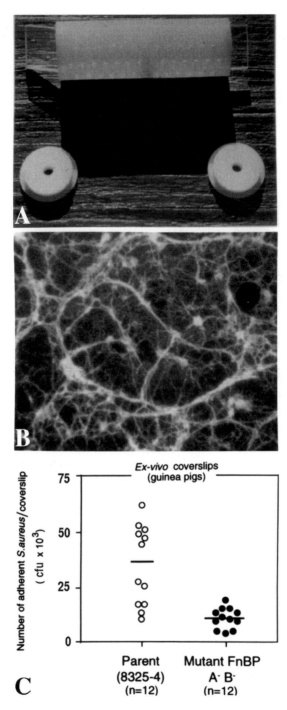

**Figure 1.** (A): View of a multiperforated Teflon tissue cage, with two inserted PMMA cover slips and two sealing caps, before implantation. (B): Immunofluorescent staining of fibronectin recovered on PMMA cover slips subcutaneously implanted into guinea pigs for 4 wk. (Reprinted with permission.[60]) (C): Adhesion of the parental strain 8325-4 of *S. aureus* and its fibronectin adhesin-defective mutant strain DU5883 which does not produce either fibronectin-binding protein (FnBP) A or B. (Reprinted with permission.[15])

## II. GUINEA PIG MODEL
## OF SUBCUTANEOUS FOREIGN BODY INFECTIONS

To analyze the role of microbial and local host factors in foreign body infections, we developed an experimental model suitable for analysis of the various microbiological, immunologic, and cellular events preceding, or associated with, a foreign body infection.[66]

In this model, rigid polytetrafluoroethylene (Teflon) tubes (internal and external diameters, 10 and 12 mm, respectively; length 32 mm) are perforated by 250 regularly spaced holes (diameter, 1 mm) and sealed at each end with a cap of identical material (Fig. 1A). Guinea pigs weighing 500 to 600 g each, which are allowed free access to food, receive an intramuscular induction of a neuroleptanalgesia (0.1 mL/100 g of body weight; Hypnorm®; Veterinary Drug Company, Dunnington, UK), composed of 10 mg/mL fluanisol and 0.2 mg/mL fentanyl, per 100 g body weight. Then a 4 cm incision is made with an aseptic technique, and the subcutaneous space dissected bluntly. Four steam-sterilized tissue cages are implanted in each flank, then the skin is closed with metal clips, which are removed 1 wk after surgery. The animals are used for experimentation 3 to 6 wk after implantation of the tissue cages, after full healing of the incision.

An important characteristic of this and other types of subcutaneous implants with a dead space is the presence of a sterile inflammatory exudate which accumulates inside the tissue cages within the 2 to 4 wk after their implantation.[66] This tissue cage fluid can be easily aspirated for analysis of its humoral and cellular components, and also to exclude occasional, spontaneous bacterial contamination.[66]

The minimal infective dose of *S. aureus* was found to be very low in the guinea pig tissue cage model, since inoculation of $10^3$ cfu produced infections in all tissue cages tested.[66] In contrast, no infection could be produced by either subcutaneous or intraperitoneal injection of $>10^6$ cfu of *S. aureus* in the absence of tissue cages,[66] thus confirming that *S. aureus* or *S. epidermidis* exhibit increased virulence only when associated with foreign implants. A few hours after inoculation of *S. aureus*, an abundant influx of neutrophils occurred into tissue cages resulting in abscess formation.[66] Despite this intense inflammatory response, formation of spontaneous fistula with purulent discharge and subsequent spontaneous shedding of the foreign bodies occurs in most cases, followed by spontaneous wound healing. No signs of bacteremic spread could be demonstrated and bacteriological cultures of other organs were uniformly negative."[66] An additional characteristic of this tissue cage model was the response to parenteral antistaphylococcal antibiotics, which could prevent or eradicate tissue cage infections only if treatment was initiated before or during the first 6 to 12 h after inoculation of *S. aureus*[52] but were ineffective if initiated >12 h after inoculation.[52] Such inefficacy of antibiotic therapy, initiated after infection has developed, is commonly observed in the clinical context of staphylococcal foreign body infections.

## III. FACTORS INFLUENCING STAPHYLOCOCCAL ADHESION
## IN VITRO AND IN VIVO

In vitro studies aiming to describe mechanisms of bacterial attachment relevant to the colonization of orthopedic biomaterial implants are frequently performed in the absence of any host-relevant factors. This omission is unfortunate when we know that exposure of any biomaterial to blood results in the immediate formation of a conditioning layer of blood proteins and cells on its surface.[6] These protein layers may influence the staphylococcal attachment in two different ways: 1) by masking nonspecifically any direct interaction between bacterial adhesive surface components and artificial surfaces,[24,46] and 2) by increasing more specifically the attractive properties of biomaterials for staphylococci. The blood components and tissue proteins containing specific binding sites for staphylococci are evidently ideal candidates for such a phenomenon, if they can be adsorbed in adequate amounts on the surface of biomaterials.

A useful way to evaluate the role of different proteins in promoting or preventing staphylococcal adhesion to indwelling devices may be the comparison between unimplanted and implanted foreign surfaces. For this purpose, our laboratory designed a specific in vitro bacterial attachment assay that originally made use of (1×1 cm) polymethylmethacrylate (PMMA) surfaces called cover slips.[60,63] Subsequently, metallic surfaces also presenting as coverslips were tested in the same assay with minor modifications.[4]

### A. Bacterial Adhesion Assay

The cover slips are first cleaned with 100% ethanol and sterilized by heating at 120°C for 30 min. Then they are either coated in vitro with pure proteins (see below) or inserted into the tissue cages which are implanted into guinea pigs (Fig. 1A).

In vitro protein-coated cover slips are incubated with indicated concentrations of the selected purified protein for 60 min at 37°C, then rinsed twice with phosphate-buffered saline (PBS) solution. Tissue cage-implanted cover slips are removed 1 mo after surgery and rinsed twice with PBS.

The bacterial adhesion assay is performed by incubating the in vitro or in vivo coated coverslips with $4 \times 10^6$ colony-forming units (cfu) of radiolabeled bacteria, in PBS supplemented with 1 mM $Ca^{2+}$, 0.5 mM $Mg^{2+}$, and 5 mg/mL of purified human albumin. Radiolabeled bacteria are prepared by incubating $1-2 \times 10^7$ cfu from overnight cultures in Mueller-Hinton broth with 25 µCi of $^3$H-thymidine in 1 mL of Mueller-Hinton broth. After 3 h of exponential growth at 37°C, the bacterial cultures containing $1-2 \times 10^8$ cfu are rinsed from unbound radioactivity by two centrifugations and suspended in 1 mL 0.9% NaCl. Albumin is added to the assay buffer to minimize direct, nonprotein-mediated adhesion of the staphylococci to the cover slips.

At the end of the attachment period, which takes place for 60 min at 37°C in a shaking waterbath, the fluids containing unbound bacteria are drained and the cover slips are transferred into new tubes containing 1 mL of 0.9% NaCl. This transfer procedure minimizes the carry-over of fluid contaminated with unbound bacteria. After 5 min at 20°C, a second wash is performed with 1 mL of fresh 0.9% NaCl for 30 min at 20°C. Thereafter, polymer cover slips may be directly immersed into 5 mL scintillation fluid (Ultima Gold™, Packard) and the radioactivity of attached bacteria directly estimated in a liquid scintillation counter. When metallic cover slips are used, they cannot be counted directly in the liquid scintillation. Therefore, the bacteria attached to metallic cover slips are first released from theses surfaces by incubation with 100 µg/mL trypsin for 15 min at 37°C. The radioactivity of each trypsin-extracted sample is then estimated by radioactivity counts.

Results of bacterial adhesion assays can be expressed in different ways. One of them is the percentage of the total bacterial output, which itself is expressed either as radioactive units (cpm or dpm) or viable units (cfu). Attachment can also be normalized as the number of adherent cfu (or cpm) per surface unit (cm$^2$).

## B. Comparison of S. aureus Attachment to Either In Vitro Protein-Coated or Explanted Cover Slips

Incubation of different types of unimplanted artificial surfaces, such as PMMA,[22,60,61] or Teflon,[55] with either whole plasma,[24] serum,[46,60] or purified serum albumin[4,22,55,60,61] was shown to prevent bacterial adhesion for a vast majority of clinical and laboratory staphylococcal strains and species. We have shown a similar inhibition by albumin of *S. aureus* and *S. epidermidis* adhesion to metallic surfaces of orthopedic use, such as stainless steel, pure titanium and titanium alloy.[4] In view of these results, many studies of bacterial adhesion on protein-coated surfaces now include this protein in the fluid phase of bacterial suspensions to reduce the contribution of nonspecific physicochemical forces to the process of staphylococcal attachment.

In contrast to unimplanted surfaces, polymeric or metallic cover slips implanted subcutaneously into guinea pigs inside the tissue cages and subsequently excised after 4 wk showed entirely different characteristics towards *S. aureus*[60] or *S. epidermidis*[4] adhesion: 1) they allowed a significant bacterial attachment even in the presence of serum, albumin, or tissue cage fluid:[4,60] 2) the new attachment characteristics of either PMMA[60] or metallic[4] cover slips, acquired during their implantation, were due to the presence of trypsin-sensitive adhesins on the surface of implants; and 3) these acquired adhesins were likely to be host conditioned. Microscopic examination revealed that explanted PMMA cover slips were coated with fibers and cellular materials that stained intensely with antifibronectin antibodies[60] (Fig. 1B). The presence of fibronectin was linked to the development of connective tissue[19] on the surface of PMMA, as evidenced by the presence of numerous fibroblasts and collagen fibers.

To demonstrate that fibronectin deposited on PMMA cover slips during in vivo exposure was a major factor mediating the adhesion of *S. aureus* Wood 46, the potential binding sites of *S. aureus* were blocked by specific antibodies to fibronectin. Treatment of explanted cover slips with such antibodies strongly reduced staphylococcal adhesion.[4,60] A similar inhibition was observed on unimplanted cover slips coated in vitro with purified fibronectin. Antibody-mediated inhibition of *S. aureus* adhesion could be demonstrated because strain Wood 46 used for this study is devoid of protein A; indeed the presence of protein A, which is an abundant surface component of most *S. aureus* strains known to attract large amounts of immunoglobulin G (IgG) by binding their Fc portion, would have masked the specific Fab-mediated blocking effect of antifibronectin antibodies on *S. aureus* adhesion.

## IV. MOLECULAR MECHANISMS OF S. AUREUS ATTACHMENT TO PROTEIN-COATED IMPLANTS

In vitro, and to some extent in vivo, studies identified host proteins that could promote *S. aureus* attachment to polymeric and metallic surfaces when immobilized on such substrates. These proteins are either plasma or extracellular matrix components and

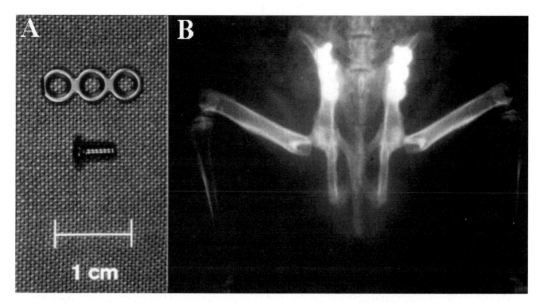

**Figure 2.** Size of titanium miniplates and miniscrews (A) and radiographic appearance (B) at 6 wk after insertion onto the iliac bones of guinea pigs. (Reprinted with permission.[7])

include fibrinogen or fibrin,[22,39] fibronectin,[22,57,60,61] collagen,[48] laminin,[22,35] bone sialoprotein,[49,65] elastin,[45] vitronectin,[1,34] thrombospondin,[21] and von Willebrand factor.[20]

Progress has also been made in elucidating cell-wall associated surface components of *S. aureus*, called bacterial adhesins, which promote specific interactions with individual host proteins. A recently introduced acronym used to describe such adhesins is MSCRAMMs (for microbial surface components recognizing adhesive matrix molecules).[47]

These studies allowed identification and characterization of the genes for two fibrinogen-binding protein components,[39,40,43] a collagen adhesin,[48,51] an elastin-binding protein,[45] and two distinct but related fibronectin-binding proteins.[8,29] Site-specific mutants of *S. aureus* specifically defective in adhesion to a single host protein such as fibrinogen,[39,43] fibronectin,[15,32] or collagen[3,23,48] have been described and used in various in vitro and in vivo studies. In particular, a knockout mutant of *S. aureus* 8325-4 that was defective in the expression of both fibronectin-binding proteins was severely impaired in its ability to attach to coverslips explanted from guinea pigs (Fig. 1C).[15]

## V. GUINEA PIG MODEL OF BONE-IMPLANTED METALLIC DEVICES

More recently, a novel experimental model has been developed to reproduce conditions of internal fixation devices and evaluate the interaction of various host and microbial factors contributing to bacterial adhesion and colonization of bone-implanted materials.[7] This novel animal model was used to evaluate the contribution of fibronectin in the attachment of *S. aureus* to the previously implanted orthopedic devices. This contribution could be determined by comparing the adhesion of a fibronectin adhesin-defective mutant of *S. aureus* with adhesion of its isogenic parental strain.

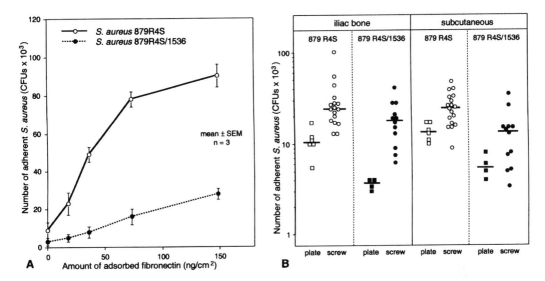

**Figure 3.** Adhesion of the parental strain 879R4S of *S. aureus* and its fibronectin adhesin-defective mutant strain 879R4S/1536 to either gelatin-PMMA cover slips coated in vitro with increasing amounts of fibronectin (A) or to titanium miniplates or miniscrews explanted from either iliac bones or the subcutaneous space of guinea pigs (B). (Reprinted with permission.[7])

## A. Surgical Procedure

This surgical procedure has been approved by official authorities of the University and State of Geneva. The orthopedic devices (Synthes®, Stratec Medical, Waldenburg, Switzerland) implanted into guinea pigs are metallic miniplates and miniscrews (6×2 mm) made of pure titanium (Fig. 2A). The day before surgery, guinea pigs are carefully shaved, depilated and kept starved overnight. Guinea pigs are anesthetized by intramuscular administration of 0.2 mL Hypnorm®. Each guinea pig is placed in the prone position with its back legs in extension. The skin is carefully disinfected with Merfen® (Zyma, Gland, Switzerland) and draped with a self adhesive sheet. A median skin incision is made, starting 4 cm proximal to the cloaca and extending 5 cm cranially. The subcutaneous tissue is dissected down to the aponeurosis, which is incised 1 cm laterally to the spinal processes. The gluteal muscles are then atraumatically dissected to expose the proximal parts of the iliac wings over a distance of 2 cm. Following positioning of a self-retaining retractor, a hole of 1.5 mm in diameter is then drilled to allow provisional fixation of each plate on the iliac bone. Two additional screws are then inserted on each side of the plate to complete the fixation (Fig. 2B). Thereafter, the surgical field is rinsed with isotonic saline and the aponeurosis sutured with 5-0 Vicryl®. Before closing the skin incision with skin staples, additional miniplates and miniscrews, identical to those fixed on the iliac bones, are subcutaneously implanted as controls. Anesthesia is antagonized by an intramuscular injection of 0.025 mL of Narcan® (Dupont Pharma GmbH, Bad Hamburg, Germany) per 100 g of body weight. No further postoperative procedure is performed and animals are left in their normal surroundings for 6 wk, with daily surveillance during the first week.

### B. Outcome of the Surgical Procedure

After initial occurrence of immediate postoperative paraplegic complications in six guinea pigs in an initial group of 18 animals, the surgical procedure was improved to further reduce the risk of nervous damage. In a second group of twelve operated animals, none of them showed any wound repair defect or infective complication. In a further series of 28 animals, only one died of postoperative complications. Thus, the novel animal model of bone-implanted metallic plates has a low rate (<5%) of complications during the 6 wk postoperative period.[7]

### C. S. aureus *Attachment to Bone-Implanted Metallic Devices*

Five to 6 wk after surgery, the miniplates and miniscrews are explanted from previously killed animals, carefully rinsed in PBS, and tested in the in vitro assay of *S. aureus* adhesion to metallic surfaces as described above. To evaluate the role of fibronectin in staphylococcal adhesion to explanted plates and screws, a mutant of *S. aureus* specifically defective in fibronectin adhesion because of decreased expression of the fibronectin adhesin was compared with its isogenic parental strain (Fig. 3A). A significant reduction in adhesion of the fibronectin adhesin-defective mutant compared to the parental strain occurred on both the subcutaneously and bone-implanted metallic plates[7] (Fig. 3B). These data suggested that fibronectin was present on bone-implanted metallic devices and promoted attachment of *S. aureus* to their surface. This novel experimental model should help to characterize several parameters of bacterial adhesion to orthopedic metallic devices and to develop novel anti-adhesive strategies for preventing such infections.

## VI. DEVELOPMENT OF IN VITRO ASSAYS MORE RELEVANT TO IN VIVO CONDITIONS

To study in more detail the interaction of *S. aureus* and *S. epidermidis* with specific host proteins deposited on foreign surfaces, several investigators have used simplified in vitro bacterial adhesion assays. A common feature of these assays was the in vitro coating of artificial surfaces with individual blood or matrix protein or sometimes a mixture of them.[9,62] A frequent limitation of many studies has been the empirical choice of protein concentrations used for coating the artificial surfaces. To compare in vitro assays with in vivo conditions, it is important to determine in vitro protein coating conditions yielding levels which on the average may approximate those recovered in vivo on biomaterial implant surfaces. Of course this estimation cannot be very accurate since host proteins frequently form complex organized networks on biomaterial implants in vivo.

In vitro adsorption of most proteins is usually quantified by using either radiolabeled preparations of these macromolecules or by appropriate immunoassays. In various studies performed with purified fibronectin radiolabeled with either $^{125}I$ or $^3H$, we found that protein adsorption on either PMMA,[61] Teflon,[55] polyvinyl chloride tissue culture plates[64] or polyurethane[10,44,59] was dose-dependent and saturable. Similar data were also obtained with cover slips of various metallic composition.[4,11,54] Quantitative dose-dependent and saturable adsorption on artificial surfaces was also observed with radio-labeled fibrinogen. "[58]

Promotion of *S. aureus* adhesion[4,21,59,61,64] by either surface-bound fibronectin or fibrinogen was dose-dependent, being a linear function of both the quantity in solution and the quantity adsorbed on each type of artificial surfaces at intermediate protein concentrations. We have repeatedly observed that fibrinogen and fibronectin were reliably adsorbed on either PMMA cover slips[41,53] or polyurethane catheter segments[59] from solutions containing submicrogram amounts of each protein per mL. In contrast to fibrinogen which could be directly adsorbed on native polymer surfaces from such highly diluted protein solutions,[41,59] precoating of the materials with gelatin was required for fibronectin.[53,59] This precoating step significantly improved the wettability of the different polymer surfaces. Of interest was the observation that optimal adhesion of *S. aureus* was promoted by levels of fibrinogen or fibronectin much lower than those leading to a monolayer coating of the surfaces.[41,53,59]

Fibronectin adsorption onto PMMA was also studied with purified radiolabeled fibronectin in complex protein mixtures such as human serum, after sequential depletion of endogenous plasma fibronectin and serum reconstitution. Under these conditions, fibronectin adsorption on PMMA cover slips was inhibited by 98%, when compared to that occurring in the absence of serum proteins.[61] Other studies have confirmed the decreased affinity of fibronectin for various hydrophobic surfaces in the presence of serum proteins.[16,17,30]

The quantity of fibronectin adsorbed on PMMA in the presence of serum proteins was indeed too low to promote a significant adhesion of *S. aureus* Wood 46 to PMMA cover-slips. These observations explain why native PMMA or Teflon cover slips preincubated with 10% whole serum, containing approx 30 µg/mL of fibronectin, were unable to promote subsequent adhesion of either *S. aureus* Wood 46[55,60,61] or other strains of *S. aureus* or *S. epidermidis*.[22,56]

Precoating of either PMMA[61] or Teflon[55] with gelatin or heat-denatured collagen restored the ability of the foreign material to adsorb fibronectin from the mixture of serum proteins. Both denatured collagen and gelatin can interact with the specific collagen-binding site[42] of fibronectin by a specific domain,[26,31] containing the mammalian collagenase cleavage site. The quantity of fibronectin adsorbed on either PMMA[61] or Teflon[55] in the presence of serum proteins was at least 25% of the quantity adsorbed in the absence of serum proteins. Interestingly, equivalent amounts of fibronectin had a much higher adhesion-promoting activity on *S. aureus* when adsorbed from serum mixtures on collagen-precoated PMMA[61] or Teflon[55] than when adsorbed on native cover slips. These data suggest a potential role for collagen as a cofactor contributing to *S. aureus* adhesion onto fibronectin-coated surfaces. Collagen-fibronectin interactions could favor conformational changes in the fibronectin molecules, such as exposure of a maximal number of binding sites to *S. aureus*.

To summarize the characteristics of host-mediated staphylococcal adhesion to artificial surfaces measured under in vitro and in vivo condition:

1. Fibronectin is by far the best characterized protein adhesin interacting with both coagulase-positive and -negative staphylococci in vitro and in vivo;
2. Except for fibrin (or fibrinogen) and collagen,[48] the presence and contribution of additional blood or matrix adhesins to bacterial adhesion and colonization of indwelling devices still needs to be fully evaluated;
3. Serum limits considerably the adsorption of fibronectin onto freshly implanted plastic materials;
4. The presence of collagen on the surface of biomaterial implants may circumvent the serum-mediated inhibition of fibronectin adsorption;

5. Biomaterials implanted in the subcutaneous space are progressively colonized by cellular and fibrillar connective tissue components, and fibroblasts may contribute by their own protein synthesis machinery to the deposition of fibronectin on the extracellular matrix coating the artificial material. This cellular form of fibronectin is closely related to the plasma form of fibronectin[25,26] and can be deposited on artificial surfaces despite the presence of serum components;[17]

6. The presence of fibrin clots on the surface of blood-exposed biomaterials may also contribute to fibronectin deposition covalently crosslinked by factor $XIII_a$.

## VII. CONCLUDING REMARKS

The first steps of in vivo bacterial attachment for *S. aureus* and other microbial organisms seem to be strongly influenced by extracellular and cellular elements coating implanted biomaterials. While further steps of bacterial colonization may involve the contribution of cellular and extracellular microbial factors leading to extensive bacterial biofilms, it is important to define the contribution of biomaterial-adsorbed host factors on early bacterial attachment. This type of study should help to characterize clinically-relevant parameters of bacterial adhesion to biomaterial implants and may contribute to the development of novel anti-adhesive strategies for preventing implant-associated infections.

**Acknowledgments:** The author is grateful to the following colleagues for their essential contribution to the development of in vivo models, namely M. Bento, M. Delmi, E. Huggler, D. Lew, M. Magnin, G. Monney, H. Vasey, and W. Zimmerli, from the Division of Infectious Diseases and the Orthopaedic Clinic, University Hospital, Geneva. Financial support: Swiss National Foundation for Scientific Research (grant no. 3200-045810.95/1).

## REFERENCES

1. Chhatwal GS, Preissner G, Müller-Berghaus, et al: Specific binding of the human S protein (vitronectin) to streptococci, *Staphylococcus aureus*, and *Escherichia coli*. *Infect Immun* 55:1878–83, 1987

2. Dankert J, Hogt AH, Feijen J: Biomedical polymers: bacterial adhesion, colonization and infection. *CRC Critical Rev Biocompatib* 2:219–301, 1986

3. Darouiche RO, Landon GC, Patti JM, et al: Role of *Staphylococcus aureus* surface adhesins in orthopaedic device infections: Are results model-dependent? *J Med Microbiol* 46:75–9, 1997

4. Delmi M, Vaudaux P, Lew DP, et al: Role of fibronectin on staphylococcal adhesion to metallic surfaces used as models of orthopaedic devices. *J Orthop Res* 12:432–8, 1994

5. Elek SD, Conen PE: The virulence of *Staphylococcus pyogenes* for man: a study of the problems of wound infection. *Br J Exp Pathol* 38:573–86, 1957

6. Feijen J, Beugeling T, Bantjes A, et al: Biomaterials and interfacial phenomena. *Adv Cardiovasc Phys* 3:100–34, 1979

7. Fischer B, Vaudaux P, Magnin M, et al: Novel animal model for studying the molecular mechanisms of bacterial adhesion to bone-implanted metallic devices: Role of fibronectin in *Staphylococcus aureus* adhesion. *J Orthop Res* 14:914–20, 1996

8. Flock JI, Froman G, Jonsson K, et al: Cloning and expression of the gene for a fibronectin-binding protein from *Staphylococcus aureus*. *EMBO J* 6:2351–7, 1987

9. François P, Vaudaux P, Foster TJ, et al: Host-bacteria interactions in foreign body infections. *Infect Control Hosp Epidemiol* 17:514–20, 1996

10. François P, Vaudaux P, Nurdin N, et al: Physical and biological effects of a surface coating procedure on polyurethane catheters. *Biomaterials* 17:667–78, 1996

11. François P, Vaudaux P, Taborelli M, et al: Influence of surface treatments developed for oral implants on the physical and biological properties of titanium. (II) Adsorption isotherms and biological activity of immobilized fibronectin. *Clin Oral Impl Res* 8:217–25, 1997

12. Galante JO, Lemons J, Spector M, et al: The biologic effects of implant materials. *J Orthop Res* 9:760–75, 1991

13. Georgiade NG, King EH, Harris WA, et al: Effect of three proteinaceous foreign materials on infected and subinfected wound models. *Surgery* 77:569–76, 1975

14. Gillespie W: Mini-symposium: Bone and joint infection. (ii) Prophylaxis against infection in orthopaedic practice. *Curr Orthop* 8:220–5, 1994

15. Greene C, McDevitt D, François P, et al: Adhesion properties of mutants of *Staphylococcus aureus* defective in fibronectin binding proteins and studies on the expression of *fnb* genes. *Mol Microbiol* 17:1143–52, 1995

16. Grinnell F, Feld MK: Adsorption characteristics of plasma fibronectin in relationship to biological activity. *J Biomed Mater Res* 15:363–81, 1981

17. Grinnell F, Feld MK: Fibronectin adsorption on hydrophilic and hydrophobic surfaces detected by antibody binding and analysed during cell adhesion in serum-containing medium. *J Biol Chem* 257:4888–93, 1982

18. Gristina AG, Costerton JW: Bacterial adherence to biomaterials and tissues. *J Bone Joint Surg* 67A:264–73, 1985

19. Hench LL, Wilson J: Surface-active biomaterials. *Science* 226:630–6, 1984

20. Herrmann M, Hartleib J, Kehrel B, et al: Interaction of von Willebrand factor with *Staphylococcus aureus*. *J Infect Dis* 176:984–91, 1997

21. Herrmann M, Suchard SJ, Boxer LA, et al: Thrombospondin binds to *Staphylococcus aureus* and promotes staphylococcal adherence to surfaces. *Infect Immun* 59:279–88, 1991

22. Herrmann M, Vaudaux P, Pittet D, et al: Fibronectin, fibrinogen and laminin act as mediators of adherence of clinical staphylococcal isolates to foreign material. *J Infect Dis* 158:693–701, 1988

23. Hienz SA, Schennings T, Heimdahl A, et al: Collagen binding of *Staphylococcus aureus* is a virulence factor in experimental endocarditis. *J Infect Dis* 174:83–8, 1996

24. Hogt A, Dankert J, Feijen J: Adhesion of *Staphylococcus epidermidis* and *Staphylococcus saprophyticus* to a hydrophobic material. *J Gen Microbiol* 131:2485–591, 1985

25. Hynes RO: Fibronectins. *Sci Am* 254:32–41, 1986

26. Hynes RO, Yamada KM: Fibronectins: multifunctional modular glycoproteins. *J Cell Biol* 95:369–77, 1982

27. Inman RD, Gallegos KV, Brause BD, et al: Clinical and microbial features of prosthetic joint infection. *Am J Med* 77:47–53, 1984

28. James PJ, Butcher IA, Gardner ER, et al: Methicillin-resistant *Staphylococcus epidermidis* in infection of hip arthroplasties. *J Bone Joint Surg* 76B:725–7, 1994

29. Jonsson K, Signäs C, Muller HP, et al: Two different genes encode fibronectin binding proteins in *Staphylococcus aureus*. The complete nucleotide sequence and characterization of the second gene. *Eur J Biochem* 202:1041–8, 1991

30. Klebe RJ, Bentley KL, Schoen RC: Adhesive substrates for fibronectin. *J Cell Physiol* 109:481–8, 1981

31. Kleinman HK, McGoodwin EB, Martin GR, et al: Localization of the binding site for cell attachment in the alpha(I) chain of collagen. *J Biol Chem* 253:5642–6, 1978

32. Kuypers JM, Proctor RA: Reduced adherence to traumatized heart valves by a fibronectin low-binding mutant of *Staphylococcus aureus*. *Infect Immun* 57:2306–12, 1989

33. Lew DP, Waldvogel FA: Osteomyelitis. *N Engl J Med* 336: 999–1007, 1997
34. Liang OD, Maccarana M, Flock J-I, et al: Multiple interactions between human vitronectin and *Staphylococcus aureus. Biochim Biophys Acta* 1225:57–63, 1993
35. Lopez JD, Dos Reis M, Bretani RR: Presence of laminin receptors in *Staphylococcus aureus. Science* 229:275–7, 1985
36. Lowy FD, Hammer SM: *Staphylococcus epidermidis* infections. *Ann Intern Med* 99:834–9, 1983
37. Maderazo EG, Judson S, Pasternak H: Late infections of total joint prostheses: a review and recommendations for prevention. *Clin Orthop* 229:131–42, 1988
38. Mauerhan DR, Nelson CL, Smith DL, et al: Prophylaxis against infection in total joint arthroplasty. One day of cefuroxime compared with three days of cefazolin. *J Bone Joint Surg* 76A:39–45, 1994
39. McDevitt D, François P, Vaudaux P, et al: Molecular characterization of the clumping factor (fibrinogen receptor) of *Staphylococcus aureus. Mol Microbiol* 11:237–48, 1994
40. McDevitt D, François P, Vaudaux P, et al: Identification of the ligand-binding domain of the surface-located fibrinogen receptor (clumping factor) of *Staphylococcus aureus. Mol Microbiol* 16:895–907, 1995
41. McDevitt D, Vaudaux P, Foster TJ: Genetic evidence that bound coagulase of *Staphylococcus aureus* is not clumping factor. *Infect Immun* 60:1514–23, 1992
42. McDonald JA, Broekelmann TJ, Kelley DG, et al: Gelatin-binding domain-specific anti-human plasma fibronectin Fab inhibits fibronectin-mediated gelatin binding but not cell spreading. *J Biol Chem* 256:5583–7, 1981
43. Ni Eidhin D, Perkins S, Francois P, et al: Clumping factor B (ClfB), a new surface-located fibrinogen-binding adhesin of *Staphylococcus aureus. Mol Microbiol* 30:245–57, 1998
44. Nurdin N, François P, Magnani A, et al: Effect of toluene extraction on Biomer™ surface: I-Esca, ATR/FTIR, contact angle analysis and biological properties. *J Biomater Sci Polym Ed* 7:49–60, 1995
45. Park PW, Rosenbloom J, Abrams WR, et al: Molecular cloning and expression of the gene for elastin-binding protein (*ebpS*) in *Staphylococcus aureus. J Biol Chem* 271:15803–9, 1996
46. Pascual A, Fleer A, Westerdaal NAC, et al: Modulation of adherence of coagulase-negative staphylococci to Teflon catheters *in vitro. Eur J Clin Microbiol Infect Dis* 5:518–22, 1986
47. Patti JM, Allen BL, McGavin MJ, et al: MSCRAMM-mediated adherence of microorganisms to host tissues. *Annu Rev Microbiol* 48:585–617, 1994
48. Patti JM, Bremell T, Krajewska-Pietrasik D, et al: The *Staphylococcus aureus* collagen adhesin is a virulence determinant in experimental septic arthritis. *Infect Immun* 62:152–61, 1994
49. Ryden C, Yacoub AI, Maxe I, et al: Specific binding of bone sialoprotein to *Staphylococcus aureus* isolated from patients with osteomyelitis. *Eur J Biochem* 184:331–6, 1989
50. Steckelberg JM, Osmon DR: Prosthetic joint infections. In: Bisno AL, Waldvogel FA, eds. *Infections Associated with Indwelling Medical Devices.* American Society of Microbiology, Washington, DC, 1994: 259–90
51. Switalski LM, Patti JM, Butcher W, et al: A collagen receptor on *Staphylococcus aureus* strains isolated from patients with septic arthritis mediates adhesion to cartilage. *Mol Microbiol* 7: 99–107, 1993
52. Tshefu K, Zimmerli W, Waldvogel FA: Short-term administration of rifampin in the prevention or eradication of infection due to foreign bodies. *Rev Infect Dis* 5: S474–S480, 1983
53. Vaudaux P: Interaction of *Staphylococcus aureus* with implanted artificial surfaces used as biomaterials. In: Möllby R, Flock JI, Nord CE, et al., eds. *Proceeding of the VIIth International Symposium on Staphylococci and Staphylococcal Infections.* Stockholm, June 29-July 3, 1992. Gustav Fischer Verlag, Stuttgart-Jena-New York, 1994:337–42

54. Vaudaux P, Clivaz X, Emch R, et al: Heterogeneity of antigenic and proadhesive activity of fibronectin adsorbed on various metallic or polymeric surfaces. In: Heimke G, Soltösz U, Lee AJC, eds. *Clinical Implant Materials. Advances in Biomaterials.* Elsevier, Amsterdam, 1990:31–6

55. Vaudaux P, Lerch P, Velazco MI, et al: Role of fibronectin in the susceptibility of biomaterial implants to bacterial infections. *Adv Biomater* 6:355–60, 1986

56. Vaudaux P, Lew DP, Waldvogel FA: Host-dependent pathogenic factors in foreign body infection: a comparison between *Staphylococcus epidermidis* and *Staphylococcus aureus*. In: Pulverer G, Quie PG, Peters G, eds. *Pathogenicity and Clinical Significance of Coagulase-Negative Staphylococci.* Gustav-Fischer-Verlag, Stuttgart, 1987:183–93

57. Vaudaux P, Lew DP, Waldvogel FA: Host factors predisposing to and influencing therapy of foreign body infections. In: Bisno AL, Waldvogel FA, eds. *Infections Associated with Indwelling Medical Devices.* American Society of Microbiology, Washington DC, 1994: 1–29

58. Vaudaux P, Pittet D, Haeberli A, et al: Host factors selectively increase staphylococcal adherence on inserted catheters: a role for fibronectin and fibrinogen/fibrin. *J Infect Dis* 160:865–75, 1989

59. Vaudaux P, Pittet D, Haeberli A, et al: Fibronectin is more active than fibrin or fibrinogen in promoting *Staphylococcus aureus* adherence to inserted intravascular catheters. *J Infect Dis* 167:633–41, 1993

60. Vaudaux P, Suzuki R, Waldvogel FA, et al: Foreign body infection: role of fibronectin as a ligand for the adherence of *Staphylococcus aureus*. *J Infect Dis* 150:546–53, 1984

61. Vaudaux P, Waldvogel FA, Morgenthaler JJ, et al: Adsorption of fibronectin onto polymethylmethacrylate and promotion of *Staphylococcus aureus* adherence. *Infect Immun* 45:768–74, 1984

62. Vaudaux P, Yasuda H, Velazco MI, et al: Role of host and bacterial factors in modulating staphylococcal adhesion to implanted polymer surfaces. *J Biomater Appl* 5:134–53, 1990

63. Vaudaux P, Zulian G, Huggler E, et al: Attachment of *Staphylococcus aureus* to polymethylmethacrylate increases its resistance to phagocytosis in foreign body infection. *Infect Immun* 50:472–7, 1985

64. Velazco MI, Waldvogel FA: Monosaccharide inhibition of *Staphylococcus aureus* adherence to human solid-phase fibronectin. *J Infect Dis* 155:1069–72, 1987

65. Yacoub A, Lindahl P, Rubin K, et al: Purification of a bone sialoprotein-binding protein from *Staphylococcus aureus*. *Eur J Biochem* 222:919–25, 1994

66. Zimmerli W, Waldvogel FA, Vaudaux P, et al: Pathogenesis of foreign body infection: Description and characteristics of an animal model. *J Infect Dis* 146:487–97, 1982

# PART V

## STUDYING MICROBIAL ADHESION TO HOST TISSUE

# 27

# Characterization of Staphylococcal Adhesins for Adherence to Host Tissues

**Mark S. Smeltzer**

*Department of Microbiology and Immunology,*
*University of Arkansas for Medical Sciences, Little Rock, AR, USA*

## I. INTRODUCTION

Studies assessing DNA–DNA reassociation kinetics have been used to define 32 staphylococcal species, approximately half of which are indigenous to humans.[59] Although all staphylococci are opportunistic pathogens, the vast majority of human infections can be attributed to one of only three species.[59,60] *Staphylococcus epidermidis* is a coagulase-negative species that can be universally found on the skin. Although relatively avirulent, it has a remarkable ability to colonize biomaterials and is a leading cause of infections centered on in-dwelling medical devices.[60] In fact, *S. epidermidis* has been referred to as a "pathogen of medical progress" based on the almost direct correlation between its increasing incidence of infection and the increasing use of medical implants.[109] *Staphylococcus aureus* is clearly the most prominent staphylococcal pathogen. In addition to its production of coagulase, *S. aureus* is distinguished from other staphy-lococcal species by its ability to produce a diverse array of virulence factors and the concomitant ability to cause a diverse array of infections.[72,105] These range from relatively benign infections of the skin (e.g., folliculitis) to debilitating and even life-threatening disease (e.g., osteomyelitis, endocarditis). *S. aureus* also has a remarkable capacity to colonize biomaterials and is probably less prominent than *S. epidermidis* as a biomaterial-related pathogen only because it is a less common inhabitant of humans. *Staphylococcus saprophyticus* is a less prevalent pathogen than *S. epidermidis* or *S. aureus* but is a frequent cause of urinary tract infections, particularly in young, sexually active women.[60] The different disease syndromes associated with each of these staphylococcal species is largely a function of their respective abilities to produce specific adhesive molecules that promote their adherence to biomaterials and/or host tissues. Because *S. epidermidis* infections are most often associated with in-dwelling medical devices, and because staphylococcal adherence to biomaterials is eloquently addressed elsewhere in this book (Chapter 20), the focus of this chapter will be on the *S. aureus* adhesins that promote the colonization of host tissues.

*Handbook of Bacterial Adhesion: Principles, Methods, and Applications*
Edited by: Y. H. An and R. J. Friedman © Humana Press Inc., Totowa, NJ

## II. COLONIZATION IN HEALTH AND DISEASE

### A. *Colonization of Healthy Adults*

A primary predisposing factor for *S. aureus* infection is colonization of the anterior nasopharynx.[151] Approximately 20% of the population are persistently colonized and an additional 30% are colonized on a more transient basis.[72] Based on the correlation between colonization with *S. aureus* and the likelihood of infection, patients at risk of infection are often treated with topical antibiotics (e.g., mupirocin) in an effort to eliminate the bacterium from the nasopharynx.[151] This approach is generally effective but only on a temporary basis. Moreover, the continued emergence of mupirocin-resistant strains[65,142] suggests that its utility will decline over time. An alternative is to reduce nasal carriage by disrupting the interactions between *S. aureus* and the nasal mucosa. This interaction appears to be primarily dependent on the mucus layer rather than the respiratory epithelium.[115] There is evidence to suggest that cell wall teichoic acids are involved in the adherence of *S. aureus* to the nasal mucosa.[2] However, recent studies suggest that *S. aureus* also produces surface proteins that specifically bind mucin.[123,137,141] For example, Shuter et al.[123] demonstrated that *S. aureus* produces 127 and 138-kDa surface proteins that bind human nasal mucin. On the other hand, some clinical isolates did not adhere to mucin-coated surfaces despite the fact that Western blot analysis confirmed the presence of both proteins.[123] These results suggest that the contribution of the mucin-binding proteins to colonization of the nasal mucosa is dependent on the context of their presentation. Indeed, Trivier et al.[141] demonstrated that nonmucoid strains of *S. aureus* bind mucin much more efficiently than mucoid strains. To date, the identity of the 127 and 138-kDa proteins and their relationship to other *S. aureus* adhesins has not been established.

### B. *Colonization in Disease*

From the anterior nasopharynx *S. aureus* is easily transferred to the skin. Its presence in either location is largely irrelevant in healthy adults; however, any compromise of the innate defense systems of the host can lead to disease. The primary initiating event is a break in the mechanical barriers of the skin or nasal mucosa,[151] with the most common manifestation being a wound infection. However, *S. aureus* can invade into deeper tissues and eventually reach the bloodstream. Once there, it can colonize the vascular endothelium either directly or by virtue of platelets and host proteins (e.g., fibronectin, fibrinogen) deposited at sites of inflammation.[72] *S. aureus* is also capable of binding a variety of host proteins found within the extracellular matrix (ECM).[98] The ability to bind plasma and ECM proteins appears to be a major contributing factor in the development of several forms of staphylococcal disease including wound infections, endocarditis and septic arthritis.[44,64,84,94,100,119] For that reason, the remainder of this chapter is devoted to a discussion of the adhesins that promote the interaction between *S. aureus* and host plasma and ECM proteins.

## III. GENERAL PROPERTIES OF *STAPHYLOCOCCUS AUREUS* ADHESINS

### A. *MSCRAMM Adhesins*

Patti et al.[98] introduced the acronym "MSCRAMM" to denote microbial surface components recognizing adhesive matrix molecules. To qualify as an MSCRAMM, the molecule must be localized to the cell surface and must recognize and bind a macro-molecular ligand found within the host ECM.[98] Like other Gram-positive pathogens, *S. aureus* produces a diverse array of proteins that qualify as MSCRAMMs. The host protein targets of the *S. aureus* MSCRAMMs include collagen,[132] fibronectin,[37] fibrinogen,[74,88] elastin,[96] laminin,[70] von Willebrand factor,[40] vitronectin,[20,155] thrombospondin[41] and bone sialoprotein.[112] The *S. aureus* MSCRAMMs that bind collagen (CNA), fibronectin (FnBPA and FnBPB), fibrinogen (ClfA and ClfB) and elastin (EbpS) have been identified and their corresponding genes cloned.[27,54,74,88,97,102] Recent data also suggests that staphylococcal protein A (Spa) is responsible for the bind-ing of von Willebrand factor (Hartlieb J, et al.: Abstract #B-080, 38th Interscience Conference on Antimicrobial Agents and Chemotherapy, 1998, San Diego, CA). With the exception of Spa, which is best known for its ability to bind immunoglobulin G (IgG), all of these MSCRAMM adhesins appear to bind a single host protein although they some-times do so by virtue of multiple binding domains.[51,69]

Although their host protein ligands have not been described, Josefsson et al.[55] recently described the characterization of three genes (*sdrC*, *sdrD* and *sdrE*) that also appear to encode MSCRAMMs. These genes were identified and cloned based on a conserved region of serine-aspartate repeats (Sdr) like those found in the ClfA and ClfB MSCRAMMs.[88] Although SdrC, SdrD and SdrE have additional structural features in common with each other (Table 1), the putative ligand-binding domains exhibit little similarity.[55] While that suggests that the Sdr adhesins may bind different host proteins, the ligand-binding domains of ClfA and ClfB also exhibit relatively little similarity despite the fact that they both bind fibrinogen.[88] Based on that, it is unclear whether the different A domains in SdrC, SdrD and SdrE reflect a binding specificity for different host proteins or for different regions of the same protein.

The *S. aureus* MSCRAMMs that bind laminin, vitronectin, thrombospondin and bone sialoprotein (BSP) have not been clearly defined. McGavin et al.[76] identified a surface protein that binds a variety of host proteins including BSP, fibrinogen, fibronectin, vitronectin, thrombospondin and, to a limited extent, collagen. Based on sequence homology with a segment of the peptide binding groove of the β chain of MHC class II proteins, this protein was designated the MHC analogous protein (Map).[53] It was originally suggested that Map binding involved lectin-like activity;[76] however, subsequent studies demonstrated that the interaction between Map and at least some of its target proteins involves a direct, protein–protein interaction.[53] Phenotypic characterization of *clfA*, *clfB*, *fnbA*, *fnbB* and *cna* mutants suggests that Map makes a relatively minor contribution to the binding of fibrinogen, fibronectin and collagen.[37,88,100] Also, because *map* is highly conserved among clinical isolates of *S. aureus*,[125] the observation that the ability to bind thrombospondin, vitronectin and BSP is a strain-dependent characteristic[20,41,104,111] suggests that adhesins other than Map are involved in the binding of these proteins. In the case of BSP, Yacoub et al.[156] isolated a 97-kDa cell-associated protein that binds a nonapeptide sequence (LKRFPVQGG) present in the amino terminus of BSP.[110] The fact that binding was not inhibited in the presence of rat chondrosarcoma proteoglycan, *N*-linked oligosaccharides purified from BSP, fibronectin, fibrinogen or collagen would appear to distinguish this protein from Map;[112,157] however, to date, the gene encoding the 97-kDa adhesin has not been identified. There are no reports describing adhesins that specifically bind laminin, thrombospondin or vitronectin.

**Table 1. Comparison of *S. aureus* MSCRAMM Adhesins**

| MSCRAMM | Ligand | Size (kDa)[a] | Signal Seq. (S) | Non-repetitive domain (A or C) | Repetitive domains[b] B (# of repeats) | D | R (SD) | Sorting Signal[c] W | LPXTG | M | C | Comments |
|---|---|---|---|---|---|---|---|---|---|---|---|---|
| CNA[102] | Collagen (all types) | 85-160 | 29 | 505 | 187 (1-4) | None | None | 64 (lysine/proline) | LPKTG | 26 | 6 | Ligand-binding A domain contains a 19 kDa collagen-binding subdomain. The B domains serve no obvious function. |
| FnBPA[54] | Fibronectin (type I modules) | 108 (200) | 36 | 508 (A) 140 (C) | 30 (2) | 38 (3) | None | 14 amino acid repeats (Wr) + 45 amino acid non-repetitive region (Wc) | LPETG | 16 | 9 | Either FnBPA or FnBPB is sufficient for binding. Binding is a function of the D repeats. |
| FnBPB[54] | Fibronectin (type I modules) | 98 (165) | 36 | 450 (A) 140 (C) | None | 38 (3) | None | | LPETG | 16 | 9 | |
| ClfA[74] | Fibrinogen (γ chain) | 97 (130) | 39 | 520[d] | None | None | 308[e] | 37 | LPDTG | 19 | 10 | The A domain in ClfA contains an EF hand and a MIDAS (DYSNS) motif. The A domain in ClfB contains a MIDAS motif but only a partial EF hand. |
| ClfB[88] | Fibrinogen (α and β chains) | ~97 (124) | 44 | 540 | None | None | 272 | 37 | LPDTG | 19 | 10 | |
| Sdr[55] | Unknown | 98-140[f] | 50-52 | 445-554 | 110-113 (2-5) | None | 132-170 | Not directly reported but includes all conserved motifs (e.g. LPXTG). Wall anchoring motifs of Sdr adhesins are more similar to each other than to Clf adhesins. | | | | Each B domain contains an EF hand motif. |
| EbpS[97] | Elastin | 23 (25-40) | None | 202 | None | None | None | ???[g] | None | ??? | ??? | Processing to cell surface may involve carboxy-terminal signal sequence. |
| Map[53] | Broad-specificity incl. BSP | 77 (72) | 30 | 19 | ~110 (6) | None | None | ??? | None | ??? | ??? | Map consists almost entirely of repetitive B domains. Each B domain contains a 31 amino acid subdomain similar to MHC class II molecules |

[a]In those cases in which size is hyphenated (e.g., CNA), the numbers represent the range defined by the number of repetitive domains. In other cases, the predicted size derived from the nucleotide sequence is given first followed parenthetically by the size observed in cell lysates. Because size determinations in *S. aureus* lysates are complicated by the need to use cell wall lytic enzymes (e.g., lysostaphin) that release the MSCRAMM along with peptidoglycan fragments, size was most often determined in *E. coli* using clones of the corresponding MSCRAMM gene. Even in *E. coli*, the presence of proline-rich regions in the W domain is also thought to result in anomalous migration patterns in SDS-PAGE gels.

[b]With the exception of the FnBPA B domains, the repetitive domains are located between the nonrepetitive A domains and the carboxy-terminal sorting signal. The FnBPA B domains are in the middle of the nonrepetitive region between the A and C domains.

[c]The sorting signal consists of a wall-spanning domain (W), an anchoring motif (LPXTG), a membrane-spanning domain (M) and a carboxy-terminal tail containing charged amino acids. The beginning of the cytoplasmic tail is generally defined as the first positively-charged amino acid (K or R) after the M domain.

[d]The ClfA and ClfB exhibit only 26% identity. The Sdr adhesins exhibit a similar level of identity both with respect to each other and with respect to the Clf adhesins.

[e]The R region in ClfA (and presumably ClfB) is highly variable ranging from 193 to 440 amino acids. Deletion analysis suggests that 80 amino acids represents a functional minimum. Whether the same size range and functional limits apply to ClfB is unknown.

[f]The primary sequence of the Sdr proteins has not been reported. Predicted sizes are based on comparisons with ClfA.

[g]The signals responsible for localization of the EbpS and Map adhesins to the cell surface are unknown. Neither contains the characteristic sorting signal and neither is covalently linked to the cell wall peptidoglycan.

**Figure 1.** Structural characteristics of *S. aureus* MSCRAMMs. A long nonrepetitive domain (A) that is often responsible for binding follows the amino-terminal signal sequence (S). The nonrepetitive region in the FnBPs is subdivided into a long region with little similarity (A) and a shorter, almost identical region (C). In FnBPA, these regions are separated by a short repetitive region (B) that is absent in FnBPB. The repetitive domains consist of relatively large (38–187 amino acid) repetitive elements (B or D) or an extended series of serine-aspartate (SD) repeats (R). The Sdr adhesins contain both forms. In the case of CNA, the repetitive domains serve no obvious function. In other cases, they are required for binding (FnBPs) or for presentation of the ligand-binding A domain (ClfA). The wall-spanning domain (W), membrane-spanning domain (M) and cytoplasmic tail (C) constitute a sorting signal that facilitates incorporation of the MSCRAMM into the *S. aureus* cell wall. The hallmark of the sorting signal is an LPXTG motif that serves as the substrate for sortase-mediated linkage of the MSCRAMM to cell wall peptidoglycan. Because the MSCRAMM becomes covalently linked to peptidoglycan, removal of LPXTG-anchored MSCRAM requires digestion with cell-wall lytic enzymes (e.g., lysostaphin).

The possibility that structural variants of Map or differences in the level of *map* expression account for strain-dependent differences in the binding of at least some of these host proteins cannot be ruled out.

### B. Structural Characteristics of MSCRAMMs

Most *S. aureus* MSCRAMMs share a number of structural characteristics (Fig. 1). The amino-terminus contains a relatively long (>29 amino acids) signal sequence that is required for *sec*-dependent processing. That is followed by a relatively large, non-repetitive region that is often responsible for the binding specificity of the MSCRAMM and a series of repeated domains that may be required for functional exposure of the ligand-binding domain or, more rarely, directly involved in binding. The carboxy-terminal region contains a wall-spanning domain that is either rich in prolines or contains a series of serine-aspartate (SD) repeats, a hydrophobic region thought to span the cell membrane, and a positively-charged hydrophilic tail that extends into the cytoplasm.[30,98] The carboxy-terminal domains constitute a sorting signal that characteristically includes an LPXTG motif at the junction of the wall and membrane-spanning domains.[86,121] The LPXTG motif is cleaved by a "sortase" during the process of anchoring the adhesin to the staphylococcal cell wall.[121] Analysis of the products released after digestion with cell wall lytic enzymes (e.g., lysostaphin) has confirmed that this cleavage results in a covalent linkage between the carboxyl group of the LPXTG threonine residue and the pentaglycine cross-link in the cell wall peptidoglycan.[87,140] The elastin-binding MSCRAMM (EbpS) is an exception in that it does not contain an amino-terminal signal sequence or the carboxy-terminal sorting signal.[97] The broad-specificity Map adhesin also lacks an LPXTG motif and appears to remain only loosely associated with the cell surface.[53,76] Additionally, despite the presence of an LPXTG motif, studies assessing the cellular compartment-alization of fusion proteins linked to sorting signals derived from different *S. aureus*

MSCRAMMs suggest that the collagen-binding adhesin (CNA) may remain anchored in the cell membrane.[122] The difference between CNA and LPXTG-anchored MSCRAMMs may be due to a reduced number of amino acids between the leucine in the LPXTG motif and the first positively charged amino acid in the cytoplasmic tail.[122] It should be noted, however, that the alternative processing of CNA is not universally accepted.[30]

### C. Extracellular Binding Proteins

In addition to the MSCRAMM adhesins, *S. aureus* produces at least three extracellular proteins that bind fibrinogen.[9,93] One of these is coagulase, which is responsible for the clotting reaction that is often used to distinguish *S. aureus* from other staphylococcal species. There is genetic evidence[125] to suggest that a fibrinogen-binding protein designated FbpA[18] is an allelic variant of the *coa* gene. Although it does not contain a carboxy-terminal hydrophobic domain or hydrophilic tail, FbpA appears to remain anchored to the cell wall by virtue of an LPSITG motif very similar to the consensus LPXTG anchoring motif.[18] The observation that different strains of *S. aureus* encode either *fbpA* or *coa*[125] suggests that some strains produce the cell-bound FbpA coagulase while others produce the extracellular form. Although the production of coagulase is the primary distinction used to distinguish *S. aureus* from the less pathogenic, coagulase-negative staphylococcal species, studies assessing the role of coagulase in the pathogenesis of *S. aureus* infection have been inconclusive.[84] Interestingly, Projan and Novick[105] recently suggested that the cell-bound form of coagulase may be a more definitive virulence factor by virtue of its ability to promote attachment to host tissues while the extracellular form may be an "evolutionary misstep" that has little or no impact on the ability to colonize the host. With that in mind, it is perhaps worth noting that the *fbpA* gene encoding the cell-associated form appears to be relatively rare.[125] On the other hand, the ClfA and ClfB fibrinogen-binding MSCRAMMs are highly conserved (*see below*), which suggests that the absence of FbpA may not be phenotypically apparent.

Boden and Flock[9,10] described a second 19-kDa extracellular fibrinogen-binding protein (Fib) that exhibits partial homology with staphylococcal coagulase. This protein was more recently designated Efb to clearly denote its role as an extracellular fibrinogen-binding protein.[151] Although a role for Efb as a virulence factor in wound infections has been established, its contribution to disease does not appear to involve binding host tissues.[94] More directly, mutation of Efb does not alter the ability of *S. aureus* to bind fibrinogen or fibronectin, and it has been suggested that it may bind soluble fibrinogen in a manner that delays clot formation and thereby delays wound healing.[94] In contrast, a third extracellular fibrinogen-binding protein (Eap) was recently shown to promote the aggregation of *S. aureus* and to promote binding of the bacterium to fibroblasts and epithelial cells.[93] Eap can bind both to itself and to *S. aureus* and probably facilitates adherence by acting as a bridging molecule between the bacterium and host tissues.[93] Eap can also bind other plasma proteins including prothrombin and fibronectin.[93] That is consistent with the observation that Eap is similar to the broad-specificity Map adhesin[53,76] and to a second binding protein designated p70.[32,159] Because studies leading to the identification of Map and Eap were done using different strains of *S. aureus*, it is unclear whether they are distinct proteins or are different variants of the same protein.[93] Importantly, only 70% of Eap is exported into the extracellular environment, which implies that a significant proportion of the protein remains at least transiently associated with the cell.[93] Since Map does not contain the carboxy-terminal sorting signals characteristic of other *S. aureus* MSCRAMMs and appears to remain only loosely associated with the staphylococcal cell surface,[53,76] the cellular localization of Eap and Map cannot be used as a definitive distinguishing characteristic.

### D. Prevalence of Staphylococcal Adhesins

The *cna* gene encoding the only well-defined collagen-binding MSCRAMM is present in approximately half of all *S. aureus* strains.[114,125] That is consistent with the observation that the ability to bind collagen is a variable characteristic.[33,114,136] Collagen-binding in coagulase-negative staphylococci has been reported but at an even lower frequency than in *S. aureus*.[104] The *epbS*, *clfA* and *clfB* genes appear to be highly conserved.[88,125] All strains also encode at least one of the two FnBP genes (*fnbA* and/or *fnbB*).[37,125] These results are consistent with the observation that most *S. aureus* isolates bind fibronectin, fibrinogen and elastin.[42,96] In contrast, the ability to bind laminin is a strain-dependent characteristic.[42] Like collagen binding, the ability to bind fibronectin, fibrinogen, elastin and laminin is a variable characteristic of the coagulase-negative staphylococci.[42,90,96] However, it was recently demonstrated that most *S. epidermidis* isolates produce a fibrinogen-binding protein (Fbe) that contains an SD repeat region like that in the *S. aureus* Clf and Sdr MSCRAMMs.[90] The Fbe and ClfA ligand-binding A domains exhibit a moderate degree of homology but appear to bind fibrinogen by different mechanisms as evidenced by the fact that *S. epidermidis* does not clump in the presence of soluble fibrinogen.[90] The genes encoding Map and Efb are highly conserved among clinical isolates of *S. aureus*.[125,152] The production of Efb appears to be limited to *S. aureus*.[153] The prevalence of the broad-specificity Eap protein has not been assessed. All *S. aureus* strains also encode at least two of the three *sdr* genes.[55] Approximately half of all strains tested bind vitronectin, thrombospondin and bone sialoprotein.[20,41,104,111] Although the ability to bind BSP is not highly conserved, there is some evidence to suggest that it is a common characteristic of strains that cause musculoskeletal infection.[111] The same correlation has been suggested for strains that bind collagen;[46,132] however, the validity of that correlation is in dispute (see below).

While some *S. aureus* MSCRAMM genes are encoded within genetic elements that are not present in all strains, they all appear to be encoded within the *S. aureus* chromosome. In general, the MSCRAMM genes are distributed throughout the approx 2800 kbp chromosome.[49] The *ebpS* gene is located on a chromosomal fragment that corresponds to SmaI fragment A in the prototypical phage group III strain 8325-4.[125] The *clfA* and *efb* (*fib*) genes are both located on SmaI fragment B, however, their location within this 361 kbp fragment is unknown.[125] Although the chromosomal location of *clfB* has not been determined, *clfA* and *clfB* are not closely linked and are probably transcribed as monocistronic mRNAs.[55] However, the *fnbA* and *fnbB* genes are also expressed under the control of independent promoters[37] but are separated by only 682 bp.[54] Although it was previously suggested that they map to SmaI fragment F,[49] more recent data indicates that the *fnb* genes are located in SmaI fragment C.[125] The *fpbA* and *coa* genes map to SmaI fragment E, which is consistent with the suggestion that they are allelic variants of the same gene.[125] The *map* gene is located in SmaI fragment F.[125] Whether the *eap* gene maps to the same location has not been determined. The chromosomal location of the *sdr* locus is also unknown. Although the *sdr* genes are closely linked, they are separated by a relatively large intergenic regions *(see below)* and are probably transcribed as monocistronic mRNAs.[55] When present, *cna* is located in a region of the chromosome corresponding to SmaI fragment G.[125] The presence of *cna* seems to be associated with certain subpopulations of *S. aureus*. For example, all of the *cna*-positive strains we have examined have a SmaI restriction pattern that is clearly distinct by comparison to phage group III strains.[125] Additionally, the common bacteriophage used to transduce genes between phage group III strains (e.g., f11) plaque poorly on *cna*-positive strains. Ryding et al[114] demonstrated a clear although not exclusive correlation between the presence of *cna* and the production serotype 8 capsular polysaccharides. Importantly, serotypes 5 and 8 are the two serotypes that account for the vast majority of human infection.[3]

## III. BINDING OF *S. AUREUS* TO SPECIFIC HOST PROTEINS

### A. Collagen Binding in S. aureus

We compared 25 strains of *S. aureus* and found an almost direct correlation between the presence of *cna* and the ability to bind collagen.[33] The only exceptions were one strain that encoded but did not express *cna* and two heavily-encapsulated strains that expressed *cna* but bound only minimal amounts of collagen. Comparisons between capsule mutants and their isogenic parent strains confirmed that the failure of heavily-encapsulated strains to bind collagen is due to masking of the adhesin by capsular polysaccharides.[33] Because both of the exceptions involved *cna*-positive strains that did not bind collagen, these results are consistent with the hypothesis that *cna* encodes the primary *S. aureus* collagen-binding MSCRAMM. On the other hand, Ryding et al.[114] examined 216 *S. aureus* isolates and found eight *cna*-negative strains that bound collagen. That clearly suggests the existence of a second collagen-binding adhesin. Because Ryding et al.[114] did not define the relative levels of collagen binding observed with *cna*-positive and *cna*-negative strains, it is difficult to assess the contribution of this alternative adhesin relative to CNA. It was noted, however, that four of the *cna*-negative strains bound collagen at levels barely above what was defined as the background level of their assay. Based on that, it was suggested that the broad-specificity Map adhesin might be responsible for this binding.[114] The more pertinent question is whether this low level binding contributes to the colonization of host tissues. Studies in which the adherence of *S. aureus* to a collagen-coated substrate was assessed under shear forces similar to those observed in the blood showed a clear correlation between the presence of CNA and the ability to adhere to immobilized collagen.[68,83,129] Moreover, a recombinant fragment corresponding to the CNA ligand-binding A domain, as well as antibodies directed against the recombinant fragment, inhibits the adherence of *S. aureus*.[83] Because collagen is not found as a soluble protein in vivo, these results suggest that the *cna*-encoded adhesin is the only *S. aureus* MSCRAMM that facilitates collagen binding under physiologically-relevant circumstances.

CNA is among the most extensively characterized of all *S. aureus* MSCRAMMs.[99,101,108,133,134] The amino-terminal signal sequence is followed by a 55-kDa A domain and a 187 amino acid B domain that may be repeated up to four times.[34] The carboxy-terminal sorting signal includes a 64 amino acid wall-spanning domain rich in lysine and proline residues.[134] The molecular weight of the CNA adhesin ranges from 85 to 160 kDa depending on the number of B domains.[133] Within the 55-kDa A domain, the collagen-binding domain (CBD) has been localized to a 19-kDa region that spans amino acids 151 through 318.[99] The region between residues 209 and 233 appears to be particularly important.[101] The 19 kDa region binds at least eight sites on the type II collagen triple helix.[99] The crystal structure of the 19-kDa CBD has been determined and appears to consist of a "jelly roll" composed of two anti-parallel β-sheets and two short α-helices.[134] One of the β-sheets forms a trench that accommodates the collagen triple helix.[30,134] This model is consistent with mutational analysis demonstrating that alteration of amino acids within this region results in a reduced ability to bind collagen.[101] Additionally, synthetic peptides corresponding to the 19-kDa CBD are capable of blocking collagen binding.[99,101] However, the fact that the full-length A domain binds collagen more specifically and with higher affinity than the 19-kDa CBD suggests that amino acids within the 55-kDa A domain, but outside the 19-kDa CBD, contribute to the conformation of the CBD in the intact adhesin molecule.[30]

The presence or absence of a B domain does not appear to have any impact on the conformational integrity of the ligand-binding A domain.[108] Rich et al.[108] demonstrated that the CNA adhesin has a mosaic architecture in that the A and B domains fold independently of each other. Additionally, the B domains are not required for processing of the collagen adhesin to the cell surface or exposure of the A domain after the adhesin is incorporated into the cell envelope.[129] In contrast, the repetitive domains of other MSCRAMM adhesins are either required for functional exposure of the ligand-binding domain [38] or are directly responsible for binding.[30] The apparent dispensability of the B domains seems odd given the fact that the B domains are very large (187 amino acids) and may be repeated up to four times.[34] Also, all recognized CNA variants include at least one B domain. Nevertheless, it seems clear that any function attributable to the B domains must either be unrelated to collagen binding or related to collagen binding only under certain circumstances. The B domains do have some similarity with the B domains of the Sdr MSCRAMMs although the significance of that observation is unknown.[55] Individual B domains also have some similarity with at least three streptococcal fibronectin proteins; however, neither the presence of a B domain or the number of B domains has any impact on the ability to bind fibronectin.[129] The possibility that multiple B domains might extend the ligand-binding A domain away from the cell surface and thereby allow the bacterium to overcome the inhibitory effects of the capsule has also been investigated.[33] However, we were unable to demonstrate any circumstance in which multiple B domains were associated with an enhanced capacity to bind collagen.[129] It should be noted that the strains in which the capsular inhibition of CNA was observed are heavily-encapsulated stereotype 1 and 2 strains that are not representative of the microencapsulated serotype 5 and 8 strains responsible for most human infections.[3] It is therefore possible that the capsular masking observed in these strains is not a biologically relevant phenomenon. Indeed, we could not demonstrate significant inhibition when we introduced even the smallest CNA variant into serotype 5 and 8 strains.[33]

Because it is not produced by all *S. aureus* strains, the CNA MSCRAMM is a less attractive therapeutic target than other, more highly conserved MSCRAMMs. On the other hand, CNA is present on the surface of *S. aureus* cells growing in bone,[35] and infection with a *cna*-positive strain elicits an anti-CNA antibody response.[113] These observations confirm that CNA is expressed in vivo during the course of *S. aureus* infection. Additionally, Nilsson et al.[89] demonstrated that active immunization with recombinant

CNA and passive immunization with CNA-specific antibodies protected mice against intravenous challenge with *S. aureus*. These results support the hypothesis that CNA is a valid candidate for inclusion in a vaccine directed toward *S. aureus* MSCRAMMs. Moreover, while the ability to bind collagen is relatively rare, there are reports suggesting that it is a conserved characteristic of strains that cause bone and joint infection.[46,132,149] The suggestion that a rare phenotype is conserved among strains that cause a specific kind of infection clearly implies that the phenotype is an important virulence factor in the pathogenesis of the infection. Support for that hypothesis comes from studies demonstrating that *S. aureus* binds directly to collagen fibrils in cartilage[149] and that the CNA adhesin is both necessary and sufficient for this binding.[102] On the other hand, there are conflicting reports that cast doubt on the correlation between collagen binding and musculoskeletal disease. Specifically, several studies have concluded that the ability to bind collagen is no more prominent among isolates from patients with bone and joint infection than it is among isolates from patients with other forms of staphylococcal disease.[113,114,136] That implies that collagen binding is not a necessary prerequisite of musculoskeletal infection, however, it should be emphasized that survey studies attempting to correlate collagen binding with etiology are complicated by a number of factors, not the least of which are the basis used to define etiology and the methods used to define collagen binding. For instance, Ryding et al.[113] found that some strains agglutinated in the presence of collagen-coated latex beads but did not bind collagen in an assay using soluble $^{125}$I-labeled collagen. These results are surprising since agglutination is generally the least sensitive of the two assays. The same authors also found that the serum of patients infected with *S. aureus* strains that were negative in $^{125}$I-collagen binding assays had anti-CNA antibody. These results suggest that some strains that bind collagen in vivo may not bind collagen when assayed in vitro.

Clearly, the more direct approach to defining the correlation between collagen binding and musculoskeletal disease is to compare *cna*-positive wild-type strains with isogenic mutants in which the ability to bind collagen has been specifically eliminated. Indeed, a mutant in which the *cna* gene was inactivated by allele replacement was less virulent than its isogenic parent strain in animal models of staphylococcal septic arthritis[100] and endocarditis.[44] Additionally, the introduction of *cna* into a *cna*-negative strain enhanced virulence in the septic arthritis model.[100] While that clearly indicates that CNA is an important contributing factor in the pathogenesis of musculoskeletal disease, it should be noted that these studies are limited to a single strain (Phillips) and that mutation of *cna* in that strain appears to have a pleiotropic effect that also results in a reduced capacity to bind fibronectin.[43] The basis for this is unclear. While the *cna* mutant produced a reduced amount of cell-associated FnBPs and had a reduced ability to aggregate in the presence of soluble fibronectin, its ability to bind immobilized fibronectin was not altered.[43,44] The fact that the reduced amount of FnBPs produced by the mutant did not limit its adherence to immobilized fibronectin suggests that its ability to bind ECM proteins was not altered in any respect other than the ability to bind collagen. Nevertheless, the reduced production of FnBPs is difficult to explain. The possibility that *cna* is linked to a regulatory element that modulates expression of the *fnb* genes can be discounted since *cna* is encoded within a discrete genetic element that extends only 202 bp upstream of the *cna* start codon and 100 bp downstream of the *cna* stop codon.[34] This element does not include any open-reading frames other than *cna* and its presence does not disrupt an open-reading frame present in strains that do not encode *cna*.[34] The possibility that the insertional inactivation of *cna* has a polar effect on a downstream regulatory element also seems unlikely since the gene immediately downstream of *cna* (*pcp*) is separated from *cna* by a 740 bp intergenic region and is transcribed in the opposite direction.[103] It is possible that the absence of *cna* somehow alters the cell surface in a fashion that negatively impacts on processing and exposure of the FnBPs. However, among *S. aureus* clinical isolates, there is no obvious correlation between the presence or absence of *cna* and the ability to bind fibronectin.[43] Taken together, these results suggest that the only consistent difference between *cna*-positive and *cna*-negative strains is the ability to bind collagen. That supports the hypothesis that the reduced virulence observed in *cna* mutants is, in fact, due to the inability to bind collagen. Nevertheless, confirmation of that hypothesis will require the analysis of additional *cna* mutants in appropriate animal models of staphylococcal disease.

### B. Fibronectin Binding in S. aureus

Unlike collagen binding, the ability to bind fibronectin is a highly conserved characteristic of *S. aureus*.[42] It is a function of two MSCRAMM adhesins encoded by closely linked but independently expressed genes designated *fnbA* and *fnbB*.[27,37,54] Most strains appear to express both genes, and expression of either gene is sufficient to confer a level of fibronectin binding comparable to that observed in those strains that encode and express both genes.[37] The *fnbA* and *fnbB* genes are very similar and presumably arose by gene duplication.[54] The corresponding MSCRAMMs (FnBPA and FnBPB) contain a 36 amino acid signal sequence followed by a relatively dissimilar nonrepetitive A domain.[54] In FnBPA, the A domain is followed by a pair of 30 amino acid repeats designated the B domain.[27] The primary difference between the FnBPA and FnBPB MSCRAMMs is that the latter lacks the repetitive B domains.[54] The remainder of the FnBPA and FnBPB adhesins is very similar and consists of 1) a second, 140 amino acid nonrepetitive region (C), 2) a series of 38 amino acid domains that are repeated three times in their entirety (D1-D3) and partially a fourth time (D4), and 3) a highly conserved sorting signal. The defining characteristic of the sorting signal is a proline-rich, wall-spanning domain that contains either four (FnBPB) or five (FnBPA) 14 amino acid repeats and a 45 amino acid nonrepetitive region that includes the LPETG anchoring motif (Table 1).

The FnBPs are unique by comparison to other *S. aureus* MSCRAMMs in that the ligand-binding domain is not contained within the nonrepetitive A or C domains. Rather, the ligand-binding domains have been localized to the repetitive D domains located immediately adjacent to the wall-spanning domain.[124] Although each individual D domain is capable of binding fibronectin, the combined D1-D3 region exhibits much higher binding affinity.[48,78,106] Also, the individual D domains are not equally efficient either with respect to binding fibronectin or inhibiting the interaction between *S. aureus* and fibronectin.[106,124] McGavin et al.[78] suggested that the interaction between the FnBPs and fibronectin was dependent on an EEDT motif that occurs once in D1 and D2 and twice in D3. The suggestion that the EEDT motif plays a pivotal role is consistent with the observation that a peptide corresponding to the D3 region is the most efficient inhibitor of the interaction between *S. aureus* and fibronectin.[78] However, amino acids flanking the EEDT motif are also important, presumably because they alter the conformation of the D domain in a manner that optimizes the interaction between the EEDT motif and fibronectin.[78] Sun et al.[131] confirmed that a distinct motif [GG(X$_{3-4}$)(I/V)DF] located within the carboxy-terminal 20 amino acids of each D domain also plays a critical role.[131] Comparisons between the FnBPs produced by various Gram-positive bacteria suggest that the peptide sequence EEDT(X$_{9-10}$)GG(X$_{3-4}$)(I/V)DF represents a consensus fibronectin-binding domain.[30,52,77] In the *S. aureus* FnBPs, this motif is located in the carboxy-terminal half of each D domain[47,48,78] and preferentially binds the 29-kDa amino-terminal region of fibronectin.[31] Binding involves all five of the repeated type I modules in the fibronectin molecule although there is evidence to suggest that the binding domains do not interact equally with all modules.[48,130] Specifically, there is an apparent preference for type I modules 4 and 5.[48,51] However, a recent report confirmed the presence of an additional FnBPA binding domain that is located in the amino-terminal half of the D3 domain and appears to interact preferentially with type I module pairs 1-2 and 2-3.[51] Although it has not been clearly defined, there is also evidence to suggest the existence of a fibronectin-binding domain located outside the D repeat region.[51]

The FnBPs are highly conserved and appear to contribute to various forms of staphylococcal disease (see below). Because their contribution to virulence presumably involves the ability to attach to host tissues and/or fibronectin-coated biomaterials, a number of studies have addressed the possibility of inhibiting the FnBP-mediated attachment of *S. aureus* using either peptide analogs of the ligand-binding D domains or specific, anti-FnBP antibodies. Signas et al.[124] demonstrated that individual D domains were capable of inhibiting the interaction between *S. aureus* and soluble fibronectin. Although the degree of inhibition was not significantly enhanced when all three full-length D domains were mixed in an equimolar ratio, a fusion protein containing all three D repeats and most of the wall-spanning region was far more inhibitory.[124] Similar results were obtained when the inhibition to plasma clots formed in vitro was assessed.[106] However, there was no appreciable difference between the inhibition observed with individual D domains and with the fusion protein when activity was assessed as a function of the attachment of *S. aureus* to fibronectin-coated microtiter plates.[106] These results imply that soluble fibronectin and fibronectin incorporated into blood clots and/or the extracellular matrix may be conformationally different and that the binding of *S. aureus* to each form of fibronectin probably involves distinct interactions between the ligand and the FnBP D domains.

Attempts to generate antibodies that block the interaction between the *S. aureus* FnBPs and fibronectin have met with limited success.[23] Sun et al.[131] immunized rabbits with a recombinant D1-D3 fragment and with a glutathione S-transferase fusion protein containing the D1-D3 fragment. Antibodies purified from the D1-D3 immunized rabbits did not inhibit fibronectin binding, apparently because the antibody response was directed primarily toward the amino-terminal region of each D domain rather than the carboxy-terminal region that contains the primary ligand-binding site.[131] Antibodies generated with the fusion protein were more effective but were not completely inhibitory because the antibody response was directed preferentially toward residues in the D1 and D2 repeats rather than the functionally-dominant D3 domain.[131] Using synthetic peptides corresponding to specific regions of the D1 and D3 domains, it was possible to generate antibodies that were relatively effective inhibitors. However, these antibodies failed to achieve complete inhibition of fibronectin binding even when they were used in combination with each other.[131] In contrast, Brennan et al.[11] recently reported that immunization with truncated or full-length forms of the D2 domain exposed on the surface of cowpea mosaic virus or rod-shaped potato virus X elicited antibodies that could both block fibronectin binding and inhibit attachment of *S. aureus* to a fibronectin-coated surface. Because the truncated D2 domain was not capable of binding fibronectin in vitro, these results suggest that folding or presentation of the fibronectin-binding domain in vivo may be critical for the development of blocking antibodies. That is consistent with the observation that the ligand-binding D domains have little secondary structure in the absence of fibronectin, which induces a conformational shift to a structure dominated by β-sheets.[47] Casolini et al.[12] subsequently demonstrated that the sera of patients with staphylococcal infection contain antibodies that recognize epitopes that are present in the fibronectin-FnBP complex but are absent in the unbound FnBPs.[12] These epitopes have been referred to as ligand-induced binding sites or LIBS.[12,30] The presence of antibodies that recognize LIBS epitopes clearly indicates that the *S. aureus* FnBPs are expressed in vivo and that they bind fibronectin sometime during the course of infection.

Fibronectin is a large glycoprotein found in soluble form in plasma and other body fluids and in a less soluble form in the extracellular matrix. Fibronectin promotes clot formation and wound healing by binding to fibrin clots and promoting the adherence of platelets and fibroblasts to sites of inflammation. The binding of soluble fibronectin could provide the bacterium with a means to escape immune recognition or, based on the interaction between bound fibronectin and other host proteins found within the extracellular matrix, could serve as a bridge between the bacterium and host tissues.[106] The ability to bind fibronectin present in the extracellular matrix or deposited on the surface of an in-dwelling medical device has obvious implications.[145] Indeed, Scheld et al.[119] reported a correlation between the ability to bind fibronectin and the propensity to cause endocarditis. Kuypers and Proctor[64] subsequently confirmed that *S. aureus* mutants with a reduced capacity to bind fibronectin also have a reduced capacity to cause endocarditis. Electron microscopic studies indicate that fibronectin is distributed evenly over the surface of *S. aureus* cells in suspension but is localized to the interface between *S. aureus* and endothelial cells when the two are mixed in culture.[144] On the other hand, Flock et al.[28] could not confirm a critical role for fibronectin binding in the pathogenesis of catheter-induced endocarditis. These contradictory results emphasize the complex nature of the interaction between *S. aureus* and ECM proteins and strongly suggest that any effective anti-adherence strategy will probably require a multivalent approach targeting a number of *S. aureus* adhesins.

## C. *Fibrinogen Binding in* S. aureus

*S. aureus* also encodes at least two fibrinogen binding MSCRAMMs.[88] Because they promote the clumping of *S. aureus* in the presence of soluble fibrinogen, these MSCRAMMs are referred to as clumping factor A (ClfA) and clumping factor B (ClfB).[74,88] ClfA and ClfB are structurally similar in that both contain a 39 amino acid signal sequence, a ligand-binding A domain, a region of SD repeats (R) and a highly-conserved carboxy-terminal sorting signal (Table 1). The SD repeats are encoded by a repeated 18 bp consensus sequence (GAYTCN GAYTCN GAYAGY where Y corresponds to either of the pyrimidines and N corresponds to any nucleotide).[88] The overall size of the R region in ClfA varies from 193 to 440 amino acids.[73] The observation that the R region diverges from the consensus sequence in the outermost repeats is consistent with the hypothesis that variations in the length of the R region have arisen by intragenic recombination.[73] Nevertheless, the *clfA* genes encoded by individual strains of *S. aureus* appear to be stable.[73] Comparison of strains producing different ClfA variants indicate that the length of the R region is not correlated to the ability to clump in the presence of soluble fibrinogen.[73] However, there is a minimum size since artificially-constructed strains containing fewer than 80 R region amino acids are defective both with respect to clumping in the presence of soluble fibrinogen and adherence to fibrinogen-coated surfaces.[38] Because the ability to form clumps was more drastically reduced than the ability to adhere to immobilized fibrinogen,[38] it was suggested that an extended R region may confer enough flexibility on the ClfA molecule to facilitate the interaction with multiple fibrinogen molecules.[38] Also, the observation that the shortest, naturally-occurring R region contains over twice the minimum number of SD repeats has led to the suggestion that an extended R domain may be required to project the ligand-binding A domain away from the cell surface and facilitate its exposure in the presence of capsular polysaccharides and/or other surface proteins.[38]

Unlike the FnBP ligand-binding domains, the ligand-binding A domains of ClfA and ClfB are very dissimilar (~26% identity). In fact, the only conserved characteristic is a TYTFTDYVD motif that is also present in the Sdr adhesins (see below) and the *S. epidermidis* Fbe fibrinogen-binding protein.[55] The function of the TYTFTDYVD motif is unknown. In fact, the dissimilarity between other regions of the ClfA and ClfB A domains is consistent with the observation that these MSCRAMMs do not bind the same region of the fibrinogen molecule. Specifically, the A domain of ClfA binds the fibrinogen $\gamma$ chains while the A domain of ClfB binds sites in the $\alpha$ and $\beta$ chains.[75,88] ClfA binds two different sites in the fibrinogen $\gamma$ chains by a mechanism similar to two different mammalian integrins.[75,117] One of these involves the same carboxy-terminal residues that serve as the recognition site for the platelet integrin $\alpha_{11b}\beta_3$.[39,58,75,88,117] The other is the recognition site for the leucocyte integrin $\alpha M\beta_2$.[1] Recognition of the $\alpha_{11b}\beta_3$ site involves an EF-hand motif while recognition of the $\alpha M\beta_2$ site involves a metal ion dependent adhesion site or MIDAS motif.[88] ClfB has a MIDAS motif (DXSXS) but contains only a partial EF hand.[88] Nevertheless, the interaction between ClfA and ClfB and their respective target sites on the fibrinogen molecule is regulated by $Ca^{++}$ and $Mn^{++}$ cations.[88] Specifically, the interaction between the ClfA ligand-binding A domains and the carboxy terminus of fibrinogen $\gamma$ chains is progressively inhibited at $Ca^{++}$ concentrations in the 1-10 mM range.[92] The interaction between the ClfB ligand-binding A domain and its $\alpha$ and $\beta$ chain targets is inhibited at even lower concentrations.[88] Since these concentrations approximate those observed in vivo, it has been suggested that the differential, calcium-responsive nature of the two adhesins may allow *S. aureus* to maintain a certain proportion of unoccupied fibrinogen receptors.[88] Presumably, these unoccupied receptors facilitate binding of the bacterium to blood clots and/or fibrinogen deposited on the surface of biomaterials even in the presence of soluble fibrinogen.

Soluble fibrinogen promotes the adherence of *S. aureus* to human endothelial cells, presumably by acting as a bridge between the bacterium and receptors present on the surface of the target cells.[17] Fibrinogen also promotes the adherence of *S. aureus* to keratinocytes and to the horny layer of the skin.[57,81] *S. aureus* also induces platelet aggregation via a fibrinogen-dependent mechanism.[5] The observation that a recombinant form of the ClfA ligand-binding A domain inhibits platelet aggregation clearly implies that ClfA is more important the ClfB in that regard.[75] ClfA has also been shown to facilitate the adherence to immobilized fibrinogen deposited on the surface of implanted medical devices.[16,42,146,148] Mutation of *clfA* results in a reduced capacity to adhere to platelet/fibrin clots and a reduced capacity to cause endocarditis.[84]

While comparison of *clfA* and *clfB* mutants has confirmed that ClfB can promote cell clumping and the adherence to immobilized fibrinogen even in the absence of ClfA, the contribution made by ClfB appears to be minor by comparison to ClfA.[88] The most direct evidence for that is the observation that *clfA* mutants clump less efficiently and have a reduced capacity to bind immobilized fibrinogen despite the presence of an intact *clfB* gene.[84] However, the production of two fibrinogen-binding proteins may also reflect the fundamental nature of the need to bind fibrinogen. For instance, in vitro studies have demonstrated that *clfB* is expressed during the early exponential growth phase while *clfA* is expressed preferentially (but not exclusively) during the post-exponential growth phase.[88,154] This differential regulation could provide the bacterium with a means to bind fibrinogen even when it is growing under conditions that do not warrant expression of one or the other *clf* gene. Alternatively, since expression of *clfA* is not limited to the post-exponential growth phase,[154] ClfA and ClfB are presumably produced simultaneously at least under some circumstances. In that case, ClfA and ClfB may act synergistically to promote adherence to thrombi even under the shear forces present in the bloodstream. Indeed, Dickenson et al.[25] demonstrated that *clfA, coa* double mutants were not displaced from a fibrinogen-coated surface even under shear forces simulating blood flow. An equally attractive hypothesis is that the production of two different fibrinogen-binding adhesins allows the bacterium to adhere to fibrin clots even in the presence of antibodies that block the activity of one of the adhesins.[88] However, the ClfB MSCRAMM was discovered relatively recently[88], and the definitive assessment of the relative contribution of the ClfA and ClfB adhesins to the colonization of host tissues and/or biomaterials will have to await a more detailed comparison of *clfA* and *clfB* mutants.

### D. The Sdr Adhesins

The Sdr proteins were recently identified based on the presence of a serine-aspartate repeat (Sdr) region like that present in ClfA and ClfB.[55] At present, it can only be assumed that they qualify as MSCRAMMs because their binding specificity, or whether they bind any host protein at all, has not been established. The *sdr* locus consists of three open-reading frames (ORFs) designated *sdrC*, *sdrD* and *sdrE*.[55] These ORFs encode 947, 1,315 and 1,166 amino acids respectively. Each Sdr has an N-terminal signal sequence containing 50 to 52 amino acids followed by a nonrepetitive region (A) containing between 445 and 554 amino acids.[55] They are distinguished from ClfA and ClfB by the presence of repeated domains (B) located between the ligand-binding A domain and the R region SD repeats. The B domains contain 110 to 113 amino acids and are repeated two, five and three times in SdrC, SdrD and SdrE respectively (Table 1). The SD repeats vary from 132 to 170 amino acids. The carboxy-terminal sorting signal in the Sdr proteins contains an LPXTG motif and is otherwise similar to the sorting signal in other *S. aureus* MSCRAMM adhesins.[55] The *sdr* genes are separated by relatively large intergenic regions (*sdrC* and *sdrD* are separated by 369 bp while *sdrD* and *sdrE* are separated by 397 bp) and are probably transcribed as monocistronic mRNAs. The regions upstream and downstream of each *sdr* gene are very similar suggesting that expression of all three genes is regulated in a similar fashion.[55]

What are presumed to be the ligand-binding A domains of the Sdr proteins are dissimilar both with respect to each other and with respect to the A domains of ClfA and ClfB.[55] In fact, the only common characteristic is the TYTFTDYVD motif that is conserved at eight of nine amino acids in at least four of the five proteins.[30] The MIDAS motifs found in the ligand-binding A domains of ClfA and ClfB are not present in any of the Sdr adhesins.[55] Because the degree of dissimilarity between the Sdr proteins and between the Sdr proteins and the Clf proteins is similar to the degree of dissimilarity between ClfA and ClfB,[30] it is not possible to speculate whether the Sdr MSCRAMMs bind different host proteins or different regions of the same protein. Also, by analogy with the FnBPs, which contain carboxy-terminal repetitive domains that are responsible for binding the host ligand, it is certainly possible that the Sdr B domains are the ligand-binding domains. However, the B domain repeats in the Sdr proteins also exhibit a surprising degree of dissimilarity. The exception are the B domains immediately adjacent to the R region, which are 95–96% identical.[55] Also, all adjacent B domains are separated by a proline residue that is absent in the junction between the carboxy-terminal B domain and the R region. Based on these distinctions, it has been suggested that the presence of at least one B domain of each type may be required for functional expression of the Sdr proteins.[55] That would imply that SdrC, which contains only two B domains, represents the functional minimum for the Sdr adhesins. All B domains contain a conserved, 29 amino acid EF-hand motif that is also present in ClfA.[55,74] In SdrD, the five B domains contain 14 $Ca^{2+}$-binding sites that fall into one of two classes based on their relative binding affinities.[56] Binding of calcium converts the B domains from a molten, unstructured state to a compact globular form.[55,56] Since the overall binding affinity is very high, the Sdr B domains are presumably fully occupied in vivo.[30,56] It is therefore unlikely that the function of the Sdr proteins is modulated by calcium availability in vivo.[55] The fact that the EF hand in ClfA is within the ligand-binding A domain is also consistent with the suggestion that the Sdr B domains may have a ligand-binding function. Alternatively, the B domains may be required for surface display of the Sdr adhesins.[55,56] The possibility that the B domains modulate the distance between region A and the bacterial cell surface[56] is an attractive hypothesis since the R regions in the Sdr proteins are smaller than the smallest R region in ClfA.[73] However, the R region in all three Sdr proteins is larger than the apparent minimum required for optimal ClfA function.[38]

### E. Elastin Binding in S. aureus

Although the level of binding varies, the ability to bind elastin appears to be a conserved characteristic of *S. aureus*.[96] Some coagulase-negative staphylococci also bind elastin.[96] The ability to bind elastin may facilitate the entry and exit of *S. aureus* from the vasculature, particularly since the bacterium also produces an elastase.[105] Elastin binding is mediated by a 25-kDa protein (EbpS) that does not contain an amino-terminal signal sequence, the carboxy-terminal sorting signals, or any of the repetitive domains characteristic of other *S. aureus* MSCRAMMs.[97] Importantly, the predicted size of EbpS based on nucleotide sequence (23 kDa) corresponds well with the size of the protein as determined by mass spectrometry (25 kDa) but not with the size of the protein as determined by Western blot of cell lysates (40 kDa).[97] The 40-kDa protein appears to represent an anomalous migration pattern rather than a dimer of the 25-kDa protein.[96] Nevertheless, it has been suggested that the larger form is an intracellular precursor while the smaller is the final, surface-exposed form of the adhesin.[96] Because both forms have an identical amino-terminal sequence, processing of the adhesin must involve modifications to the carboxy terminus, perhaps by virtue of a unique but not unprecedented carboxy-terminal signal sequence.[97] However, because the size of the protein observed in Western blots appears to reflect a migration artifact rather than the actual size of an intracellular precursor, the modification is presumably less drastic than the difference in molecular weights would suggest. The lack of an amino-terminal signal sequence is also consistent with the observation that the elastin-binding motif consists of a hexameric sequence (TNSHQD) located in the extreme amino-terminal region of EbpS.[95] Binding must also be dependent on amino acids flanking this motif since a synthetic TNSHQD peptide does not inhibit elastin binding.[95] EbpS binds a target in the amino terminus of the elastin molecule.[96] The role of EbpS in the colonization of host tissues has not been defined, however, the observation that elastin is present in most mammalian tissues and is particularly abundant in the skin suggests that elastin binding may contribute both to the ability of *S. aureus* to persist as a commensal bacterium and to its ability to invade to deeper tissues.

### F. Interaction of S. aureus with Host Cells

*S. aureus* is considered a prototypical extracellular pathogen. Based on the preceding discussion, it is evident that its role as an extracellular pathogen is facilitated by its ability to bind host proteins present in plasma and in the extracellular matrix. However, *S. aureus* is also capable of directly binding a variety of host cells including endothelial and epithelial cells, osteoblasts and keratinocytes.[6,50,81,82] Tompkins et al.[138] demonstrated that the ability to bind endothelial cells was associated with the enhanced expression of four surface proteins, one of which appeared to correspond to protein A. Conversely, endothelial cells produce a 50-kDa membrane protein that specifically mediates the binding of *S. aureus*.[139] Subsequent studies have demonstrated that the interaction between *S. aureus* and host cells often results in internalization of the bacterium. Specifically, internalization of *S. aureus* has been confirmed using human endothelial cells,[82] bovine mammary epithelial cells[6] and osteoblasts.[50,107] There is some evidence to suggest that internalization requires *de novo* protein synthesis in the host cell but not in the bacterium.[50] Internalized *S. aureus* exists both within membrane-bound vacuoles and free in the cytoplasm.[6,82] Bacteria remain viable for at least 72 h.[82] The end result of invasion is the induction of programmed cell death (apoptosis), which may provide the bacterium with a means by which it can persist in the host without inducing an acute inflammatory response.[6,82] It may also help the bacterium avoid immune recognition and exposure to antimicrobial agents. Although some *S. aureus* exoproteins can induce apoptotic death, several studies have confirmed that the induction of apoptosis by intact *S. aureus* cells is dependent on internalization.[6,136] For example, Menzies and Kourteva[82] demonstrated that UV-killed bacteria could attach to endothelial cells but that attachment did not induce apoptosis. The analysis of adhesin mutants indicates that several MSCRAMMs promote the binding of host cells[81] but only the FnBPs are required for internalization.[26] The fact that the D3 fibronectin-binding domain and soluble fibronectin are capable of inhibiting internalization emphasizes the specificity of this interaction.[26] Moreover, the observation that fibronectin inhibits rather than promotes binding suggests that the *S. aureus* FnBPs bind directly to host cells rather than indirectly via a fibronectin bridge.[26]

## G. Regulation of S. aureus MSCRAMMs

The virulence factors of *S. aureus* can be categorized into those that remain exposed on the cell surface and those that exported into the extracellular environment. In vitro, the two groups are coordinately and inversely regulated, with the extracellular virulence factors being produced at the expense of surface proteins as cultures enter the post-exponential growth phase.[105] It has been suggested that the regulatory events observed in vitro have an in vivo corollary that corresponds to before and after abscess formation.[105] More directly, the corollary suggests that *S. aureus* produces its array of surface proteins early during the course of infection when the most important considerations are avoiding host defenses and colonizing an appropriate tissue. The extracellular virulence factors are turned on only after growth in the localized environment of an abscess results in limited nutrient availability and the need to invade adjacent tissues. This scenario implies that the MSCRAMMs would be expressed during the early stages of infection, which should translate to the exponential phase of in vitro growth. While that is the case with the FnBPs,[79,105] CNA[8,35] and ClfB,[88] *clfA* is expressed preferentially during the post-exponential growth phase.[154] One interpretation of this apparent inconsistency is that the differential production of the Clf MSCRAMMs may allow *S. aureus* to bind fibrinogen under a more diverse set of environmental conditions. The need to produce the MSCRAMMs during the early stages of infection is somewhat intuitive. However, it may also be important to turn these genes off and to remove them from the cell surface as a means of avoiding immune recognition. Indeed, it could be argued that their protease sensitivity *(see below)* reflects that need.

The primary regulatory system controlling expression of *S. aureus* virulence factors is the accessory gene regulator (*agr*), which encodes a two-component quorum sensing system that responds to the accumulation of an octapeptide pheromone and modulates the transition between expression of surface proteins and expression of extracellular toxins and enzymes.[105] Optimal expression of *agr* is also dependent on a DNA-binding protein (SarA) encoded within a second locus called the staphylococcal accessory regulator or *sar*.[21,22,85] While mutation of *agr* results in over-expression of protein A and the fibronectin-binding proteins,[105] it has little impact on transcription of *clfA*[154] or *cna*.[33,35] In the case of *cna*, SarA appears to modulate transcription directly by binding *cis* elements upstream of the *cna* structural gene and repressing transcription.[8] The *sar* locus also affects production of MSCRAMMs via an indirect mechanism arising from its impact on the production of several proteases. Specifically, Chan and Foster[14] demonstrated that SarA is a repressor of several proteases including V8 serine protease. McGavin et al.[79] subsequently demonstrated that the FnBPs are particularly sensitive to V8 protease degradation. That is consistent with the observation that *sar* mutants have a reduced capacity to bind fibronectin.[19] ClfB also appears to be particularly sensitive to protease degradation. In fact, ClfB cannot be detected on the surface of *S. aureus* cells after the exponential growth phase.[88] Although certain MSCRAMMS (e.g., CNA) appear to be more stable,[35] the observation that it has been difficult to confirm the size of several *S. aureus* MSCRAMMs because they are often found in cell lysates in multiple forms that include protease-degraded fragments[31,37,54,88] supports the hypothesis that protease production makes an important contribution to *S. aureus* MSCRAMM function.

## IV. METHODS FOR CHARACTERIZING THE INTERACTION BETWEEN *S. AUREUS* AND HOST TISSUES

The methods used to characterize *S. aureus* MSCRAMMs and their interaction with host proteins are not unique and are eloquently addressed elsewhere in this book. For that reason, this section will focus on those methods that pertain specifically to the study of the staphylococci.

### A. *Growth of* S. aureus

*S. aureus* is typically grown in complex nutrient media like tryptic soy or brain-heart infusion broth. A chemically-defined medium has been described.[153] The staphylococci are facultative anaerobes but are generally grown under aerobic conditions. Under such conditions, *S. aureus* grows rapidly and reaches the postexponential growth phase within a matter of hours. Because the interaction of *S. aureus* with host cells[138] and with different matrix proteins *(see above)* is growth-phase dependent, it is imperative that a standard growth curve be established and that any results be reported in the context of that growth curve. For example, ClfB was not identified during the course of studies leading to the identification and characterization of ClfA because it is produced only during the early exponential growth phase and is rapidly degraded as cultures enter postexponential growth.[88]

The ability of *S. aureus* to adhere to plastic and to host cells also varies depending on growth medium.[143] Krajewska-Pietrasik et al.[62] recently demonstrated that growth in iron-limited medium is associated with an increased capacity to bind collagen. Because collagen binding was assessed by enzyme-linked immunosorbent assay (ELISA) using an anti-CNA antiserum, this effect was presumably mediated either by a direct effect on *cna* transcription by an indirect effect that resulted in enhanced exposure of the CNA adhesin. Growth on agar is also known to affect the production of several *S. aureus* surface proteins and the production of capsular polysaccharides.[15,67] The latter raises the possibility that MSCRAMM-mediated adherence of host proteins will be reduced when cells grown on solid media are assayed.[129] These observations clearly emphasize the need to consider alternative growth conditions when evaluating the ability of *S. aureus* to bind host proteins.

The inclusion of antibiotics can also have an impact on adherence. For example, Vaudaux et al.[147] recently demonstrated that the introduction of *mecA* into the chromosome results in a reduced capacity to bind fibrinogen and fibronectin despite the fact that the production of the FnBP and Clf adhesins is unaltered. It was suggested that this paradox could be explained either by aberrant processing resulting in exposure of an inactive form of the MSCRAMM or the presence of additional surface proteins that interfere with the interaction between the adhesins and their ligands. Support for the latter hypothesis comes from the observation that some methicillin-resistant strains produce a 230-kDa surface protein that is absent in methicillin-sensitive strains.[45] The 230-kDa protein is sensitive to plasmin degradation, and its removal is associated with an increased capacity to bind fibronectin, fibrinogen and IgG.[45] In fact, the presence of the 230-kDa protein has been associated with a false-negative reaction in commercial agglutination tests used to identify *S. aureus* based on its ability to bind fibrinogen and/or protein A.[63] In contrast, subinhibitory levels of ciprofloxacin have been associated with an increased capacity to bind fibronectin in fluoroquinolone-resistant strains of *S. aureus*.[7] Because it enhances the ability to bind fibronectin, and because the ability to bind fibronectin contributes to the ability to colonize the host *(see above)*, it was suggested that this correlation may actually be related to the emergence of fluoroquinolone-resistant strains.[7] Whether or not that is the case, antibiotics should be used only when required to maintain plasmids. Chromosomal resistance markers in *S. aureus* are stable and do not require selection, however, it remains important to verify the purity of unselected broth cultures by plating on solid medium with and without selection at the completion of each experiment. Given the complex and interactive nature of *S. aureus* virulence factors, it is also important to verify that genetic manipulation did not affect any phenotype other than the phenotype under study. The possibility that mutation of *cna* has an impact on production of the FnBPs has already been discussed.

## B. Genetic Manipulations of S. aureus

The "K12" of *S. aureus* is the phage group III strain 8325-4.[153] It was generated by curing the prophage from a strain (8325) known to carry at least three lysogens (f11, f12 and f13). Because the *att* site for one of these phage (f13) is within the gene (*hlb*) encoding b-toxin, only the prophage-cured strain 8325-4 is hemolytic on sheep blood.[127] Historically, 8325-4 has been identified by other designations including RN450, RN6390 and ISP479C.[33,126,153] It encodes *fnbA, fnbB, clfA, sdrC, sdrD, ebpS, map* and *efb* (*fib*).[37,55,125] It does not encode *sdrE*[55] or *cna*[34]. Greene et al.[37] has described 8325-4 derivatives carrying mutations in *fnbA* and/or *fnbB*. Studies leading to the identification of ClfA, ClfB and Eap were done with *S. aureus* strain Newman.[74,88,93] Hartford et al.[38] generated an extensive set of Newman derivatives that differ in the length of the ClfA SD repeat (R) region. Newman derivatives carrying mutations inactivating *clfA* and/or *clfB* have also been described.[88] Newman encodes all three *sdr* genes but does not encode *cna*.[55,133] CNA was originally purified from *S. aureus* strain Cowan.[133] The gene was subsequently cloned from FDA574 by screening a lgt11 library with anti-CNA antibody.[102] The genetic element encoding *cna* has been characterized and is essentially identical in strains encoding each of the four *cna* variants.[34] All four *cna* variants, as well as an artificially-constructed variant that does not contain a B domain, have also been cloned.[129] To date, the mutagenesis of *cna* has only been done in *S. aureus* strain Phillips, which encodes the *cna* variant with two B domains.[100]

Mutation of *S. aureus* MSCRAMM genes has been done by random insertion of transposons (e.g., Tn917) and by allele replacement.[37,74,88,100] Foster[29] has compiled a detailed and comprehensive summary of the specific methods used for both protocols. The most common approach is to use a temperature-sensitive delivery vector. The permissive and nonpermissive temperatures are typically 30 and 43°C respectively.[29] It is also possible to generate chromosomal mutants using *E. coli* plasmids as suicide vectors[116] and by directed plasmid integration.[71] Although both of these have the potential to eliminate the requirement for a selectable marker within the target gene, the plasmid integration approach is preferable because the plasmid can be established and verified in *S. aureus* prior to undertaking the mutagenesis experiments. In this case, a fragment of the target gene that is truncated at both the 5' and 3' ends is generated by PCR and then cloned into an appropriate delivery vector. Once the construct is in the desired strain *(see below)*, it is grown at the permissive temperature with selection and then shifted to the nonpermissive temperature without selection. After at least two overnight cultures, integrants are selected by plating at the nonpermissive temperature on medium containing appropriate antibiotics. Lowe et al.[71] used this approach to successfully knockout 11 of 15 genes identified using an in vivo expression technology (IVET) protocol. It was suggested that the failure to knockout the remaining four genes was due to the fact that they were probably essential genes rather than an inherent limitation of the mutagenesis protocol. The vector used in that study was pAUL-A,[118] which is an attractive alternative because the same antibiotic (erythromycin) can be used for selection in both *E. coli* (125 μg/mL) and *S. aureus* (10 μg/mL).[71]

The two most common methods used to move genes into *S. aureus* are transformation and transduction. Natural competence has not been demonstrated in *S. aureus* but it can be transformed either by protoplast fusion or, more commonly, by electroporation.[4,61,120] Electrocompetent cells can be prepared by harvesting cells from early to mid-exponential phase cultures and washing in decreasing volumes of 500 mM ice-cold sucrose. The final suspension should be at least a 100-fold concentration of the original culture volume. It is possible to transform *S. aureus* directly but the frequency of transformation is very low, particularly when transforming with a ligation mixture. The most common alternative is to employ an *E. coli-S. aureus* shuttle vector (e.g., pLI50).[129] Whether attempting to transform *S. aureus* directly with a ligation mixture or after transforming *E. coli* and isolating plasmid DNA, it is absolutely essential to utilize the restriction-deficient strain RN4220 as an intermediate host.[29] One of the most critical parameters in the electroporation of *S. aureus* is the immediate recovery of cells from the electroporation cuvette.[4] The recovery medium we use is SMMP, which consists of equal parts of 4× Penassay Broth (Antibiotic Medium #3, Difco) and 2× SMM (1 M sucrose, 0.04 M maleic acid, 0.04 M $MgCl_2$, pH 6.5). Other formulations include small amounts of bovine serum albumin.[4] Recovery should be continued at 37°C with gentle agitation for at least one hour. It is sometimes necessary to add small amounts of antibiotic (e.g., 0.5 mg/mL erythromycin for *ermC*) to the recovery medium in order to induce expression of the resistance markers.[29] Many *E. coli–S. aureus* shuttle vectors were constructed using origins from *S. aureus* plasmids that replicate by the rolling-circle mechanism. Since such plasmids become less stable as their size increases,[91] the need to include multiple components (e.g., resistance markers for both *E. coli* and *S. aureus*) often results in decreased plasmid stability. For that reason, we have found it prudent to verify plasmids by selection and by restriction analysis at both the *E. coli* and RN4220 stages. For example, in experiments in which we cloned each of the *cna* structural variants into a shuttle vector (pLI50) and introduced each variant into *E. coli*, less than 0.4% of the ampicillin-resistant transformants contained the intact plasmid (pLI50:*cna*). That proportion was increased to an acceptable level (~10%) by growing in minimal media at a reduced temperature (30°C).

Once the appropriate constructs are verified in RN4220, they are relatively easy to move into the target strain either by electroporation or transduction. Transduction of plasmids is much more efficient than electroporation although that level of efficiency is not necessarily required. In our experience, the frequency of transducing a given chromosomal marker is on the order of $10^{-8}$. Unlike electroporation, transduction efficiency does not seem to be dependent on the use of exponentially-growing cells. In fact, we routinely use bacteria harvested from fresh plates and resuspended at high density ($>10^{10}$ colony forming units per mL) in tryptic soy broth. Transduction protocols take advantage of the requirement for $Ca^{2+}$ to promote phage infection. Specifically, phages (e.g., f11, 80a) are propagated on the donor strain and then mixed with the recipient strain in the presence of 5 mM calcium chloride. After incubating for exactly 20 min at 37°C, ice-cold sodium citrate is added to chelate the calcium and prevent further phage infection. Cells are then harvested, resuspended in 20 m$M$ sodium citrate and plated on selective media containing sodium citrate.[29] Transductants are screened either by phenotypic analysis or by PCR and then verified by Southern blot analysis using probes for both the selected marker and, when possible, an additional unselected chromosomal marker. Useful resistance genes for selection of single-copy chromosomal markers include *ermC* (erythromycin), *tetK* (tetracylince) and *tetM* (tetracyline and minocycline).[118] Although aminoglycoside-resistance genes have been successfully used to generate knockout mutants,[14,100] these genes are sometimes difficult to transduce because the level of spontaneous mutation leading to resistance is relatively high.[118] In fact, we have encountered tremendous difficulty in transducing the *cna* mutation in PH100, which is marked with a gentamicin-resistance determinant,[100] into other *cna*-positive strains of *S. aureus*. Once chromosomal mutations are established, it is usually not necessary to maintain antibiotic selection.

Complementation in *S. aureus* is most often done using plasmids. In most cases, that is entirely appropriate. However, all of the *S. aureus* MSCRAMM genes identified to date are encoded within the chromosome, and the possibility that the introduction of a chromosomal gene on a multicopy plasmid will misrepresent the situation observed in nature should always be considered. Also, it is sometimes difficult to extend these experiments to animal studies because it is difficult to maintain plasmid selection in vivo. A remarkably well-used alternative is the series of integration vectors developed by Chia Y. Lee at the University of Kansas Medical Center.[66] These vectors (e.g., pCL84) take advantage of the site-specificity of phage insertion. Specifically, they include selectable markers for both *E. coli* and *S. aureus* but do not include an *S. aureus* replication origin. After cloning the target gene in *E. coli*, the construct is transformed into an RN4220 derivative (CYL316) that includes a second plasmid (pYL112D19) that contains phage L54a integrase (*int*) gene. Since these vectors also include the L54a *att* site, the presence of the second plasmid drives integration of the construct into a specific target site within the lipase (*geh*) gene. Inclusion of the integrase gene within a resident plasmid rather than the integration vector itself greatly enhances the efficiency of integration.[66] This system therefore provides a reproducible way to introduce single-copy genes into the chromosome of *S. aureus*. The fact that the integration site is in the lipase gene also provides a convenient method of confirmation since integrants will be lipase-negative. However, if the intention is to introduce an MSCRAMM gene into the chromosome using this system, two things must be kept in mind. First, it is certainly possible that lipase contributes to pathogenesis, which means that any strains generated with this system would carry a potentially abrogating mutation. Indeed, one of the genes identified by Lowe et al.[71] in their IVET selection was the *geh* gene that encodes lipase (aka glycerol ester hydrolase). Second, it is not necessarily true that the introduction of a gene into *geh* will result in wild-type transcriptional levels. Indeed, when we introduced *cna* into the *geh* locus of 8325-4, we observed a relatively low level of *cna* transcription by comparison to *cna*-positive strains.[8,33,35]

## C. Animal Models of Staphylococcal Disease

Several studies have attempted to assess the contribution of different MSCRAMMs to human infection by correlating binding phenotype with the etiology of disease or the prevalence of antibody in patients suffering from staphylococcal infection.[12,46,113,114,136] These studies are certainly informative, but in the end it is necessary to test any apparent correlation in appropriate animal models. That can be a difficult task given the opportunistic nature of *S. aureus* infection. Although obvious, it is also important to emphasize that all animal models are not the same. That is perhaps most evident in the report of Coulter et al.,[24] who used a signature-tagged mutagenesis IVET system to identify genes specifically induced in vivo and then compared the pool of target genes obtained using each of three animal models (bacteremia, abscess and wound infection). Only 10% (23 of 237) of the induced genes were identified in all three models. Another 63 were identified in two of three models; however, the vast majority (151) were identified in only one model. To date, three different IVET systems have been applied to *S. aureus*.[24,71,80] None of these have identified any of the MSCRAMM genes. However, there could be several explanations for that. For instance, the signature-tagged mutagenesis (STM) protocols[24,80] are functional assays (i.e., does the loss of a given function attenuate the bacterium with respect to survival in vivo?), and the redundancy in the MSCRAMMs may simply mean that mutation of any one gene does not attenuate the bacterium enough to facilitate its detection. In that regard, it is of some interest to note that, while CNA is an apparent exception to the functional redundancy of *S. aureus* MSCRAMMs, all of the IVET protocols have been done with a strain (8325-4) that does not encode *cna*. The protocol used by Lowe et al.[71] avoided the problem of functional redundancy by more directly assessing whether a given gene is expressed in vivo. Since this model should allow detection of any highly expressed gene regardless of when that gene is expressed during the course of infection, the failure to detect MSCRAMM genes would suggest that these genes are not strongly induced in vivo at least by comparison to other genes. Nevertheless, the more direct method of comparing wild-type strains with their isogenic MSCRAMM mutants clearly suggests that the MSCRAMMs make an important contribution to at least some forms of staphylococcal disease. Included among the models used to assess that contribution are animal models of wound infection,[94] bacteremia,[89,135] endocarditis,[44,64,84,119] septic arthritis[100] and osteomyelitis (*see below*). Because some of these models rely heavily on the use of in-dwelling medical devices, it is perhaps worth reiterating that several *S. aureus* MSCRAMMs have also been shown to promote the colonization of biomaterials (Chapter 20).

### 1. Wound Infection

Palma et al.[94] described a model in which an incision was made in the shoulder of male Wistar rats and the submuscular space inoculated with $10^4$ colony-forming units (cfu) of *S. aureus*. This dose was empirically chosen from a range of $10^3$ to $10^8$. Control animals were either infected subcutaneously without prior incision or injected with sterile saline. Rats infected with a wild-type strain (FDA486) or an isogenic *efb* (*fib*) mutant were monitored for 1 wk and evaluated based on the absence of infection vs. mild to severe disease. None of the control animals developed signs of infection while the infection rate was 80% in rats infected with the wild-type strain. The infection rate with the Efb mutant was 57.2%. However, the number of animals with severe signs of disease was decreased from 66.7 to 28.6% by mutation of *efb*. In a parallel comparison using a catheter-induced endocarditis model, there was no difference between the wild-type strain and the *efb* mutant.[94]

## 2. Bacteremia

Bacteremia models have been used for selection in IVET systems[24,71] and to assess the efficacy of immunization.[89] For instance, intravenous injection of *S. aureus* in mice was used to confirm that immunization with CNA conferred protection against septic death.[89] Bacteremia models are generally done by injecting $10^6$–$10^7$ cfu of bacteria into the tail vein of mice. It can also be done by intraperitoneal injection, however, in that case, it is generally necessary to increase the dose dramatically ($10^8$–$10^{10}$ cfu). Tissues used to evaluate the infection include the spleen, liver and kidneys. Hematogenous seeding in mice has also been used in murine models of septic arthritis[100] and, more recently, osteomyelitis *(see below)*.[13,158] Importantly, the septic arthritis studies confirmed that CNA is a relevant virulence factor.

## 3. Endocarditis

Endocarditis studies have been done in rabbits[28,64,84,119] and rats.[44] In all cases, a catheter is inserted through the right carotid artery, across the aortic valve and into the left ventricle. The catheter is usually left in place for 24–30 h.[44,64] However, catheterization for as little as 1 h results in the development of nonbacterial thrombotic endocarditis (NBTE) characterized by the deposition of platelets and fibrin on the damage heart valve.[44,119] In some cases, the catheter is left in place throughout the experiment.[28,84] Animals are infected intravenously with $10^6$–$10^8$ cfu.[44,64] Results are assessed both as infection rate and the density of bacterial vegetations.[28,44] Flock et al.[28] derived an adherence rate based on the number of bound bacteria vs. the number of bacteria injected. There is conflicting data concerning the role of the FnBPs in endocarditis.[28,64,119] Other studies suggest that CNA[44] and ClfA[84] play an important role.

## 4. Septic Arthritis

The only MSCRAMM that has been specifically evaluated using a septic arthritis model is CNA. Patti et al.[100] generated a *cna* mutant (PH100) in Phillips by allele replacement and then compared the wild-type and mutant strains using a murine model. They also did the corresponding experiment of introducing *cna* into the chromosome of the *cna*-negative strain CYL316. Neither the mutation of *cna* or the introduction of *cna* had any significant impact on the ability to bind fibronectin, fibrinogen or thrombospondin.[100] To evaluate the impact on the pathogenesis of septic arthritis, mice were inoculated intravenously with $10^7$ cfu and then evaluated based both on infection rate and joint histopathology. When *cna* was mutated, the infection rate was reduced from 70 to 27.2%. When it was introduced, the infection rate was increased from 33.3 to 76.5%. Histological analysis of the CYL316 derivative (CYL574) was not done. However, 100% of the mice infected with Phillips had histological signs of arthritis while that was true of only 50% of the mice infected with PH100.

## 5. Osteomyelitis

It is extremely difficult to cause experimental osteomyelitis. A number of models have been described, most of which employ the use of schlerosing agents and/or implants.[128] In almost all cases, these models have been used to characterize the disease process or to evaluate therapeutic protocols rather than to make comparisons between strains of *S. aureus*. In such cases, it is often sufficient to utilize the direct approach of introducing bacteria directly into the bone. For instance, Gracia et al.[36] drilled a hole in the tibia and then inoculated the cavity directly with either free bacteria or adherent bacteria attached to a metal rod. In our case, we use a rabbit model in which a 1 cm midradial segment is removed, inoculated with bacteria and then replaced. Although our model does not employ sclerosing agents or biomaterial implants, it does involve the direct introduction of bacteria into devascularized bone. Nevertheless, we have demonstrated the time and dose-dependent nature of this model and have demonstrated strain-dependent differences in infectivity.[128] We have also used the model to confirm that CNA is present on the surface of *S. aureus* cells growing in bone.[35] To date, however, we have been unable to demonstrate a reproducible difference between Phillips and its *cna*-negative derivative PH100. On the other hand, we have done experiments in which we immunized rabbits with the CNA ligand-binding A domain and then challenged by intravenous infection. In an effort to facilitate the development of osteomyelitis, we did a sham surgery immediately prior to infection. Importantly, four of six unimmunized rabbits died within 24 h of hematogenous infection while all but one of the immunized rabbits survived. Bacteria were also isolated from the bone defect in half of the unimmunized rabbits but only 17% of the immunized group (data not shown). A mouse model of hematogenous osteomyelitis was recently described in which an incomplete cartilaginous fracture of the right proximal tibial growth plate was used to facilitate infection.[13] This model is relatively unique and may prove particularly useful since it was possible to cause osteomyelitis in a significant number of animals via hematogenous delivery of bacteria. The use of mice also has the advantage of allowing for more detailed immunological studies given the availability of murine reagents.[158]

**Acknowledgments:** Every effort has been made to recognize the people responsible for the work summarized in this chapter. I sincerely apologize to anyone who thinks that effort has fallen short of its objective. I would also like to express my appreciation to past and present members of my laboratory including Jim Snodgrass, Jon Blevins, Karen Beenken and Frankie Pratt. Dr. Allison Gillaspy deserves special recognition for her contribution to the study of CNA. Thanks are also due Marcella Gardner, whose help with formatting is the only reason I still have an intact computer. Dr. Carl L. Nelson, who chairs the Department of Orthopaedic Surgery at the University of Arkansas for Medical Sciences (UAMS) deserves special recognition as does Robert A. Skinner, Charles Stewart and other members of the UAMS Orthopaedic Research Section. Very special thanks are due Dr. Joseph M. Patti (Inhibitex, Alpharetta, GA) for his careful review of the manuscript and his helpful comments, almost all of which were taken to heart. The willingness of Drs. Greg Bohach and Ken Bayles (University of Idaho, Moscow, ID) to share unpublished data is also greatly appreciated. Funding for the author's laboratory has been provided by the National Institute of Allergy and Infectious Disease, the Orthopaedic Research Society and the Arkansas Science and Technology Association.

# REFERENCES

1. Altieri DC, Plescia J, Plow EF: The structural motif glycine 190-valine 202 of the fibrinogen γ chain interacts with CD11b/CD18 integrin ($\alpha_M\beta_2$, Mac-1) and promotes leukocyte adhesion. *J Biol Chem* 268:1847–53, 1993

2. Aly R, Shinefield HR, Litz C, et al: Role of teichoic acid in the binding of *Staphylococcus aureus* to nasal epithelial cells. *J Infect Dis* 141:464–5, 1979

3. Arbeit RD, Karakawa WW, Vann WF, et al: Predominance of two newly described capsular polysaccharide types among clinical isolates of *Staphylococcus aureus*. *Diagn Microbiol Infect Dis* 2:85–91, 1984

4. Augustin J, Gotz F: Transformation of *Staphylococcus epidermidis* and other staphylococcal species with plasmid DNA by electroporation. *FEMS Microbiol Lett* 66:203–8, 1990

5. Bayer AS, Sullam PM, Ramos M, et al: *Staphylococcus aureus* induces platelet aggregation via a fibrinogen-dependent mechanism which is independent of principal platelet glycoprotein independent of principal platelet glycoprotein IIb/IIIa fibrinogen-binding domains. *Infect Immun* 63:3634–41, 1995

6. Bayles KW, Wesson CA, Liou LE, et al: Intracellular *Staphylococcus aureus* escapes the endosome and induces apoptosis in epithelial cells. *Infect Immun* 66:336–42, 1998

7. Bisognano C, Vaudaux PE, Lew DP, et al: Increased expression of fibronectin-binding proteins by fluoroquinolone-resistant *Staphylococcus aureus* exposed to subinhibitory levels of ciprofloxin. *Antimicrob Agents Chemotherapy* 41:906–13, 1997

8. Blevins JS, Gillaspy AF, Rechtin TM, et al: The staphylococcal accessory regulator (*sar*) represses transcription of the *Staphylococcus aureus* collagen adhesin gene (*agr*) in an *agr*-independent manner. *Mol Microbiol* 33:317–26, 1999

9. Boden MK, Flock JI: Evidence for three different fibrinogen-binding proteins with unique properties from *Staphylococcus aureus* strain Newman. *Microb Pathog* 12:289–98, 1992

10. Boden MK, Flock JI: Cloning and characterization of a gene for a 19 kDa fibrinogen-binding protein from *Staphylococcus aureus*. *Mol Microbiol* 12:599–606, 1994

11. Brennan FR, Jones TD, Longstaff M, et al: Immunogenicity of peptides derived from a different fibronectin-binding protein of *Staphylococcus aureus* expressed on two different plant viruses. *Vaccine* 17:1846–57, 1999

12. Casolini F, Visai L, Joh D, et al: Antibody response to fibronectin-binding adhesin FnbpA in patients with *Staphylococcus aureus* infections. *Infect Immun* 66:5433–42, 1998

13. Chadha HS, Fitzgerald RH, Jr, Wiater P, et al: Experimental acute hematogenous osteomyelitis in mice. I. Histopathological and immunological findings. *J Orthop Res* 17: 376–81, 1999

14. Chan PF, Foster SJ: Role of SarA in virulence determinant production and environmental signal transduction in *Staphylococcus aureus*. *J Bacteriol* 180:6232–1, 1998

15. Cheung AL, Fischetti VA: Variation in the expression of cell wall proteins of *Staphylococcus aureus* grown on solid and liquid media. *Infect Immun* 56:1061–5, 1988

16. Cheung AL, Fischetti VA: The role of fibrinogen in mediating staphylococcal adherence to fibers. *J Surg Res* 50:150–5, 1991

17. Cheung AL, Krishnan M, Jaffe EA, et al: Fibrinogen acts as a bridging molecule in the adherence of *Staphylococcus aureus* to cultured human endothelial cells. *J Clin Invest* 87:2236–45, 1991

18. Cheung AL, Projan SJ, Edmiston CE Jr, et al: Cloning, expression, and nucleotide sequence of a *Staphylococcus aureus* gene (*fbpA*) encoding a fibrinogen-binding protein. *Infect Immun* 63:1914–20, 1995

19. Cheung AL, Ying P: Regulation of α- and β-hemolysins by the *sar* locus of *Staphylococcus aureus*. *J Bacteriol* 176:580–5, 1994

20. Chhatwal GS, Preissner KT, Muller-Berghaus KT, et al: Specific binding of the human S protein (vitronectin) to streptococci, *Staphylococcus aureus*, and *Escherichia coli*. *Infect Immun* 55:1878–83, 1987

21. Chien Y, Cheung AL: Molecular interactions between two global regulators, *sar* and *agr*, in *Staphylococcus aureus*. *J Biol Chem* 273:2645–52, 1998

22. Chien Y, Manna AC, Cheung, AL: SarA level is a determinant of *agr* activation in *Staphylococcus aureus*. *Mol Microbiol* 30:991–1001, 1998

23. Ciborowski P, Flock JI, Wadstrom T: Immunological response to a *Staphylococcus aureus* fibronectin-binding protein. *J Med Microbiol* 37:376–81, 1992

24. Coulter SN, Schwan WR, Ng EY, et al: *Staphylococcus aureus* genetic loci impacting growth and survival in multiple infection environments. *Mol Microbiol* 30:393–404, 1998

25. Dickinson RB, Nagel JA, McDevitt D, et al: Quantitative comparison of clumping factor- and coagulase-mediated *Staphylococcus aureus* adhesion surface-bound fibrinogen under flow. *Infect Immun* 63:3143–50, 1995

26. Dziewanowska K, Patti JM, Deobald CF, et al: Fibronectin binding protein and host cell tyrosine kinase are required for internalization of *Staphylococcus aureus* by epithelial cells. *Infect Immun* 67:4673–4678

27. Flock JI, Froman G, Jonsson K, et al: Cloning and expression of the gene for fibronectin-binding protein from *Staphylococcus aureus*. *EMBO J* 6:2351–7, 1987

28. Flock JI, Hienz SA, Heimdahl A, et al: Reconsideration of the role of fibronectin binding in endocarditis caused by *Staphylococcus aureus*. *Infect Immun* 64:1876–8, 1996

29. Foster TJ: Molecular genetic analysis of staphylococcal virulence. *Methods Microbiol* 27:433–54, 1998

30. Foster TJ, Hook M: Surface protein adhesins of *Staphylococcus aureus*. *Trends Micro* 6:484–8, 1998

31. Fröman G, Switalski LM, Speziale P, et al: Isolation and characterization of a fibronectin receptor from *Staphylococcus aureus*. *J Biol Chem* 262:6564–6571, 1987

32. Fujigaki Y, Froman G, Jonsson K, et al: Glomerular injury induced by cationic 70kD staphylococcal protein; specific immune response is not involved in early phase in rats. *J Bacteriol* 181:2840–5, 1996

33. Gillaspy AF, Lee CY, Sau S, et al: Factors affecting the collagen binding capacity of *Staphylococcus aureus*. *Infect Immun* 66:3170–8, 1998

34. Gillaspy AF, Patti JM, Pratt FL Jr, et al: The *Staphylococcus aureus* collagen adhesin-encoding gene (*cna*) is within a discrete genetic element. *Gene* 196:239–48, 1997

35. Gillaspy AF, Patti JM, Smeltzer MS: Transcriptional regulation of the *Staphylococcus aureus* collagen adhesin gene, *cna*. *Infect Immun* 65:1536–40, 1997

36. Gracia E, Lacleriga A, Monzon M, et al: Application of a rat osteomyelitis model to compare *in vivo* and *in vitro* the antibiotic efficacy against bacteria with high capacity to form biofilms. *J Surg Res* 79:146–153, 1998

37. Greene C, McDevitt D, Francois P, et al: Adhesion properties of mutants of *Staphylococcus aureus* defective in fibronectin-binding proteins and studies on the expression of fnb genes. *Mol Microbiol* 17:1143–52, 1995

38. Hartford O, Francois P, Vaudaux P, et al: The dipeptide repeat region of the fibrinogen-binding protein (clumping factor) is required for functional expression of the fibrinogen-binding domain on the *Staphylococcus aureus* cell surface. *Mol Microbiol* 25:1065–76, 1997

39. Hawiger J: Adhesive ends of fibrinogen and its antiadhesive peptides: the end of a saga? *Semin Hematol* 32:99–109, 1995

40. Herrmann M, Hartleib J, Kehrel B, et al: Interaction of von Willebrand factor with *Staphylococcus aureus*. *J Infect Dis* 176:984–91, 1997

41. Herrmann M, Suchard SJ, Boxer LA, et al: Thrombospondin binds to *Staphylococcus aureus* and promotes staphyloccal adherence to surfaces. *Infect Immun* 59:279–88, 1991

42. Herrmann M, Vaudaux PE, Pittet D, et al: Fibronectin, fibrinogen, and laminin act as mediators of adherence of clinical staphylococcal isolates to foreign material. *J Infect Dis* 158:693–701, 1988

43. Hienz SA, Palma M, Flock JI: Insertional inactivation of the gene for collagen-binding protein has a pleiotropic effect on the phenotype of *Staphylococcus aureus. J Bacteriol* 178:5327–9, 1996

44. Hienz SA, Schennings T, Heimdahl A, et al: Collagen binding of *Staphylococcus aureus* is a virulence factor in experimental endocarditis. *J Infect Dis* 174:83–8, 1996

45. Hildén P, Savolainen K, Tyynelä J, et al: Purification and characterization of a plasmin-sensitive surface protein of *Staphylococcus aureus. Eur J Biochem* 236:904-10, 1996

46. Holderbaum D, Spech T, Ehrhart LA, et al: Collagen binding in clinical isolates of *Staphylococcus aureus. J Clin Microbiol* 25:2258–61, 1987

47. House-Pompeo K, Xu Y, Joh D, et al: Conformational changes in the fibronectin binding of MSCRAMMs are induced by ligand binding. *J Biol Chem* 271:1379–84, 1996

48. Huff S, Matsuka YV, McGavin MJ, et al: Interaction of N-terminal fragments of fibronectin with synthetic and recombinant D motifs from its binding protein on *Staphylococcus aureus* studied using fluorescence anisotropy. *J Biol Chem* 269:15563–70, 1994

49. Iandolo JJ, Bannantine JP, Stewart GC: Genetic and physical map of the chromosome of *Staphylococcus aureus*. In: Crossley KB, Archer GL, eds: *The Staphylococci in Human Disease,* New York, NY, 1997:39–53

50. Jevon M, Guo C, Ma B, et al: Mechanism of internalization of *Staphylococcus aureus* by cultured human osteoblasts. *Infect Immun* 67:2677–81, 1999

51. Joh D, Speziale P, Gurusiddappa S, et al: Multiple specificities of the staphylococcal and streptococcal fibronectin-binding microbial surface components recognizing adhesive matrix molecules. *Eur J Biochem* 258:897–905, 1998

52. Joh HJ, House-Pompeo K, Patti JM, et al: Fibronectin receptors from Gram-positive bacteria: comparison of active sites. *Biochemistry* 33:6086–92, 1999

53. Jonsson K, McDevitt D, McGavin MH, et al: *Staphylococcus aureus* expresses a major histocompatibility complex class II analog. *J Biol Chem* 270:21457–60, 1995

54. Jonsson K, Signas C, Muller HP, et al: Two different genes encode fibronectin binding proteins in *Staphylococcus aureus*: The complete nucleotide sequence and characterization of the second gene. *Eur J Biochem* 202:1041–8, 1991

55. Josefsson E, McCrea KW, Eidhin DN, et al: Three new members of the serine-aspartate repeat protein multigene family of *Staphylococcus aureus. Microbiology* 144:3387–95, 1998

56. Josefsson E, O'Connell D, Foster TJ, et al: The binding of calcium to the B-repeat segment of SdrD, a cell surface protein of *Staphylococcus aureus. J Biol Chem* 273:31145–52, 1998

57. Kanzaki H, Morishita Y, Akiyama H, et al: Adhesion of *Staphylococcus aureus* to horny layer: role of fibrinogen. *J Dermatol Sci* 12:132–9, 1996

58. Kloczewiak M, Timmons S, Lukas TJ, et al: Platelet receptor recognition site on human fibrinogen. Synthesis and structure-function relationship of peptides corresponding to the carboxy-terminal segment of the γ chain. *Biochemistry* 23:1767–74, 1984

59. Kloos W: Taxonomy and systematics of staphylococci indigenous to humans. In: Crossley KB, Archer GL, eds: *The Staphylococci in Human Disease,* New York, NY, 1997:113–37

60. Kloos WE, Bannerman TL: Update on clinical significance of coagulase-negative staphylocci. *Clin Microbiol Rev* 7:117–40, 1994

61. Kraemer GR, Iandolo JJ: High-frequency transformation of *Staphylococcus aureus* by electroporation. *Curr Microbiol* 21:373–6, 1995

62. Krajewska-Pietrasik D, Sobis-Glinkowska M, Sidorczyk Z, et al: The influence of iron occuring in the growth medium of *Staphylococcus aureus* on the bacterial adhesion to collagen. *Acta Microbiol Polonica* 47:349–56, 1997

63. Kuusela P, Hildén P, Savolainen K, et al: Rapid detection of methicillin-resistant *Staphylococcus aureus* strains not identified by slide agglutination tests. *J Clin Microbiol* 32:143–7, 1994

64. Kuypers JM, Proctor RA: Reduced adherence to traumatized rat heart valves by a low-fibronectin-binding mutant of *Staphylococcus aureus. Infect Immun* 57:2306–12, 1989

65. Layton MC, Patterson JE: Mupirocin resistance among consecutive isolates of oxacillin-resistant and borderline oxacillin-resistant *Staphylococcus aureus* at a university hospital. *Antimicrob Agents Chemother* 38:1664–7, 1994

66. Lee CY, Buranen SL, Ye ZH: Construction of single-copy integration vectors for *Staphylococcus aureus*. *Gene* 103:101–5, 1991

67. Lee JC, Takeda S, Livolsi PJ, et al: Effects of *in vitro* and *in vivo* growth conditions on expression of type 8 capsular polysaccharide by *Staphylococcus aureus*. *Infect Immun* 61:1853–8, 1993

68. Li Z, Hook M, Patti JM, et al: The effect of shear stress on the adhesion of *Staphylococcus aureus* to collagen I, II, IV and VI. *Annals Biomed Eng* 24:S–55, 1996

69. Liang OD, Flock JI, Wadstrom T: Evidence that the heparin-binding consensus sequence of vitronectin is recognized by *Staphylococcus aureus*. *J Biochem* 116:457–63, 1994

70. Lopes JD, Dos Reis M, Brentani RR: Presence of laminin receptors in *Staphylococcus aureus*. *Science* 229:275–7, 1985

71. Lowe AM, Beattie DT, Deresiewicz RL: Identification of novel staphylococcal virulence genes by *in vivo* expression technology. *Mol Microbiol* 27:967–76, 1998

72. Lowy FD: Medical progress - *Staphylococcus aureus* infections. *N Engl J Med* 339:520–32, 1998

73. McDevitt D, Foster TJ: Variation in the size of the repeat region of the fibrinogen receptor (clumping factor) of *Staphylococcus aureus* strains. *Mol Microbiol* 141:937–43, 1995

74. McDevitt D, Francois P, Vaudaux P, et al: Molecular characterization of the clumping factor (fibrinogen receptor) of *Staphylococcus aureus*. *Mol Microbiol* 11:237–48, 1994

75. McDevitt D, Nanavaty T, House-Pompeo K, et al: Characterization of the interaction between the *Staphylococcus aureus* clumping factor (ClfA) and fibrinogen. *Eur J Biochem* 247:416–24, 1997

76. McGavin MH, Krajewska-Pietrasik D, Ryden C, et al: Identification of a *Staphylococcus aureus* extracellular matrix-binding protein with broad specificity. *Infect Immun* 61:2479–85, 1993

77. McGavin MJ, Gurusiddappa S, Lindgren P-E, et al: Fibronectin receptors from *Streptococcus dysgalactiae* and *Staphylococcus aureus*. *J Biol Chem* 268:23946–53, 1993

78. McGavin MJ, Raucci G, Gurusiddappa S, et al: Fibronectin binding determinants of the *Staphylococcus aureus* fibronectin receptor. *J Biol Chem* 266:8343–7, 1991

79. McGavin MJ, Zahradka C, Rice K, et al: Modification of the *Staphylococcus aureus* fibronectin binding phenotype by V8 protease. *Infect Immun* 65:2621–8, 1997

80. Mei J-M, Nourbakhsh F, Ford CW, et al: Identification of *Staphylococcus aureus* virulence gene in a murine model of bacteraemia using signature-tagged mutagenesis. *Mol Microbiol* 26:399–407, 1997

81. Mempel M, Schmidt T, Weidinger S, et al: Role of *Staphylococcus aureus* surface-associated proteins in the attachment to cultured HaCaT keratinocytes in a new adhesion assay. *J Invest Dermatol* 111:452–6, 1999

82. Menzies BE, Kourteva I: Internalization of *Staphylococcus aureus* by endothelial cells induces apoptosis. *Infect Immun* 66:5994–8, 1999

83. Mohamed N, Teeters MA, Patti JM, et al: Inhibition of *Staphylococcus aureus* adherence to collagen under dynamic conditions. *Infect Immun* 67:589–94, 1999

84. Moreillon P, Entenza JM, Francioli P, et al: Role of *Staphylococcus aureus* coagulase and clumping factor in pathogenesis of experimental endocarditis. *Infect Immun* 63:4738–43, 1995

85. Morfeldt E, Tegmark K, Arvidson S: Transcriptional control of the *agr*-dependent virulence gene regulator, RNAIII, in *Staphylococcus aureus*. *Mol Microbiol* 21:1227–37, 1996

86. Navarre WW, Schneewind O: Surface proteins of gram-positive bacteria and mechanisms of their targeting to the cell wall envelope. *Microbiol Mol Biol Rev* 63:174–229, 1999

87. Navarre WW, Ton-That H, Faull KF, et al: Anchor structure of staphylococcal surface proteins. *J Biol Chem* 273:29135–42, 1998

88. Ni Eldhin D, Perkins S, Francois P, et al: Clumping factor B (ClfB), a new surface-located fibrinogen-binding adhesin of *Staphylococcus aureus*. *Mol Microbiol* 30:245–57, 1998

89. Nilsson I-M, Patti JM, Bremell T, et al: Vaccination with a recombinant fragment of collagen adhesin provides protection against *Staphylococcus aureus*-mediated septic death. *J Clin Invest* 101:2640–9, 1998

90. Nilsson M, Frykberg L, Flock J-I, et al: A fibrinogen-binding protein of *Staphylococcus epidermidis*. *Infect Immun* 66:2666–73, 1998

91. Novick RP: Staphylococcal plasmids and their replication. *Ann Rev Microbiol* 43:537–65, 1989

92. O'Connel DP, Natavaty T, McDevitt D, et al: The fibrinogen-binding MSCRAMM (clumping factor) of *Staphylococcus aureus* has a $Ca^{2+}$-dependent inhibitory site. *J Biol Chem* 273:6821–9, 1998

93. Palma M, Haggar A, Flock J-I: Adherence of *Staphylococcus aureus* is enhanced by an endogeous secreted protein with broad binding activity. *J Bacteriol* 181:2840–5, 1999

94. Palma M, Nozohoor S, Schennings T, et al: Lack of the extracellular 19-kilodalton fibrinogen-binding protein from *Staphylococcus aureus* decreases virulence in experimental wound infection. *Infect Immun* 64:5284–9, 1996

95. Park PW, Broekelmann TJ, Mecham BR, et al: Characterization of the elastin binding domain in the cell-surface 25-kDa elastin-binding protein of staphylococcus aureus. *J Biol Chem* 274:2845–50, 1999

96. Park PW, Roberts DD, Grosso LE, et al: Binding of elastin to *Staphylococcus aureus*. *J Biol Chem* 266:23399–406, 1991

97. Park PW, Rosenbloom J, Abrams WR, et al: Molecular cloning and expression of the gene for elastin-binding protein (*ebpS*) in *Staphylococcus aureus*. *J Biol Chem* 271:15803–9, 1996

98. Patti JM, Allen BL, McGavin MJ, et al: MSCRAMM-mediated adherence of microorganisms to host tissue. *Ann Rev Microbiol* 48:585–617, 1994

99. Patti JM, Boles JO, Hook M: Identification and biochemical characterization of the ligand binding domain of the collagen adhesin from *Staphylococcus aureus*. *Biochemistry* 32:11428–35, 1993

100. Patti JM, Bremell T, Krajewska-Pietrasik D, et al: The *Staphylococcus aureus* collagen adhesin is a virulence determinant in experimental septic arthritis. *Infect Immun* 62:152–61, 1994

101. Patti JM, House-Pompeo K, Boles JO, et al: Critical residues in the ligand-binding site of the *Staphylococcus aureus* collagen-binding adhesin (MSCRAMM). *J Biol Chem* 270: 12005–11, 1995

102. Patti JM, Jonsson H, Guss B, et al: Molecular characterization and expression of a gene encoding a *Staphylococcus aureus* collagen adhesin. *J Biol Chem* 267:4766–72, 1992

103. Patti JM, Schneider A, Garza N, et al: Isolation and characterization of *pcp*, a gene encoding a pyrrolidone carboxyl peptidase in *Staphylococcus aureus*. *Gene* 166:95–9, 1995

104. Paulsson M, Wadstrom T: Vitronectin and type-I collagen binding by *Staphylococcus aureus* and coagulase-negative staphylococci. *FEMS Microbiol Immun* 65:55–62, 1990

105. Projan SJ, Novick RP: The molecular basis of pathogenicity. In: Crossley KB, Archer GL: *The Staphylococci in Human Disease*, New York, NY, 1997:55–81

106. Raja RH, Raucci G, Hook M: Peptide analogs to a fibronectin receptor inhibit attachment of *Staphylococcus aureus* to fibronectin-containing substrates. *Infect Immun* 58:2593–8, 1990

107. Reilly SS, Ramp WK, Zane SF, et al: Internalization of *Staphylococcus aureus* by embryonic chicken osteoblasts *in vivo*. *J Bone Miner Res* 12:S231, 1997

108. Rich RL, Demeler B, Ashby K, et al: Domain structure of the *Staphylococcus aureus* collagen adhesin. *Biochemistry* 37:15423–33, 1998

109. Rupp ME, Archer GL: Coagulase-negative staphylococci: Pathogens associated with medical progress. *Clin Infect Dis* 19:231–245, 1994

110. Ryden C, Tung HS, Nikolaev V, et al: *Staphylococcus aureus* causing osteomyelitis binds to a nonapeptide sequence in bone sialoprotein. *Biochem J* 327:825–9, 1997

111. Ryden C, Yacoub AI, Maxe I, et al: Selective binding of bone matrix sialoprotein to *Staphylococcus aureus* in osteomyelitis. *Lancet* 2:515–515, 1987

112. Ryden C, Yacoub AI, Maxe I, et al: Specific binding of bone sialoprotein to *Staphylococcus aureus* isolated from patients with osteomyelitis. *Eur J Biochem* 184:331–6, 1989

113. Ryding U, Christensson B, Soderquist B, et al: Antibody response to *Staphylococcus aureus* collagen binding protein in patients with *S. aureus* septicaemia and collagen binding properties of corresponding strains. *J Med Microbiol* 43:328–34, 1995

114. Ryding U, Flock JI, Flock M, et al: Expression of collagen-binding protein and types 5 and 8 capsular polysaccharide in clinical isolates of *Staphylococcus aureus*. *J Infect Dis* 176:1096–9, 1997

115. Sanford BA, Thomas VL, Ramsay MA: Binding of staphylococci to mucus *in vivo* and *in vitro*. *Infect Immun* 57:3735–42, 1989

116. Sau S, Bhasin N, Wann ER, et al: The *Staphylococcus aureus* allelic genetic loci for serotype 5 and serotype 8 capsule expression contain the type-specific genes flanked by common genes. *Microbiology* 143:2395–405, 1997

117. Savage B, Bottini E, Ruggeri ZM: Interaction of integrin $\alpha_{IIb}\beta_3$ with multiple fibrinogen domains during platelet adhesion. *J Biol Chem* 270:28812–7, 1995

118. Schaferkordt S, Domann E, Chakraborty T: Molecular approaches for the study of *Listeria*. *Methods Microbiol* 27:421–31, 1998

119. Scheld WM, Strunk RW, Balian G, et al: Microbial adhesion to fibronectin *in vitro* correlates with production of endocarditis in rabbits. *Proc Soc Exp Biol Med* 180:474–82, 1985

120. Schenk S, Laddaga RA: Improved method for electroporation of *Staphylococcus aureus*. *FEMS Microbiol Lett* 94:133–8, 1992

121. Schneewind O, Fowler A, Faull KF: Structure of the cell wall anchor of surface proteins in *Staphylococcus aureus*. *Science* 268:103–6, 1995

122. Schneewind O, Mihaylora-Petkov D, Model P: Cell wall sorting signals in surface proteins of Gram-positive bacteria. *EMBO J* 12:4803–11, 1993

123. Shuter J, Hatcher VB, Lowy FD: *Staphylococcus aureus* binding to human nasal mucin. *Infect Immun* 64:310–8, 1996

124. Signas C, Raucci G, Jonsson K, et al: Nucleotide sequence of the gene for a fibronectin-binding protein from *Staphylococcus aureus*: Use of this peptide sequence in the synthesis of biologically active peptides. *Proc Nat Acad Sci USA* 86:699–703, 1989

125. Smeltzer MS, Gillaspy AF, Pratt FL Jr, et al: Prevalence and chromosomal map location of *Staphylococcus aureus* adhesin genes. *Gene* 196:249–59, 1997

126. Smeltzer MS, Hart ME, Iandolo JJ: Phenotypic characterization of *xpr*, a global regulator of extracellular virulence factors in *Staphylococcus aureus*. *Infect Immun* 61:919–25, 1993

127. Smeltzer MS, Hart ME, Iandolo JJ: The effect of lysogeny on the genomic organization of *Staphylococcus aureus*. *Gene* 138:51–57, 1994

128. Smeltzer MS, Thomas JR, Hickmon SG, et al: Characterization of a rabbit model of staphylococcal osteomyelitis. *J Orthop Res* 15:414–21, 1997

129. Snodgrass JL, Mohamed N, Ross JM, et al: Functional analysis of the *Staphylococcus aureus* collagen adhesin B domain. *Infect Immun* 67:3952–9, 1999

130. Sottile J, Schwarzbauer J, Selegue J, et al: Five type I modules of fibronectin form a functional unit that binds to fibroblast and *Staphylococcus aureus*. *J Biol Chem* 266:12840–3, 1991

131. Sun Q, Smith GM, Zahradka C, et al: Identification of D motif epitopes in *Staphylococcus aureus* fibronectin-binding protein for the production of antibody inhibitors of fibronectin binding. *Infect Immun* 65:537–43, 1997

132. Switalski LM, Patti JM, Butcher WG, et al: A collagen receptor on *Staphylococcus aureus* strains isolated from patients with septic arthritis mediates adhesin to cartilage. *Mol Microbiol* 7:99–107, 1993

133. Switalski LM, Speziale P, Hook M: Isolation and characterization of a putative collagen receptor from *Staphylococcus aureus* strain Cowan 1. *J Bacteriol* 264:21080–6, 1989

134. Symersky J, Patti JM, Carson M, et al: Structure of the collagen-binding domain from a *Staphylococcus aureus* adhesin. *Nature Struct Biol* 4:833–8, 1997

135. Thakker M, Park J-S, Carey V, et al: *Staphylococcus aureus* serotype 5 capsular polysaccharide is antiphagocytic and enhances bacterial virulence in a murine bacteremia model. *Infect Immun* 66:5183–9, 1998

136. Thomas MG, Peacock S, Daenke S, et al: Adhesion of *Staphylococcus aureus* to collagen is not a major virulence determinant for septic arthritis, osteomyelitis, or endocarditis. *J Infect Dis* 179:291–3, 1999

137. Thomas VL, Sanford BA, Ramsay MA: Calcium- and mucin-binding proteins of staphylococci. *J Gen Microbiol* 139:623–9, 1993

138. Tompkins DC, Blackwell LJ, Hatcher VB, et al: *Staphylococcus aureus* proteins that bind to human endothelial cells. *Infect Immun* 60:965–9, 1992

139. Tompkins DC, Hatcher VB, Patel D, et al: A human endothelial cell membrane protein that binds *Staphylococcus aureus in vitro*. *J Clin Invest* 85:1248–54, 1990

140. Ton-That H, Faull K F, Schneewind O: Anchor structure of staphylococcal surface proteins. *J Biol Chem* 272:22285–92, 1997

141. Trivier D, Houdret N, Courcol RJ, et al: The binding of surface proteins from *Staphylococcus aureus* to human bronchial mucins. *Eur Resp J* 10:804–10, 1997

142. Udo EE, Jacob LE, Mokadas EM: Conjugative transfer of high-level mupirocin resistance from *Staphylococcus haemolyticus* to other staphylococci. *Antimicrob Agents Chemother* 41:693–5, 1997

143. van Wamel WJ, Vandenbrucke-Grauls CM, Verhoef J, et al: The effect of culture conditions on the in vitro adherence of methicillin-resistant *Staphylococcus aureus*. *J Med Microbiol* 47:705–9, 1998

144. Vann JM, Hamill RJ, Albrecht RM, et al: Immunoelectron microscopic localization of fibronectin in adherence of *Staphylococcus aureus* to cultured bovine endothelial cells. *J Infect Dis* 160:538–42, 1989

145. Vaudaux P, Pittet D, Haeberli A, et al: Fibronectin is more active than fibrin or fibrinogen in promoting *Staphylococcus aureus* adherence to inserted intravascular catheters. *J Infect Dis* 167:633–41, 1993

146. Vaudaux PE, Francois P, Proctor RA, et al: Use of adhesin-defective mutants of *Staphylococcus aureus* to define the role of specific plasma proteins in promoting bacterial adhesion to canine arteriovenous shunts. *Infect Immun* 63:585–90, 1995

147. Vaudaux PE, Monzillo V, Francois P, et al: Introduction of the *mec* element (methicillin resistance) into *Staphylococcus aureus* alters *in vitro* functional activities of fibrinogen and fibronectin adhesins. *Antimicrob Agents Chemother* 42:564–70, 1999

148. Vaudaux PE, Pittet D, Haeberli A, et al: Host factors selectively increase staphylococcal adherence on inserted catheters: A role for fibronectin and fibrinogen or fibrin. *J Infect Dis* 160:865–75, 1989

149. Voytek A, Gristina AG, Barth E, et al: Staphylococcal adhesion to collagen in intra-articular sepsis. *Biomaterials* 9:107–10, 1988

150. Wade D, Palma M, Lofving-Arvholm I, et al: Identification of functional domains in Efb, a fibrinogen binding protein of *Staphylococcus aureus*. *Biochem Biophys Res Commun* 248:690–5, 1998

151. Waldvogel FA: *Staphylococcus aureus* (including toxic shock syndrome). In: Mandell GL, Douglas RG Jr, Bennett JE, eds. *Principles and Practice of Infectious Diseases,* New York, NY, 1990:1489–1510

152. Wastfelt MK, Flock J-I: Incidence of the highly conserved *fib* gene and expression of the fibrinogen-binding protein (Fib) among clinical isolates of *Staphylococcus aureus*. *J Clin Microbiol* 33:2347–52, 1995

153. Wilkinson BJ: Biology. In: Crossley KB, Archer GL: *The Staphylococci in Human Disease,* New York, NY, 1997:1–38
154. Wolz C, McDevitt D, Foster TJ, et al: Influence of *agr* on fibrinogen binding in *Staphylococcus aureus* Newman. *Infect Immun* 64:3142–7, 1996
155. Wong AC, Bergdoll MS: Effect of environmental conditions on production of toxic shock syndrome toxin 1 by *Staphylococcus aureus. Infect Immun* 58:1026–9, 1990
156. Yacoub A, Lindahl P, Rubin K, et al: Purification of a bone sialoprotein-binding protein from *Staphylococcus aureus. Eur J Biochem* 222:919–25, 1994
157. Yoon KS, Fitzgerald RH Jr, Sud S, et al: Experimental acute hematogenous osteomyelitis in mice. II. Influence of Staphylococcus aureus infection on T-cell immunity. *J Orthop Res* 17:382–91, 1999
158. Yousif Y, Drager R, Schlitz M, et al: Nucleotide sequence of a *S. aureus* gene encoding outer surface binding 70 kD protein. *National Center for Biotechnology Information Database.* Accession No. Y10419:1997

# 28

# Studying Bacterial Adhesion to Tooth Surfaces

**Shigetada Kawabata,[1] Taku Fujiwara,[2] and Shigeyuki Hamada[1]**

*[1]Department of Oral Microbiology, and [2]Department of Pedodontics, Osaka University Faculty of Dentistry, Osaka, Japan*

## I. INTRODUCTION

Mutans streptococci (MS) have been implicated as primary causative agents of dental caries in humans and experimental animals.[14,15] Among these organisms, *Streptococcus mutans* (serotypes c, e, and f) and *Streptococcus sobrinus* (serotypes d and g) are frequently isolated from human dental plaque.[26] Dental plaque is a complex but typical bacterial biofilm, which contains MS and other oral bacteria and their products. Dental plaque is formed in two different stages; the initial and reversible attachment of oral bacteria (mainly streptococci) to the tooth surface, and subsequent sucrose-dependent firm and irreversible adhesion of MS and other organisms.

The former step is initiated by adsorption of organisms to acquired pellicle on the tooth surface. Salivary receptor molecules are likely to promote the bacterial adhesion. These receptors can influence the adhesion in several ways. They can induce aggregation of oral bacteria:[6,7] such aggregation promotes adhesion of MS to other bacteria, but can also facilitate removal of oral bacteria by swallowing or flushing. Fimbrial cell surface proteins such as PA,[36] I/II,[38] P1,[8] and SR[34] of MS are presumed to participate in the initial attachment of the organisms to the tooth surface via acquired pellicle. In this regard, surface protein is considered to be a protective antigen for dental caries. [20,27]

Glucosyltransferases (GTFs; EC2.4.1.5) are strongly involved in the latter step. These enzymes utilize dietary sucrose as a substrate, and yield free fructose and glucan with predominant a($1 \rightarrow 3$) and a($1 \rightarrow 6$) bonds. Several species of oral streptococci produce GTFs. However, only those of MS cooperatively synthesize adhesive and water-insoluble glucan, resulting in firm adhesion of MS to the tooth surface. It is reported that *S. mutans* and *S. sobrinus* produce three and four kinds of GTFs, respectively.[12,13,48] On the other hand, fructose is used as an energy source for these organisms, and it is metabolized to release lactic acid and other organic acids that serve as erosive agents in the cariogenic process. The biofilm may keep acids inside the structure, which results in localized decalcification of the enamel surfaces.

Molecular biological approaches to the study of MS have produced interesting results. PA-deficient mutants with a disrupted PA gene displayed a low level of hydrophobicity and saliva-mediated adsorption in comparison to their parent strains.[23,36] When the *gtfB*

*Handbook of Bacterial Adhesion: Principles, Methods, and Applications*
Edited by: Y. H. An and R. J. Friedman © Humana Press Inc., Totowa, NJ

and *gtfC* genes coding for GTFs were inactivated by allelic replacement, the resultant mutant strains were less virulent than the wild type strain in the development of experimental dental caries in rat.[30,47] These findings are briefly reviewed in the following sections.

## II. BIOFILM-FORMING FACTORS OF MUTANS STREPTOCOCCI

### A. Molecular Characterization of Cell Surface Components Involved in Adhesion

The hydrophobicity of the bacterial cell surface has been considered to affect several biological phenomena such as the attachment of bacteria to host tissues, adhesion of bacteria to solid surfaces, and interactions between bacteria and phagocytes. *S. mutans* possesses a cell surface protein with a molecular mass of 190 kDa known as PAc (protein antigen serotype *c*). *S. sobrinus* produces a protein antigen with a molecular mass of 210 kDa named PAg (protein antigen serotype g).[35] The PA of mutans streptococci is highly involved in cell surface hydrophobicity. PA-deficient mutants showed less hydrophobicity and were less adsorptive on saliva-coated hydroxyapatite beads (S-HA) than their parent strains.[22,23] Furthermore, isogenic mutants, which synthesized larger amounts of cell-associated PA, exhibited increased hydrophobicity as compared with their wild type strains.[22,42]

The antigenicity, immunogenicity, and protective efficiency of several synthetic peptides derived from the deduced amino acid sequence of the PAc protein have been studied extensively.[41,43] PAc peptide corresponding to amino acid residues 301-319 of the alanine-rich region was strongly reactive with anti-PAc antibody and induced proliferation of T cells from BALB/c mice immunized with the PAc protein. Furthermore, intranasal immunization of mice with the peptide PAc (301-319) coupled to the nontoxic subunit of the cholera toxin molecule elicited strong serum IgG antibody responses to the peptide and suppressed the colonization of *S. mutans* on murine teeth. These results suggest that the alanine-rich region containing residues 301-319 of the PAc protein may be a good immunogen to elaborate antibodies that protect against dental caries.

### B. Preparation of Native GTFs

*S. mutans* produces GTFs in cell-free (CF) and cell-associated (CA) states. CF-GTF synthesizes water-soluble glucan from sucrose. CA-GTF was released by 8 M urea extraction and chromatographically purified. It exclusively forms water-insoluble glucan.[13] CA-GTF is considered to be the most important GTF in the pathogenesis of *S. mutans*, because mutants lacking it do not induce experimental dental caries.[21] Molecular biological studies[9,40,45] revealed that CA-GTF contains GTF-B-synthesizing insoluble glucans as well as GTF-C-synthesizing both insoluble and soluble glucans. Sequencing of the genes encoding GTF-B and GTF-C reveals that they are very similar in terms of nucleotide sequence.[10]

On the other hand, *S. sobrinus* produces at least three kinds of GTFs; one GTF synthesizes water-insoluble glucan from sucrose, the other two synthesize water-soluble glucan.[12] The organisms aggregate upon addition of high molecular weight dextran. *S. sobrinus* grown in sucrose-containing media binds GTFs, while this is not the case when sucrose-free media are used for cultivation of the organisms. GTFs can be isolated and purified from sucrose-free culture supernatant by chromatofocusing with PBE74 (Amersham Pharmacia Biotech, Tokyo, Japan).[12] In the following method section, we focus on the purification of CA-GTF from *S. mutans*.

## C. Function and Role of GTFs in Biofilm Development

The ability of *S. mutans* and other species of MS to adhere firmly to the tooth surface in the presence of sucrose depends on the GTFs of the organisms. Although the initial attachment of MS to saliva-coated enamel surfaces occurs through the surface components of the organisms, the synthesis of water-insoluble, adhesive glucan from sucrose by GTFs is essential for biofilm and dental caries development. In recent years, a variety of methods for purifying GTFs have been reported and several GTFs were purified from mutans streptococci. *S. mutans* possesses three GTF genes, *gtfB*, *gtfC*, *gtfD*.[18,40] GTF-B (GTF-I; 162 kDa) and GTF-C (GTF-SI; 149 kDa) primarily synthesize water-insoluble glucan, while GTF-D (GTF-S; 155 kDa) exclusively synthesizes water-soluble glucan.

The sucrose-binding domain of GTF that functions in the binding and hydrolysis of sucrose is located in the N-terminal two-thirds of the molecule. Conserved amino acid sequences of the catalytic region are DSIRVDAVD for GTF-I and DGIRVDAED for GTF-S.[29] Site-directed mutagenesis of the corresponding Asp residue in the GTF-B enzyme completely inactivated the enzyme.[19] On the other hand, a carboxyl-terminal region constitutes the glucan-binding domain (GBD). Deletion studies of this region showed that GBD was essential for glucan synthesis but not for sucrase activity.[25,32]

## III. METHODS FOR ANALYZING BIOFILMS

### A. Hydrophobicity Assay Using Hydrocarbon

For the estimation of the hydrophobic nature of bacteria, a hexadecane method[37] has been developed. Lyophilized bacterial cells can be used, and experimental time may be saved.

MS organisms are grown at 37°C for 18 h in brain heart infusion broth (Difco Laboratories, Detroit, MI, USA). The organisms are washed twice with phosphate-buffered saline (PBS; pH 7.4) and lyophilized. When live cells are used, this lyophilization step can be omitted. The organisms are suspended in PUM buffer (pH 7.1; 17.0 g $K_2HPO_4$, 7.3 g $KH_2PO_4$, 1.8 g urea, 0.2 g $MgSO_4 \cdot 7H_2O$/1000 mL distilled water) to an optical density of 0.6 at 550 nm. Aliquots (3 mL) of bacterial suspension are transferred into glass tubes (13×100 mm). Hexadecane (0.3 mL) is mixed to a suspension using a Vortex mixer for 1 min. The reaction mixtures are allowed to stand for 15 min at room temperature. The optical density of the lower, aqueous phase is measured. Adsorption is calculated as the percentage loss in optical density relative to that of the initial cell suspension.

### B. Adsorption of MS by Salivary Glycoproteins

Bacterial adhesion is often the result of a specific interaction between the carbohydrate portions of receptor glycoproteins and proteinaceous complexes termed adhesins on bacterial cell surfaces.[33] An example is the interaction between human salivary proteins and the A-repeat portion of PA of MS.[4,31] The assay for bacterial adsorption by S-HA described below is very sensitive.

Whole saliva is collected in an ice-cooled container and clarified by centrifugation (44,000 $g$, 30 min, 4°C). The collected samples are aliquoted and frozen at –20°C until use.[2] To prepare S-HA, spheroidal hydroxyapatite beads (20 mg, BDH Chemicals, Poole, UK) are coated with 1 mL of KCl buffer (pH 6.0; 50 mM KCl in 2 m$M$ potassium phosphate buffer) or diluted whole saliva (1 part saliva to 5 parts KCl buffer) for 1 h at room temperature and washed three times with KCl buffer. Organisms are grown at 37°C for 18 h in a chemically defined medium[44] containing [*methyl*-³H]thymidine (2.22 TBq [60 Ci]/mmol; ICN Biomedicals, Costa Mesa, CA, USA) at a final concentration of 370 kBq (10 µCi)/mL. [³H]thymidine-labeled bacteria ($5\times10^7$) are incubated with S-HA in 1 mL of KCl buffer for 2 h at 37°C with continuous rotation. The S-HA beads are washed twice with KCl buffer to remove unattached cells. The radioactivity associated with the S-HA beads is estimated using liquid scintillation spectroscopy. The number of bacteria adsorbed by the S-HA is estimated from the calculated specific radioactivity of the bacteria.

## C. Measurement of GTF Activity

GTF activity is estimated by [¹⁴C]glucan synthesis from [¹⁴C]sucrose using [¹⁴C-*glucose*] sucrose (1.85 MBq [50 µCi]/mmol, Amersham Pharmacia Biotech).[13,21] The standard reaction mixture contains 10 mM [¹⁴C-*glucose*] sucrose with or without 20 µM dextran T10 in 20 µL of 0.1 M potassium phosphate buffer (pH 6.0).

The standard mixture and GTF samples are mixed and incubated for 1 h at 37°C using a round-bottomed 96-well plate. They are spotted on a filter-paper square (1.5×1.0 cm) and dried in air. The squares are washed three times by stirring with methanol and dried in air. To determine the amount of [¹⁴C]glucan synthesized, the filters are immersed in scintillation fluid, and are quantitated by a liquid scintillation counter. One unit of GTF activity is defined as the amount of enzyme incorporating 1 µmole of glucose from a sucrose molecule into glucan/min under the conditions described above.

## D. Sucrose-Dependent Cell Adhesion to Glass Surfaces

The number of adherent MS cells was determined turbidimetrically, and expressed as a percentage of the total cell mass (percentage cell adherence) as follows.[16]

MS are grown at 37°C at a 30° angle to the horizontal for 18 h in 3 mL of BHI broth containing 1% sucrose. The use of 13×100 mm disposable glass tubes (Corning Inc, Corning, NY) is recommended for this assay. Organisms that form a biofilm on the surface of the tube are removed by rubber scraper and dispersed by ultrasonication (total cell mass). The sample tubes are vigorously vibrated with a Vortex mixer for 3 s. Supernatant containing detached glucans and bacteria (unadhered cell mass) is ultrasonicated, and the absorbance at 550 nm is measured in a spectrophotometer. Following measurement of $A_{550}$ using BHI broth as background, the percentage of adhesion is defined as $100 \times A_{550}$ (total cell mass – unadhered cell mass)/$A_{550}$ (total cell mass).

## E. Purification of GTFs from Cell Extract and Culture Supernatant

*S. mutans* was grown in 8 L of TTY broth culture for 16 h at 37°C.[13,21] When necessary, CF-GTF can be purified from the culture supernatant by 50% saturated ammonium sulfate precipitation, chromatofocusing on a Polybuffer exchanger PBE94 (Amersham

**Table 1. Summary of the Purification of Cell-Associated GTF from *S. mutans* MT8148**

| Purification step | Total protein (mg) | Total activity (U) | Specific activity (U/mg) | Recovery (%) | Purification (fold) |
|---|---|---|---|---|---|
| 8 M urea extract | 207 | 417 | 2.01 | 100 | 1.0 |
| Ammonium sulfate (60% saturation) | 119 | 260 | 2.18 | 62 | 1.1 |
| DEAE-Sephacel | 37.4 | 85.9 | 2.30 | 21 | 1.1 |
| Hydroxyapatite | 4.7 | 32.8 | 6.96 | 8 | 3.5 |

Cells recovered from TTY culture (81) were used for the extraction of cell-associated GTF.

Pharmacia Biotech), and subsequent HA chromatography (Bio-Rad, Hercules, CA, USA) as described previously. [39]

The organisms are collected by centrifugation and washed three times with 50 mM sodium phosphate buffer (NaPB, pH 6.0). The organisms are stirred with 8 M urea solution for 1 h at room temperature to release CA-GTF. The centrifuged supernatant is concentrated by 60% saturated ammonium sulfate precipitation, dialyzed against 50 mM NaPB (pH 7.5), and loaded onto a DEAE-Sephacel column (2.5×13 cm, Amersham Pharmacia Biotech). After washing with 50 mM NaPB (pH 7.5), bound proteins are eluted with a linear gradient of 0 to 1.0 M NaCl. Fractions (15 mL) are collected and GTF activity is determined as described above. Fractions containing GTF eluted at about 0.6 M NaCl are pooled, dialyzed against 50 mM NaPB (pH6.0), and the GTF solution is applied onto a hydroxyapatite column (1.0×13 cm, Bio-Rad). The column is first eluted with 225 mL of 50 mM NaPB (pH 6.0), and then eluted step-wise with 50 mM (90 mL), 0.2 M (165 mL), 0.26 M (135 mL), and 0.5 M (255 mL) NaPB (pH 6.0) at a flow rate of 2 mL/min. The absorbance at 280 nm and GTF activity of each fraction (15 mL) are measured. CA-GTF should be obtained by elution with 0.5 M NaPB. The method for the purification of GTF is summarized in Table 1. Long-term storage is best achieved at –80°C after the enzyme solution has been frozen in liquid nitrogen.

## F. Genetic Manipulation of the gtf Genes in *S. mutans*

To examine the role of each GTF in sucrose-dependent adhesion, the construction of *gtf*-isogenic mutants is essential. Assays for studying the *gtf* gene functions were reported by several research groups. For example, recombinant GTFs were expressed in *Escherichia coli* or *Streptococcus milleri*, purified, and characterized.[9,11,17] Sucrase and glucan binding activities of GTF of MS were also determined by independent groups.[1,19,46] In this section, we describe the inactivation and reconstruction of the *gtf* genes of *S. mutans* based on our recent report.

Transformation of *S. mutans* is performed as previously described[24] with some modification. *S. mutans* is cultured in Todd-Hewitt broth (Difco) supplemented with 10% heat-inactivated horse serum (Life Technologies, Rockville, MD, USA) for 18 h at 37°C. The overnight culture is diluted 1:40 into the broth (10 mL) and incubated for 1.5 h at 37°C, after which donor DNA is added to a final concentration of 25 µg/mL. It is further incubated for 2 h at 37°C, concentrated approx 10-fold by centrifugation, and

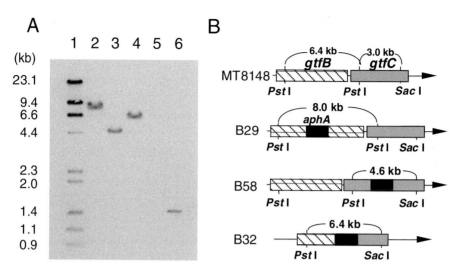

**Figure 1.** Southern blot analysis of *S. mutans* MT8148 and its GTF-deficient mutants (A) and the location of the inserted *aphA* gene in the GTF-deficient mutants (B). (A) Chromosomal DNA of the test organisms was digested and separated in a 0.8% agarose gel, and then transferred onto a nylon membrane. The blotted membrane was hybridized stringently with the *aphA* gene fragment. Lanes: 1, DNA size marker; 2, mutant B29; 3, B58; 4, B32; 5, parent strain MT8148; 6, the *aphA* gene. (B) The location of the inserted *aphA* gene (■) was determined by Southern blot analysis and the restriction enzyme mapping of the *gtjB* (▧) and *gtfC* (▨) genes.

spread on Mitis-Salivarius agar (Difco) plates containing appropriate antibiotics. The plates are incubated at 37°C for 2 to 3 d, and possible transformants are harvested for further examination.

The 2-kb central portion of both *gtfB* and *gtfC* shows 98% identity. Therefore, *gtfB-*, *gtfC-*, and *gtfBC*-inactivated isogenic mutants may be obtained simultaneously when a unique *MluI* site in the central region of the genes is inserted by an antibiotic marker, and the resultant plasmid is transformed into *S. mutans*. According to this strategy, the *gtfB* harboring plasmid, pSK6,[9] is cleaved with *MluI*, and ligated with a kanamycin resistance gene (*aphA*) from transposon Tn*1545* [3] to generate pTF55. Following linearization at a unique *KpnI* site of the plasmid, the DNA fragment (10 µg) is introduced into *S. mutans* strain MT8148 by transformation to allow allelic replacement. Transformants are selected on MS agar containing 500 µg/mL kanamycin, and are examined for their colony morphology on MS-agar under a dissecting microscope.

*S. mutans* grown on MS-agar shows a rough colonial morphology. Transformants are screened based on their colonial appearances, and three kinds of mutant strains are obtained. Southern blot analysis using the 1.6-kb *aphA* gene as probe reveals that strains B29, B58, and B32 are inactivated *gtfB*, *gtfC*, and *gtfBC*, respectively (Fig. 1).

## G. Reconstruction of gtfB and gtfC, and Reintroduction of GTF

A *Streptococcus-E. coli* shuttle vector pVA838 [28] is digested with *XbaI* and *SalI* to generate a shuttle plasmid carrying *gtfB* (Fig. 2). A 5.7-kb DNA fragment containing the genes of the origin of replication in *Streptococcus* and erythromycin-resistant gene (*erm*) is isolated from agarose gel. The fragment is ligated into pTF41, which is derivative of

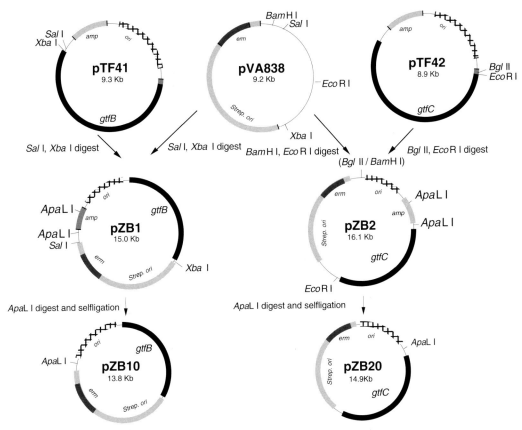

**Figure 2.** Construction of *E. coli-S. mutans* shuttle plasmids carrying the *gtfB* or *gtfC* gene.

pSK6 harboring the *gtfB* gene cleaved with the same enzymes, generating pZB1. To remove the *amp* gene, pZB1 is digested with *Apa*LI and self-ligated, resulting in pZB10.

The shuttle vector containing the *gtfC* gene, pZB2, is constructed by digesting pVA838 with *Eco*RI and *Bam*HI (Fig. 2). A 7.2-kb fragment is ligated with pTF42 digested with *Eco*RI and *Bgl*II. To remove the *amp* gene, pZB2 is digested with *Apa*LI and self-ligated, resulting in pZB20. Both *E. coli* and *S. mutans* harboring pZB10 or pZB20 should be confirmed to be sensitive to ampicillin. To reintroduce the deficient GTFs, pZB10 and pZB20 are transformed into the GTF-deficient mutants B29 and B58, respectively (Fig. 3).

An examination of sucrose dependent adherence of the parent strain and the mutants verified the predicted mutations (Table 2). The sucrose dependent cellular adherence of these mutants was significantly decreased as compared with that of parent strain MT8148. Mutant B32 showed the lowest adherence, while the cellular adherence of mutant B58 was lower than that of B29. The difference in the adherence ability among these mutants was significant as revealed by ANOVA ($P < 0.05$). The degree of sucrose dependent adherence of the GTF-I reintroduced transformant B29(pZB10) was greater than that of the recipient strain B29, but did not reach the level of that of the parent strain MT8148. On the other hand, the sucrose-dependent adherence of GTF-SI reintroduced transformants B58(pZB20) was almost equal to that of the parent strain MT8148 (Table 2).

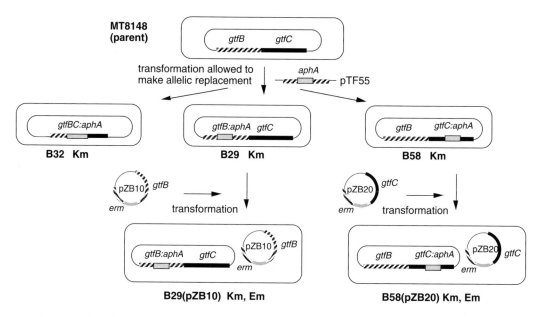

**Figure 3.** Deletion and reconstruction of GTFs of S. *mutans* MT8148. The *gtfB* or *gtfC*, or both genes were inactivated by insertional mutagenesis. Then the shuttle plasmids harboring the inactivated *gtf* genes was transformed into the GTF deficient mutant.

**Table 2. Sucrose Dependent Adherence of GTF Reintroduced Mutants**

| Strain | Antibiotic resistance | GFT | Adherence (%) |
|---|---|---|---|
| MT8148 | None | I/SI/S | 72.8±2.6 |
| B29 | Km | /SI/S | 16.3±1.0 |
| B29 (pZB10) | Km, Em | I/SI/S | 46.9±5.9 |
| B58 | Km | I/ /S | 9.6±1.0 |
| B58 (pZB20) | Km, Em | I/SI/S | 69.9±1.8 |
| B32 | Km | / /S | 1.4±0.4 |

## IV. CONCLUDING REMARKS

This chapter outlines how MS adhere to the tooth surface and, together with other oral bacterial species, form dental plaque, a biofilm. GTFs contribute to the adherence capability of these organisms, and should be recognized as critical virulence factors of these organisms. Traditional and novel molecular biological techniques are required to understand the mechanism of the biofilm formation in the oral cavity. S. *mutans* and S. *sobrinus* are apparently favored among oral bacterial flora for the study of biofilms. It should be noted that similar biofilms are occasionally formed on the heart valves, which eventually leads to bacterial endocarditis. In such cases, MS as well as other oral and enteric streptococcal species are reported to be etiologically important.

**Acknowledgments:** This work was supported by grants from the Japan Society for the Promotion of Science, and the Ministry of Health and Welfare.

# REFERENCES

1. Abo H, Matsumura T, Kodama T, et al: Peptide sequences for sucrose splitting and glucan binding within *Streptococcus sobrinus* glucosyltransferase (water-insoluble glucan synthetase). *J Bacteriol* 173:989–96, 1991

2. Appelbaum B, Golub E, Holt SC: *In vitro* studies of dental plaque formation: adsorption of oral streptococci to hydroxyapatite. *Infect Immun* 25:717–28, 1979

3. Caillaud F, Carlier C, Courvalin P: Physical analysis of the conjugative shuttle transposon Tn*1545*. *Plasmid* 17:58–60, 1987

4. Crowley PJ, Brady LJ, Piacentini DL, et al: Identification of a salivary agglutinin-binding domain within cell surface adhesin P1 of *Streptococcus mutans*. *Infect Immun* 61:1547–52, 1993

5. Eifert R, Rosan B, Golub E: Optimization of an hydroxyapatite adhesion assay for *Streptococcus sanguis*. *Infect Immun* 44:287–91, 1984

6. Emilson CG, Ciardi JE, Olsson J, et al: The influence of saliva on infection of the human mouth by mutans streptococci. *Arch Oral Biol* 34:335–40, 1989

7. Ericson T, Rundegren J: Characterization of a salivary agglutinin reacting with a serotype c strain of *Streptococcus mutans*. *Eur J Biochem* 133:255–61, 1983

8. Forester H, Hunter N, Knox KW: Characteristics of a high molecular weight extracellular protein of *Streptococcus mutans*. *J Gen Microbiol* 129:2779–88, 1983

9. Fujiwara T, Kawabata S, Hamada, S: Molecular characterization and expression of the cell-associated glucosyltransferase gene from *Streptococcus mutans*. *Biochem Biophys Res Commun* 187:1432–8, 1992

10. Fujiwara T, Terao Y, Hoshino T, et al: Molecular analyses of glucosyltransferase genes among strains of *Streptococcus mutans*. *FEMS Microbiol Lett* 161:331–6, 1998

11. Fukushima K, Ikeda T, Kuramitsu HK: Expression of *Streptococcus mutans gtf* genes in *Streptococcus milleri*. *Infect Immun* 60:2815–22, 1992

12. Furuta T, Koga T, Nishizawa T, et al: Purification and characterization of glucosyltransferase from *Streptococcus mutans* 6715. *J Gen Microbiol* 131:285–93, 1985

13. Hamada S, Horikoshi T, Minami T, et al: Purification and characterization of cell-associated glucosyltransferase synthesizing water-insoluble glucan from serotype *c Streptococcus mutans*. *J Gen Microbiol* 135:335–44, 1989

14. Hamada S, Koga T, Ooshima T: Virulence factors of *Streptococcus mutans* and dental caries prevention. *J Dent Res* 63:407–11, 1984

15. Hamada S, Slade HD: Biology, immunology and cariogenicity of *Streptococcus mutans*. *Microbiol Rev* 44:331–84, 1980

16. Hamada S, Torii M: Effect of sucrose in culture media on adherence to glass surfaces. *Infect Immun* 20:592–9, 1978

17. Hanada N, Kuramitsu HK: Isolation and characterization of the *Streptococcus mutans gtfC* gene, coding for synthesis of both soluble and insoluble glucans. *Infect Immun* 56:1999–2007, 1988

18. Honda O, Kato C, Kuramitsu HK: Nucleotide sequence of the *Streptococcus mutans gtfD* gene encoding the glucosyltransferase-S enzyme. *J Gen Microbiol* 136:2099–105, 1990

19. Kato C, Nakano Y, Lis M, et al: Molecular genetic analysis of the catalytic site of *Streptococcus mutans* glucosyltransferase. *Biochem Biophys Res Commun* 189:1184–8, 1992

20. Katz J, Harmon CC, Buckner GP, et al: Protective salivary immunoglobulin A responses against *Streptococcus mutans* infection after intranasal immunization with *S. mutans* antigen I/II coupled to the subunit of cholera toxin. *Infect Imuun* 61:1964–71, 1993

21. Koga T, Asakawa H, Okahashi N, et al: Sucrose dependent cell adherence and cariogenicity of serotype *c Streptococcus mutans*. *J Gen Microbiol* 132:2873–83, 1986

22. Koga T, Okahashi N, Takahashi I, et al: Surface hydrophobicity, adherence, and aggregation of cell surface protein antigen mutants of *Streptococcus mutans* serotype c. *Infect Immun* 58:289–96, 1990

23. Lee SF, Progulske-Fox A, Erdos GW, et al: Construction and characterization of isogenic mutants of *Streptococcus mutans* deficient in major surface protein antigen P1 (I/II). *Infect Immun* 57:3306–13, 1989

24. Lindler L, Macrina FL: Characterization of genetic transformation in *Streptococcus mutans* by using a novel high-efficiency plasmid marker rescue system. *J Bacteriol* 166:658–65, 1986

25. Lis M, Shiroza T, Kuramitsu HK: Role of the C-terminal direct repeating units of the *Streptococcus mutans* glucosyltransferase-S in glucan binding. *Appl Environ Microbiol* 61:2040–2, 1995

26. Loesche WJ: Role of *Streptococcus mutans* in human dental decay. *Microbiol Rev* 50:353–80, 1986

27. Ma JK, Hunjan M, Smith R, et al: An investigation into the mechanism of protection by local passive immunization with monoclonal antibodies against *Streptococcus mutans*. *Infect Imuun* 58:3407–14, 1990

28. Macrina FL, Evans RP, Tobian JA, et al: Novel shuttle plasmid vehicles for *Escherichia–Streptococcus* transgenetic cloning. *Gene* 25:145–50, 1983

29. Mooser G, Hefta SA, Paxton RJ, et al: Isolation and sequence of an active-site peptide containing a catalytic aspartic acid from *Streptococcus sobrinus* glucosyltransferases. *J Biol Chem* 266:8916–22, 1991

30. Munro C, Michalek SM, Macrina FL: Cariogenicity of *Streptococcus mutans* V403 glucosyltransferase and fructosyltransferase mutants constructed by allelic exchange. *Infect Immun* 59:2316–23, 1991

31. Nakai M, Okahashi N, Ohta H, et al: Saliva-binding region of *Streptococcus mutans* surface protein antigen. *Infect Immun* 61:4344–9, 1993

32. Nakano Y, Kuramitsu HK: Mechanism of *Streptococcus sobrinus* glucosyltransferases: hybrid-enzyme analysis. *J Bacteriol* 174:5639–46, 1992

33. Ofek I, Doyle RJ: Relationship between bacterial cell surfaces and adhesins. In: Ofek I, Doyle RJ, eds. *Bacterial Adhesion to Cells and Tissues*. Chapman & Hall, New York, NY, 1994: 54–93.

34. Ogier JA, Scholler M, Leproivre Y, et al: Complete nucleotide sequence of the *sr* gene from *Streptococcus mutans* OMZ175. *FEMS Microbiol Lett* 68:223–8, 1990

35. Okahashi N, Koga T, Hamada S: Purification and immunochemical properties of a protein antigen from serotype g *Streptococcus mutans*. *Microbiol Immunol* 30:35–47, 1986

36. Okahashi N, Sasakawa C, Yoshikawa M, et al: Molecular characterization of a surface protein antigen gene from serotype c *Streptococcus mutans*, implicated in dental caries. *Mol Microbiol* 3:673–8, 1989

37. Rosenberg MD, Gutnick D, Rosenberg E: Adherence of bacteria to hydrocarbons: a simple method for measuring cell surface hydrophobicity. *FEMS Microbiol Lett* 9:29–33, 1980

38. Russell MW, Bermeier LA, Zanders ED, et al: Protein antigens of *Streptococcus mutans*: purification and properties of a double antigen and its protease-resistant component. *Infect Immun* 28:486–93, 1980

39. Sato S, Koga T, Inoue M: Isolation and some properties of extracellular D-glucosyltransferases and D-fructosyltransferases from *Streptococcus mutans* serotype c, e, and f. *Carbohyd Res* 134:293–304, 1984

40. Shiroza T, Ueda S, Kuramitsu HK: Sequence analysis of the *gtfB* gene from *Streptococcus mutans*. *J Bacteriol* 169:4263–70, 1987

41. Takahashi I, Matsushita K, Nishizawa T, et al: Genetic control of immune responses in mice to synthetic peptides of a *Streptococcus mutans* surface protein antigen. *Infect Immun* 60:623–9, 1992

42. Takahashi I, Okahashi N, Hamada S: Molecular characterization of a negative regulator of *Streptococcus sobrinus* surface protein antigen gene. *J Bacteriol* 175:4345–53, 1993

43. Takahashi I, Okahashi N, Matsushita K, et al: Immunogenicity and protective effect against oral colonization by *Streptococcus mutans* of synthetic peptides of a streptococcal surface protein antigen. *J Immunol* 146:332–6, 1991

44. Terleckyj B, Willett NP, Shockman GD: Growth of several cariogenic strains of oral streptococci in a chemically defined medium. *Infect Immun* 11:649–55, 1975

45. Ueda N, Shiroza T, Kuramitsu HK: Sequence analysis of the *gtfC* gene from *Streptococcus mutans* GS-5. *Gene* 69:101–9, 1988

46. Wong C, Hefta SA, Paxton RJ, et al: Size and subdomain architecture of the glucan-binding domain of sucrose: 3-α-D-glucosyltransferase from *Streptococcus mutans*. *Infect Immun* 58:2165–70, 1990

47. Yamashita Y, Bowen WH, Kuramitsu HK: Molecular analysis of a *Streptococcus mutans* strain exhibiting polymorphism in the tandem *gtfB* and *gtfC* genes. *Infect Immun* 60:1618–24, 1992

48. Yamashita Y, Hanada N, Takehara T: Purification of a fourth glucosyltransferase from *Streptococcus sobrinus*. *J Bacteriol* 171:6265–70, 1989

# Studying Bacterial Adhesion to Respiratory Mucosa

**Maria Cristina Plotkowski,[1] Sophie de Bentzmann,[2] and Edith Puchelle[2]**

[1]*Department of Microbiology and Immunology, State University of Rio de Janeiro (UERJ), Rio de Janeiro, Brazil and [2]INSERM U 514, Reims University IFR 53, Reims, France*

## I. INTRODUCTION

Adhesion to host tissues enables human pathogens to withstand host defense mechanisms such as removal by fluid flow, mucociliary clearance, and other physical processes. Adhesion is therefore an essential prerequisite for successful colonization of epithelial surfaces and is recognized as a virulence factor for bacterial, viral and fungal pathogens. However, adhesion alone is rarely, if ever, responsible for inducing disease. Most frequently, the combination of adhesion, pathogen growth in the lining epithelial cells, and toxin production or adhesion, penetration, and growth within mucosal epithelial cells determines the course of human diseases.

Bacterial adhesion may result from relatively nonspecific hydrophobic and ionic interactions between the pathogen and host tissues. However, the available evidence strongly suggests that adhesion is a highly specific process, the result of the interaction between complementary chemical and conformational structures on the surface of the microorganisms, called adhesins,[49] and protein or carbohydrate epitopes present on host cell plasma membranes.[64] In the simplest cases, adhesion is mediated by binding of a unique bacterial adhesin to host receptors. However, unimodal adhesive pathway seems to be uncommon among successful pathogens. Most frequently, multiple ligands on the bacterial surface, recognizing more than one carbohydrate sequence or both host proteins and glycoconjugates, serve to increase the strength of adhesion, when these ligands are engaged in concert.[91]

Since bacterial adhesins have very specific requirements for the recognition of eukaryotic cell epitopes, adhesion is usually restricted to a set of cell populations carrying their optimal receptors. In vivo, this partly determines the localization of pathogens in selected tissue sites and the set of infections by which a bacterial pathogen is characterized. Experimentally, this restriction of host cells for a given bacterium should be the guiding factor in the choice of models selected to study its adhesiveness.

In healthy conditions, the lower respiratory tract is sterile. That is so, despite the continuous inhalation of pathogens because the respiratory mucosa is equipped with efficient mechanisms to clean up inhaled microorganisms.[90] On the other hand, once

*Handbook of Bacterial Adhesion: Principles, Methods, and Applications*
Edited by: Y. H. An and R. J. Friedman © Humana Press Inc., Totowa, NJ

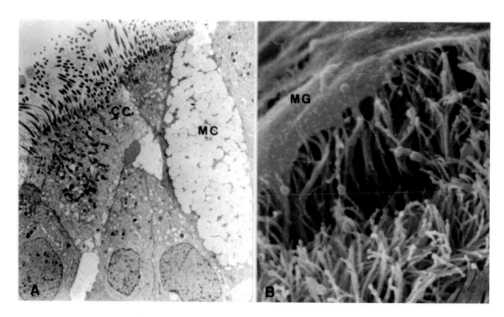

**Figure 1.** A. Transmission electron micrograph of the human tracheal mucosa showing ciliated (CC) and surface mucus secretory cells (MC) with electron-lucent secretory granules; B. Scanning electron micrograph of the human tracheal mucosa showing the mucus gel (MG) at the tips of cilia.

defense mechanisms such as the mucociliary clearance or the integrity of the epithelial barrier are overcome by physicochemical or viral agents, or even by inflammatory mechanisms, different pathogenic microorganisms can colonize the respiratory mucosa.[74,95] Therefore, bacterial colonization and infection of the airways are most frequently the result of a preliminary damage to local defense mechanisms, rather than of bacterial virulence factors.

## II. STRUCTURE AND FUNCTIONS OF AIRWAY EPITHELIUM

The lining of the human respiratory airways is predominantly a pseudostratified, columnar epithelium, the main cell types seen in surface epithelium being ciliated, mucous-secreting (goblet), intermediate and basal cells (Fig. 1A). In the smaller distal airways, the epithelium is composed of a single layer of ciliated and Clara cells and basal cells are absent. Ciliated cells predominate also in terminal and respiratory bronchioles, where they are adjacent to the alveolar lining cells.[35,47]

The epithelial airway cells are held together by specialized cell junctions which are essential for the barrier function of the epithelium against nocive agents.[58] Tight junctions seal cells together in a way that inhibits even small molecules from leaking from one side of the sheet to the other.[3,12] Circumferential apical tight junctions also separate apical and basolateral domains of cell membranes, assisting in maintenance of epithelial cell polarity. By isolating basolaterally restricted proteins, tight junctions also restrict the repertoire of potential interactions that may occur between luminal molecules and epithelial cell surface receptors. Anchoring junctions (adherens junctions and desmosomes) mechanically attach cells (and their cytoskeletons) to their neighbors or to the extracellular matrix. In response to environmental aggression, the cell junctions may be dramatically altered. The respiratory epithelium may therefore become "leaky" and permeable to noxious elements (pollutant gases, allergens, as well as infectious agents) present in the airway lumen.[42,78]

The respiratory epithelium is covered by mucus which forms a continuous film at the tips of the cilia of ciliated cells (Fig. 1B). This mucus film is made of two phases: an aqueous phase (called the periciliary layer), with low viscosity, in which the cilia beat, and a superficial (or gel layer), characterized by a high viscosity and elasticity. The film of mucus is permanently kept in motion by the action of ciliary beating. Inhaled exogenous microorganisms and particles are trapped by the gel mucus and are transported by the ciliary beating up to the pharynx where they are constantly swallowed. The entrapment of bacteria by the respiratory mucus, and the elimination of the mucus-embedded bacteria by the ciliary activity, actually represents the first stage in the defense of the respiratory epithelium.[95] The efficacy of the mucociliary escalator depends both on the integrity of the ciliary apparatus and on the rheological properties of respiratory mucus.[37,110] The respiratory mucus also has antioxidant, antiprotease and other antibacterial functions.

From a biochemical point of view, bronchial mucus is a water rich gel (over 97%) in which bathe numerous macromolecules (proteins, glycoproteins, lipids) and inorganic salts. The hydration of the mucus, the osmolarity of the periciliary layer, the relative thickness and the physical properties of the gel layer are directly regulated by the transepithelial movement of ions and water across the apical cell membranes, the ATPase pumps providing the necessary energy.[122] In cystic fibrosis (CF), the abnormalities in the Cl$^-$ channel due to the abnormal protein cystic fibrosis transmembrane conductance regulator (CFTR) cause significant dehydration of the mucus. Together with considerable problems of hypersecretion and hyperviscosity, mucus dehydration contributes to the deficiency of the mucociliary transport which in turn leads to the recurrent airway infections detected in CF patients.

Mucins, the most important component of the mucus gel, are a broad family of highly glycosylated molecules with a molecular mass ranging from about $3 \times 10^5$ to more than $10^6$ daltons.[93,94] Several mucin genes have been identified so far and the peptide part of the mucin, called apomucin, constitutes a family of diverse molecules.[5] The expression of different apomucin genes certainly contributes to the diversity of human respiratory mucins but this diversity is tremendously increased by glycosylation, a major post-transcriptional phenomenon responsible for about 70% of the weight of the mucin molecules. Some mucins may present several hundred different carbohydrate chains,[93] having 1-20 sugar residues. The majority of the carbohydrate chains are attached to the peptide by O-glycosidic bonds between the hydroxyl group of a serine or threonine residue of the apomucin and a *N*-acetylgalactosamine (GalNAc) residue at the reducing end of the carbohydrate chain. In addition to GalNAc, mucins may also contain *N*-acetylglucosamine (GlcNAc), galactose (Gal), fucose (Fuc), *N*-acetylneuraminic acid (NeuAc), and sulfate.

Since many microbial adhesins recognize and bind to host glycoconjugates,[95,105] it is likely that the carbohydrate diversity of the respiratory mucins represents a mosaic of potential sites for the attachment of inhaled microorganisms, leading to their trapping, and ultimately to their elimination. In normal conditions, therefore, mucin carbohydrate diversity represents a protection for the underlying mucosa against bacterial colonization. However, in pathological conditions, the mucociliary transport may be severely reduced due to a decrease in the number and activity of ciliated cells, changes in the biochemical and rheological properties of the mucus gel, or modifications in the surface properties of the respiratory mucosa.[37] As soon as this protective mechanism is impaired, the specific recognition and attachment of bacteria to mucins represents a critical pathway for bacterial colonization and infection of the underlying epithelial cells, generally associated with local inflammatory reaction.

## III. RESPONSE OF AIRWAY EPITHELIUM TO ENVIRONMENTAL INJURY

Generally speaking, the acute response of the respiratory epithelium to injury begins by an increase in mucus secretion. At the initial stage, this hypersecretion represents a phenomenon of protection by three principal mechanisms: 1) an increase in the physical protective barrier of the epithelium (the thickness of the gel phase increases); 2) an increase in the pool of secretory cells, essential for the rapid regeneration of ciliated cells which are particularly sensitive to the aggressive agents; and 3) an increase in the pool of biochemical molecules intervening in the defense of the epithelium.[78,79]

If there is an imbalance between the defense mechanisms and the factors of aggression, this will progressively give rise to the migration of polymorphonuclear leukocytes towards the surface epithelium. Lysis of the mucus gel associated with the injury of ciliated cells by oxidant molecules and proteases released by inflammatory cells are very quickly followed by the decrease of mucociliary clearance. Moreover, in response to aggression, the airway epithelium may suffer from a succession of cellular events which culminate with the loss of surface epithelial impermeability normally ensured by the tight junctions, or the shedding of the epithelium (only basal cells still being attached to the basal lamina), or even the complete denudation of the basement membrane.[79]

After acute injury, the respiratory epithelium initiates a healing process to recover its integrity. This process involves such mechanisms as internalization of cilia, and loss of the polarity of epithelial respiratory cells which then flatten, spread and migrate to recover denuded basement membranes.[51,52,59] Proliferation of the cells located behind the wound begins later. Cells in mitosis are generally secretory cells and not basal cells.

In vivo studies have shown that the wound space is colonized by different cell types, the sequential processes by which these cells interact with each other being orchestrated by the release of signal substances, such as cytokines and inflammatory mediators. However, in vitro studies have shown that airway epithelial cells have the capacity by themselves, without the contribution of any other nonepithelial cells, to migrate at a speed ranging from 10 to 30 $\mu$m/h, and to re-establish a barrier junction within 48–72 h, according to the size of the wound.[126]

Whereas intact respiratory mucosa is relatively resistant to bacterial adhesion and colonization, an extensive literature has shown that injured mucosa is highly susceptible to bacterial infection.[31,74,88,128] Mechanisms by which tissue injury predisposes respiratory airways to bacterial infection may be related to the uncovering of potential receptors for bacterial adhesins. Desquamated cells[68,81,88,118] and different components of the denuded extracellular matrix (ECM)[20,75] represent preferential targets for bacterial adhesion. Moreover, the remodeling of airway epithelium during injury and repair seems to represent a critical event which may further favor the recurrence of infections. By using an in vitro model of the repair process of the airway epithelium,[125] we have shown that spreading epithelial cells located at the margins of the wounds (regenerating cells) are particularly susceptible to the adhesion of *Pseudomonas aeruginosa*.[19,72]

When attempting to study bacterial adhesion to the respiratory mucosa, it is important to bear in mind the characteristics of the infections caused in vivo by the bacterial species of interest. Answers to questions such as "Can these microorganisms infect healthy people, like *Mycoplasma pneumoniae* do?"[56] or "Are these microorganisms mainly opportunistic bacteria, as is the case of *P. aeruginosa*?" will guide the choice of the in vitro model which would better simulate the conditions which the different bacterial pathogens find in vivo.

## IV. STUDYING BACTERIAL ADHESION TO RESPIRATORY MUCUS

*Streptococcus pneumoniae*,[70,87] *Haemophilus influenzae*,[7,17,31,46,89] *Staphylococcus aureus*,[101,119] *Burkholderia cepacia*,[97,99] and *P. aeruginosa*[68,81-83,98] are human respiratory pathogens that seldom, if ever, adhere to normal ciliated cells but bind avidly to respiratory mucus.

Bacterial adhesion to mucus has been assessed by scanning electron microscopic (SEM) observation of experimentally infected animal airways,[83,101] organotypic cultures,[31,118] and cell culture models.[73] The effect of mucus hypersecretion on *S. aureus* adhesion was recently assessed by treating epithelial respiratory cells in primary culture with human neutrophil elastase (HNE) at 1 µg/mL for 30 min. At this concentration, HNE was not toxic for ciliated cells and was shown to significantly increase the number of mucus granules released by the respiratory cells.[119] In another study in which cultured cells were treated with HNE at higher concentration (250 µg/mL), a substantial increase in *P. aeruginosa* adhesion to respiratory mucosa was observed and bacteria were seen to adhere both to mucus granules and to desquamated respiratory cells (Fig. 2).[70]

Although SEM observation of the respiratory mucosa allows the qualitative assessment of bacterial adhesion to mucus, the definition of the precise mucus receptor (mucin or other mucus glycoconjugate; epitopes on these molecules)[84,86] and of the bacterial ligands accounting for this adhesion[11,85,106] require the use of other in vitro assays.

### A. Bacterial Binding to Mucin Immobilized onto Solid Phases

#### 1. Microtiter Plate Assays

Different concentrations of purified mucin or mucin glycopeptides in deionized water[84,97,121] or in 0.1 M carbonate buffer, pH 9.6,[101] are added to 96-well microtiter plates and the plates are left at 37°C overnight to allow mucin to coat the wells. Just before adding bacterial suspensions, the wells are washed with sterile phosphate-buffered saline (PBS; pH 7.0) and the plates are incubated with bacteria at 37°C for different periods. Alternatively, before the addition of the bacterial suspension, the remaining receptor sites in microplate wells can be blocked with 3% bovine serum albumin (BSA) for 1 h.[97,98] After the incubation with bacteria, the wells are washed with sterile PBS to remove unbound bacteria. Adherent microorganisms can then be lifted with 0.5% Triton X-100,[84,121] with 0.5% Tween 80[7] or with 0.05% trypsin - 0.5 mM EDTA in Hanks'

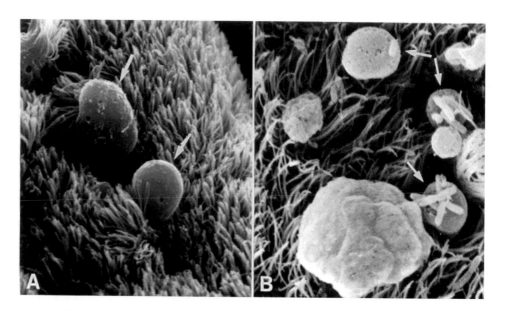

**Figure 2.** Scanning electron micrograph of the respiratory mucosa treated with human neutrophil elastase (HNE) at 250 µg/mL for 1 h (A) and of HNE-treated mucosa exposed to *P. aeruginosa* suspension. Note in (A) the secretion of mucus granules (arrows) and in (B), the presence of bacteria adherent to secreted mucus (arrows).

balanced salt solution without calcium or magnesium.[109] Quantification of adherent bacteria is obtained by serial diluting and culturing the content of each well. The viable count of the starting inoculum should also be quantified by dilution of a sample and plating.

*2. Slot Dot Assay*

Different concentrations of purified mucin or mucin glycopeptides are blotted onto nitrocellulose membranes which are washed with PBS containing 0.05% Tween 20 (PBS-Tween), blocked with 1% BSA[17] or 3% gelatin[97] for 30 min to 1 h and incubated with suspensions of [35]S-labeled bacteria in 1% BSA-PBS-Tween for different periods. The membranes are then washed in PBS-Tween, and air dried. The presence of radio-activity can be detected by autoradiography and quantitated with a scanning densitometer[17,97] or by the counting of the associated radioactivity with scintillation counters.

*3. Thin Layer Chromatography (TLC) Assay*

Chloroform:methanol extracts of mucin or glycolipids are separated by thin layer chromatography on silica gel.[86,97,98] The silica plates are then either stained with Orcinol and visualized, or blocked with 1% gelatin or BSA overnight at 37°C, incubated with [35]S-labeled bacteria for 1 h, washed, air dried and processed for autoradiography. By using TLC and the neoglycolipid technology, Ramphal et al.[86] have determined some mucin carbohydrate receptors for *P. aeruginosa* adhesion. They have also gathered evidence that adhesins that are different from pilin account for this bacterium affinity for respiratory mucus.

## B. Bacterial Aggregation by Mucin

Purified mucins at 1 mg/mL (10 μL) are mixed with bacterial suspensions at $2\times10^9$ cfu/mL (50 μL), shaken for 2 h at room temperature, and examined by interference microscopy. Aggregates are graded as present (+, ++, or +++) or absent. If present, the mucins are serially diluted in PBS and the bacterium-mucin interaction is quantitated as the lowest concentration of mucin that caused aggregation.[17]

## C. Interaction of Mucin with Bacterial Fractions

To identify the ligands accounting for adhesion to mucus, bacterial outer membrane proteins (OMP) can be extracted and separated by sodium dodecyl sulfate-polyacrylamide gel electrophoresis (SDS-PAGE), blotted onto nitrocellulose replicas, blocked with 2% BSA, and probed with human bronchial mucins (or other mucus components) labeled with [125]I. The blots are then washed with PBS (4 times for 15 min each), and with 0.05% Tween 20-PBS (4 times for 15 min each), dried, and processed for autoradiography. By using this method Carnoy et al.[11] identified several *P. aeruginosa* OMPs recognizing both respiratory mucin and lactotransferrin. More recently, Scharfman et al.[103] demonstrated that the expression of mucin-binding proteins in OMPs of *P. aeruginosa* is affected by the iron content of the medium in which the bacteria are grown.

## V. STUDYING BACTERIAL ADHESION TO RESPIRATORY CELLS

### A. Animal Models

Ideally, the animal model used to study a bacterial disease should have the following characteristics: the bacterium should infect the animal by the same route it uses to infect humans, have the same tissue distribution in animals as in humans and cause the same symptons. Strains known to be more virulent than others in humans should also be more virulent in the animal model. Such an ideal animal model has rarely been found.[100]

Different animal species have been used for investigations of bacterial adhesion to respiratory mucosa.[69,80,115,128] At different periods after bacterial inoculation, animals were sacrified, and the respiratory tissues removed to determine the extent of adhesion by SEM or by the counting of viable bacteria associated with tissues.

The in vivo system using the inoculation of experimental animals for investigations of bacterial virulence factors has long been accepted as a substitute for using human subjects and considered to be the gold standard of research on bacterial virulence.[100] However, there are many disadvantages in using animal models to study bacterial adhesion to host tissues. 1) They represent a complex system in which many variables can not be controlled. In in vivo models, there is always a cascade of epithelial, mesenchymal and inflammatory cell-derived molecules and it is often difficult to distinguish which cell phenotype and molecules are specifically responsible for bacterial adhesion. 2) Since there may exist significant differences in the expression of surface carbohydrates between animal species, receptors for bacterial adhesins exhibited by animal cells may not be the same exhibited by human cells. 3) It is necessary to have a large enough animal population in order to provide statistical validation of the data. 4) Experimental animal models are not adequate to attain a molecular understanding of bacterial adhesion.

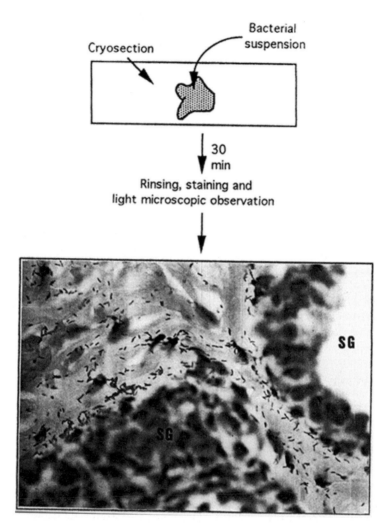

**Figure 3**. Schematic representation of the tissue section model to assess bacterial adhesion to respiratory mucosa. Cryosections of bovine tracheal mucosa were exposed to *P. aeruginosa* suspension for 30 min and stained with methylene blue. Note the presence of many bacteria adherent to the submucosal extracellular matrix, located preferentially at the basal side of epithelial cells from submucosal glands (SG), suggesting the affinity of bacteria for cell basement membranes.

Although animal models are usually inadequate as the initial approach to investigate bacterial adhesion to respiratory mucosa, they remain excellent tools to confirm data obtained from other experimental models in which less complex targets are used.

## B. Tissue Section Models

*In situ* screening of human tissue sections exposed to bacterial suspensions has been used to assess adhesion to respiratory mucosa. The advantage of this model is that it allows direct and specific evaluation of the distribution of receptor molecules among different cell populations in human target tissues. Moreover, it can be used to further characterize the receptor directly on the tissues by performing double labeling of adherent bacteria and tissue components, biochemical modifications of the tissues, and inhibition assays.[28]

Sterk et al.[113] described a technique for using paraffin-embedded sections of adenoid human tissue to assess the role of fimbriae in *H. influenzae* adhesion to respiratory cells. Briefly, tissue sections were transferred to poly-L-lysine-coated glass slides and incubated with human plasma for 4 h at 37°C, to improve adhesion, and stored. Before using, the sections were deparaffinized in 100% xylene for 10 min, rehydrated in graded ethanols, washed in slowly running distilled water, and pre-incubated for 15 min at 37°C with 0.5% pepsin in diluted HCl (pH 2.0) and then with 10% normal swine serum for 15 min at 22°C, to reduce nonspecific binding. Tissue sections were overlaid with bacterial suspensions, incubated for 1 h in a humid atmosphere, and washed three times with PBS containing 0.05% Tween 20. Sections were then stained immunochemically with anti-bacterial antibodies to detect adherent *H. influenzae*. The slides were counterstained with hematoxylin and the sections were examined by light microscopy. Alternatively, bacteria can be labeled with fluorochromes, such as fluorescein isothiocyanate (FITC) or tetramethyl-rhodamine isothiocyanate (TRITC) and the bacterial adhesion assessed by fluorescence microscopy.[29]

Tissue section models can also be used in inhibition assays. In the study carried out by Sterk et al.[113] to ascertain the role of fimbriae in *H. influenzae* adhesion to respiratory cells, bacteria were pre-incubated with anti-fimbriae monoclonal antibodies prior to their incubation with tissue sections.

*In situ* screening of frozen human tissue sections exposed to bacterial suspensions was described by Beuth et al.[10] Frozen sections are expected to give origin to more reliable results because cell reactivity is not hampered by chemical treatments. Briefly, tissue fragments were embedded in Optimum Cutting Temperature compound (OCT, Tissue Tek, Elkhart, IN), immersed in liquid nitrogen for 5 min, and kept at −80°C. Thin cryotome sections were deposited onto gelatin-coated slides and air-dried. Sections were rehydrated with PBS containing 1% BSA for 10 min and then exposed to either bacterial suspensions or bacterial suspensions containing different carbohydrates. Sections were then rinsed in ice-cold PBS, fixed, stained with methylene blue and microscopic examination followed.

By using the technique described by Beuth et al.[10] applied to bovine airway epithelium, we could observe the adhesion of *P. aeruginosa* to respiratory mucus as well as to the extracellular matrix (Fig. 3; unpublished data).

## C. Isolated Cell Model

In this model, human epithelial respiratory cells obtained from buccal,[23,44,124] nasal[63] or tracheal mucosa[23,33,44,63] were mixed with suspensions of radiolabeled bacteria in conical bottom tubes. After incubation, the mixtures were centrifuged, and the supernatants, containing nonadherent bacteria, decanted.[63] The cell pellets were then resuspended in PBS and any nonadherent bacteria still remaining were eliminated by filtration in polycarbonate filters. After washing, the filter-associated radioactivity was counted in scintillation counters. In other studies, in which nonradiolabeled bacteria were used instead, the adhesion was assessed by light microscopy, after cell fixation and staining.

Although the isolated cell model was the model used in early studies on bacterial adhesion, nowadays its merit is recognized as being mainly historical. That is the case because in vivo, epithelial cells are held together by cell junction complexes that separate apical and basolateral domains of cell membranes.[3,12] Membranes of these two domains differ structurally, biochemically and physiologically. Thus, in healthy conditions, only receptors from apical membranes are exposed to bacteria and the in vitro finding of bacteria adherent to cell lateral sides may have no physiological significance. It may, however, be representative of conditions found in early phases of epithelial injury, when many exfoliated cells are found lying on the respiratory mucosa. In fact, several authors have observed the adhesion of respiratory pathogens, as *P. aeruginosa*[68,70,81,118] and *H. influenzae,*[46,88,89] to exfoliated epithelial respiratory cells.

### D. Cultured Cell Model

Cultured mammalian cells provide a simple and easily controlled model for investigating the host–bacterium interaction. Cells can be grown in defined culture medium under reproducible conditions and only one cell type is represented. Another advantage of using cell cultures to assess bacterial adhesion, that has been underlined by Salyers and Whitt,[100] is that "cultured cells cost less per day to house than laboratory animals, do not fight with each other and have seldom been known to escape from their cages."

Due to the variation in expression of cell surface epitopes known to exist between species and tissues/organs, as well as among cells originating from the same tissue as a function of developmental and differentiation stages,[29,30] careful consideration must be made to the type of cultured cell selected for the adhesion assays. The cultured cell chosen for analysis should resemble as closely as possible the naturally colonized cells with which the organism interacts with in tissues.

It may be possible to prepare and maintain cell populations from explants of normal tissue in primary culture. In fact, cells in primary cultures should be used whenever possible, since they should reflect the cells found in nondiseased tissues in vivo. For tissues whose cells cannot be conveniently obtained in primary culture, it is often possible to identify a transformed cell line that is derived from the tissue of interest.

A number of important limitations should be considered while attempting to compare results obtained with cultured cells to host tissue.

1. The state of differentiation that a given cell type can be maintained at in culture may not reflect the state that is present in the tissue. This is particularly important in studies on bacterial adhesion to respiratory mucosa. As previously mentioned, the respiratory airways, from the nose to respiratory bronchioles, are covered by ciliated cells. The beating activity of cilia has been shown to protect cells from bacterial adhesion.[72,118]

2. Cultured cells frequently lose many of the traits that are characteristic of the cells in vivo, especially because most cell lines have been established from malignant tissue. Also, if the cells have been transformed, the repertoire of receptors that they will express may be considerably altered as the process of transformation may produce numerous mutations and rearrangements in the genome of the cell. Moreover, cell lines that are passed repeatedly in culture will continue to develop new mutations and rearrangements.[29] To make the transition from a differentiated, nondividing cell found in vivo to a rapidly dividing cell, cultured cells must be stripped of many of the properties that made them the type of cell they were. One feature that

can be lost in the process of cell passage is tissue-specific cell surface molecules that normally function as bacterial receptors, unmasking low affinity receptors.[29] Such an occurrence may explain the surprising fact that many bacterial pathogens that are highly specific for a particular tissue when causing an infection in an intact animal are able to adhere to and invade cultured cells derived from tissue they do not infect normally. In addition, most cultured cells lose their normal shape and distribution of surface antigens, and, as they do not form tight junctions, are unable to form impermeable barriers.[77] These disadvantages may also be circumvented by culturing cells on porous membranes[40,66] or on supporting thick gels of ECM components, such as type I collagen[13] or EHS gels.[9] We have recently shown that when human epithelial respiratory cells, in primary culture and from the cell line, 1 HAEo-,[16] were cultured on thick collagen gels, cells were polarized, tight and impermeable to the penetration of lanthanum nitrate. In contrast, these same cells cultured on thin collagen films or on plastic supports gave origin to cultures that did not exclude lanthanum nitrate and that exhibited receptors from cell basolateral membranes, such as $\beta 1$ integrins, also at their upper membranes.[76] These cells differed also in their susceptibility to *P. aeruginosa* adhesion and internalization, the non-polarized cells being highly susceptible to bacterial infection.

3. Another problem concerning cultured cells as representative of human mucosa is that real mucosal surfaces are covered with mucus and bathed in solutions that are difficult to mimic in an in vitro system.

The fact that pitfalls may occur with cultured cells does not mean that they have not been extremely useful for investigating the bacterial adhesion to respiratory mucosa. If their limitations are kept in mind, cultured cells may be important for generating hypothesis that may be validated in animal models.

Different methods have been used to investigate bacterial adhesion to cultured cells. In all these methods, cultured cells are exposed to bacterial suspensions for different periods and rinsed to eliminate nonadherent microorganisms. Different approaches are then adopted, depending on the method selected.

## 1. Light Microscopy Assays

Cells with attached bacteria are fixed, stained with Giemsa or Gram dyes and observed under light microscopy to determine the percentage of cells with adherent bacteria as well as the mean number of adherent bacteria per cell. Alternatively, light microscopes equipped with ocular grids can be used to determine the mean number of bacteria per optical field.[8,67] Although this method can be carried out with cells cultured on thick gels of ECM, the best accuracy is obtained when cells have been cultured on rigid supports.

## 2. Electron Microscopy

Bacterial adhesion can also be assessed by scanning or by transmission electron microscopy (SEM and TEM, respectively). In SEM assays, cells with adherent microorganisms are fixed, submitted to critical point drying, and coated with gold palladium particles, before qualitative or qualitative assessment of bacterial adhesion.[54] By using image analysis workstations connected to the SEM, it is possible to obtain topographic information, such as the localization of the object within a region of interest drawn at any magnification of the microscope or quantitative data. In a study carried out by Colliot et al.,[14] a relationship between *P. aeruginosa* adhesion to respiratory cells and the labeling of respiratory cells with the lectin RCA II adsorbed to colloidal gold granules could be demonstrated. Computer-assisted SEM was also very useful in the determination of the number of aggregated bacteria adherent to regenerating epithelial respiratory cells.[18] SEM studies can be carried out on cells cultured on rigid supports and, most interestingly, also with cells cultured on substrates allowing a better differentiation of cells, such as thick gels of ECM components and porous membranes.

TEM of sectioned material has primarily been used to provide high-resolution information about the mechanisms of interaction of bacteria with target cell surfaces. The combined use of immunolabeling can provide information on the nature of the epitopes of both the bacterium and the target cell involved in adhesion.[54] The major disadvantages of the method are: 1) only very small areas of cell surfaces are examined in each ultrathin section, and 2) the specimen preparation requires a high level of skill and the technique is time-consuming. TEM should therefore be used for studies where alternative simpler techniques are inappropriate.

### 3. Viable Count Assays

Cells with adherent bacteria are treated with sterile detergents such as 0.1% Triton X-100 for 5 to 10 min to detach bound microorganisms. Cell lysates are then diluted and cultured.[67,76] Results are presented both as the number of bacteria per ml of cell lysate and as the percentage of the starting bacterial inoculum. This method can be carried out with cells cultured on rigid supports, on thick gels of ECM components or on porous membranes.

### 4. Enzyme Linked Immunosorbent Assay (ELISA) Based Adhesion System

Cell cultures are fixed in order to minimize any detachment from the plastic surfaces. Adherent bacteria are then treated with anti-bacterial antibodies followed by secondary antibody complexed with horseradish peroxidase or alkaline phosphatase. The development of color after adding an appropriate chromogenic substratum for the enzymes reflects the amount of bound bacteria and can be quantified with a microplate reader.[4,65] However, this method can be carried out only when cells have been cultured on 96-well microtiter plates and it determines total adhesion to a given population of animal cells. It presents, as an advantage over other methods, the possibility of screening large numbers of potential inhibitors with high sensitivity. Moreover, when the target cells are capable of internalizing the adherent bacteria, the ELISA-based system allows discrimination between the attachment and internalization stages of the endocytic process.[4] Fixatives that do not permeabilize cell membranes (such as 4% paraformaldehyde) will only allow the detection of adherent bacteria whereas fixatives permeabilizing cell membranes (such as methanol) allow the detection of both adherent and internalized microorganisms.

### 5. Fluorescence Flow Cytometry

In a technique recently described by Jiang et al.[48] to ascertain whether respiratory syncytial virus infection (RSV) would enhance the adhesion of *H. influenzae* to respiratory cells, virus-infected and noninfected cultured cells were detached from the culture wells with 0.1 M EDTA, pH 7.5, for 10 min, washed with PBS, and labeled with FITC-conjugated anti-RSV antibody. Cells were then incubated, at various ratios, with suspensions of bacteria previously labeled with a red fluorescent dye, washed by differential centrifugation to remove unattached bacteria, and fixed in 1% paraformalde-hyde prior to analysis by two-colour flow cytometry. This technique has also been used to assess the role of virulence factors in *B. pertussis* adhesion to respiratory cells.[120]

### 6. Radioactivity Count Assay

In this technique, bacteria are labeled with different isotopes[53,71] and exposed to cells cultured on rigid supports or on porous membranes. After the incubation period, cells are rinsed and adherent bacterial are quantified by scintillation counting of solubilized cells. Results are usually expressed as the percentage remaining of the initial radioactivity in-oculum added to the respiratory cells, according to the formula: % of adherent bacteria = counted dpm in solubilized cells/counted dpm in bacterial inoculum × 100.

Due to due to its simplicity, this technique has been used in inhibitory assays to assess the role of both bacterial[57] and host cell components[8,112] implicated in bacterial adhesion as well as the effect of antibacterial drugs.[123]

All these experimental approaches mentioned above have been used with success in studies of bacterial adhesion to respiratory cells. However, it is highly recommended that the initial assessment of adhesion should be conducted by microscopy to ascertain that bacteria are actually binding to the cells and not to supporting matrices or dishes. This recommendation is justified, for instance, by our finding during studies on *P. aeruginosa* interaction with human epithelial respiratory cells in primary culture. In those studies we have worked on the culture model described by Chevillard et al.[13] in which explants from human nasal polyps were cultured on a thick type I collagen matrix in defined culture medium. Under these conditions, cells from the explants migrated over the collagen gel, giving origin to an outgrowth area around the explants (Fig. 4A). When the cultures were exposed to *P. aeruginosa*, we observed that bacterial adhesion to the supporting matrix was significantly higher than to the epithelial cells (Fig. 4B, C[72]).

We also observed large aggregates of bacteria trapped at the extremities of cilia (Fig. 5A). By TEM and immunocytochemical labeling techniques, we demonstrated the presence of fibronectin in the fibrillar material surrounding aggregated microorganisms (Fig. 5B–D[72]). Had we worked on radiolabeled bacteria, all the radioactivity associated with the collagen matrix would have been attributed to bacterial adhesion to the epithelial cells. Moreover, we would not have detected the capability of fibronectin to agglutinate *P. aeruginosa* cells. Another advantage of microscopic methods to assess bacterial adhesion is that they may reveal that the organisms are binding nonuniformly to cultured cells. Bacteria may be found binding to specific area of the culture that may correlate with distribution of cellular receptors.

Although microscopic methods may give origin to precious information about bacterial adhesiveness, they also present a few disadvantages: 1) they are tedious and time consuming and this is particularly true for SEM methods and, 2) research on bacterial adhesion by light microscopy is often hampered by the inability to distinguish between bacteria that are attached to the surface of the cells from those that have been internalized by them.

### 7. Distinguishing Extracellular from Intracellular Bacteria

Different methods have been proposed to distinguish between attached and internalized bacteria. Some of these methods take advantage of the impermeability of

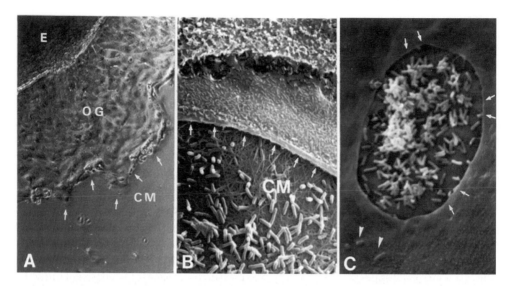

**Figure 4.** (A). Phase contrast micrograph of human nasal polyp explant (E) cultured on type I collagen matrix (CM) in defined culture medium. Note the presence of the epithelial outgrowth (OG) surrounding the explant, due to cell migration from the explant and to cell proliferation. Arrows point to the edges of the outgrowth. (B, C). Scanning electron micrographs of the cell culture illustrating the affinity of *P. aeruginosa* for the collagen matrix (CM) on which cells have been seeded. Arrows in B point to the edges of the outgrowth and in C to an area of retraction of the epithelial cells. The uncovered collagen matrix exhibit many adherent bacteria, in sharp contrast with the paucity of bacteria adherent to the cell membrane (arrowheads).

eukaryotic cell membranes to large molecules, such as antibodies. While investigating the interaction of *P. aeruginosa* with Caco-2 cells, Pereira et al.[67] fixed cell monolayers previously exposed to bacterial suspensions with 4% paraformaldehyde in PBS, a fixative solution that does not permeabilize cell membranes. Thereafter, cell cultures were treated with anti-bacterial antibodies complexed to horseradish peroxidase and with a solution containing diaminobenzidine (DAB), imidazol and hydrogen peroxide, to label attached microorganisms, and were counterstained with the Giemsa dye. By light microscopy, extracellular (EC) bacteria were seen stained in brown, while epithelial cells and intracellular (IC) microorganisms appeared stained in blue.

Another approach taking advantage of the impermeability of cell membranes to large molecules depends on the quenching of the fluorescence of EC bacteria by crystal violet[41] or by propidium iodide.[25] In the first method, following the incubation of eukaryotic cells cultured on glass coverslips with bacteria previously labeled with FITC, cells are submitted to fluorescence microscopy to determine the percentage of cells with associated microorganisms as well as the mean number of adherent bacteria per cell. In parallel, other coverslips are treated with crystal violet and re-submitted to fluorescence microscopy. EC bacteria have their fluorescence quenched by the dye and appear dark and nonfluorescent while IC microorganisms remain brightly fluorescent. In the second method, FITC-labeled IC bacteria remain green fluorescent whereas EC bacteria exhibit red or orange fluorescence.

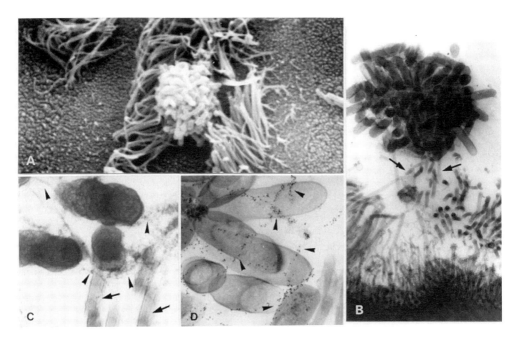

**Figure 5.** (A). Scanning electron micrograph of epithelial respiratory cells in primary culture showing the presence of a large aggregate of *P. aeruginosa* associated with ciliated cells. (B–D). Transmission electron micrographs of thick sections of a ciliated respiratory cell exposed to *P. aeruginosa* suspension showing in B, aggregated bacteria trapped at the extremities of cilia (arrows); in C, the presence of a fibrillar material (arrowheads) surrounding aggregated bacteria, which appears to establish the interaction between bacteria and cilia (arrows); in D, the labeling of the fibrillar material (arrowheads) which surround aggregated bacteria with colloidal gold granules, revealing the presence of fibronectin secreted by the respiratory cells in culture.

Although the methods mentioned may allow the distinction of adherent from internalized bacteria, the most current test to quantify IC bacteria is the gentamicin exclusion assay in which cultured cells previously exposed to bacteria are treated with gentamicin to eliminate EC bacteria. Internalized bacteria are not usually affected by the antibiotic. Cells are then treated with detergents, such as 0.1% Triton X-100. Dilution and plating of cell lysates allow precise quantification of IC microorganisms.[114]

### E. Tissue Culture Models

Small respiratory tissue fragments, or explants, may be maintained immersed in cell culture medium. Under these conditions, ciliated cells maintain their morphological and functional characteristics for up to 7 d.[88] In tissue culture models, the cut edges of the explants should be sealed with molten agar. Explants are then incubated with bacterial suspensions for different periods, and rinsed to eliminate nonadherent microorganisms. Bacterial adhesion to respiratory cells can be assessed by SEM, TEM, and by viable count assays, after cell treatment with detergent to detach adherent bacteria, as described above. Alternatively, radiolabeled microorganisms can be used. In this case, bacterial adhesion will be assessed by counting the radioactivity associated with the tissues. Cultures of

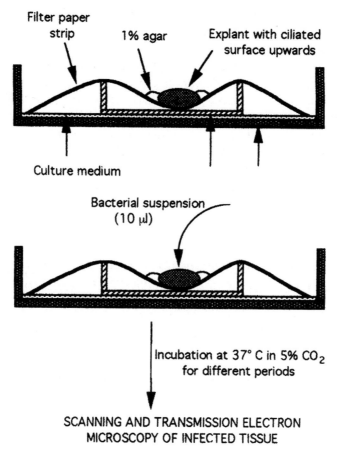

**Figure 6.** Schematic representation of the tissue culture model with an air–mucosal interphase. The respiratory tissue was placed on a strip of filter paper whose edges were immersed in culture medium. Bacterial inoculum was minimal. After infection, the tissue fragments were processed for electron microscopic observation.

tracheal rings[6,34,61,104,115] or of nasopharyngeal tissue[1,31,88] have been used to assess the bacterial adhesion to epithelial respiratory cells.

A major disadvantage of the in vitro models used as tools for studying bacterial adhesion to respiratory cells is that in vivo the interplay between these cells occurs at an air interface. Recently, Tsang et al.[118] have described a novel tissue culture model to study the interaction of *P. aeruginosa* with adenoid tissue in which bacterial interaction with respiratory mucosa occurs in an air–mucosal interphase. This model is more physiological than others in which cells are immersed in culture medium. Moreover, the mucociliary system is expected to function as it does in vivo. Briefly, a 3 cm petri dish was placed in the center of a 5 cm petri dish (Fig. 6). Cell culture medium was added to the 5 cm petri dish. A strip of paper measuring 50×5 mm was soaked in sterile cell culture medium and then placed aseptically onto and across the diameters of the two petri dishes. The two ends of the filter paper remained immersed in the culture medium. One piece of adenoid tissue was placed with its ciliated surface upwards onto the filter paper strip at the center of the smaller inner petri dish. Semimolten 1% agar was pipetted around the edge of the adenoid tissue in order to seal its cut edges. Ten microliters of bacterial suspension was dropped directly onto the center of the tissue which was incubated in 5% $CO_2$ in a humidified atmosphere. After different incubation periods, control noninfected and infected tissues were fixed and submitted to SEM and to TEM. By using this model the authors confirmed 1) the adhesion of *P. aeruginosa* to respiratory mucus; 2) the inability of bacteria to adhere to functionally intact ciliated cells; 3) the bacterial affinity for damaged epithelia and for basement membranes; and 4) the bacterial ability to induce epithelial damage, characterized by loss of epithelial junctions and cell exfoliation.

Recently, Jackson et al.[46] compared the interaction of *H. Influenzae* type b with adenoid tissue cultures maintained with an air interface or immersed in medium. They observed that bacterial adhesion to mucus and to damaged epithelium was significantly higher in the mucosa-air interface model than in immersed tissue cultures. They concluded that immersion of respiratory tissue during experimental infection can substantially influence the results obtained.

## VI. STUDYING BACTERIAL ADHESION TO INJURED AND REPAIRING RESPIRATORY EPITHELIUM

As underlined before, most bacteria rarely adhere to intact and functionally active respiratory ciliated cells. In contrast, airway epithelial injury and repair represent key moments which favour bacterial implantation within airways. Apart from the bacterial products themselves, which may be determinant factors of epithelial damage, bacteria may find, during epithelial injury, local conditions that are optimal for adhesion and replication. This is particularly the case for opportunistic bacteria such as *P. aeruginosa*. Therefore, when attempting to investigate the pathogenesis of infections by microorganisms that only infect respiratory mucosa previously damaged in some way, it is necessary to use an adequate experimental model. Both animal and cell culture models have been used to study bacterial adhesion to injured and repairing respiratory epithelium.

### A. Animal Models

In in vivo animal models, the surface epithelium has to be previously damaged in order to allow bacterial adhesion. This has been obtained by acid treatment,[81] by endotracheal intubation,[80] by viral infection[69,80] or even by $SO_2$ inhalation.[128] After injury, animal tracheas were filled with bacterial suspensions and adhesion was assessed by the counting of the radioactivity associated with tracheal mucosa, when radiolabeled bacteria were used, or by treating tracheas with detergents, to detach bound bacteria. Then, bacterial adhesion was quantitated by dilution and plating of cell lysates. However, by using these two methods to assess bacterial adhesion in animal models of epithelial injury, it is impossible to identify, among the many epithelial, mesenchymal, and inflammatory cells participating in the wound repair process, which cell phenotypes and molecules are specifically responsible for bacterial adhesion. Therefore, bacterial adhesion to animal mucosa should rather be assessed by SEM.

### B. Cell Culture Models

Another approach to studying bacterial adhesion to injured mucosa is the in vitro model of the wound repair process described by Zahm et al.[125] In this model, airway cells

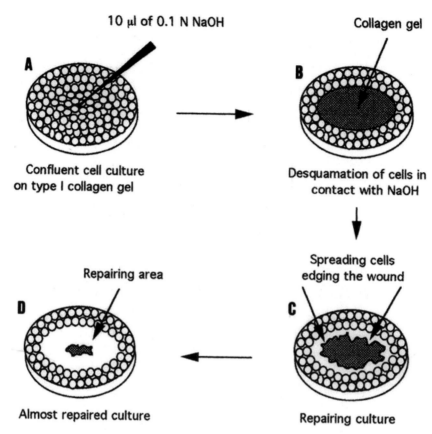

**Figure 7.** Schematic representation of the wound repair model described by Zahm et al.[126] and modified by Herard et al.[42] Dissociated epithelial respiratory cells were cultured on thick type I collagen matrix. (A) When cells reached confluence, a 1 μL drop of 0.1 N NaOH was deposited in the cell monolayer and immediately diluted in 1 mL of culture medium. (B) The cells which were directly in contact with NaOH desquamated from the collagen gel, creating a circular wound of about 30 mm². The wounded cultures were then rinsed and incubated with culture medium. (C) Cells at the edges of the wound spread over the collagen gel, to repair the wound. (D) The diameter of the epithelial wound has diminished due to cell spreading and migration.

dissociated from human nasal polyps or from human bronchial tissue are seeded on a thick type I collagen gel. When cells reach confluence, a circular mechanical or chemical wound (0.1 N NaOH) is made in the center of the culture (Fig. 7, Fig. 8A). Cell cultures are then rinsed and reincubated with culture medium. About 2 h after the wound has been made, cells edging the wound begin to extrude prominent lamellipodia in the forward direction of cellular motion (Fig. 8B), to spread, and migrate over the collagen matrix, leading to the progressive closure of the wounded area, which occurs in about 72 h. Within 24 h after injury, cell proliferation takes place, with a mitotic activity which peaks at 48 h after injury and involves 24% of the cells in the repaired area.[79]

To identify the factors that determine the *P. aeruginosa* adhesion to injured respiratory mucosa, we have worked on the wound repair model at 24 h after injury has been performed. By SEM, we observed the high affinity of bacteria for cells with the

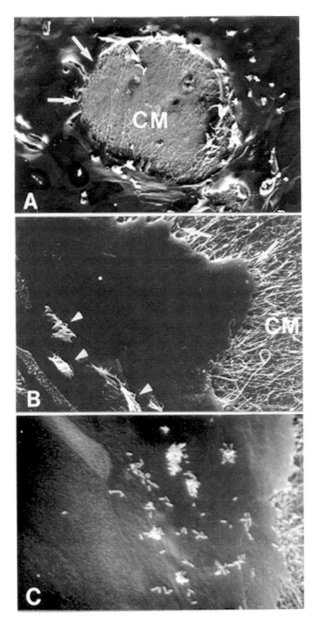

**Figure 8.** Scanning electron micrographs of a wounded epithelial respiratory cell culture. (A). Cells in contact with NaOH desquamated, creating a circular wound and uncovering the underlying collagen matrix (CM). Arrows point to the edges of the wound. (B). Cells edging the wound spread over the collagen matrix (CM). Lamellipodia (arrows) are seen running on the matrix. Microvilli usually detected at the cell surface progressively disappeared. Arrowheads point to ciliated cells. (C). Many isolated and aggregated bacteria are seen adherent to the spreading cells edging the wound.

spreading phenotype from the wound borders.[72-74](Fig. 8C) It has been shown by de Bentzmann et al.[19] that the upper membranes of these spreading cells are particularly rich in asialo GM1 residues, a major receptor for *P. aeuginosa* adhesin whereas Roger et al.[92]

have very recently shown that besides asialo GM1 residues, also fibronectin and α5β1 integrins mediate *P. aeruginosa* adhesion to the repairing epithelial respiratory cells.

## C. Xenograft Models

A denuded tracheal xenograft model was developed to analyze the regeneration potential of different animal airway cell subpopulations.[55] In this model, rat tracheas denuded of their surface epithelium by repeated cycles of freezing and thawing were seeded with adult airway epithelial cells and were subcutaneously implanted into immunodeficient nude mice. Seeded cells adhered to the tracheal ECM, grew, differentiated and progressively developed an epithelium similar in morphology and function to epithelium from normal proximal conducting airways. With this xenograft model it was shown that the regenerative process described in rat tracheas is similar to the regenerative process observed after a focal denuding mechanical injury of the respiratory mucosa.[107,108]

The xenograft model was later adapted to human bronchial epithelial cells,[27] offering a unique opportunity to perform studies on regenerating or on fully-differentiated human airway epithelium within an in vivo experimental environment. Using the humanized model it was shown that regenerating human airway epithelium is more prone than fully differentiated airway epithelium to be transduced by viral gene vectors.[38,39] Very recently, Dupuit et al.[26] modified further the tracheal xenograft model by using human nasal epithelial cells to mimic the regeneration of human proximal airway epithelium up to the fully-differentiated state.

In a similar model, human fetal tracheas were implanted under the skin of severe combined immunodeficiency (SCID) mice.[116] In these conditions, immature fetal human tracheas reached full histological maturation.

Both tracheal xenograft models described above seem to be very promising in studies on bacterial adhesion to normal and to regenerating human respiratory epithelium. Tirouvanziam et al.[117] have recently used the human fetal trachea xenograft model to compare the response of respiratory mucosa from cystic fibrosis (CF) patients and from non-CF individuals to *P. aeruginosa* primoinfection. They showed that leucocytes transmigrated more rapidly and massively through CF surface epithelium, exacerbating exfoliation of surface epithelial cells, thereby facilitating bacterial access to preferential sites of adhesion to basal cells and the basement membranes.

## VII. IDENTIFIATION OF CELL RECEPTORS AND BACTERIAL ADHESINS

After the establishment of the adhesion capability of a given bacterium, the characterization of the molecular nature of the host cell receptors and of bacterial adhesins should take place. Many different approaches can be adopted:

## A. Inhibition Assays

The chemical nature of receptors can be assessed by enzyme or chemical treatment of cultured cell monolayers, after fixing with 2% glutaraldehyde in PBS at 4°C, as described by St. Geme.[112] Briefly, after fixation, cells were rinsed and treated with 10 mM sodium metaperiodate-50 mM sodium acetate and later with 50 mM sodium borohydride in PBS, or with 100 μm of neuraminidase, or with 50 μg/mL of proteinase K, or with 5 U of peptide-*N*-glycosidase F. Control untreated and treated cells were then incubated with bacterial suspensions and the adhesion was quantitated by the viable count and the radio-activity assays. However, before treating the cells, one should ascertain whether cell fixation does not interfere with bacterial adhesion, due to the possibility of crosslinking of cell receptors by glutaraldehyde, as observed by St Geme and Falkow.[111]

Receptor active molecules can be further characterized indirectly by inhibition experiments in which cultured cells are incubated with defined antibodies[19] or lectins[112] prior to the incubation with bacterial suspensions. Alternatively, bacteria can be treated with oligosaccharides[8] or glycoconjugates[96] to prevent attachment in a competitive fashion. Inhibition by free ligand and structural analogs, besides defining the receptor specificity, may also define the avidity of binding (for review see ref[15]).

The nature of the adhesins can be assessed by bacterial treatment with antibodies raised against potential ligands, such as pili,[24] or by treatment with components purified from the bacterial cell.[57]

Although adhesion of bacteria to host cells may depend on the binding of a unique bacterial adhesin to a unique host receptor, most frequently bacterial adhesion depends on the simultaneous interaction between multiple adhesins with different cell receptors. The inhibition of a single pathway therefore may have no effect on overall adhesion or at best may produce a partial inhibition. In order to characterize the potential multiple receptor-ligands interactions between a pathogen and its host, it is often necessary to reduce the complexity by using purified ligands or receptors from one cell and study its binding to or its ability to promote adhesion to another cell type.

### B. Bacterial Binding to Immobilized Receptors

Host cell fractions with potential activity of receptor for bacterial adhesion can be extracted from cell membranes, separated in a solid phase and then exposed either to purified bacterial adhesins or to bacterial suspensions. Host cell proteins and glyco-proteins can be immobilized in nitrocellulose membranes in Western blotting assays,[92] whereas lipids and glycolipids can be separated on silica thin-layer chromatograms.[28,127] Potential cell receptors can also be immobilized onto plastic 96-well plates.[39] Identification of cellular receptors can be assessed by different approaches:

1. Overlaying of nitrocellulose replica or chromatograms with radiolabeled bacteria. After the incubation period, solid supports are rinsed, to eliminate nonadherent microorganisms, and submitted to autoradiography and to densitometric analysis.[28]
2. Overlaying of solid supports successively with unlabeled bacteria followed by treatment with primary antibacterial antibodies and with secondary antibodies complexed to horseradish peroxidase or alkaline phosphatase. Adherent microorganisms are assessed either by incubating the solid supports with chromogenic substrates or by the enhanced chemiluminescence (ECL) technique, this latter technique being more sensitive than the former.
3. Overlaying of solid supports with solutions of purified potential adhesins, such as pilin or OMPs, which may be unlabeled or labeled with radioisotopes or with biotin. Detection of bound adhesins will then be assessed as described above.

The role of purified potential receptors from host cell membranes in bacterial adhesion can also be assessed by ELISA assays.[36,127] Briefly, host cell proteins, glycoproteins or glycolipids are immobilized on wells from 96-well micotiter plates. After blocking with

PBS containing BSA at 1 or 2%, wells are exposed successively to bacterial suspensions (or with potential bacterial ligands), to primary antibacteria antibodies and to secondary antibodies complexed to horseradish peroxidase or alkaline phosphatase. After addition of a chromogenic substrate, bacterial adhesion is quantitated in a microplate reader. Alternatively, one can use radiolabeled microorganisms. Quantification of adherent bacteria is performed by cutting away each well and placing them in scintillation counters.

Finally, Amano et al.[2] have applied the recently developed biomolecular interaction analysis (BIAcore) system, involving the use of surface plasmon resonance (SPR), to investigate the interaction between bacterial fimbriae and host proteins immobilized onto a carboxymethyldextran chip. In the BIAcore system, one interactant is covalently immobilized onto a sensory chip whereas the other interactant is kept in solution and flows over the sensory chip. This system can detect small changes on or near the chip surface by measuring refractive index and can specify which ligands are immobilized. The benefits of SPR assay are: 1) direct and real-time observation of the interactions without any labeling of the proteins, 2) kinetic analysis to provide rate and affinity constants, and 3) screening of unknown interactants in crude samples.[2,62]

## VIII. CONCLUDING REMARKS

The ultimate goal of studies on bacterial adhesion is to understand what molecular interactions between host and bacteria occur in vivo and the impact of these interactions on disease processes.

Adhesion of bacterial pathogens to a mammalian cell is usually not a static process but rather elicits a response in the targeted cells, which depends ultimately on the kind of receptor that is bound.

Each cell receptor has a limited repertoire of responses, which depend on their associated signal transduction systems. Adhesion to plasma membrane receptors may be a prerequisite for invasion of an intracellular pathogen.[45] For other bacteria, the expression of genes for virulence factors may be triggered by the process of contact with and adhesion to specific host derived factors.[32,60] Adhesion may also elicit epithelial cells to produce cytokines,[21,22] with the possibility of initiating and regulating the mucosal inflammatory and immune response.[43] It may as well favor the activity of bacterial toxins, the intimate contact between bacteria and host cells allowing toxic products to reach concentrations sufficient to damage them. Moreover, adhesins may even be toxins which may mediate cellular toxicity by presentation of the enzymatically active subunit directly to host cell receptors.[43]

Many bacteria bind to specific carbohydrate structures. Within a given cell, these carbohydrate structures determine which of the their many glycoproteins or glyco-lipids[50] will interact with the bacteria. On the other hand, the noncarbohydrate parts of the epithelial cell receptors (either a protein or a lipid), is usually inserted in the membrane, and involved directly in specific cell physiological functions (at least in the case of glycoproteins). Although the nature of these noncarbohydrate moieties of cell receptors is likely to be most important for understanding the effects bacteria have on the cell, so far they have seldom been adequately characterized.

Increased knowledge of cellular response triggered by bacterial adhesion has the potential of leading to the development of new generations of antimicrobial drugs that could be more specifically directed towards the molecular events that lead to disease, thereby circumventing many of the undesirable side effects encountered with broad-spectrum antibiotics.[29] Finally, because antibodies to bacterial adhesins have been associated with protection from colonization, the identification of relevant adhesins for respiratory pathogens raises the possibility they can be used as components of vaccines.

# REFERENCES

1. Almagor M, Kahane I, Wiesel JM, et al: Human ciliated epithelial cells from nasal polyps as an experimental model for *Mycoplasma pneumoniae* infection. *Infect Immun* 48:552–5, 1985
2. Amano A, Nakamura T, Kimura S, et al: Molecular interactions of *Porphyromonas gingivalis* fimbriae with host proteins: kinetic analyses based on surface plasmon resonance. *Infect Immun* 67:2399–2405, 1999
3. Anderson JM, Balda MS, Fanning AS: The structure and regulation of tight junctions. *Curr Opin Cell Biol* 5:772–8, 1993
4. Athama A, Ofek I: Enzyme-linked immunosorbent assay for quantitation of attachment and ingestion stages of bacterial phagocytosis. *Infect Immun* 26:62–6, 1988
5. Audié JP, Janin A, Porchet N, et al: Expression of human mucin genes in respiratory, digestive, and reproductive tracts ascertained by *in situ* hybridization. *J Histochem Cytochem* 41:1479–85, 1993
6. Baker NR, Marcus H: Adherence of clinical isolates of *Pseudomonas aeruginosa* to hamster tracheal epithelium *in vitro*. *Curr Microbiol* 7:35–40, 1982
7. Barsum W, Wilson R, Read RC, et al: Interaction of fimbriated and nonfimbriated strains of unencapsulated *Haemophikus influenzae* with human respiratory tract mucus *in vitro*. *Eur Respir J* 8:709–14, 1995
8. Barthelson R, Mobasseri A, Zope D, et al: Adherence of *Streptococcus pneumoniae* to respiratory epithelial cells is inhibited by syalylated oligosaccharides. *Infect Immun* 66:1439–44, 1998
9. Benali R, Dupuit F, Jacquot J, et al: Growth and characterization of isolated bovine tracheal gland cells in culture. Influence of a reconstituted basement membrane matrix. *Biol Cell* 66:263–70, 1989
10. Beuth J, Ko HL, Schroten H, et al: Lectin mediated adhesion of *Streptococcus pneumoniae* and its specific inhibition *in vitro* and *in vivo*. *Zbl Bakt Hyg A* 265:160–8, 1987
11. Carnoy C, Scharfman A, Van Brussel E, et al: *Pseudomonas aeruginosa* outer membrane adhesins for human respiratory mucus glycoprotein. *Infect Immun* 62:1896–1900, 1994
12. Cereijido M, Gonzales-Mariscal L, Contreras RG, et al: Epithelial tight junctions. *Am Rev Respir Dis* 138:S17–21, 1988
13. Chevillard M, Hinnrasky J, Zahm JM, et al: Proliferation, differentiation and ciliary beating frequency of human respiratory ciliated cells in different conditions of primary culture. *Cell Tiss Res* 264:49–55, 1991
14. Colliot G, de Bentzmann S, Plotkowski MC, et al: Quantitative analysis and cartography in scanning electron microscopy: application to the study of bacterial adhesion to respiratory epithelium. *Microsc Res Tech* 24:527–36, 1993
15. Cowan MM: Kinetic analysis of microbial adhesion. *Method Enzymol* 253:179–89, 1995
16. Cozens AL, Yezzi MJ, Yamada M, et al: Transformed human epithelial cell line that retains tight junctions post crisis. *In Vitro Cell Dev Biol* 28A:735–744, 1992.
17. Davies J, Carlstedt I, Nilsson AK, et al: Binding of *Haemophilus influenzae* to purified mucins from the human respiratory tract. *Infect Immun* 63:2485–92, 1995.
18. De Bentzmann S, Bajolet-Laudinat O, Plotkowski MC, et al. Digital stereology to quantify the filling rate of bacterial aggregates of *Pseudomonas aeruginosa*. *J Microbiol Meth* 17:193–8, 1993

19. De Bentzmann S, Roger P, Bajolet-Laudinat O, et al: Asialo GM1 is a receptor for *Pseudomonas aeruginosa* adherence to regenerating respiratory epithelium. *Infect Immun* 64:1582–8, 1996

20. De Bentzmann S, Plotkowski MC, Puchelle E: Receptors in the *Pseudomonas aeruginosa* adherence to injured and repairing airway epithelium. *Am J Respir Crit Care Med* 154:S155–62, 1996

21. DiMango E, Zar HJ, Bryan R, et al: Diverse *Pseudomonas aeruginosa* gene products stimulate respiratory epithelial cells to produce interleukin-8. *J Clin Invest* 96:2204–10, 1995

22. DiMango E, Ratner AJ, Bryan R, et al: Activation of NF-kB by adherent *Pseudomonas aeruginosa* in normal and cystic fibrosis respiratory epithelial cells. *J Clin Invest* 101:2598–2606, 1998

23. Doig P, Todd T, Sastry PA, et al: Role of pili in adhesion of *Pseudomonas aeruginosa* to human respiratory epithelial cells. *Infect Immun* 56:1641–6, 1988

24. Doig P, Sastry PA, Hodges RS, et al: Inhibition of pili-mediated adhesion of *Pseudomonas aeruginosa* to human buccal epithelial cells by monoclonal antibodies directed against pili. *Infect Immun* 58:124–30, 1990

25. Drevets D, Campbell PA: Macrophage phagocytosis: use of fluorescence to distinguish between extracellular and intracellular bacteria. *J Immunol Meth* 142:31–8, 1991

26. Dupuit F, Gaillard D, Hinnrasky J, et al: Differentiated and functional human airway epithelium regeneration in tracheal xenografts. *Am J Physiol* Submitted, 1999

27. Engelhardt JF, Yankaskas JR, Wilson JM: *In vitro* retroviral gene transfer into human bronchial epithelia xenografts. *J Clin Invest* 90:2598–2607, 1992

28. Fakih MG, Murphy TF, Pattoli MA, et al: Specific binding of *Haemophilus influenzae* to minor gangliosides of human respiratory epithelial cells. *Infect Immun* 65:1695–700, 1997

29. Falk P, Bóren T, Haslam D et al: Bacterial adhesion and colonization assays. In: Russell DG, ed. *Microbes As Tools For Cell Biology*. Academic Press, San Diego, CA, 1994:165–92

30. Falk P, Roth KA, Gordon JI: Lectins are sensitive tools for defining the differentiation programs of epithelial cell lineages in the development of adult mouse gastrointestinal tract. *Am J Physiol* 266:G987–1003, 1994

31. Farley MM, Stephens DS, Mulks MH, et al: Pathogenesis of IgA, protease-producing and non-producing *Haemophilus influenzae* in human nasopharyngeal organ cultures. *J Infect Dis* 154:752–9, 1986

32. Finlay BB, Heffron F, Falkow S: Epithelial cell surfaces induce *Salmonella* protein required for bacterial adherence and invasion. *Science* 243:940–3, 1989

33. Franklin AL, Todd T, Gurman G, et al: Adherence of *Pseudomonas aeruginosa* to cilia of human tracheal epithelial cells. *Infect Immun* 55:1523–5, 1987

34. Funnell SGP, Robinson A: A novel adherence assay for *Bordetella pertussis* using tracheal organ cultures. *FEMS Microbiol Lett* 110:197–204, 1993

35. Gaillard D, Plotkowski MC: Changes in airway structure after airway infection. In: Chrétien J, Dusser D, eds: *Environmental Impact on the Airways*. M. Dekker, New York, 1996: 471–505

36. Geuijen CAW, Willems RJL, Mooi FR: The major fimbrial subunit of *Bordetella pertussis* binds to sulfated sugars. *Infect Immun* 64:2657–65, 1996

37. Girod S, Zahm JM, Plotkowski MC, et al: Role of the physicochemical properties of mucus in the protection of the respiratory epithelium. *Eur Respir J* 5:477–87, 1992

38. Goldman MJ, Wilson JM: Expression of avb5 integrin is necessary for efficient adenovirus-mediated gene transfer in the human airway. *J Virol* 69:5951–8, 1995

39. Goldman MJ, Lee PS, Yang JS, et al: Lentiviral vectors for gene therapy of cystic fibrosis. *Hum Gene Ther* 8:2261–8, 1997

40. Gruenert DC, Finkbeiner WE, Widdicombe JH: Culture and transformation of human airway epithelial cells. *Am J Physiol* 268:L347–60, 1995

41. Hed J: The extinction of fluorescence by crystal violet and its use to differentiate between attached and ingested microorganisms in phagocytosis. *FEMS Lett* 1:357–61, 1977
42. Herard AL, Zahm JM, Pierrot D, et al: Epithelial barrier integrity during *in vitro* wound repair of the airway epithelium. *Am J Respir Cell Mol Biol* 15:624–32, 1996
43. Hoepelman AIM, Tuomanen EI: Consequences of microbial attachment: directing host cell functions with adhesins. *Infect Immun* 60:1729–33, 1992
44. Irvin RT, Doig P, Lee KK, et al: Characterization of the *Pseudomonas aeruginosa* pilus adhesin: confirmation that the pilin structural protein subunit contains a human epithelial cell-binding domain. *Infect Immun* 57:3720–6, 1989
45. Isberg R: Discrimination between intracellular uptake and surface adhesion of bacterial pathogens. *Science* 252:934–8, 1991
46. Jackson AD, Cole PJ, Wilson R: Comparison of *Haemophilus influenza* type b interaction with respiratory mucosa organ cultures maintained with an a air interface or immersed in medium. *Infect Immun* 64: 2353–5, 1996
47. Jeffery PK: Morphology of airway surface epithelial cells and glands. *Am Rev Respir Dis* 128:S14–20, 1983
48. Jiang Z, Nagata N, Molina E, et al: Fimbria-mediated enhanced attachment of nontypable *Haemophilus influenzae* to respiratory syncytial virus-infected respiratory epithelial cells. *Infect Immun* 67:187–92, 1999
49. Jones GW, Isaacson RE. Proteinaceous bacterial adhesins and their receptors. *CRC Crit Rev Microbiol* 10:229–60, 1983
50. Karlsson KA: Animal glycosphingolipids as membrane attachment sites for bacteria. *Annu Rev Biochem* 58:309–50, 1988
51. Keenan KP, Combs JW, Mc Dowell EM: Regeneration of hamster tracheal epithelium after mechanical injury. *Virchows Arch B Cell Pathol* 41:193–214, 1982
52. Keenan KP, Wilson TS, McDowell EM: Regeneration of hamster tracheal epithelium after mechanical injury. IV: Histochemical, immunocytochemical and ultrastructural studies. *Virchows Arch B Cell Pathol* 43:213–40, 1983
53. Kishore R: Radiolabeled microorganisms: comparison of different radioisotopic labels. *Rev Infect Dis* 3:1179–85, 1981
54. Knutton S: Electron microscopical methods in adhesion. *Meth Enzymol* 253:145–58, 1995
55. Liu j, Nettesheim P, Randell SH: Growth and differentiation of tracheal progenitor cells. *Am J Physiol* 266 (*Lung Cell Mol Physiol*) 10:L296–307, 1994
56. Loveless RW, Feizi T: Sialo-oligosaccharide receptors for *Mycoplasma pneumoniae* and related oligosaccharides of poly-N-acetyllactosamine series are polarized at the cilia and apical-microvillar domains of the ciliated cells in human bronchial epithelium. *Infect Immun* 57:1285–9, 1989
57. Marty N, Pasquier C, Dournes JL, et al: Effects of characterised *Pseudomonas aeruginosa* exopolysaccharide on adherence to human tracheal cells. *J Med Microbiol* 47:129–34, 1998
58. Matsumara H, Setoguti T: Freeze-fracture replica studies of tight junctions in normal human bronchial epithelium. *Acta Anat* 134:219–26, 1989
59. McDowell EM, Becci PJ, Schurch W, et al: The respiratory epithelium: Epidermoid metaplasia of hamster tracheal epithelium during regeneration following mechanical injury. *J Natl Cancer Inst* 62:995–1008, 1979
60. Miller JF, Mekalanos JJ, Falkow S: Coordinate regulation and sensory transduction in the control of bacterial virulence. *Science* 243:916–22, 1989
61. Muse KE, Collier AM, Baseman JB: Scanning electron microscopic study of hamster tracheal organ cultures infected with *Bordetella pertussis*. *J Infect Dis* 136:786–77, 1977
62. Myszka DG: Kinetic analysis of macromolecular interactions using surface plasmon resonance biosensors. *Curr Opin Biotechnol* 8:50–7, 1997
63. Niederman MS, Rafferty TD, Sasaki CT, et al: Comparison of bacterial adherence to ciliated and squamous epithelial cells obtained from the human respiratory tract. *Am Rev Respir Dis* 127:85–90, 1983

64. Ofek I, Doyle RJ, eds: *Bacterial Adhesion To Cells And Tissues*. Chapman & Hall, New York, 1994

65. Ofek I: Enzyme-linked immunosorbent-based adhesion assays. *Meth Enzymol* 253: 528–36, 1995

66. Patrone LM, Cook JR, Crute BE, et al: Differentiation of epithelial cells on microporous membranes. *J Tiss Cult Meth* 14:225–34, 1992

67. Pereira SHM, Cervante MP, de Bentzmann S, et al: *Pseudomonas aeruginosa* entry into Caco-2 cells is enhanced in repairing wounded monolayers. *Microb Path* 23:249–55, 1997

68. Philippon S, Streckert HJ, Morgenroth K: *In vitro* study of the bronchial mucosa during *Pseudomonas aeruginosa* infection. *Virchows Archiv A Pathol Anat* 423:39–43, 1993

69. Plotkowski MC, Puchelle E, Beck G, et al: Adherence of type I *Sreptococcus pneumoniae* to tracheal epithelium of mice infected with influenza A/PR8 virus. *Am Rev Respir Dis* 134: 1040–4, 1986

70. Plotkowski MC, Beck G, Tournier JM, et al: Adherence of *Pseudomonas aeruginosa* to respiratory epithelium and the effect of leococyte elastase. *J Med Microbiol* 30:285–93, 1989

71. Plotkowski MC, Beck G, Bernardo Filho M, et al: Evaluation of the 99m Technetium labelling effect on *Pseudomonas aeruginosa* surface properties. *Ann Inst Pasteur/Microbiol* 138:415–26, 1987

72. Plotkowski MC, Chevillard M, Pierrot D, et al: Differential adhesion of *Pseudomonas aeruginosa* to human epithelial respiratory cells in primary culture. *J Clin Invest* 87: 2018–28, 1991

73. Plotkowski MC, Chevillard M, Pierrot D, et al: Epithelial respiratory cells from cystic fibrosis patients do not possess specific *Pseudomonas aeruginosa* adhesive properties. *J Med Microbiol* 36:104–11, 1992

74. Plotkowski MC, Bajolet-Laudinat O, Puchelle E: Cellular and molecular mechanisms of bacterial adhesion to respiratory mucosa. *Eur Respir J* 6:903–16, 1993

75. Plotkowski MC, Tournier JM, Puchelle E: *Pseudomonas aeruginosa* possess specific adhesins for laminin. *Infec Immun* 64:600–5, 1996

76. Plotkowski MC, de Bentzmann S, Pereira SHM et al: *Pseudomonas aeruginosa* internalization by human epithelial respiratory cells depends on cell differentiation, polarity, and junctional complex integrity. *Am J Respir Cell Mol Biol* 20:880–90, 1999

77. Pucciarelli MG, Finlay BB: Polarized epithelial monolayers: model systems to study bacterial interaction with host epithelial cells. *Methods Enzymol* 236:438–47, 1994

78. Puchelle E, Zahm JM: Repair processes of the airway epithelium. In: Chrétien J, Dusser D, eds. *Environmental Impact on the Airways. From Injury To Repair*. M. Dekker, New York, 1996:157–82

79. Puchelle E, Zahm JM, Tournier JM, et al: Airway epithelial injury and repair. *Eur Respir Rev* 7:136–41, 1997.

80. Ramphal R, Small PM, Shands Jr JW, et al: Adherence of *Pseudomonas aeruginosa* to tracheal cells injured by influenza infection or by endotracheal intubation. *Infect Immun* 27:614–9, 1980

81. Ramphal R, Pyle M: Adherence of mucoid and nonmucoid *Pseudomonas aeruginosa* to acid-injured tracheal epithelium. *Infect Immun* 41:345–51, 1983

82. Ramphal R, Pyle M: Evidences for mucins and sialic acid as receptors for *Pseudomonas aeruginosa* in the lower respiratory tract. *Infect Immun* 41:339–44, 1983

83. Ramphal R, Guay C, Pier G: *Pseudomonas aeruginosa* adhesins for tracheobronchial mucin. *Infect Immun* 55:600–3, 1987

84. Ramphal R, Houdret N, Koo L, et al: Differences in adhesion of *Pseudomonas aeruginosa* to mucin glycopeptides from sputa of patients with cystic fibrosis and chronic bronchitis. *Infect Immun* 57:66–71, 1989

85. Rampahl R, Koo L, Ishimoto K, et al: Adhesion of *Pseudomonas aeruginosa* pilin-deficient mutants to mucin. *Infect Immun* 59:1307–11, 1991

86. Ramphal R, Carnoy C, Fievre E, et al: *Pseudomonas aeruginosa* recognizes carbohydrate chains containing type 1 (Gal β1-3GlCNAc) or type 2 (Gal β1-4GlcNAc) disaccharide units. *Infect Immun* 59:700–4, 1991

87. Rayner CFJ, Jackson AD, Rutman A, et al: Interaction of pneumolysin-sufficient and –deficient isogenic variants of *Streptococcus pneumoniae* with human respiratory mucosa. *Infect Immun* 63:442–7, 1995

88. Read RC, Wilson W, Rutman A, et al: Interaction of nontypable *Haemophilus influenzae* with human respiratory mucosa *in vitro*. *J Infect Dis* 163:549–58, 1991

89. Read RC, Rutman A, Jeffery PK, et al: Interaction of capsulated *Haemophilus influenzae* with human airway mucosa *in vitro*. *Infect Immun* 60:3244–52, 1992

90. Reynolds HY: Respiratory host defense-surface immunity. *Immunobiol* 191:402–12, 1994

91. Roberts DD: Interactions of respiratory pathogens with host cell surface and extracellular matrix components. *Am J Respir Cell Mol Biol* 3:181–6, 1990

92. Roger P, Puchelle E, Bajolet-Laudinat O, et al: Fibronectin and α5β1 integrin mediate binding of *Pseudomonas aeruginosa* to repairing airway epithelium. *Eur Respir J* 13:1301–9, 1999

93. Rose MC: Mucins: structure, function and role in pulmonary diseases. *Am J Physiol* 263: L413–29, 1992

94. Roussel P, Lamblin G, Lhermitte M, et al: The complexity of mucins. *Biochimie* 70:1471–82, 1988

95. Roussel P, Ramphal R, Lamblin G: Bacterial infection and airways epithelium. In: Chrétien J, Dusser D, eds. *Environmental Impact on the Airways. From Injury To Repair*. M. Dekker, New York, 1996:437–9

96. Saiman L, Prince A: *Pseudomonas aeruginosa* pili bind to asialo GM1 which is increased on the surface of cystic fibrosis epithelial cells. *J Clin Invest* 92:1875–0, 1993

97. Sajjan US, Coey M, Karmali MA, et al: Binding of *Pseudomonas cepacia* to normal human intestinal mucin and reapiratory mucin from patients with cystic fibrosis. *J Clin Invest* 89:648–56, 1992

98. Sajjan U, Doig RP, Irvin RT, et al: Binding of nonmucoid *Pseudomonas aeruginosa* to normal human intestinal mucin and respiratory mucin from patients with cystic fibrosis. *J Clin Invest* 89:657–65, 1992

99. Sajjan US, Forstner JF, et al: Identification of the mucin-binding adhesin of *Pseudomonas cepacia* isolated from patients with cystic fibrosis. *Infect Immun* 60:1434–40, 1992

100. Salyers AA, Whitt DD: Experimental approaches to investigating the host-bacterium interaction. In: Salyers AA, Whitt DD, eds. *Bacterial Pathogenesis. A Molecular Approach*. ASM Press, Washington, DC, 1994, 73–89

101. Sanford BA, Thomas VL, Ramsay M: Binding of staphylococci to mucus *in vivo* and *in vitro*. *Infect Immun* 57:3735–42, 1989

102. Scharfman A, van Brussel E, Houdret N, et al: Interactions between glycoconjugates from human respiratory airways and *Pseudomonas aeruginosa*. *Am J Respir Crit Care Med* 154:S163–9, 1996

103. Scharfman A, Kroczynski H, Carnoy C, et al: Adhesion of *Pseudomonas aeruginosa* to respiratory mucins and expression of mucin-binding proteins are increased by limiting iron during growth. *Infect Immun* 64:5417–20, 1996

104. Sekiya K, Flutaesaku Y, Nakase Y: Electron microscopic observations on ciliated epithelium of tracheal organ cultures infected with *Bordetella bronchiseptica. Microbiol Immunol* 33:111–21, 1989

105. Sharon N. Bacterial lectins, cell-cell recognition and infectious disease. *FEBS Lett* 217:145–57, 1987

106. Simpson DA, Ramphal R, Lory S: Genetic analysis of *Pseudomonas aeruginosa* adherence: distinct genetic loci control attachment to epithelial cells and mucins. *Infect Immun* 60:3771–9, 1992

107. Shimizu T, Nettesheim P, Ramaekers FCS, et al: Expression of "cell-type-specific" markers during rat tracheal epithelial regeneration. *Am J Respir Cell Mol Biol* 7:30–41, 1992

108. Shimizu T, Nishihara M, Kawaguchi S, et al: Expression of phenotypic markers during regeneration of rat tracheal epithelium following mechanical injury. *Am J Respir Cell Mol Biol* 11: 85–94, 1994

109. Shuter J, Hatcher VB, Lowy FD: *Staphylococcus aureus* binding to human nasal mucin. *Infect Immun* 64:310–8, 1996

110. Sleigh MA, Blake JR, Liron N: The propulsion of mucus by cilia. *Am Rev Respir Dis* 137: 726–41, 1988

111. St. Geme JW III, Falkow S: *Haemophilus influenzae* adheres to and enters cultured human epithelial cells. *Infect Immun* 58:403–44, 1990

112. St. Geme JW III: The HMW1 adhesin of nontypable *Haemophilus influenzae* recognizes sialylated glycoprotein receptors on cultured human epithelial cells. *Infect Immun* 62:3881–9, 1994

113. Sterk LM, Alphen LV, Den Broek GV, et al: Differential binding of *Haemophilus influenzae* to human tissue by fimbriae. *J Med Microbiol* 35:129–38, 1991

114. Tang P, Foubister V, Pucciarelli MC, et al: Methods to study bacterial invasion. *J Microbiol Methods* 18:227–40, 1993

115. Temple LM, Weiss AA, Walker KE, et al: *Bordetella avium* virulence measured *in vivo* and *in vitro. Infect Immun* 66:5244–51, 1998

116. Tirouvanziam R, Desternes M, Saari A, et al: Bioelectric properties of human cystic fibrosis and non-cystic fibrosis fetal tracheal xenografts in SCID mice. *Am J Physiol Cell Physiol* 43:C875–82, 1998

117. Tirouvanziam R, De Bentzmann S, Hinnraski J, et al: Native inflammatory imbalance in CF human airway xenografts leads to exacerbation of *Pseudomonas aeruginosa* primoinfection. *Abstract from the Inflammation Meeting,* Paris, June 27–30, 1999

118. Tsang KWT, Rutman A, Tanaka E, et al: Interaction of *Pseudomonas aeruginosa* with human respiratory mucosa *in vitro. Eur Respir J* 7:1746–53, 1994

119. Ulrich M, Herbert S, Berger J, et al: Localization of *Staphylococcus aureus* in infected airways of patients with cystic fibrosis in a cell culture model of *S. aureus* adherence. *Am J Respir Cell Mol Biol* 19:83–91, 1998

120. Van den Berg BM, Beekhuizen H, Willems RJL, et al: Role of *Bordetella pertussis* factors in adherence to epithelial cell lines derived from the human respiratory tract. *Infect Immun* 67:1056–62, 1999

121. Vishwanath S, Ramphal R: Adherence of *Pseudomonas aeruginosa* to human tracheobronchial mucin. *Infect Immun* 45:197–02, 1984

122. Welsh MJ: Electrolyte transport by airway epithelia. *Physiol Rev* 67:1143–84, 1987

123. Wolter JM, McCormack JG: The effect of subinhibitory concentrations of antibiotics on adherence of *Pseudomonas aeruginosa* to cystic fibrosis (CF) and non-CF affected tracheal epithelial cells. *J Infect* 37:217–23, 1998

124. Woods DE, Starus DC, Johanson WG, et al: Role of pili in adherence of *Pseudomonas aeruginosa* to mammalian buccal epithelial cells. *Infect Immun* 29:1146–51, 1980

125. Zahm JM, Puchelle E: Wound repair of human surface respiratory epithelium. *Am Rev Respir Cell Mol Biol* 5:242–9, 1991

126. Zahm JM, Kaplan H, Herard AL, et al: Cell migration and proliferation during the *in vitro* wound repair of the respiratory epithelium. *Cell Motil Cytoskeleton* 37:33–43, 1997

127. Zhang Q, Young TF, Ross RF: Glycolipid receptors for attachment of *Mycoplasma hyopneumoniae* to porcine respiratory ciliated cells. *Infect Immun* 62:4367–73, 1994

128. Zoutman DE, Hulbert WC, Pasloske BL, et al: The role of polar pili in the adherence of *Pseudomonas aeruginosa* to injured canine tracheal cells: a semiquantitative morphologic study. *Scanning Microsc* 5:109–26, 1991

# 30

# Studying Bacterial Adherence to Endothelial Cells

**David A. Elliott[1] and Franklin D. Lowy[2]**

[1]*Program in Cellular and Molecular Medicine, Department of Pathology,*
*Johns Hopkins School of Medicine, Baltimore, MD, USA*
[2]*Department of Medicine, Microbiology and Immunology,*
*Albert Einstein College of Medicine, Bronx, NY, USA*

## I. INTRODUCTION

Bacterial adhesion to host surfaces is the critical first step in the initiation of infection. The ability of different bacterial species to establish infections at particular sites is in large part a consequence of their ability to colonize these tissue sites. Selective colonization of host surfaces is often a result of specific bacterial adhesin–host cell surface receptor interactions. Binding may occur to the surface directly via specific or non-specific interactions, or via bridging ligands. These bridging ligands, usually serum or cellular components, contain multiple binding sites that allow them to simultaneously bind both bacterial and tissue molecules thus mediating adherence to host surfaces.

Bacterial adherence to endovascular tissue has become an area of intense interest. The ability of bacteria to adhere to and invade endovascular tissue is a major pathogenic virulence mechanism that enables these bacterial species to cause a multitude of different, potentially life-threatening diseases. There is growing evidence that the endothelium has an active role both in the initiation and the progression of the inflammatory response to infection. Endothelial cells contribute to regulation of leukocyte transmigration to foci of infection, changes in vascular permeability and tone, the release of cellular cytokines and the expression of surface molecules such as Fc receptors.[25] As a result, alterations of the endothelium contribute to the clinical manifestations of such diverse syndromes as sepsis, vasculitis and meningitis. The present discussion reviews studies describing the interaction of different bacterial species with endothelial cells and the methods used to investigate these interactions. Because of our particular interest in staphylococcal–human endothelial cell interactions, many of the methods discussed in the later sections are based on studies performed with this pathogen.

## II. THE ADHERENCE OF DIFFERENT BACTERIAL SPECIES TO ENDOTHELIAL CELLS, IN VITRO

The ability of different bacterial species to adhere to endothelial cells varies. Those pathogens capable of adhering, are also the ones associated with endovascular infections

*Handbook of Bacterial Adhesion: Principles, Methods, and Applications*
Edited by: Y. H. An and R. J. Friedman © Humana Press Inc., Totowa, NJ

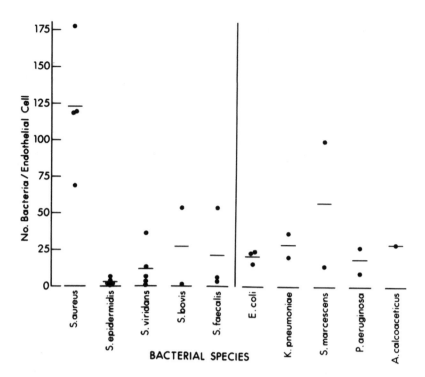

**Figure 1.** Adherence of bacterial blood culture isolates to human umbilical vein endothelial cells after a 2 h incubation at 37°C. The average initial bacterial inoculum was $2.2 \times 10^8$ colony forming units/mL.[29] Reprinted with permission of the American Society for Microbiology.

such as infective endocarditis. Ogawa et al.[29] demonstrated that the clinical blood culture isolates most commonly associated with acute infective endocarditis, such as *Staphylococcus aureus*, were also the ones that demonstrated the highest adherence to human endothelial cells in an adherence assay (Fig. 1). This section reviews in vitro studies that examine the ability of different bacterial species to adhere to and invade endothelial cells. These species utilize a variety of ligands to attach to the endothelial cell surface. In many instances binding involves more than one adhesin-receptor interaction. This redundancy of adhesins ensures the ability of these strains to adhere. There is presently limited understanding of the molecular mechanisms involved in bacterial adhesion to and invasion of these cells.

*Bartonella henselae* colonization of vascular surfaces may cause tumorlike growth with endothelial cell proliferation. This fastidious Gram-negative bacillus has been associated with bacillary angiomatosis, parenchymal bacillary peliosis, cat scratch disease and bacterial endocarditis. The cell surface aggregation and subsequent engulfment of this organism apparently utilizes a unique mechanism that involves bacterial clumping followed by internalization by an endothelial cell structure termed an "invasome." This process occurs over 24 h rather than minutes and appears to be an actin-dependent process. The bacterial adhesin and the cellular receptor mediating internalization are unknown.[11]

*Borrelia burgdorferi* is the causative agent of lyme disease which is the most prevalent vector-borne illness in the United States. *B. burgdorferi* produces extracellular vesicles that contain different subsets of the outer surface proteins on their surfaces. Lipoproteins OspA and OspB are contained in these blebs, and mediate attachment to endothelial cells. These blebs can competitively inhibit *B. burgdorferi* from binding to endothelial cells.[33] Coburn et al.[8] report that $\alpha(v)\beta3$ and $\alpha5\beta1$ integrins are the endothelial cell receptors for *B. burgdorferi*.

*Escherichia coli* is the most common cause of Gram-negative bacterial meningitis during the neonatal period. *E. coli* K1 does not invade systemic endothelial cells, but does bind to and invade brain microvascular endothelial cells suggesting that they have a specific affinity for this endothelial cell type. The *E. coli* Outer Membrane Protein A (OmpA) specifically interacts with GlcNAcβ1-4GlcNAc epitopes on brain micro-vascular endothelial cells.[31] S fimbriae also appear to facilitate bacterial adherence.[30]

*Listeria monocytogenes* is one of the major pathogens that causes bacterial meningitis in infants and the elderly. Immunocompromised individuals can become infected and undergo bacteremia, sepsis, abortion, meningitis, and encephalitis. Greiffenberg et al.[18] reported that InlB (a 65-kDa protein) is responsible for binding to and invasion of microvascular endothelial cells. The endothelial cell receptor is unknown.

*Mycobacterium tuberculosis* infects one third of the world's population, and is the cause of devastating disease world wide. *M tuberculosis* and *M. smegmatis* both have a polar phosphatidylinositol mannoside on their surfaces. Hoppe et al.[19] suggest that this component is coated by a serum opsonin, the mannose-binding protein that mediates adherence to endothelial cells.

*Neisseria meningitidis* is a Gram-negative extracellular pathogen responsible for meningitis and septicemia. Virgi et al.[41,42] have suggested that the PilC-expressing *N. meningitidis* binds a serum component that in turn binds to $\alpha(v)\beta3$ integrin on the endothelial cells. Thus a trimolecular complex may drive neisserial invasion. The corresponding endothelial receptor is not known.

*Porphyromonas gingivalis* is associated with periodontitis, a major cause of tooth loss in the adult population. It has also recently been associated with cardiovascular disease and preterm delivery of low birth weight infants. The major fimbriae protein is required for both adherence to and invasion of endothelial cells. The host cell receptor is not known.[12]

*Rickettsia rickettsii* are responsible for the systemic disease Rocky Mountain Spotted Fever. A major clinical manifestation of this disease is vasculitis. In vitro studies have demonstrated that the rickettsia entry of endothelial cells requires both cellular and bacterial participation.[43]

*Staphylococcus aureus* causes a diversity of diseases that range from minor skin and soft tissue to life-threatening systemic infections including endocarditis and sepsis. *S. aureus* adheres to endovascular tissue and endothelial cells more avidly than other bacterial species (Fig. 1).[17,29] Several candidate staphylococcal surface proteins that bind endothelial cells have been identified, although their functional role has not been further defined.[26,36] Endothelial cell receptors for staphylococci have also not been well characterized. A 50 kDa endothelial cell membrane protein that binds staphylococci has been partially characterized.[37] Cheung et al.[6] have found that fibrinogen appears to act as a bridging ligand enhancing staphylococcal adherence to endothelial cells.

*Streptococcus agalactiae* (group B streptococci) are the most common cause of meningitis in human newborns, with an incidence of 1.8 to 3.2 cases per 1000 live births, and are associated with significant morbidity and mortality. These bacteria gain access to the central nervous system by crossing the blood-brain barrier. Serotype 3, the type most commonly associated with meningitis, is also the most efficient invader of human brain microvascular endothelial cells.[28] The bacterial adhesin is not known, but the endothelial cell receptor has been proposed to be complement S protein (vitronectin).[39]

*Streptococcus gordonii* gains access to the microcirculation through breaks in the oral mucosa. Once circulating in the blood, the bacteria may adhere to and damage the endothelium causing endocarditis. A single adhesin protein of 153 kDa (a surface-localized glucosyltransferase) has been identified as the adhesin responsible for streptococcal adherence to human umbilical vein endothelial cells. Antibodies against the purified protein inhibited adherence.[38]

*Streptococcus pneumoniae* colonizes the nasopharynx of up to 40% of healthy adults, can cause pneumonia, otitis media, sepsis, meningitis, and in developing countries is the most frequent cause of bacterial pneumonia among children and adults.[2,22] Pneumococci bind both to resting and activated endothelial cells. Pneumococci bind to resting endothelial cells by attaching to GalNAcβ1-4Gal and GalNAcβ1-3Gal. The associated bacterial adhesin is not known. The activation of endothelial cells results in a dramatic increase in pneumococcal binding. Activated endothelial cells upregulate expression of platelet-activating factor (PAF) and PAF receptor. Phosphorylcholine on the pneumococcal surface then binds to the PAF receptor on the activated host cells.[10]

## III. ENDOTHELIAL CELL TYPES

Primary tissue culture systems have been used to investigate bacterial adherence to endothelial cells. Human tissue culture monolayers are easy to work with, can be maintained under controlled conditions, and are relevant to human disease. Several different systems will be discussed. In tissue culture, endothelial cells exhibit Weibel-Palade bodies, and grow in monolayers with a typical cobblestone morphology. Endothelial cell differentiation is in large part determined by the local tissue microenvironment. Therefore cells derived from different sites can vary dramatically in both structural and functional properties. The nature of the intercellular junctions vary depending on the source of the endothelial cell as do the synthesis of cytokines, proteoglycans and surface receptors.[15,44] Traditionally endothelial cells have been cultured from umbilical vein endothelial cells, pulmonary arteries, saphenous veins as well as other tissue sites. Cells can be maintained in tissue culture for multiple passages, however with long-term passage there is a tendency for these cells to undergo dedifferentiation with loss of many of their characteristic properties.[15]

The growth conditions of the endothelial cells also has an effect on the susceptibility of these cells to bacterial infection. Blumberg et al.[5] demonstrated that the absence of acidic fibroblast growth factor increased cellular susceptibility to infection. Others have shown that the presence of serum components or stimulation of cells with cytokines such as tumor necrosis factor also alters susceptibility to infection.[6,7]

Stably transformed endothelial cells are also available. These cells maintain acceptable levels of differentiation while removing the variability of the differing genetic makeup of

**Table 1. Endothelial Cells Currently Used in In Vitro Studies**

1. Bovine Aortic Endothelial Cells (BAEC) — can be acquired from the NIA. Aging Cell Culture Repository, Coriell Institute for Medical Research.
2. Brain Microvascular Endothelial cells (BMEC) — can be isolated by the method of Stins et al.[34,35] They can also be immortalized by transfection with simian virus 40 large T antigen while maintaining their morphologic and functional characteristics.
3. EA-hy926 — endothelial cells are the result of a fusion of human umbilical vein endothelial cells (HUVEC) and a human adenocarcinoma cell line. They are reported to maintain acceptable differentiation.
4. Fetal Bovine Heart Endothelial Cells (FBHEC) — are available from ATCC.
5. Human Saphenous Vein Endothelial Cells (HSVEC) — can be isolated by the method of Klein-Soyer et al.[23]
6. Primary Human Umbilical Vein Endothelial Cells (PHUVEC) — can be isolated by the method of Gordon et al.[16] These cells are a mainstay of tissue culture systems since they can be easily isolated from discard tissue. They can be grown easily and can be reliably used for up to 6 passages (*see* Table 2). These cells are now available from the ATCC.
7. Lung Microvascular Endothelial Cells (LmvEC) — can be isolated by the method of Meyrick.[27] Pulmonary Artery Endothelial Cells (PAEC) — may be isolated by the method of Schwartz.[32]

the primary culture cells' donors. Transformation may, however, alter other cellular characteristics. A listing of different endothelial cell types currently used in different adherence assays is shown in Table 1. Table 2 outlines a method for the isolation and passage of human umbilical vein endothelial cells in tissue culture. Although monolayer systems are immensely useful, they have limitations for the study of host–bacterial interactions. When infecting a human endothelial cell, bacteria must interact with multiple layers of cells. This is not the case with most in vitro adherence assays. There are, however, several models that permit the investigation of more complex systems. Biegel et al.[3] developed a system of growing brain microvessel endothelial cells on collagen gels over transwell filters, that permits study of blood–brain barrier physiology and central nervous system inflammation. Birkness et al.[4] developed a system that permits investigation of bacterial interactions with multiple cell layers and types. Alston et al.[1] demonstrated that endothelial cells grown on extracellular matrix synthesized by *S. aureus*-infected endothelial cells become more susceptible to subsequent staphylococcal infection.

## IV. METHODS OF MEASURING BACTERIAL ADHERENCE TO ENDOTHELIAL CELLS

This section will review in vitro methods for measuring bacterial adherence to endothelial cell monolayers. The specificity of the binding interaction is often determined by measuring whether adherence is saturable in a dose and time dependent manner. As noted above variations in the growth conditions of the endothelial cells can have significant effects on bacterial adherence. It is therefore important to maintain standard culture conditions for the performance of these assays. In addition, alterations of the bacterial growth conditions will also affect adherence. For example, *S. aureus* in the logarithmic growth phase are more adherent than bacteria in the stationary phase of growth. Finally the conditions of the adherence assay are also critical. Elliott et al.[13] found

**Table 2. Method to Isolate and Passage Human Umbilical
Vein Endothelial Cells (HUVEC)**

A. ISOLATION OF HUVEC FROM UMBILICAL CORDS

1. Coat 3.5 cm tissue culture plates with 0.2% gelatin. Leave at room temperature for 15 min and then remove the gelatin.
2. Wipe blood off the outside of the umbilical cord with sterile gauze. Cut off the ends the of cord with a razor blade. Attach a female cannula adapter to one end of the umbilical vein and secure with a tie. Flush the vein once with 20 mL 37°C sterile saline.
3. Attach a male cannula adapter to the other end of the umbilical vein and secure with a tie. Flush the vein once again with 20 mL 37°C sterile saline.
4. Fill the vein with as much collagenase (37°C) as the cord can hold. Clamp both ends with forceps. Place in a beaker containing warm saline and incubate at 37°C for 10 min.
5. Flush the vein once with 20 mL (37°C) saline to collect the endothelial cells from the vein.
6. Spin cells at 1000 rpm for 5 min. Discard the supernatant and resuspend the pellet in 3 mL of 37°C complete medium.
7. Transfer to a gelatin coated plate. Change the medium every day until the cells become confluent.

B. PASSAGE OF HUVEC

1. Coat 10 cm tissue culture plates with 0.2% gelatin as outlined above.
2. Wash the HUVEC 3 times with Hank's Balanced Salt Solution without $Ca^{++}$ or $Mg^{++}$.
3. Add 37°C trypsin (0.05% + 0.53 M EDTA) to the cells. Incubate at 37°C for 10 min.
4. Add complete medium to inactivate trypsin and then transfer the suspension to a 15 mL conical tube. Spin the cells at 1000 rpm for 5 min. Discard the supernatant.
5. Resuspend the cell pellet in 10 mL 37°C complete medium. Plate 5 mL of cells into gelatin-coated 100×20 mm plates.

C. REAGENTS

1. Collagenase (2 mg/mL)
   45 mL distilled $H_2O$
   5 mL 10X HEPES buffer, pH 7.5
   Add 100 mg collagenase (stir @ 4°C until in solution).
   Filter through a 0.22 μm filter
2. Complete Medium: per 100 mL
   74 mL M199
   0.1 mL Ascorbic acid (50 mg/mL)
   0.1 mL Heparin (25 mg/mL)
   0.8 mL Glutamine (Gibco)
   1 mL pen/strep (Gibco)
   0.25 mL Sigma growth factor (15 mg in 5 mL M199)
   5 mL Human serum (Biocell)
   20 mL Newborn calf serum (Gibco BRL)
3. M199: per 1 L
   1 package of M199 powder (Gibco BRL + 900 mL distilled $H_2O$)
   2.2 g $NaHCO_3$ +15 mL 1 M HEPES (pH 7.4)
   pH to 7.0–7.2
   Add distilled $H_2O$ to 1 L
   Filter through 0.22 μm.

**Table 3. Bacterial Infection Assay[29]**

1. Seed endothelial cells into tissue culture plates (4 to 96 well sets) and allow the cells to grow to confluence. Prior to the experiment determine the number of endothelial cells in several wells from each plate. There should be limited variability in cell counts between wells.
2. Select a single bacterial colony, inoculate in appropriate broth, and grow overnight. Wash bacteria and adjust to appropriate density by optical density.
3. Inoculate bacteria resuspended in the endothelial cell growth media minus antibiotics (e.g., Medium 199) to the confluent endothelial cellmonolayers. Incubate for 30 to 120 min at 37°C in 5.5% $CO_2$.
4. Following incubation, the nonadherent bacteria are removed and the cell surfaces washed three times with endothelial cell growth media.
5. The cells and adherent bacteria are lifted with trypsin (or with a solution of 0.5% triton X-100 in distilled water). For staphylococci, the endothelial cells are ruptured by incubation in distilled water for 5 min.
6. The bacteria are then serially diluted and plated into appropriate agar. The number of colonies are counted after 48 h.
7. Results are expressed as the number of bacteria per well or the number of bacteria per endothelial cell.

in one assay that a glycoprotein in the bacterial growth media (Casamino Acids–Yeast Extract) passively adsorbed to the bacteria and significantly reduced adherence to endothelial cell monolayers.

## A. Bacterial Adherence Assay

A standard adherence assay first described by Ogawa et al.[29] is outlined in Table 3. Radioactively-labeled bacteria ([³H]-thymidine) have also been used as an alternative method to measure adherence.[29] The latter assay is less labor intensive, however it is also less sensitive than the method outlined in Table 3. A third method for counting bacteria is to make the bacteria visible and then microscopically count adherent bacteria. Both Giemsa stain and acridine orange have been used to identify adherent bacteria. This is a fast technique that, in the case of some stains (e.g., acridine orange) may be used without killing the endothelial cells or the bacteria. Thus time-dependent measurements may be made. The disadvantages include the labor intensive nature of obtaining sufficiently large sample sizes. This type of assay provides more qualitative than quantitative data and as such is often used as a screening assay.

## B. Scanning Electron Microscopy (SEM)

The techniques of SEM are beyond the scope of this chapter (Hunter et al.[20] is recommended for further reading), but this is a common approach. The limitations of SEM are similar to those described above for other microscopic evaluations. However, SEM observations can provide important insights into the nature of early events of infection and allow visualization of the different steps in the invasion pathway.

## C. Bacterial Internalization Assay

In addition to adhesion, many bacteria are internalized by endothelial cells. This has, to date, been an in vitro observation for many bacterial species, however it is likely that a similar process occurs in vivo as well. The pathway used by bacteria to traverse the endothelium and gain access to adjacent tissues remains poorly defined. Measurement of the rates of internalization versus adherence of bacteria relies on the use of agents that are bactericidal but are not internalized by endothelial cells. One method to measure the rate of endothelial cell internalization of bacteria is described below.

The initial steps in the assay are as outlined in Table 3. After nonadherent bacteria are removed and the cell surfaces washed, gentamicin (100 μg/mL) is added for 30 min at 37°C. This antimicrobial agent will kill susceptible extracellular bacteria. Other antimicrobial agents may be used if the bacterial strain is not gentamicin-susceptible. Intracellular bacteria are unaffected because gentamicin achieves minimal intracellular concentrations. Following this step, the remainder of the assay is unchanged. By measuring the number of total and internalized bacteria one can also calculate the number of adherent bacteria.

### D. Identification of Bacterial Adhesins

Techniques to identify the specific bacterial adhesins responsible for attachment to eukaryotic cells include the use of molecular techniques such as transposon mutagenesis or biochemical methods such as protein purification followed by inhibition assays using purified protein or antibody. A further discussion of these techniques is beyond the scope of this chapter.

**Acknowledgment:** Supported by grants from the American Heart Association and the National Institute on Drug Abuse (DA09656 and DA11868).

## REFERENCES

1. Alston WK, Elliott DA, Epstein ME, et al: Extracellular matrix heparin sulfate modulates endothelial cell susceptibility to *Staphylococcus aureus*. *J Cell Physiol* 173:102–9, 1997
2. Austrian R: Some aspects of the pneumococcal carrier state. *J Antimicrob Chemother* 18 (Suppl A):35–45, 1986
3. Biegel D, Pachter FS: Growth of brain microvessel endothelial cells on collagen gels: applications to the study of blood-brain barrier physiology and CNS inflammation. *In Vitro Cell Dev Biol Anim* 30A:581–8, 1994
4. Birkness KA, Swisher BL, White EH, et al: A tissue culture bilayer model to study the passage of *Neisseria meningitidis. Infect Immun* 63:402–9, 1995
5. Blumberg EA, Hatcher VB, Lowy FD: Acidic fibroblast growth factor modulates *Staphylococcus aureus* adherence to human endothelial cells. *Infect Immun* 56:1470–4, 1988
6. Cheung AL, Krishnan M, Jaffe EA, et al: Fibrinogen acts as a bridging molecule in the adherence of *Staphylococcus aureus* to cultured human endothelial cells. *J Clin Invest* 87:2236–45, 1991
7. Cheung AL, Koomey JM, Lee S, et al: Recombinant human tumor necrosis factor alpha promotes adherence of *Staphylococcus aureus* to cultured human endothelial cells. *Infect Immun* 59:3827–31, 1991
8. Coburn J, Magoun L, Bodary SC: Integrins $\alpha(v)\beta3$ and $\alpha5\beta1$ mediate attachment of lyme disease spirochetes to human cells. *Infect Immun* 66:1946–52, 1998
9. Comstock LE, Thomas DD: Penetration of endothelial cell monolayers by *Borrelia burgdorferi. Infect Immun* 57:1626–8, 1989
10. Cundell D, Masure HR, Tuomanen EI: The molecular basis of pneumococcal infection: a hypothesis. *Clin Infect Dis* 21(Suppl 3):S204–12, 1995

11. Dehio C, Meyer M, Berger J, et al: Interaction of *Bartonella henselae* with endothelial cells results in bacterial aggregation on the cell surface and the subsequent engulfment and internalization of the bacterial aggregate by a unique structure, the invasome. *J Cell Science* 110:2141–54, 1997

12. Deshpande RG, Khan MB, Genco CA: Invasion of aortic and heart endothelial cells by *Porphyromonas gingivalis*. *Infect Immun* 66:5337–43, 1998

13. Elliott DA, Hatcher VB, Lowy FD: A 220-kilodalton glycoprotein in yeast extract inhibits *Staphylococcus aureus* adherence to human endothelial cells. *Infect Immun* 59:2222–3, 1991

14. Gaillard JL, Berche P, Frehel C, et al: Entry of *L. monocytogenes* into cells is mediated by internalin, a repeat protein reminiscent of surface antigens from Gram-positive cocci. *Cell* 65:1127–1141, 1991

15. Garlanda C, Dejana E: Heterogeneity of endothelial cells. *Arteriosclerosis Thrombovascular Biol* 17:1193-202, 1997

16. Gordon PB, Sussman II, Hatcher VB: Long-term culture of human endothelial cells. *In Vitro* 11:661–71, 1983

17. Gould K, Ramirez-Ronda CH, Holmes RK, et al: Adherence of bacteria to heart valves *in vitro*. *J Clin Invest* 56:1364–1370, 1975

18. Greiffenberg L, Goebel W, Kim KS, et al: Interaction of *Listeria monocytogenes* with human brain microvascular endothelial cells: In1B-dependent invasion, long-term intracellular growth, and spread from macrophages to endothelial cells. *Infect Immun* 66:5260–7, 1998

19. Hoppe HC, DeWet BJ, Cywes C, et al: Identification of phosphatidylinositol mannoside as a mycobacterial adhesin mediating both direct and opsonic binding to nonphagocytic mammalian cells. *Infect Immun* 65:3896–905, 1997

20. Hunter E, Maloney P, Bendayan M, et al: *Practical Electron Microscopy: A Beginner's Illustrated Guide*. 2nd Ed. Cambridge University Press, Cambridge, UK, 1993

21. Johnson AR: Human pulmonary endothelial cells in culture. *J Clin Invest* 65:841–50, 1980

22. Johnston RB: Pathogenesis of pneumococcal pneumonia. *Rev Infect Dis* 13(Suppl 6): S509–17, 1991

23. Klein-Soyer C, Beretz A, Millon-Collard R, et al: A simple *in vitro* model of mechanical injury of confluent culture endothelial cells to study quantitatively the repair process. *Thromb Haemostasis* 56:232–235, 1986

24. Levine EM, Mueller SN, Grinspan JB, et al: The limited life-span of bovine endothelial cells. In: Jaffe EA, ed. *Biology of Endothelial Cells*. Martinus Nijhoff Publishers, Boston, MA, 1984:109–17

25. Lowy FD: *Staphylococcus aureus* infections. *N Engl J Med* 339:520–32, 1998

26. McIntire-Campbell K, Johnson CM: Identification of *Staphylococcus aureus* binding proteins on isolated porcine cardiac valve cells. *J Lab Clin Med* 115:217–23, 1990

27. Meyrick B, Hoover R, Jones MR, et al: *In vitro* effects of endotoxin on bovine and sheep lung microvascular and pulmonary artery endothelial cells. *J Cell Physiol* 138:165–74, 1989

28. Nizet V, Kim KS, Stins M, et al: Invasion of brain microvascular endothelial cells by group B streptococci. *Infect Immun* 65:5074–81, 1997

29. Ogawa SK, Yurberg ER, Hatcher VB, et al: Bacterial adherence to human endothelial cells *in vitro*. *Infect Immun* 50:218–24, 1985

30. Parkkinen J, Ristimäki A, Westerlund B: Binding of *Escherichia coli* S fimbriae to cultured human endothelial cells. *Infect Immun* 57:2256–9, 1989

31. Prasadarao NV, Wass CA, Kim KS: Endothelial cell GlcNAcβ1-4GlcNAc epitopes for outer membrane protein A enhance traversal of *Escherichia coli* across the blood-brain barrier. *Infect Immun* 64:154–60, 1996

32. Schwartz SM: Selection and characterization of bovine aortic endothelial cells. *In Vitro* 14:966–80, 1978

33. Shoberg RJ, Thomas DD: Specific adherence of *Borrelia burgdorferi* extracellular vesicles to human endothelial cells in culture. *Infect Immun* 61:3892–3900, 1993

34. Stins MF, Prasadarao NV, Zhow J, et al: Bovine brain microvascular endothelial cells transfected with SV40-large T antigen: Development of an immortalized cell line to study pathophysiology of CNS disease. *In Vitro Cell Dev Biol* 33:243–7, 1997

35. Stins MF, Prasadarao NV, Ibric LV, et al: Binding of S-fimbriated *E. coli* to brain microvascular endothelial cells. *Am J Pathol* 145:1228–36, 1994

36. Tompkins DC, Blackwell LJ, Hatcher VB, et al: *Staphylococcus aureus* proteins that bind to human endothelial cells. *Infect Immun* 60:965–9, 1992

37. Tompkins DC, Hatcher VB, Patel D, et al: A human endothelial cell membrane protein that binds *Staphylococcus aureus in vitro. J Clin Invest* 85:1248–54, 1990

38. Vacca-Smith AM, Jones CA, Levine MJ, et al: Glucosyltransferase mediates adhesion of *Streptococcus gordonii* to human endothelial cells *in vitro. Infect Immun* 62:2187–94, 1994

39. Valentin-Weigand P, Grulich-Henn J, Chhatwal GS, et al: Preissner. Mediation of adherence of streptococci to human endothelial cells by complement S protein (vitronectin). *Infect Immun* 56:2851–5, 1988

40. Vernier A, Diab M, Soell M, et al: Cytokine production by human epithelial and endothelial cells following exposure to oral *Viridans streptococci* involves lectin interactions between bacteria and cell surface receptors. *Infect Immun* 64:3016–22, 1996

41. Virgi M, Makepeace K, Peak IR, et al: Opc- and pilus-dependent interactions of meningococci with human endothelial cells: molecular mechanisms and modulation by surface polysaccharides. *Molec Microbiol* 18:741–54, 1995

42. Virgi M, Makepeace K, Moxon ER: Distinct mechanisms of interactions of Opc-expressing meningococci at apical and basolateral surfaces of human endothelial cells; the role of integrins in apical interactions. *Molec Microbiol* 14:173–84, 1994

43. Walker TS: Rickettsial interactions with human endothelial cells *in vitro:* adherence and entry. *Infect Immun* 44:205–10, 1984

44. Watson A, Camera-Benson L, Palmer-Crocker R, et al: Variability among human umbilical vein endothelial cultures. *Science* 268:448, 1995

# 31

# Studying Bacterial Adhesion to Gastric Epithelium

**Shunji Hayashi,[1] Yoshikazu Hirai,[1] Toshiro Sugiyama,[2]**
**Masahiro Asaka,[2] Kenji Yokota,[3] and Keiji Oguma[3]**

*[1]Department of Microbiology, Jichi Medical School, Tochigi-ken, Japan*
*[2]Department of Internal Medicine, Hokkaido University School of Medicine, Sapporo, Japan*
*[3]Department of Bacteriology, Okayama University Medical School, Okayama, Japan*

## I. INTRODUCTION

It had long been held that the acidic gastric environment is deadly to microorganisms, and that a stomach with normal acid secretion is sterile. Thus, gastric microbiology had unfortunately been neglected until Warren and Marshall cultured and identified *Helicobacter pylori* (initially called gastric *Campylobacter*-like organisms, *Campylobacter pyloridis* and then *Campylobacter pylori*).[74] It is now confirmed that two *Helicobacter* species, *H. pylori* and *H. heilmannii* (originally named *Gastrospirillum hominis*) colonize the human gastric epithelium and are associated with gastritis and other gastric diseases. However, *H. heilmannii* has not been cultured.[78] Thus, the adhesive properties of *H. heilmannii* are unknown. Other bacteria and fungi are occasionally isolated from gastric specimens. However, it is not confirmed whether these microorganisms colonize the human gastric epithelium. Accordingly, we will review the adhesive properties of *H. pylori* to human gastric epithelium in this chapter.

## II. BIOLOGICAL AND PATHOLOGICAL SIGNIFICANCE OF *H. PYLORI* ADHESION

*H. pylori* is a Gram-negative, spiral-shaped rod that colonizes human gastric epithelium. In humans, *H. pylori* plays a causal role in chronic gastritis and peptic ulcers,[21,76] and is an important factor in the occurrence of gastric cancer and gastric mucosa-associated lymphoid tissue lymphoma.[18,75] *H. pylori* is isolated with high frequency from gastric biopsy specimens obtained from the patients with these diseases, but is rarely isolated from other specimens. This suggests natural *H. pylori* infection is specific to human gastric mucosa.

The adhesion of *H. pylori* to human gastric epithelium is the initial step of *H. pylori* colonization, and is mediated through the interactions between *H. pylori* adhesins and host cell receptors. *H. pylori* that is unable to adhere to gastric mucosa tends to be rapidly removed by shedding of surface epithelial cells and the mucous layer. Thus, adhesion is

*Handbook of Bacterial Adhesion: Principles, Methods, and Applications*
Edited by: Y. H. An and R. J. Friedman © Humana Press Inc., Totowa, NJ

essential for the maintenance of *H. pylori* colonization. In mucosal infections, the epithelium that the bacteria colonize is generally dependent on their adhesive properties. For example, *Escherichia coli* isolates from urine of patients with urinary tract infections are highly adhesive to urinary tract epithelial cells.[68] Thus, the specificity of *H. pylori* infection to human gastric mucosa could be determined by the adhesive properties of *H. pylori*.

The adhesion of *H. pylori* is closely related to its pathogenicity. *H. pylori* adherent to epithelial cells induces mucosal injury by direct and indirect mechanisms. *H. pylori* scarcely invades the epithelial barrier,[41] but produces a vacuolating cytotoxin (VacA).[5] Furthermore, *H. pylori* adhesion induces mucosal immune response, which causes chronic inflammation and mucosal injury.[64,65] Aihara et al. have reported that the adhesion of live *H. pylori* to gastric epithelial cells induces the production of interleukin-8, a cytokine which is a potent activator and chemotactic agent for neutrophils.[1] Segal et al. have reported that *H. pylori* adhesion to gastric epithelial cells induces effacement of microvilli, pedestal formation, cytoskeletal rearrangement, and tyrosine phosphorylation of host cell proteins at the site of adherence.[54] This indicates that the effect of *H. pylori* adhesion on gastric epithelial cells is similar to that of enteropathogenic *E. coli*.

Accordingly, in developing strategies against *H. pylori* infection, it is important to analyze the mechanism of *H. pylori* adhesion, especially the interaction between *H. pylori* adhesins and receptors. Inhibition of adhesion would be an ideal target for the prevention of *H. pylori* colonization.

## III. ANALYSIS OF *H. PYLORI* ADHESION IN VITRO

It is difficult to analyze the adhesion of *H. pylori* to gastric epithelial cells in vivo. Thus, several methods to analyze *H. pylori* adhesion in vitro have been developed.[17] Here, we review the results obtained by in vitro experiments.

### A. H. pylori *Adhesion to Erythrocytes*

Many microorganisms adhere to erythrocytes, which can be conveniently obtained from a variety of animals. The bacterial adhesion to erythrocytes can be estimated semi-quantitatively with the hemagglutination (HA) test. This test is widely used for analysis of bacterial adhesion and was initially used on *H. pylori* adhesion.[13] Most *H. pylori* strains adhere to human, guinea-pig, rabbit and sheep erythrocytes. However, *H. pylori* adhesion mechanisms differ in erythrocytes and gastric epithelial cells. Thus, the HA test is not suitable for the analysis of *H. pylori* adhesion to human gastric epithelial cells.

### B. H. pylori *Adhesion to Tissue Sections*

A histological method is also used to analyze bacterial adhesion to tissue.[7] In the analysis of *H. pylori* adhesion, tissue sections of gastric biopsies are used as targets. Fluorescein-isothiocyanate-labeled *H. pylori* is added to the section. After incubating 1 h at room temperature, nonadherent bacteria are removed by washing. Binding of *H. pylori* to gastric epithelial cells can be observed with a fluorescence microscope. This method is suitable for distribution analysis of bacterial receptor molecules in host tissue. It is, however, difficult to accurately quantify the bacteria adhering to tissue by using this method.

## C. H. pylori *Adhesion to Primary Cultured Cells*

Primary gastric epithelial cells, isolated from humans and other animal species, are occasionally used for the analysis of *H. pylori* adhesion. *H. pylori* can adhere to gastric epithelial cells in primary cultures obtained from mice, rats, Mongolian gerbils, guinea pigs, pigs, and cynomolgus monkeys.[34] However, *H. pylori* adhesion to primary cultured gastric epithelial cells varies greatly among these animals. *H. pylori* adheres well to epithelial cells from monkey and pig gastric antra. Conversely, *H. pylori* only weakly adheres to fundic epithelial cells from monkeys and pigs and to gastric epithelial cells from the other animals.

*H. pylori* also adheres to primary gastric epithelial cells isolated from humans.[9] Endoscopic examinations are necessary to obtain human gastric epithelial cells, therefore it is difficult for the laboratory analyst to use primary cultured human gastric epithelial cells.

## D. H. pylori *Adhesion to Cell Lines*

Established cell lines in tissue culture have been used for the analysis of bacterial adhesion. MKN-28, MKN-45, KATO III and AGS cells, derived from human gastric carcinomas, are used as target cells for the analysis of *H. pylori* adhesion.[12,24,29] The mechanisms of adhesion may differ between these cells and normal human gastric mucosal cells but as normal human gastric mucosal cells are not available for laboratory adhesion assays, these cell lines are commonly used for the analysis of *H. pylori* adhesion.

*H. pylori* also adheres to human cervical epithelial HeLa cells, human laryngeal epithelial HEp-2 cells,[15] human embryonic intestine INT-407 cells,[8] and mouse adrenal Y-1 cells.[14] These adhesive properties suggest that *H. pylori* adhesion is not specific to human gastric epithelium.

## IV. METHODS TO QUANTIFY *H. PYLORI* ADHESION

It is necessary for further analysis to estimate quantitatively *H. pylori* adhesion. Several different methods of quantifying *H. pylori* adherence to target cells have been developed. Here, we review the utilities of these methods.

### A. Microscopic Method

The standard method to quantify bacterial adhesion to target cells is performed by counting the number of bacteria binding to target cells under a light microscope.[17,34] Bacteria are visualized with Giemsa staining or immunostaining. However, this method demands much effort of investigators and thus it is not suitable for screening the adhesion activities of many *H. pylori* strains.

### B. Viable Count

This method counts the number of viable bacteria adherent to target cells.[17,29] Target cells and adherent *H. pylori* are harvested from a tissue culture plate by treatment with trypsin. Cells and bacteria are washed and collected by centrifugation, suspended in sterile distilled water, and vortexed vigorously to break down the cells and disperse the bacteria. The suspension is 10-fold serially diluted in sterile distilled water, and each

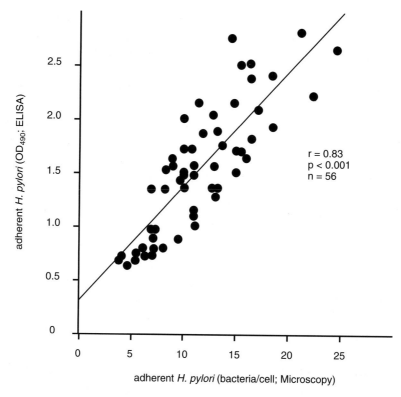

**Figure 1.** The correlation between ELISA and microscopic method. Fifty six *H. pylori* strains were assayed by using both methods. The correlation between the results obtained by ELISA and microscopic method was evaluated by Spearman's rank correlation.

dilution is inoculated onto a 5% horse serum-supplemented brain-heart-infusion agar plate. Plates are incubated at 37°C under microaerophilic conditions for 3 to 4 d, and then the numbers of *H. pylori* colonies on the plates are counted. In this method, adherent bacteria are scored as colony-forming units. However, this method is time-consuming and complicated. Furthermore, some of the organisms would become nonculturable during the procedure under the usual atmosphere so this method is rarely used for the analysis of *H. pylori* adhesion.

### C. Flow Cytometry

Flow cytometry is also used for the analysis of *H. pylori* adhesion.[12,47] This method can analyze quantitatively and exactly *H. pylori* adhesion to target cells. When analyzing *H. pylori* adhesion by flow cytometry, target cells must be dispersed and suspended in sample fluid. However, normal gastric epithelial cells form cell–cell junctions and construct gastric epithelium in vivo, which indicates the manner of *H. pylori* adhesion to the dispersed cells may be different from that to normal human gastric epithelium.

### D. ELISA

To solve the problems of the methods described above, we developed a rapid and simple method to quantitatively analyze *H. pylori* adhesion by utilizing enzyme-linked immunosorbent assay (ELISA).[24] This method used gastric epithelial cells grown in a 96-well tissue culture plate as target cells. The amount of *H. pylori* adhering to target cells was quantified per well through use of indirect enzyme-labeled antibody technique. This method can estimate adhesion activities of many *H. pylori* strains at one time.

Details of our ELISA are described here. Target cells are washed often in this procedure so cells which easily detach from tissue culture plates during washing are not suitable for this method. MKN-28 and MKN-45 cells are relatively resistant to washing thus were selected as target cells. The target cells were suspended at a concentration of $3 \times 10^5$ cells/mL in RPMI-1640 medium (ICN Biomedicals, Costa Mesa, CA, USA) containing 10% fetal calf serum, penicillin G (100 units/mL) and streptomycin (0.1 mg/mL). For the assay, 100 μL of cell suspension was placed in each well of a flat-bottomed 96-well tissue culture plate (FALCON 3072; Becton Dickinson, Lincoln Park, NJ, USA), and the plate was incubated at 37°C under 5% $CO_2$ for 2 d. After the target cells formed confluent monolayers, the medium was decanted from the microplates. The plates were then washed three times with 10 mM phosphate-buffered saline (PBS; pH 7.4), 100 μL of *H. pylori* suspension ($10^9$ bacteria/mL) was added to each well, and the plates were incubated at 37°C under 8% $CO_2$ for 90 min. The plates were then washed three times to remove nonadherent *H. pylori*, 100 μL of 8% paraformaldehyde was added to each well, and adherent *H. pylori* and cells were fixed at 4°C for 60 min. After washing, 100 μL of 1% $H_2O_2$ in methanol was added to each well and the plates were incubated at room temperature for 10 min, inactivating the endogenous peroxidase. After washing, 100 μL of rabbit anti-*H. pylori* polyclonal antibody (10 μg/mL) was added to each well and the plates were incubated for 2 h at 37°C. After washing, 100 μL of peroxidase-conjugated goat anti-rabbit immunoglobulins (Wako Chemicals, Osaka, Japan) diluted 1:1000 in PBS was added to each well and the plates were incubated for 2 h at 37°C. After the final wash, 100 μL of *o*-phenylenediamine (0.4 mg/mL) in 100 mM citrate-phosphate buffer (pH 5.0) containing 0.02% $H_2O_2$ was added to each well and the plates were incubated at room temperature for 15 min. The reaction was terminated by adding 50 μL of 2 M $H_2SO_4$. The optical density (OD) of the reaction was measured at 490 nm with a microplate reader (Model 3550 EIA Reader; Bio-Rad, Richmond, CA, USA). The OD indicates the amount of *H. pylori* adhering to the target cells.

### E. Comparison Between ELISA and Microscopic Method

We compared our established ELISA to the standard microscopic method. In this experiment, MKN-28 cells were used as target cells. A strong correlation between our ELISA and the microscopic method is shown in Figure 1. These results indicate that the value obtained by ELISA represents the amount of *H. pylori* adherent to target cells. The comparison between reproducibility of ELISA and microscopic method is shown in Figure 2. The ELISA method was consistently reproducible, whereas the results obtained by microscopic method varied widely. These results suggest that our established ELISA method is well suited for the analysis of *H. pylori* adhesion.

## V. DIVERSITY OF *H. PYLORI* ADHESION

We assayed adhesion activities of 56 *H. pylori* strains, obtained from 19 patients with chronic gastritis, 18 patients with gastric ulcer, 9 patients with duodenal ulcer, and

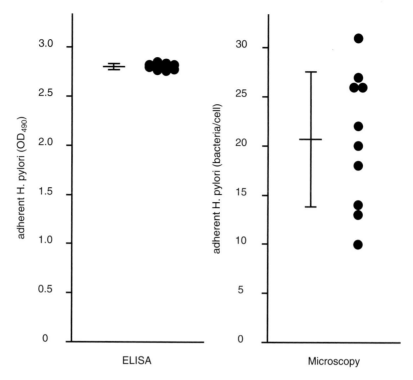

**Figure 2.** Reproducibility of ELISA and microscopic method. These results were obtained with the strain that had the highest adhesion activity. This strain was tested ten times, and each point indicates the result of each test. Bars indicate the mean±SD of ten tests.

10 patients with gastric cancer, by ELISA. MKN-28 cells were used as target cells. The results are shown in Figure 3. *H. pylori* is known to have a strain diversity in its pathogenetic factors, such as the production of VacA and the induction of inflammatory cytokines.[5,10] Our results indicate that *H. pylori* has a large strain diversity in adhesion activity as well. However, there were no significant differences in *H. pylori* adhesion activity between chronic gastritis, gastric ulcer, duodenal ulcer, and gastric cancer strains. These results suggest that the specific relation was not observed between adhesion activity and *H. pylori*-related clinical results.

## VI. *H. PYLORI* ADHESINS AND RECEPTORS

Several adhesins of *H. pylori* and their host tissue receptors, have been identified (Table 1). The receptor–adhesin interaction plays an important role in *H. pylori* adhesion to human gastric mucosa. The representative adhesins are described below.

### A. Hemagglutinins

*H. pylori* has at least two hemagglutinins and shows hemagglutinating activity in vitro. One is the 25-kDa protein (HpaA) identified by Evans et al.[16] This protein forms a fibrillar structure and recognizes *N*-acetylneuraminyllactose in erythrocytes.[13] The gene *hpaA*, which codes for this hemagglutinin, is found in all clinical isolates. However, a

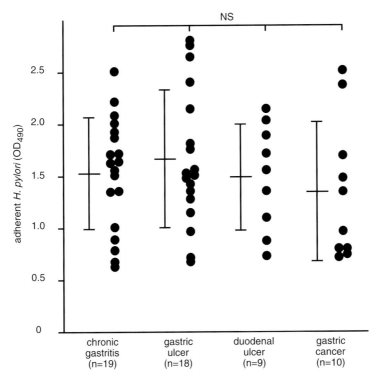

**Figure 3.** Adhesion activities of 56 *H. pylori* strains to MKN-28 cells. Bars indicate the mean±SD of each group. Differences among chronic gastritis, gastric ulcer, duodenal ulcer and gastric cancer were evaluated by two-tailed unpaired Student's t-test. NS, Not significant.

**Table 1.** *Helicobacter pylori* **Adhesins and Receptors**

| Adhesins | Receptors | References |
|---|---|---|
| 25-kDa hemagglutinin (HpaA) | N-acetylneuraminyllactose | Evans et al.[13,15,16] |
| 63-kDa exoenzyme S-like adhesin | Phosphatidylethanolamine Gangliotriaosylceramide Gangliotetraosylceramide | Lingwood et al.[39,40] and Gold et al.[20] |
| 19.6-kDa protein | Laminin | Doig et al.[11] |
| 25-kDa protein | Laminin | Valkonen et al.[72] |
| 75-kDa Lewis[b]-binding adhesin (BabA) | H-1, Lewis[b] blood group antigens | Boren et al.[7] and Ilver et al.[31] |
| 61-kDa protein | H-2, Lewis[a], Lewis[b] blood group antigens | Alkout et al.[2] |
| 16-kDa protein | Lewis[x] blood group antigen, mucin | Namavar et al.[45] |
| 59-kDa hemagglutinin | | Huang et al.[30] |
| 60-kDa heat shock protein | | Yamaguchi et al.[77] |
| | GM3, lactosylceramide sulfate | Slomiany et al.[60] |
| | GM3, sulfatide | Saitoh et al.[52] |
| | Heparan sulphate | Ascencio et al.[4] |

mutant defective for this gene can adhere to gastric epithelial cells. Thus, HpaA may not be essential for *H. pylori* adhesion to gastric epithelium.[16] Another hemagglutinin is the 59-kDa protein, but the receptor specific to this protein is unknown.[30]

### B. Exoenzyme S-Like Adhesin

Lingwood et al. have reported that *H. pylori* binds to phosphatidylethanolamine, gangliotriaosylceramide and gangliotetraosylceramide.[20,39] The binding specificity of *H. pylori* to these lipids is similar to that of *Pseudomonas aeruginosa*. Exoenzyme S is an important adhesin of *P. aeruginosa*. Monoclonal antibodies to exoenzyme S react with a 63-kDa protein of *H. pylori*. This 63-kDa exoenzyme S-like protein inhibits competitively *H. pylori* binding to phosphatidylethanolamine in vitro.[40] This suggests that the 63-kDa exoenzyme S-like protein is the adhesin which is responsible for the lipid-binding specificity of *H. pylori*.

### C. Le^b-Binding Adhesin

Borén and colleagues found that the fucosylated blood group antigens Lewis$^b$ (Le$^b$) and H-1 mediate *H. pylori* adhesion to human gastric epithelial cells.[7] Furthermore, they identified Le$^b$-binding adhesin (BabA).[31] The fucosylated blood group antigens are associated with blood group O phenotype. Borén et al. speculated that this finding explains epidemiological observations that individuals of blood group O phenotype run a greater risk for developing gastric ulcers.[7] However, this hypothesis is still controversial.

### D. Lipopolysaccharide

Lipopolysaccharide (LPS) is an outer membrane constituent of Gram-negative bacteria. Valkonen et al. have reported that *H. pylori* LPS binds to laminin and plays an important role in adhesion.[71] However, endotoxic activity of *H. pylori* LPS is very low.[42]

### E. Perspectives of H. pylori Adhesins and Receptors

*H. pylori* has several adhesins which have respective receptors in host tissue. Thus, *H. pylori* adhesion to human gastric epithelium will be due to several combinations between *H. pylori* adhesins and their receptors. However, the role of each molecule in *H. pylori* adhesion has not been analyzed. Furthermore, host receptor molecules (Table 1) are expressed in various tissues besides gastric epithelium, which suggests that *H. pylori* adhesion is not specific to gastric epithelium.

*H. pylori* adhesins and receptors might competitively inhibit adhesion. For example, hemagglutinins described above can inhibit *H. pylori* adhesion to erythrocytes.[13,30] *H. pylori* pretreated with Le$^b$ antigen cannot adhere to human mucosal tissue *in situ*, in which the terminal fucose of Le$^b$ antigen is essential.[7] However, adhesion-related epitopes of other molecules are not analyzed enough.

It is very important to analyze the molecular structures of *H. pylori* adhesins and receptors. On the basis of these molecular structures, new therapeutic agents could be designed and developed to prevent the adhesion of *H. pylori* to gastric mucosa.

## VII. ANTIBODIES TO INHIBIT *H. PYLORI* ADHESION

In general, bacterial adhesion to host cells is blocked by antibody to their adhesins. For example, the component pertussis vaccine consists of pertussis toxin and filamentous hemagglutinin (FHA) which mediates *Bordetella pertussis* adhesion to the ciliated epithelial cells of the upper respiratory tract. Anti-FHA antibody induced by the vaccine inhibits adhesion.[53]

Yamaguchi et al. have reported that *H. pylori* adhesion to MKN-45 cells is reduced by pretreating *H. pylori* with a monoclonal antibody recognizing 60-kDa heat shock protein, one of the *H. pylori* adhesins.[77] Osaki et al. have established a monoclonal antibody to inhibit *H. pylori* adhesion in vitro.[48] The monoclonal antibody recognizes *H. pylori* LPS which plays an important role in *H. pylori* adhesion. These experiments suggest that antibodies to *H. pylori* adhesins can inhibit the adhesion of *H. pylori* to gastric epithelial cells.

However, *H. pylori* infection occurs in gastric mucosa. Serum immunoglobulins cannot make contact with *H. pylori* on gastric epithelium. Secretory immunoglobulin A (S-IgA) is the main immunoglobulin in gastric juice.[23,28] S-IgA antibodies to *H. pylori* adhesins may inhibit *H. pylori* adhesion in human gastric mucosa. Oral immunization can induce S-IgA in the alimentary tract.[38,50] Thus, oral immunization with *H. pylori* adhesins may have potential vaccine applications.

## VIII. ANTI-*H. PYLORI* ADHESION AGENTS

It is reported that some mucoprotective antiulcer agents can inhibit the adhesion of *H. pylori* to human gastric mucosa.[27] Inhibition of adhesion is an ideal way to prevent *H. pylori* colonization.

### A. Sucralfate

Sucralfate is recognized as an effective antiulcer agent.[69] Slomiany et al. have conducted a series of in vitro experiments to investigate the effect of sucralfate on *H. pylori*. In vitro experiments suggest that sucralfate competes with lactosylceramide sulfate and GM3 ganglioside, which are receptors for *H. pylori*, and inhibits *H. pylori* attachment to the epithelium.[60,62] Furthermore, sucralfate inhibits *H. pylori* enzyme activities,[61] and enhances the susceptibility of *H. pylori* to antibiotics.[63] In clinical trials, sucralfate monotherapy reduced the density of *H. pylori*, and sucralfate combined with antibiotics and proton pump inhibitors enhanced the *H. pylori* eradication rate.[36]

### B. Sofalcone

Sofalcone is an antiulcer agent that has multiple effects against *H. pylori*.[19] It has antibacterial activity, induces morphological changes and inhibits lipolytic activity. Sunairi et al. have reported that sofalcone also has anti-*H. pylori* adhesion effect.[66] The adhesion of *H. pylori* to gastric mucin was strongly inhibited in a dose-dependent manner by the in vitro administration of sofalcone at a concentration of more than 15 μmol/L. Several clinical trials have demonstrated that sofalcone enhances the *H. pylori* eradication rate when combined with antibiotics and proton pump inhibitors.[35]

### C. Rebamipide

Rebamipide is an antiulcer agent that has antioxidant and free radical scavenging activities.[22,43,67] Rebamipide itself does not have antibacterial activity. However, we have reported that rebamipide inhibits the adhesion of *H. pylori* to gastric epithelial cells in vitro.[26] Furthermore, Kato et al. have reported that triple therapy with lansoprazole, amoxicillin, and rebamipide combined showed a high eradication rate in a clinical trial.[32]

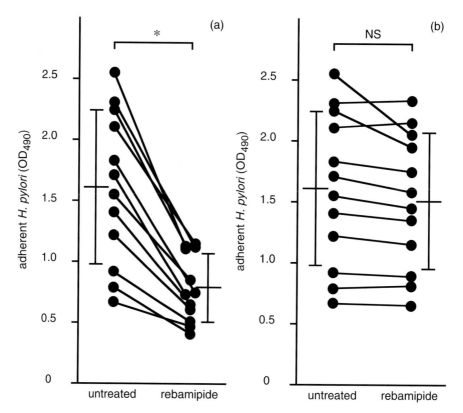

**Figure 4.** Effect of rebamipide on *H. pylori* adhesion. Twelve *H. pylori* strains were used for the evaluation of rebamipide. MKN-28 cells were used as target cells. Each point indicates the adhesion activity of each strain. (a) MKN-28 cells were pretreated with 100 µg/mL rebamipide for 90 min. (b) *H. pylori* was pretreated with 100 µg/mL rebamipide for 90 min. The difference between rebamipide treatment and no treatment was evaluated by two-tailed paired Student's t-test. *P < 0.0001; NS, Not significant.

We evaluated the effect of rebamipide on *H. pylori* adhesion by ELISA. The adhesion of *H. pylori* to target cells was reduced to approximately half by pretreatment of target cells with 100 µg/mL rebamipide for 90 min (Fig. 4a). The inhibitory activity reached a maximum at this condition, and this concentration can be achieved in the gastric mucous layer by the recommended clinical dose of rebamipide.[44] These results suggest that rebamipide has potential as a therapeutic agent against *H. pylori* infection. Conversely, *H. pylori* adhesion was not affected by pretreating *H. pylori* with the same concentration of rebamipide (Fig. 4b). These results indicate that rebamipide directly affects the gastric epithelial cells and does not act on *H. pylori*.

### D. Ecabet Sodium

Ecabet sodium is a mucoprotective antiulcer agent.[58] Ecabet sodium has urease-inhibiting activity,[55] but does not have direct bactericidal activity against *H. pylori*. We have reported that ecabet sodium has an anti-*H. pylori* adhesion effect in vitro.[25] Furthermore, triple therapy of lansoprazole, amoxicillin, and ecabet sodium showed a high eradication rate in a clinical trial.[46]

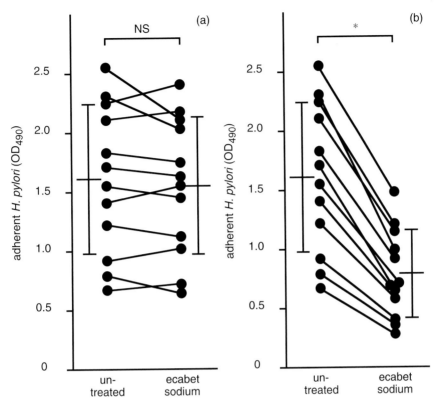

**Figure 5.** Effect of ecabet sodium on *H. pylori* adhesion. Twelve *H. pylori* strains were used for the evaluation of ecabet sodium. MKN-28 cells were used as target cells. Each point indicates the adhesion activity of a strain. (a) MKN-28 cells were pretreated with 1 mg/mL ecabet sodium for 90 min. (b) *H. pylori* was pretreated with 1 mg/mL ecabet sodium for 90 min. The difference between ecabet sodium treatment and no treatment was evaluated by two-tailed paired Student's t-test. *P < 0.0001; NS, Not significant.

We estimated the effect of ecabet sodium on *H. pylori* adhesion by ELISA. *H. pylori* adhesion to target cells was reduced to approximately half by pretreating *H. pylori* with 1 mg/mL ecabet sodium for 90 min (Fig. 5b). The inhibitory activity reached a maximum at this condition, and this concentration can be achieved in gastric mucous layer by the recommended clinical dose of ecabet sodium.[3] However, the adhesion was not affected by pretreating target cells with the same concentration of ecabet sodium (Fig. 5a). This indicates that ecabet sodium acts on *H. pylori* and does not affect the gastric epithelial cells.

## E. Perspectives of Anti-**H. pylori** Adhesion Agents

In the gastric mucosa, *H. pylori* localizes on the surface of the epithelial cells as well as in the mucous layer.[56,57] *H. pylori* colonized on epithelial cells can induce mucosal injury by direct and indirect mechanisms. *H. pylori* in the mucous layer may survive after an insufficient eradication therapy, and the organisms adhere again to gastric epithelial cells and recolonize. Anti-adhesion effects can contribute to the prevention of *H. pylori* recolonization. Several clinical trials have indicated that the anti-adhesion agents described above enhance *H. pylori* eradication rates when they are combined with antibiotics and proton pump inhibitors.[32,35,36,46] Thus, the anti-adhesion agents are promising for *H. pylori* eradication regimens.

Furthermore, antiadhesion agents may have potential as probes for the analysis of molecular mechanisms of *H. pylori* adhesion. The adhesion of *H. pylori* to gastric epithelial cells is performed by binding between *H. pylori* adhesins and their receptors. The targets of antiadhesion agents will be these molecules. Several *H. pylori* adhesins and receptors in host tissue have been identified. Anti-adhesion agents are a useful tool to define the fine structure and the essential epitope of adhesins or receptors.

We evaluated the anti-adhesion effects of rebamipide and ecabet sodium quantitatively by ELISA. Both significantly inhibited *H. pylori* adhesion. However, their anti-adhesion effects were incomplete (Fig. 4, Fig. 5). The adhesion of *H. pylori* to gastric epithelial cells will be due to some combinations of *H. pylori* adhesins and their receptors. Each anti-adhesion agent may only partially inhibit the combinations. It may be difficult to inhibit the adhesion completely using only one anti-adhesion agent.

## IX. FUTURE WORK

The adhesion of *H. pylori* to human gastric epithelium has been analyzed in detail by using cellular and molecular biological methods. Several molecules related to adhesion have been identified. However, many problems remain to be resolved. At the end of this chapter, we describe the future work we must perform in the study on *H. pylori* adhesion.

### A. *Analysis of* **H. pylori** *Adhesion In Vivo*

Most experiments to analyze *H. pylori* adhesion to gastric epithelium have been performed in vitro. In vitro experiments are suitable for molecular analysis. However, there are discrepancies between the results obtained by in vitro experiments and *H. pylori* infection in human gastric mucosa. For example, in vitro experiments suggest that *H. pylori* adhesion is not specific to gastric epithelial cells, though natural *H. pylori* infection is very specific to human gastric mucosa. In the early study on *H. pylori*, we did not have a good animal model of *H. pylori* infection. However, we can now use some animal models; mice,[37] Mongolian gerbils,[79] piglets,[33] Japanese monkeys,[59] etc. In vivo experiments using these animal models will be necessary for the further study on *H. pylori* adhesion to gastric epithelium.

### B. *Adhesion of* **H. pylori** *to Mucous Layer*

The study on *H. pylori* adhesion has been mainly performed by analyzing the adhesion to gastric epithelial cells. However, Shimizu et al. have reported that *H. pylori* in the gastric mucous layer are more abundant than those attached to epithelial cells.[56,57] The adhesion to the mucous layer may be as important as that to the epithelial cells. Thus, it will be necessary to analyze *H. pylori* adhesion to the gastric mucous layer as well as that to the epithelial cells. Tzouvelekis et al. have reported that *H. pylori* binds to gastric mucin which is a major constituent of the gastric mucous layer.[70] Furthermore, Namavar et al. identified a 16-kDa surface protein which binds to mucin.[45] However, many questions about *H. pylori* adhesion to the gastric mucous layer remain.

### C. H. pylori *Vaccine*

In the development of *H. pylori* vaccine, the effect of oral immunization with *H. pylori* urease is now being investigated.[38,50] Urease is an essential enzyme for *H. pylori* colonization.[49] Oral immunization with urease prevented *Helicobacter* infections in animal models, but has not succeeded in human clinical trials.

Oral immunization with *H. pylori* adhesins has potential as another approach to *H. pylori* vaccine. S-IgA antibodies to the adhesins, induced by oral immunization, may block *H. pylori* adhesion to the gastric epithelium. Thus, oral immunization with adhesins, as well as that with urease, should be investigated.

### D. *Eradication of* H. pylori

Combination therapies with antimicrobial drugs (clarithromycin, amoxicillin, metronidazole) and proton pump inhibitors (omeprazole, lansoprazole, rabeprazole) show high *H. pylori* eradication rates. However, *H. pylori* strains resistant to these antimicrobial drugs have been reported.[6,51,73] In the near future, resistance to antimicrobial drugs will be an important problem in the eradication of *H. pylori*. Thus, another regimen to eradicate *H. pylori* will be necessary.

Anti-*H. pylori* adhesion agents described in this chapter have potential as new therapeutic agents.[27] *H. pylori* cannot be eradicated by a single anti-adhesion agent alone but when they are combined with antibiotics and proton pump inhibitors anti-adhesion agents can enhance eradication rates.[32,35,36,46] Thus, effects of these anti-adhesion agents on *H. pylori* infection should be evaluated. In addition, it is important to discover novel anti-adhesion agents. Our established ELISA method is well suited for screening anti-adhesion agents.[24]

## X. CONCLUSION

The adhesion of *H. pylori* to human gastric epithelium has been analyzed in detail. However, the results obtained by the analyses are not widely applied in clinical medicine. In the strategies against *H. pylori* infection, we must apply these results for the prevention and the treatment of *H. pylori* infection.

## REFERENCES

1. Aihara M, Tsuchimoto D, Takizawa H, et al: Mechanisms involved in *Helicobacter pylori*-induced interleukin-8 production by a gastric cancer cell line, MKN45. *Infect Immun* 65:3218–24, 1997
2. Alkout AM, Blackwell CC, Weir DM, et al: Isolation of a cell surface component of *Helicobacter pylori* that binds H type 2, Lewis[a], and Lewis[b] antigens. *Gastroenterology* 112:1179–87, 1997
3. Asada S, Miyoshi H, Takiuchi H, et al: Effect of TA-2711, a new anti-ulcer agent on human gastric juice. *Gastroenterology* 100:A27, 1991
4. Ascencio F, Fransson LA, Wadstrom T: Affinity of the gastric pathogen *Helicobacter pylori* for the N-sulphated glycosaminoglycan heparan sulfate. *J Med Microbiol* 38:240–4, 1993
5. Atherton JC, Cao P, Peek RM Jr, et al: Mosaicism in vacuolating cytotoxin alleles of *Helicobacter pylori*. Association of specific vacA types with cytotoxin production and peptic ulceration. *J Biol Chem* 270:17771–7, 1995
6. Becx MC, Janssen AJ, Clasener HA, et al: Metronidazole-resistant *Helicobacter pylori*. *Lancet* 335:539–40, 1990

7. Borén T, Falk P, Roth KA, et al: Attachment of *Helicobacter pylori* to human gastric epithelium mediated by blood group antigens. *Science* 262:1892–5, 1993

8. Chmiela M, Lawnik M, Czkwianianc E, et al: Attachment of *Helicobacter pylori* strains to human epithelial cells. *J Physiol Pharmacol* 48:393–404, 1997

9. Clyne M, Drumm B: Adherence of *Helicobacter pylori* to primary human gastrointestinal cells. *Infect Immun* 61:4051–7, 1993

10. Crabtree JE, Farmery SM, Lindley IJ, et al: CagA/cytotoxic strains of *Helicobacter pylori* and interleukin-8 in gastric epithelial cell lines. *J Clin Pathol* 47:945–50, 1994

11. Doig P, Austin JW, Kostrzynska M, et al: Production of a conserved adhesin by the human gastroduodenal pathogen *Helicobacter pylori*. *J Bacteriol* 174:2539–7, 1992

12. Dunn BE, Altmann M, Campbell GP: Adherence of *Helicobacter pylori* to gastric carcinoma cells: analysis by flow cytometry. *Rev Infect Dis* 13(Suppl 8):S657–64, 1991

13. Evans DG, Evans DJ Jr, Moulds JJ, et al: *N*-acetylneuraminyllactose-binding fibrillar hemagglutinin of *Campylobacter pylori*: a putative colonization factor antigen. *Infect Immun* 56:2896–906, 1988

14. Evans DG, Evans DJ Jr, Graham DY: Receptor-mediated adherence of *Campylobacter pylori* to mouse Y-1 adrenal cell monolayers. *Infect Immun* 57:2272–8, 1989

15. Evans DG, Evans DJ Jr, Graham DY: Adherence and internalization of *Helicobacter pylori* by HEp-2 cells. *Gastroenterology* 102:1557–67, 1992

16. Evans DG, Karjalainen TK, Evans DJ Jr, et al: Cloning, nucleotide sequence, and expression of a gene encoding an adhesin subunit protein of *Helicobacter pylori*. *J Bacteriol* 175:674–83, 1993

17. Evans DG, Evans DJ Jr: Adhesion properties of *Helicobacter pylori*. *Methods Enzymol* 253:336–60, 1995

18. Forman D and the Eurogast Study Group: An international association between *Helicobacter pylori* infection and gastric cancer. *Lancet* 341:1359–62, 1993

19. Fujioka T, Murakami K, Kubota T, et al: Sofalcone for treatment of *Helicobacter pylori* infection. *J Gastroenterol* 31(Suppl 9):56–8, 1996

20. Gold BD, Huesca M, Sherman PM, et al: *Helicobacter mustelae* and *Helicobacter pylori* bind to common lipid receptors *in vitro*. *Infect Immun* 61:2632–8, 1993

21. Graham DY: *Helicobacter pylori*: its epidemiology and its role in duodenal ulcer disease. *J Gastroenterol Hepatol* 6:105–13, 1991

22. Han BG, Kim HS, Rhee KH et al: Effects of rebamipide on gastric cell damage by *Helicobacter pylori*-stimulated human neutrophils. *Pharmacol Res* 32:201–7, 1995

23. Hayashi S, Sugiyama T, Hisano K, et al: Quantitative detection of secretory immunoglobulin A to *Helicobacter pylori* in gastric juice: antibody capture enzyme-linked immunosorbent assay. *J Clin Lab Anal* 10:74–7, 1996

24. Hayashi S, Sugiyama T, Yachi A, et al: A rapid and simple method to quantify *Helicobacter pylori* adhesion to human gastric MKN-28 cells. *J Gastroenterol Hepatol* 12:373–5, 1997

25. Hayashi S, Sugiyama T, Yachi A, et al: Effect of ecabet sodium on *Helicobacter pylori* adhesion to gastric epithelial cells. *J Gastroenterol* 32:593–7, 1997

26. Hayashi S, Sugiyama T, Amano K, et al: Effect of rebamipide, a novel antiulcer agent, on *Helicobacter pylori* adhesion to gastric epithelial cells. *Antimicrob Agents Chemother* 42:1895–9, 1998

27. Hayashi S, Sugiyama T, Asaka M, et al: Modification of *Helicobacter pylori* adhesion to human gastric epithelial cells by antiadhesion agents. *Dig Dis Sci* 43(Suppl):56–60S, 1998

28. Hayashi S, Sugiyama T, Yokota K, et al: Analysis of immunoglobulin A antibodies to *Helicobacter pylori* in serum and gastric juice in relation to mucosal inflammation. *Clin Diagn Lab Immunol* 5:617–21, 1998

29. Hemalatha SG, Drumm B, Sherman P: Adherence of *Helicobacter pylori* to human gastric epithelial cells *in vitro*. *J Med Microbiol* 35:197–202, 1991

30. Huang J, Keeling PW, Smyth CJ: Identification of erythrocyte-binding antigens in *Helicobacter pylori*. *J Gen Microbiol* 138:1503–13, 1992

31. Ilver D, Arnqvist A, Ogren J, et al: *Helicobacter pylori* adhesin binding fucosylated histo-blood group antigens revealed by retagging. *Science* 279:373–7, 1998

32. Kato M, Asaka M, Sugiyama T, et al: Effects of rebamipide in combination with lansoprazole and amoxicillin on *Helicobacter pylori*-infected gastric ulcer patients. *Dig Dis Sci* 43(Suppl):198–202S, 1998

33. Krakowka S, Morgan DR, Kraft WG, et al: Establishment of gastric *Campylobacter pylori* infection in the neonatal gnotobiotic piglet. *Infect Immun* 55:2789–96, 1987

34. Kobayashi Y, Okazaki K, Murakami K: Adhesion of *Helicobacter pylori* to gastric epithelial cells in primary cultures obtained from stomachs of various animals. *Infect Immun* 61:4058–63, 1993

35. Kodama R, Fujioka T, Fujiyama K, et al: Combination therapy with clarithromycin and sofalcone for eradication of *Helicobacter pylori*. *Eur J Gastroenterol Hepatol* 6(Suppl 1):S125–8, 1994

36. Lam SK, Hu WHC, Ching CK: Sucralfate in *Helicobacter pylori* eradication strategies. *Scand J Gastroenterol* 30(Suppl 210):89–91, 1995

37. Lee A, Fox JG, Otto G, et al: A small animal model of human *Helicobacter pylori* active chronic gastritis. *Gastroenterology* 99:1315–23, 1990

38. Lee CK, Weltzin R, Thomas WD, et al: Oral immunization with recombinant *Helicobacter pylori* urease induces secretory IgA antibodies and protects mice from challenge with *Helicobacter felis*. *J Infect Dis* 172:161–72, 1995

39. Lingwood CA, Huesca M, Kuksis A. The glycerolipid receptor for *Helicobacter pylori* (and exoenzyme S) is phosphatidylethanolamine. *Infect Immun* 60:2470–4, 1992

40. Lingwood CA, Wasfy G, Han H, et al: Receptor affinity purification of a lipid-binding adhesin from *Helicobacter pylori*. *Infect Immun* 61:2474–8, 1993

41. Mai UE, Pérez-Pérez GI, Allen JB, et al: Surface proteins from *Helicobacter pylori* exhibit chemotactic activity for human leukocytes and are present in gastric mucosa. *J Exp Med* 175:517–25, 1992

42. Muotiala A, Helander IM, Pyhala L, et al: Low biological activity of *Helicobacter pylori* lipopolysaccharide. *Infect Immun* 60:1714–6, 1992

43. Naito Y, Yoshikawa T, Tanigawa T, et al: Hydroxyl radical scavenging by rebamipide and related compounds: electron paramagnetic resonance study. *Free Radic Biol Med* 18:117–23, 1995

44. Naito Y, Yoshikawa T, Iinuma S, et al: Local gastric and serum concentrations of rebamipide following oral administration to patients with chronic gastritis. *Arzneimittelforschung* 46:698–700, 1996

45. Namavar F, Sparrius M, Veerman EC, et al: Neutrophil-activating protein mediates adhesion of *Helicobacter pylori* to sulfated carbohydrates on high-molecular-weight salivary mucin. *Infect Immun* 66:444–7, 1998

46. Ohkusa T, Takashimizu I, Fujiki K, et al. Prospective evaluation of a new anti-ulcer agent, ecabet sodium, for the treatment of *Helicobacter pylori* infection. *Aliment Pharmacol Ther* 12:457–1, 1998

47. Osaki T, Yamaguchi H, Taguchi H, et al: Studies on the relationship between adhesive activity and hemagglutination by *Helicobacter pylori*. *J Med Microbiol* 46:117–21, 1997

48. Osaki T, Yamaguchi H, Taguchi H, et al: Establishment and characterisation of a monoclonal antibody to inhibit adhesion of *Helicobacter pylori* to gastric epithelial cells. *J Med Microbiol* 47:505–12, 1998

49. Owen RJ, Martin SR, Borman P: Rapid urea hydrolysis by gastric *Campylobacters*. *Lancet* 1:111, 1985

50. Pappo J, Thomas WD Jr, Kabok Z, et al: Effect of oral immunization with recombinant urease on murine *Helicobacter felis* gastritis. *Infect Immun* 63:1246–52, 1995

51. Parasakthi N, Goh KL: Primary and acquired resistance to clarithromycin among *Helicobacter pylori* strains in Malaysia. *Am J Gastroenterol* 90:519, 1995

52. Saitoh T, Natomi H, Zhao W, et al: Identification of glycolipid receptors for *Helicobacter pylori* by TLC-immunostaining. *FEBS Lett* 282:385–7, 1991

53. Sato Y, Kimura M, Fukumi H: Development of a pertussis component vaccine in Japan. *Lancet* 1:122–6, 1984

54. Segal ED, Falkow S, Tompkins LS: *Helicobacter pylori* attachment to gastric cells induces cytoskeletal rearrangements and tyrosine phosphorylation of host cell proteins. *Proc Natl Acad Sci USA* 93:1259–64, 1996

55. Shibata K, Ito Y, Hongo A, et al: Bactericidal activity of a new antiulcer agent, ecabet sodium, against *Helicobacter pylori* under acidic conditions. *Antimicrob Agents Chemother* 39:1295–9, 1995

56. Shimizu T, Akamatsu T, Ota H, et al: Immunohistochemical detection of *Helicobacter pylori* in the surface mucous gel layer and its clinicopathological significance. *Helicobacter* 1:197–206, 1996

57. Shimizu T, Akamatsu T, Sugiyama A, et al: *Helicobacter pylori* and the surface mucous gel layer of the human stomach. *Helicobacter* 1:207–18, 1996

58. Shimoyama T, Fukuda Y, Fukuda S, et al: Ecabet sodium eradicates *Helicobacter pylori* infection in gastric ulcer patients. *J Gastroenterol* 31(Suppl 9):59–62, 1996

59. Shuto R, Fujioka T, Kubota T, et al: Experimental gastritis induced by *Helicobacter pylori* in Japanese monkeys. *Infect Immun* 61:933–9, 1993

60. Slomiany BL, Piotrowski J, Samanta A, et al: *Campylobacter pylori* colonization factor shows specificity for lactosylceramide sulfate and GM3 ganglioside. *Biochem Int* 19:929–36, 1989

61. Slomiany BL, Murty VL, Piotrowski J, et al: Glycosulfatase activity of *H. pylori* toward human gastric mucin: Effect of sucralfate. *Am J Gastroenterol* 87:1132–7, 1992

62. Slomiany BL, Murty VL, Piotrowski J, et al: Gastroprotective agents in mucosal defense against *Helicobacter pylori*. *Gen Pharmacol* 25:833–41, 1994

63. Slomiany BL, Piotrowski J, Majka J, et al: Sucralfate affects the susceptibility of *Helicobacter pylori* to antimicrobial agents. *Scand J Gastroenterol* 30(Suppl 210):82–4, 1995

64. Sugiyama T, Awakawa T, Hayashi S, et al: The effect of the immune response to *Helicobacter pylori* in the development of intestinal metaplasia. *Eur J Gastroenterol Hepatol* 6(Suppl 1):S89–92, 1994

65. Sugiyama T, Yabana T, Yachi A: Mucosal immune response to *Helicobacter pylori* and cytotoxic mechanism. *Scand J Gastroenterol* 30(Suppl 208):30–2, 1995

66. Sunairi M, Watanabe K, Suzuki T, et al: Effects of anti-ulcer agents on antibiotic activity against *Helicobacter pylori*. *Eur J Gastroenterol Hepatol* 6(Suppl 1):S121–4, 1994

67. Suzuki M, Miura S, Mori M, et al: Rebamipide, a novel antiulcer agent, attenuates *Helicobacter pylori* induced gastric mucosal cell injury associated with neutrophil derived oxidants. *Gut* 35:1375–8, 1994

68. Svanborg-Eden C, Hagberg L, Hanson LA, et al: Bacterial adherence—a pathogenetic mechanism in urinary tract infections caused by *Escherichia coli*. *Prog Allergy* 33:175–88, 1983

69. Terano A, Shimada T, Hiraishi H: Sucralfate and *Helicobacter pylori*. *J Gastroenterol* 31(Suppl 9):63–5, 1996

70. Tzouvelekis LS, Mentis AF, Makris AM, et al: *In vitro* binding of *Helicobacter pylori* to human gastric mucin. *Infect Immun* 59:4252–4, 1991

71. Valkonen KH, Wadström T, Moran AP: Interaction of lipopolysaccharides of *Helicobacter pylori* with basement membrane protein laminin. *Infect Immun* 62:3640–8, 1994

72. Valkonen KH, Wadström T, Moran AP: Identification of the *N*-acetylneuraminyllactose-specific laminin-binding protein of *Helicobacter pylori*. *Infect Immun* 65:916–23, 1997

73. van Zwet AA, Vandenbrouke-Grauls CM, Thijs JC, et al: Stable amoxicillin resistance in *Helicobacter pylori*. *Lancet* 352:1595, 1998

74. Warren JR, Marshall BJ: Unidentified curved bacilli on gastric epithelium in active chronic gastritis. *Lancet* 1:1273–5, 1983
75. Wotherspoon AC, Ortiz-Hidalgo C, Falzon MR, et al: *Helicobacter pylori*-associated gastritis and primary B-cell gastric lymphoma. *Lancet* 338:1175–6, 1991
76. Wyatt JI, Dixon MF: Chronic gastritis - a pathogenetic approach. *J Pathol* 154:113–24, 1988
77. Yamaguchi H, Osaki T, Kurihara N, et al: Heat-shock protein 60 homologue of *Helicobacter pylori* is associated with adhesion of *H. pylori* to human gastric epithelial cells. *J Med Microbiol* 46:825–31, 1997
78. Yeomans ND, Kolt SD: *Helicobacter heilmannii* (formerly *Gastrospirillum*): association with pig and human gastric pathology. *Gastroenterology* 111:244–7, 1996
79. Yokota K, Kurebayashi Y, Takayama Y, et al: Colonization of *Helicobacter pylori* in the gastric mucosa of Mongolian gerbils. *Microbiol Immunol* 35:475–80, 1991

# 32

# Studying Bacterial Adhesion in the Urinary Tract

**James A. Roberts and M. Bernice Kaack**

*Tulane Regional Primate Research Center, Covington, LA, USA*

## I. INTRODUCTION

Bacterial adhesion is necessary for a bacterial infection to occur in either the urinary, respiratory or gastrointestinal tract. Adhesion to mucous membranes is the initial event in any infection other than those associated with urethral instrumentation or catheterization. The surface energy theory of bacterial adhesion, DLVO, attempts to explain the mechanism of bacterial adhesion, and the necessity of surface appendages for adhesion to occur in an energy efficient manner.[7]

The net negative surface charge of both tissue cells and bacteria as well as diffuse ion clouds in the area repulse adhesion. While at 15 nm there is very little repulsion, as bacteria approach 10 nm, maximum repulsion occurs. Since the magnitude of both attractive as well as repulsive forces depends on the diameter of the approaching body, bacterial fimbriae or polymers, being of a much smaller diameter, allow bacterial adherence that might not otherwise occur, reaching cell surface receptors for firm adherence to the cell surface. Thus adherent, the normal flow of body fluids such as mucous or in the case of the urinary tract, urine, does not wash the bacteria away. This leads to bacterial multiplication reaching the critical mass necessary for a clinical infection.

Bacterial surface factors which predispose to urinary tract infection include lipopolysaccharides, capsular antigens, and adhesive factors.[6] The adhesins in urinary tract infections include type-1 fimbriae which adhere to mucins containing mannose residues,[16] P-fimbriae which adhere to glycolipids on cell surfaces,[11] and the afimbrial adhesins such as the Dr adhesins.[15] The ability of both fimbrial and nonfimbrial adhesins to colonize the gut, the perineum of females and prepuce of uncircumcised males, followed by the urethra, bladder, ureter and kidney, is the most important initiating factor in the development of urinary tract infection.[13] While the nonfimbrial adhesins may be important in colonization of the perineum and urethra, it has been suggested that their effect in the urinary tract may be minimal, although they can produce pyelonephritis.[14] S fimbriae, type-1 fimbria and 075X, while recognizing urothelial receptors, are inhibited by substances in urine. This appears to be Tamm Horsfall protein in the case of S fimbriae, as well as lower molecular weight compounds which bind and prevent adherence by both type-1 fimbria and the 075X adhesin.[13] The further effects of

*Handbook of Bacterial Adhesion: Principles, Methods, and Applications*
Edited by: Y. H. An and R. J. Friedman © Humana Press Inc., Totowa, NJ

endotoxin, hemolysin and other factors such as aerobactin 3 (which allow the bacteria to access the iron needed for growth and further colonization) are all important. Bacteria which have been shown to grow well in urine are the bacteria usually found associated with urinary tract infections. This includes the enterobacteriaceae, enterococci and *Pseudomonas*.[23]

Phase variation is another factor important in the ability of bacteria to colonize surfaces. Such variation can change fimbriated bacteria to a nonfimbriated state and appears to be advantageous to the bacteria. One experiment showed that heavily-fimbriated *Proteus* when given intravenously to the experimental animal led to rapid eradication of urinary tract infection. These fimbriae adhere well to leukocytes and seem to be rapidly phagocytosed and killed.[20] The same fimbriated bacteria when introduced via an ascending route led to ascending pyelonephritis because of the ability of the fimbriae to adhere to urothelial cells and establish infection.[21] The same factor may well be important in phase variation of type-1 fimbriae which adhere well to leukocytes. Thus, while type-1 fimbriae may be important in colonization of the vagina and perineum by adherence to vaginal mucus,[26] once phagocytosis begins in the bladder, phase variation to the non-type-1 fimbriated state may well occur, decreasing adherence to and activation of leukocytes.[12] P-fimbriae, however, do not adhere well to leukocytes, thus their presence in an ongoing infection will lead to continuing bacterial adherence to urothelial cells, without an increase in their phagocytic killing. Our studies have involved both in vitro and in vivo measures of bacterial adhesion using a primate model. They have been very successful because Maqaque monkey species such as the rhesus monkey appear to have the same receptors for fimbriated *Escherichia coli* as those causing UTI in humans.

## II. METHODS OF STUDY

### A. *In Vitro Methods*

#### 1. Adherence to Urothelial Cells

Adherence to urothelial cells was first reported by Svanborg-Eden[24] who studied urothelial cells from normal patients voided urine to which she added bacteria usually found in urinary tract infections. Soon after, Kallenius and Winberg[10] examined urothelial cells collected from patients with urinary tract infections. Both reported bacterial adhesion to the urothelial cell wall.

Urothelial cells can be collected from bladder or vaginal tissue at sacrifice by scraping the cells of the epithelium with the edge of a microscopic slide into phosphate-buffered saline (PBS). These cells are immediately ready to use. Urothelial cells can also be collected from urine by centrifugation (3000 rpm for 10 min). These cells must then be washed with PBS twice before they are ready for use.

Most of the work done in this lab has utilized *E. coli*. Bacteria are grown on blood agar or, if P-fimbrial growth is wanted, on colony forming antigen agar (CFA agar, see appendix for reagents).[1,9] They are harvested in saline, washed, and diluted to $1 \times 10^{10}$ by spectrophotometry of the density of the solution. Using our machine in lab, this density is produced by 10% transmission of light through the bacterial solution.

Adhesion is observed by mixing equal amounts of cells and bacteria (usually 100 µL) on a slide or mirror and rocking them until agglutination is evident by white clumps. This

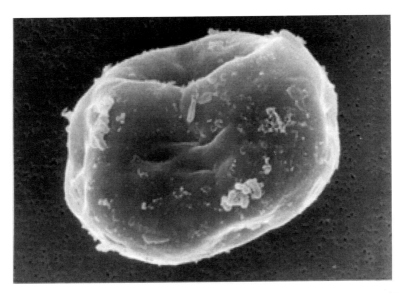

**Figure 1.** Scanning electron micrograph of urothelial cell with adherent bacteria. Negative stain with uranyl acetate. Reduced from ×35,000.

agglutination is rated from 1+ through 4+. A control slide is set up by using the same amount of cells (100 µL) with 100 µL of saline to control for volume.

Agglutination can be inhibited by several chemicals depending on the type of fimbriae found on the bacteria.[19] Inhibition studies can be done in vitro either by incubation of the bacteria with its receptor or incubation of the urothelial cells with the antibody to the receptor. The minimal receptor of the P-fimbriae of *E. coli* responsible for adherence is the disaccharide α-Gal (1-4) β-Gal of the globoseries of glycolipids. We obtained Gala 1-4Gal-β-O-Et (ethanol) from Pierce Chemical Company (Rockford, IL) and 4-0-α-D-galactopyranosyl-D-galactopyranose from Sigma Chemical (St. Louis, MO). A 30 mM solution of either inhibited the binding of P-fimbriae to urothelial cells. Equal amounts (100 µL) of the inhibitor and the bacteria are mixed on a mirror and allowed to combine for 15 min. 100 µL of cells are added and the mirror is rocked until agglutination is seen in the control, but not in the inhibited mixture. The control consists of bacteria with saline to which are added the cells. There are several epitopes of the P-fimbrial adhesin: Class I which has not been associated with disease; Class II, the epitope associated with pyelo-nephritis and whose receptor is the Gal-Gal containing glycolipid; and the Class III epitope associated with cystitis.

The Forssman antigen acts as a receptor for the Class III bacterial adhesin. Rabbit Mab antihuman Forssman was obtained from Accurate Chemical & Scientific, Westbury, NY. It was supplied in solution and was used without dilution.

Fibronectin on epithelial cells can act as a receptor for P-fimbriae by the PapE protein, which binds fibronectin. To test for adhesion by this mechanism, we studied inhibition of adhesion using anti-human fibronectin (Cappel/Organon Teknika, Durham, NC). The reagent was diluted according to the directions and used without further dilution. These antibodies are incubated with the epithelial cells and the bacteria is added after incubation.

A 5% solution of methyl α-D-mannopyranoside (Sigma) will inhibit the binding of Type-1 fimbriae to cells. For bacteria that possess both P and Type-1 fimbriae, a mixture of 60 m$M$ and 10% mannose gives the proper concentration for inhibition.

Adhesion can also be observed on individual cells by scanning electron microscopy (SEM) (Fig. 1). The method of fixation is described in detail under the section Electron Microscopy (EM).

### B. Studies Using the Aggregometer

The Chrono-Log Aggregometer (Havertown, PA) records agglutination as change in resistance between two electrodes. It is used to measure the aggregation of platelets in whole blood. We, however, used it to measure agglutination of red blood cells and epithelial cells by bacteria. For this procedure the machine is calibrated to allow 5 cm change to equal 20 ohms resistance. All materials are warmed to 37°C. Approximately 1 mL of heparinized whole blood or 1 mL urothelial cells is pipetted into a vial and the electrodes are inserted. Allow equilibration for a few minutes and set the baseline at 90%. Introduce the bacteria ($1\times10^{10}$) close to the electrodes without agitating the blood or cells. As the bacteria agglutinate the rbc's or the cells, the curve goes downward denoting an increase in resistance between the electrodes. The unit change/minute (based on the #cm travelled on the chart which translates into minutes and the 5 cm/20 ohms calibration denotes the amount of agglutination. If the amount of inhibition is to be measured using Gal–Gal, mannose, etc., a baseline of agglutination (cells + bacteria) has to be established.[8]

### C. Studies of Fluorescein Isothiocyanate (FITC)-Labeled Bacteria

Bacterial adherence to cells is often easier to measure if the bacteria are stained. For this, a saturated solution of FITC (Sigma) in phosphate-buffered saline (PBS) pH 8.5 is used. The stain is centrifuged and the supernatant is used for staining. Grow bacteria overnight on blood agar or CFA. Harvest with PBS, and wash two times. Dilute the bacteria to $1\times10^{10}$ bacteria/mL and add 2 mL of FITC to 10 mL of bacteria. Agitate slowly at room temperature for 6 h. Wash gently three times with PBS and one time with Solution X (*see* appendix). Let stand overnight in Solution X. Wash two times more with Solution X and dilute to $1\times10^{10}$.

### D. Flow Cytometry

To prepare cells for flow cytometry collect urine and centrifuge it at 2000 rpm for 15 min. Discard the supernatant and resuspend the percipitate in PBS, pH 7.5. Wash the cells twice with PBS and once with Solution X. The amount of Solution X used to resuspend the cells depends on how many different bacteria you are testing. Allow 0.5 mL for each sample and 0.5 mL for the control. Do nothing more to the control cells. 0.5 mL is enough to establish your gates and run for autofluorescence. Incubate the sample urothelial cells with 0.5 mL FITC-labeled bacteria ($1\times10^{9}$) for 1 h at room temperature. Keep the samples gently moving as on a tube rocker. Wash two times with Solution X. Between washings, centrifuge at a very low speed (about 1000 rpm is maximum). This spins down the urothelial cells and gets rid of the excess bacteria which remain in the supernatant. Resuspend the cells/bacteria in 0.5 mL Solution X.[17,25] The computer-generated curves from the control and experimental samples are shown in Figure 2.

## E. Fluorescence Microscopy

The fluorescence microscope can be used to examine slides of cells collected from urine or bladder as well as tissue that has been processed, embedded in paraffin and sectioned. Cells are prepared as described above for flow cytometry. After incubation with bacteria a drop of material on a slide with a cover slip can be examined under the fluorescence microscope. The number of cells with adherent bacteria can be counted. The average number of bacteria adhering to each cell can also be ascertained.

Unstained tissue slides of bladder or kidney are deparaffinized and processed as described by Falk et al.[2] Tissue sections are overlaid with 150 μL of FITC-labeled bacteria and incubated at 4°C for 1 h. Slides are washed extensively with PBS and examined under the fluorescence microscope (Fig. 3).[18]

## F. Fluorometry

The Fluoreskan fluorometer is made by LabSystems, Temecula, CA, and utilizes Ascent software. Adherence of FITC-bacteria to cells is measured in a black EIA plate, with a flat, clear bottom. A 96-well plate can be used but if more sample is needed, larger wells can be used and the machine will read specified sections of the well. Scraped bladder or vaginal cells, or cells from spun urine are washed with Solution X as described above. A control sample of cells (100 μL) is removed and diluted to 600 μL with Solution X. A control (100 μL) of the FITC-labeled bacteria is also removed and diluted to 600 μL with Solution X. Nothing more is done to these two control samples. 100 μL samples of cells are incubated with 100 μL FITC-labeled bacteria for 1 h at 37°C. They are washed one time with Solution X, the supernatant is discarded and they are resuspended to 600 μL with Solution X. 250 μL duplicates of each sample and controls are pipetted into the wells, processed in the Fluoreskan, and averaged. The bacteria alone gives a value for 100% fluorescence. The cells alone give a correction factor for autofluorescence of Solution X which is subtracted from the cells + bacteria values. This amount divided by the 100% fluorescence value for the bacteria alone gives the % adherence.

## G. Electron Microscopy (EM)

Bacterial adherence can be visualized vividly with both scanning and transmission electron microscopy as shown in our studies of experimental infection. For in vitro studies, tissue specimens are incubated with bacteria ($1\times10^9$ cells/mL saline) at 37°C with constant rotation for 1 to 2 h. They are then fixed in modified Karnovsky's fixative (*see* appendix).

Samples for scanning EM are washed three times in 0.2 M sodium cacodylate buffer, three times in distilled water, dehydrated in ascending ethanols and critical point dried in absolute ethanol. They are then sputter coated with gold, examined and photographed in a JEOL-T300 scanning electron microscope.

Tissues for transmission EM are washed six times with 0.2 M sodium cacodylate buffer, post-fixed in 1% osmium tetroxide, dehydrated in ascending concentrations of ethanol and embedded in Spurr's epoxy resin.[22] A Siemens Elmiskop 101 electron microscope is used to observe the sectioned samples. To compare bacterial adherence among strains, the

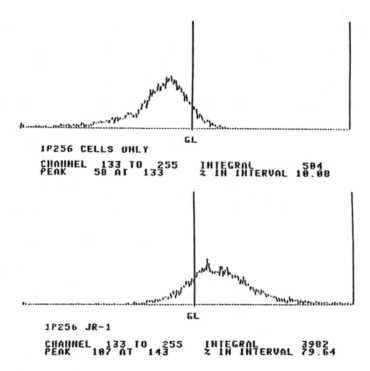

1P256 CELLS ONLY

CHANNEL    133 TO    255      INTEGRAL         504
PEAK       50 AT    133       % IN INTERVAL 10.08

1P256  JR-1

CHANNEL    133 TO    255      INTEGRAL         3902
PEAK      107 AT   143        % IN INTERVAL 79.64

**Figure 2.** Computer generated curves of flow cytometry data from human urothelial cells incubated with FITC-labeled *E. coli*. The curve represents every particle with fluorescence. The upper curve shows autofluorescence of untreated cells. The bottom curve shows cells to which FITC-labeled bacteria are attached. The lower gate(GL) line indicates the point where the two curves intersect. The area under the curve to the right of the GL on the top curve is subtracted from the area under the curve to the right of the GL on the bottom curve to yield 69%, the percentage of cells to which FITC-labeled bacteria are attached.

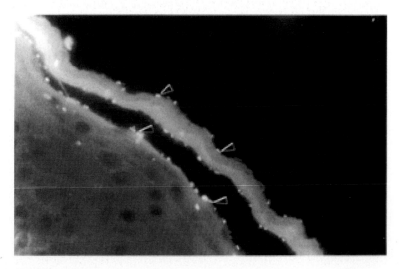

**Figure 3.** Fluorescence microscopy of squamous epithelium from the bladder trigone with marked adherence of FITC-labeled *E. coli*.

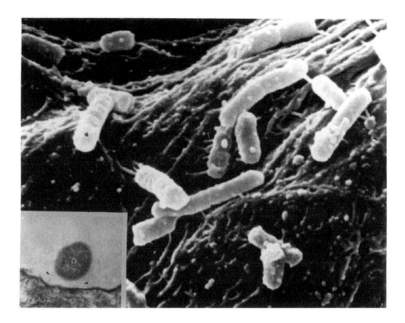

**Figure 4.** SEM image of the mucosal surface of human foreskin showing attachment of *E. coli* by fimbriae. Reduced from ×20,000. Inset, some bacteria (*) appear to attach without fimbriae. Distance (arrowheads) between bacteria and epithelial cellular membrane is 550-600 Å. Reduced from ×32,000. (Reprinted with permission: Fussell EN, et al: Adherence of bacteria to human foreskins. *J Urol* 140: 997-1001, 1988)

**Figure 5.** SEM image of massive colonization of *P. mirabilis* on foreskin mucosal epithelium. Fimbriae are shown by small arrowheads. Reduced from ×15,000. Inset: attachment of bacteria (*) to mucosal epithelium (arrowheads). Distance between bacteria and epithelium is 550 to 600 Å. Reduced from 38,000×. (Reproduced with permission—*see* Fig. 4.)

**Figure 6.** Scanning EM of bladder from monkey 48 h after infection with *E. coli* showing fimbrial attachment of bacteria to mucosal cells of the bladder. Reduced from ×5000.

number of bacteria adhering to each ×7500 field are counted. Adherence by P or Type-1 fimbriae can be determined by incubating some bacteria with 30 mM Gal-Gal or 5% mannose before introducing the tissue.

Adherence of bacteria to foreskins can be seen in Figures 4 and 5 showing both scanning and transmission EM.[5]

### H. In Vivo Methods

Perhaps the most salient proof that adherence of bacteria is a factor in urinary tract infections comes from studies of the production of experimental pyelonephritis in the monkey with subsequent identification of bacteria adhering to the ureter and kidney tubular epithelium after sacrifice.[4] *E. coli* strain JR1 (P+), used extensively in this laboratory, is introduced into one kidney by ureteral catheter leaving the other kidney as a control. The animals are sacrificed at various times after inoculation and the tissues are processed as indicated above for scanning and transmission EM. Bacterial adherence is seen in Figures 6 and 7 from monkey tissue after infection.

**Acknowledgment:** Supported by USPHS Grants: RR-00164 and DK-14681.

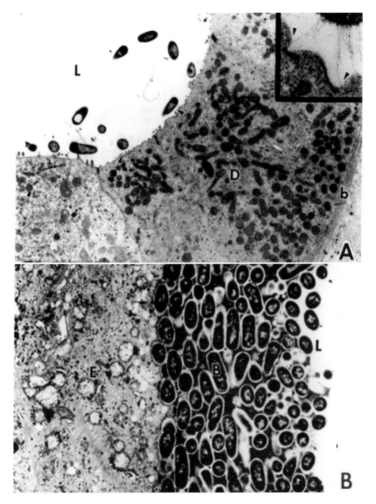

**Figure 7.** Transmission electron micrograph of tissue from monkey infected with the JR1 strain of *E. coli*. (A) Kidney showing adherence of bacteria to the renal tubular cell (D) by fimbriae (arrows); lumen (L); basement membrane (b). Reduced from ×15,000. Insert reduced from ×100,000. (B) Numerous bacteria adhering to ureteral epithelium (E) forming a biomass (B); lumen of ureter (L). Reduced from ×25,000. (Reproduced with permission—Roberts JA, Suarez GM, Kaack B, et al: Experimental pyelonephritis in the monkey. VII. Ascending pyelonephritis in the absence of vesicoureteral reflux. *J Urol* 133: 1068–75, 1985).

## REFERENCES

1. Evans DG, Evans DJ Jr: New surface-associated heat-labile colonization factor antigen (CFAII) produced by enterotoxigenic *Excherichia coli* of serographs 06 and 08. *Infect Immun* 21:638–47, 1978
2. Falk P, Roth PK, Boren T, et al: An *in vitro* adherence assay reveals that *Helicobacter pylori* exhibits cell lineage-specific tropism in the human gastric epithelium. *Proc Nat Acad Sci* 90:2035–9, 1993
3. Finkelstein RA, Scirotino CU, McIntosh MA: Role of iron in microbe-host interactions. *Rev Infect Dis* 5:S759–S77, 1983

4. Fussell EN, Roberts JA: The ultrastructure of acute pyelonephritis in the monkey. *J Urol* 133:179–83, 1984

5. Fussell EN, Kaack MB, Cherry R, Roberts JA: Adherence of bacteria to human foreskins. *J Urol* 140:997–1001, 1988

6. Johnson JR: Virulence factors in *Escherichia coli* urinary tract infection. *Clin Microbiol Rev* 4:80–128, 1991

7. Jones GW: The attachment of bacteria to the surfaces of animal cells. In: Rissing JL, ed. *Microbial Interaction.* Ellis Harwook, Chichester, UK, 1980:79

8. Kaack MB, Dowling KJ, Patterson GM, et al: Immunity of pyelonephritis. VIII. *E. coli* causes granulocytic aggregation and renal ischemia. *J Urol* 136:1117–22, 1986

9. Kaack MB, Pere A, Korhonen TK, et al: P-fimbriae vaccines. l. Cross-reactive antibodies to heterologous P-fimbriae. *Pediat Nephrol* 3:386–90, 1989

10. Kalleanius G, Winberg J: Bacterial adherence to periurethral epithelial cells in girls prone to urinary-tract infecion. *Lancet* 2:540, 1978

11. Kallenius G, Mollby R, Svenson SB, et al: The $P^k$ antigen as receptor for the haemagglutinin of pyelonephritic *Escherichia coli. FEMS Microbiol Lett* 7:297–302, 1980

12. Kisielius P, Schwan W, Amundsen S, et al: *In vivo* expression and phase variation of type 2 pili by *Escherichia coli* in urine of adults with acute urinary tract infection. *Infect Immun* 57:1656–62, 1989

13. Korhonen TK, Virkola R, Westerlund B, et al: Tissue tropism of *Escherichia coli* adhesins in human extraintestinal infections. *Curr Top Microbiol Immunol* 151:115–27, 1990

14. Labigne-Roussel A, Lark D, Schoolnik G, et al: Cloning and expression of an afimbrial adhesin (AFA-I) responsible for P-blood-group independent mannose resistant hemagglutination from a pyelonephritic *Escherichia coli* strain. *Infect Immun* 46:251–9, 1984

15 Nowicki B, Labigne A, Moseley S, et al: The Dr hemagglutinin, afimbrial adhesins AFA-I and AFA-III, and F1845 fimbriae of uropathogenic and diarrhea-associated *Escherichia coli* belong to a family of hemagglutinins with DR receptor recognition. *Infect Immun* 58:279–81, 1990

16. Ofek I, Beachey EH: Mannose binding and epithelial cell adherence of *Escherichia coli. Infect Immun* 22:247–54, 1975

17. Roberts JR, GM Suarez, B Kaack, et al: Experimental pyelonephritis in the monkey. VII. Ascending pyelonephritis in the absence of vesicoureteral reflux. *J Urol* 133:1068–75, 1985

18. Roberts JA, Marklund BI, Ilver D, et al: The Gal (α 1-4) Gal-specific tip adhesin of *Escherichia coli* P-fimbriae is needed for pyelonephritis to occur in the normal urinary tract. *Proc Nat Acad Sci* 91:11889–93, 1994

19. Roberts JA, Kaack MB, Baskin G, et al: Epitopes of the P-fimbrial adhesin of *E. coli* cause different urnary tract infections. *J Urol* 158:1610–3, 1997

20. Silverblatt FJ, Ofek I: Influence of pili on the virulence of *Proteus mirabilis* in experimental hematogenous pyelonephritis. *J Infect Dis* 138:664–7, 1978

21. Silverblatt FJ, Cohen LS: Antipili antibody affords protection against experimental ascending pyelonephritis. *J Clin Invest* 64:333–6, 1979

22. Spurr, AR: A low-viscosity epoxy resin embedding medium for electron microscopy. *J Ultrastruct Res* 26:31–, 1969

23. Stamey TA, Mihara G: Observations on the growth ofurethral and vaginal bacteria in sterile urine. *J Urol* 124:461–3, 1980

24. Svanborg-Eden C, Hanson LA, Jordal U: Variable adherence to normal human urinary tract epithelial cells of *E. coli* strains associated with various forms of urinary tract infections. *Lancet* 2:490, 1976.

25. Svenson SB, Kallenius G, Korhonen TK et al: Bacterial Infection 20. Initiation of clinical pyeloenphritis — The role of P-fimbriae-mediated bacterial adhesion. *Contributions Nephrology* 39:252–72, 1984

26. Venegas MF, Navas EL, Gaffney RA et al: Binding of type 1-piliated *Escherichia coli* to vaginal mucus. *Infect Immun* 63:416–22, 1995

# APPENDIX

## CFA Agar:

1 L Distilled water
10 g Casaminio acid (1%)
1.0 g Yeast extract (0.15%)
2.22 mL $MgSO_4$ (0.005%)
0.005 g $MnCl_2$ (0.0005%)
20 g Granulated agar (2%)
Autoclave, cool, pour into sterile petri dishes

## Solution X:

1% Glycerol
5% Bovine serum albumin
0.5% Sodium azide
Distilled water

## Karnovsky's Fixative:

2.5% Glutaraldehyde
2% Paraformaldehyde
in 0.1 M Na-cacodylate buffer

0.2 M Na-cacodylate buffer (mol wt 214.03)
42.8 g/L Water
Bring to pH 7.2–7.4 with HCl
For 0.1 M dilute 1:1

# 33

# Studying *Candida albicans* Adhesion

**Lakshman P. Samaranayake[1] and Arjuna N. B. Ellepola[2]**

*[1]Oral Biosciences, Faculty of Dentistry, The University of Hong Kong, Hong Kong*
*[2]Department of Oral Medicine and Periodontology, University of Peradeniya, Sri Lanka*

## 1. INTRODUCTION

Increasing recognition of the ecological, medical and economic significance of microbial adhesion has resulted in a vast escalation of research effort in this area. Although studies on bacterial adhesion have predominated, with an extensive body of data, investigations on candidal adherence are comparatively limited.[23] Nevertheless, a rapidly expanding literature on candidal adherence attests to the potential importance of understanding the behavior of this ubiquitous yeast and the pathogenesis of infections which it causes in the human host.[37,59] Adhesion of *Candida* to epithelial cells has been investigated to define parameters relevant to the pathogenesis of oral, gastrointestinal, vaginal and urinary candidiasis.[24,48,57,58] Further, the attachment of the organism to fibrin, fibrin-platelet matrices and to vascular endothelial cells have been examined to elucidate initial events leading to candidal endocarditis and hematogenously disseminated infection.[48,49,60] There is also a growing body of information on the adhesion of *Candida* to inert/nonbiological surfaces such as denture prostheses, intravascular and urinary catheters, and prosthetic cardiac valves.[12,31-33,48,50,51,54,55]

*C. albicans* is one of several *Candida* species isolated from humans and is responsible for the majority of superficial and systemic fungal infections.[38] Thus, most of the foregoing studies pertain to this isolate although a number of workers have studied the adhesion of emerging pathogens such as *C. krusei*[61,62] and *C. parapsilosis*.[40] In this report we outline briefly the variety of laboratory methods available for quantification of *C. albicans* to both biological and inert surfaces. For further details of most of the studies given below and the parameters which guide their use the reader is referred to reviews such as Kennedy,[23] Douglas,[12] Samaranayake and MacFarlane,[59] and Odds.[37]

The term "adhesin" and "receptor" are widely used to describe surface components that mediate attachment of microorganisms to animal cells. Adhesins are the adhesive structures on microbial surfaces, whereas receptors are complementary adhesive structures on the surfaces of the host cells. Throughout this chapter the terms "adhesion," "adherence" and "attachment" will be used synonymously to describe associations between *Candida* and surfaces.

*Handbook of Bacterial Adhesion: Principles, Methods, and Applications*
Edited by: Y. H. An and R. J. Friedman © Humana Press Inc., Totowa, NJ

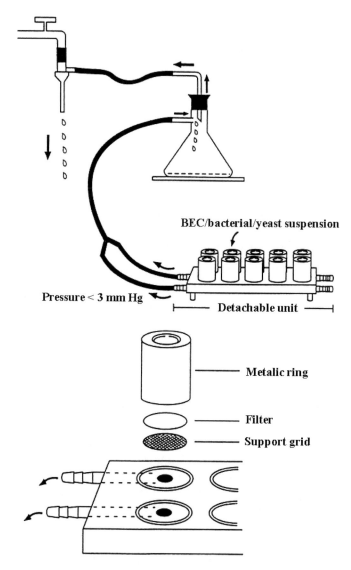

BEC/bacterial/yeast suspension

Pressure < 3 mm Hg

Detachable unit

Metalic ring

Filter

Support grid

**Figure 1.** A schematic representation of the filter manifold used in buccal epithelial cell adhesion assays and its mode of action (Courtesy Dr. RG Nair).

## II.  ADHESION TO EXFOLIATED EPITHELIAL CELLS

### A. *Microscopic Analysis*

A quantitative method to  determine yeast adhesion to epithelial cells in vitro was first reported by Liljemark and Gibbons,[28] which was subsequently modified by Kimura and Pearsall.[24] The procedure involved incubation of equal volumes of standardized suspensions of yeasts and epithelial cells, subsequent removal of unattached yeasts, and recovering the epithelial cells with adherent yeasts from the incubation mixtures using polycarbonate filters. The filters had a pore size of 14 μm that allowed passage of unattached yeast cells, but not the epithelial cells. The filters with retained epithelial cells and the adherent yeasts were dried, fixed, and stained with Gram's method. The filters

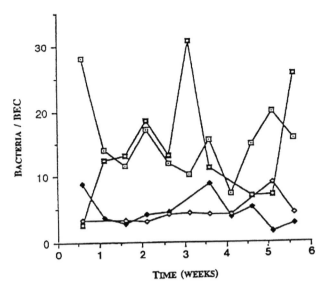

**Figure 2.** Variation in indigenous bacterial counts on buccal epithelial cells in four individuals over a period of 6 wk. [53]

were then placed directly on glass slides, mounted on oil and the adherent yeasts counted under light microscope at ×40 mag. At least 100 epithelial cells with adherent yeasts were quantified. This method, with further minor modifications, has been used by other investigators to determine yeast adhesion to buccal and vaginal epithelial cells as described below.

A number of workers have investigated candidal adhesion to buccal epithelial cells (BEC) and also parameters which affect adhesion by using the following popular technique.[8-10,14,15,18,34,53,56] In this method equal volumes of the BEC and the yeast suspension are mixed gently in Bijoux bottles and incubated in a rotary incubator with gentle agitation at 37°C for 1 h. The yeast/BEC suspension is thereafter diluted in 4 mL of sterile phosphate-buffered saline (PBS). The BEC are then harvested under negative pressure using a manifold filter (Sartorius, SM16547) onto a 12 mm pore size poly-carbonate filters (Millipore, UK), and washed twice with 50 mL of sterile PBS to remove unattached fungi (Fig. 1). Thereafter each filter is removed carefully with forceps and placed firmly on a glass slide with the BEC against the glass surface. After 10 s, the filter is removed gently, leaving the BEC adherent on to the glass slide. This process results in transfer of ample number of individual BEC on to the surface of the slide for observation and quantification. The preparations are then air dried, fixed with methanol and stained with Gram's stain. The number of *Candida* cells attached to 50 BEC is quantified using a light microscope at ×40. In a further modification to this method, a color video camera (TK-1080E, JVC, Japan) is attached to the microscope and the images of the cells fed into an image analyzer (Videoplan Image Processing System, Kontron Image Analysis Division, Germany) and the number of adherent yeasts per BEC recorded semi-automatically using a computer mouse to flag each individual yeast.[14,15,18] The following criteria have been used in quantifying the yeasts: a yeast with a daughter cell smaller than the mother cell is counted as one unit, overlapping and folded BEC are not counted and, only a single BEC in isolation is counted.

**Table 1. Factors Affecting the Adhesion of *Candida***

**Yeast factors**
    Concentration and viability
    Phase and temperature of growth
    Growth-medium composition
    Species and strain
    Germ tube formation
    Hydrophobicity

**Epithelial cell factors**
    Cell type
    Mucosal cell size and viability
    Sex hormones
    Yeast carriers vs patients with overt candidosis

**Environmental factors**
    Temperature and period of contact
    Hydrogen-ion concentration
    Bacteria
    Antibodies
    Antibacterial drugs
    Sugars (especially dietary carbohydrates)

Braga et al.[2] investigated the ability of various *Candida* species to adhere to vaginal epithelial cells (VEC) by mixing 1:1 volumes of *Candida* with VEC in polystyrene tubes rotated end-over-end at 24 rpm for 1 h at 37°C. The epithelial cells were then separated from nonadherent *Candida* by centrifuging three times in sterile phosphate-buffered saline (PBS). The final epithelial cell pellet is resuspended in PBS, placed on a round microscope cover slip and dried. The round cover slip with cells is then fixed in 2.5% glutaraldehyde in 0.1 mol/L cacodylate buffer, for 1 h at 4°C. After several dehydration steps in alcohol, the coverslips were critical-point dried and coated with 200 Å gold and counted in a scanning electron microscope (SEM). This method is complex, time consuming and is dependant on the availability of a SEM. Hence, the more simpler methods described above have proven to be popular.

The quality of the epithelial cells used is critical in conducting the foregoing assays. Buccal or vaginal cell preparations can be readily obtained from human volunteers by gently swabbing or scraping the mucosal surface. Cells from the cheek are preferred to other sites as they harbor a lower a number of indigent bacteria.[53] The major drawback of experiments using BEC or VEC is that the cells derived from donors differ from person to person and also in the same person at different times. Hence, the collection of BEC has to be strictly standardized due to the latter problem as well as the complexity of the oral environment. Early morning (fasting) BEC are essentially devoid of food debris and are preferred by some workers. Also, in females the quality of cells vary depending on the hormonal levels of the menstrual cycle.[37] Due to such extreme variations between individuals it is proposed that pooled BEC collected from at least three volunteers should be used to obtain consistent, globally reproducible data. Qualitative variations are also seen due to other factors that effect *Candida* adhesion of to epithelial cells as shown in Table 1.[12,63]

As adhesion increases with increasing yeast concentration, yeast to epithelial cell ratio of 500:1 or 1000:1 are commonly employed.[12] It is customary to count the total number of adherent yeasts on 100 to 200 epithelial cells retained on the filter and, to prevent bias, the latter should be coded and read "blind." Although this visual method can be rather tedious, it does have the advantage that adhesion to individual epithelial cells as well as yeast-to-yeast co-adherence can be monitored.

Similar experimental procedures have been utilized for measuring yeast adherence to human uro-epithelial cells[5] and epidermal corneocytes.[6,46]

## B. Radioisotope Analysis

King et al.[25] used radioactively-labeled *Candida* to measure its adherence to buccal and vaginal epithelial cells. The yeasts were grown in phytone–peptone broth supplemented with [U-$^{14}$C] glucose. Standardized suspensions of labeled organisms were incubated with suspensions of epithelial cells, and the latter with adherent yeasts were collected on polypropylene filters. The filters were treated with a solubilizing agent and their radio-activity counted in a liquid scintillation spectrometer. Control assays, containing yeasts only, showed that 2% of the organisms adsorbed to the filters and all results were corrected for this background count.

The adherence of *C. albicans* isolates to different types of filters can vary. Zierdt[74] showed that *C. albicans* adsorbs in higher numbers to polypropylene filters than to polycarbonate filters. As an alternative to filtration, differential centrifugation can be used to separate epithelial cells from unattached yeasts.[65] These protocols with slight modifications have been used by other investigators.[3,42,70] For instance Vargas et al.[68] studied the adhesion of *C. albicans* to buccal epithelium and stratum corneum, and Pereiro et al.[42] the adhesion of *C. albicans* from patients with and without HIV infection to oral mucosal cells, using such radiometric assays.

Radiolabeling of yeasts provides a potentially attractive means of evaluating adhesion to mucosal surfaces in vitro. It is relatively rapid and considerably less laborious than the visual microscopic method. However, during the assay, intact or lysing yeast could release radioactive label which might be subsequently bound or incorporated onto the epithelial cells. The possibility of yielding spuriously high adherence values due to this phenomenon should therefore be kept in mind when using radiometric methods.

## C. Fluorescent Labeled Yeasts and Flow Cytometry

As discussed, the methods based on optical microscopic evaluations and radiometric analysis are popular and frequently used in determining candidal adhesion to epithelial cells. However, the optical adherence assay is usually conducted with a large excess of yeasts (100-fold) over epithelial cells in order to obtain microscopically quantifiable data. On the other hand, the radiometric analysis fails to measure individual cells. Further the phenomenon known as "co-adherence" where individual yeast cells aggregate to yield potentially spurious results cannot be ruled out in radiometric assays.

A method for measuring yeast adherence that overcomes some of these problems has been developed by Polacheck et al.[43] It is based on labeling the yeasts with a fluorogenic marker (2', 7'-bis-[2-carboxyethyl]-5[6]-carboxyfluorescein acetoxymethyl ester) prior to the assay, and analyzing epithelial cells with adherent yeasts by flow cytometry, while non-bound yeasts are excluded by "gating." The increase in the fluoroscence intensities of the BEC is considered a quantitative measure of the extent of yeast adherence. The possibility of studying reliably low yeast: epithelial cell ratios, which mimic adhesion as it occurs in vivo, appear to be an important advantage of this assay.

A similar method, but based on photometric quantification of yeast adherence to epithelial cells has recently been described by Borg-von Zepelin and Wagner.[1] Here the adherent *Candida* are detected by staining with the fluorescent dye Calcofluor white (CFW). The assay comprises the following steps. Coating of a microtest plate with epithelial cells and inoculation with a standard concentration of *Candida*, staining *Candida* with CFW, rinsing to remove nonadherent cells and unbound dye and, detection of adherent fluorescent yeasts using photometry.

It is noteworthy that although CFW is well known to differentially stain the yeast cells in preference to epithelial cells, previous workers have had little success with this technique when studying candidal adhesion.

## III. ADHESION TO CULTURED EPITHELIAL CELLS

Although readily obtained and widely used in investigating microbial adhesion to mucosal surfaces, exfoliated epithelial cell preparations have several disadvantages. They invariably consist of a heterogeneous mixture of viable and nonviable cells with a multitude of adherent bacteria (Fig. 2). Studies of both bacterial and yeast adhesion to such preparations have demonstrated substantial cell-to-cell variation in the number of adherent microbes.[53] Further, as stated earlier exfoliated cell preparations may vary according to the donor, the time of sampling, and the degree of exposure to various secretions such as saliva, serum and food.

To avoid some of these problems, methods for assaying yeast adhesion to more uniform cell populations have been developed. Samaranayake and MacFarlane[56] were the first to describe the adhesion of *C. albicans* to cultured human epithelial cells (i.e., HeLa cells and human embryonic kidney epithelial cells). They incubated standard yeast suspensions with confluent cell monolayers grown on a cover slip. Following removal of unattached yeasts, the number of adherent organisms per unit area of the monolayer was determined by direct microscopy after air drying, Gram staining, and mounting on glass slides. This technique has been used to determine the adhesion and colonization of other *Candida* species as well.[61]

Another in vitro adherence model in a primary culture of human keratinocytes, which allows molecular study of mechanisms responsible for *C. albicans* adherence in cutaneous candidosis, has also been developed.[39] But the disadvantage of such primary cultures as opposed to continuous cultures (e.g., HeLa) is that the latter could be propagated almost indefinitely, and hence, replicate experiments performed easily in many laboratories. However, only a limited amount of information could be obtained using primary culture systems as they can be passaged only for a few generations.

## IV. ADHESION TO ENDOTHELIAL CELLS

A method for studying adhesion of *Candida* to vascular endothelium was devised by Klotz et al.[26] Segments of freshly obtained pig blood vessels were secured between two sheets of Lucite; the upper sheet contained 12 mm perforations to create wells, each of which had the endothelial surface as its base. Standardized yeast suspensions were incubated in the wells and, adherence to endothelium evaluated by enumerating the number of nonadherent yeasts that could be removed from each well by washing. Multiple wells were provided to obtain several measurements using the same vascular segment.

Adhesion to cultured vascular endothelial cells was described by Rotrosen et al.[49] who used both human umbilical vein endothelial cells and rabbit aortic cells. Confluent monolayers of either cell type were incubated with yeast suspensions, and number of organisms adherent to each monolayer determined by viable counts using an agar-overlay technique. Separate experiments demonstrating the absence of yeast clumping indicated that each colony resulted from a single adherent organism, although experience with optical assay techniques indicate that this may not be always true. However, this method with slight modifications is widely used. For instance, it has been used to investigate the potential mechanisms by which g interferon protects endothelial cells from candidal injury.[21] Also the role of iron in endothelial cell injury caused by *C. albicans* and, the impact of the new triazole, voriconazole, on the interactions between *Candida* species and endothelial cells has also been evaluated using this method.[19,20]

Radiometric analysis has also been performed to study adhesion between human dermal microvascular endothelial cells (HDMEC) and *C. albicans*.[27] HDMEC were plated in gelatin-coated 96-well flat-bottomed culture plates. A standard concentration of *C. albicans* suspension labeled with [$^{35}$S] methionine was incubated with HDMEC, after which the supernatant was removed from each well and the monolayer was washed twice to remove non-adherent yeasts. The latter were then counted in a scintillation counter. Adherent yeast cells and HDMEC monolayers were removed subsequently, solubilized, counted in a scintillation counter and the percentage binding of *C. albicans* calculated.

## V. ADHESION TO ANIMAL TISSUES

Many have studied the adherence of *C. albicans* using epithelial cells and this undoubtedly remains the most popular technique, probably due to the ready availability of the substrate. On the contrary, adherence of *C. albicans* to internal organs is less studied and not well understood. Nonetheless, during candidamia, especially in compromised hosts a number of target organs such as the spleen and kidney are profoundly affected.[37]

A method first described by Cutler et al.[7] and later modified by Riesselman et al.[47] is commonly used to study the adherence of *C. albicans* to internal organs. In their method, mice are sacrificed and various organs are removed, rapidly frozen on dry ice, and sectioned. Standard concentrations of *C. albicans* suspensions are then inoculated onto the tissue sections and incubated for 15 min with or without agitation at 4 to 6°C. At the end of 15 min, the majority of unattached *C. albicans* are decanted, the section blotted onto plastic backed absorbent paper, fixed in glutaraldehyde, rinsed by dipping repeatedly in cold tap water, and air dried for crystal violet staining.[47] The stained sections are dehydrated by dipping in ethanol and allowed to air dry. The adherent *C. albicans* are quantified microscopically either by manual counting or by computerised image analysis techniques.

The above method has been subsequently modified to study the adherence of *C. albicans* germ tubes to murine tissues, the latter exposed to cytotoxic drugs,[29,30] candidal adhesion to brain tissues of the primate *Macaca mulata*[11] and to tissues from immunocompromised mice.[70]

## VI. ADHESION TO FIBRIN CLOTS

Infective endocarditis is characterised by the colonization of platelet–fibrin thrombi on heart valves by microorganisms which subsequently produce an endothelial vegetation. *C. albicans* is recognized as an important cause of infective endocarditis in patients with pre-existing cardiac disease, particularly after open-heart surgery.[22] Thus, in vitro candidal adhesion to surfaces such as fibrin and fibrin–platelet matrices have been investigated by a few researchers[4,60] due to the critical importance of this primary event in the pathogenesis of infective endocarditis.

The methodology used by Samaranayake et al.[60] is as follows. The fibrin clots are prepared using a modification of the method described by Toy et al.[67] Bovine plasma is mixed with a clotting reagent (bovine thrombin and 0.125 M calcium chloride in 0.15 M sodium chloride) and deposited in a well of a leucocyte migration plate. The well containing the clotting mixture is placed within multi-well tissue culture plates and incubated to ensure complete solidification. Afterwards the surface of the clot is washed to remove excess plasma and immediately used for the adhesion assay.

Adhesion assay is carried out by introducing a standard yeast inoculum into each well, and incubating the wells on a rotary shaker for 30 min. The yeast suspension is then aspirated from the wells and the clot washed. The washed fibrin clots are removed from the leucocyte migration wells into sterile PBS and vortex mixed. A standard volume of resultant yeast suspension is removed, inoculated onto Sabouraud's dextrose agar plates using a Spiral Plater (Spiral Systems, Cincinnati, OH), incubated for 2 d and the number of colony forming units (cfu) estimated.

This method which quantifies yeast adhesion to fibrin clots in terms of cfu's is simple, sensitive and versatile, as compared with other methods such as radiometry.

## VII. ADHESION TO DENTURE ACRYLIC

Undoubtedly the study of candidal adhesion to soft tissue surfaces or their derivatives is critical in understanding the pathogenesis of yeast infections and the means of their prevention. Nonetheless, it is well recognized that inert surfaces which are either implanted or in superficial contact with the human host frequently act as a conduit of infection transmission or a reservoir of infection. For these reasons a number of researchers have studied the adherence of *Candida* to a variety of materials found in medical devices such as catheters and oral prosthesis.

*Candida*-associated denture stomatitis is the most common form of yeast infection seen amongst the denture wearing elderly.[52] The major causative agent of this disease is *C. albicans,* and acrylic dentures which may be ill-fitting, with suboptimal hygiene, act as reservoirs of infection. As the ability of *Candida* to adhere to acrylic surfaces plays an important role in the pathogenesis of this condition many have investigated yeast adhesion to denture acrylic or similar inert polymeric material.[16,17,35,40,61]

Samaranayake and MacFarlane[55] first described an assay system to evaluate *Candida* adhesion to acrylic and since then it has been used by several workers with or without minor modifications. In this method, transparent self-polymerizing acrylic powder is spread on an aluminum foil-covered glass slide and the monomer liquid is poured onto the surface of the slide and immediately a second slide placed on top of the polymerizing mixture, and the slides firmly secured. After bench curing for 30 min the glass slides are separated, the resultant acrylic strips cut into 5×5 mm squares, water cleaned, disinfected by dipping in 70% alcohol, and washed with sterile distilled water. The strips are then used for the adhesion assay, after checking the sterility.

The acrylic strips are placed vertically in wells of sterile serology plates (Dynatech Immunlon, USA) and a standard yeast cell suspension added to each well, and the whole assembly incubated for one hour at 37°C with gentle agitation. The strips are recovered, washed, dried, and stained using a modified Gram stain, without the counter stain. After air drying the strips are mounted on glass slides and the adherent yeast quantified by light microscopy by counting up to 100 fields.

Recently a computerized semi-automated program (IBAS 2000, Kontron, Berlin, Germany) has been used to quantify the adherent yeasts to denture acrylic surfaces.[16,17,40,41] This system allows for semi-automated, rapid detection of adherent yeasts by scanning the surface area occupied by the adherent cells. The method, in brief, is as follows.

Once the image of a specific region of the acrylic strip is captured, the margins of the individual yeasts and the clumped cells are contoured. However, as the semi-automated system contours the cells which are adherent to the edges of the strips as well as other dark artifacts, subsequent manual editing is required to eliminate these from the counting field. Thereafter, the total area of the adherent yeasts in a given field is recorded. The area occupied by a single adherent yeast is derived using a simple formula.

$$\text{Area occupied by a single yeast (pixels)} = \frac{\text{Total area of adherent cells in a given field (pixels)}}{\text{Number of yeasts in the given field (visual counting)}}$$

Subsequently, the readings from 40 fields are divided by the area of a unitary yeast to yield the total number of yeasts attached to a unit area of the acrylic strip. The results obtained from the image analyzer correlate well with the adhesion data obtained using visual counting with light microscopy.[41] This protocol with slight modifications has also been used to study the adherence of *C. albicans* to denture-base materials[44] with different surface finishes[45] and to denture soft lining materials.[72]

Due to the opaque nature of the acrylic resin material, counting stained yeast with normal light microscopy is difficult if the fabricated test pieces are too thick. Hence, some investigators have stained the adherent yeasts with acridine orange for 2 min to enumerate attached yeasts using fluorescent microscopy.[69,72] Radiolabeled *Candida* has also been used to study the adhesion to polymethylmethacrylate[13] and this suffers from low sensitivity due to the high background counts of radioactivity. Others have used scanning electron microscopy to study the adhesion of *Candida* to polymethylmethacrylate,[71] which is a laborious and expensive method that cannot be recommended for novice workers.

Recently Nikawa et al.[36] quantified *Candida* adhesion and biofilm formation on denture acrylic surfaces by using luciferin-luciferase ATP (adenosine triphosphate) assay. The acrylic strips with adherent yeasts are immersed in an extraction-reagent (benzalkonium) after the conventional assay procedure. The resultant reagent solution is then filtered to clarify, and subjected to ATP-measurement using a bioluminescent apparatus (ATPA-1000, T6A Electronics, Tokyo, Japan). This apparatus makes use of the firefly luciferase system to determine the concentration of cellular ATP in live *Candida*. In essence, the machine quantifies the light emitted during oxidation of luciferine by molecular oxygen in the presence of ATP and magnesium ions. The intensity of light emitted is directly proportional to the ATP concentration in adherent yeasts.[36] These workers have shown that the results obtained with ATP assays are consistent with those of conventional viable counts or radiolabeling methods. The excellent correlation between yeast cells and the ATP content in this method is not surprising as the assay is based on the fundamental principle that the amount of cellular ATP correlates with the dry weight, the volume and the number of viable cells.[64] Although this method appears to be a promising technique for accurate measurement of adherent yeasts, the cost of the bioluminescent apparatus is a major drawback.

A novel approach for the assessment of candidal adherence to translucent acrylic material was described recently.[73] The method uses the inverted microscope to visualize yeast adhering to acrylic surfaces while the test material remains immersed in buffer. The authors claim that as the adherent cells are not subjected to surface tension forces, which operate during the drying process, an even distribution of yeasts with no aggregation occurs when the experiment is conducted, mimicking the in vivo situation. Nonetheless, until further experiments are performed in parallel with conventional techniques, the veracity of this new method remains to be determined.

## VIII. CONCLUSIONS

Although a rather bewildering number of methods are available to evaluate candidal adhesion to biological and non-biological surfaces it is important to recognize the limitations of these methods. Most of the currently available in vitro methods suffer from the inability to reproduce the in vivo environment, and hence, the data emanating are approximate at best. For instance, the adhesion of microbes within the oral environment is highly regulated by the continuous salivary flow and the surface coating of saliva which is a complex fluid comprising numerous organic and inorganic constituents and many individual variations. Though some have attempted to simulate the oral environment in vitro by coating the substrate surface with pure, mixed, stimulated and unstimulated saliva or buffers mimicking the composition of saliva, with some degree of success, further studies are required to substantiate these findings.[59]

We have alluded above to the quality of the substrate in conducting adhesion experiments. In the case of the epithelial cells, this is critical in deriving worthwhile data. The variables which should be considered include: the number of cells in the suspension mixture, whether the epithelial cells are pooled or not, the time of collecting epithelial cells (as the diurnal and menstrual/hormonal rhythm affect quality) and the processing of epithelial cells prior to the assay. As for the yeasts in the assay suspension, their quality is as important as their quantity. For instance, wild type *Candida* are far better than domesticated, reference strains which are less likely to adhere than the former. However, it is salutary to include at least one reference strain within an assay system to yield globally comparable data between laboratories. The source of the isolate including the clinical condition, and concurrent drug therapy affects the attributes of the yeast populations and should be kept in mind when interpreting data. Further, when intraspecies variations are compared it is important to use more than six isolates from each species to obtain statistically assessable data.

The above account illustrates the pitfalls of *Candida* adhesion assays for the unwary. Nevertheless, the assay systems currently available have served us well to understand the adherence phenomena in *C. albicans*. However, the researchers who seek further into this arena should be in for many surprising and interesting findings, as the journey is yet incomplete.

# REFERENCES

1. Borg-von Zepelin M, Wagner T: Fluorescence assay for the detection of adherent *Candida* yeasts to target cells in microtest plates. *Mycoses* 38:339–47, 1995
2. Braga PC, Maci S, Dal Sasso M, et al: Experimental evidences for a role of subinhibitory concentrations of rilopirox, nystatin and fluconazole on adherence of *Candida* species to vaginal epithelial cells. *Chemotherapy* 42:259–65, 1996
3. Brassart D, Woltz A, Golliard M, et al: *In vitro* inhibition of adhesion of *Candida albicans* clinical isolates to human buccal epithelial cells by Fuc α1→2 Gal β-bearing complex carbohydrates. *Infect Immun* 59:1605–13, 1991
4. Calderone RA, Rotondo MF, Sande MA: *Candida albicans* endocarditis: ultrastructural studies of vegetation formation. *Infect Immun* 20:279–89, 1978
5. Centeno A, Davis CP, Cohen MS, et al: Modulation of *Candida albicans* attachment to human epithelial cells by bacteria and carbohydrates. *Infect Immun* 39:1354–60, 1983
6. Collins-Lech C, Kalbfleisch JH, Franson TR, et al: Inhibition by sugars of *Candida albicans* adherence to human buccal mucosal cells and corneocytes *in vitro*. *Infect Immun* 46:831–4, 1984
7. Cutler JE, Brawner DL, Hazen KC, et al: Characteristics of *Candida albicans* adherence to mouse tissue. *Infect Immun* 58:1902–8, 1990
8. Darwazeh AM, Lamey PJ, Samaranayake LP, et al: The relationship between colonisation, secretor status and *in vitro* adhesion of *Candida albicans* to buccal epithelial cells from diabetics. *J Med Microbiol* 33:43–9, 1990
9. Darwazeh AM, Lamey PJ, Lewis MA, et al: Systemic fluconazole therapy and *in vitro* adhesion of *Candida albicans* to human buccal epithelial cells. *J Oral Pathol Med* 20:17–9, 1991
10. Darwazeh AM, MacFarlane TW, Lamey PJ: The *in vitro* adhesion of *Candida albicans* to buccal epithelial cells (BEC) from diabetic and non-diabetic individuals after *in vivo* and *in vitro* application of nystatin. *J Oral Pathol Med* 26:233–6, 1997
11. Denaro FJ, Lopez-Ribot JL, LaJean Chaffin W: Adhesion of *Candida albicans* to brain tissues of *Macaca mulata* in an *ex vivo* assay. *Infect Immun* 63:3438–41, 1995
12. Douglas LJ: Adhesion to surfaces. In: Rose AH, Harrison JS, Eds. *The yeasts*. Academic Press, London, 1987:239–80
13. Edgerton M, Scannapieco FA, Reddy MS, et al: Human submandibular-sublingual saliva promotes adhesion of *Candida albicans* to polymethylmethacrylate. *Infect Immun* 61:2644–52, 1993
14. Ellepola ANB, Samaranayake LP: Adhesion of oral *Candida albicans* to human epithelial cells following limited exposure to antifungal agents. *J Oral Pathol Med* 27:325–32, 1998
15. Ellepola ANB, Samaranayake LP: The effect of limited exposure to antimycotics on the relative cell surface hydrophobicity and adhesion of oral *Candida albicans* to buccal epithelial cells. *Arch Oral Biol* 43:879–87, 1998
16. Ellepola ANB, Samaranayake LP: The post-antifungal effect (PAFE) of antimycotics on oral *Candida albicans* isolates and its impact on candidal adhesion. *Oral Dis* 4:260–7, 1998

17. Ellepola ANB, Samaranayake LP: Adhesion of oral *Candida albicans* isolates to denture acrylic following limited exposure to antifungal agents. *Arch Oral Biol* 43: 999–1007, 1998

18. Ellepola ANB, Panagoda GJ, Samaranayake LP: Adhesion of oral *Candida* species to human buccal epithelial cells following brief exposure to nystatin. *Oral Microbiol Immunol* 14:in press, 1999

19. Fratti RA, Belanger PH, Ghannoum MA, et al: Endothelial cell injury caused by *Candida albicans* is dependent on iron. *Infect Immun* 66:191–6, 1998

20. Fratti RA, Belanger PH, Sanati H, et al: The effect of the new triazole, voriconazole (UK-109,496), on the interactions of *Candida albicans* and *Candida krusei* with endothelial cells. *J Chemother* 10:7–16, 1998

21. Fratti RA, Ghannoum MA, Edwards JE Jr, et al: Gamma interferon protects endothelial cells from damage by *Candida albicans* by inhibiting endothelial cell phagocytosis. *Infect Immun* 64:4714–8, 1996

22. Hurley R, de Louvois J, Mulhall A: Pathogenic attributes of *Candida*. In: Rose AH, Harrison JS, Eds. *The Yeasts*. Academic Press, London, 1986:207–81

23. Kennedy MJ: Adhesion and association mechanisms of *Candida albicans*. *Cur Topics Med Mycol* 2:73–169, 1988

24. Kimura LH, Pearsall NN: Adherence of *Candida albicans* to human buccal epithelial cells. *Infect Immun* 21:64–8, 1978

25. King RD, Lee JC, Morris AL: Adherence of *Candida albicans* and other *Candida* species to mucosal epithelial cells. *Infect Immun* 27:667–74, 1980

26. Klotz SA, Drutz DJ, Harrison JL, et al: Adherence and penetration of vascular endothelium by *Candida* yeasts. *Infect Immun* 42:374–84, 1983

27. Lee KH, Yoon MS, Chun WH: The effects of monoclonal antibodies against iC3b receptors in mice with experimentally induced disseminated candidiasis. *Immunology* 92:104–10, 1997

28. Liljemark WF, Gibbons RJ: Suppression of *Candida albicans* by human oral streptococci in gnotobiotic mice. *Infect Immun* 8: 846–9, 1973

29. Lopez-Ribot JL, McVay CS, LaJean Chaffin W: Murine tissues exposed to cytotoxic drugs display altered patterns of *Candida albicans* adhesion. *Infect Immun* 62:4226–32, 1994

30. Lopez-Ribot JL, Vespa MV, LaJean Chaffin W: Adherence of *Candida albicans* germ tubes to murine tissues in an *ex vivo* assay. *Can J Microbiol* 40:77–81, 1994

31. McCourtie J, MacFarlane TW, Samaranayake LP: Effect of chlorhexidine gluconate on the adherence of *Candida* species to acrylic surfaces. *J Med Microbiol* 20:97–104, 1985

32. McCourtie J, MacFarlane TW, Samaranayake LP: Effect of saliva and serum on the adherence of *Candida* species to chlorhexidine treated denture acrylic. *J Med Microbiol* 21:209–13, 1986

33. McCourtie J, MacFarlane TW, Samaranayake LP: A comparison of the effects of chlorhexidine gluconate, amphotericin B and nystatin on the adherence of *Candida* species to acrylic. *J Antimicrobial Chmother* 17:575–82, 1986

34. Nair RG, Samaranayake LP: The effect of oral commensal bacteria on candidal adhesion to human buccal epithelial cells *in vitro*. *J Med Microbiol* 45:179–85, 1996

35. Nair RG, Samaranayake LP: The effect of oral commensal bacteria on candidal adhesion to denture acrylic surfaces. An *in vitro* study. *APMIS*: 104:339–49, 1996

36. Nikawa H, Nishimura H, Yamamoto T, et al: The role of saliva and serum in *Candida albicans* biofilm formation on denture acrylic surfaces. *Microbial Ecol Health Dis* 9:35–48, 1996.

37. Odds FC: *Candida and candidosis: a review and bibliography*. Bailliere Tindall, London, 1994

38. Odds FC: Pathogenesis of *Candida* infections. *J Am Acad Dermatol* 31:S2–S5, 1994

39. Ollert MW, Sohnchen R, Korting HC, et al: Mechanisms of adherence of *Candida albicans* to cultured human epidermal keratinocytes. *Infection and Immunity*: 61:4560–8, 1993

40. Panagoda GJ, Ellepola ANB, Samaranayake LP: Adhesion to denture acrylic surfaces and relative cell surface hydrophobicity of *Candida parapsilosis* and *Candida albicans*. *APMIS*: 106:736–42, 1998

41. Panagoda GJ, Samaranayake LP: A new, semi-automated technique for quantification of *Candida* adherence to denture acrylic surfaces. *Mycoses* 42:265–7, 1999

42. Pereiro M Jr, Losada A, Toribio J: Adherence of *Candida albicans* strains isolated from AIDS patients. Comparison with pathogenic yeasts isolated from patients without HIV infection. *Brit J Dermatol* 137:76–80, 1997

43. Polacheck I, Antman A, Barth I, et al: Adherence of *Candida albicans* to epithelial cells: studies using fluorescently labelled yeasts and flow cytometry. *Microbiology* 141:1523–33, 1995

44. Radford DR, Challacombe SJ, Walter JD: Adherence of phenotypically switched *Candida albicans* to denture base materials. *Int J Prosthodont* 11:75-81, 1998

45. Radford DR, Sweet SP, Challacombe SJ, et al: Adherence of *Candida albicans* to denture base materials with different surface finishes. *J Dent* 26:577–83, 1998

46. Ray TL, Digre KB, Payne CD: Adherence of *Candida* species to human epidermal corneocytes and buccal mucosal cells: correlation with cutaneous pathogenicity. *J Invest Dermatol* 83:37–41, 1984

47. Riesselman MH, Kanbe T, Cutler JE: Improvements and important considerations of an *ex vivo* assay to study *Candida albicans* - splenic tissue interactions. *J Immunol Meth* 145:153–60, 1991

48. Rotrosen D, Calderone RA, Edwards JE Jr: Adherence of *Candida* species to host tissues and plastic surfaces. *Rev Infect Dis* 8:73–85, 1986

49. Rotrosen D, Edwards JE Jr, Gibson TR, et al: Adherence of *Candida* to cultured vascular endothelial cells: mechanisms of attachment in endothelial cell penetration. *J Infect Dis* 152:1264–74, 1985

50. Rotrosen D, Gibson TR, Edwards JE Jr: Adhesion of *Candida* species to intravenous catheters. *J Infect Dis* 147:594, 1983

51. Ruechel R: Virulence factors of *Candida* species. In: Samaranayake LP, MacFarlane TW, Eds. *Oral candidosis*. Wright, London, 1990:47–65

52. Samaranayake LP: *Essential Microbiology for Dentistry*. Churchill Livingstone, Edinburgh, 1996

53. Samaranayake LP, Hamilton D, MacFarlane TW: The effect of indigenous bacterial populations on buccal epithelial cells on subsequent microbial adhesion *in vitro*. *Oral Microbiol Immunol* 9:236–40, 1994

54. Samaranayake LP, McCourtie J, MacFarlane TW: Factors affecting the adhesion of *Candida albicans* to acrylic surfaces. *Arch Oral Biol* 25:611–5, 1980

55. Samaranayake LP, MacFarlane TW: An *in vitro* study of the adherence of *Candida albicans* to acrylic surfaces. *Arch Oral Biol* 25:603–9, 1980

56. Samaranayake LP, MacFarlane TW: The adhesion of the yeast *Candida albicans* to epithelial cells of human origin *in vitro*. *Arch Oral Biol* 26:815–20, 1981

57. Samaranayake LP, MacFarlane TW: Factors affecting the *in vitro* adhesion of the oral fungal pathogen *Candida albicans* to epithelial cells of human origin. *Arch Oral Biol* 27:869–73, 1982

58. Samaranayake LP, MacFarlane TW: The effect of dietary carbohydrates on the *in vitro* adhesion of *Candida albicans* to human epithelial cells. *J Med Microbiol* 15:511–7, 1982

59. Samaranayake LP, MacFarlane TW: *Oral candidosis*. Wright, Butterworth, London, 1990

60. Samaranayake LP, McLaughlin L, MacFarlane TW: Adherence of *Candida* species to fibrin clots *in vitro*. *Mycopathologia* 102:135–8, 1988

61. Samaranayake YH, Wu PC, Samaranayake LP, et al: Adhesion and colonisation of *Candida krusei* on host surfaces. *J Med Microbiol* 41:250–8, 1994

62. Samaranayake YH, Wu PC, Samaranayake LP, et al: Relationship between the cell surface hydrophobicity and adhesion of *Candida krusei* and *Candida albicans* to epithelial and denture acrylic surfaces. *APMIS* 103:707–13, 1995

63. Scully C, El-Kabir M, Samaranayake LP: *Candida* and oral candidosis: a review. *Critical Reviews in Oral Biol Med* 5:125–7, 1994

64. Siro RM, Romar H, Lovgren T: Continuous flow method for extraction and bioluminescence assay of ATP in baker's yeast. *Eur J Appl Microbiol Biotechnol* 15:258–64, 1982

65. Sobel JD, Obedeanu N: Effects of subinhibitory concentrations of ketoconazole on *in vitro* adherence of *Candida albicans* to vaginal epithelial cells. *Eur J Clin Microbiol* 2:445–2, 1983

66. Tobgi RS, Samaranayake LP, MacFarlane TW: The adhesion of *Candida albicans* to buccal epithelial cells exposed to chlorhexidine gluconate. *J Med Vet Mycol* 25:335–8, 1987

67. Toy PT, Lai L, Drake TA, et al: Effect of fibronectin on the adherence of *Staphylococcus aureus* to fibrin thrombi *in vitro*. *Infect Immun* 48:83–6, 1985

68. Vargas K, Wertz PW, Drake D, et al: Differences in adhesion of *Candida albicans* 315A cells exhibiting switch phenotypes to buccal epithelium and stratum corneum. *Infect Immun* 62:1328–35, 1994

69. Verran J, Maryan CJ: Retention of *Candida albicans* on acrylic resin and silicone of different surface topography. *J Prosth Dent* 77:535–9, 1997

70. Vespa MN, Lopez-Ribot JL, LaJean Chaffin W: Adherence of germ tubes of *Candida albicans* to tissues from immunocompromised mice. *FEMS Immunol Med Microbiol* 11:57–64, 1995

71. Waltimo T, Tanner J, Vallittu P, et al: Adherence of *Candida albicans* to the surface of polymethylmethacrylate – E glass fiber composite used in dentures. *Int J Prosthodont* 12:83–6, 1999

72. Waters MGJ, Williams DW, Jagger RG, et al: Adherence of *Candida albicans* to experimental denture soft lining materials. *J Prosth Dent* 77:306–12, 1997

73. Williams DW, Waters MG, Potts AJ, et al: A novel technique for assessment of adherence of *Candida albicans* to solid surfaces. *J Clin Pathol* 51:390–1, 1998

74. Zierdt CH: Adherence of bacteria, yeast, blood cells, and latex spheres to large-porosity membrane filters. *Appl Environ Microbiol* 38:1166–72, 1979

# 34

# Studying Bacterial Adhesion to Cultured Cells

## M. John Albert,[1] Travis Grant,[2] and Roy Robins-Browne[2]

[1]*International Centre for Diarrhoeal Disease Research, Bangladesh (ICDDR, B), Dhaka, Bangladesh.* [2]*Department of Microbiology and Infectious Disease, Royal Children's Hospital, Parkville, Victoria, Australia*

## *Abbreviations*

| | |
|---|---|
| A/E | attaching and effacing |
| BFP | bundle forming pilus |
| CFs | colonization factors |
| DAEC | diffuse adherent *E. coli* |
| EAEC | enteroaggregative *E. coli* |
| EAF | enteropathogenic *E. coli* adherence factor |
| EHEC | enterohemorrhagic *E. coli* |
| EIEC | enteroinvasive *E. coli* |
| ELISA | enzyme-linked immunosorbent assay |
| EPEC | enteropathogenic *E. coli* |
| Esp | *E. coli* secreted protein |
| ETEC | enterotoxigenic *E. coli* |
| F-actin | filamentous actin |
| FAS | fluorescent actin staining |
| FITC | fluorescein isothiocyanate |
| IL-8 | interleukin 8 |
| IP | ionositol phosphate |
| LT | heat-labile enterotoxin |
| PKC | protein kinase C |
| PLC | phospholipase C |
| PMN | polymorphonuclear leukocyte |
| ST | heat-stable enterotoxin |
| Stx | Shiga toxin |

## I. INTRODUCTION

Adherence of bacteria to the body surface is the first step in the pathogenesis of most bacterial infections. Adherence reflects a specific interaction between a ligand expressed on the bacterial surface and a receptor on the epithelial cell surface. This adherence process can be studied in vitro using cultured mammalian cells to provide a simple model

*Handbook of Bacterial Adhesion: Principles, Methods, and Applications*
Edited by: Y. H. An and R. J. Friedman © Humana Press Inc., Totowa, NJ

for investigating host–bacterium interactions. These interactions include the mechanisms of adhesion of bacteria to tissue culture cells, identification of bacteria by investigation of their adherence phenotype, and the events that occur after the bacteria associate with the host cell, such as the capacity to invade these cells.

Cultured cells, which represent a single cell type, can be grown in defined media under reproducible conditions. However, there are certain limitations which may affect interpretation of experimental data. Cultured mammalian cells are derived either from a tumor, in which many genetic changes have already occurred, or by a process of immortalization that produces numerous mutations. During the process of immortalization, cell lines lose many traits of the original tissue from which they were derived. One feature that can be lost in this process is tissue-specific surface molecules that normally function as receptors for bacterial adhesins. This may explain the fact that many bacterial pathogens that are highly specific for a particular tissue of the host are frequently able to adhere to cultured cells derived from tissue that they do not normally infect. Another problem is that most cultured cells exhibit changes to their normal morphology and distribution of surface antigens. Cells in intact animals are polarized, whereby different parts of the cell are exposed to different environments. For example, the apical surface of normal mucosal cells is exposed to the external environment, whereas the basal and lateral surfaces are in contact with extracellular matrix. In addition, cells of the mucosa are joined by impermeable tight junctions. By contrast, most tissue culture cells do not have differentiated surfaces, but this has been achieved for some cell lines. Another limitation of cultured cells as representatives of human mucosal surfaces is that mucosal surfaces in vivo are coated by mucus and bathed in solutions that are difficult to mimic in an in vitro system. Finally, real tissues consist of multiple cell types, not of a single cell type as seen in most tissue culture models.

In spite of the numerous limitations of existing cell lines, they have been extremely useful when investigating bacterium–host cell interaction, and, if their limitations are kept in mind, cultured cells will continue to be invaluable models. Once a new phenomenon has been found with cultured cell lines, experiments can be designed with animal models to test the importance of the phenomenon in vivo.

## II. TISSUE CULTURE CELL LINES, COMMON BACTERIA, AND COMMON APPROACHES

### A. Commonly Used Cell Lines

Some of the commonly used nonpolarized cell-lines for studying the bacteria–host interaction are: HeLa cells (derived from human cervical epithelial carcinoma), HEp-2 cells (derived from human laryngeal epidermoid carcinoma), and INT-407 cells (derived from human embryonic intestinal cells). The commonly used polarized tissue culture cell-lines are Caco-2 cells and HT-29 cells (both derived from human colonic adeno-carcinoma). These cells undergo morphologic and functional differentiation into mature enterocytes with a well developed microvilli and brush border enzymes. Another polarized cell line is T84 cells (derived from human colonic carcinoma). These cells grow to confluence as monolayers, exhibit compartmentalization of organelles and form tight junctions and desmosomes, and an apical surface with microvilli. Polarized T84 cells are morphologically similar to colonic crypt cells.

## B. Commonly Studied Bacteria

A variety of bacteria including those causing infections of the gastrointestinal tract (e.g., diarrheagenic *Escherichia coli*), genitourinary tract (e.g., *Neisseria gonorrhoeae* and uropathogenic *E. coli*), respiratory tract (e.g., *Streptococcus* sp. and *Hemophilus influenzae*), and systemic infections (e.g., *Salmonella* sp.) have been shown to adhere to tissue culture cells. Some of the consequences of bacterial adherence to tissue culture cells include changes in cell morphology without structural damage (e.g., morphological changes in Y1 mouse adrenal tumor cells and Chinese hamster ovary cells due to the cytotonic cholera toxin of *Vibrio cholerae*),[7,13] damage to the cellular structure (e.g., enteropathogenic *E. coli*),[17] internalization of bacteria (e.g., *Yersinia enterocolitica*),[20] replication of internalized bacteria and their spread to adjacent cells (for example *Shigella* sp.),[28] cytotoxicity (e.g., Shiga toxins of enterohemorrhagic *E. coli*),[25] stimulation of cytokine production (e.g., enteroaggregative *E. coli*),[30] and necrosis (e.g., cytotoxic necrotizing factor producing *E. coli*).[6]

## C. Bacterial Adherence to Formalin-Fixed Cells

Conventional bacterial adherence assays are performed using viable cells. The cells are harvested from stock culture and the monolayer is prepared at the time of use; however, for some assays, the monolayer can be prepared in advance. This is achieved by fixing the cells in formalin, which enables the cells to be stored for up to 3 wk. There are several advantages of this method. Cover slips with fixed cells can be sent to laboratories where tissue culture facilities are not available. Moreover, the fixed cells are resistant to the cytotoxic action of some bacteria, which enables the adherence characteristics of these bacteria to be easily studied.[29]

## D. Bacterial Adherence to Cell Culture Suspension

The use of cells in suspension is particularly useful for bacteria which cause damage to or sloughing of the cell monolayer. Tissue culture cells in suspension can be mixed with bacteria and incubated in a rotator. Nonadherent bacteria are then removed by centrifugation of the tissue culture cells and repeated washing. Adherent bacteria can be counted by colony count, or by light microscopy with or without staining.[3]

## E. Bacterial Adherence to Differentiated Human Cells

The adherence of bacteria can be studied using differentiated human intestinal cells such as Caco-2, T84 or HT-29 cells as described for nondifferentiated cells.[32-34] However, unlike nondifferentiated tissue culture cells, differentiated cells possess a number of characteristics which enable more complex bacterial–host cell interactions to be investigated. These include the study of bacterial transcytosis, which can be investigated by seeding epithelial cells on 3 μm pore-size polycarbonate filter membranes. The filter units are then incubated in wells of tissue culture plates containing culture medium until a confluent monolayer is formed. The epithelial cell surface adherent to the filter forms the basolateral surface, while the nonadherent surface forms the apical surface. Separating these surfaces are tight junctions and polarity which give rise to transepithelial electrical resistance. Bacteria are added to the apical surface and transcytosis is measured by culturing the bacteria from the basolateral medium at varying times. Transepithelial electrical resistance can also be determined. Ultrastructural changes in the monolayer can be studied by transmission and scanning electron microscopy.[11]

### III. METHODS FOR STUDYING BACTERIAL ADHERENCE TO TISSUE CULTURE CELLS

Important considerations before performing bacterial adhesion assays are the bacterial growth conditions, as they will influence the bacteria–cell interaction. These conditions include the bacterial growth phase, temperature of incubation, pH and osmolarity of culture media, and the presence or absence of $O_2$ and $CO_2$. In general, growth conditions approximating the in vivo milieu have been found to produce the best results.[16] Tissue culture cells are routinely grown in the wells of tissue culture plates (on glass coverslips for some methods) until they are at least semi-confluent. An appropriate concentration of bacteria (approximate multiplicity of infection: 10 bacteria per cell) is added to the cell monolayer to interact with the tissue culture cells for an appropriate time, most commonly being 3 h. Nonadherent bacteria are removed by washing the monolayer with phosphate-buffered saline (PBS).[5] The capacity of bacteria to adhere to cells can then be studied by a variety of methods.

### A. Colony Count

The cell monolayer can be lysed with an appropriate detergent, such as 1% (v/v) digitonin or Triton X-100, which does not affect bacterial viability. The volume in each well is then increased to 1 mL with culture media, followed by vigorously pipetting the lysate to ensure cell disruption. Appropriate dilutions are then plated on agar plates for enumeration of cell-associated bacteria by colony count.[27] The number of cell-associated bacteria is most commonly expressed as a percentage of the original bacterial inoculum.

### B. Radioactivity Count

Bacteria can be labeled with a radioactive isotope, such as tritiated thymidine, before investigation by the adhesion assay. The number of cell-associated bacteria can then be measured by lysing the cell monolayer, and measuring the radioactivity of the lysate.[8]

### C. Enzyme-Linked Immunosorbent Assay

The number of cell-associated bacteria can be quantitated by ELISA, by either using specific antibodies to the adherent bacteria followed by enzyme-conjugated secondary antibody, or by using biotinylated bacteria and avidin-enzyme as the detecting agent.[23]

When interpreting the results of quantitative methods, a number of limitations must be considered. For example, some bacteria, such as *Y. enterocolitica* and enteroinvasive *E. coli*, may invade tissue culture cells in large numbers. Thus, when considering the number of adhesive bacteria, the number of bacteria that have been internalized by the cell may need to be determined independently. This is usually quantified by incubating infected tissue culture cells in tissue culture medium containing bactericidal antibiotics which do not penetrate the tissue culture cell membrane (e.g., gentamicin), and thus, kill extra-cellular bacteria while leaving intracellular bacteria intact. After removal of the antibiotic from the cells by washing with PBS, the cells are lysed, and the number of intracellular bacteria are determined as described for the adhesion assay. In addition, quantification of adherence by colony count measures the number of colony-forming units (cfu) and not individual bacteria, which can dramatically underestimate the number of adherent bacteria if they exhibit hydrophobic or aggregative adherent properties.

### D. Light and Electron Microscopy

Tissue culture cells that have been grown on glass cover slips can be infected with bacteria as described previously, and then processed for analysis by light microscopy. This is achieved by fixing the cells with methanol, followed by staining with Giemsa, and then examining the cells by light microscopy. This method can be used to determine the pattern of adherence of bacteria, the proportion of cells with adherent bacteria, and an estimation of the number of bacteria adherent to each cell. Changes in cell morphology as a result of bacterial adherence can also be examined.[22,36]

The ultrastructural characteristics of adherence can be studied by visualizing infected cells using transmission or scanning electron microscopy.[10]

## IV. TISSUE CULTURE ADHERENCE ASSAYS FOR STUDYING DIARRHEAGENIC *E. COLI*

Diarrheagenic *E. coli* provides a good example of how bacterial adherence assays using tissue culture cells have advanced our understanding of infections. Six categories of diarrheagenic *E. coli* exist, these being enterotoxigenic *E. coli* (ETEC), enteroinvasive *E. coli* (EIEC), enterohemorrhagic *E. coli* (EHEC), enteropathogenic *E. coli* (EPEC), enteroaggregative *E. coli* (EAEC), and diffuse adherent *E. coli* (DAEC).[21]

ETEC strains adhere to differentiated colonic epithelial cell lines such as Caco-2 and HT-29 by means of colonization factors (CFs). Diarrhea due to ETEC is mediated by heat-labile enteroxin (LT) and heat-stable enterotoxin (ST). The mechanisms of diarrhea have been elucidated using a variety of systems. LT exerts its secretory effect via the adenyl cyclase-cyclic AMP pathway, whereas ST acts via the guanylate cyclase-cyclic GMP pathway. ETEC can be internalized by tissue culture cells; however, since ETEC diarrhea is a prototype of secretory diarrhea, the significance of this observation is not known.[9]

EIEC and *Shigella* sp. are genetically similar, and share identical pathogenic mechanisms. EIEC are engulfed upon contact with tissue culture cells. This is achieved by signal transduction events, whereby bacteria send signals to the cell to induce dramatic membrane ruffling and cytoskeletal rearrangements that result in macropinocytosis and virtually passive entry of bacteria. The current model of *Shigella* and EIEC pathogenesis comprises epithelial cell penetration, lysis of the endocytic vacuole, intracellular multiplication, directional movement through the cytoplasm, and extension into adjacent cells.[24]

EPEC strains produce a characteristic histopathology, known as attaching and effacing (A/E) lesions, in the intestines of infected humans and animals. This phenotype is characterized by effacement of microvilli and intimate adherence between the bacterium and the epithelial cell membrane.[21] EPEC adhere to cultured mammalian cell-lines, such as HeLa and HEp-2 cells, in a characteristic pattern known as localized adherence (Fig. 1). Marked cytoskeletal changes including accumulation of polymerized actin are seen directly beneath the adherent bacteria.[12] This observation led to the development of the fluorescent-actin staining (FAS) test. In this test, fluorescein isothiocyanate (FITC)-

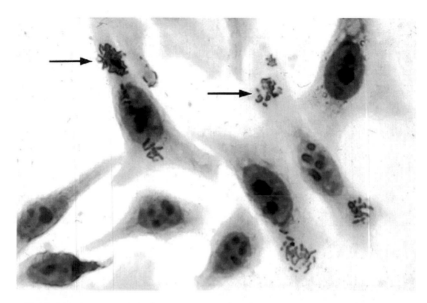

**Figure 1.** Light micrograph of adherence to HEp-2 epithelial cells by enteropathogenic *E. coli* (EPEC), showing characteristic localized adherence (arrows). Mag ×400

labeled phalloidin binds specifically to filamentous actin in cultured epithelial cells directly beneath the adherent bacteria. The fluorescence can be visualized by fluorescence microscopy (Fig. 2). Other cytoskeletal components, including alpha actinin, talin, erzin, and villin, also accumulate directly under the bacteria, and are involved in crosslinking of the actin microfilaments.

The interactions between EPEC and host cells have been divided into three stages. The initial adherence to cultured epithelial cells is mediated by bundle forming pili (BFP), a type IV fimbriae encoded by the EPEC adherence factor (EAF) plasmid. BFP is also responsible for binding of bacteria into aggregates seen as localized adherence to tissue culture cells. While BFP is not essential for forming the characteristic A/E lesions, it is believed that BFP mediates close contact between the bacteria and the host cell. The proteins which mediate the formation of A/E lesions are encoded by a 35-kilobase pathogenicity island, termed the locus of enterocyte effacement (LEE). The genes required for this process include *esps* (*E. coli*-secreted protein), *escs* (*E. coli* secretion), *sep* (secretion of *E. coli* proteins), *eae* (*E. coli* attaching and effacing, encoding intimin), and *tir* (translocated intimin receptor).

The second stage of EPEC pathogenesis involves the secretion of bacterial proteins including EspA, EspB and EspD into the host cell. The translocation of these proteins is essential for activating a number of signal tranduction pathways. EspB is the critical protein involved in the host cell's signaling pathways. All of these proteins are secreted by a type-III secretion system encoded by the *esc* and *sep* genes of LEE.

The third stage of EPEC pathogenesis is characterized by the intimate attachment of the bacteria to the host cell. Intimin, a 94-kDa outer membrane protein encoded by the *eae* gene binds to its receptor on the host cell membrane. This receptor, *tir*, is of bacterial

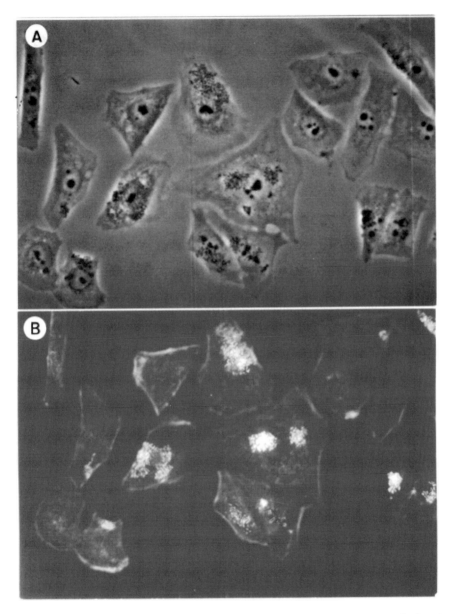

**Figure 2.** Phase contrast (A), and fluorescent micrographs (B) of HEp-2 cells incubated with EPEC, showing fluorescent-actin staining (FAS) corresponding to areas of bacterial adherence. Mag ×400.

origin and is translocated from the bacterial cell into the host membrane, where it becomes phosphorylated. Purified intimin also binds β1 integrins, which suggests that intimin may be binding more than one receptor on the epithelial cell. The resultant association is accompanied by the formation of actin pedestals on which the pathogen resides (Fig. 3). Some of these bacteria become internalized. In response to intimate attachment, signal transduction events are activated, including activation of phospho-

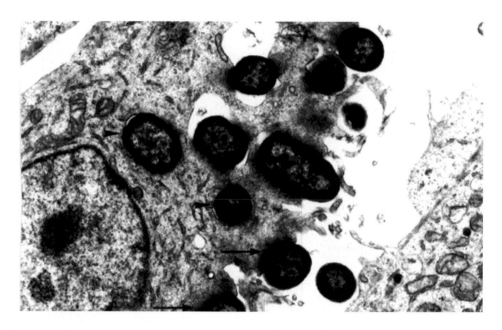

**Figure 3.** Transmission electron micrograph of HEp-2 cells incubated with EPEC. Some bacteria adhere to the host cell membrane before inducing attachment effacement lesions and pedestal formation (arrows), while other bacteria are internalized by the cell (arrowheads). Mag ×8000.

lipase C (PLC) and protein kinase C (PKC), inositol triphosphate (IP3) fluxes, $\alpha v^d$ $Ca^{2+}$ release from internal stores. The signal transduction response to EPEC also results in migration of polymorphonuclear leukocytes (PMN). It is suggested that the binding of EPEC to epithelial cells activates a eukaryotic trancription factor which in turn upregulates the expression of the pro-inflammatory cytokine, IL-8, which attracts PMN. Multiple mechanisms could be involved in diarrhea due to EPEC. These include active chloride secretion involving PKC, malabsorption due to loss of microvilli and the local inflammatory response, and increased intestinal permeability.

Enterohemorrhagic *E. coli* are the causative agents of hemorrhagic colitis and hemolytic uremic syndrome. EHEC adhere to cultured epithelial cells and produce an A/E lesion similar to EPEC. However, they do not produce BFP. All strains of the major serotype O157:H7 contain a plasmid, designated pO157, but the role of this plasmid in the pathogenesis of disease, including its role in adherence to epithelial cells, is controversial. A major virulence factor of EHEC is the Shiga toxin (Stx), of which there are two major types: Stx1 and Stx2. Stx production can be determined by using tissue culture cells, as they produce characteristic cytopathic effect on HeLa and Vero cells.[21,25]

Enteroaggregative *E. coli* have been associated with persistent diarrhea in children in developing countries. EAEC adhere to tissue culture cells in a characteristic pattern of layered bacteria known as "stacked-brick" configuration (Fig. 4). EAEC are also internalized by tissue culture cells. As with invasion of epithelial cells by many other bacterial pathogens, the invasion of host cells by EAEC can be inhibited by chemicals that interfere

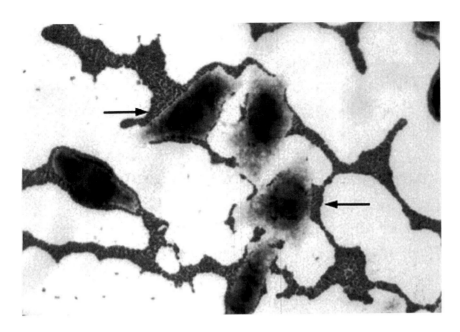

**Figure 4.** Light micrograph of adherence to HEp-2 epithelial cells by enteroaggregative *E. coli* (EAEC), showing characteristic 'stacked-brick' phenotype (arrows). Mag ×400.

with signal transduction and actin polymerization.[2] Another consequence of adherence of EAEC to tissue culture cells is the production of IL-8. This characteristic is clinically relevant, as children infected with EAEC have evidence of inflammation in the intestine.[30]

Diffuse adherent *E. coli* adhere to cells in a distinctive pattern, whereby bacteria are evenly distributed over the surface of the cells without forming microcolonies or clumps (Fig. 5). DAEC strains induce the production of finger-like projections extending from the surface of tissue culture cells, which "embed" bacteria without complete internalization.[4] Some strains of adherent bacteria induce cytoskeletal and cytochemical changes, but the majority of DAEC strains appear to have no effect on mammalian cells. Some strains of DAEC may also produce a lesion similar to the A/E lesion of EPEC in cultured cells. This is accompanied by accumulation of actin and phosphorylation of proteins at the site of bacterial attachment. Furthermore, these strains secrete proteins homologous to EspA, EspB, and EspD of EPEC that are necessary for signal transduction leading to A/E lesions.[1] By contrast, some other strains of DAEC induced F-actin disassembly by piracy of decay-accelerating factor signal transduction.[26] Another category of DAEC adheres to mammalian cells in a clustered pattern and then invades the cells as a cluster causing capsule-like actin accumulation around each cluster.[35]

Although we have emphasized how in vitro adherence studies with tissue culture cells have helped in the identification and characterization of diarrheagenic *E. coli*, the same principles of investigation are applicable to other pathogenic bacteria. In fact, cell culture adherence studies have been used to investigate the pathogenesis of infection with a number of other pathogens, including *Bordetella pertussis*,[18] *Neisseria gonorrhoeae*,[19] uropathogenic *E. coli*,[15] and *Yersinia* sp.[14]

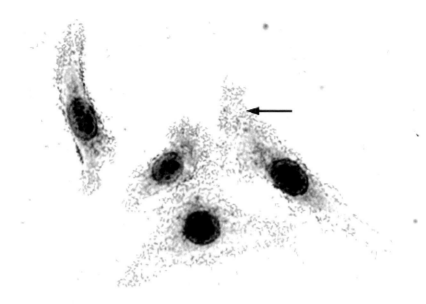

**Figure 5.** Light micrograph of adherence to HEp-2 epithelial cells by diffuse adherent *E. coli* (DAEC), showing characteristic even distribution of bacteria on epithelial cells (arrow). Mag ×400.

## V. CONCLUDING REMARKS

Mucosal colonization of the host by adhesion to host cells is the first step in the pathogenesis of many infectious diseases. In vitro adherence studies of pathogenic bacteria to tissue culture cells have advanced our understanding of bacteria–host cell interactions. In some cases the bacterial ligands mediating adhesion to the receptors in host cell membrane have been identified, which has contributed to progress in the preparation of anti-adhesive vaccines.[31] However, the adhesion molecules and corresponding receptors for many pathogenic bacteria are yet to be discovered. In vitro adherence studies of bacteria to tissue culture cell should continue to be an area of intensive research because they can provide practical information concerning targets for preventing or containing disease. In addition, much can be learned about the biology of the cell itself, including transmembrane signaling, cytoskeletal reorganization and programmed cell death.

## REFERENCES

1. Beinke C, Laarmann S, Wachter C, et al: Diffusively adhering *Escherichia coli* strains induce attaching and effacing phenotypes and secrete homologs of Esp proteins. *Infect Immun* 66:528–39, 1998
2. Benjamin P, Federman M, Wanke CA: Characterization of an invasive phenotype associated with enteroaggregative *Escherichia coli*. *Infect Immun* 63:3417–21, 1995
3. Carruthers MM: In vitro adherence of Kanagawa-positive *Vibrio parahaemolyticus* to epithelial cells. *J Infect Dis* 136:588–92, 1977
4. Cookson ST, Nataro JP: Characterization of HEp-2 cell projection formation induced by diffusively adherent *Escherichia coli*. *Microb Pathog* 21:421–34, 1996

5. Cravioto A, Gross RJ, Scotland SM, et al: An adhesive factor found in strains of *Escherichia coli* belonging to the traditional infantile enteropathogenic serotypes. *Curr Microbiol* 3:95–9, 1979

6. De Rycke J, Gonzalez EA, Blanco J, et al: Evidence for two types of cytotoxic necrotizing factor in human and animal clinical isolates of *Escherichia coli*. *J Clin Microbiol* 28:694–9, 1990

7. Donta ST, Moon HW, Whipp SC: Detection of heat-labile *Escherichia coli* enterotoxin with the use of adrenal cells in tissue culture. *Science* 183:334–6, 1974

8. Dunkle LM, Blair LL, Fortune KP: Transformation of a plasmid encoding an adhesin of *Staphylococcus aureus* into a nonadherent staphylococcal strain. *J Infect Dis* 153:670–5, 1986

9. Elsinghorst EA, Kopecko DJ: Molecular cloning of epithelial cell invasion determinants from enterotoxigenic *Escherichia coli*. *Infect Immun* 60:2409–17, 1992

10. Favre-Bonte S, Darfeuille-Michaud A, Forestier C: Aggregative adherence of *Klebsiella pneumoniae* to human intestine-407 cells. *Infect Immun* 63:1318–28, 1995

11. Finlay BB, Falkow S: Salmonella interactions with polarized human intestinal Caco-2 epithelial cells. *J Infect Dis* 162:1096–106, 1990

12. Goosney DL, Knoechel DG, Finlay BB: Enteropathogenic *E. coli*, *Salmonella*, and *Shigella*: Masters of host cell cytoskeletal exploitation. *Emerg Infect Dis* 5:216–23, 1999

13. Guerrant RL, Brunton LL, Schnaitman TC, et al: Cyclic adenosine monophosphate and alteration of Chinese hamster ovary cell morphology: a rapid, sensitive in vitro assay for the enterotoxins of *Vibrio cholerae* and *Escherichia coli*. *Infect Immun* 10:320–7, 1974

14. Iriarte M, Vanooteghem J-C, Delor I, et al: The Myf fibrillae of *Yersinia enterocolitica*. *Mol Microbiol* 9:507–520, 1993

15. Johnson JR: Virulence factors in *Escherichia coli* urinary tract infection. *Clin Microbiol Rev* 4:80–128, 1991

16. Kenny B, Abe A, Stein M, et al: Enteropathogenic *Escherichia coli* protein secretion is induced in response to conditions similar to those in the gastrointestinal tract. *Infect Immun* 65:2606–12, 1997

17. Knutton S, Baldwin T, Williams PH, et al: Actin accumulation at sites of bacterial adhesion to tissue culture cells: basis of a new diagnostic test for enteropathogenic and enterohemorrhagic *Escherichia coli*. *Infect Immun* 57:1290–8, 1989

18. Leininger E, Roberts M, Kenimer JG, et al: Pertactin, an Arg-Gly-Asp containing *Bordetella pertussis* surface protein that promotes adherence to mammalian cells. *Proc Natl Acad Sci USA* 88:345–9, 1991

19. Meyer TF: Pathogenic *Neisseriae*: Complexity of pathogen-host cell interplay. *Clin Infect Dis* 28:433–41, 1999

20. Miller VL, Falkow S: Evidence for two genetic loci in *Yersinia enterocolitica* that can promote invasion of epithelial cells. *Infect Immun* 56:1242–8, 1988

21. Nataro JP, Kaper JB: Diarrheagenic *Escherichia coli*. *Clin Microbiol Rev* 11:142–201, 1998

22. Nataro P, Kaper JB, Robins-Browne R, et al: Patterns of adherence of diarrheagenic *Escherichia coli* to HEp-2 cells. *Pediatr Infect Dis J* 6:829–31, 1987

23. Ofek I, Courtney HS, Schifferli DM, et al: Enzyme-linked immunosorbent assay for adherence of bacteria to animal cells. *J Clin Microbiol* 24:512–6, 1986

24. Parsot C, Sansonetti PJ: Invasion and the pathogenesis of *Shigella* infections. *Curr Top Microbiol* 209:25–42, 1996

25. Paton JC, Paton AW: Pathogenesis and diagnosis of Shiga toxin-producing *Escherichia coli* infections. *Clin Microbiol Rev* 11:450–79, 1998

26. Peiffer I, Servin AL, Bernet-Camard M-F: Piracy of decay-accelerating factor (CD55) signal transduction by the diffusively adhering strain *Escherichia coli* C1845 promotes cytoskeletal F-actin rearrangements in cultured human intestinal INT407 cells. *Infect Immun* 66:4036–42, 1998

27. Robins-Browne R, Bennett-Wood V: Quantitative assessment of the ability of *Escherichia coli* to invade cultured animal cells. *Microb Pathog* 12:159–64, 1992

28. Sansonetti PJ. Bacterial pathogens, from adherence to invasion: comparative strategies. *Med Microbiol Immunol* 182:223–32, 1993

29. Spencer J, Chart H, Smith HR, et al: Improved detection of enteroaggregative *Escherichia coli* using formalin-fixed HEp-2 cells. *Lett Appl Microbiol* 25:325–6, 1997

30. Steiner TS, Lima AM, Nataro JP, et al: Enteroaggregative *Escherichia coli* produce intestinal inflammation and growth impairment and cause interleukin-8 release from intestinal epithelial cells. *J Infect Dis* 177:88–96, 1998

31. Tacket CO, Levine MM: Vaccines against enterotoxigenic *Escherichia coli* infections. II. Live oral vaccines and subunit (purified fimbriae and toxin subunit) vaccines. In: Levine MM, Woodrow GC, Kaper JB, et al., eds. *New Generation Vaccines*. 2nd edition. Marcel Decker Inc., New York, NY, 1997:875–83

32. Tartera C, Metcalf ES: Osmolarity and growth phase overlap in regulation of *Salmonella typhi* adherence to and invasion of human intestinal cells. *Infect Immun* 61:3084–9, 1993

33. Viboud GI, McConnell MM, Helander A, et al: Binding of enterotoxigenic *Escherichia coli* expressing different colonization factors to tissue-cultured Caco-2 cells and to isolated human enterocytes. *Microb Pathog* 21:139–47, 1996

34. Winsor Jr. DK, Ashkenazi S, Chiovetti R, et al: Adherence of enterohemorrhagic *Escherichia coli* strains to a human colonic epithelial cell line (T84). *Infect Immun* 60:1613–7, 1992

35. Yamamoto T, Kaneko M, Changchawalit S, et al: Actin accumulation associated with clustered and localized adherence in *Escherichia coli* isolated from patients with diarrhea. *Infect Immun* 62:2917–29, 1994

36. Zafriri D, Ofek I, Adar R, et al: Inhibitory activity of cranberry juice on adherence of type 1 and type P fimbriated *Escherichia coli* to eukaryotic cells. *Antimicrob Agents Chemother* 33:92–8, 1989

# PART VI

## STRATEGIES FOR PREVENTION OF MICROBIAL ADHESION

# 35

# Strategies for Preventing Group A Streptococcal Adhesion and Infection

**Harry S. Courtney,**[1,2] **James B. Dale,**[1,2] **and David L. Hasty**[1,3]

[1]*The Veterans Affairs Medical Center,* [2]*The Departments of Medicine,*
*and* [3]*Anatomy and Neurobiology, University of Tennessee, Memphis, TN, USA*

## I. INTRODUCTION

Group A streptococci are responsible for a number of clinical syndromes including pharyngitis, impetigo, pneumonia, puerperal sepsis, and myositis. Recently, an increase in the incidence of streptococcal toxic shock syndrome and necrotizing fasciitis have been noted which have a high morbidity and mortality rate.[25,140] Two nonsuppurative sequelae of group A streptococcal infections, acute rheumatic fever (ARF) and acute glomerulo-nephritis (AGN), are the most significant health problems stemming from group A streptococcal infections worldwide. Development of ARF is usually preceded by a streptococcal infection of the pharynx, whereas AGN may be caused by infections of either the skin or oral mucosa, but only by certain nephritogenic strains.[20,153] Exactly how an antecedent streptococcal infection leads to ARF or AGN is unknown, but autoimmune humoral and cellular responses to infection are suspected of contributing to these diseases.

Although group A streptococci remain sensitive to penicillin, antibiotic therapy is not always available or effective. In some cases, group A streptococcal infections progress so rapidly that antibiotic therapy may have little or no impact on the outcome. Some patients have developed ARF without realizing a need for medical attention. In an outbreak of ARF in Utah, a large number of patients did not even recall having had a sore throat before the onset of ARF symptoms.[146]

Thus, there is a clear need for alternatives to antibiotic therapy. Other forms of treatment being considered are vaccines that protect the host against colonization and invasion, adhesin or receptor analogs that block adhesion, neutralization of toxins and superantigens by intravenous immunoglobulins, and replacement therapy with commensal organisms. One requirement for developing effective new therapies is a detailed knowledge of the molecular mechanisms utilized by group A streptococci to adhere to and/or invade host cells, how they survive after invasion, and how elaboration of virulence factors leads to tissue damage. In this chapter, we will review some of the currently known mechanisms of adhesion and virulence and how this information is being

*Handbook of Bacterial Adhesion: Principles, Methods, and Applications*
Edited by: Y. H. An and R. J. Friedman © Humana Press Inc., Totowa, NJ

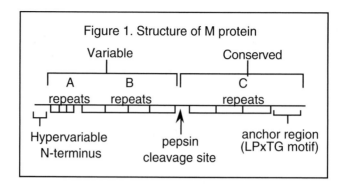

**Figure 1.** Structure of M protein.

used to develop strategies for preventing streptococcal adhesion and subsequent infections.

## II. SURFACE STRUCTURES AND CLASSIFICATION OF GROUP A STREPTOCOCCI

Before discussing the pathogenic mechanisms of *Streptococcus pyogenes*, it will be helpful to describe some of the surface structures used for classification. Group A streptococci are Gram-positive cocci that demonstrate β-hemolysis on blood agar plates. These organisms are classified as group A based on serological reactivity with antibodies to the cell wall carbohydrate (C-carbohydrate). The C-carbohydrate is composed of polymers of L-rhamnose and N-acetyl glucosamine. The latter sugar contains the antigenic epitope.

M proteins are alpha-helical coiled-coil proteins that emanate from the surface of the bacteria.[62,135] The general structure of M proteins is shown schematically in Figure 1. The N-terminal half of the molecule varies considerably between different types of M proteins, while the C-terminal half is highly conserved. The N-terminal 30-50 amino acids of the M protein are hypervariable and serve as the basis for serotyping group A streptococci into more than 90 different M types.

M proteins have been classically defined by functional characteristics, namely the ability to confer resistance to phagocytosis and to evoke opsonic antibodies. However, these definitions have begun to blur as new information has become available. In some serotypes, M-like or M-related proteins are also required for optimal resistance to phagocytosis, but these proteins have not yet been found to evoke opsonic antibodies.[114] The nomenclature for these proteins was derived from the observation that there is a high degree of conservation between the C-terminal halves of these proteins and those of M proteins. The genes for M-related proteins (Mrp), M proteins, and M-like proteins (Mlp) are linked tandemly in the same gene cluster under control of the positive regulator, Mga (Fig. 2).[94,110] The genes for these proteins are *mrp*, *emm*, and *enn* respectively. Mrp may also be referred to as FcRA and Mlp may be referred to as Enn. Mrp and Enn usually bind the Fc domain of IgG and IgA respectively. Mga may also control expression of several other genes.[113] A global regulator was recently identified that negatively regulates expression of the genes for Mga, protein F2, and a collagen-binding protein (Cpa).[112]

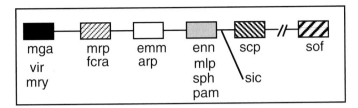

**Figure 2.** Generalized scheme for the Mga regulon. The positive regulator Mga controls the expression of several different genes. These genes are linked tandemly and are flanked on the 5' end by *mga* and on the 3' end by *scp*. Mga also controls expression of SOF.[94] The *sof* gene is located at a different chromosomal locus as indicated by the gap. All group A streptococci tested so far contain *mga, emm,* and *scp*. Other serotypes may contain one or more of the remaining genes. Alternate designations are listed below each gene. The *sic* gene has been found in only M1 and M57 strains and is located between *sph* and *scp*. There may be an additional gene between *sic* and *scp²*. Note that the protective antigen in M type 4, in which *arp* was described, has not been defined and *arp* does not confer resistance to phagocytosis.[75]

Serum opacity factor (SOF), so named for its ability to opacify mammalian sera, is another group A streptococcal protein that is used to categorize and serotype *S. pyogenes*. SOF contains type-specific determinants that covary with the determinants of M protein.[161] Thus, the M serotype can be determined by typing with SOF antisera, assuming that the strain is SOF-positive. SOF typing sera are mainly used in those cases where M protein typing sera are not available or difficult to prepare.

Group A streptococci have also been divided into two classes based on reactivity with monoclonal antibodies that distinguish differences in the conserved C-repeat region of M proteins.[16] Class I organisms are generally SOF-negative, are associated with upper respiratory tract infections, and do not generally express Mrp or Enn proteins. Class II organisms are SOF-positive, express Mrp or Enn or both, and are associated with skin or pharyngeal infections.

Another typing scheme is based on expression of T proteins and R proteins.[162,163] The function of these proteins is not known and they have not yet been shown to be involved in the virulence of group A streptococci and will not be discussed further.

## III. ADHESION

Most bacterial infections are preceded by adhesion of bacteria to the tissues of the skin or mucosal surfaces. This initial adhesion is required for the bacteria to avoid the cleansing mechanisms of the host, such as the muco-ciliary blanket and the rapid shedding of epithelial cells. The organisms must also be able to resist bactericidal and/or bacteriostatic agents, such as lysozyme and hydrogen peroxide in oral secretions, and the defensins and fatty acids of the skin. Adhesion also provides a survival advantage in the competition for nutrients and may effectively increase concentrations of bacterial toxins in the vicinity of the host cell.[166]

The attachment of bacteria to the epithelium of the host requires an interaction between ligands on the surface of the bacteria (adhesins) and surface structure(s) of the host cell (receptors). In the case of group A streptococci, fibronectin (Fn) was the first host protein identified as a receptor.[138] (For a review of the general role of Fn in bacterial adhesion see references.[69–71]) Succinctly, the evidence that Fn is a receptor is: 1) Fn and antisera against Fn blocked adhesion of group A streptococci to human buccal epithelial cells;[42] 2) adhesion of streptococci to buccal cells correlated with the amount of Fn on the surface of these cells;[1] and 3) the streptococcal-binding domain of Fn was identified as an N-terminal 28 kDa fragment and this fragment also blocked streptococcal adhesion to human buccal cells.[55,37] Other host proteins suggested to be receptors for group A streptococci are listed in Table 1. Undoubtedly, additional host cell receptors will be discovered as different streptococcal strains and host cells are tested.

Lipoteichoic acid (LTA) was the first purified adhesin found to block attachment of group A streptococci to a wide variety of host cells.[7,34] LTA reacts with Fn, inhibits Fn-binding to streptococci, and reacts with the N-terminal streptococcal-binding domain of Fn. Anti-LTA blocked adhesion to host cells and colonization of the oral cavity of mice by group A streptococci. Penicillin induced the release of LTA complexed with Fn from the surface of streptococci and the released Fn was precipitated by anti-LTA serum but not by antiserum directed against other surface antigens of streptococci.[103] Since the initial report on LTA, at least 15 additional surface components have been suggested to play a role in adhesion of group A streptococci to host cells (Table 1). The 28-kDa protein, glyceraldehyde-3-phosphate dehydrogenase (G3PDH), PBFP, and SOF can bind to Fn, but no direct role in adhesion has been demonstrated for these putative adhesins. The vitronectin-binding protein is presumed to bind to vitronectin on host cells and mediate adhesion, but this has not been demonstrated directly. The streptococcal binding domain of vitronectin has been localized.[87] There is substantial data indicating that LTA, M protein, protein F/Sfb, protein F2, and FBP54 mediate adhesion to host cells. Recent evidence also suggests that the hyaluronate capsule may mediate adhesion to certain types of cells by interaction with CD44.[133]

M protein was initially thought to be involved in adhesion because M-positive strains attached in greater numbers to host cells than did non-isogenic, M-negative strains. Subsequently, M protein was purified and found not to block streptococcal adhesion to human buccal epithelial cells.[5,34,69-71] Recently, the role of M protein in adhesion was investigated by comparing the adhesion of M-negative mutants to their wild type parents.[27,39,40] There were no differences between the adhesion of M-negative mutants and wild type parents to human buccal cells, which agrees with earlier findings. However, the M-negative mutants failed to attach to HEp-2 tissue culture cells or to human keratinocytes, whereas the wild type parents readily attached to these cells. Purified M proteins and LTA were shown to block adhesion to HEp-2 cells.[33] These data suggested that M proteins play a role in adhesion but only to certain types of host cells. M proteins may also bind to influenza A-infected cells by interactions with fibrinogen bound to viral particles on the surface of host cells.[127] It was suggested that this finding may help to explain the association between certain viral infections and those of group A streptococci.

Several Fn-binding proteins of group A streptococci have been identified and some of these were shown to be involved in adhesion (Table 1). Protein F (or Sfb, an allelic variant) was discovered by two independent groups.[68,144] Protein F/Sfb contains two different domains that react with Fn.[136] Protein F2 is another Fn-binding protein that reacts with Fn via two distinct domains.[76] These domains contain repeated sequences that bear striking homology to repeats in the Fn-binding proteins of other bacteria. Purified protein F/Sfb was found to block adhesion to HEp-2 cells. Inactivation of the gene for

**Table 1. List of Putative Streptococcal Adhesins and Receptors**

| Adhesin | Receptor | Reference |
|---|---|---|
| Lipoteichoic acid | Fn, macrophage scavenger receptor | 7,33,36,38,59,61, 66,138 |
| M protein | Fgn, CD46, galactose, fucose/fucosylated glycoprotein | 33,39,40,105, 106,148,151,152 |
| | Fn, laminin | 47 |
| Protein F1/Sfb1 | Fn, integrins | 68, 98,108,144 |
| Protein F2 | Fn | 76,132 |
| PFBP | Fn | 122 |
| FBP54 | Fn, Fgn | 37, 43 |
| 28 kDa protein | Fn | 35 |
| G3PDH | Fn, Fgn | 109,164 |
| Vn-binding protein | Vn | 87,145 |
| Galactose-binding protein | ? | 65 |
| Hyaluronic acid | CD44 | 133 |
| C-carbohydrate | ? | 22 |
| SOF | Fn | 32,85,120 |
| Collagen-binding protein | Collagen | 112,147 |
| SpeB | Integrin | 141 |
| ? | Laminin | 143 |

Fn, fibronectin; Fgn, fibrinogen; Vn, vitronectin.

protein F/Sfb resulted in a dramatic decrease in streptococcal adhesion to HEp-2 cells and Langerhans cells. It was proposed that Fn acts as a bridge between the Fn-binding proteins on the streptococcal surface and integrins on the surface of host cells.[108] Southern blot analysis indicated that approximately 48% of the strains tested did not contain the gene for protein F/Sfb.[102] Most of those found to be negative were ARF strains, suggesting that protein F does not have a role in the pathogenesis of ARF and may be more important in the pathogenesis of skin infections.

FBP54 is another Fn-binding protein that is expressed on the surface of group A streptococci.[37,43] The N-terminal domain of Fn binds to streptococci, interacts with FBP54, and blocks streptococcal adhesion to human buccal cells. The Fn-binding domain of FBP54 was localized to the first 89 residues of the N-terminus of the molecule. A degenerate repeat motif was identified which had little homology with other Fn-binding repeats. Analysis of serum from patients with poststreptococcal ARF or AGN indicated that FBP54 was expressed and immunogenic in the human host. Antiserum to FBP54 reacted with 5 out of 15 strains tested, indicating that FBP54 is expressed by some but not all group A streptococci. It is not yet known if FBP54 is preferentially expressed by skin or pharyngeal strains of group A streptococci. Confirmation of the role of FBP54 in adhesion will require experiments with FBP54-negative mutants.

The hyaluronate capsule appears to contribute to the ability of group A streptococci to colonize the oral cavity of mice.[155-156] Inactivation of the *hasA* gene in an encapsulated strain of group A streptococci resulted in loss of capsule expression, and a reduction in their ability to colonize the oral cavity of mice and to resist phagocytosis. The hyaluronate capsule has recently been suggested to mediate streptococcal adhesion to human keratinocytes by interaction with CD44.[133] These same investigators have also reported

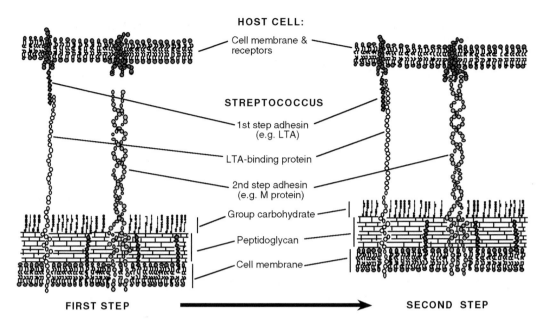

**Figure 3.** Hypothetical two-step model for adhesion of *S. pyogenes* to host receptors. It is proposed that adhesion occurs in two distinct steps. The first step serves to overcome repulsion that occurs between two negatively charged particles such as bacteria and host cells. It is suggested that LTA mediates this first step by hydrophobic interactions between its lipid moiety and receptors on host surfaces. This step is nonspecific and reversible. Completion of the first step facilitates the interaction of a second step adhesin with its receptor. The second step adhesin confers tissue specificity and leads to high affinity adhesion. M protein is shown as a second step adhesin in the model, but other adhesins such as protein F/Sfb, FBP54, etc., can also serve as a second adhesin. (Reproduced with permission.[69])

that expression of the hyaluronic acid capsule is inversely related to internalization of streptococci by human keratinocytes.[134] It is not clear how the capsule can initiate adhesion to keratinocytes and block internalization at the same time. Our data indicate that expression of capsule decreases the ability of streptococci to adhere to a variety of host cells.[44] Perhaps the hyaluronate capsule has an indirect role in adhesion. Hyaluronate has been shown to stimulate cytokine expression in host cells,[93] and it may be possible that the hyaluronate capsule upregulates expression of host cell receptors for other streptococcal adhesins.

At first glance, the data on the potential roles of the various adhesins appear to be contradictory. In order to account for these diverse and seemingly inconsistent findings, we proposed that group A streptococci utilize multiple adhesins to adhere to host cells and that adhesion occurs by a two-step mechanism.[69] In the first step, the streptococci must overcome the electrostatic repulsion separating its surface from that of the substratum (Fig. 3). We proposed that LTA mediates this initial step for most, if not all, group A streptococci. Completion of the first step would provide for relatively weak, reversible adhesion which is not sufficiently specific to provide for tissue tropism. A second step dependent upon another tier of adhesins (e.g., protein F, FBP54, M protein, etc.) must then occur in order to achieve firm or perhaps functionally irreversible adhesion that is tissue specific. If either the first step or the second step adhesin is missing or blocked, firm adhesion will not occur. The validity of this hypothesis is supported by the findings[44] that: 1) streptococci in which exposure of LTA is masked by the hyaluronate capsule are rapidly removed by washing; 2) streptococci in which LTA is exposed on the surface, but M protein is not expressed, exhibited only loose adhesion; 3) streptococci in which both LTA and M protein are exposed remained firmly attached even after six washes; 4) both LTA and M protein inhibited streptococcal attachment to HEp-2 cells;[33] and 5) antiserum to both LTA and M protein blocked colonization of the oral cavity of mice.[26,58]

Recent findings from several laboratories have also provided support for the concept that group A streptococci utilize multiple adhesins to mediate adhesion and that the type of second-step adhesin expressed will determine what specific host tissues are targeted. M protein mediates adhesion to HEp-2 cells and keratinocytes, but not to human buccal cells or Langerhans cells.[33,39,105,106] Protein F mediates adhesion to Langerhans cells, but not to keratinocytes.[105] FBP54 mediates adhesion to human buccal cells, but not to HEp-2 cells.[33] Although our model offers an explanation for some of the data discussed above, it does not resolve all of the issues. The model will undoubtedly be modified as new information becomes available.

Adhesion of group A streptococci to host cells may also lead to invasion of these cells.[46,47,86,134] Streptococci expressing M1 protein bound Fn and invaded human lung cell cultures, whereas an M-negative mutant had a reduced capacity to bind Fn and to invade these cells. It was suggested that one mechanism of invasion was mediated by interactions between M protein, Fn, and integrin. Laminin was also found to promote invasion of the M1 positive strain but not the M-negative strain. Laminin-stimulated invasion also appears to be mediated by an integrin. Protein F1 is another adhesin that promotes invasion of host cells. Similar to the mechanism used by M1 protein, Fn serves to link protein F1 and integrins on Hela cells resulting in uptake of streptococci.[108]

While attachment of bacteria to host surfaces is considered a prerequisite for infection, attachment can also induce subsequent changes in both the host cells and the adherent bacteria. One consequence of adhesion of bacteria to epithelial cells is an increase in the secretion of proinflammatory cytokines by host cells. Attachment of group A streptococci stimulated the release of cytokines from HEp-2 cells and human keratinocytes.[44,149] M protein was found to be the major adhesin involved in attachment to both of these cell types. In the case of HEp-2 cells, M protein was required for firm adhesion necessary for an increase in IL-6 secretion, but it is not known whether M protein directly initiated this increase or if some other streptococcal factor is involved after attachment occurs. Although M protein-mediated adhesion to keratinocytes stimulated cytokine production, other streptococcal factors also appear to be involved.[149]

A number of group A streptococcal products other than M proteins can elicit cytokine production. Peptidoglycan-polysaccharide complexes and superantigens from streptococci induced secretion of cytokines from a variety of host cells. It has been proposed that the cytokines IL-1, IL-6, tumor necrosis factor, and interferon play a central role in tissue destruction and death due to shock triggered by superantigens.[84,129,140] Adhesion of streptococci to host cells would serve to increase the effective concentrations of these streptococcal products in the milieu immediately surrounding host cells and may help to initiate an inflammatory cascade.

**Table 2. Plasma Proteins That Bind to *S. pyogenes***

| Host protein | Streptococcal receptor | Reference |
|---|---|---|
| Fgn | M protein, Mrp, T protein, G3PDH, FBP54 | 114,130,132 43,109 |
| Fn | M protein, G3PDH, protein F/Sfb, FBP54, SOF, LTA | 109,132 32,36,43,68 |
| Plasminogen | Plr (G3PDH)* | 89,164 |
|  | PAM (Mlp), M protein | 14,30,150 |
| Kininogens | M protein | 12,13 |
| FHL-1 | M protein | 79 |
| C4BP | Arp, Sir, protein H | 81 |
| Albumin | M protein | 3,109 |
| Factor H | M protein | 72,73 |
| IgA | Enn, Arp, Sir | 19,114 |
| IgG | Mrp, M protein, Sir, protein H | 3,114,119,139 |
| IgM | Mrp | 114 |
| Clusterin | Protein SIC | 2 |
| HRG | Protein SIC | 2 |
| α2 macroglobulin | Protein GRAB | 121 |

*Proteins in parenthesis indicate alternative nomenclature.
Abbreviations:
  G3PDH = glyceraldehyde-3-phosphate dehydrogenase
  Mrp = M-related protein
  Mlp = M-like protein
  HRG = histidine rich glycoprotein
  GRAB = protein G related α2-macroglobulin binding protein
  FHL-1 = factor H like protein 1.

## IV. VIRULENCE FACTORS

### A. M Proteins

The interaction between group A streptococci and a number of plasma proteins has been linked to increased resistance to phagocytosis, survival in deep tissues, and increased virulence in animals (Table 2). It has long been recognized that expression of M protein is linked to virulence of group A streptococci and several mechanisms have been proposed to account for this association. One proposal suggests that the binding of fibrinogen to M protein confers resistance to phagocytosis. Whitnack et al.[158-160] found that M type 24 *S. pyogenes* were effectively killed in serum but not in plasma. Addition of fibrinogen to serum restored the ability of streptococci to survive. Immunofluorescence assays of C3 binding to streptococci indicated that fibrinogen effectively reduced deposition of complement on the surface of the bacteria. These results have been confirmed using M-negative and M-positive isogenic pairs of streptococci.[39,40,115] The M types used in these assays, M5, M6, M18, and M24, were all class I organisms which do not generally express Mrp or Enn proteins. Recently, Mrp from an M type 49 strain was found to bind fibrinogen and to be required for resistance to phagocytosis.[114] Although the role of fibrinogen was not investigated in this study, it remains a possibility that resistance mediated by Mrp49 was due to interaction with fibrinogen.

The interaction of fibrinogen with M proteins may have other consequences as well. Plasminogen can interact with fibrinogen that is bound to M proteins or Mrp. Once bound, plasminogen is activated by streptokinase[30] which may contribute to the invasive properties of the streptococci.[88] Other proteins such as G3PDH (also called Plr) and PAM (also called Mlp or Enn) bind plasminogen directly,[14,164,165] providing group A streptococci with multiple mechanisms for binding plasminogen.

M proteins also react with Factor H,[73] a regulatory component of the alternative complement pathway that inhibits soluble C3b and helps control substrate bound C3b. Binding of Factor H to M protein was suggested to inhibit complement deposition on the bacterial surface. Factor H can also bind to fibrinogen,[72] and may help in limiting complement deposition by interacting with fibrinogen complexed with M protein. However, binding of factor H is not required for resistance to phagocytosis. An *S. pyogenes* mutant expressing an M6 protein from which the C-repeat was deleted did not bind factor H, but was able to multiply in human blood.[111] The role of fibrinogen-binding in the growth of this mutant in blood was not investigated.

The hypervariable domain of M proteins and related molecules may also bind C4BP, which down regulates activation of complement by the classical pathway.[81] Arp proteins from M types 4 and 60, Sir protein from M type 22, and protein H from M type 1 were all capable of binding C4BP. The binding sites in these proteins were localized to their N-terminal residues. The N-termini of these proteins have little homology, but key residues were proposed as necessary for binding. It was suggested that the degree to which members of the M protein family formed coiled-coil dimers was inversely related to their ability to bind C4BP. Interestingly, M5 protein did not bind C4BP, indicating that not all members of the M protein family bind C4BP. However, the N-termini of M5 and M6 proteins were found to bind the complement inhibitor FHL-1.[79] Thus, M proteins have evolved multiple mechanisms for limiting complement deposition.

Another plasma protein that binds to M proteins is human kininogen.[12,13] A large number of M-positive serotypes bound kininogen whereas M-negative strains did not. Purified M proteins also bound kininogen. Binding of kininogen to streptococci in human plasma resulted in the release of bradykinin, a potent vasoactive peptide. It was suggested that the binding of kininogen to streptococci could have other biological consequences. Since kininogens are inhibitors of cysteine proteases, the binding of kininogen may modulate the effects of the streptococcal cysteine protease (SPEB). Kininogen also binds to several types of host cells and could potentiate adhesion of streptococci to these cell types.

It should be stressed that M proteins do not exhibit precisely the same structures or functions. For example, M3 protein binds Fn,[132] whereas M5 and M24 proteins do not.[39] Some M proteins bind immunoglobulins by interacting with their Fc domain, whereas other M proteins do not.[31,119] M proteins that vary in structure may also vary in their functions or may achieve the same function by a different mechanism. The contributions of M proteins to a particular function very likely depend upon the genetic background of the strain, co-expression of other virulence factors, and the environment.

## B. Serum Opacity Factor

SOF is a large protein of ~100 kDa that is found in a cell-bound and extracellular form. SOF specifically cleaves apolipoprotein A1 in high density lipoproteins, which may account for the opalescence in serum caused by SOF.[128] Five *sof* genes from *S. pyogenes* have been sequenced.[32,85,120] All contained a highly conserved Fn-binding region that consisted of tandemly repeated peptides. In addition, the leader sequences and the membrane anchor regions were highly conserved. Short spans of conserved sequences were interspersed throughout the remaining parts of the proteins. The enzymatic domain was contained between amino acid residues 148 and 843. The role of SOF in virulence was investigated by insertional inactivation of *sof2*.[32] The SOF-negative mutant did not opacify serum and Fn binding was reduced by 70%. Complementation with *sof28* fully restored the ability to opacify serum and bind Fn. Only 7% of mice challenged with the SOF-positive parent survived, whereas 80% of mice challenged with the SOF-negative mutant survived. None of the mice challenged with the SOF-negative mutant that was complemented with *sof28* survived. These data indicate that SOF is a virulence factor of group A streptococci.

### C. Immunoglobulin Binding Proteins

The binding of immunoglobulins to the streptococcal surface has also been suggested to have a role in virulence. Group A streptococci express a number of different proteins that bind the Fc domain of immunoglobulins by non-immune mechanisms (Table 2). Mrp proteins, Mlp (or Enn) proteins and some serotypes of M proteins can bind immuno-globulins. Mrp proteins usually bind IgG molecules whereas Mlp proteins usually bind IgA molecules. These proteins discriminate between immunoglobulins of different mammalian species and preferentially bind immunoglobulins of a particular isotype.[23, 117-119]

The binding of immunoglobulins by group A streptococci has been found to be associated with the invasive potential of the organisms.[98,99] Boyle and coworkers have classified IgG-binding proteins that react with $IgG_1$, $IgG_2$, and $IgG_4$ as type IIa, those that react with $IgG_{1-4}$ as type IIo, and those react with only $IgG_3$ as type IIb.[117-119] It was shown that serial passage of a strain of group A streptococci in murine skin air sacs or human blood resulted in a dramatic increase in IgG binding activity. Recently, expression of the IgG-binding protein, Mrp, was found to be required for growth in human blood, a clear indication of virulence.[114] Protein H is a member of the M protein family that binds IgG and activates complement.[3,15] Protein H can be cleaved by SCPA (C5a peptidase) and released. Intravenous administration of protein H was found to be lethal in rabbits. It was speculated that the ability of soluble protein H to form complexes with immunoglobulins and activate complement caused these deaths and may contribute to AGN in humans.[15] Inactivation of the protein H gene, *sph*, resulted in decreased deposition of complement on the surface of the mutant. Whether the mutant was able to resist phagocytosis was not reported. Thus, soluble protein H activates complement, whereas surface bound protein H prevents activation of complement. It was suggested that other streptococcal IgG-binding proteins may function in a similar manner.

It has been suggested that coating of bacteria with immunoglobulins or other host proteins may accomplish several other functions.[31] This could be a form of molecular mimicry in which the coated bacteria looks like the host to effector cells. Binding of host proteins may also serve to signal the bacteria that they are in a particular niche of the host and that relevant genes need to be turned on or off. The immunoglobulin-binding proteins may also act as adhesins by reacting with immunoglobulin-like receptor molecules on host cells. IgG-binding proteins were found to be expressed primarily by those strains associated with skin infections, suggesting a role in tissue tropism.[18] The immunoglobulin-binding proteins may serve multiple, overlapping functions such as avoiding immune surveillance, adhesion, and resistance to phagocytosis.

### D. Protein SIC

Protein SIC (streptococcal inhibitor of complement-mediated lysis) has been found in only M types 1 and 57.[2] Protein SIC has several unusual features. First, it does not have any homology with other streptococcal proteins, either within or outside the Mga regulon. Second, it does not contain the LPxTG anchoring motif and is an extracellular protein. Protein SIC binds to two different plasma regulators of the membrane attack complex of complement, incorporates into C5b-C9 complexes, and inhibits complement-mediated lysis.[2]

### E. C5a Peptidase

C5a peptidase (SCPA) is an endopeptidase that is expressed on the surface of all group A streptococci that have been examined and some also release SCPA into culture supernatants. The peptidase is highly specific for the complement-derived chemotaxin C5a and it has been proposed that cleavage of this chemotaxin by SCPA inhibits recruitment of phagocytes into the area of infection.[157] To determine the role of SCPA, the *scp* gene was inactivated and the mutant tested in a mouse air sac model of infection.[77] There were no significant differences between the rate of mortality in mice receiving SCPA-negative mutant and those receiving the wild type parent. However, SCPA-negative mutants were cleared more rapidly than the wild type parent and the total number of neutrophils infiltrating the site of infection was greater in mice challenged with wild type streptococci than with the mutant. The route of dissemination of these organisms was also different. The mutant streptococci were transported to lymph nodes first, whereas the wild-type parent spread by a hematogenous route. It was speculated that SCPA-negative streptococci were more efficiently phagocytized by dendritic cells and macrophages and delivered to the lymph nodes, whereas the wild-type strain was more resistant to phagocytosis and invaded the blood stream.

The role of SCPA in colonization of the oral cavity of mice was also tested. SCPA-negative mutants were cleared more rapidly from the oral cavity of mice than wild type strains.[77] Up to 50% of the mice were colonized after intranasal challenge with wild type streptococci, whereas only 11% of mice were colonized after challenge with the SCPA-negative mutant. It was suggested that expression of SCPA prevented rapid clearance of streptococci by inactivating C5a and reducing infiltration of phagocytes.

### F. Protein GRAB

Protein GRAB (protein G related $\alpha$2-macroglobulin binding protein) is a streptococcal surface protein that binds $\alpha$2-macroglobulin, a protease inhibitor in plasma, and inhibits proteolytic activity of both *S. pyogenes* and host proteases.[121] The binding domain was localized to the *N*-terminal region. The *grab* gene was found in most group A streptococci. An isogenic GRAB mutant was less virulent in mice than the parental strain when injected intraperitoneally.

### G. Hyaluronic Acid Capsule

The hyaluronate capsule is composed of polymers of the disaccharide glucoronic-$\beta$-1,3-N-acetylglucosamine. The expression of hyaluronate capsule has long been thought to contribute to the ability of group A streptococci to survive in blood. Streptococcal strains that have caused the recent outbreaks of severe infections were heavily encapsulated. The role of capsule has been previously investigated by treating the streptococci with hyaluronidase, which enzymatically removes the capsule. Such treatments were found to increase phagocytosis of streptococci, suggesting that the capsule contributed to resistance to phagocytosis. Recently, data have been presented that confirms this concept. Inactivation of the *hasA* gene, which encodes for hyaluronate synthase, resulted in loss of capsule expression, reduced the ability to multiply in blood, and reduced virulence in an animal model.[101,156,131] Capsule negative mutants were defective in their ability to colonize the oral cavity of mice.[155] The hyaluronate capsule may also mediate adhesion to certain types of host cells, as mentioned above. Although the capsule is clearly a virulence factor, it is not a vaccine candidate because hyaluronate is expressed by mammalian cells and is poorly immunogenic.

While the capsule is clearly a virulence factor, it should not be construed that it is important for virulence in all strains of group A streptococci. In a recent survey of 1,100 isolates of *S. pyogenes*, 3% of isolates from uncomplicated pharyngitis were mucoid, 21% of isolates from serious infections were mucoid, and 42% of isolates from rheumatic fever patients were mucoid.[80] The relative contribution of the hyaluronate capsule to virulence may depend upon co-expression of other virulence factors.[74,101,156]

The hyaluronate capsule has been thought to be relatively inert with its main function being to impede phagocytosis by providing a relatively large barrier that sterically hindered interactions with phagocytes. However, recent studies suggest that fragments of hyaluronate may interact with CD44 on host cells and stimulate secretion of cytokines.[93] These studies utilized fragments of human hyaluronic acid and streptococcal forms of hyaluronic acid have not been tested. Although both human and streptococcal hyaluronic acid are composed of identical subunits, only certain sized fragments were capable of stimulating cytokines and it is not known if streptococci generate such sizes. These studies suggest that it is possible that the hyaluronic acid capsule may have a more active role in the pathogenesis of streptococcal infections than previously thought.

### H. Toxins

Group A streptococci elaborate a number of extracellular substances that may contribute to the virulence of the organism by acting directly on host cells or by generating by-products detrimental to the host. Among these are streptokinase, DNAse, hyaluronidase, LTA, streptolysin O, streptolysin S, NADase, streptococcal pyrogenic exotoxins, and other less well characterized substances.

Streptokinase catalyzes the conversion of plasminogen to plasmin. There are different antigenic forms of streptokinase, but all appear to perform the same function. It has been speculated that streptokinase aids streptococci in penetrating the fibrin barrier that may form around focal lesions. Streptokinase also activates plasminogen that is bound to fibrinogen complexed with M or M-like proteins on the surface of streptococci.[150] Inactivation of the streptokinase gene, *ska*, resulted in a dramatic decrease in the binding of plasminogen and streptococcal-bound enzyme activity.[30]

DNAse degrades DNA and may aid in liquefying the highly viscous pus that forms in lesions. Streptolysin O is an oxygen-labile hemolysin that is cytotoxic for a number of host cell types. Streptolysin S is stable in the presence of oxygen but generally requires some form of carrier for activity. It is thought to be a small peptide that may be generated from cleavage of a another protein and is toxic for host cells. LTA is cytotoxic for a number of cell types and LTA bound to cells can activate complement.[41,99,137]

The streptococcal pyrogenic exotoxins (SPEs), also known as erythrogenic toxins, have been implicated as central contributors to the streptococcal toxic shock syndrome (STSS). The SPEs are superantigens capable of stimulating T cell proliferation and induction of cytokines.[84,129,140] It has also been proposed that SPEs may deplete certain types of T cells.[154] The induction of cytokines such as TNFα/β, IL-1, and IL-6 has been proposed as one mechanism contributing to tissue injury.

The role of SPEB in virulence was investigated by inactivation of the *speB* gene in M types 3 and 49 *S. pyogenes*.[90] Approximately 93% of mice challenged with the M3 SPEB-negative mutant survived, whereas only 7% of mice challenged with the wild type parent survived by day 5. Similarly, 47% of mice challenged with the M49 SPEB-negative mutant survived, whereas only 3% survived challenge with the wild type parent. Recently, a SPEB variant was found to contain the cell binding motif, RGD.[141] It was suggested that this variant may act as adhesin since this variant interacted with human integrins. SPEB may also contribute to virulence by degrading host proteins such as Fn and vitronectin or by activating metalloproteases of the host.

A number of other toxins and mitogenic factors have been described but their role in virulence is less clear. There may be as many as 20 different mitogens produced by a single strain of group A streptococci.[84]

## V. VACCINES FOR PREVENTING STREPTOCOCCAL INFECTIONS

Antibodies to surface components of group A streptococci can provide protection against infections by several different mechanisms: 1) antibodies may opsonize the streptococci and promote phagocytic killing; 2) antibodies may block colonization of the host; and 3) antibodies may neutralize the activity of a toxin. Specific examples of each type of mechanism and the targeted virulence factor are presented.

### A. Vaccines to Hypervariable Domains of M Proteins

Current efforts to produce a group A streptococcal vaccine have focused on M proteins and were stimulated by the finding that M proteins evoked bactericidal antibodies that lasted for as long as 30 years following natural infections. Early vaccine trials indicated M proteins evoked protective antibodies in humans.[60,92] However, a significant number of the vaccinated people developed toxic reactions and some actually developed ARF,[91] suggesting that these preparations were not of sufficient purity. Subsequently, it was found that even highly purified preparations of M proteins may induce antibodies that cross-react with human tissues.[24,52,54] The epitopes that induced these cross-reactive antibodies have been identified for many types of M proteins and appear to reside within the A, B, and C repeat regions of M proteins.[24,48,52]

The potential problems associated with M protein epitopes that induce cross-reactive antibodies were avoided by focusing on the N-terminal, hypervariable domain of M proteins. It was demonstrated that the hypervariable domain contains the type-specific epitopes of M proteins and that protective antibodies were evoked by synthetic peptides copying the amino acid residues in the hypervariable N-terminus of several serotypes of

M proteins.[6,8-11,54,56] Importantly, antisera raised against these synthetic peptides did not cross-react with human tissues.

A second obstacle to an M protein based vaccine is that antisera to M protein usually provide type-specific protection and will not protect against infections from heterologous serotypes. Presently there are more than 90 different serotypes of group A streptococci, indicating the difficulty of preparing a vaccine effective against all known serotypes. However, not all serotypes are associated with ARF or AGN,[20] and the majority of life-threatening infections are caused by a limited number of serotypes.[80] Thus, it is theoretically possible to prepare a multivalent vaccine against a limited number of serotypes that would provide protection against a majority of the serotypes causing serious illness.

The first multivalent vaccine consisted of a hybrid synthetic peptide containing type-specific, N-terminal domains of M5, M6, and M24 proteins in tandem.[9] The trivalent vaccine evoked opsonic antibodies against all three serotypes of group A streptococci and none of the antisera cross-reacted with human tissues. However, the synthetic peptide approach becomes cumbersome or impossible as more serotypes are added. Thus, recombinant DNA techniques were used to create a tetravalent vaccine.[55] The approach entails amplifying the 5' termini of *emm* genes, splicing these *emm* sequences, and ligating the hybrid gene into an expression vector. The first construct consisted of the 5' termini of *emm24*, *emm5*, *emm6*, and *emm19*. The hybrid, recombinant protein evoked opsonic antibodies against all four serotypes of group A streptococci, although antibody titers were significantly lower against the C-terminal M19 protein.

An octavalent, recombinant, hybrid construct was prepared as described above consisting of N-terminal peptides of M24, M5, M6, M19, M1, M3, M18, and M2 proteins linked in tandem.[56] This octavalent vaccine evoked antibodies in rabbits that reacted with all 8 serotypes of group A streptococci. Opsonic antibodies were produced only against the first six M proteins; none of the rabbit antisera opsonized M type 18 or M type 2 group A streptococci. This finding was similar to that with the tetravalent vaccine in which the C-terminal M19 peptide evoked only low levels of opsonic antibody. Whether the poor responses to the C-terminal peptide was due to an unfavorable conformation of the C-terminus or excessive proteolytic processing is not known. This problem was overcome by reiterating the M24 peptide at the C-terminus.[50] Adding the M24 peptide to the C-terminus led to high titered antisera against the penultimate peptide that opsonized strains of the corresponding serotype.

There are several advantages in using a vaccine constructed as described above. First, such a vaccine can incorporate peptides from many M types and offer protection against a wide variety of serotypes. Second, the construction of a single molecule containing multiple protective epitopes should help to simplify purification and testing of vaccines. Incorporation of a purification tag such as a polyhistidine leader sequence would also aid in purification. Third, by preparing a number of such hybrid proteins it would be possible to tailor the vaccine to target relevant serotypes as warranted by new epidemiological evidence. Fourth, hybrid, multivalent constructs do not need carrier proteins to elicit protective responses.[50]

### B. Vaccine Against Conserved Epitopes of M Proteins

Another approach that is being pursued is a vaccine against conserved domains of M proteins. The rationale for this approach is that antisera to a common domain of

M proteins may provide protection against a wide variety of serotypes. A logical choice of antigen is the C-repeat domain of M proteins which is highly conserved among different serotypes. Since secretory IgA is thought to play a significant role in blocking adhesion of bacteria to mucosal surfaces of the host, it seems reasonable to focus on vaccine conjugates that stimulate a specific IgA response. Intranasal immunization of mice with synthetic peptides copying residues of the C-repeat domain of M6 protein conjugated to the B subunit of the cholera toxin was found to increase levels of salivary IgA and serum IgG that reacted with M6 protein.[17] These antibodies reacted with multiple serotypes of group A streptococci indicating the presence of shared epitopes. Mice immunized with the conjugate had a reduced level of pharyngeal colonization when challenged with M type 6 *S. pyogenes* as compared to mice immunized only with the cholera toxin subunit. However, this C-repeat vaccine did not protect against systemic infections since the rate of mortality was the same in both groups of mice. The failure of the vaccine to protect against systemic infections was suggested to be due to the inability of serum antibodies to the C-repeat domain to opsonize group A streptococci.

We also found that intranasal immunization of mice with synthetic peptides from the C-repeat domain of M5 protein conjugated to the cholera toxin subunit B protected mice against pharyngeal colonization and death after challenge with M type 24 group A streptococci.[24] Antisera to the conjugate vaccine did not opsonize streptococci, in agreement with other findings. Thus, protection against death was presumed to be due to stimulation of antibodies that block adhesion of the streptococci to oral tissues.

One concern regarding the use of C-repeat peptides in a vaccine is the possibility of stimulating antibodies that cross-react with human tissues. Antisera to the synthetic peptides from the C-repeat domain of M5 protein were found to cross-react with a protein in human kidneys and myosin,[11,48] In addition, the level of antibodies reacting with C-repeat domains are elevated in patients with ARF and it was suggested that these antibodies may play a role in development of ARF through cross-reactions with myosin in heart tissues.[100] Although these findings are discouraging, it may be possible to avoid epitopes within the C-repeat domains that elicit antibodies that cross-react with human tissues. For example, antisera from ARF patients reacted with a limited number of epitopes within the C-repeats, suggesting that there may be epitopes that do not induce cross-reactive antibodies.[100]

## C. Vaccine to Group A C-Carbohydrate

The group A streptococcal cell wall carbohydrate has been proposed as a potential vaccine, based on the findings that high titered antisera to its antigenic epitope, *N*-acetyl glucosamine, were able to opsonize group A streptococci.[126] These antisera also opsonized multiple serotypes indicating cross-protection against heterologous serotypes. It was suggested that an increase in antibodies to group A C-carbohydrate might explain the observation that upper airway infections due to group A streptococci usually diminish with age. While this antigen may ultimately prove useful in development of a multi-component vaccine, antisera to the group A cell wall carbohydrate have not been tested for cross-reactions with human tissues. In addition, protection required high titers that may be difficult to achieve in humans.

### D. Vaccine to C5a Peptidase (SCPA)

The structure of SCPA does not appear to vary significantly among serotypes of group A streptococci, suggesting that such a vaccine construct might provide protection against multiple serotypes. To determine if this was the case, mice were immunized with recombinant SCPA from M type 49 *S. pyogenes*.[78] Antisera from the immunized mice neutralized the peptidase activity of SCPA from 4 different serotypes, demonstrating a lack of serotype specificity. Furthermore, the immunized mice were protected against intranasal challenge from M type 49 *S. pyogenes*. Immunized mice also had a reduced rate of infection when challenged with M types 1, 2, 11, and 6 *S. pyogenes*. These data suggested that a C5a peptidase vaccine may be able to evoke cross-protective antibodies that neutralized the peptidase activity of SCPA. It would be interesting to ascertain the relative frequency and titers of antibodies to C5a peptidase in patients and controls to determine if the development of antibodies to C5a peptidase correlates with protection.

### E. Vaccines to Adhesins

Of all the putative adhesins listed in Table 2, only LTA, M proteins, and Sfb1 have been shown to evoke protective antibodies in an animal model of infection. Passive immunization of mice with rabbit antisera to LTA blocked colonization of the oral cavity. Instillation of LTA in the nares of mice prior to challenge with group A streptococci dramatically reduced colonization of the oral cavity. Deacylated LTA had no effect on colonization. These studies suggested that LTA might be useful as a vaccine component. However, LTA is not a viable vaccine candidate because it is a common surface antigen of most Gram-positive organisms and an LTA vaccine might alter the microbial flora. In addition, antibodies to LTA can react with human antigens. Although LTA may not be useful as a vaccine, it may have use in certain limited circumstances. Topical application of LTA to teats prevented infection with group B streptococci in mice neonates.[45]

Immunization with M proteins also evokes antibodies that block colonization of the oral cavity of mice by group A streptococci (see sections V.A. and V.B. above). Whether a vaccine will evoke adhesion-blocking antibodies or opsonic antibodies or both depends on the vaccine construct and route of immunization. Vaccines containing the hyper-variable domain of M proteins can evoke both types of antibodies when administered intranasally, whereas vaccines to conserved domains contained in C-repeats only evoke antibodies that block colonization.

We have tested rabbit antisera to FBP54 for opsonic activity in whole human blood and found no effect (unpublished data). Although antibodies to FBP54 did not opsonize group A streptococci, it still remains to be determined if antibodies to FBP54 can block adhesion in the oral cavity of mice.

Intranasal vaccination of mice with the Fn-binding protein, SfbI (also termed protein F1), protected mice against lethal, intranasal challenges with homologous and heterologous serotypes.[67] The cholera toxin subunit B was used as an adjuvant and conjugated to Sfb1. The degree of protection correlated with the levels of the specific antibodies in serum and lung lavages. However, intranasal vaccination did not provide protection against an intraperitoneal challenge, suggesting that Sfb1 does not evoke opsonic antibodies. The authors concluded that an IgA or IgG response that blocks adhesion was critical in protecting against intranasal challenges. Interestingly, Sfb1 was found to be an effective adjuvant and the Fn-binding domain stimulated B cell activation.[95,96]

In screening for antigens that have the potential to induce antibodies that block adhesion, it is logical to first investigate those antigens known to be involved in adhesion. However, it is possible that antibodies that react with a surface antigen that is not involved in adhesion may also block adhesion by steric hindrance. We are not aware of a single case where antibodies that reacted with the surface of streptococci failed to block adhesion.

### F. Vaccine to SPEs

Since many of the features of STSS are thought to be due to SPEA, the effect of anti-serum to SPEA on streptococcal infections in rabbits was investigated.[129] Rabbits were immunized with SPEA and then challenged subcutaneously with either M type 3 or M type 1 *S. pyogenes*. Only 20% of mice receiving placebo survived a challenge with M type 3 *S. pyogenes*, whereas 84% of immunized mice survived. Approximately 53% of mice receiving placebo survived a challenge with M type 1 *S. pyogenes*, whereas 87% of immunized mice survived.

Passive immunization of mice with antibodies to cysteine protease (SPEB) enhanced survival of mice challenged with group A streptococci.[82] Immunization of mice with SPEB significantly prolonged survival time in mice challenged with heterologous strains of group A streptococci. These data suggests that immunization with SPEB can elicit non-type specific immunity to group A streptococci.

### G. Vaccine to a New Protective Antigen (Spa)

Recently, a protein from an M type 18 strain was described that evokes antibodies that opsonized the streptococci and protected mice against challenge infections of type 18 group A streptococci.[49] The protein was termed streptococcal protective antigen (Spa). Antisera against Spa was also found to opsonize M types 3 and 28 group A streptococci. Serotypes that were not opsonized included M types 1, 2, 5, 6, 13, 14, 19, and 24. This is the first time that a group A streptococcal protein has been identified, other than M protein, that evokes opsonic antibodies. Whether or not Spa has a role in resistance to phagocytosis remains to be determined.

## VI. NONVACCINE BASED STRATEGIES

### A. Competition with Commensal Organisms

Recurrent infections are not uncommon after treatment of group A streptococcal pharyngitis with antibiotics. Recurrences can be due to failure to complete the prescribed regimen or to intracellular reservoirs of the organism,[107] but other factors may also be involved. Antibiotic treatment may disturb the composition of the normal flora, especially those capable of interfering with growth of group A streptococci. It has been reported that the presence of bacteria in the oral cavity that are bactericidal for group A streptococci increases with age, whereas bacteria that are bacteriostatic for group A streptococci are more prevalent in the young individuals. It was suggested that this disparity may contribute to greater resistance of adults to group A streptococcal infections. Therefore, restoration of normal α-streptococci in the oral cavity may help to prevent recurrent infections. In a double blind, placebo-controlled study, patients with recurrent group A streptococcal infections were treated with antibiotics for 10 d followed by either a spray of a pool of selected α-streptococci or placebo for 5 d.

Twenty-three percent of patients receiving the placebo had recurrent infections whereas only two percent of the patients receiving α-streptococcal treatment had a recurrent infection.[123]

### B. Desorption of Adherent Bacteria

The adhesion of many bacteria including group A streptococci has been found to be related to the surface hydrophobicity of the organism.[34] An emulsan that prevents the interaction of bacteria with hydrocarbons and desorbs bacteria attached to a variety of surfaces has been described. Recently clinical trials were performed with an oil-water, two-phase mixture containing cetylpyridinium chloride.[124] Subjects rinsed with the two-phase mouthwash, Listerine, or placebo in the morning and at bedtime. Rinsing with the two-phase mouthwash consistently reduced the levels of oral bacteria, plaque accumulation, and halitosis, when compared to placebo treatments. Listerine was slightly less effective in controlling malodor as compared to the two-phase mouthwash.

### C. Intravenous Immunoglobulin Therapy (IVIG)

Some streptococcal infections progress so rapidly that antibiotic therapy has little to no effect on the outcome. This is particularly relevant to streptococcal toxic shock syndrome (STSS) and necrotizing fasciitis. One of the major factors believed to be responsible for such a rapid deterioration is the elaboration of toxins by group A streptococci. The superantigens SPEA, SPEB, and SPEC have all been linked to STSS. Other mitogenic factors may also play a role.[84] It is thought that these superantigens initiate a cascade of inflammatory responses that results in extremely high levels of TNF and other cytokines. Limited studies have suggested that pooled human immunoglobulins may be beneficial in treatments of STSS. In a matched case-control study, IVIG treatment reduced mortality from 80% to 40%, a significant rate of protection.[83] Analysis of IVIG indicated the presence of antibodies that neutralize the mitogenic and cytokine-inducing activities of streptococcal superantigens.[104] Plasma from patients treated with IVIG had increased levels of neutralizing antibodies. Interestingly, the level of protection did not correlate with antibody titers to superantigens but did correlate with the level of neutralizing antibodies to superantigens. Based on these findings it was proposed that IVIG treatment be administered in conjunction with antibiotic treatment.

### D. IL-12 Therapy

IL-12 enhances cell-mediated immunity by increasing cytolytic activity of macrophages, NK cells, and T-cells.[29] Since IL-12 can increase resistance of the host to a variety of bacteria, it was tested for its effect on group A streptococcal skin infections in mice.[97] Mice were given IL-12 intraperitoneally and then inoculated with group A streptococci in skin air sacs. By day four, only 20% of mice receiving placebo survived, whereas 60% of mice receiving IL-12 survived.

### E. Antibiotics

The primary effect of antibiotics on bacteria is their ability to inhibit growth or to kill the organism. Due to the possibility of acquisition of resistance to antibiotics by bacteria, new forms of antibiotics are being constantly researched. One novel antibiotic was described by Bjorck et al.[21] A peptide mimicking the binding domain of a cysteine protease was synthesized and found to prevent growth of group A streptococci. Moreover, the synthetic peptide protected mice against lethal infections of group A streptococci. It was suggested that the primary target of the peptide inhibitor was the streptococcal cysteine protease (or SPEB) and that inactivation of SPEB inhibited growth.

Antibiotics may have other effects in addition to bacteriostatic and bactericidal actions. Penicillin was found to induce the release of LTA from group A streptococci in the stationary phase of growth, to reduce binding of Fn to streptococci, and to reduce adhesion to host cells.[4,103] Berberine sulfate was also found to induce the release of LTA from group A streptococci, to reduce binding of Fn to streptococci, and to reduce adhesion to host cells.[142]

## VII. GENERAL COMMENTS

A general theme that can be found concerning the pathogenicity of group A streptococci is that these organisms can use multiple mechanisms for adhesion, survival, and toxicity. The types of mechanisms used will depend upon the genetic background of the strain, the type of tissues involved, and environmental factors. Many of the surface proteins can serve multiple functions and the functions of different proteins may overlap. This provides the organisms with a safety net if one protective mechanism fails to function in a particular environment.

From a perusal of Tables 1 and 2 it can readily be seen that only a few of the adhesive and virulence factors of group A streptococci have been tested for their potential to prevent streptococcal infections. While M proteins are an obvious choice for a vaccine, other possibilities should not be overlooked. It is likely that as other surface proteins are investigated, one or more will be found to evoke protective antibodies. Likely candidates are the Fn-binding proteins since these proteins contain conserved Fn-binding repeats and antisera against one will likely react with another. Antisera to the Fn-binding domains of *Staphylococcus aureus* have been shown to provide protection in an animal model.[125] Since the domains of the Fn-binding proteins of group A streptococci bear remarkable homology with those of *S. aureus*, it seems reasonable to investigate whether antisera to these domains may also protect against infections from *S. pyogenes*. Other likely vaccine candidates are the SPEs. Identification of common domains within a particular SPE could lead to a vaccine that neutralizes its activity.

Many of the virulence factors of group A streptococci have been identified. This knowledge has been used to develop several constructs that may be useful as vaccines. Further investigations will provide new information leading to the development of more complex vaccines. One could visualize a multivalent vaccine construct that induces protective immunity against a broad range of human pathogens. It has been shown that conjugates of the diphtheria toxin, hepatitis surface antigen, tetanus toxoid, and M protein can evoke immunity to all 4 antigens.[28] Incorporation of M protein in the measles vaccinia virus stimulated antibodies to both the virus and M proteins.[64] A recombinant form of *Streptococcus gordinii* has been engineered to express hybrid proteins anchored on the surface by the C-terminal domain of the M6 protein.[63,116] Intranasal and oral inoculations in mice with live recombinants resulted in colonization and stimulation of salivary and serum antibodies. Hopefully, such constructs are only the harbinger of vaccines to come.

# REFERENCES

1. Abraham SN, Beachey EH, Simpson WA: Adherence of *Streptococcus pyogenes*, *Escherichia coli*, and *Pseudomonas aeruginosa* to fibronectin-coated and uncoated epithelial cells. *Infect Immun* 41:1261–8, 1983
2. Akesson P, Sjoholm A, Bjorck L: Protein SIC, a novel extracellular protein of *Streptococcus pyogenes* interfering with complement activity. *J Biol Chem* 271:1081–8, 1996
3. Akesson P, Schmidt K, Cooney J, et al: M1 protein and protein H: IgGFc- and albumin-binding streptococcal surface proteins encoded by adjacent genes. *Biochem J* 300:877–86, 1994
4. Alkan M, Beachey EH: Excretion of lipoteichoic acid by group A streptococci: influence of penicillin on excretion and loss of ability to adhere to human oral mucosal cells. *J Clin Invest* 61:671–7, 1978
5. Beachey EH, Courtney HS: Bacterial adherence: the attachment of group A streptococci to mucosal surfaces. *Rev Infect Dis* 9:S475–81, 1987
6. Beachey EH, Seyer JM: Protective and non-protective epitopes of chemically synthesized peptides of the NH2-terminal region of type 6 streptococcal M protein. *J Immunol* 136: 2287–92, 1986
7. Beachey EH, Ofek I: Epithelial cell binding of group A streptococci by lipoteichoic acid on fimbriae denuded of M protein. *J Exp Med* 143:759–71, 1976
8. Beachey EH, Stollerman G, Johnson R, et al: Human immune response to immunization wtih a structurally defined polypeptide fragment of streptococcal M protein. *J Exp Med* 150: 862–77, 1979
9. Beachey EH, Seyer JM, Dale JB: Protective immunogenicity and T lymphocyte specificity of a trivalent hybrid peptide containing NH2-terminal sequences of types 5, 6, and 24 M proteins synthesized in tandem. *J Exp Med* 166:647–56, 1987
10. Beachey EH, Seyer JM, Dale JB, et al: Type specific protective immunity evoked by synthetic peptide of *Streptococcus pyogenes* M protein. *Nature* 292:457–9, 1981
11. Beachey EH, Bronze MS, Dale JB, et al: Protective and autoimmune epitopes of streptococcal M protein vaccine. *Vaccine* 6:192–6, 1988
12. Ben Nasr A, Herwald H, Sjobring U, et al: Absorption of kininogen from human plasma by *Streptococcus pyogenes* is followed by release of bradykinin. *Biochem J* 326:657–60, 1997
13. Ben Nasr A, Herwald H, Muller-Esterl W, et al: Human kininogens interact with M protein, a bacterial surface protein and virulence determinant. *Biochem J* 305:173–80, 1995
14. Berge A, Sjobring U: PAM, a novel plasminogen-binding protein from *Streptococcus pyogenes*. *J Biol Chem* 34:25417–24, 1993
15. Berge A, Kihlberg B, Sjoholm A, et al: Streptococcal protein H forms soluble complement-activating complexes with IgG, but inhibits complement activation by IgG-coated targets. *J Biol Chem* 272:20774–81, 1987
16. Bessen D, Fischetti VA: Differentiation between two biologically distinct classes of group A streptococci by limited substitution of amino acids with the shared region of M protein-like molecules. *J Exp Med* 172:1757–64, 1990
17. Bessen D, Fischetti VA: Influence of intranasal immunization with synthetic peptides corresponsding to conserved epitopes of M protein on mucosal colonization by group A streptococci. *Infect Immun* 56:2666–72, 1988
18. Bessen D, Sotir C, Readdy T, et al: Genetic correlates of throat and skin isolates of group A streptococci. *J Infect Dis* 173:896–900, 1996
19. Bessen DE: Localization of immunoglobulin A-binding sites within M or M-like proteins. *Infect Immun* 62:1968–74, 1994
20. Bisno AL, Pearce IA, Wall HP, et al: Contrasting epidemiology of acute rheumatic fever and acute glomerulonephritis. *N Eng J Med* 283:561–5, 1970
21. Bjorck L, Akesson P, Bohus M, et al: Bacterial growth blocked by a synthetic peptide based on the structure of a human proteinase inhibitor. *Nature* 337:385–6, 1989

22. Botta G: Surface components in adhesion of group A streptococci to pharyngeal epithelial cells. *Curr Microbiol* 6:101–4, 1981

23. Boyle M, Raeder R: Analysis of heterogeneity of IgG-binding protein expressed by group A streptococci. *Immunomethods* 2:41–53, 1993.

24. Bronze MS, Dale JB: Epitopes of streptococcal M proteins that evoke antibodies that cross-react with human brain. *J Immunol* 151:2820–8, 1993

25. Bronze MS, Dale JB: The reemergence of serious group A streptococcal infections and acute rheumatic fever. *Amer J Med Sci* 311:41–54, 1996

26. Bronze MS, Courtney HS, Dale JB: Epitopes of group A streptococcal M protein that evoke cross-protective local immune response. *J Immunol* 148:888–93, 1992

27. Caparon M, Stevens D, Olsen A, et al: Role of M protein in adherence of group A streptococci. *Infect Immun* 59:1811–7, 1991

28. Chedid L, Jolivet M, Audibert F, et al: Antibody responses elicited by a polyvalent vaccine containing synthetic diptheric, streptococcal, and hepatitis peptides coupled to the same carrier. *Biochem Biophysic Res Comm* 117:908–15, 1983

29. Chehimi J, Trinchieri G: Interleukin-12: a bridge between innate resistance and adaptive immunity with a role in infection and acquired immunodeficiency. *J Clin Immunol* 14:149–61, 1994

30. Christner R, Li Z, Raeder R, et al: Identification of key gene products required for acquisition of plasmin-like enzymatic activity by group A streptococci. *J Infect Dis* 175:1115–20, 1997

31. Cleary P, Retnoningrum D: Group A streptococcal immunoglobulin-binding proteins: adhesins, molecular mimicry or sensory proteins. *Trends Microbiol* 2:131–6, 1994

32. Courtney H, Hasty D, Li Y, et al: Serum opacity factor is a major fibronectin-binding protein and a virulence determinant of M type 2 *Streptococcus pyogenes*. *Mol Microbiol* 32:89–98, 1999

33. Courtney HS, von Hunolstein C, Dale JB, et al: Lipoteichoic acid and M protein: dual adhesins of group A streptococci. *Microb Path* 12:199–208, 1992

34. Courtney HS, Hasty DL, Ofek I: Hydrophobicity of group A streptococci and its relationship to adhesion of streptococci to host cells. In: Doyle RJ, Rosenberg M, eds. *Microbial Cell Surface Hydrophobicity*. ASM Press, Washington, DC, 1990

35. Courtney HS, Hasty DL, Dale JB, et al: A 28 kilodalton fibronectin-binding protein of group A streptococci. *Curr Microbiol* 25:245–50, 1992

36. Courtney HS, Ofek I, Simpson WA, et al: Binding of lipoteichoic acid to fatty acid binding sites on human plasma fibronectin. *J Bacteriol* 153:763–70, 1983

37. Courtney HS, Dale JB, Hasty DL: Differential effects of the streptococcal fibronectin-binding protein, FBP54, on adhesion of group A streptococci to human buccal cells and HEp-2 tissue culture cells. *Infect Immun* 64:2415–9, 1996

38. Courtney HS, Stanislawski L, Ofek I, et al: Localization of a lipoteichoic acid binding site to a 24 kDa N-terminal fragment of fibronectin. *Rev Infect Dis* 10:S360–S2, 1988

39. Courtney HS, Bronze MS, Dale JB, et al: Analysis of the role of M24 protein in group A streptococcal adhesion and colonization by use of Ω-interposon mutagenesis. *Infect Immun* 62:4868–73, 1994

40. Courtney HS, Liu S, Dale JB, et al: Conversion of M serotype 24 of *Streptococcous pyogenes* to M serotypes 5 and 18: Effect on resistisance to phagocytosis and adhesion to host cells. *Infect Immun* 65:2472–4, 1997

41. Courtney HS, Simpson WA, Beachey EH: Relationship of critical micelle concentrations of bacterial lipoteichoic acids to biological activity. *Infect Immun* 51:414–8, 1986

42. Courtney HS, Simpson WA, Hasty DL, et al: Binding of *Streptococcus pyogenes* to soluble and insoluble fibronectin. *Infect Immun* 53:454–9, 1983

43. Courtney HS, Li Y, Dale JB, et al: Cloning, sequencing, and expression of a fibronectin/fibrinogen-binding protein from group A streptococci. *Infect Immun* 62: 3937–46, 1994

44. Courtney HS, Ofek I, Hasty DL: M protein mediated adhesion of M type 24 *Streptococcus pyogenes* stimulates release of interleukin-6 by HEp-2 tissue culture cells. *FEMS Microbiol Lett* 151:65–70, 1997

45. Cox. F: Prevention of group B streptococcal colonization with topically applied lipoteichoic acid in a maternal-newborn mouse model. *Pediatr Res* 16:816–9, 1982

46. Cue D, Cleary P: High-frequency invasion of epithelial cells by *Streptococcus pyogenes* can be activated by fibrinogen and peptides containing RGD. *Infect Immun* 65:2759–64, 1997

47. Cue D, Dombekl P, Lam H, et al: *Streptococcus pyogenes* serotype M1 encodes multiple pathways for entry into human epithelial cells. *Infect Immun* 66:4593–601, 1998

48. Cunningham M, Antone S, Smart M, Kosane S: Molecular analysis of human cardiac myosin-cross-reactive- B- and T- cell epitopes of the group A streptococcal M5 protein. *Infect Immun* 65:3913–23, 1997

49. Dale J, Chiang E, Liu S, et al: New protective antigen of group A streptococci. *J Clin Invest* 103:1261–8, 1999

50. Dale JB: Multivalent group A streptococcal vaccine designed to optimize the immunogenicity of six tandem M protein fragments. *Vaccine* 17:193–200, 1999

51. Dale JB, Chiang EC: Intranasal immunization with recombinant group A streptococcal M protein fragment fused to the B subunit of *Escherchia coli* labile toxin protects mice against systemic infecions. *J Infect Dis* 171:1038–41, 1995

52. Dale JB, Beachey EH: Epitopes of streptococcal M protein shared with cardiac myosin. *J Exp Med* 162:583–91, 1985

53. Dale JB, Beachey EH: Localization of protective epitopes of the amino terminus of type 5 streptococcal M protein. *J Exp Med* 163:1191–202, 1986

54. Dale JB, Beachey EH: Human cytotoxic T lymphocytes evoked by group A streptococcal M proteins. *J Exp Med* 166:1825–35, 1987

55. Dale JB, Chiang E, Lederer JW: Recombinant tetravalent group A streptococcal M protein vaccine. *J Immunol* 151:2188–94, 1993

56. Dale JB, Simmons M, Chiang EC, et al: Recombinant, octavalent group A streptococcal M protein vaccine. *Vaccine* 14:944–948, 1996

57. Dale JB,Washburn RG, Marques MB, et al: Hyaluronate capsule and surface M protein in resistance to opsonization of group A streptococci. *Infect Immun* 64:1495–501, 1996

58. Dale JB, Baird R, Courtney H, et al: Passive protection of mice against group A streptococcal pharyngeal infection by lipoteichoic acid. *J Infect Dis* 169:319–23, 1994

59. Dale JB, Beachey EH: Protective antigenic determinant of streptococcal M protein shared with sarcolemmal membrane protein of human heart. *J Exp Med* 156:1165–76, 1982

60. D'Alessandri R, Plotkin G, Kluge RM, et al: Protective studies with group A streptococcal M protein vaccine. III. Challenge of volunteers after systemic or intranasal immunization with type 3 or type 12 group A streptococcus. *J Infect Dis* 138:712–8, 1978

61. Dunne DW, Resnick D, Greenberg DJ, et al: The type 1 macrophage scavenger receptor binds to Gram-positive bacteria and recognizes lipoteichoic acid. *Proc Natl Acad Sci USA* 91:1863–7, 1994

62. Fischetti VA: Streptococcal M protein: molecular design and biological behaviour. *Clin Microbiol* 2:285–314, 1989

63. Fischetti VA, Medaglini D, Oggioni M, et al: Expression of foreign proteins on Gram-positive commensal bacteria for mucosal vaccine delivery. *Curr Opinion Biotech* 4: 603–10, 1993

64. Fischetti VA, Hodges W, Hyrby DE: Protection against streptococcal pharyngeal coloni-zation with a vaccinia: M protein recombinant. *Science* 244:1487–90, 1990

65. Gerlach D, Schalen C, Tigyi Z, et al: Identification of a novel lectin in *Streptococcus pyogenes* and its possible role in bacterial adherence to pharyngeal cells. *Curr Microbiol* 28:331–8, 1994

66. Grabovskaya KB, Totolian A, Ryc M, et al: Adherence of group A streptococci to epithelial cells in tissue culture. *Zentral Bakteriol Mikrobiol Hyg* [A] 247:303–14, 1980

67. Guzman C, Talay S, Molinari G, et al: Protective immune response against *Streptococcus pyogenes* in mice after intranasal vaccination with the fibronectin-binding protein SfbI. *J Infect Dis* 179:901–6, 1999

68. Hanski E, Caparon M: Protein F, a fibronectin-binding protein, is an adhesin of the group A streptococcus, *Streptococcus pyogenes. Proc Natl Acad Sci USA* 89:6172–6, 1992

69. Hasty DL, Ofek I, Courtney HS, et al: Multiple adhesins of streptococci. *Infect Immun* 60: 2147–52, 1992

70. Hasty DL, Courtney HS: Group A streptococcal adhesion: all of the theories are correct. *Adv Exp Med Biol* 408:81–94, 1996

71. Hasty DL, Beachey EH, Courtney HS, et al: Interactions between fibronectin and bacteria. In: Carsons SE, ed. *Fibronectin in Health and Disease.* CRC Press, Boca Raton, FL, 1989:89–112

72. Horstmann R, Sievertsen H, Leippe M, et al: Role of fibrinogen in complement inhibition by streptococcal M protein. *Infect Immun* 60:5036–41, 1992

73. Horstmann R, Sievertsen J, Knobloch J, et al: Antiphagocytic activity of streptococcal M protein: selective binding of complement control protein factor H. *Proc Natl Acad Sci USA* 85:1657–61, 1988

74. Husmann L, Yung D, Hollingshead S, et al: Role of putative virulence factors of *Streptococcus pyogenes* in mouse models of long-term throat colonization and pneumonia. *Infect Immun* 65:1422–30, 1997

75. Husmann L, Scott J, Lindahl G, et al: Expression of the Arp protein, a member of the M protein family, is not sufficient to inhibit phagocytosis. *Infect Immun* 63:345–8, 1995

76. Jaffe J, Natanson-Yaron S, Caparon MG, et al: Protein F2, a novel fibronectin-binding protein from *Streptococcus pyogenes*, possesses two binding domains. *Mol Microbiol* 21: 373–84, 1996

77. Ji Y, McLandsborough A, Kondagunta A, et al: C5a peptidase alters clearance and trafficking of group A streptococci by infected mice. *Infect Immun* 64:503–10, 1996

78. Ji Y, Carlson B, Kondagunta A, et al: Intranasal immunization with C5a peptidase prevents nasopharyngeal colonization of mice by the group A streptococcus. *Infect Immun* 65:2080–7, 1997

79. Johnnson E, Berggard K, Kotarsky H, et al: Role of the hypervariable region in streptococcal M proteins: binding of a human complement inhibitor. *J Immunol* 161:4894–01, 1998

80. Johnson DR, Stevens D, Kaplan EL: Epidemiologic analysis of group A streptococcal serotypes associated with severe systemic infections, rheumatic fever, or uncomplicated pharyngitis. *J Infect Dis* 166:374–82, 1992

81. Johnsson E, Thern A, Dahlback B, et al: A highly variable region in members of the streptococcal M protein family binds the human complement regulator C4BP. *J Immunol* 157:3021–9, 1996

82. Kapur V, Maffei J, Greer G, et al: Vaccination with streptococcal extracellualr cysteine protease (interleukin-1 convertase) protects mice against challenge with heterologous group A streptococci. *Microb Pathog* 16:443–50, 1994

83. Kaul R, McGeer A: Ontario Streptococcal Study Group, Norby-Teglund A, Kotb M, Low D: Intravenous immunoglobulin (IVIG) therapy in streptococcal toxic shock syndrome (STSS): results of a matched case-control study. Abstract LM68. *Abstracts of the 35th ICACC.* ASM, Washington, DC, 1995

84. Kotb M: Bacterial pyrogenic exotoxins as superantigens. *Clin Microbiol Rev* 8:411–26, 1995

85. Kreikmeyer B, Talay SR, Chhatwal GS: Characterization of a novel fibronectin-binding surface protein in group A streptococci. *Mol Microbiol* 17:137–145, 1995

86. LaPenta D, Rubens C, Chi E, et al: Group A streptococci efficiently invade human respiratory epithelial cells. *Proc Natl Acad Sci USA* 91:12115–9, 1994

87. Liang O, Preissner K, Chhatwal G: The hemopexin-type repeats of human vitronectin are recognized by *Streptococcus pyogenes*. *Biochem Biophys Res Comm* 234:445–9, 1997

88. Lottenberg R, Minning-Wenz D, Boyle M: Capturing host plasmin(ogen): a common mechanism for invasive pathogens. *Trends Microbiol* 2:20–24, 1994

89. Lottenburg R, Broder C, Boyle M, et al: Cloning, sequence analysis, and expression in *Escherichia coli* of a streptococcal plasmin receptor. *J Bacteriol* 174:5204–10, 1992

90. Lukomski S, Sreevatsan S, Amberg C, et al: Inactivation of *Streptococcus pyogenes* extracellular cysteine protease significantly decreases mouse lethality of serotype M3 and M49 strains. *J Clin Invest* 99:2574–80, 1997

91. Massel B, Honikman L, Amezcua J: Rhematic fever following streptococcal vaccination. *J Amer Med Assoc* 207:1115–9, 1969

92. Massel BF, et al: Secondary and apparent primary antibody responses after group A streptococcal vaccination of 21 children. *Appl Microbiol* 16:509–18, 1968

93. McKee C, Penno M, Cowman M, et al: Hyaluronan (HA) fragments induce chemokine gene expression in alveolar macrophages. *J Clin Invest* 98:2403–13, 1996

94. McLandsborough L, Cleary P: Insertional inactivation of virR in *Streptococcus pyogenes* M49 demonstrates that VirR functions as a positive regulator of ScpA, FcRA, OF, and M protein. *FEMS Microbiol Lett* 128:45–51, 1995

95. Medina E, Talay S, Chhatwal G, et al: Fibronectin-binding protein I of *Streptococus pyogenes* is a promising adjuvant for antigens delivered by mucosal route. *Eur J Immunol* 28:1069–77, 1998

96. Medina E, Talay S, Chhatwal G, et al: Fibronectin-binding protein I of *Streptococcus pyogenes* promotes T cell-independent proliferation of murine B lymphocytes and enhances the expression of MHC class II molecules on antigen-presenting cells. *Int Immunol* 10:1657–64, 1998

97. Metzger D, Raeder R, Van Cleave V, et al: Protection of mice from group A streptococcal skin infections by Interleukin-12. *J Infect Dis* 171:1643–5, 1995

98. Molinari G, Talay S, Valentin-Weigand P, et al: The fibronectin-binding protein of *Streptococcus pyogenes* is involved in internalization of group A streptococci by epithelial cells. *Infect Immun* 65:1357–63, 1997

99. Monefeldt K, Tollefesen T: Effects of streptococcal lipoteichoic acid on complement activation *in vitro*. *J Clin Periodontol* 20:186–192, 1993

100. Mori K, Kamikawaji N, Sasazuki T: Persistent elevation of immunoglobulin G titer against the C region of recombinant group A streptococcal M protein in patients with rhematic fever. *Ped Res* 39:336–42, 1996

101. Moses A, Wessels M, Zalcman K, et al: Relative contributions of hyaluronic acid capsule and M protein to virulence in a mucoid strain of the group A streptococcus. *Infect Immun* 65:64–71, 1997

102. Natanson S, Sela S, Musser J, et al: Distribution of fibronectin-binding proteins among group A streptococci of different M types. *J Infect Dis* 171:871–8, 1995

103. Nealon TJ, Beachey EH, Courtney HS, et al: Release of fibronectin-lipoteichoic acid complexes from group A streptococci with penicillin. *Infect Immun* 51:529–35, 1986

104. Norrby-Teglund A, Kaul R, Low D, et al: Evidence for the presence of streptococcal-superantigen-neutralizing antibodies in normal polyspecific immunoglobulin G. *Infect Immun* 64:5395–8, 1996

105. Okada N, Pentland A, Falk P, et al: M protein and protein F act as important determinants of cell-specific tropism of *Streptococcus pyogenes* in skin tissue. *J Clin Invest* 94:965–77, 1994

106. Okada N, Liszewski M, Atkinson J, et al: Membrane cofactor protein (CD46) is a keratinocyte receptor for the M protein of the group A streptococcus. *Proc Natl Acad Sci USA* 92:2489–93, 1995

107. Osterlund A, Popa R, Nikkila T, et al: Intracellular reservoir of *Streptococcus pyogenes in vivo*: a possible explanation for recurrent pharyngotonsillitis. *Laryngoscope* 107:640–7, 1997

108. Ozeri V, Rosenshinie I, Mosher D, et al: Roles of integrins and fibronectin in the entry of *Streptococcus pyogenes* into cells via protein F1. *Mol Microbiol* 30:625–37, 1998

109. Pancholi V, Fischetti VA: A major surface protein on group A streptococci is glyceraldehyde-3-phosphate-dehydrogenase with multiple binding activity. *J Exp Med* 176:415–26, 1992

110. Perez-Casal J, Caparon M, Scott J: Mry, a trans-acting positive regulator of the M protein gene with similarity to the receptor proteins of two-component regulatory systems. *J Bacteriol* 173:2617–24, 1991

111. Perez-Casal J, Okada N, Caparon M, et al: Role of the conserved C-repeat region of the M protein of *Streptococcus pyogenes*. *Mol Microbiol* 15:907–16, 1995

112. Podbielski A,Woischnik M, Leonard B, et al: Characterization of *nra*, a global negative regulator gene in group A streptococci. *Mol Microbiol* 31:1051–64, 1999

113. Podbielski A, Woischnik M, Pohl B, et al: What is the size of the group A streptococcal *vir* regulon? the *mga* regulator affects expression of secreted and surface virulence factors. *Med Microbiol* 185:171–81, 1996

114. Podbielski A, Schnitzler N, Beyhs P, et al: M-related protein (Mrp) contributes to group A streptococcal resistance to phagocytosis by human granulocytes. *Mol Microbiol* 19:429–41, 1996

115. Poirier TP, Kehoe M, Whitnack E, et al: Fibrinogen binding and resistance to phagocytosis of *Streptococcus sanguis* expressing cloned M protein of *Streptococcus pyogenes*. *Infect Immun* 57:29–35, 1989

116. Pozzi G, Contorni M, Oggioni M, et al: Delivery and expression of a heterologous antigen on the surface of streptococci. *Infect Immun* 60:1902–7, 1992

117. Raeder R, Faulmann E, Boyle M: Evidence for functional heterogeneity in IgG Fc-binding proteins associated wtih group A streptococci. *J Immnol* 146:1247–53, 1991

118. Raeder R, Boyle M: Assocation between expression of immunoglobulin G-binding proteins by group A streptococci and virulence in a mouse skin infection model. *Infect Immun* 61:1378–84, 1993

119. Raeder R, Boyle M: Properties of IgG-binding proteins expressed by *Streptococcus pyogenes* isolates are predictive of invasive potential. *J Infect Dis* 173:888–95, 1996

120. Rakonjac JV, Robbins JC, Fischetti VA: DNA sequence of the serum opacity factor of group A streptococci: identification of a fibronectin-binding repeat domain. *Infect Immun* 63:622–31, 1995

121. Rasmussen M, Muller H, Bjorck L: Protein GRAB of *Streptococcus pyogenes* regulates proteolysis at the bacterial surface by binding alpha2-macroglobulin. *J Biol Chem* 274:15336–44, 1999

122. Rocha C, Fischetti V: Identification and characterization of a novel fibronectin-binding protein on the surface of group A streptococci. *Infect Immun* 67:2720–8, 1999

123. Roos K, Holm S, Grahn E, Lind L: Alpha-streptococci as supplementary treatment of recurrent streptococcal tonsillitis: a randomized placebo-controlled study. *Scand J Infect Dis* 25:31–35, 1993

124. Rosenberg M, Greenstein R, Barki M, et al: Hydrophobic interactions as a basis for interferring with microbial adhesion. In: Kahane I, Ofek I, eds. *Toward Anti-Adhesion Therapy for Microbial Diseases*. Plenum Press, New York, 1996:241–8

125. Rozalska B, Wadstrom T: Protective opsonic activity of antibodies against fibronectin-binding proteins (FnBPs) of *Staphylococcus aureus*. *Scand J Immunol* 37:575–80, 1993

126. Salvadori LG, Blake MS, McCarty M, et al: Group A streptococcus-liposome ELISA titers to group A polysaccharide and opsonic capabilities of the antibodies. *J Infect Dis* 171:593–600, 1995

127. Sanford BA, Davison VE, Ramsay MA: Fibrinogen-mediated adherence of group A streptococci to influenza A virus-infected cell cultures. *Infect Immun* 38:513–20, 1982

128. Saravani G, Martin D: Opacity factor from group A streptococci is an apolipoproteinase. *FEMS Microbiol Lett* 68:35–40, 1990

129. Schlievert P, Assimacopoulos A, Cleary P: Severe invasive group A streptococcal disease: clinical description and mechanisms of pathogenesis. *J Lab Clin Med* 127:13–22, 1996

130. Schmidt KH, Kohler W: T-proteins of *Streptococcus pyogenes* IV. Communication: isolation of T1-protein by affinity chromatography on immobilized fibrinogen. *Zbl Bakt Hyg A.* 258:449–56, 1984

131. Schmidt KH, Gunther E, Courtney HS: Expresson of both M protein and hyaluronic acid capsule by group A streptcoccal strains results in high virulence for chicken embryos. *Med Microbiol Immunol* 184:169–73, 1996

132. Schmidt KH, Mann K, Cooney J, et al: Multiple binding of type 3 streptococcal M protein to human fibrinogen, albumin, and fibronectin. *FEMS Immunol Med Microbiol* 7:135–44, 1993

133. Schrager HM, Dougherty G, Wessels MR: CD44 is a major receptor for the group A streptococcal hyaluronic acid capsule on skin and pharyngeal keratinocytes. *Proc Lancefield Society*, San Francisco, CA, 1997

134. Schrager HM, Rheinwald J, Wessels MR: Hyaluronic acid capsule and the role of streptococcal entry into keratinocytes in invasive skin infections. *J Clin Invest* 98:1954–8, 1996

135. Scott JR: The M protein of group A streptococcus: evolution and regulation. In: Inglewski B and Clark V, eds., *The Bacteria.* Academic Press, San Diego, CA. 1990

136. Sela S, Aviv A, Tovi A, et al: Protein F: an adhesin of *Streptococcus pyogenes* binds fibronectin via two distinct domains. *Mol Microbiol* 10:1049–55, 1993

137. Simpson WA, Beachey EH: Cytotoxicity of the glycolipid region of streptococcal lipoteichoic acid for cultures of human heart cells. *J Lab Clin Med* 99:118–26, 1982

138. Simpson WA, Beachey EH: Adherence of group A streptococci to fibronectin on oral epithelial cells. *Infect Immun* 39:275–9, 1983

139. Stenberg L, O'Toole P, Lindahl G: Many group A streptococcal strains express two different immunoglobulin-binding proteins encoded by closely linked genes: characterization of the proteins expressed by four strains of different M-type. *Mol Microbiol* 6:1185–94, 1992

140. Stevens DL: Invasive group A streptococcal infections: past, present and future. *Pediatr Infect Dis J* 13:561–6, 1994

141. Stockbauer K, et al: A natural variant of the cysteine protease virulence factor of group A streptococcus with an arginine-glycine-aspartic acid (RGD) motif preferentially binds human integrins avb3 and aIIbbe. *Proc Natl Acad Sci USA* 96:242–7, 1999

142. Sun D, Courtney HS, Beachey EH: Berberine sulfate blocks adherence of *Streptococcus pyogenes* to epithelial cells, fibronectin, and hexadecane. *Antimicrob Agents Chemother* 32:1370–4, 1988

143. Switalski LM, Speziale P, Hook M, et al: Binding of *Streptococcus pyogenes* to laminin. *J Biol Chem* 259:3734–8, 1984

144. Talay SR, Valentin-Weigand P, Jerlstrom PG, et al: Fibronectin-binding protein of *Streptococcus pyogenes*: Sequence of the binding domain involved in adherence of streptococci to epithelial cells. *Infect Immun* 60:3837–44, 1993

145. Valentin-Wiegand P, Grulich-Henn J, Chhatwal GS, et al: Mediation of adherence of streptococci to human endothelial cells by complement S protein (vitronectin). *Infect Immun* 56:2851–5, 1988

146. Veasey L, Wiedmeir S, Orsmond G, et al: Resurgence of acute rheumatic fever in the intermountain area of the United States. *New Eng J Med* 316:421–3, 1987

147. Visai L, Bozzini S, Raucci G, et al: Isolation and characterization of a novel collagen-binding protein from *Streptococcus pyogenes* strain 6414. *J Biol Chem* 270:347–53, 1995

148. Wadstrom T, Tylewska S: Glycoconjugates as possible receptors for *Streptococcus pyogenes*. *Curr Microb* 7:343–6, 1982

149. Wang B, Ruiz N, Pentland A, Caparon M: Keratinocytes proinflammatory responses to adherent and nonadherent group A streptococci. *Infect Immun* 65:2119–26, 1997

150. Wang H, Lottenberg R, Boyle M: A role for fibrinogen in the streptokinase-dependent acquisition of plasmin(ogen) by group A streptococci. *J Infect Dis* 171:85–92, 1995

151. Wang J, Stinson M: M protein mediates streptococcal adhesion to HEp-2 cells. *Infect Immun* 62:442–8, 1994

152. Wang J, Stinson M: Streptococcal M6 protein binds to fucose-containing glycoproteins on cultured human epithelial cells. *Infect Immun* 62:1268–74, 1994

153. Wannamaker LW: Differences between streptococcal infections of the throat and of the skin. *New Engl J Med* 282:23–31, 1970

154. Watanable-Ohnishi R, Low D, McGeer A, et al: Selective depletion of VB-bearing T cells in patients with severe invasive group A streptococcal infections and streptococcal shock syndrome. *J Infect Dis* 171:74–9, 1995

155. Wessels M, Bronze M: Critical role of the group A streptococcal capsule in pharyngeal colonizaton and infection in mice. *Proc Natl Acad Sci USA* 91:12238–42, 1994

156. Wessels MR, Moses A, Goldberg J, et al: Hyaluronic acid capsule is a virulence factor for mucoid group A streptococci. *Proc Natl Acad Sci USA* 88:8317–21, 1991

157. Wexler D, Chenoweth E, Cleary P: Mechanism of action of the group A streptococcal C5a inactivator. *Proc Natl Acad Sci USA* 82:8144–8, 1985

158. Whitnack E, Beachey EH: Antiopsonic activity of fibrinogen bound to M protein on the surface of group A streptococci. *J Clin Invest* 69:1042–5, 1982

159. Whitnack E, Beachey EH: Biochemical and biological properties of the binding of human fibrinogen to M protein in group A streptococci. *J Bacteriol* 164:350–8, 1985

160. Whitnack E, Dale JB, Beachey EH: Common protective antigens of group A streptococcal M proteins masked by fibrinogen. *J Exp Med* 159:1201–2, 1984

161. Widdowson J, Maxted W, Grant D: The production of opacity in serum by group A streptococci and its relation with the presence of M antigen. *J Gen Microbiol* 61:343–53, 1970

162. Wilkinson H: Comparison of streptococcal R antigens. *Appl Microbiol* 24:669–670, 1972

163. Wilson E, Zimmerman R, Moody M: Value of T-agglutination typing of group A streptococci in epidemiologic investigatons. *Health Lab Sci* 5:199–207, 1968

164. Winram SB, Lottenberg R: The plasmin-binding protein Plr of group A streptococci is identified as glyceraldehyde-3-phosphate-dehydrogenase. *Microbiology* 142:2311–20, 1996

165. Wistedt AC, Ringdahl U, Mueller-Esterl W, et al: Identification of a plasminogen-binding motif in PAM, a bacterial surface protein. *Mol Microbiol* 18:569–78, 1995

166. Zafiri D, Eisenstein B, Ofek I: Growth advantage and enhanced toxicity of *Escherichia coli* adherent to tissue culture cells due to restricted diffusion of products secreted by cells. *J Clin Invest* 79:1210–6, 1987

# 36

# Changing Material Surface Chemistry for Preventing Bacterial Adhesion

**Wolfgang Kohnen and Bernd Jansen**

*Department of Hygiene and Environmental Medicine,
Johannes Gutenberg University, Mainz, Germany*

## I. INTRODUCTION

Since the initial step in the pathogenesis of catheter related infections is the adhesion of bacteria to the polymer, prevention of bacterial adherence should be an ideal way to avoid foreign body infections.[21] A scientific understanding of the interactions between microorganisms and commonly used polymers is the basis of any effective development of devices with antiinfective surfaces. Thus Ferreirós et al. studied the in vitro adhesion of twenty nine *Staphylococcus epidermidis* strains to Teflon, polyethylene and polycarbonate.[11] It was found that all strains showed a high adhesion to polymers with a high surface free energy. Reid et al. investigated the adhesion of microorganisms to urinary catheter surfaces.[40] They found *Lactobacillus acidophilus* adhesion correlated with substratum surface tension, whereas adherence of a *S. epidermidis* strain did not. *Escherichia coli* adhered very poorly to all polymers tested. A correlation between surface tension of different synthetic polymers used for medical purposes and staphylococci was also observed in a study conducted by us:[33] bacterial attachment decreased with increasing surface tension of synthetic materials.

As most bacteria in an aqueous environment exhibit a negative surface charge, it had been expected that negatively charged polymers would lead to a reduction in bacterial adhesion. Carballo et al. found a correlation between bacterial charge of several *S. edidermidis* strains and the adhesion to seven synthetic polymers.[5] Adherence of *E. coli* to an anion exchange resin was affected by the pH of the bacterial suspension, suggesting that ionic groups were involved in the adhesion process.[17] Harkes et al. investigated the adhesion of three *E. coli* strains to poly(methacrylates).[16] They found a correlation between bacterial adhesion and the zeta potentials of the polymers, indicating an influence of surface charge. We observed a reduction of *S. epidermidis* adhesion to polyurethane which had been surface-grafted with acrylic acid (resulting in a negatively charged surface) when compared with control polyurethane.[28]

*Handbook of Bacterial Adhesion: Principles, Methods, and Applications*
Edited by: Y. H. An and R. J. Friedman © Humana Press Inc., Totowa, NJ

### Table 1. Surface Modification by Physicochemical Methods

| Modified material | Modification | 1st Author[Ref.] |
|---|---|---|
| Titanium | Crosslinked ovine serum albumin | An[1] |
| Polyurethane | Glycerophosphorylcholine | Baumgartner[2] |
| Polystyrene | Poly(ethylene oxide) Poly(propylene oxide) | Bridgett[4] |
| Polyethylene terephthalate, Latex | Poly(ethylene oxide) | Desai[9] |
| Polycarbonate, Polyethylene, Polymethylmethacrylate, Polypropylene, Polystyrene, Polysulfone, Polyurethane, Poly(vinyl chloride), Silicone | Photochemical coating | Dunkirk[10] |
| Polyurethane | Sulfonated poly(ethylene oxide) | Han[15] |
| Gold film | Self-assembled monolayers of substituted alkanethiols | Ista[20] |
| Polyurethane | Radiation techniques | Jansen[24] |
| Polyurethane, Polyethylene, Polypropylene, Polyethylene terephthalate | Glow discharge technique | Jansen[22] |
| Polyurethane | Hydrophilic coated catheter (Hydrocath®) | Tebbs[42] |
| Poly(vinyl chloride) | Heparin | Zdanowski[48] |

## II. SURFACE MODIFICATION BY PHYSICOCHEMICAL METHODS

On the basis of the above results, several groups have tried to change the surface chemistry of polymer materials in order to reduce the force of attraction between microorganisms and the biomaterial (Table 1). Special interest has been set on the alteration of surface hydrophilicity, i.e., surface tension or surface energy.

Tebbs et al.[42] compared the adherence of five *S. epidermidis* strains to a polyurethane catheter and to a commercially available hydrophilic, poly(vinyl pyrrolidone)-coated polyurethane catheter (Hydrocath®, British Viggo, Swindon, UK). Adhesion of three strains was considerably reduced by the coated catheters. Our own approaches to developing anti-infective materials comprised the modification of polymer surfaces by radiation or glow discharge techniques, e.g., the hydrophilic 2-hydroxyethylmethacrylate (HEMA) was covalently bonded to a polyurethane surface by means of radiation grafting, leading to a reduced in vitro adhesion of *S. epidermidis*.[24] A photochemical coating of polymers was used by Dunkirk et al.[10] demonstrating that the coating reduced adhesion of a variety of bacterial strains.

One method of achieving hydrophilic surfaces is by modification of polymer materials with poly(alkylene oxides) or substituted poly(alkylene oxides). Thus Bridgett et al.[4] studied the adherence of three isolates of *S. epidermidis* to polystyrene surfaces which were modified with a copolymer of poly(ethylene oxide) and poly(propylene oxide). A substantial reduction in bacterial adhesion was achieved in vitro with all surfactants tested, independent of the poly(ethylene oxide) or poly(propylene oxide) block lengths. Similiar results were found by Desai et al.[9] who investigated the adhesion of *S. epidermidis*, *S. aureus* and *Pseudomonas aeruginosa* to polyethylene terephthalate films which were surface-modified with poly(ethylene oxide). They observed reductions between 70 and 95% in adherent bacteria as compared to the untreated polymer.

Han et al.[15] improved the effectiveness of a polyurethane copolymer by using sulfonated poly(ethylene oxide) as surfactant. The sulfonated poly(ethylene oxide) surface showed higher water uptake than poly(ethylene oxide) or the polyurethane itself, indicative of increased hydrophilicity. The sulfonated material showed significantly lower adhesion of *S. epidermidis* than both the polyurethane control and a sample which was modified with poly(ethylene oxide). Baumgartner et al.[2] incorporated glycerophosphorylcholine (GPC) as a chain extender in a series of poly(tetramethylene oxide)-based polyurethane block copolymers. Water absorption was increased with GPC content. In a radial flow chamber, utilizing automated video microscopy, decreased bacterial adhesion was found on the GPC-containing materials compared to other functionalized polyurethanes both in the absence of and after pre-adsorption with plasma proteins.

Ista et al.[20] studied the attachment of *S. epidermidis* and a marine organism (*Deleya marina*) to the surface of self-assembled monolayers firmed by the adsorption of substituted alkanethiols on gold films. This technique allows the generation of surfaces with a high density of the surfactant. The alkanethiol chains were terminated with hexa(ethylene glycol), methyl, carboxylic acid or fluorocarbon groups. The adhesion of both test organisms was studied by phase contrast microscopy under flow conditions. Again, the poly(ethylene oxide) group – the hexa(ethylene glycol) – showed a very low adhesion with a 99.7% reduction of attachment for both organisms when compared to the most fouled surface for each microorganism. On the other surfaces, *S. epidermidis* and *D. marina* were shown to exhibit very different attachment which responses to the wettability of the substratum. While the attachment of *S. epidermidis* correlated positively with surface hydrophilicity, *D. marina* showed a preference for hydrophobic surfaces. These results make clear that not only the hydrophilicity of the polymer device influences the adhesion of microorganisms but also the surface properties of the bacterium itself.

A parameter which considers the surface tension (or hydrophilicity) of the polymer material, the bacterium, and the surrounding medium is the free enthalpy of adhesion. We used this parameter in a study where the adherence of *S. epidermidis* to a variety of polymers with different surface properties, generated by means of the glow discharge technique, was investigated.[22] Although no influence of the hydrophilicity of the surface modified materials on bacterial adherence was noticed, a strong correlation between the free enthalpy of adhesion and adherence was observed. We found that adhesion of the bacterium to the modified materials decreased with increasing negative free enthalpy values. We could prove a certain minimum number of the adherent *S. epidermidis* strain if the free enthalpy of adhesion had positive values. These results suggest that in vitro there seems to exist a certain minimum number of adherent bacteria independent of the nature of the polymer, so that in reality a "zero adherence" seems impossible. However, with the described materials a reduction in bacterial adhesion could be achieved. Modified polymers with negative surface charge allow for bacterial adherence close to the adherence minimum.

**Table 2. Modification with Antimicrobial Substances Different from Antibiotics**

| Device | Modification | 1st Author[Ref.] |
|---|---|---|
| Polyurethane, Silicone | Silver | Boeswald[3] |
| Central venous catheter (Hydrocath®) | Silver | Gatter[13] |
| Polyurethane | Parabens | Golomb[14] |
| Modified polyurethane | Silver | Jansen[22] |
| Central venous catheter (Hydrocath®) | Iodine | Jansen[23] |
| Silicone urinary catheter | Silver oxide | Johnson[25] |
| Ethylvinyl acetate, Polyethylene, Polypropylene, Poly(4-methyl-1-pentene) | IRGASAN® | Kingston[27] |
| Modified poly(vinyl fluoride) | Iodine | Kristinsson[29] |
| Latex urinary catheter | Silver | Liedberg[30] |
| Collagen cuff | Silver | Maki[34] |
| Polyurethane catheter | Chlorhexidine and silver sulfadiazine | Maki[35] |
| Silicone, Poly(vinyl chloride), Teflon, Butyl rubber | Sputter coating | McLean[36] |
| Swan-Ganz pulmonary artery catheter | Benzalkonium chloride Oligodynamic iontophoresis-enhanced material | Mermel[37] Milder[38] |
| Silicone catheter | Electrically generated silver | Raad[39] |
| Silicone catheter | Ion implantation | Sioshansi[41] |

Also, heparin-coated poly(vinyl chloride) decreases the adherence of *S. epidermidis*, *S. aureus* and *E. coli* compared to the uncoated material.[48] A precoating with human plasma reduced the adhesion of the tested species to plain poly(vinyl chloride) but did not affect the binding to modified poly(vinyl chloride). However, after precoating with plasma, the heparin-coated material showed a higher binding of *S. aureus* than the unmodified polymer, which might possibly be due to bridging effects of fibronectin or other plasma proteins. The study emphasizes that the adhesion to a biomaterial is also influenced by plasma proteins. An et al.[1] coated titanium surfaces with bovine serum albumin using carbodiimide, a crosslinking agent. Only 10% of the crosslinked albumin decayed off the surface during a 20-d incubation period and throughout the experiment bacterial adherence was reduced.

## III. MODIFICATION WITH CHEMICAL SUBSTANCES EXHIBITING ANTIMICROBIAL PROPERTIES

To reduce foreign body infections substances with antimicrobial properties different from those of antibiotics were used in developing new catheter materials (Table 2). Golomb et al. embedded parabens in polyurethane by a solvent cast method. The incorporated drug decreased the number of *S. epidermidis* colony forming units on the polymer surface considerably.[14] The disinfectant IRGASAN® incorporated into several polymers showed a reduction in polymer-associated infections in rabbits as well.[27] Most Swan–Ganz pulmonary artery catheters have heparin bonded to the surface with benzalkonium chloride to reduce thrombosis. Because the benzalkonium chloride has intrinsic anti-microbial activity, heparin-bonded catheters exhibit activity against a wide range of microbial pathogens, including *Candida albicans*.[37]

We developed a technique to load a central venous catheters with iodine. Adherence of various microorganisms (*Staphylococcus* sp., *E. coli*, or *Candida* sp.) to the catheter was completely inhibited for the duration of iodine release.[23] Also, iodine was complexed to polyvinylfluoride films which were grafted with *N*-vinylpyrrolidone.[29] In in vitro experiments no viable microorganisms on these materials could be detected at least for 5 d.

Among metals with antimicrobial activity silver has raised the interest of many investigators because of its good antimicrobial action and low toxicity. Sioshansi et al.[41] used the technique of ion implantation to deposit silver-based coatings on a silicone rubber which thereafter demonstrated antimicrobial activity. Also silver–copper surface films, sputter-coated onto catheter materials, showed antibacterial activity against *Pseudomonas aeruginosa* biofilm formation.[36] We developed an antimicrobial polymer by binding silver ions to an acid-modified, negatively-charged polyurethane surface.[22] The number of adherent *S. epidermidis* cells in vitro was reduced compared to the native modified polymer. Also a hydrophilic central venous catheter was loaded with silver.[13] The device showed good antimicrobial efficacy in a stationary and a dynamic model with different microorganisms. Boeswald et al.[3] have described the impregnation of poly-urethane and silicone with low concentrations of silver by two different methods. In in vitro tests these catheters led to a reduction of colonization with several microorganisms and they are now being tested in a clinical trial.

Liedberg et al. investigated the interaction between silver alloy-coated urinary catheters and *Pseudomonas aeruginosa*. The in vitro biofilm formation was suppressed,[30] and in a randomized clinical study the incidence of catheter-associated urinary tract infection was reduced.[31] In a prospective clinical trial involving 482 hospitalized patients, Johnson et al.[25] used a silver oxide-coated urinary catheter. The incidence of catheter-associated urinary tract infection was similar in recipients of a silver-catheter or a control silicone catheter. However, the coated catheter reduced catheter-related infections in the subgroup of women not receiving other antimicrobials.

Maki et al. have developed a silver-impregnated cuff attached to the subcutaneous segment of central venous catheters. This cuff acts as tissue-interface barrier preventing the migration of microorganisms along the catheter from the surface to the intradermal part of the catheter. Clinical trials have shown the effectiveness of this commercially available device.[34,12] Because of the biodegradable nature of the collagen cuff, the antimicrobial activity is short-lived.

Raad et al. used a catheter in which silver ions are electrically generated at the subcutaneous segment of a catheter in order to have a long durability in preventing the migration of the microorganisms along the catheter. In a rabbit model this iontophoretic silver catheter showed its efficacy in preventing colonization with *S. aureus*.[39]

A newly developed polymer containing silver is the so-called "oligodynamic ionto-phoresis-enhanced" material.[38] The polymer is impregnated with silver, platinum and carbon particles. The silver and platinum particles act as electrodes in a battery-like chemistry, releasing a steady flow of silver which provides an effective colonization resistance. By controlling the particle size and concentration different silver release profiles can be achieved which could be successful in the prevention of catheter-related infections.

One of the first commercially available catheters impregnated with silver was the ARROWgard blue™ (Arrow International, Reading, PA, USA). This central venous catheter is coated with a combination of chlorhexidine and silver sulfadiazine. There are several clinical studies dealing with this device. In comparison to an uncoated device, Maki et al.[35] could prove the effectiveness of the chlorhexidine/silver sulfadiazine catheter in a randomized, controlled clinical trial with 158 participants. The impregnated materials were less likely to be colonized than the control catheter (13.5% vs 24.1%) and blood-stream infection also was reduced (1.0% vs 4.6%). In another randomized controlled trial in the surgical intensive care unit of a university hospital, 157 uncoated triple-lumen catheters and 151 devices coated with chlorhexidine/silver sulfadiazine were investigated.[18] Impregnated catheters were effective in reducing the rate of bacterial growth on either the tip or the intradermal segment as compared with control catheter (40% vs 52%, p = 0.04). However, there was no difference in the incidence of catheter related bacteremia (3.8% vs 3.3%). There are other studies which could not prove the effectiveness of the impregnated catheter to prevent bloodstream infections. Logghe et al.[32] performed a prospective double-blind randomized controlled trial in order to investigate the effectiveness of the ARROWgard blue™ in patients suffering from hematologic malignancy treated by chemotherapy through a central venous catheter. A total of 680 catheters were inserted, of which 338 were antiseptic impregnated. There was no statistically significant difference between the overall rates of bloodstream infection for impregnated and control catheters (14.5% vs 16.3%). The incidence of catheter-related infection was also similiar in both groups (5% vs 4.4%). Moreover, there is a case of a 28-yr-old male patient reported who developed anaphylactic shock possibly due to the contact with the chlorhexidine/silver sulfadiazine impregnated central venous catheter.[43] The chlorhexidine was confirmed as the causative agent of the anaphylactic shock. Nevertheless systems impregnated with chemical substances exhibiting animicrobial properties seem to be effective in the prevention of foreign body infections.

## IV. CONCLUSIONS

The initial step in the pathogenesis of foreign-body infections is the adhesion of micro-organisms to the medical device. Several studies showed that modification of polymer surfaces leads to a reduction of bacterial adhesion, but a "zero adhesion" could not be reached.

Nevertheless, the adhesion of bacteria to a biomaterial in vivo is also influenced by factors not only depending on the nature of the synthetic material. It must be assumed that specific interactions occur between bacteria and a polymer surface, e.g., mediated by bacterial adhesins.[7,44,45] Further, preadsorption of albumin decreases bacterial adhesion while fibronectin promotes adherence.[6,8,19,26,47] Contact-activated platelets[46] also increase bacterial adhesion. Therefore, polymers which favour the adsorption of albumin or prevent platelet adhesion seem to be useful in the prevention of foreign-body infections.

Modification of polymers by physicochemical methods should be an effective tool to create materials showing altered interactions with proteins or host cells, resulting in a further reduction of foreign body infections. This goal has not been reached satisfactorily. So far none of the modified polymers have been used in clinical applications with the exception of the polyvinylpyrrolidone-coated Hydrocath.

Incorporation of chemical substances with antimicrobial properties into the catheter material results in polymeric drug release systems which primarily reduce or prevent the colonization of the polymer surface. Due to the limited action of most of these release systems, this approach seems to be particularly useful for short-term or mid-line catheters and implants, preventing "early onset infections." The emergence of resistant micro-organisms and the risk of side effects like anaphylactic shock might be a disadvantage of these systems; however, in contrast to antibiotics the antimicrobial substances incorporated in the discussed systems are normally not used for the therapy of bacterial or fungal infections. The investigations mentioned above show that there are promising approaches in the development of antiinfective polymer materials, but more detailed studies are necessary to evaluate the efficiency of the new polymers in the clinical application.

## REFERENCES

1. An YH, Stuart GW, McDowell SJ, et al: Prevention of bacterial adherence to implant surfaces with a crosslinked albumin coating *in vitro*. *J Orthop Res* 14:846–9, 1996
2. Baumgartner JN, Yang CZ, Cooper SL: Physical property analysis and bacterial adhesion on a series of phosphonated polyurethanes. *Biomaterials* 18:831–7, 1997
3. Boeswald M, Girisch M, Greil J, et al: Antimicrobial activity and biocompatibility of poly-urethane and silicone catheters containing low concentrations of silver: a new perspective in prevention of polymer-associated foreign-body-infections. *Zentralbl Bakteriol* 283:187–200, 1995
4. Bridgett MJ, Davies MC, Deneyer SP: Control of staphylococcal adhesion to polystyrene surfaces by polymer surface modification with surfactants. *Biomaterials* 13:411–6, 1992
5. Carballo J, Ferreirós CM, Criado MT: Factor analysis in the evaluation of the relationship between bacterial adherence to biomaterials and changes in free energy. *J Biomat Appl* 7: 130–1, 1992
6. Cheung AL, Fischetti VA: The role of fibrinogen in staphylococcal adherence to catheters *in vitro*. *J Infect Dis* 161:1177–86, 1990
7. Christensen GD, Simpson WA, Bisno AL, et al: Adherence of slime-producing strains of *Staphylococcus epidermidis* to smooth surfaces. *Infect Immun* 37:318–26, 1982
8. Delmi M, Vaudaux P, Lew DP, et al: Role of fibronectin in staphylococcal adhesion to metallic surfaces used as models for orthopedic devices. *J Orthop Res* 12:432–8, 1994
9. Desai NP, Hossainy SF, Hubbell JA: Surface-immobilized polyethylene oxide for bacterial repellence. *Biomaterials* 13:417–20, 1992
10. Dunkirk SG, Gregg SL, Duran LW, et al: Photochemical coatings for the prevention of bacterial colonization. *J Biomat Appl* 6:131–55, 1991
11. Ferreirós CM, Carballo J, Criado MT, et al: Surface free energy and interaction of *Staphylococcus epidermidis* with biomaterials. *FEMS Microbiol Lett* 35:89–94, 1989
12. Flowers RH, Schwenzer KJ, Kopel RF, et al.: Efficacy of an attachable subcutaneous cuff for the prevention of intravascular catheter related infection. A randomized, controlled trial. *JAMA* 261:878–83, 1989
13. Gatter N, Kohnen W, Jansen B: *In vitro* efficacy of a hydrophilic central venous catheter loaded with silver to prevent microbial colonization. *Zentralbl Bakteriol* 287:157–69, 1998

14. Golomb G, Shpigelman A: Prevention of bacterial colonization on polyurethane *in vitro* by incorporated antibacterial agent. *J Biomed Mater Res* 25:937–52, 1991

15. Han DK, Park KD, Kim YH: Sulfonated poly(ethylene oxide) – grafted polyurethane copolymer for biomedical applications. *J Biomater Sci Polym Ed* 9:163–74, 1998

16. Harkes G, Feijen J, Dankert J: Adhesion of *Escherichia coli* on to a series of poly-(methacrylates) differing in charge and hydrophobicity. *Biomaterials* 12:853–60, 1991

17. Hattori R, Hattori T: Adsorptive phenomena involving bacterial cells and an anion exchange resin. *J Gen Appl Microbiol* 31:147–63, 1989

18. Heard SO, Wagle M, Vijayakumar E, et al: Influence of triple-lumen central venous catheters coated with chlorhexidine and silver sulfadiazine on the incidence of catheter-related bacteremia. *Arch Intern Med* 158:81–7, 1998

19. Herrmann M, Suchard SJ, Boxer LA, et al: Thrompospondin binds *Staphylococcus aureus* and promotes staphylococcal adherence to surfaces. *Infect Immun* 59:279–88, 1991

20. Ista LK, Fan H, Baca O, et al: Attachment of bacteria to model solid surfaces: oligo (ethylene glycol) surfaces inhibit bacterial attachment. *FEMS Microbiol Lett* 142:59–63, 1996

21. Jansen B: Current approaches to the prevention of catheter-related infections. In: Seifert H, Jansen B, Farr BM, eds. *Catheter-Related Infections*. Marcel Dekker Inc., New York, NY, 1997:411–36

22. Jansen B, Kohnen W: Prevention of biofilm formation by polymer modification. *J Ind Microb* 15: 391–6,1995

23. Jansen B, Kristinsson KG, Jansen S, et al: *In vitro* efficacy of a central venous catheter complexed with iodine to prevent bacterial colonization. *J Antimicrob Chemother* 30:135–9, 1992

24. Jansen B, Schareina S, Steinhauser H, et al: Development of polymers with anti-infective properties. *Polymer Mater Sci Eng* 5743–6, 1987

25. Johnson JR, Roberts PL, Olson RJ, et al: Prevention of urinary tract infection with a silver oxide-coated urinary catheter: clinical and microbiologic correlates. *J Infect Dis* 162:1145–50, 1990

26. Keogh JR, Eaton JW: Albumin binding surfaces for biomaterials. *J Lab Clin Med* 124:537–45, 1994

27. Kingston D, Seal DV, Hill ID: Self-disinfecting plastics for intravenous catheters and prosthetic inserts. *J Hyg Camb* 96:185–98, 1986

28. Kohnen W, Jansen B, Ruiten D, et al: Novel anti-infective biomaterials by polymer modification. In: Gebelein CG, Carraher CE Jr, eds. *Biotechnology and Bioactive*. Plenum, New York, NY, 1993:317–26

29. Kristinsson KG, Jansen B, Treitz U, et al: Antimicrobial activity of polymers coated with iodine-complexed polyvinylpyrrolidone. *J Biomat Appl* 5:173–84, 1991

30. Liedberg H, Ekman P, Lundeberg T: *Pseudomonas aeruginosa*: adherence to and growth on different urinary catheter coatings. *Int Urol Nephrol* 22: 87–92, 1990

31. Liedberg H, Lundeberg T: Silver alloy coated catheters reduce catheter-associated bacteriuria. *Br J Urol* 65:379–81, 1990

32. Logghe C, van Ossel C, D'Hoore W, et al: Evaluation of chlorhexidine and silver-sulfadiazine impregnated central venous catheters for the prevention of bloodstream infection in leukemic patients: a randomized controlled trial. *J Hosp Infect* 37:145–56, 1997

33. Ludwicka A, Jansen B, Wadström T, et al: Attachment of staphylococci to various synthetic polymers. *Zbl Bakt Hyg A* 256:479–89, 1984

34. Maki DG, Cobb L, Garman JK, et al: An attachable silver-impregnated cuff for prevention of infection with central venous catheters. *Am J Med* 85: 07–14, 1988

35. Maki DG, Stolz SM, Wheeler S, et al: Prevention of central venous catheter-related bloodstream infection by use of an antiseptic-impregnated catheter. A randomized, controlled trial. *Ann Intern Med* 127:257–66, 1997

36. McLean RJ, Hussain AA, Sayer M, et al: Antibacterial activity of multilayer silver-copper surface films on catheter material. *Can J Microbiol* 39:895–9, 1993

37. Mermel LA, Stolz SM, Maki DG: Surface antimicrobial activity of heparin-bonded and antiseptic-impregnated vascular catheters. *J Inf Dis* 167:920–4, 1993

38. Milder F: Device-related nosocomial infection – reducing infection with antimicrobial materials and coatings. *Medical Device Technology* 10:34–9, 1999

39. Raad I, Hachem R, Zermeno A, et al: Silver iontophoretic catheter: a prototype of long-term anti-infective vascular access device. *J Inf Dis* 173:495–8, 1996

40. Reid G, Hawthorn LA, Eisen A, et al.: Adhesion of *Lactobacillus acidophilus, Escherichia coli* and *Staphylococcus epidermidis* to polymer and urinary catheter surfaces. *Coll Surf* 42:299–311, 1989

41. Sioshansi P: New processes for surface treatment of catheters. *Artif Org* 18:266–71, 1994

42. Tebbs SE, Elliott TSJ: Modification of central venous catheter polymers to prevent *in vitro* microbial colonisation. *Eur J Clin Microbiol Infect Dis* 13:111–7, 1994

43. Terazawa E, Nagase K, Masue T, et al: [Anaphylactic shock associated with a central venoous catheter impregnated with chlorhexidine and silver sulfadiazine.] *Masui* 47:556–561, 1998

44. Timmermann CP, Fleer A, Besnier JM, et al: Characterization of a proteinaceous adhesin of *Staphylococcus epidermidis* which mediates attachment to polystyrene. *Infect Immun* 59:4187–92, 1991

45. Tojo M, Yamashita N, Goldmann DA, et al: Isolation and characterization of a capsular polysaccharide adhesin from *Staphylococcus epidermidis. J Infect Dis* 157:13–22, 1988

46. Wang I, Anderson JM, Marchant RE: *Staphylococcus epidermidis* adhesion to hydrophobic biomedical polymer is mediated by platelets. *J Infect Dis* 167:329–6, 1993

47. Yu JL, Ljungh A, Andersson R, et al: Promotion of *Escherichia coli* adherence to rubber slices by adsorbed fibronectin. *J Med Microbiol* 41:133–8, 1994

48. Zdanowski Z, Koul B, Hallberg E, et al: Influence of heparin coating on *in vitro* bacterial adherence to poly(vinyl chloride) segments. *J Biomater Sci Polym Ed* 8:825–32, 1997

# 37

# Antimicrobial Agent Incorporation
# for Preventing Bacterial Adhesion

## Rabih O. Darouiche[1] and Issam I. Raad[2]

*[1]Center for Prostheses Infection, Department of Physical Medicine and Rehabilitation,
and Infectious Disease Section, Dept of Medicine, Baylor College of Medicine and VAMC,
Houston, TX, USA, [2]Section of Infectious Diseases, Department of Medical Subspecialties,
The University of Texas M.D. Anderson Cancer Center, Houston, TX, USA*

## I. INTRODUCTION

Medical devices are indispensable in the modern care of patients. Despite adherence to sterile guidelines for the insertion and maintenance of medical devices, infection remains the most common serious complication of indwelling medical devices. Infections related to medical devices account for nearly half of all nosocomial infections.[35] The contribution of medical devices to nosocomial infection is particularly prominent with certain devices, such as vascular and urinary catheters. For instance, vascular catheters account for most cases of nosocomial bloodstream infections,[24] and catheter-related urinary tract infection is the most common nosocomial infection in health care institutions.[35] In addition to causing serious medical complications, infection of medical devices is very expensive to manage. For instance, the extra cost of treating one episode of catheter-related bloodsteam infection in a critically ill patient was estimated to be $28,690 per survivor, and each such episode resulted in an additional average stay of 6.5 d in the intensive care unit.[23]

The major medical and economic burdens emanating from infectious complications of medical devices have propelled the development of novel antimicrobial devices in an effort to reduce the risk of device-related infection. Although numerous antimicrobial devices have been suggested to guard against device-related infection, only few have been reported to be protective in vivo. Since the majority of cases of device-related infection emanate from the use of vascular and urinary catheters, most technologies that incorporate antimicrobial agents onto the medical device have focused on the prevention of infection of these two types of catheters. In this chapter, we will discuss only the antimicrobial vascular and urinary catheters that had been shown in animal and/or clinical studies to protect against catheter-related infection (Table 1).

## II. ANTIMICROBIAL CENTRAL VENOUS CATHETERS

We have witnessed over the last decade a new era in the prevention of infections associated with vascular catheters. This era of heightened interest in developing novel

*Handbook of Bacterial Adhesion: Principles, Methods, and Applications*
Edited by: Y. H. An and R. J. Friedman © Humana Press Inc., Totowa, NJ

**Table 1. Antimicrobial Catheters with Reported Efficacy In Vivo**

| Central venous catheters | Bladder catheters |
|---|---|
| Dipping surfactant-treated catheters in antibiotics | Catheters coated with silver |
| Catheters coated with chloerhexidine | Catheters coated with nitrofurazone |
| Catheters coated with minocycline and rifampin | |
| Catheters coated with silver | |

antimicrobial catheters was accurately predicted by Maki and colleagues who reported in 1988 that "binding of a nontoxic antiseptic or antimicrobial to the entire catheter surface, or incorporation of such substance into the catheter material itself, may ultimately prove to be the most effective technologic innovation for reducing the risk of device-related infections."[18] Sherertz and colleagues established a solid foundation for our current understanding of the antiinfective mechanism of antimicrobial-coated vascular catheters by demonstrating a correlation between in vitro and in vivo findings.[31,32] Using a modified Kirby–Bauer technique, they initially examined in vitro the zones of inhibition by catheters coated with a variety of antimicrobial agents, including chlorhexidine, dicloxacillin, clindamycin, and fusidic acid. Later, they demonstrated the quantitative relationship between the size of the zone of inhibition in vitro and the concentration of bacteria cultured from the subcutaneous segments of the antimicrobial-coated catheters in a rabbit model of *Staphylococcus aureus* infection of percutaneously inserted catheters.[31] They concluded that antimicrobial-coated catheters with zones of inhibition ≥10–15 mm were likely to be efficacious in preventing colonization of indwelling catheters in the rabbit model.

### A. Dipping Surfactant-Treated Vascular Catheters in Antibiotic Solutions

This approach of noncovalent bonding of negatively charged antibiotics to cationic surfactants (such as benzalkonium chloride) on the surface of the device was initially designed by Harvey and colleagues to render vascular grafts anti-infective.[9] The same principle was applied later to bind negatively charged antibiotics to a cationic surfactant (tridodecyl methyl ammonium chloride: TDMAC) on the surface of treated vascular catheters.[37] The clinical efficacy of this antimicrobial coating approach was initially evaluated in a prospective, randomized clinical trial that entailed immersing TDMAC-treated central venous catheter (CVC) in a solution of cefazolin just prior to catheter insertion.[15] Although cefazolin-immersed catheters were about 7-fold less likely to be colonized than untreated catheters, the efficacy of this approach in protecting against catheter-related bloodstream infection could not be directly assessed in that clinical trial because there were no cases of bloodstream infection in either group of patients.

Using the same principle, bedside dipping of TDMAC-treated CVC in a solution of vancomycin was reportedly associated with about 25% reduction in the rate of catheter colonization, as compared with untreated catheters.[36] As with cefazolin-treated catheters, the ability of vancomycin-treated catheters to prevent catheter-related bloodstream infection was not clinically assessed. Furthermore, vancomycin-treated catheters were more likely than untreated catheters to be colonized with Gram-negative bacteria and

*Candida* organisms. A somewhat similar approach was used to bind another glycopeptide antibiotic, namely teicoplanin, to polyvinylpyrrolidone on the surface of a hydrophilic catheter (Hydrocath) which was shown to prevent percutaneous staphylococcal infection in a mouse model.[30] In addition to the potential for fungal superinfection made posible by dipping catheters in a solution of antibiotic that lacks any antifungal activity, this bedside approach is time consuming, rather impractical and provides only short-lived antimicrobial activity (up to few days). This explains the much larger interest in developing catheters that are precoated with appropriate antimicrobial agent(s).

### B. Vascular Catheters Coated with Chlorhexidine

Polyurethane central venous catheters coated with a low concentration of chlorhexidine then sterilized by gamma irradiation provided only small (<10 mm) zones of inhibition in vitro.[34] Not unexpectedly, such chlorhexidine-coated catheters failed to demonstrate anti-infective efficacy in an animal study using the established rabbit model and in a prospective, randomized, multicenter clinical trial.[34] The results of the rabbit study suggested that higher concentrations of chlorhexidine on the surface of CVC could be associated with better efficacy.[34]

Although coating of catheters with chlorhexidine alone was found to be ineffective,[34] central venous catheters coated with the combination of chlorhexidine and silver sulfadiazine was shown in in vitro and animal studies to reduce bacterial adherence.[8] The largest prospective, randomized, multicenter study of short-term (mean duration of placement of about one week) CVC coated with the combination of chlorhexidine and silver sulfadiazine demonstrated that these coated catheters are two-fold less likely to become colonized and at least fourfold less likely to cause bloodstream infection, as compared with uncoated.[19] Although several smaller clinical trials showed only a nonsignificant trend toward lower rates of catheter-related bloodstream infection among catheters coated with chlorhexidine and silver sulfadiazine, none had sufficient power to examine differences in the rates of catheter-related bloodstream infection.[2,11,22] It is important to note that although the combination of chlorhexidine and silver sulfadiazine may be synergistic in vitro, the advantage of adding silver sulfadiazine to chlorhexidine has not been demonstrated in vivo. In fact, catheters coated with silver sulfadiazine were ineffective in reducing catheter-related infection in the rabbit model when compared with uncoated catheters.[33]

### C. Vascular Catheters Coated with Minocycline and Rifampin

Unlike other antibiotics, such as the glycopeptides and cephalosporins which are considered as drugs of choice for treating established systemic infections, the two antibiotics minocycline and rifampin are rarely used therapeutically as such and, therefore, are appropriate for prophylactic use on coated catheters.[4] The approach of coating central venous catheters with this novel combination of minocycline and rifampin has been very successful. In vitro studies demonstrated that catheters coated with minocycline and rifampin provide a broad-spectrum inhibitory activity against Gram-positive bacteria, Gram-negative bacteria and *C. albicans*.[25] Using the established rabbit model of subcutaneous infection by *S. aureus*,[32] catheters coated with minocycline and rifampin were shown to protect against infection, as compared with uncoated catheters.[26] A prospective, randomized, multicenter clinical trial showed that short-term

CVC coated with minocycline and rifampin are threefold less likely to be colonized than uncoated catheters and prevent the occurrence of catheter-related bloodstream.[28]

When examined both in vitro and in animals, catheters coated with minocycline and rifampin exhibited superior antimicrobial activity, as compared with catheters coated with chlorhexidine and silver sulfadiazine.[25,26] Rifampin is more active than other antibiotics against the slowly growing bacteria within the biofilm surrounding the device.[39] Furthermore, catheters coated with minocycline and rifampin provide antimicrobial activity along both the external (to protect against migration of skin organisms) and internal catheter surfaces (to protect against migration of bacteria causing contamination of catheter hub), whereas the antimicrobial activity of catheters coated with chlorhexidine and silver sulfadiazine is limited to only the external surface of the catheter. These differences and others may help explain the results of a recently completed large, prospective, randomized, multicenter clinical trial which demonstrated that short-term CVC coated with minocycline and rifampin are threefold less likely to be colonized and 12-fold less likely to cause catheter-related bloodstream infection than catheters coated with chlorhexidine and silver sulfadiazine.[5] Not unexpectedly,[4] clinical trials have demonstrated no evidence for developing antibiotic resistance among bacteria recovered from patients who had received catheters coated with minocycline and rifampin.[5,28]

### D. Vascular Catheters Coated with Silver

Silver can be applied either to the subcutaneous cuff of the tunneled CVC or to the surface of the catheter. The use of the silver-impregnated subcutaneous cuff is intended to provide both an antimicrobial deterrent (due to the silver) and a mechanical barrier (due to the subcutaneously placed collagen cuff) to the migration of bacteria along the external surface of the catheter. Although the use of the silver-impregnated subcutaneous cuff was reported in two prospective, randomized clinical trials[6,18] to reduce the incidence of infection among critically ill patients with indwelling short-term CVC (duration of placement = 5.6–9.1 d), a recent prospective but nonrandomized study failed to demonstrate a beneficial effect.[10] Owing to the biodegradable nature of the collagen cuff to which the silver ions are chelated, the antimicrobial activity of the silver-impregnated subcutaneous cuff is short-lived. Moreover, the use of the silver-impregnated subcutaneous cuff offers no protection against contamination of the catheter hub, a major source of organisms causing infection of long-term CVC. These factors help explain why the use of the silver-impregnated subcutaneous cuff failed to protect against infection of CVC with longer duration of placement (mean = 20 d)[3] or the long-term, tunneled Hickman (Bard Access Systems, Salt Lake City, Utah) catheters.[7]

Silver-containing catheters may also be constructed by depositing silver ions on the external surface of the catheter. The results of in vitro evaluation and a nonrandomized clinical study suggested that such silver-coated catheters are less prone to bacterial colonization.[1] However, the clinical efficacy of silver-coated CVC was not confirmed in prospective randomized studies. In fact, a recent prospective, randomized clinical trial demonstrated a nonsignificant trend for higher rates of catheter-related infection when using silver-coated tunneled hemodialysis catheters vs uncoated catheters.[38] Because of the strong adhesion of silver molecules to the surface of such coated catheters, the silver ions are not appreciably released from the surface of the catheter to produce zones of inhibition and, therefore, are unlikely to provide antimicrobial activity against bacteria embedded in the biofilm. The production of an effective zone of inhibition by antimicrobial-coated catheters serves to inhibit adherence of organisms not only to the surface of the catheter but to the biofilm layer around the indwelling catheter which contains a variety of host-derived adhesins, such as fibronectin, fibrinogen, fibrin, etc.[12] Unlike catheters coated with silver alone, the silver iontophoretic catheter which allows leaching of silver ions from the surface of the catheter produces zones of inhibition against most potential pathogens.[27] In the rabbit model, the silver iontophoretic catheter was shown to be significantly more protective against infection than catheters coated with chlorhexidine and silver sulfadiazine.[27]

## III. ANTIMICROBIAL BLADDER CATHETERS

### A. Bladder Catheters Coated with Silver

As with silver-coated vascular catheters, the antiinfective efficacy of bladder catheters coated with silver alone remains controversial. Earlier small-sized clinical trials had indicated that silver-coated bladder catheters significantly reduce the rate of catheter-associated bacteriuria when compared with uncoated catheters.[16,17] However, more recent, larger-sized clinical trials demonstrated similar overall rates of catheter-associated bacteriuria in patients who received silver-coated bladder catheters vs. uncoated catheters.[13,29] Since these silver-coated bladder catheters do not produce zones of inhibition, they may not provide antimicrobial activity against bacteria present in the aqueous micromilieu adjacent to the surface of the indwelling catheter.

Because the hydrophobic nature of bacteria and the surfaces of most catheters enhance the adsorption of bacteria onto the catheter surface, it is generally thought that rendering the catheter surface hydrophilic may decrease bacterial colonization of the catheter. In that regard, a preliminary report from a prospective, randomized, double-blind study indicated that a novel silver hydrogel-coated bladder catheter with a mean duration of catheterization of 6.5 d reduces the rate of catheter-associated bacteriuria by 30%, as compared with uncoated catheters.[21] This reduction in the overall rate of catheter-associated bacteriuria was primarily due to protection against Gram-positive bacteria and, to a lesser extent, against yeast. However, there was a nonsignificant trend for higher rates of Gram-negative bacteriuria in patients who received the silver-hydrogel-coated bladder catheter.

### B. Bladder Catheters Coated with Nitrofurazone

Coating of bladder catheters with nitrofurazone, a nitrofuran derivative chemically related to nitrofurantoin, results in some zones of inhibition in vitro against a variety of potential urinary pathogens.[14] However, nitrofurazone-coated bladder catheters produce no zones of inhibition against organisms such as *Pseudomonas, Serratia, Proteus,* and *Candida* species. A preliminary report of a recently completed prospective, randomized clinical trial indicated that the use of nitrofurazone-coated bladder catheters in newly catheterized patients who received a catheter within 7 d of admission to the hospital was associated with a fivefold decrease in the rate of bacterial catheter-associated urinary tract infection, as compared with uncoated catheters.[20] However, the decrease in the overall rate of catheter-related urinary tract infection was statistically insignificant. Moreover, the efficacy of the nitrofurazone-coated catheter was demonstrated neither in patients beyond 7 d nor in long-term hospitalized patients who are likely to be colonized by multiresistant bacteria and yeast.

## IV. CONCLUSIONS AND FUTURE WORK

Although several antimicrobial catheters have been proposed to reduce the rate of catheter-related infection, only few have been proven to be protective in vivo. At present, short-term, polyurethane central venous catheters coated with minocycline and rifampin along both the external and internal surfaces appear to be the most clinically protective catheters against catheter colonization and catheter-related bloodstream infection. In contrast to the remarkable progress made in the prevention of vascular catheter-related infection, the clinical results obtained with various antimicrobial bladder catheters have not been very promising, particularly in those who require bladder catheterization for more than 1 wk.

So far, most antimicrobial coating measures have focused on short-term vascular and urinary catheters. There is a pressing need to explore the clinical efficacy of antimicrobial-coated long-term silicone vascular catheters. The development of an anti-infective long-term central venous catheter may obviate the need for the expensive and time-consuming practice of subcutaneous tunneling of vascular catheters. As with vascular catheters, the ultimate goal for bladder catheters would be to develop an antimicrobial catheter that can protect against infection in patients who require long-term catheterization, such as spinal cord-injured patients, nursing home residents and elderly persons.

## REFERENCES

1. Bambauer P, Mestres P, Schiel R, et al: New surface-treatment technologies for catheters used for extracorporeal detoxification methods. *Dialysis Transplantation* 24:228–37, 1995
2. Ciresi D, Albrecht RM, Volkers PA, et al: Failure of an antiseptic bonding to prevent central venous catheter-related infection and sepsis. *Am Surgeon* 62:641–6, 1996
3. Clementi E, Mario O, Arlet, G, et al: Usefulness of an attachable silver-impregnated cuff for the prevention of catheter-related sepsis (CRS). Abstract #460. *Program and Abstracts of the 31st Interscience Conference on Antimicrobial Agents and Chemotherapy*, Chicago, IL,1990
4. Darouiche RO, Raad II, Bodey GP, et al: Antibiotic susceptibility of staphylococcal isolates from patients with vascular catheter-related bacteremia: potential role of the combination of minocycline and rifampin. *Int J Antimicrob Agents* 6:31–6, 1995
5. Darouiche RO, Raad II, Heard SO, et al: A comparison of two antimicrobial-impregnated central venous catheters. *N Engl J Med* 340:1–8, 1999
6. Flowers RH III, Schwenzer KJ , Kopel RF, et al: Efficacy of an attachable subcutaneous cuff for the prevention of intravascular catheter-related infection. *JAMA* 261:878–83, 1989
7. Groeger JS, Lucas AB, Coit D, et al: A prospective randomized evaluation of silver-impregnated subcutaneous cuffs for preventing tunneled chronic venous access catheter infections in cancer patients. *Ann Surg* 218:206–10, 1993
8. Greenfield JI, Sampath L, Popilskis SJ, et al: Decreased bacterial adherence and biofilm formation on chlorhexidine and silver sulfadiazine-impregnated central venous catheters implanted in swine. *Crit Care Med* 23:894–900, 1995
9. Harvey RA, Greco RS: The noncovalent bonding of antibiotics to a polytetrafluoroethylene-benzalkonium graft. *Ann Surg* 194:642–7, 1981

10. Hasaniya NW, Angelis M, Brown MR, et al: Efficacy of subcutaneous silver-impregnated cuffs in preventing central venous catheter infections. *Chest* 109:1030–2, 1996

11. Heard SO, Wagle M, Vijayakumar E, et al: The influence of triple-lumen central venous catheters coated with chlorhexidine/silver sulfadiazine on the incidence of catheter-related bacteremia: a randomized, controlled clinical trial. *Arch Intern Med* 158:81–7, 1998

12. Hermann M, Vaudaux PE, Pittet D, et al: Fibronectin, fibrinogen, and laminin act as mediators of adherence of clinical staphylococcal isolates to foreign material. *J Infect Dis* 158:693–710, 1988

13. Johnson JR, Roberts PL, Olson RJ, et al: Prevention of catheter-associated urinary tract infection with silver oxide-coated urinary catheter: clinical and microbiologic correlates. *J Infect Dis* 162:1145–50, 1990

14. Johnson JR, Berggren T, Conway AJ: Activity of a nitrofurazone matrix urinary catheter against catheter-associated uropathogens. *Antimicrob Agents Chemother* 37:2033–6, 1993

15. Kamal GD, Pfaller MA, Rempe LE, et al: Reduced intravascular catheter infection by antibiotic bonding. *JAMA* 265:2364–8, 1991

16. Liedberg H, Lundberg T, Ekman P: Refinement in the coating of urethral catheters reduces the incidence of catheter-associated bacteriuria: an experimental and clinical study. *Eur Urol* 17:236–40, 1990

17. Liedberg H, Lundeberg T: Silver alloy coated catheters reduce catheter-associated bacteriuria. *Brit J Urol* 65:379–81, 1990

18. Maki DG, Cobb L, Garman JK, et al: An attachable silver-impregnated cuff for prevention of infection with central venous catheters: a prospective randomized multicenter trial. *Am J Med* 85:307–14, 1988

19. Maki DG, Stolz SM, Wheeler S, et al: Prevention of central venous catheter-related bloodstream infection by use of an antiseptic-impregnated catheter: a randomized, controlled study. *Ann Intern Med* 127:257–66, 1997

20. Maki DG, Knasinski V, Tambyah PA: A prospective investigator-blinded trial of a novel nitrofurazone-impregnated indwelling urinary catheter. Abstract #M49. *The 7th Annual Meeting of the Society for Healthcare Epidemiology of America*, St. Louis, MO, 1997

21. Maki DG, Knasinski V, Halverson K, et al: A novel silver-hydrogel-impregnated indwelling urinary catheter reduces CAUTIs: a prospective double-blind trial. Abstract #10. *The 8th Annual Meeting of the Society for Healthcare Epidemiology of America*, Orlando, FL, 1998

22. Pemberton LB, Ross V, Cuddy P, et al: No difference in catheter sepsis between standard and antiseptic central venous catheters: A prospective randomized trial. *Arch Surg* 131:986–9, 1996

23. Pittet D, Tarara D, Wenzel RP: Nosocomial bloodstream infection in critically ill patients: excess length of stay, extra costs, and attributable mortality. *JAMA* 271:1598–1601, 1994

24. Raad II, Bodey GP: Infectious complications of indwelling vascular catheters. *Clin Infect Dis* 15:197–210, 1992

25. Raad I, Darouiche R, Hachem R, et al: Antibiotics and prevention of microbial colonization of catheters. *Antimicrob Agents Chemother* 39:2397–400, 1995

26. Raad I, Darouiche R, Hachem R, et al: The broad spectrum activity and efficacy of catheters coated with minocycline and rifampin. *J Infect Dis* 173:418–24, 1996

27. Raad I, Hachem R, Zermeno A, et al: Silver iontophoretic catheter: A prototype of a long-term anti-infective vascular access device. *J Infect Dis* 173:495–8, 1996

28. Raad I, Darouiche R, Dupuis J, et al: Central venous catheters coated with minocycline and rifampin for the prevention of catheter-related colonization and bloodstream infections: a randomized, double-blind trial. *Ann Intern Med* 127:267–74, 1997

29. Riley DK, Classen DC, Stevens LE, et al: A large randomized clinical trial of a silver impregnated urinary catheter: lack of efficacy and staphylococcal superinfection. *Am J Med* 98:349–56, 1995

30. Romano E, Berti M, Goldstein BP, et al: Efficacy of a central venous catheter (Hydrocath) loaded with teicoplanin in preventing subcutaneous staphylococcal infections in mouse. *Zbl Bakt* 279:426–3, 1993

31. Sherertz RJ, Carruth A, Hampton AA, et al: Efficacy of antibiotic-coated catheters in preventing subcutanous *Staphylococcus aureus* infection in rabbits. *J Infect Dis* 167:98–106, 1993

32. Sherertz RJ, Forman DM, Solomon DD: Efficacy of dicloxacillin-coated polyurethane catheters in preventing subcutaneous *Staphylococcus aureus* infection in mice. *Antimicrob Agents Chemother* 33:1174–8, 1989

33. Sherertz R, Hu Q, Clarkson L, et al: The chlorhexidine (CH) on Arrow catheters (C) may be more important than AG sulfadiazine (AGSD) at preventing C-related infection. Abstract #1622. *Program and Abstracts of the 33rd Interscience Conference on Antimicrobial Agents and Chemotherapy*, New Orleans, LA, 1993

34. Sherertz RJ, Heard SO, Raad II, et al: Gamma radiation-sterilized, triple-lumen catheters with a low concentration of chlorhexidine were not efficacious at preventing catheter infections in intensive care unit patients. *Antimicrob Agents Chemother* 40:1995–7, 1996

35. Stamm WE: II. Prevention of infections. Infections related to medical devices. *Ann Intern Med* 89:764–9, 1978

36. Thornton J, Todd NJ, Webster NR: Central venous line sepsis in the intensive care unit. A study comparing antibiotic coated catheters with plain catheters. *Anesthesia* 51:1018–20, 1996

37. Trooskin SZ, Donetz AP, Harvey RA, et al: Prevention of catheter sepsis by antibiotic bonding. *Surgery* 97:547–51, 1985

38. Trerotola SO, Johnson MS, Shah H, et al: Tunneled hemodialysis catheters. Use of a silver-coated catheter for prevention of infection — a randomized study. *Radiology* 207:491–6, 1998

39. Widmer AF, Frei R, Zimmerli W: Correlation between *in vivo* and *in vitro* efficacy of antimicrobial agents against foreign body infections. *J Infect Dis* 162:96–102, 1990

# 38

# Studying Bacterial Adhesion
# to Antibiotic Impregnated Polymethylmethacrylate

**Charles E. Edmiston, Jr.[1] and Michael P. Goheen[2]**

[1]*Surgical Microbiology Research Laboratory, Department of Surgery,*
*Medical College of Wisconsin, Milwaukee, WI USA,*
[2]*Department of Pathology, Indiana University School of Medicine, Indianapolis, IN, USA*

## I. INTRODUCTION

### A. Bacterial Adherence to Bioinert and Bioactive Materials

When infection of an implantable orthopedic device occurs, its impact is often catastrophic, often necessitating additional operative procedures and associated with significant patient morbidity. The Gram-positive staphylococci are the most frequent isolates recovered from orthopedic device-related infections. Several investigators have documented that both *Staphylococcus aureus* and *Staphylococcus epidermidis* will adhere to a myriad of bioinert and bioactive materials. In some studies, *S. aureus* has been shown to adhere tenaciously to metal surfaces such as steel and titanium.[7,21] However, other studies have shown that *S. epidermidis* demonstrates a preferential adherence depending upon the type of material and charge characteristics of the orthopedic substrate.[16] Metals such as titanium tend to exhibit a negative surface charge, similar to the bacterial cell and therefore, fewer cells adhere to this surface than to a bioactive substrate such as hydroxyapatite. Hydroxyapatite has a surface that contains positive charges originating from free calcium and phosphate compounds. Hydroxyapatite, therefore, tends to exhibits greater staphylococcal adherence compared to negatively charged metals. Metal surfaces that are oxidized will alsdo exhibit variable binding characteristics that may influence microbial adherence.[9] While surface charge characteristics can influence microbial adherence to a biomedical device, other factors promote microbial persistence and involve the elaboration of a bacterial exopolysaccharide substance (biofilm). Staphylococcal strains that produce a biofilm are sheltered from the host cellular and humoral immune defense mechanisms, in addition these strains also exhibit a recalcitrance to conventional antimicrobial therapy.[4,7,20]

### B. Inhibiting Microbial Adherence

Efforts to prevent microbial adherence to the surface of biomedical devices have been an area of active if not controversial research. Previous studies have demonstrated that

*Handbook of Bacterial Adhesion: Principles, Methods, and Applications*
Edited by: Y. H. An and R. J. Friedman © Humana Press Inc., Totowa, NJ

metals such as titanium (bioinert) support fewer adherent staphylococcal cells than bioactive substances such as hydroxyapatite. However, the use of bioactive materials in implant surgery is highly desirable since they enhance biocompatibility of the device, reducing host efforts to biodegrade (acute phase process) or sequester (fibrous encapsulation) the implant from the adjacent tissues. Coating a bioinert substrate such as titanium (net [−] charges) with hydroxyapatite (net [+] charge) enhances tissue biocompatibility while discouraging microbial adherence. Titanium surfaces coated with hydroxyapatite demonstrate an adherence index (1.4) that is similar to titanium alone (1.0) but significantly less than hydroxyapatite (29.0).[16] Alternatively, coating an implantable material with host tissue proteins often demonstrates a variable impact on microbial adherence. When titanium is coated with crosslinked (carbodiimide) albumin there was a significant reduction (85%) in the adherence of S. epidermidis over a 20 d test period.[2] However, coating vascular prostheses (Dacron) with albumin does not result in uniform inhibition of staphylococcal adherence.[19] Albumin inhibition of staphylococcal adherence appears to be influenced by strain selection and composition of the coated biomaterial.

Over the past 30 yr a large body of literature has accumulated on the use of antibiotic impregnated materials in orthopedic surgery. The first reported use of antibiotic incorporation into orthopaedic bone cement occurred in Europe in 1970 and was used to prevent deep organ space infection after total hip or knee arthroplasty.[5] Presently, this technique is also used to treat chronic osteomyelitis and other bone and deep tissue infections.[13,17] Over the past 20 yr, the use of antibiotic-impregnated polymethylmethacrylate (PMMA) has gained wide acceptance among orthopaedic surgeons in the United States with the aminoglycosides and various cephalosporins being some of the most commonly used agents.[22]

### C. Studies Using Polymethylmethacrylate (PMMA)

Polymethylmethacrylate (PMMA) is an adhesive material that is available in various commercial preparations. To incorporate antimicrobials into this material, the drug powder (parenteral formulation) is added to the powdered cement and thoroughly mixed before the addition of liquid methylmethacrylate. Since this procedure generates significant heat (100°C) the antibiotic must be stable during the polymerization process or anti-infective activity will be lost. Previous studies have shown that if prepared properly the addition of antibiotic to bone cement has no effect on color, polymerization time, or tensile strength.[15] Furthermore, several studies have demonstrated that a wide variety of antimicrobial agents can be effectively incorporated into PMMA, eluting from the surface of the bone cement at concentrations exceeding the minimal inhibitory concentration (MIC) of most anticipated pathogens.

The elution of various antibiotics has been measure under both in vitro and in vivo (animal model) conditions. The elution of antibiotic from bone cement is dependent upon solvent conditions, pH and stability of the drug following polymerization. In the tissues, drug concentrations above the MIC have been demonstrated for cefazolin, ciprofloxacin, clindamycin, tobramycin and vancomycin.[1] The following discussion focuses on an in vitro model for studying antibiotic impregnated PMMA as a strategy to prevent microbial adherence.

## II. MATERIAL AND METHODS

### A. Microbial Populations and Handling

While the staphylococci are the predominant microbial pathogens in orthopedic device-related infections, Gram-negative bacteria such as *E. coli* may also be associated with implant-associated infections. Studies conducted in our laboratory have used *Staphylococcus epidermidis* RP62A a copious slime producer, *Staphylococcus epidermidis* LDE2, *Staphylococcus epidermidis* MCW8A (both clinical laboratory strains: weakly slime-positive and slime-negative, respectively, by alcian blue staining), and *Escherichia coli* (ATCC 25922) a laboratory reference strain. If the focus of the study involves using an exopolysaccharide (slime) producing strain then efforts must be made to validate the slime producing capabilities of the desired test isolate. Alcian blue staining is a quick and dependable method for ascertaining slime production in clinical isolates. Reference strains recovered from cold storage (–70°C) must be thawed, plated to Trypticase Soy Agar with 5% sheep's blood and incubated for 24 h at 35°C to check for viability and purity.

### B. Drug Susceptibility and Preparation of Bone Cement

The minimal inhibitory concentration (MIC) and minimal bactericidal concentration (MBC) for all test isolates must first be determined by standard methods (NCCLS, 1988). In the present model, five antibiotics were selected for testing in PMMA: cefazolin (Sigma-Aldrich, St. Louis, MO), ciprofloxacin (Bayer, West Haven, CT), clindamycin (Upjohn, Kalamazoo, MI), tobramycin (Sigma-Aldrich, St. Louis, MO) and vancomycin (Sigma-Aldrich, St. Louis, MO). The antibiotic powder was thoroughly mixed with low viscosity methylmethacrylate powder (Howmedica, Houston, TX) in the following concentration: 4 g/40 g powder — cefazolin; 4 g/40 g powder — ciprofloxacin; 5 g/40 g powder — clindamycin; 8 g/40 g powder — tobramycin; and 4 g/40 g powder — vancomycin. Liquid copolymer PMMA was slowly added to the antibiotic-cement mixture and the suspension distributed to silastic molds to produce a test surface, 5 mm$^2$. The antibiotic impregnated blocks are allowed to cure for 4 h, individually packaged and sterilized in an autoclave to remove any potential surface contamination. Steam sterilization for 30 min does not appear to significantly reduce the inhibitory activity of the test anti-infectives. Following sterilization, the antibiotic-impregnated blocks can be stored aseptically at –70°C until needed.

### C. In Vitro Assay of Antimicrobial-Impregnated Bone Cement

It is desirable to validate the antimicrobial activity of the antibiotic impregnated blocks prior to forming the adherence assay. The polymerization process is associated with heat generation (100°C) and the in vitro assay of antimicrobial activity assures that active agent is present in the PMMA blocks. One block each of the antibiotic-impregnated cement is placed in 1 mL of physiological buffered saline (PBS) and incubated at 35°C. After 24 h, the original PBS is decanted and following two buffered washings, 1 mL of fresh PBS is added to the tube and placed back in the incubator. A total of 3 PBS changes are made over a 4-d test interval. At 2, 8, 24, 48, and 96 h (4 d), an aliquot of the PBS (200 μL) is saved for analysis. Antimicrobial quantitation is determined by biological assay.[1] Five samples are tested at each time interval to determine mean elution

**Table 1. Minimal Inhibitory Concentration (MIC) and Minimal Bacteriocidal Concentration (MBC) of Selected Anti-Infectives Against Staphylococcal and Gram-Negative Test Strains**

| Strains | Antibiotic | MIC | MBC |
|---|---|---|---|
| *S. epidermidis* RP62A | Cefazolin | 2.0 | 16 |
| | Ciprofloxacin | 0.5 | 1 |
| | Clindamycin | 0.25 | 0.5 |
| | Tobramycin | 1 | 4 |
| | Vancomycin | 2 | 16 |
| *S. epidermidis* LDE2 | Cefazolin | 0.5 | 2 |
| | Ciprofloxacin | 0.25 | 0.5 |
| | Clindamycin | 0.5 | 1 |
| | Tobramycin | 1 | 8 |
| | Vancomycin | 1 | 4 |
| *S. epidermidis* MCW8A | Cefazolin | 0.25 | 1 |
| | Ciprofloxacin | 0.12 | 0.25 |
| | Clindamycin | 0.5 | 2 |
| | Tobrmycin | 0.5 | 2 |
| | Vancomycin | 1 | 4 |
| *E. coli* ATCC 25922 | Ciprofloxacin | 0.12 | 0.12 |
| | Tobramycin | 0.06 | 0.12 |

concentrations. In addition, control samples obtained from antibiotic free cement are incubated in PBS and run with all test samples to validate that PMMA is free of inhibitory activity.

### D. Bacterial Adherence to PMMA Blocks

The microbial test strains are incubated in Trypticase Soy Broth for 18 h at 35°C, washed twice in PBS and resuspended in PBS containing 1.0% dextrose. The final test inoculum is adjusted to 7.0 $\log_{10}$ colony forming units/mL spectrophotometrically and 2 mL of this inoculum added to a series of tubes containing the antibiotic impregnated blocks. The blocks are incubated for 2 h followed by aspiration of the test inoculum. The blocks are gently washed (2 times) in the tubes and 2 mL of fresh PBS with 1.0% dextrose added. At 2, 8, 24, 48, and 96 h three blocks each are removed, sonicated (20 kHz) for 10 min and the sonicate serially diluted in PBS prior to plating on TSA.[6] Quantitative recovery is determined after 48 h incubation at 35°C and final counts reported as $\log_{10}$ cfu/mm$^2$ bone cement. Antibiotic free control blocks are prepared as per impregnated blocks and tested for comparison of microbial recovery.

### E. SEM and TEM

At selected time intervals, blocks are removed and prefixed in 2.5% glutaraldehyde with 0.1 M sodium cacodylate buffer. An alternatively prefixation involves using 75 mM lysine, 0.075% ruthenium red, 2.5% glutaraldehyde in 0.1 M sodium cacodylate buffer to enhance visualization of the exopolysaccharide (slime) layer of selected staphylococcal

**Table 2. In Vitro Analysis of Cefazolin, Ciprofloxacin, Clindamycin,
Tobramycin and Vancomycin Mean Antimicrobial Activity
from Polymethylmethacrylate-Impregnated Blocks**

| Antibiotic | Concentration ± S.D. (µg/mL)[a] | | | | |
|---|---|---|---|---|---|
| | 2 h | 8 h | 24 h | 48 h | 96 h |
| Cefazolin | 67.5 ± 19.2 | 95.5 ± 17.2 | 119.2 ± 21.1 | 64.6 ± 6.5 | 18.6 ± 5.5 |
| Ciprofloxacin | 44.8 ± 8.1 | 87.4 ± 11.1 | 71.5 ± 7.5 | 37.4 ± 11.9 | 15.6 ± 7.1 |
| Clindamycin | 34.5 ± 7.5 | 89.8 ± 9.7 | 140.6 ± 25.7 | 81.3 ± 13.5 | 42.1 ± 10.1 |
| Tobramycin | 81.8 ± 11.1 | 112.6 ± 16.8 | 175.8 ± 30.9 | 110.5 ± 17.3 | 69.9 ± 8.8 |
| Vancomycin | 56.9 ± 14.6 | 99.3 ± 10.4 | 197.9 ± 27.6 | 100.1 ± 9.9 | 27.6 ± 5.5 |

[a]Five 0.1 mL samples were analyzed at each time interval.

isolates.[8] Specimens for TEM are post-fixed in $OsO_4$ for 4 h followed by several buffer washes (three times), dehydrated through a graded alcohol series, infiltrated overnight with Spurrs/Polybed resin under vacuum, sectioned, stained with uranyl acetate and lead citrate followed by TEM observation. Specimens for scanning electron microscopy are processed using the osmium thiocarbohydrazide osmium (OTO) technique.[12] Total OTO fixation time is approximately 6 h (4.5 h 1% $OsOm_4$; 15 min 1% thiocarbohydrazide at 45°C; 45 min final $O_sOm_4$) with 10 buffer rinses between each of the three steps. Following dehydration the PMMA blocks are critical point dried in $CO_2$, mounted on stubs with silver paint and examined by SEM.

## III. OBSERVATIONS AND DISCUSSION

### A. *Microbial Adherence to Antibiotic-Impregnated PMMA*

Table 1 demonstrated the MIC and MBC values for four test microbial strains. The values reported for all five agents reflect concentrations, which are clinically achievable in the host. Table 2 demonstrates the in vitro diffusion from antibiotic-impregnated polymethylmethacrylate blocks. The eluted concentrations of all 5 antibiotics at 24 h represented levels that are 4 to 20 times the therapeutic level routinely achieved clinically following delivery of the maximal parenteral dose in man. The values reported at 96-h represent concentrations higher than the minimal bactericidal concentration reported for the 4 test strains in Table 3. The antibiotic concentrations eluted for ciprofloxacin and tobramycin at 96 h represent a value 50 to 100 times the MBC for *E. coli*. Table 3 documents the mean microbial recovery from antibiotic impregnated PMMA following 2-h incubation in a standardized inoculum of the test strains. Microbial recovery from control (antibiotic free PMMA) at 8 h ranged from 3.7 to 4.5 $log_{10}$ CFU/mm². No significant decrease in mean microbial recovery was observed with antibiotic-impregnated PMMA colonized with *Staphylococcus epidermidis* RP62A, a well-documented slime producing strain. Mean microbial recovery at 96 h was similar to 8 and 96 h control values. However, antibiotic-impregnated blocks colonized with LDE2 demonstrate a lower mean microbial recovery at 96 h compared to RP62A. At 96 h, surface colonization was significantly ($p \leq 0.05$) reduced on ciprofloxacin, clindamycin and tobramycin-impregnated PMMA blocks compared to control (antibiotic free PMMA).

**Table 3. Mean Microbial Recovery
from Antibiotic-Impregnated Polymethylmethacrylate (PMMA)**[a]

| Organism | Drug[b] | Recovery ± S.D. ($\log_{10}$ cfu/mm$^2$) | | | |
|---|---|---|---|---|---|
| | | 8 h | 24 h | 48 h | 94 h |
| *S. epidermidis* RP62A[c] | Cef | 3.2 ± 1.2 | 4.2 ± 1.6 | 3.8 ± 2.1 | 3.5 ± 1.0 |
| | Cip | 2.4 ± 0.9 | 3.1 ± 1.1 | 2.9 ± 1.9 | 2.6 ± 0.8 |
| | Cld | 4.1 ± 1.5 | 3.8 ± 0.8 | 3.3 ± 1.9 | 3.0 ± 1.2 |
| | Tob | 3.1 ± 0.5 | 2.7 ± 0.5 | 2.5 ± 0.7 | 2.9 ± 1.0 |
| | Van | 3.7 ± 1.7 | 3.0 ± 1.2 | 3.5 ± 1.3 | 3.9 ± 1.6 |
| *S. epidermidis* LDE2[d] | Cef | 3.5 ± 1.3 | 3.9 ± 0.8 | 2.5 ± 1.1 | 1.2 ± 0.5 |
| | Cip | 3.8 ± 1.0 | 2.9 ± 1.2 | 1.6 ± 0.5 | 0.7 ± 0.2[g] |
| | Cld | 4.1 ± 2.0 | 3.6 ± 0.7 | 2.2 ± 1.0 | 1.1 ± 0.4[g] |
| | Tob | 2.6 ± 0.7 | 2.4 ± 1.0 | 1.5 ± 0.4 | 0.5 ± 0.2[g] |
| | Van | 3.1 ± 0.9 | 3.4 ± 0.9 | 2.3 ± 0.9 | 1.5 ± 1.0 |
| *S. epidermidis* MCW8A[e] | Cef | 3.3 ± 1.0 | 3.1 ± 1.2 | 1.9 ± 0.5 | 0.2 ± 0.1[g] |
| | Cip | 4.1 ± 1.6 | 3.8 ± 1.6 | 2.9 ± 0.9 | 1.8 ± 0.5 |
| | Cld | 3.0 ± 1.3 | 2.8 ± 0.9 | 2.1 ± 1.2 | 1.1 ± 0.4[g] |
| | Tob | 3.7 ± 0.9 | 3.3 ± 1.9 | 2.6 ± 1.0 | 0.9 ± 0.5[g] |
| | Van | 4.2 ± 1.9 | 3.7 ± 1.5 | 1.8 ± 0.9 | - [g] |
| *E.coli* ATCC 25922[f] | Cip | 4.3 ± 2.2 | 2.5 ± 0.5 | 1.3 ± 0.4[g] | - [g] |
| | Tob | 3.3 ± 1.5 | 3.5 ± 0.8 | 2.1 ± 0.8 | 1.1 ± 0.3[g] |

[a] five blocks were analyzed at each time interval
[b] cefazolin = Cef; ciprofloxacin = Cip; clindamycin = Cld; tobramycin = Tob; vancomycin = Van
[c] PR62A controls: 4.2 ± 1.7 (8 h) and 3.7 ± 1.9 (96 h)
[d] LDE2 controls: 3.9 ± 1.5 (8 h) and 3.5 ± 1.0 (96 h)
[e] MCW8A controls: 3.8 ± 1.9 (8 h) and 3.6 ± 2.0 (96 h)
[f] ATCC 25922 controls: 4.2 ± 2.1 (8 h) and 4.1 ± 2.3 (96 h)
[g] $p \leq 0.05$.

This effect was more pronounced with MCW8A colonized blocks. At 96 h all antibiotic-impregnated blocks revealed a significant ($p \leq 0.05$) decrease in surface contamination compared to 96 h controls. At 96 h, vancomycin-impregnated PMMA blocks were culture negative following sonication. Similar findings were observed with *E. coli* inoculated PMMA. Mean microbial recovery from both ciprofloxacin and tobramycin-impregnated blocks was significantly ($p \leq 0.05$) reduced at 96 h. In addition, *E. coli* recovery from ciprofloxacin-impregnated PMMA was significantly reduced at 48 h compared to control blocks.

### B. TEM and SEM Observations

Figure 1A documents the colonization of cefazolin-impregnated PMMA at 24 h post-inoculation. Figure 1B demonstrates the gelatinous colonies on the surface of the cefazolin-impregnated cement at 48 h postinoculation. The high magnification of RP62A (cefazolin-impregnated PMMA) at 96 h reveals exopolysaccharide (fuzzy) substance embedded between individual cocci (Fig. 1C). The transmission electron micrograph in

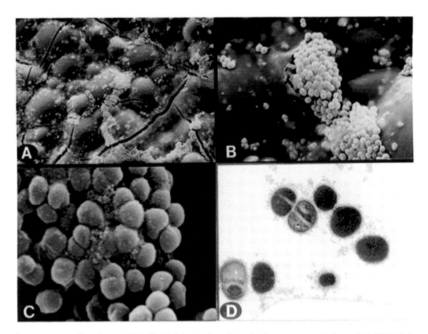

**Figure 1.** (A) colonization of PMMA block by *Staphylococcus epidermidis* RP62A at 24 h postinoculation, cefazolin-impregnated PMMA (Mag. ×640). (B) Strain RP62A (slime+) producing large gelatinous colonies on surface of cefazolin-impregnated PMMA at 48 h postinoculation (Mag. ×2500). (C) High magnification of RP62A at 96 h (cefazolin-impregnated PMMA) demonstrates exopolysaccharide matrix between individual bacterial cells (Mag. ×10,000). (D) TEM image of RP62A colonizing the surface of ciprofloxacin impregnated resin block at 24 h. Cells were stained with L-lysine-ruthenium red-glutaraldehyde to enhance visualization of exopolysaccharide (slime) layer (Mag. ×18,000).

Figure 1D demonstrates the glycocalyx substance associated with individual RP62A cells and adherent to the surface of the ciprofloxacin-impregnated PMMA blocks at 24 h. Figure 2A documents the abundant surface colonization of vancomycin-impregnated PMMA (4.2 $\log_{10}$ cfu/mm$^2$) at 24 h post-inoculation. Figure 2B is a lower magnification taken at 96 h postinoculation, showing few cells contaminating the surface of the vancomycin-impregnated PMMA. Figure 3A demonstrates abundant *E. coli* adherent to control cement at 8 h. Figure 3B reveals the surface of tobramycin-impregnated PMMA at 48 h demonstrating a reduction in *E. coli* adherence compared to control blocks (Fig. 3A).

## C. Previous Studies with Antibiotic-Impregnated Cement

Several studies have measured the diffusion of antibiotics from orthopedic bone cement (polymethylmethacrylate) in a variety of diluents including distilled water, serum, buffered PBS, and synovial fluid.[10,11] The values obtained in the present protocol represent values within the midrange of published studies. The levels obtained with all test antimicrobials exceeded the breakpoint for all drugs tested. However, the data suggests that staphylococcal strains that produce copious amounts of slime, are less susceptible to antibiotic impregnated PMMA than poor slime (or negative) producers.

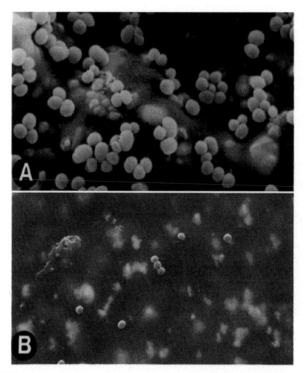

**Figure 2.** (A) SEM image of staphylococcal strain LDE2 (slime) colonizing surface of vancomycin-impregnated PMMA at 24 h post-inoculation (Mag. ×5000). (B) Lower magnification of vancomycin-impregnated PMMA at 96 h post-inoculation demonstrating fewer cells adherent to the cement surface (Mag. ×2100).

This verifies recent studies suggesting that bacterial slime or biofilm promotes phenotypic resistance compared to biofilm-deficient strains.[18,20] It was apparent that nonslime-producing strains were much more susceptible to test antimicrobials than RP62A. This also appears to be the case with other microbial populations such as the Gram-negative cocci (*E. coli*). The production of copious amounts of exopolysaccharide may physically prevent antibiotics from entering the contaminated site. In addition, bacteria existing within the biofilm matrix express varying levels of metabolic competence, which may appear to mimic bacterial resistance.[3]

### D. Problems with Current Antibiotic-Impregnated Protocols

There are two potential problems associated with the present protocol and previous investigations of microbial adherence to antibiotic impregnated orthopaedic cement. First, lack of standardization has resulted in the use of various dosing (antibiotic loading) schedules that have been derived empirically. Therefore, it is difficult to make a comparative evaluation of this technique with other published studies without careful consideration of all selected variables (drug, dosing and surgical procedure). Second, antibiotic selection often focuses upon agents that cover anticipated microbial pathogens and little thought is given to the metabolic competence of these organisms once adherent to PMMA. Following adherence to an inert biomaterial surface, microorganisms undergo a down-regulation of their metabolism, increasing the generational turnover time.[4,7]

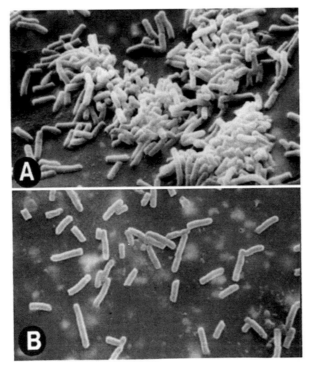

**Figure 3.** (A) *E. coli* colonizing surface of control PMMA block at 8 h (Mag. ×3100). (B) SEM image of *E. coli* on surface of tobramycin-impregnated PMMA demonstrating fewer adherent cells at 48 h postinoculation compared to controls (Mag. ×3100).

Therefore, in designing an in vitro model of microbial adherence to antibiotic-impregnated materials, there must be a sufficient period of exposure to the impregnated substrate to compensate for the diminished metabolic capacity of the test micro-organisms. We have found a time frame of 4 to 7 d adequate for testing the inhibitory impact of impregnated antibiotics on microbial adherence to PMMA.

Preparation of PMMA samples for transmission or scanning electron microscopy is relatively straightforward. The advantages of using lysine for improved visualization of staphylococcal glycocalyx are discussed in Chapter 15 (Fassel and Edmiston). The principle advantage of the OTO (SEM) technique by Kelley is enhanced sample preservation without the subsequent heat damage that often occurs when applying traditional metal (Au/Pd) coatings. The OTO technique does not reduce sample durability, allowing for long observation times under the electron beam. In addition, the OTO coating does not reduce resolution of staphylococcal exopolysaccharide (glycocalyx) and is an appropriate technique for visualizing microbial adherence to a broad-range of biomaterial surfaces.

## REFERENCES

1. Adams K, Couch L, Cierny G et al: *In vitro* and *in vivo* evaluation of antibiotic diffusion from antibiotic-impregnated polymethylmethacrylate bead. *Clin Orthop* 278:244–52, 1992
2. An YH, Stuart GW, McDowell SJ, et al: Prevention of bacterial adherence to implant surfaces with crosslinked albumin coating *in vitro*. *J Orthop Res* 14:846–9, 1996

3. Anwar H, Dasgupta K, Lam K et al: Tobramycin resistance of mucoid *Pseudomonas aeruginosa* biofilm grown under iron limitations. *J Antimicrob Chemother* 24:647–55, 1989

4. Anwar H, Strap JL, Costerton JW: Establishment of aging biofilms: possible mechanism of bacterial resistance to antimicrobial therapy. *Antimicrob Agents Chemother* 36:1347–51, 1992

5. Buchholz HW, Engelbrecht H: Uber die depotwirkung einiger antibiotica bei vermischung mit dem kunstharz palacos. *Chirurg* 41:511–5, 1970

6. Edmiston CE, Schmitt DD, Seabrook GR: Etiology and microbial pathogenesis of acute and late-onset vascular graft infections. In: Wadstrom T, Eliasson I, Holder I, et al., eds. *Pathogenesis of Wound and Biomaterial-Associated Infections.* Springer-Verlag, Berlin, Germany, 1990:465–78

7. Edmiston CE: Prosthetic devices infections in surgery. In: Nichols RL, Nyhus LM, eds. *Problems in General Surgery: Surgical Sepsis 1993 and Beyond.* Lippincott, Philadelphia, PA, 1993:115–137

8. Fassel TA, Sanger JR, Edmiston CE: Lysine effect on ruthenium red and alcian blue preservation and staining of the staphylococcal gylcocalyx. *Cells Mater* 3:327–36, 1993

9. Gabriel BL, Gold J, Gristina AG, et al: Site-specific adhesion of *Staphylococcus epidermidis* (PR12) in Ti-Al-V metal systems. *Biomaterials* 15:628–34, 1994

10. Goodell JA, Flick AB, Hebert JC et al: Preparation and release characteristics of tobramycin-impregnated polymethylmethacrylate beads. *J Bone Joint Surg* 43:1454–61, 1986

11. Hill J, Klenerman L, Trustey S, et al: Diffusion of antibiotics from acrylic bone-cement *in vitro. J Bone Joint Surg* 59B:197–203, 1976

12. Kelley RO, Dekker RA, Bluemink JD: Ligand-mediated osmium binding: its application in coating biological specimens for SEM. *J Ultrast Res* 45:245–58, 1973

13. Majid SA, Lindberg LT, Gunterbert B, et al: Gentamicin-PMMA bead in the treatment of chronic osteomyelitis. *Acta Orthop Scand* 56:265–8, 1985

14. National Committee for Clinical Laboratory Standards: Methods for dilution antimicrobial susceptibility test for bacteria that grow aerobically, p.895. Document M7-T2, 2nd ed. National Committee for Clinical Laboratory Standards, Villanova, PA, 1988

15. Nelson CL, Griffin FM, Harrison BH, et al: *In vitro* elution characteristics of commercially and noncommercially prepared antibiotic PMMA beads. *Clin Orthop* 284:303–9, 1992

16. Oga M, Arizono T, Sugioka Y: Bacterial adherence to bioinert and bioactive materials studies *in vitro. Acta Orthop Scand* 64:273–6, 1993

17. Shapiro SA: Cranioplasty, ventral body replacement, and spinal fusion with tobramycin-impregnated methylmethacrylate. *Neurosurgery* 28:789–91, 1991

18. Schwank S, Rajacic Z, Zimmerli W, et al: Impact of bacterial biofilm formation on *in vitro* and *in vivo* activities of antibiotics. *Antimicrobial Agents Chemother* 42:895–8, 1998

19. Silverhus DJ, Schmitt DD, Edmiston CE, et al: Adherence of mucin on non-mucin producing staphylococci to preclotted and albumin coated velour knitted vascular grafts. *Surgery* 10:613–9, 1990

20. Souli M, Giamarellou H: Effects of slime produced by clinical isolates of coagulase-negative staphylococci on activities of various antimicrobial agents. *Antimicrob Agents Chemother* 42:939–41, 1998

21. Verheyen CC, Dhert WJ, de Blieck-Hogervorst JM, et al: Adherence to a metal, polymer and composite by *Stapylococcus aureus* and *Staphylococcus epidermidis. Biomaterials* 14:383–91, 1993

22. Wininger DA, Fass RJ: Antibiotic-impregnated cemnet and beads for orthopaedic infections. *Antimicrob Agents Chemother* 40:2675–9, 1996

# 39

# Macromolecule Surface Coating for Preventing Bacterial Adhesion

**Yuehuei H. An, Brian K. Blair, Kylie L. Martin, and Richard J. Friedman**

*Department of Orthopaedic Surgery, Medical University of South Carolina,
Charleston, SC, USA*

## I. INTRODUCTION

One of the most serious complications in implant surgeries is infection, which may lead to complete failure of the implanted device.[3,6] Bacterial contamination of prostheses or bacterial adherence to biomaterial surfaces during implantation surgery (through air or direct contact) or during the postoperative period (hematogenously) is an important step in the pathogenesis of prosthetic infection.[4] It is thought that certain strains of bacteria, particularly coagulase-negative staphylococci (CNS, a major bacteria for prosthetic infection) secrete bacterial slime which forms a biofilm on implants. Once the adhesion occurs on the implant surface, the bacteria make themselves less accessible to the human defense system and significantly increase antibiotic resistance.

Over the years, the implant infection rate has decreased due to improved operating room techniques, such as the ultra clean air system (in the case of total joint replacement surgery) and the use of prophylactic antibiotics.[3] Also, much attention has been paid to surface modification or surface coating of biomaterials in order to reduce the chance of bacterial adherence and subsequent prosthetic infection. Examples of these efforts include material surfaces modified by anodal polarization or photochemical immobilization of antimicrobial peptide or coated with protein, silver, heparin, or salicylates, although these approaches are still in the experimental stage.[6]

Among the various methods available for changing the surface of biocompatible materials, one approach is to attach or coat the surface of the material with appropriate macromolecules of biological origin, such as albumin or heparin. The classical rationale for using albumin to coat blood contacting biomaterial surface is that it neither initiates coagulation nor attracts platelets, thus leading to improved blood compatibility. Many proteins (serum or tissue proteins) have been studied for their effects on bacterial adhesion to material surfaces, including albumin, heparin, fibronectin, fibrinogen, laminin, denatured collagen, and more.[5,7,53,88] They promote or inhibit bacterial adhesion through binding to substrata surfaces, binding to bacterial surfaces, or being present in the liquid medium during the adhesion period. For the latter situation most of the

*Handbook of Bacterial Adhesion: Principles, Methods, and Applications*
Edited by: Y. H. An and R. J. Friedman © Humana Press Inc., Totowa, NJ

proteins inhibited bacteria adhesion,[17,28] possibly affecting bacterial adhesion by their association with the bacterial cell surface, the material surface, or both. Most of the bindings between bacteria and proteins are specific ligand/receptor-like interactions. Proteins may also change the adherent behavior of bacteria by changing bacterial surface physicochemical characteristics.[55,75,76]

So far, the most promising proteins having bacterial repellant ability include heparin and albumin, which are the foci of this article. Several other proteins with potential anti-adhesion properties, such as serum and bacterial surface proteins, are also briefly mentioned. Several proteins showing the effect of antibacterial adhesion are listed in Table 1.

## II. HEPARIN

Heparin contains (1→4)-linked 2-amino-2-deoxy-α-D-glucopyranosyl, α-L-idopyranosyluronic acid, and a small proportion of β-D-glucopyranosyluronic acid residue. The hexosamine and hexuronic acid residues are linked alternately and are partially O-sulfated. Most of the 2-amino-2-deoxy-D-glucosyl residues are N-sulfated, the remainder being N-acetylated Heparin exerts its main blood-anticoagulant activity by binding, and thereby potentiating the inhibitory effect of, the plasma protein anti thrombin.[77] The concept of coating surfaces with heparin began with Gott et al.[32] Since this research began, extensive efforts have been carried out to explore the effects of heparin coating of various biomaterials. Much of the early research focused on developing various methods in which to bind heparin to thrombogenic and nonthrombogenic surfaces alike.[21,22,38,49] Once this hurdle had been accomplished, further studies were done in hopes of developing a new technique of inhibiting bacterial adhesion to biomaterial surfaces using heparin coating.

### A. Heparin Bonding to Biomaterial Surfaces

Various efforts have aimed at achieving the goal of finding the most efficient way in which to bind heparin to the surface of biomaterials. The initial data obtained by Gott suggested that the mechanism of binding by heparin was via the negatively charged sulfate groups of heparin complexed with the quaternary groups contained on the surfaces of materials.[32] Data from this research developed the idea that surfaces coated with heparin are generally and relatively thromboresistant. This concept was the basis for continued research by various other groups. The concept of introducing quaternary groups began to be applied to a variety of surfaces, such as polyethylene and silicone rubber. The work by Yen and Renbaum further developed this method by using polyurethanes containing amino groups which were then quaternized with HCl.[91] Merker et al. used 3-aminopropyltriethoxysilane to react with silicone rubber, and then the amino groups were then quaternized. Much simpler procedures were then developed by Grode whereby treatment of polymers with a solution of tridodecyl-methyl ammonium chloride (TDMAC).[35] Merrill et al. also prepared a cross-linked material in which heparin is covalently bound to a biomaterial surface first by heating a water solution of heparin, polyvinyl alcohol, glutaraldehyde, and an acid catalyst.[54]

Early work by Chang and associates further studied the platelet–surface interaction on a heparin complexes coated thrombogenic surface (heparin-benzokonium-cellulose

## Table 1. Surface-Bond Proteins Having Antibacterial Adhesion Abilities

| Protein | Specific protein | Material coated | Bacteria studied | References |
|---|---|---|---|---|
| Albumin | BSA | Polystyrene | *Pseudomonas* sp. | Fletcher 1976[28] |
| | BSA | HA disks | *Strep. mutans* | Reynolds 1983[75] |
| | BSA | HA beads | *S. sanguis, S. mutans* | Gibbons 1985[31] |
| | BSA | HA beads | *S. mutans, S. sanguis, S. mitis* | Yen 1987[90] |
| | BSA | FEP, CA, glass | *S. mutans, S. sanguis, S. mitis* | Pratt-Terpstra1987[72] |
| | HSA | Silicone catheters | *S. aureus, S. epidermidis* | Espersen 1990[27] |
| | HAS | Teflon, PC, PE | *S. epidermidis* | Carballo 1991[19] |
| | HSA | PTFE | *S. aureus, S. epidermidis* | Zdanowski 1993[93] |
| | | Dacron | *S. aureus, S. epidermidis, E coli* | |
| | HSA | Polyetherurethane | *S. epidermidis* | Keogh 1994[45] |
| | HSA | Ti, Ti-alloy | *Actinomyces, Actinobacillus* | Steinberg 1998[84] |
| Bacterial proteins | Protein A | Silicone | *S. aureus* | Espersen 1990[27] |
| | Clumping factor | Silicone | *S. aureus* | Espersen 1990[27] |
| | Cell membrane | Silicone | *S. aureus, S. epidermidis* | Espersen 1990[27] |
| | SA I/II | HA Beads | *S. mutans* | Munro 1993[59] |
| Heparin | - | Latex, teflon-vinyl | *E. coli* | Ruggieri 1987[78] |
| | - | PMMA IOL | *S. epidermidis* | Arciola 1994[10] |
| | - | PE, polystyrene | *S. aureus, S. epidermidis* | Paulsson 1994[69] |
| | - | PVC | *E. coli, S. aureus* | Nagaoka 1995[61] |
| | - | Silicone | *E. coli, S. aureus* | Homma 1996[42] |
| | - | Polyurethane | *S. aureus, S. epidermidis* | Appelgren 1996[8] |
| | - | PVC, silicone | *S. epidermidis* | Nomura 1997[65] |
| | - | PMMA IOL | *S. epidermidis* | Abu el-Asar 1997[1] |
| | - | PMMA IOL | *S. epidermidis* | Lundberg 1998[50] |
| | - | PMMA | *S. epidermidis* | Schmidt 1998[79] |
| | - | Polyurethane-PEG | *S. epidermidis, E. coli* | Park 1998[66] |
| Gelatin | - | Polystyrene | *Pseudomonas* sp. | Fletcher 1976[28] |
| Kininogen | Human | Polyurethane | *S. aureus* | Nagel 1996[63] |
| Pepsin | - | Polystyrene | *Pseudomonas* sp. | Fletcher 1976[28] |
| Plasma or serum | - | PU, PVC, glass | *S. epidermidis, P. aeruginosa* | Mohammad 1988[56] |
| | Human | Silicone catheters | *S. aureus, S. epidermidis* | Espersen 1990[27] |
| | Human | Teflon catheters | *S. aureus* | Muller 1991[57] |
| | Human | Stainless steel | *S. aureus, E. coli, P. aeruginosa* | Wassall 1997[89] |
| Saliva | Human | Ti, Ti-alloy | *Actinomyces, Actinobacillus* | Steinberg 1998[84] |
| Other proteins | Human IgG | Silicone catheters | *S. aureus, S. epidermidis* | Espersen 1990[27] |
| | Polyclonal IgG | Dacron | *S. epidermidis, E coli* | Zdanowski 1993[93] |
| | | PTFE | *S. epidermidis* | |
| | Poly-L-glutamate | HA disks | *Strep. mutans* | Reynolds 1983[75] |
| | Phosvitin | | | |
| | $\alpha_{s1}$-Casein | | | |
| | β-Casein | | | |
| | κ-Casein | | | |
| | β-lactoglobulin | | | |
| | α-lactalbumin. | | | |

Abbreviations:
CA = cellulose acetate
FEP = fluorethylene propylene copolymer
HA = hydroxyapatite
IOL = intraocular lens
PC = polycarbonate
PE = polyethylene

PMMA = polymethylmethacrylate
PTFE = polytetrafluoroethylene
PU = polyurethane
PVC = polyvinyl chloride
Ti = titanium

nitrate coated),[21] as well as surface radiation grafting of heparin to thrombogenic surfaces.[22] Two-hour gamma irradiated *N,N*-diethylaminoethyl cellulose acetate membranes retained the highest amount of heparin; however, there was a 25% decrease in tensile strength. In vitro studies showed that blood did not clot even after 60 min on this membrane. In vivo studies on dogs showed a significantly higher thromboresistance on the heparinized samples as compared to control samples, indicating these as a promising heparinize nonthrombogenic sufaces.[22]

A heparinized hydrophilic polymer was reported by Idezuki et al.,[43] which is H-RSD, a graftcopolymer composed of ethylene, vinyl acetate, vinyl chloride, polyethylene-glycolmethacrylate, quarternized dimethylaminoethylmethacrylate and ionically bound heparin. It continuously releases heparin from its surface at the rate of approximately 0.004 units/cm$^2$/min when placed in the plasma. It has been proven both experimentally and clinically that H-RSD has excellent antithrombogenic[43,62] and antibacterial adhesion properties.[42,61]

Since the effects of albumin have been determined to provide a bacterial resistant coating to bioactive material, Hennick et al.[37,38] developed a albumin–heparin conjugate and studied its effects on coagulation in vitro. This substance was found to be an inhibitor of coagulation formation, but more importantly, it was shown that the heparin part of the complex was found to be responsible for the neutralization of clotting factors at the blood-material interface.

Later, Larm et al.[49] and Hoffman et al.[40] from the same group developed a method for the covalent binding of heparin to a nonthrombogenic surface via a reducing terminal residue. This new method of binding heparin involved the initial degradation of heparin and then coupling the remaining fragments together via their reducing terminal ends. Later the method was named as "end-point attachment."[64] This study was performed in vitro and resulted in several important effects of heparin binding. Key to the strength of attachment of heparin to the biomaterial surface was determined to be the formation of multiple ionic bonds between amino groups on the surfaces and heparin itself. The increased strength due to the ionic interaction resulted in a decreased motility of heparin once it is bound to the surface. This heparin coating method gives a stable anti-adhesive surface with low activation of complement and decreased bacterial adhesion, resulting lowered infection rates of implantable devices.[8,9]

### B. Antibacterial Adhesion

One major purpose of using heparin surface modification is to inhibit bacterial adhesion to tubular shaped medical devices, such as urinary catheters,[78] central venous catheters,[8] ventriculoperitoneal shunts,[65] or portconnected catheter.[42]

Despite many advances in catheter design and use, the most common cause of hospital-acquired infections is catheterization of the urinary tract. Ruggieri et al. found that coating latex catheter material with TDMAC[35] without heparin resulted in 3.6-fold higher adherence by *E. coli* whereas coating with the TDMAC–heparin complex reduced adherence to less than 10% of control untreated latex. TDMAC–heparin also significantly reduced bacterial adherence to teflon coated latex (Bardex) and vinyl catheter material. Less than 30% of the original heparin was removed after wash periods of up to 1 wk. These results indicate that TDMAC–heparin coating of urethral catheters reduces bacterial adherence and thereby may delay the acquisition of catheter associated urinary tract infection.

The adhesion and growth of two pathogenic bacteria, *E. coli* and *S. aureus,* on the surface of a heparinized hydrophilic polymer (H-PSD as mentioned above) were studied by Nagaoka et al.[61] Heparinized hydrophilic polymer is composed of poly(vinyl chloride) grafted with poly(ethylene glycol) monomethacrylate, diethylaminoethyl methacrylate, and ionically bound heparin. Poly(vinyl chloride) was used as a control. Polymer films were stored in bacterial suspensions under gentle shaking at 37°C for 24 h. The results demonstrated that a large amount of bacterial adhesion and biofilm formation was found on the control surface of poly(vinyl chloride), whereas significant reductions in bacterial adhesion and no biofilm formation were observed on heparinized polymer. Later, they studied the inhibitory effects on bacterial adhesion of a catheter heparinized using a similar technique to be used in patients with malignant obstructive jaundice, a randomized controlled study of indwelling endoprostheses (implantable port-connected heparinized catheters).[42] In vitro examination of the two type of catheters exposed to suspensions of *E. coli* and *S. aureus* was performed using electron microscopy and a luminometer. The formation of a biofilm coated with glycocalyces was found in silicone catheters, but not in the heparinized catheters. In vitro experiments demonstrated little bacterial adhesion to the heparinized surface, but significant formation of biofilm on the silicone catheter surface.

Nomura et al.[65] studied the adhesion by coagulase negative staphylococci in vitro to polyvinyl chloride (PVC), silicone, and to PVC and silicone with end-point (EPA) attached heparin. Bacterial adhesion was quantitated by bioluminescence. Heparinization of silicone and PVC decreased the numbers of adhered bacteria by 23 to 54% and 0 to 43% compared to unheparinized surfaces. Among putative inhibitors tested, suramin, chondroitin sulfate, and fucoidan inhibited adhesion to 81±19, 78±22, and 64±7%, respectively. These findings indicate that hydrophobic interactions play an important role, and heparinization rendering the biomaterial surface hydrophilic is therefore effective to reduce bacterial adhesion. Heparinized polymers incubated with putative inhibitors may be the optimal way to prevent shunt infections. A similar study was reported by Zdanowski et al.[92] with similar results. They found that plain PVC as compared to EPA–PVC bound significantly more cells of all three tested species, *S. aureus, S. epidermidis,* and *E. coli.* Plasma precoating significantly decreased adherence of the tested species to plain PVC but did not affect the binding to EPA–PVC. However, after precoating with human plasma, EPA–PVC compared to plain PVC showed a higher binding of *S. aureus* which might possibly be due to bridging effects of fibronectin or other plasma proteins, interacting with *S. aureus.*

Paulsson et al. studied the adherence of CNS *S. aureus* to surfaces containing various glycosaminoglycans.[64] Results from this study showed that in general, cells of the CNS family showed a greater adherence to polyethylene surfaces than did those from the *S. aureus* family. However, when the surface of the material was heparinized, it was shown that the adherence of the *S. aureus* cells showed a decreased adherence whereas the CNS cells adhered in greater numbers than those of *S. aureus,* and heparinizing the surface seemed to have little effect on their activity. When substances such as vitronectin, laminin, fibronectin, and collagen were preabsorbed into the surfaces, this seemed to increase the binding of *S. epidermidis* to the heparinized surfaces.

The effect of surface heparinization of central venous catheters on bacterial adhesion was studied by Appelgren et al.[8] Adhesion of 17 radiolabeled clinical isolates of staphylococci to catheters was examined in vitro and the outcome of heparinized and control catheters was compared in vivo in patients receiving long-term parenteral nutrition. The results showed that CNS adhered less in vitro to heparinized catheters than to control catheters. Among 32 central venous catheters, or patients who completed the study, catheter-associated bacteremia or fungemia was observed in five patients in the control group ($n = 19$) and in no patient with a heparinized catheter ($n = 13$). Four of 13 catheters in the heparin group were colonized compared with 14 of 19 in the control group. The numbers of organisms found on colonized catheters were larger in the control group than in the heparin group. They concluded that the covalent end point surface heparinization appears to have a great impact on both in vitro and in vivo bacterial colonization of central venous catheters.

Recently, attention has been paid to the potential anti-adhesive effects of heparin coating on intraocular lenses (10 L) and contact lenses. As stated by Arciola et al.,[10] intraocular lenses implanted after cataract removal must not only be biocompatible but also be able to inhibit cell adhesion. Studies have shown that heparin surface coating can inhibit cell adhesion to IOLs thus lowering the incidence of complications and reducing the risk of inflammation and infection.[51,52] At about the same time, Portolés et al. in 1993[71] and Arciola et al. in 1994[10,11] both hypothesized that heparin surface modification of IOLs may also inhibit bacterial adhesion. Their results did show the significant effects of heparin surface treatment in hindering *S. epidermidis* and *S. aureus* adhesion. A later study by Abu el-Asrar et al.[1] with similar design testing on the same lenses also confirmed their findings.

Lundberg et al.[50] also investigated the adhesion of staphylococcal cells to IOLs coated with heparin under in vitro flow conditions, 280 µL/min at 37°C. The intraocular lenses were incubated with human cerebrospinal fluid prior to bacterial challenge. Surface coating with heparin significantly decreased bacterial adhesion of both strains after incubation with cerebrospinal fluid including 0.50% plasma for 12 h. Microscopy showed that more bacteria were present on intraocular lenses without heparin than on intraocular lenses with heparin.

Adherence of bacteria to the surface of contact lenses may play an important role in contact lens intolerance and corneal infections. To decrease the capability of bacteria to adhere to contact lenses Durán et al.[24] incubated two types of soft contact lenses with two strains of *Pseudomonas aeruginosa* for 12 h. When heparin was added to the medium at a concentration of 1000 IU/mL the numbers of bacteria adhering to the contact lenses were significantly fewer than in the controls. Portolés et al.[71] also found that in the presence of heparin, either bound to the IOL surface (heparin-PMMA + PBS) or in the incubation solution (PMMA IOLs + heparin), the adhesion of *S. epidermidis* to the surfaces of IOLs was significantly diminished. Their results suggest that heparin, either included in contact lens solutions or bonded to the surface of the contact lens or IOLs, may decrease the incidence of biomaterial-related infections.

One mechanism for the anti-adhesive effects of heparin coating on bacterial adhesion is that heparin creates hydrophilic surfaces to biomaterials or places a highly hydrated layer between bacteria and biomaterial surface, which is very effective for repelling hydrophobic bacterial strains.[61,79] In the case of *S. epidermidis* adhesion to IOLs, Arciola et al. found that some structural fatty acids of *S. aureus* and *S. epidermidis* undergo significant variations after adhesion to heparin surface modified PMMA. They hypothesized that this bacterial fatty acid modification, the percentage value variation of the fatty acid, and in particular the disappearance of the 18:0 and I-17:0 peaks, may be the mechanism through which heparin alters staphylococcal adhesion in vitro.[10]

There are two studies demonstrated the efficacy of immobilised heparin on skin wound healing[48] and vascular graft healing.[67] These studies may lead to a new concept of heparin coating on implantable devices which has triple functions, preventing blood clotting, accelerating tissue healing or ingrowth, and antibacterial adhesion.

## III. ALBUMIN

Albumin is a simple protein found in both plants and animals (cells, blood, and tissue fluids). Common forms of albumin for clinical treatment and research are commercially available. Most of them are serum albumins, such as human serum albumin (HSA) and bovine serum albumin (BSA). There are also many other commercially available albumins isolated from most research animals including goats, sheep, dogs, pigs, cats, rabbits, turkeys, rats, mice, or chicken eggs (Sigma, St. Louis, MO).

### A. Albumin Bonding to Biomaterials

Most research studies used simple adsorption method to coat experimental material surfaces with albumin.[5,27,68,75] Among earlier reports, Chang coated collodion microencapsulated charcoal in an albumin saline solution, 1 g% bovine albumin fraction V from Sigma, for 15 h at 4°C, then removing the supernatant.[20] Similar methods have been used by our laboratory[5] and others.[81] Another procedure is perfusing tubular membrane materials with albumin solution, as for polyvinyl chloride (PVC) tubings reported by Mulvihill et al[58] and for dialysis membranes (cuprophane and polymethylmethacrylate) by Remuzzi and Boccardo.[74] Uyen et al.[87] studied the adsorption of serum albumin to substrata with a broad range of wettabilities from solutions with protein concentrations between 0.03 and 3.00 mg/mL in a parallel-plate flow cell. Wall shear rates were varied between 20 and 2000/s. The amount of albumin adsorbed in a stationary state was always highest on PTFE, the most hydrophobic material employed and decreased with increasing wettability of the substrata. Increasing stationary amounts of adsorbed albumin were observed with increasing wall shear rates at the lowest protein concentration. Inverse observations were made at the highest protein concentration. Transmission electron micrographs of replicas from the albumin-coated substrata showed that proteins were mostly adsorbed in islandlike structures on the hydrophobic substrata. The study demonstrates that both the amount of adsorbed albumin as well as the surface structure of the adsorbed proteins are regulated by the substratum wettability.

For albumin crosslinking methods, the two common cross-linking agents are carbodiimide (CDI)[7,53] and glutaraldehyde.[23,36,47,81] However, the later was found by some early work to cause severe tissue reactions.[14,15] These results suggest that CDI has the potential for crosslinked albumin to coat an implant surface, inhibit bacterial adherence and reduce the possibility of prosthetic infections in vivo. For albumin crosslinking with CDI,[7,14,53,80] the cp-Ti plates were soaked in a solution consisting of 11 mL of 20% (w/v) bovine serum albumin (BSA) and 10 mL of 0.2 mol/L CDI in phosphate buffered saline (PBS, pH 7.3) for 15 min. The plates were removed prior to gelation, and air-dried in covered petri dishes for 48 h. An in vivo degradation study is needed to explore this concept further. Previous studies have shown that both albumin adsorption and coating with crosslinking on titanium surfaces inhibited bacterial adherence.[5,53] A recent study further confirmed the long-lasting effect of CDI crosslinked albumin on titanium surface.[7] During the 20 d incubation in PBS at 37°C with agitation, the coated albumin decayed only 10% compared to day 0. This means that most of the albumin molecules still remained on the surface after 20 d of incubation in PBS. This is in agreement with the work by Ben Slimane et al.,[16] in which a crosslinked albumin coating on an arterial polyester prosthesis decayed only 2% in PBS or plasma over a 6 d incubation period. An unanswered question is what are the effects of mechanical impaction during the implantation procedure on the integrity of the crosslinked albumin coating? How firmly the crosslinked albumin layer bonds to the titanium surface or how much mechanical stress the coating can sustain are unknown. If the crosslinked albumin coating cannot sustain the stresses that occur during routine implantation, another method for incorporating albumin to the implant surface may be needed.

Sipehia et al.[83] reported enhanced albumin binding to polypropylene via anhydrous ammonia gaseous plasma. The technique was used to add amino groups onto the polypropylene surface by exposing them to anhydrous ammonia plasma. Through these amino groups, albumin was attached to the polypropylene beads. Attached albumin was further stabilized by crosslinking with glutaraldehyde. The effect of washing albuminated polypropylene beads with saline and human plasma was investigated. It was found that after initial rapid removal of albumin, the concentration of attached albumin tended to reach a steady-state. After 52 h of washing, the amount of albumin retained on the beads varied between 125 and 171 µg/cm². The same procedure was also used to coat polypropylene membrane.[82]

Albumin can be covalently attached to various biomaterial surfaces such as polypropylene, polycarbonate, and poly(vinyl chloride) by γ-irradiation.[44] The amount of the grafted albumin is dependent on the γ-irradiation dose and the concentration of albumin used for adsorption. The grafted albumin molecules remained on the surface even after exposure to blood for prolonged time periods. This approach was used to graft albumin to polymeric materials of an oxygenator. The covalent grafting of functionalized albumin by γ-irradiation obviates the need for premodification of chemically inert polymer surfaces. It is useful for albumin grafting to various biomaterial surfaces.

Tseng et al.[86] reported that albumin was grafted on to polypropylene (PP) films by thermolysis of the azido groups of 4-azido-2-nitrophenyl albumin (ANP-albumin) with no pre-modification of the PP surface. The albumin-grafted surface was characterized by electron spectroscopy for chemical analysis (ESCA) and by quantitative determination of platelet adhesion and activation. The bulk concentration of ANP-albumin used for adsorption varied from 0.001 to 30 mg/mL, and the albumin-adsorbed PP films were incubated at 100°C for up to 7 h. From the same group,[85] albumin was also grafted onto dimethyldichlorosilane-coated glass (DDS-glass) by photolysis of the azido groups of ANP-albumin without any premodification of the surface. The albumin-grafted DDS-glass was characterized by determining the relative amount of nitrogen resulting from the grafted albumin on the surface using electron spectroscopy for chemical analysis (ESCA).

The amount of nitrogen increased when the concentration of ANP-albumin in the adsorption solution increased up to 0.1 mg/mL. The maximum platelet-resistant effect was observed when the ANP-albumin was adsorbed for more than 50 min at the solution concentration ranging from 0.05 to 10 mg/mL.

The early work by Hoffman et al.[39] used ε-amino caproic acid as a spacer or "arms" to chemically attach human serum albumin, heparin and streptokinase to hydrogels (composed of different ratios of hydroxyethyl methacrylate and *N*-vinyl pyrrolidone in water) radiation-grafted onto silicone rubber films.

Keogh et al.[46] attempted to produce biomaterial surfaces that would selectively bind host albumin because albumin-coated surfaces were known to diminish both coagulation and bacterial adherence. An albumin-binding high molecular weight dextran:Cibacron blue adduct was bulk incorporated into polyetherurethane (PU). The modified material bound albumin selectively and reversibly, and also showed evidence of enhanced biocompatibility. However, approximately 30% of the surface of this material was evidently unmodified and still capable of exerting the above adverse effects. In their later work,[45] they have covalently surface-modified polyetherurethane with sequential additions of acrylamide, amino-propylmethacrylamide, dextran, and Cibacron blue, which derivatized polyurethane preferentially and reversibly binds albumin, even from complex protein mixtures such as plasma. This new surface inhibits the clotting of nonanticoagulated whole human blood, perhaps by virtue of binding and activation of antithrombin III by the sulfonic acid residues on the surface-immobilized Cibacron blue. Finally, such surfaces diminish the adherence of *S. epidermidis*, a pathogen frequently associated with device-centered infections.[45]

Another kind of albumin-binding surface is the alkylated or alkyl derivatized (polyurethane, polyethylene, cellulose acetate) polymer surfaces.[25,26,60] Albumin in blood has a high affinity for circulating free fatty acids. In 1981, Munro et al. proposed the covalent binding to polymer surfaces of 16 or 18 carbon alkyl chains ($C_{16}$ and $C_{18}$), which mimic the nonpolar structrue of the saturated fatty acids and thus develop a strong hydrophobic interaction with albumin.[25,60]

### B. Antibacterial Adhesion

The early work on the effects of surface coated albumin on bacterial adhesion was reported by Fletcher.[28] She found that BSA impaired the attachment of a marine pseudomonad to polystyrene Petri dishes, apparently through adsorption on the dish surface. Albumin adsorbed or coated on material surfaces has shown obvious inhibitory effects on bacterial adhesion to polymers,[41,45,68,70,73] silicone,[27] ceramic,[30] hydroxyapatite,[75] titanium (Ti),[5,7,53,84] Ti-alloy[84] surfaces. In general, albumin coating has anti-adhesive effects on many bacterial species or strains, including *S. aureus*, *S. epidermidis*, *E coli*, *S. mutans*, *S. sanguis*, *S. mitis*, *Pseudomonas* sp., *Actinomyces* sp, *Actinobacillus* sp, or *Porphyromonas gingivalis* (Table 1),[84] Because the anti-adhesive effect of albumin is so definite, it has been used often as a control coating for the studies on bacterial adhesion promoting proteins or factors, such as fibronectin, fibrinogen, fibrin, or thrombin.[13,56]

Researchers have been trying to decrease bacterial adhesion to the inside wall of tubular shaped medical devices, such as catheters and stents, using albumin coating. Pascual et al.[68] found that adherence to Teflon catheters was significantly related to the degree of hydrophobicity of the strains. When hydrophobic groups were removed from *S. epidermidis* by pepsin treatment, adhesion was almost completely abolished. Preincubation of catheters in human serum also caused a 80 to 90% reduction of adherence. Preincubation of *S. epidermidis* in serum similarly decreased adhesion. This effect of serum was mainly due to albumin, while IgG and fibronectin were less effective.

Carballo et al[19] studied the influence of human plasma albumin on *S. epidermis* adhesion to teflon, polyethylene, and polycarbonate in an in vitro quantitative assay by scintillation counting. Bacterial adhesion was generally reduced by the presence of protein. The effect of these plasma proteins on bacterial surface properties resulted in strong increases of surface charge as measured by ion-exchange chromatography and with no effect on hydrophobicity, estimated as contact angles. In another study from the same group,[18] the adhesion of five coagulase-negative strains onto polyethylene, nylon and polyvinyl-chloride catheters, after treatment of bacteria, catheters or both with citrated human plasma and HSA was studied. Plasma and serum albumin produced a marked inhibition of bacterial adherence by means of adsorption on biomaterial surface.

In the oral and dental field, bacterial adhesion which can lead to biofilm formation or infection is a great challenge to researchers and clinicians. Attention has been paid to the effects of adsorbed albumin on oral bacterial adhesion to the surfaces of tooth, salivary pellicles,[31] hydroxyapatite,[75] and titanium.[84] It has been clear that albumin adsorbed on hydroxyapatite surfaces significantly reduces bacterial adhesion of common oral species such as *S. mutans, S. sanguis,* and *S. mitis.* Yen and Gibbons[90] found that fewer streptococci adsorbed in vitro to hydroxyapatite beads treated with bovine albumin when compared to HA treated with buffer or with saliva. Approximately 60% of adsorbed $^3$H-albumin persisted on HA when incubated for 24 h in clarified whole or parotid saliva. Also, fewer bacteria were recovered from vigorously-pumiced, molar-tooth surfaces 24 h after application of albumin compared to buffer-treated controls.[90]

In a study by Steinberg et al.,[84] the adhesion of radiolabeled *Actinomyces viscosus, Actinobacillus actinomicetemcomitans* and *Porphyromonas gingivalis* to titanium and Ti-6-Al- 4V alloy (Ti-alloy) coated with albumin or human saliva was investigated. All the tested bacteria displayed greater attachment to Ti-alloy than to Ti. *P. gingivalis* exhibited less adhesion to Ti and Ti-alloy than did the other bacterial strains. Adhesion of *A. viscosus* and *A. actinomicetemcomitans* was greatly reduced when Ti or Ti-alloy were coated with albumin or saliva. *P. gingivalis* demonstrated a lesser reduction in adhesion to albumin or saliva-coated surfaces. The results show that oral bacteria have different adhesion affinities for Ti and Ti-alloy and that both albumin and human saliva reduce bacterial adhesion.

Another challenge of bacterial adhesion is the prosthetic infection in orthopedic field. The use of albumin coating to reduce bacterial adhesion and evantually the implant site infection is still in experimental stage. HSA inhibited *S. epidermidis* adhesion to commercially pure titanium (cp-Ti) surfaces by more than 95 % after adsorption of 200 mg/ml of HSA at 37°C for 2 h.[5] In another study, titanium surfaces were coated with 10% BSA in PBS using carbodiimide (CDI) as the cross-linking agent.[7] The durability of the coated surfaces and the inhibitory effect of the albumin coating on bacterial adherence were tested in an in vitro condition (at 37°C, in PBS, with intermittent agitation) for 20 consecutive days. The results showed that only 10% of the coated BSA decayed off the surface during the 20 d incubation period. The inhibition rate of the albumin coating on bacterial adherence remained high (>85%) throughout the length of the experiment. The results suggested the potential use of this crosslinked albumin coating to reduce bacterial adherence and the possibility of subsequent prosthetic or implant infection in vivo. The durability of the crosslinked albumin coating appeared to be satisfactory for this in vitro condition, according to the theory of the "race for the surface." This is a term that signifies a contest between tissue cell integration of, and bacterial adhesion to, an available implant surface.[33,34] This knowledge is applicable not only to orthopaedic prostheses but to implantation of any foreign material, especially those with implant surfaces that have direct contact with the circulatory system where bacteria reach the implant surfaces through blood circulation.

In a recent study, a crosslinked albumin coating has been shown to reduce prosthetic infection rate in a rabbit model. Albumin coated and uncoated cylindrical implants were exposed to *S. epidermidis* (RP62A) suspension for 1 h before inplantation. Animals with albumin coated implants had a much lower infection rate (27%) than those with uncoated implants (62%). This finding may represent a new method for preventing prosthetic infection.[2]

Most proteins reduce bacterial adhesion through the adsorption to substrata surface, while serum albumin also inhibits adhesion by binding to bacterial cells.[17] Albumin may also reduce bacterial adhesion by changing substratum surface hydrophobicity, because in the presence of adsorbed albumin, the substrata surface becomes much less hydrophobic.[29,75] In one of our recent studies,[7] the SEM showed that the albumin coated titanium surfaces (using CDI as the crosslinking agent) are smoother than the uncoated ones. The results of water contact angle measurement indicated that the coated surfaces are more hydrophobic than the uncoated surfaces. Although no conclusions regarding the mechanism of the bacterial inhibiting effect of the albumin surface can be drawn from this experimental design, these results suggest that some relationship may exist between surface texture, hydrophobicity, and bacterial adherence. For example, albumin may reduce bacterial adherence by making the titanium surface smoother. Baker and Greenham[12] found that roughening the surface of either glass or polystyrene with a grindstone greatly increased the rate of bacterial colonization in a river environment. Albumin may also reduce bacterial adhesion by changing substratum surface hydrophobicity, since in the presence of adsorbed bovine serum albumin (BSA) substrata surface became much less hydrophobic.[29,72]

## IV. SERUM AND OTHER PROTEINS

### A. Serum

The adhesion of various CNS onto plasma-coated FEP was studied by Hogt et al.[41] The adhesion of all strains onto plasma FEP was much lower than onto the untreated control FEP surface. Pascual et al found that pre-incubation of Teflon catheters in human serum caused an 80 to 90% reduction of adhesion of *S. epidermidis*. Similar effects were also found when polymers were pre-incubated with plasma or albumin.[27,57,70] Like albumin, the anti-adhesive effect of serum is definite, so it has been often used as a control coating for the studies on bacterial adhesion promoting proteins or factors.[13,27,56] Pre-incubation of *S. epidermidis* in serum similarly decreased adhesion. This effect of serum was mainly due to albumin, while IgG and fibronectin were less effective.[68] However, a controversal results was reported by Zdanowski et al.[93] They found that after coating with human plasma, the binding of all three species, *S. aureus, S. epidermidis,* and *E. coli,* to PTFE was significantly enhanced, whereas the binding to Dacron was reduced, indicating that the effect of serum coating depends also on the substrata.

### B. Bacterial Surface Proteins

By pre-incubation of silicone catheters the influence of purified staphylococcal cell surface components on the binding was evaluated by Espersen et al.[27] The most potent inhibitors of the binding of *S. aureus* were the two surface proteins, clumping factor and protein A, and the cytoplasmic membrane. Surface proteins and the cell membrane of *S. epidermidis* also blocked the binding. Another report stated that streptococcal antigen (SA) I/II of *S. mutans* prevented the bacteria from adhesion to hydroxyapatite beads.[59]

### C. Other Proteins

The adherence of *S. mutans* to hydroxyapatite disks pretreated with various acidic and basic proteins was reported by Reynolds and Wong.[75] Adsorption of a basic protein, including Histone H1, Histone H3, and poly-L-lysine, onto an hydroxyapatite disk enhanced or had no effect on bacterial adherence, whereas adsorption of acidic protein reduced adherence. These acidic proteins include BSA, poly-L-glutamate, phosvitin, $\alpha_{s1}$-Casein, $\beta$-Casein, $\kappa$-Casein, $\beta$-lactoglobulin, and $\alpha$-lactalbumin.

Another comprehensive study on the effects of individual proteins on bacterial adhesion to biomaterial surfaces was reported by Zdanowski et al.[93] They studied the in vitro adhesion of *S. aureus*, *S. epidermidis* and *E coli* (one strain of each species) to commercially available microporous polytetrafluoroethylene (PTFE) and woven Dacron vascular grafts before and after coating with different proteins. They found that coating with HSA reduced the binding of all three species to Dacron and of staphylococci to PTFE. IgG decreased the binding of *S. epidermidis* and *E. coli* to Dacron and of *S. epidermidis* to PTFE. In contrast, fibrinogen enhanced the binding of *S. aureus* both to Dacron and PTFE, and that of *E. coli* to PTFE, but decreased the binding of *S. epidermidis* and *E. coli* to Dacron. Fibronectin enhanced the binding of *S. aureus* to Dacron, and of *E. coli* to PTFE, but decreased the binding of *S. aureus* to PTFE and of *S. epidermidis* both to PTFE and Dacron.

High molecular weight kininogen (HMWK) is a plasma protein that has recently been found to have an anti-adhesive effect on osteosarcoma cells, platelets, monocytes, and endothelial cells. Using a radial flow chamber, Nagel et al.[63] found the anti-adhesive effect of HMWK (Enzyme Res Lab, Southbend, IN, USA) coating on adhesion of *S. aureus* to polyurethane, especially the hydrophilic polyurethanes. When fibrinogen and HMWK were adsorbed from the same solution rather than consecutively, a significant decrease in bacterial attachment rate was observed on all material surfaces (base, sulfonated, quaternized amine, and phosphonated polyurethane)

The work by Reynolds and Wong[75] may shed some light on the understanding of mehcanisms of the effects of surface adsorbed proteins on bacterial adhesion. They studied the adherence of *S. mutans PK1* to hydroxyapatite disks pretreated with various acidic and basic proteins in imidazole buffer. Adsorption of a basic protein onto an hydroxyapatite disk enhanced or had no effect on bacterial adherence, whereas adsorption of an acidic protein reduced adherence. The effect of adsorbed protein on bacterial adherence was of both short and long range. The long-range effect of the acidic proteins in reducing the number of bacteria adhering to hydroxyapatite was related to protein adsorption causing an increase in surface net negative charge, as shown by zeta potential measurement. Basic protein produced a net positive surface charge which facilitated adherence. Within the acidic protein group, the acidic residue percentage of the adsorbed protein was negatively correlated with the number of bacteria adhering, whereas the nonpolar residue percentage was positively correlated with bacterial adherence. Within the basic protein group, the basic residue percentage was correlated with the number of cells adhering. These results indicate the involvement of short-range hydrophobic and ionic interactions in bacterial adherence to protein-coated hydroxyapatite.[75]

**Acknowledgment:** a MUSC institutional grant (1992) and a grant from the Arthritis Foundation (1993) supported this study.

## REFERENCES

1. Abu el-Asrar AM, Shibl AM, Tabbara KF, et al: Heparin and heparin-surface-modification reduce *Staphylococcus epidermidis* adhesion to intraocular lenses. *Int Ophthalmol* 21:71–4, 1997
2. An YH, Bradley J, Powers DL, et al: The prevention of prosthetic infection using a cross-linked albumin coating in a rabbit model. *J Bone Joint Surg Br* 79:816–9, 1997
3. An YH, Friedman RJ: Prevention of sepsis in total joint arthroplasty. *J Hosp Infect* 33:93–108, 1996
4. An YH, Friedman RJ: Concise review of mechanisms of bacterial adhesion to biomaterial surfaces. *J Biomed Mater Res* 43:338–48, 1998
5. An YH, Friedman RJ, Draughn RA: Rapid quantification of staphylococci adhered to titanium surfaces using image analyzed epifluorescence microscopy. *J Microbiol Meth* 24:29–40, 1995
6. An YH, Friedman RJ, Draughn RA, et al: Bacterial adhesion to biomaterial surfaces. In: Wise DE, et al., ed. *Human Biomaterials Applications.* Humana Press, Totowa, NJ, 1996: 19–57
7. An YH, Stuart GW, McDowell SJ, et al: Prevention of bacterial adherence to implant surfaces with a crosslinked albumin coating *in vitro. J Orthop Res* 14:846–9, 1996
8. Appelgren P, Ransjo U, Bindslev L, et al: Surface heparinization of central venous catheters reduces microbial colonization *in vitro* and *in vivo*: results from a prospective, randomized trial. *Crit Care Med* 24:1482–9, 1996
9. Appelgren P, Ransjo U, Bindslev L, et al: Does surface heparinization reduce bacterial colonisation of central venous catheters? *Lancet* 345:130, 1995
10. Arciola CR, Caramazza R, Pizzoferrato A: In vitro adhesion of *Staphylococcus epidermidis* on heparin-surface- modified intraocular lenses. *J Cataract Refract Surg* 20:158–61, 1994
11. Arciola CR, Radin L, Alvergna P, et al: Heparin surface treatment of poly (methylmethacrylate) alters adhesion of a *Staphylococcus aureus* strain: utility of bacterial fatty acid analysis. *Biomaterials* 14:1161–4, 1993
12. Baker AS, Greenham LW: Factors affecting the bacterial colonization of various surfaces in a river. *Can J Microbiol* 30:511–5, 1984
13. Baumgartner JN, Cooper SL: Influence of thrombus components in mediating *Staphylococcus aureus* adhesion to polyurethane surfaces. *J Biomed Mater Res* 40:660–70, 1998
14. Ben Slimane S, Guidoin R, Marceau D, et al: Characteristics of polyester arterial grafts coated with albumin: the role and importance of the cross-linking chemicals. *Eur Surg Res* 20:18–28, 1988
15. Ben Slimane S, Guidoin R, Mourad W, et al: Polyester arterial grafts impregnated with cross-linked albumin: the rate of degradation of the coating *in vivo. Eur Surg Res* 20:12–7, 1988

16. Benslimane S, Guidoin R, Roy PE, et al: Degradability of crosslinked albumin as an arterial polyester prosthesis coating in in vitro and in vivo rat studies. *Biomaterials* 7:268–72, 1986

17. Brokke P, Dankert J, Carballo J, et al: Adherence of coagulase-negative staphylococci onto polyethylene catheters in vitro and *in vivo*: a study on the influence of various plasma proteins. *J Biomater Appl* 5:204–26, 1991

18. Carballo J, Ferreiros CM, Criado MT: Importance of experimental design in the evaluation of the influence of proteins in bacterial adherence to polymers. *Med Microbiol Immunol* 180:149–55, 1991

19. Carballo J, Ferreiros CM, Criado MT: Influence of blood proteins in the *in vitro* adhesion of *Staphylococcus epidermidis* to Teflon, polycarbonate, polyethylene and bovine pericardium. *Rev Esp Fisiol* 47:201–8, 1991

20. Chang TM: Removal of endogenous and exogenous toxins by a microencapsulated absorbent. *Can J Physiol Pharmacol* 47:1043–5, 1969

21. Chang TM: Platelet-surface interaction: effect of albumin coating or heparin complexing on thrombogenic surfaces. *Can J Physiol Pharmacol* 52:275–85, 1974

22. Chawla AS, Chang TM: Nonthrombogenic surface by radiation grafting of heparin: preparation, in vitro and in vivo studies. *Biomater Med Devices Artif Organs* 2:157–69, 1974

23. Domurado D, Thomas D, Brown G: A new method for producing proteic coatings. *J Biomed Mater Res* 9:109–10, 1975

24. Duran JA, Malvar A, Rodriguez-Ares MT, et al: Heparin inhibits *Pseudomonas* adherence to soft contact lenses. *Eye* 7:152–4, 1993

25. Eberhart RC, Munro MS, Frautschi JR, et al: Influence of endogenous albumin binding on blood-material interactions. *Ann N Y Acad Sci* 516:78–95, 1987

26. Eberhart RC, Munro MS, Williams GB, et al: Albumin adsorption and retention on C18-alkyl-derivatized polyurethane vascular grafts. *Artif Organs* 11:375–82, 1987

27. Espersen F, Wilkinson BJ, Gahrn-Hansen B, et al: Attachment of staphylococci to silicone catheters in vitro. *Apmis* 98:471–8, 1990

28. Fletcher M: The effects of proteins on bacterial attachment to polystyrene. *J Gen Microbiol* 94:400–4, 1976

29. Fletcher M, Marshall KC: Bubble contact angle method for evaluating substratum interfacial characteristics and its relevance to bacterial attachment. *Appl Environ Microbiol* 44:184–92, 1982

30. Gibbons RJ, Etherden I: Comparative hydrophobicities of oral bacteria and their adherence to salivary pellicles. *Infect Immun* 41:1190–6, 1983

31. Gibbons RJ, Etherden I: Albumin as a blocking agent in studies of streptococcal adsorption to experimental salivary pellicles. *Infect Immun* 50:592–4, 1985

32. Gott VL, Whiffen JD, Dutton RC: Heparin bonding on colloidal graphite surfaces. *Science* 142:1297, 1963

33. Gristina AG, Naylor PT, Myrvik QN: Musculoskeletal infection, microbial adhesion, and antibiotic resistance. *Infect Dis Clin North Am* 4:391–408, 1990

34. Gristina AG, Naylor PT, Myrvik QN: Mechanisms of musculoskeletal sepsis. *Orthop Clin North Am* 22:363–71, 1991

35. Grode GA, Falb RD, Crowley JP: Biocompatibie materials for use in the vascular system. *J Biomed Mater Res* 6:77–84, 1972

36. Guidoin R, Martin L, Marois M, et al: Polyester prostheses as substitutes in the thoracic aorta of dogs. II. Evaluation of albuminated polyester grafts stored in ethanol. *J Biomed Mater Res* 18:1059–72, 1984

37. Hennink WE, Feijen J, Ebert CD, et al: Covalently bound conjugates of albumin and heparin: synthesis, fractionation and characterization. *Thromb Res* 29:1–13, 1983

38. Hennink WE, Kim SW, Feijen J: Inhibition of surface induced coagulation by preadsorption of albumin–heparin conjugates. *J Biomed Mater Res* 18:911–26, 1984

39. Hoffman AS, Schmer G, Harris C, et al: Covalent binding of biomolecules to radiation-grafted hydrogels on inert polymer surfaces. *Trans Am Soc Artif Intern Organs* 18:10–8, 1972

40. Hoffman J, Larm O, Scholander E: A new method for covalent coupling of heparin and other glycosaminoglycans to substances containing primary amino groups. *Carbohydr Res* 117:328–31, 1983

41. Hogt AH, Dankert J, Feijen J: Adhesion of *Staphylococcus epidermidis* and *Staphylococcus saprophyticus* to a hydrophobic biomaterial. *J Gen Microbiol* 131:2485–91, 1985

42. Homma H, Nagaoka S, Mezawa S, et al: Bacterial adhesion on hydrophilic heparinized catheters, with compared with adhesion on silicone catheters, in patients with malignant obstructive jaundice. *J Gastroenterol* 31:836–43, 1996

43. Idezuki Y, Watanabe H, Hagiwara M, et al: Mechanism of antithrombogenicity of a new heparinized hydrophilic polymer: chronic *in vivo* studies and clinical application. *Trans Am Soc Artif Intern Organs* 21:436–49, 1975

44. Kamath KR, Park K: Surface modification of polymeric biomaterials by albumin grafting using h-irradiation. *J Appl Biomater* 5:163–73, 1994

45. Keogh JR, Eaton JW: Albumin binding surfaces for biomaterials. *J Lab Clin Med* 124:537–45, 1994

46. Keogh JR, Velander FF, Eaton JW: Albumin-binding surfaces for implantable devices. *J Biomed Mater Res* 26:441–56, 1992

47. Kottke-Marchant K, Anderson JM, Umemura Y, et al: Effect of albumin coating on the *in vitro* blood compatibility of Dacron arterial prostheses. *Biomaterials* 10:147–55, 1989

48. Kratz G, Arnander C, Swedenborg J, et al: Heparin-chitosan complexes stimulate wound healing in human skin. *Scand J Plast Reconstr Surg Hand Surg* 31:119–23, 1997

49. Larm O, Larsson R, Olsson P: A new non-thrombogenic surface prepared by selective covalent binding of heparin via a modified reducing terminal residue. *Biomater Med Devices Artif Organs* 11:161–73, 1983

50. Lundberg F, Gouda I, Larm O, et al: A new model to assess staphylococcal adhesion to intraocular lenses under in vitro flow conditions. *Biomaterials* 19:1727–33, 1998

51. Lundgren B, Holst A, Tarnholm A, et al: Cellular reaction following cataract surgery with implantation of the heparin-surface-modified intraocular lens in rabbits with experimental uveitis. *J Cataract Refract Surg* 18:602–6, 1992

52. Lundgren B, Ocklind A, Holst A, et al: Inflammatory response in the rabbit eye after intra-ocular implantation with poly(methyl methacrylate) and heparin surface modified intraocular lenses. *J Cataract Refract Surg* 18:65–70, 1992

53. McDowell SG, An YH, Draughn RA, et al: Application of a fluorescent redox dye for enumeration of metabolically active bacteria on albumin-coated titanium surfaces. *Lett Appl Microbiol* 21:1–4, 1995

54. Merrill EW, Salzman EW, Lipps BJ, Jr., et al: Antithrombogenic cellulose membranes for blood dialysis. *Trans Am Soc Artif Intern Organs* 12:139–50, 1966

55. Miorner H, Myhre E, Bjorck L, et al: Effect of specific binding of human albumin, fibrinogen, and immunoglobulin G on surface characteristics of bacterial strains as revealed by partition experiments in polymer phase systems. *Infect Immun* 29:879–85, 1980

56. Mohammad SF, Topham NS, Burns GL, et al: Enhanced bacterial adhesion on surfaces pretreated with fibrinogen and fibronectin. *ASAIO Trans* 34:573–7, 1988

57. Muller E, Takeda S, Goldmann DA, et al: Blood proteins do not promote adherence of coagulase-negative staphylococci to biomaterials. *Infect Immun* 59:3323–6, 1991

58. Mulvihill JN, Faradji A, Oberling F, et al: Surface passivation by human albumin of plasma-pheresis circuits reduces platelet accumulation and thrombus formation. Experimental and clinical studies. *J Biomed Mater Res* 24:155–63, 1990

59. Munro GH, Evans P, Todryk S, et al: A protein fragment of streptococcal cell surface antigen I/II which prevents adhesion of *Streptococcus mutans*. *Infect Immun* 61:4590–8, 1993

60. Munro MS, Quattrone AJ, Ellsworth SR, et al: Alkyl substituted polymers with enhanced albumin affinity. *Trans Am Soc Artif Intern Organs* 27:499–503, 1981
61. Nagaoka S, Kawakami H: Inhibition of bacterial adhesion and biofilm formation by a heparinized hydrophilic polymer. *Asaio J* 41:M365–8, 1995
62. Nagaoka S, Mikami M, Noishiki Y: Evaluation of antithrombogenic thermodilution catheter. *J Biomater Appl* 4:22–32, 1989
63. Nagel JA, Dickinson RB, Cooper SL: Bacterial adhesion to polyurethane surfaces in the presence of pre-adsorbed high molecular weight kininogen. *J Biomater Sci Polym Ed* 7:769–80, 1996
64. Nilsson UR, Larm O, Nilsson B, et al: Modification of the complement binding properties of polystyrene: effects of end-point heparin attachment. *Scand J Immunol* 37:349–54, 1993
65. Nomura S, Lundberg F, Stollenwerk M, et al: Adhesion of staphylococci to polymers with and without immobilized heparin in cerebrospinal fluid. *J Biomed Mater Res* 38:35–42, 1997
66. Park KD, Kim YS, Han DK, et al: Bacterial adhesion on PEG modified polyurethane surfaces. *Biomaterials* 19:851–9, 1998
67. Parsson H, Jundzill W, Johansson K, et al: Healing characteristics of polymer-coated or collagen-treated Dacron grafts: an experimental porcine study. *Cardiovasc Surg* 2:242–8, 1994
68. Pascual A, Fleer A, Westerdaal NA, et al: Modulation of adherence of coagulase-negative staphylococci to Teflon catheters in vitro. *Eur J Clin Microbiol* 5:518–22, 1986
69. Paulsson M, Gouda I, Larm O, et al: Adherence of coagulase-negative staphylococci to heparin and other glycosaminoglycans immobilized on polymer surfaces. *J Biomed Mater Res* 28:311–7, 1994
70. Paulsson M, Kober M, Freij-Larsson C, et al: Adhesion of staphylococci to chemically modified and native polymers, and the influence of preadsorbed fibronectin, vitronectin and fibrinogen. *Biomaterials* 14:845–53, 1993
71. Portoles M, Refojo MF, Leong FL: Reduced bacterial adhesion to heparin-surface-modified intraocular lenses. *J Cataract Refract Surg* 19:755–9, 1993
72. Pratt-Terpstra IH, Weerkamp AH, Busscher HJ: Adhesion of oral streptococci from a flowing suspension to uncoated and albumin-coated surfaces. *J Gen Microbiol* 133:3199–206, 1987
73. Pringle JH, Fletcher M: Influence of substratum hydration and adsorbed macromolecules on bacterial attachment to surfaces. *Appl Environ Microbiol* 51:1321–5, 1986
74. Remuzzi A, Boccardo P: Albumin treatment reduces *in vitro* platelet deposition to PMMA dialysis membrane. *Int J Artif Organs* 16:128–31, 1993
75. Reynolds EC, Wong A: Effect of adsorbed protein on hydroxyapatite zeta potential and *Streptococcus mutans* adherence. *Infect Immun* 39:1285–90, 1983
76. Rosenberg M, Gutnick D, Rosenberg E: Adherence of bacteria to hydrocarbons: a simple method for measuring cell-surface hydrophobicity. *FEMS Microbiol Lett* 9:29–33, 1980
77. Rosenberg RD, Damus PS: The purification and mechanism of action of human antithrombin-heparin cofactor. *J Biol Chem* 248:6490–505, 1973
78. Ruggieri MR, Hanno PM, Levin RM: Reduction of bacterial adherence to catheter surface with heparin. *J Urol* 138:423–6, 1987
79. Schmidt H, Schloricke E, Fislage R, et al: Effect of surface modifications of intraocular lenses on the adherence of Staphylococcus epidermidis. *Zentralbl Bakteriol* 287:135–45, 1998
80. Sheehan JC, Hlavka J: The cross-linking of gelatin using a water-soluble carbodiimide. *J Am Chem Soc* 79:4528–9, 1957
81. Sigot-Luizard MF, Lanfranchi M, Duval JL, et al: The cytocompatibility of compound poly-ester-protein surfaces using an in vitro technique. *In Vitro Cell Dev Biol* 22:234–40, 1986
82. Sipehia R, Chawla AS: Albuminated polymer surfaces for biomedical application. *Biomater Med Devices Artif Organs* 10:229–46, 1983
83. Sipehia R, Chawla AS, Chang TM: Enhanced albumin binding to polypropylene beads via anhydrous ammonia gaseous plasma. *Biomaterials* 7:471–3, 1986

84. Steinberg D, Sela MN, Klinger A, et al: Adhesion of periodontal bacteria to titanium, and titanium alloy powders. *Clin Oral Implants Res* 9:67–72, 1998

85. Tseng YC, Kim J, Park K: Photografting of albumin onto dimethyldichlorosilane-coated glass. *J Biomater Appl* 7:233–49, 1993

86. Tseng YC, Mullins WM, Park K: Albumin grafting on to polypropylene by thermal activation. *Biomaterials* 14:392–400, 1993

87. Uyen HM, Schakenraad JM, Sjollema J, et al: Amount and surface structure of albumin adsorbed to solid substrata with different wettabilities in a parallel plate flow cell. *J Biomed Mater Res* 24:1599–614, 1990

88. Vaudaux PE, Waldvogel FA, Morgenthaler JJ, et al: Adsorption of fibronectin onto polymethylmethacrylate and promotion of *Staphylococcus aureus* adherence. *Infect Immun* 45:768–74, 1984

89. Wassall MA, Santin M, Isalberti C, et al: Adhesion of bacteria to stainless steel and silver-coated orthopedic external fixation pins. *J Biomed Mater Res* 36:325–30, 1997

90. Yen S, Gibbons RJ: The influence of albumin on adsorption of bacteria on hydroxyapatite beads *in vitro* and human tooth surfaces *in vivo*. *Arch Oral Biol* 32:531–3, 1987

91. Yen SP, Rembaum A: Complexes of heparin with elastomeric positive polyelectrolytes. *J Biomed Mater Res* 5:83–97, 1971

92. Zdanowski Z, Koul B, Hallberg E, et al: Influence of heparin coating on in vitro bacterial adherence to poly(vinyl chloride) segments. *J Biomater Sci Polym Ed* 8:825–32, 1997

93. Zdanowski Z, Ribbe E, Schalen C: Influence of some plasma proteins on *in vitro* bacterial adherence to PTFE and Dacron vascular prostheses. *Apmis* 101:926–32, 1993

## Appendix 1
### Basic Glossary on Bacterial Adhesion and Biofilm Studies

### Adhesion, Adherence, and Attachment

Bacterial adhesion is a process that bacteria adhere firmly to a surface by a complete interactions between them, including an initial phase of reversible, physical contact and a time-dependent phase of irreversible, chemical and cellular adherence. It is an energy involved formation of a adhesive junction between bacteria and surfaces. Sorption is an out-of-date synonym for adhesion. Adherence is a general description of bacterial adhesion, or the initial process of attachment of bacteria directly to a surface, and is a less scientific term for bacterial adhesion and is not a legitimate alternative to adhesion. Attachment can be defined as the initial stage of bacterial adhesion, refers more to physical contact than complicated chemical and cellular interactions, and is usually reversible. Adhesion, adherence, and attachment are often used interchangeably.

### Adhesin and Receptor

Adhesin is a substance (a surface macromolecule, commonly lectins or lectin-like proteins or carbohydrate, of bacteria) produced by bacteria and is thought to be a specific material for specific adhesion. But generally, any structures responsible for adhesive activities can be called adhesins. Bacteria may have multiple adhesins for different surfaces (different receptors). A receptor is a component (both known and putative) on the surfaces of biomaterials or host tissue which is bound by the active site of an adhesin during the process of specific adhesion.

### Adsorption and Deposition

The accumulation of molecules onto a solid surface at a concentration exceeding that in the bulk fluid, brought about as a result of random Brownian motion. Deposition is normally used to describe the accumulation of particles at a fluid interface brought about by the application of an external force. In most circumstances, gravitational force brings particles to deposit on the bottom of a aqueous container.

### Biofilm and Biofouling

An accumulated biomass of bacteria and extracellular materials (basically slime) on a solid surface is called a biofilm. Slime is defined as an extracellular substance (the exopolymers composed of mainly polysaccharide) produced by the bacteria which may be partially free from the bacteria after dispersion in a liquid medium (water-soluble) and can be removed from bacterial cells by washing. Biofouling is the fouling or contamination of an area, which is basically the process of biofilm formation on the surfaces of non-medical devices.

### Colony Forming Unit (CFU or cfu)

A colony forming unit (cfu) is a colony on a culture plate that is thought to have derived from a single bacterium. For many bacteria, 24–48 h of incubation is enough to obtain countable colonies.

### Flow Chambers or Flow Cells

Flow devices or reactors used to grow and observe bacterial adhesion and biofilm development.

### Glycocalyx

Extracellular polymeric material produced by some bacteria. Term initially applied to the polysaccharide matrix excreted by epithelial cells forming a coating on the surface of epithelial tissue. General term for polysaccharide compounds outside the bacterial cell wall. Also called slime layer, EPS, or matrix polymer.

### Hydrophobicity and Hydrophilicity

The structure of water in the region near any surface (such as solid material surface or bacterial surface) is perturbed over distances of up to several tens of molecular layers. Hydrophobicity or hydrophilicity are relative descriptions. Near a hydrophobic surface the water is less structured in terms of intermolecular hydrogen bonding between the water molecules, whilst near a hydrophilic surface water is more structured. Water contact angle (WCA) is a good example of hydrophobic or hydrophilic nature of a surface. A high WCA represents hydrophobicity and a low WCA represents hydrophilicity.

### Slime

Slime is defined as an extracellular substance (the exopolymers composed of mainly polysaccharide) produced by the bacteria which may be partially free from the bacteria after dispersion in a liquid medium (water-soluble) and can be removed from bacterial cells by washing. An accumulated biomass of bacteria and extracellular materials (basically slime) on a solid surface is called a biofilm.

### Substrate and Substratum (substrata)

Substrate is a material utilized by microorganisms as a source of energy, but it is often used as an alternative of substratum. Substratum is a solid surface to which a microorganism may adhere.

# INDEX